PHOTOCHEMISTRY RESEARCH PROGRESS

PHOTOCHEMISTRY RESEARCH PROGRESS

ALEJANDRO SÁNCHEZ
AND SERGIO J. GUTIERREZ
EDITORS

Nova Science Publishers, Inc.
New York

For permission to use material from this book please contact us:
Telephone 631-231-7269; Fax 631-231-8175
Web Site: http://www.novapublishers.com

LIBRARY OF CONGRESS CATALOGING-IN-PUBLICATION DATA

Sánchez, Alejandro, 1953-
Photochemistry research progress / Alejandro Sánchez and Sergio J. Gutierrez.
p. cm.
ISBN 978-1-60456-568-3 (hardcover)
1. Photochemistry--Research. I. Gutierrez, Sergio J., (Sergio Jose), 1952- II. Title.
QD708.2.S36 2008
541'.35--dc22
2008014841

Published by Nova Science Publishers, Inc. ✣ New York

CONTENTS

PREFACE

Photochemistry, a sub-discipline of chemistry, is the study of the interactions between atoms, small molecules, and light (or electromagnetic radiation). Like most scientific disciplines, photochemistry utilizes the SI or metric measurement system. Important units and constants that show up regularly include the meter (and variants such as centimeter, millimeter, micrometer, nanometer, etc.), seconds, hertz, joules, moles, the gas constant R, and the Boltzmann constant. These units and constants are also integral to the field of physical chemistry.

There are important processes based in the photochemistry principles. One case is photosynthesis, in which some plants use light to create glucose in their chloroplasts to contribute to cell metabolism. The glucose is used by the plant's mitochondria to produce adenosine triphosphate. Medicine bottles are made with darkened glass to prevent the medicine itself from reacting chemically with light. In fireflies, an enzyme in the abdomen works to produce bioluminescence. The mercaptans or thiols are produced by photochemical addition of hydrogen sulfide (H2S) to alfa olefins. Among their many uses as a chemical reagent these mercaptans are used to provide a distinctive odor (an odorant) to otherwise odorless natural gas.

This new book is dedicated to presenting the latest research from around the world in this field.

Chapter 1 - The effects of the two botryticides, fludioxonil (fdx) and fenhexamid (fhd), were investigated on grapevine leaves (Vitis vinifera L. cv. Pinot noir) following photosynthesis mechanisms. Treatments were carried out in vineyard at the end of flowering. Phytotoxicity of both fungicides was evaluated by measuring variations of leaf photosynthetic parameters and correlated expression of photosynthesis-related genes. Results demonstrated that similar perturbation of photosynthesis was caused by fdx and fhd applications. Moreover, the mechanism leading to photosynthesis alteration seems to be the same for both fungicides. Photosynthesis modification was mostly attributed to non-stomatal limitation since the reduction of photosynthesis was associated with few changes in intercellular CO_2 concentration despite stomatal conductance decline. Effective PSII quantum yield (ΦPSII) was not affected. Additionally, decreased photosynthesis was coupled with repression of rubisco small subunit (rbcS) and rubisco large subunit (rbcL) genes. To the authors' knowledge, this is the first study of photosynthesis-related gene expression following fungicide stress.

Chapter 3 - Phytochrome (phy), a chromoprotein, is the major photoreceptor in plants, determining the strategy of their growth and development throughout their lifecycle. The chain of the regulatory steps leading to the photoresponse includes (i) light signal perception via the photoisomerization of the photoreceptor, (ii) transduction of the light signal from the activated photoreceptor through the complex network of mediators, and (iii) realizations of the light stimulus via modulation of expression of light-regulated genes. In recent years, we have been witnessing great progress in our understanding of the nature of this pigment and intimate mechanisms of its action. Among the fundamental discoveries are structural and functional phytochrome polymorphism with phyA and phyB as major ones and concrete mechanisms of their action including cytoplasm-to-nucleus translocation of the photoreceptor and interaction with transduction chain partners. It was shown that phyA is unique among other phys in many respects. In particular, it mediates photoresponse modes specific only for this pigment (the so-called very low fluence responses, VLFR, and high irradience response, HIR) and also low fluence responses, LFR, characteristic of phyB and other minor phys. The nature of this uniqueness remains, however, obscure. In the group of the author, the polymorphism of phyA itself was discovered with the use of an original method of fluorescence spectroscopy and photochemistry of phytochrome in the cell, which may explain, at least partially, the specific phenomenological and functional properties of phyA. In this chapter, the data on the structural and functional polymorphism are summarized and the literature on phyA is analyzed from this point of view.

Chapter 4 - In this chapter, the primary photochemical deactivation processes and secondary thermal reactions of iron(III) compounds are reviewed. Along with the theoretical aspects and deactivation mechanisms, particular stress is paid to application importance of the processes occurring in irradiated systems of iron(III) compounds. Both the high-spin and low-spin iron(III) complexes are dealt with here.

A dominant share of the accumulated knowledge concerns photoredox reactivity involving both the photoreduction of central atom Fe(III) to Fe(II), and photooxidation of Fe(III), mainly to Fe(V). The chapter covers intramolecular photoredox decompositions as well as outer-sphere electron-transfer processes. While the majority of the results has been obtained investigating mononuclear complexes, a deep insight into the mechanisms of photochemical and photophysical deactivation processes has been reached by also studying polynuclear complexes and solid polymeric compounds.

The complexes are classified according to the bonding properties of their ligands into groups with monodentate, chelating and macrocyclic ligands. Due to their peculiar properties, native proteins containing the iron(III) central atom are discussed in a separate section. In addition to solution photochemistry, the photoredox processes occurring in microheterogeneous systems, polymer matrices and solid complexes are evaluated. In each type of complex, the current knowledge is assessed and the main directions for future photochemical research are tentatively offered.

One of the driving forces in focusing long-lasting and systematic attention to the photochemical properties of iron(III) compounds is their practical importance. Due to this fact, application aspects of the photochemistry of such complexes are documented as well, with particular attention paid to their contribution to environmental protection, the role played in biological systems, therapeutical significance, photomagnetic materials, understanding the processes occurring in the marine environment, and technological applications.

Chapter 5 - Intermolecular hydrogen bonding, as a site-specific interaction between hydrogen donor and acceptor molecules, is a very important type of solute-solvent interaction. It is central to the understanding of the microscopic structure and function in many molecular systems. It also plays an important role in the photochemistry of the hydrogen-bonded systems in their electronically excited states. In this chapter, the authors review their recent investigations about the hydrogen bonding effects on the photochemistry of chromophores in solution. The electronically excited states of chromophores were theoretically studied using the time-dependent density functional theory (TDDFT) method. At the same time, the infrared spectra in both the ground state and electronically excited states of chromophores were also calculated. By monitoring the spectral shifts of some characterized vibrational modes involved in the formation of hydrogen bonds in different electronic states, it has been demonstrated for the first time that the intermolecular hydrogen bonds between chromophores and hydrogen bonding solvents are significantly strengthened upon photoexcitation. This excited-state hydrogen bonding dynamic occurs on the ultrafast timescale mainly set by vibrational motions of the hydrogen donor and acceptor groups, and it controls the excited-state dynamics of hydrogen-bonded complexes in a significant way. The authors have demonstrated that the radiationless deactivation via internal conversion (IC) from the fluorescent state to ground state can be enhanced due to the hydrogen bond strengthening in the fluorescent state. The ultrafast solute-solvent intermolecular photoinduced electron transfer (PET) reaction, which is facilitated by the hydrogen bond strengthening, has been proposed to account for the drastic fluorescence quenching behaviors of chromophores in alcoholic solvents. Moreover, the site-specific solvation dynamics can be initiated by the intermolecular hydrogen bond strengthening in the electronically excited state.

Chapter 6 - This chapter is organized according to classes of silicon-containing compounds that give rise to silyl-centered radicals through photolytic methods, and then review their reactivity, their spectroscopic data, and their possible applications. Such diverse sources of silicon compounds include simple hydrosilanes, alkyl-, alkyl-aryl, and arylsilanes, disilanes, polysilanes, acylsilanes, silacycloalkanes, and silyl ethers.

Photochemically-induced homolytic cleavage of Si-X bonds (where X can be B, Ge, or other elements) are also described as available sources of silyl radicals. The study of the photochemistry of different organic compounds, which bear common silicon-protective groups, their photolytic de-protection reactions, and the silyl radicals generated thereafter, are presented.

It is worthwhile to emphasize that this chapter covers only silicon-centered radicals that are directly or indirectly generated by the interaction of silicon-containing compounds with light, thus excluding silyl radicals that are obtained through thermal methods. Thermal-decomposition methods for the production of silicon-centered radicals are widely distributed, and constitute the subject of future review articles. This chapter also circumvents other silicon-based intermediates, such as silylenes, silenes, and radical ions of silicon, which are also the subject of future review studies. Throughout the chapter it will become apparent that the photochemical generation of silicon-centered radicals is often accompanied by the production of other kinds of reactive silicon intermediates (silicon–centered biradicals, silylenes, silenes, etc.).

Chapter 7 - The chapter is concerned with ultraviolet (UV) and vacuum ultraviolet (VUV) excilamps and excilamp-based photoreactors and reviews the authors' efforts in UV/VUV electrodeless barrier-discharge excilamps and their applications in photochemistry, namely, for oxidation and mineralization of organic substrates, VUV-initiated photodimerization and removal of water from hydrocarbon gas mixtures, and in actinometry.

Chapter 8 - Photochemical processes play an important role in the atmospheric aqueous phase. They are involved in the transformation of organic and inorganic compounds, some of which have an impact on global climate. Atmospheric waters (cloud, fog, mist, rain), due to their elevated surface-to-volume ratio, are also a favourable environment for various kinds of interface phenomena. Photoinduced reactions at the air-water interface have been the object of much attention, because different kinds of surface effects could considerably enhance the reaction rates.

Early studies have been focused on the distribution of the light-field intensity inside disperse water droplets, starting from the assumption that the radiation intensity in the droplet could be considerably enhanced by cavity resonance effects compared to the surrounding gas phase. However, such phenomena would only involve peculiar combinations of radiation wavelength and droplet diameter and would play a minor role for polychromatic irradiation, as is the case of sunlight, and a continuous distribution of droplet diameters as is typical of the atmosphere. Multiple radiation reflections at the air-water interface would nevertheless increase the effective radiation intensity in the droplet by a factor of 1.3-1.8 compared to the gas phase. A number of additional effects are then able to substantially enhance the interface photochemical activity.

Many organic compounds are known to undergo accumulation at the air/water interface, and some of them can also interact with inorganic ions. Such an interaction, which can be approximately assessed by means of the Wagner-Onsager-Samaras approach, allows the concentration of photoactive anions near the surface to increase in the presence of organic compounds. For instance, the surface co-adsorption of nitrate and benzene considerably enhances the rate of the photochemical conversion of benzene into phenol upon nitrate photolysis. Just because of co-adsorption, in a spherical water drop of 1 μm radius around 15% of the cited reaction would take place in a surface layer of sub-nanometric thickness, which accounts for just 0.3% of the total volume.

Also very important is the enhancement of the photolysis quantum yields at the interface, where photoactive molecules are surrounded by an incomplete solvent cage. The presence of a surrounding solvent cage is able to considerably decrease the photolysis quantum yields by facilitating the recombination of photogenerated radical fragments. Molecules at interfaces would undergo the solvent-cage effect on photolysis to a very limited extent, which could considerably enhance their surface reactivity. The quantum yield of nitrate photolysis might be increased up to 6 times at the interface, and the combination of surface co-adsorption and increased quantum yield would cause over 50% of the photochemical conversion of benzene into phenol to occur in just 0.3% of the drop volume. This means that the reaction rate would be increased by a factor of two if a solution bulk is dispersed into 1 μm drops. Other photochemical reactions (e.g., the photolysis of $FeCl_2^+$) could be even more markedly enhanced. These findings could substantially modify the current knowledge concerning multiphase atmospheric chemistry.

Chapter 9 - In this chapter, besides a brief description of the matrix isolation technique and discussion of its main advantages and drawbacks when applied to photochemical studies on organic molecules, a general overview of the photochemistry of tetrazoles and the general pattern of their photoreactions is presented.

The application of tetrazoles in agriculture and medicine is widespread and well known. In many drugs and drug-candidates, replacement of carboxylic acid groups, −COOH, by tetrazolic acid fragments, $-CN_4H$, is frequent, resulting in an increase in bioavailability and potency of the pharmacophore. Tetrazoles also have many other applications, such as in photography and photoimaging, and in the automobile industry as gas generating agents for airbags. These compounds also exhibit a very rich photochemistry, which is strongly influenced by the substituents present in the tetrazolic ring.

Several representative tetrazoles were trapped in a rigid environment of solidified noble gases (argon, xenon) at cryogenic temperatures (typically 10 K) and subjected to in situ photolysis. The range of possible rearrangements of the isolated species was controlled by choice of the excitation wavelengths. FT-IR spectroscopy provided experimental frequencies and intensities of characteristic absorptions of the isolated chemical species, both for reagents and photoproducts. The analysis of experimental data was assisted by their direct comparison with the vibrational spectra theoretically calculated for the single molecule in vacuum. This comparison was facilitated because the inert matrix only slightly affects the structure of the isolated molecules, and also owing to the high resolution of infrared matrix spectra (a few tenths of cm^{-1}).

UV-excitation resulted in photofragmentation of the monomeric tetrazoles with a wide range of exit channels. Since the obtained fragments in general stay in the matrix cage where they are formed, no subsequent cross-reactions involving species resulting from photolysis of different reactant molecules can occur, strongly reducing the number of possible photoproducts in comparison with gas phase or solution studies. This fact introduced a useful simplification for the interpretation of the reaction mechanisms.

A number of relatively unusual or highly reactive molecules, such as antiaromatic azirines, azides, isocyanates or isothiocyanates, may be formed from photolysis of tetrazoles in a matrix. In some examples described in this chapter, the spectroscopic characterization of these molecular species was presented for the first time.

Chapter 10 - The carbazole family is subjected to an overview in terms of the photophysics and photochemistry aspects. In this chapter no attempts are made to include polymeric systems or macromolecules in which the carbazole nuclei are pendant groups. The photophysical behavior of carbazoles deals with the characterization of the electronic excited states (singlet and triplet states) as well as the derivation of the dynamic properties of the excited states in terms of radiative and radiationless rate constants by means of steady-state and time-resolved spectroscopy (absorption, emission fluorescence and phosphorescence spectroscopy [optical methods]; laser induced optoacoustic spectroscopy [LIOAS; optoaocousitc method]). This discussion is restricted to experiments done in solution at 298 K and in solid matrix at 77 K. The photochemistry of carbazoles can be grouped according to different classes of photoreactions in which these compounds are involved: photoinduced electron transfer process (PET), photosensitized process and photoinduced rearrangement process. In this regard, the photoreactions of carbazoles are discussed from preparative and mechanistic viewpoints.

Chapter 11 - The Paternò-Büchi reaction on furan derivatives is reviewed. The photochemical reaction of furan derivatives in the reaction with carbonyl compounds gives the corresponding monoadduct with high regio- and stereoselectivity. When the reaction is carried out in benzene, a large amount of the diadduct can be obtained. The exo stereoselectivity at C-6 was explained assuming the presence of a secondary orbital effect on spin orbit coupling. Regio- and stereoselectivity of the reaction of carbonyl compounds with 2,3-dihydrofuran is described. Applications of the reaction to the synthesis of 2- and 3-furylmethanol derivatives is reported. The synthesis of asteltoxin, avenaciolide, and oxetanocin is also reported. Asymmetric reactions on chiral glyoxylate is reported. The low diastereoselectivity of the reaction with chiral ketones and aldehydes is discussed. Finally, the high diastereoselectivity observed in the reaction with 2- and 3-furylmethanol derivatives is reported.

Chapter 12 - Plant growth and development are dynamic processes which are continuously changed by environmental conditions. Nowadays, with the availability of performance instruments, many physiological parameters, which correlate with plant growth, development and economic yield under biotic and abiotic stresses, are exploited as indicators in plant breeding programs. Chlorophyll *a* fluorescence is a fast, non-invasive, non-destructive and informative tool to study physiological state of photosynthetic apparatus of any photosynthesizing material. It gives the ability to analyze the individual life processes and to develop an understanding of the plant's response to fluctuations in environmental conditions. In this chapter, the authors discuss responses of photosynthetic apparatus of barley plants growing under various abiotic stresses: high and low temperature, high and low light, salinity, heavy metals, nutrient deficiencies and drought. In addition, measurement of basic chlorophyll fluorescence and quantitative analysis of the fluorescence transient (JIP-test) are discussed. This test helps to explain the stepwise flow of energy through PSII: absorption, trapping and electron transport. This sort of analysis enables an understanding of the relationships between the biophysical side of photosynthesis, signals of the fluorescence and its analytical expressions. The results show the differentiation in extracted and technical fluorescence parameters, quantum efficiencies, and specific and phenomenological fluxes of barely plant growing under unfavorable growth conditions. This work proves that chlorophyll *a* fluorescence is one of the vital tools to monitor PSII reaction to different abiotic stresses. Moreover, it allows stress types that plants experience in certain environmental growth conditions to be distinguished. Different chlorophyll afluorescence parameters are discussed and proposed to be used as trustworthy indicators in selection for abiotic stress tolerance in barley germplasm.

Chapter 13 - Environmental systems models now simulate major element geocycles spanning the ocean and periodic table, but development is lagging for aqueous photochemistry mechanisms which interrelate biogeochemical tracers in most surface waters. This chapter reviews knowledge of critical open marine photochemical channels for the major atom types. Potential feedbacks to climate change are described, occurring initially via the generation of small radicals but leading to low molecular weight volatiles such as carbon monoxide, carbonyl sulfide, dimethyl sulfide and hydrocarbons. Examples are provided of box and one-dimensional models developed over the last decade or so which entail relatively complete energy transfer and reaction schemes. Only a handful of large-scale, three-dimensional approaches have appeared. The available global photochemical codes are surveyed, with emphasis on studies conducted recently in the oceanic component of the Community Climate System Model. Numerical and computational issues arising are enumerated in all cases. Coupling with atmospheric chemistry and aerosol-consistent radiative transfer must be achieved within the model frameworks. Stiff reaction systems will ultimately result as the pelagic diel cycle is resolved. Multiple scattering algorithms must be incorporated to augment purely absorptive schemes within the water column. The dissolved organic matter must be segregated into tracer bins of graded lability in order to deal with bleaching and aging. Aspects of biogeochemical complexity are mentioned to conclude. It is noted that the composition of photochemical source material implies a rich reaction web involving polymers, colloids, surfactants and trace metals. Complete mechanism development will thus be challenging.

Chapter 14 - Both various organic compounds (such as surfactants, industrial intermediates, and amino acids) and heavy-metal ions are characteristic pollutants of wastewaters, due to their extended industrial and domestic use. Heavy-metal ions and many organic contaminations are generally toxic, thus their reliable and cost-effective removal or degradation is the focus of numerous investigations. Photocatalytic procedures offer efficient solutions to this problem. One of the most promising directions of research in this area is based on photoactive semiconductors. Colloidal titanium dioxide proved to be applicable for both reductive deposition of metals and (even simultaneously) mineralization of organic compounds. In this chapter, the authors' survey and interpret the results achieved in photocatalytic procedures based mostly on their experiments with titanium dioxide suspensions.

Although heavy-metal ions are mostly toxic, the corresponding metals are generally precious because of their application in the industry and everyday life as well. Thus, photoreductive deposition of these metals such as mercury, silver, and copper is also important from economic points of view. Furthermore, in several cases, photocatalytic efficiency of colloidal titanium dioxide in mineralization of organic contaminants can be significantly enhanced by silver and copper deposition. The effects of such metal depositions on the mechanism of the photoinduced degradation of various organic compounds are discussed in details.

In addition, simultaneous photoreduction of metal ions can considerably increase the photocatalytic oxidative mineralization of organic contaminants such as surfactants, and vice versa. This synergism strongly depends on the salt content and pH of the systems due to the effects of surface charge of the colloidal photocatalyst particles as well as complex and ion-pair formations. Detailed interpretation of these effects are presented in this chapter for the preparation of practical application of this heterogeneous photocatalytic method in the treatment of various wastewaters.

Chapter 15 - Room temperature ionic liquids (RTILs) have recently emerged as one of the preferred "green" solvents. These liquids are composed of ions and are fluid at room temperature, but still can dissolve a wide range of organic molecules. The use of RTILs as novel media for various photophysical and photochemical studies is a topic of recent interest since the influence of the solvent in molecular organic photochemistry is well established. This chapter comments on some aspects of the photochemistry in RTILs, including solvent relaxation dynamics, electron and energy transfer kinetics, photoisomerization, solvatochromic absorption and fluorescence of dipolar molecules, unconventional wavelength-dependent fluorescence, empirical polarity parameters based on dye molecules, transparency of RTILs in the UV-visible region and the effect of viscosity and/or polarity of RTILs on the mechanism, dynamics (rate constant) and yield of photoinduced electron transfer (PET) processes.

Chapter 16 - Photoinitiated free radical polymerization is a very rapid procedure in which a photoinitiator decomposes and then produces free radicals very quickly under UV or visible light irradiation. The disadvantage of photopolymerization is that the light source is necessary for initiating and maintaining the reaction. For this reason, photopolymerization is generally used in photo-imaging or photolithography technology and surface application such as coating and printing, but not suitable for the preparation of materials. Recently, the authors have tried to introduce the procedure of photopolymerization to the preparation of polymeric microspheres. In this chapter, they report on the photoinitiated dispersion polymerization of methyl methacrylate which was carried out in a mixture of ethanol and water as dispersion medium in the presence of poly(N-vinylpyrrolidone) (PVP) as the steric stabilizer and Darocur 1173 as photoinitiator. Within 30 mins of UV irradiation at room temperature, 93.7% of conversion was achieved, and microspheres with 0.94μm number-average diameter and 1.04 polydispersity index (PDI) were obtained. Effects of reaction parameters on particle size and distribution of the resulted microspheres have been investigated. It is shown that by photoinitiated dispersion polymerization, polymeric microspheres with narrow size distribution can be prepared in a very short time, whereas around 20 hours is normally needed in a traditional thermal initiated dispersion polymerization.

In: Photochemistry Research Progress
Editors: A. Sánchez, S. J. Gutierrez

ISBN: 978-1-60456-568-3

Chapter 1

TWO BOTRYTICIDE EFFECTS ON GRAPEVINE LEAF PHOTOSYNTHESIS

Anne-Noëlle Petit, Florence Fontaine, Christophe Clément and Nathalie Vaillant-Gaveau *

Université de Reims Champagne-Ardenne, UFR Sciences, France

ABSTRACT

The effects of the two botryticides, fludioxonil (fdx) and fenhexamid (fhd), were investigated on grapevine leaves (Vitis vinifera L. cv. Pinot noir) following photosynthesis mechanisms. Treatments were carried out in vineyard at the end of flowering. Phytotoxicity of both fungicides was evaluated by measuring variations of leaf photosynthetic parameters and correlated expression of photosynthesis-related genes. Results demonstrated that similar perturbation of photosynthesis was caused by fdx and fhd applications. Moreover, the mechanism leading to photosynthesis alteration seems to be the same for both fungicides. Photosynthesis modification was mostly attributed to non-stomatal limitation since the reduction of photosynthesis was associated with few changes in intercellular CO_2 concentration despite stomatal conductance decline. Effective PSII quantum yield (ΦPSII) was not affected. Additionally, decreased photosynthesis was coupled with repression of rubisco small subunit (rbcS) and rubisco large subunit (rbcL) genes. To our knowledge, this is the first study of photosynthesis-related gene expression following fungicide stress.

Keywords: fungicide, gene expression, grapevine, photosynthesis

* Address correspondence to: E-mail: nathalie.vaillant-gaveau@univ-reims.fr

INTRODUCTION

Grey mould, caused by the fungus Botrytis cinerea, is a serious disease in grapevine (Vitis vinifera L.) culture, which affects both the quantity of harvest and the quality of wine produced from infected grapes (Bulit and Dubos, 1988). This disease is mainly controlled in vineyards using chemical fungicides. Three preventive applications are usually recommended: at the end of flowering (BBCH 69), at bunch closure (BBCH 77) and at the beginning of berry ripening (BBCH 81). Fludioxonil (fdx) and fenhexamid (fhd) are commonly used as botryticides (Rosslenbroich and Stuebler, 2000). Their modes of action are different suggesting that their by side effects on grapevine leaves may also be different. The phenylpyrrole fdx is a non-systemic molecule which inhibits spore germination, germ-tube elongation and mycelium growth of B. cinerea. It increases the glycerol content in the fungus, leading to a perturbation of the osmoregulation potential (Pillonel and Meyer, 1997). Fhd belongs to the new hydroxyanilide category. It is a locosystemic compound which inhibits germ-tube elongation and mycelial growth (Hänβler and Pontzen, 1999). Fhd is a sterol biosynthesis inhibitor (SBI) but its target site appears to be different from the ones of other widely used SBI fungicides. It inhibits the 3-keto reductase involved in the enzymatic complex of the sterol C-4 demethylation (Debieu et al., 2001).

In grapevine, studies of fdx and fhd mostly focused on their modes of action (Hänβler and Pontzen, 1999; Rosslenbroich and Stuebler, 2000), their residues in grapes and wine (Cabras et al., 1997, 2001; Teixeira et al., 2004; Likas et al., 2007) and the induced resistance of Botrytis populations to these fungicides (Leroux et al., 1999, 2002; Baroffio et al., 2003). Concerning their effects on grapevine physiology, impacts of fdx were already followed by leaf water content, carbohydrate and nitrogenous compounds levels, CO_2 assimilation, O_2 production and photosynthetic pigment contents (Saladin et al., 2003a, b). These studies revealed transient alterations of grapevine physiology after fdx application. Nevertheless, no information is available on the molecular mechanisms that can lead to physiological changes. Concerning fhd, no study has been performed on its by side effects on grapevine at both physiological and molecular levels.

Phytotoxicity effects of fungicides can be examined through leaf photosynthesis which is a key factor in carbon metabolism and which is involved in grapevine yield and vigour. Effects of fungicides on photosynthesis have been already revealed by modifications of photosynthetic activity, chlorophyll fluorescence, pigment contents or thylakoid function and integrity (Lidon et al., 1993; Krugh and Miles, 1996; Van Iersel and Bugbee, 1996; Benton and Cobb, 1997; Saladin et al., 2003b; Tort and Türkyilmaz, 2003; Alaoui-Sossé et al., 2004; Untiedt and Blanke, 2004; Xia et al., 2006). An alteration of photosynthesis may disturb the whole carbon balance of the plant affecting both restitution of reserves to storage organs (Liu and Huang, 2000; Petit et al., 2006) and carbon nutrition in annual leaves and inflorescences (Saladin et al., 2003a, b; Lebon et al., 2005). In response to stresses, plants can react in complex ways that are reflected by various metabolic responses. It includes both facultative metabolic adaptations that may afford stress protection, and metabolic impairments or injuries (Saladin and Clément, 2005). Thus, a decrease of photosynthesis may be a part of plant "adaptive strategy" to overcome environmental stresses (Chapin et al., 1993).

This work reports phytotoxic effects of both fdx and fhd on leaf photosynthesis of vineyard-grown plants (Vitis vinifera L. cv. Pinot noir) when applied at the end of flowering.

Different aspects of photosynthesis were analysed such as gas exchange and chlorophyll a fluorescence measurements. Indeed, fluorescence combined with gas exchange measurements allows to obtain a full picture of photosynthetic machinery plant response to environmental variations (Roháček, 2002). Additionally, chlorophyll a fluorescence is a sensitive and early detection tool of damage to photosynthesis apparatus and to the plant physiology resulting from environmental stresses (Maxwell and Johnson, 2000). The physiological approach was supplemented with the expression of photosynthesis-related genes.

MATERIAL AND METHODS

Plant material

Experiments were performed in 2005 and 2006 on Pinot Meunier grapevines (Vitis vinifera L.) planted in 1986, grafted on 41B rootstock, and trained according to the Chablis method. The experimental site was located in the CIVC vineyard in Loisy-en-Brie (France). The fungicides fdx and fhd (as the formulated products Geoxe® and Teldor®, respectively) were sprayed using the recommended concentration at the end of flowering (BBCH 69). Control plants in non-treated areas were chosen as perfectly healthy and received none botryticide application. Gas exchange and chlorophyll fluorescence measurements were performed on nine leaves of different plants and twelve leaves were used for real-time RT-PCR analysis. Sunexposed and fully expanded leaves were used for the measurements.

Leaf gas exchange

Net photosynthesis (Pn), stomatal conductance (g_s) and intercellular CO_2 concentration (Ci) were determined simultaneously on leaves with a portable infrared gas analysis system (Li-Cor Model 6400, Lincoln, NE). Measurements were conducted between 9 a.m. and 11 a.m. before the "midday depression" (Chaumont et al., 1994). The gas-exchange system was equipped with a clamp-on leaf cuvette that included 6 cm^2 of leaf area. Environmental conditions during photosynthetic measurements were the following: photosynthetically active radiation = 1500 $\mu mol\ m^{-2}s^{-1}$ provided by a red-blue light emitting diode, relative humidity around 30% and chamber temperature at 20°C. Carbon dioxide concentration was maintained at a constant level of 400 $\mu mol\ l^{-1}$ using a LI-6400-01 CO_2 injector with a high-pressure liquid CO_2 cartridge source. Gas exchange measurements were performed three times per leaf. The same leaves were used during all the kinetic corresponding to 2 days before fungicide application and 1, 2, 4 and 8 days after treatment.

Chlorophyll a fluorescence

Chlorophyll a fluorescence was evaluated following each gas exchange measurement using a pulse-modulated fluorometer (FMS2, Hansatech, King's Lynn, UK) at ambient CO_2 concentration and temperature. Chlorophyll a fluorescence reflects the functionality of the photosynthetic apparatus since it results from absorbed light energy that was not used for photosynthetic reactions and heat dissipation (Oxborough, 2004). The measuring probe of the fluorometer was placed in the adaxial side in the central part of the leaf. The effective PSII quantum yield (ΦPSII) was computed as (Fm' – Fs)/Fm' according to Genty et al. (1989) where Fm' is the maximal fluorescence and Fs is the steady-state fluorescence yield in the light-adapted state (Roháček, 2002). ΦPSII represented the number of electrons transported by a PSII reaction centre per mole of quanta absorbed by PSII. It measures the proportion of the light absorbed by chlorophyll associated with PSII that is used in photochemistry (Maxwell and Jonhson, 2000). Chlorophyll a fluorescence measurements were performed on the same leaves and according to the same kinetic than gas exchange.

RNA extraction

Leaves were collected one day after fungicide spraying. They were immediately frozen in liquid nitrogen then stored at -80°C. The leaves were ground in liquid nitrogen to a fine powder. One hundred mg of powder were used for total RNA extraction and homogenized in extraction buffer (Plant Purification RNA Reagent, Invitrogen, France), according to the manufacturer's instructions. The RNA pellet was resuspended in 20 µL of RNase-free water and quantified by absorbance at 260 nm.

Real-time RT-PCR analysis

Table 1. Genes analysed by real-time RT-PCR

Gene	Encoding	Pimer sequence	Accession numbe
EF1-α	elongation factor 1-alpha	Sense 5' GAA CTG GGT GCT TGA TAG GC Antisense 5' AAC CAA AAT ATC CGG AGT AAA AGA	BQ799343
Lhca3	light-harvesting chlorophyll-binding protein of photosystem I	Sense 5' GAC ATA CAA CTA CTG GGC AG Antisense 5' TGA GCT TCA GAT CCT TGA GG	AY194366
rbcS	small subunit of ribulose-1,5-bisphosphate carboxylase	Sense 5' GTG CAA TGC ATC GCT TTC ATT Antisense 5' TCC ACA AGG GTC CTA AAC ATG AG	CB340401
rbcL	large subunit of ribulose-1,5-bisphosphate carboxylase	Sense 5' AAT TTT TCC TCC ACG GCG ATA Antisense 5' ATC TGC GCC CGC CTT TAT A	AJ635355

One hundred and fifty ng of total RNA were reverse-transcribed using M-MLV reverse-transcriptase (Invitrogen, France) according to the manufacturer's protocol. PCR conditions were described in Bézier et al. (2002). The reaction was carried out in a GeneAmp 5700 sequence detection system (Applied Biosystems, France) using the following thermal profile: 15 s at 95°C (denaturation) and 1 min at 60°C (annealing/extension) for 40 cycles. PCR efficiency of the primer sets was calculated by performing real-time PCR on several dilutions. Results were standardized to the EF1α gene expression level (Terrier et al., 2005). Control sample (non-treated leaves) was chosen to represent 1x expression of the genes of interest and treated samples were expressed relative to the corresponding control. Expression of three photosynthesis-related genes was tracked (Table 1). The entire experiment, including both the

RT and real-time PCR steps, was repeated four times with RNA from different sets of samples.

Statistical analysis

To determine whether values of untreated plants were significantly different from treated plants, analysis of variance (ANOVA) followed by a Student's t test was used. Differences at $P < 0.05$ were considered as significant.

RESULTS

Leaf gas exchanges

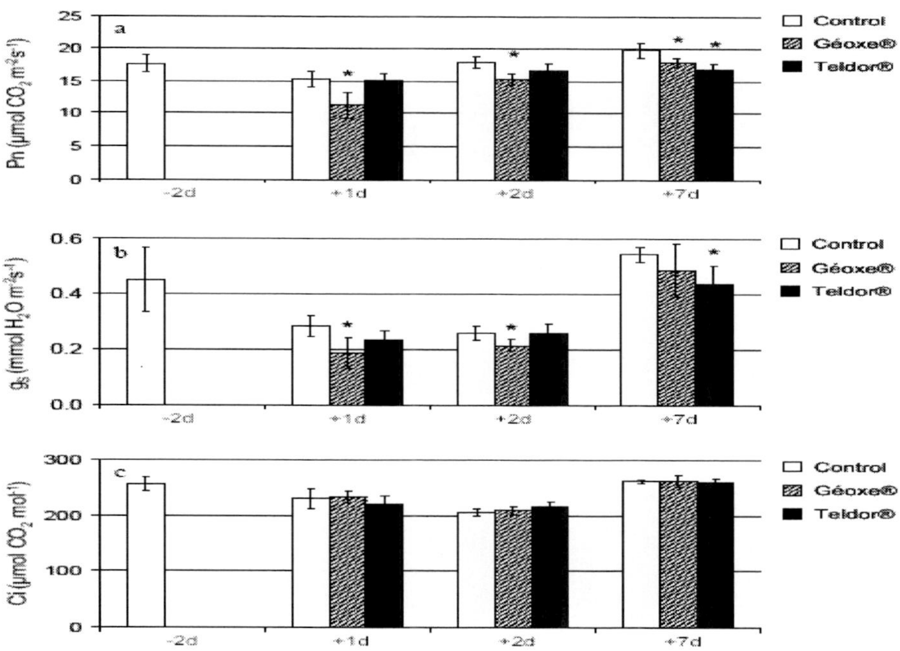

Figure 1. Changes in (a) net photosynthesis (Pn), (b) stomatal conductance (gs) and (c) intercellular CO2 concentration (Ci) in control and fdx-treated leaves of grapevine. Measurements were made 2 days before treatment (-2d) then 1 (+1d), 2 (+2d) and 7 (+7d) days after treatment Data are means ± standards errors (n=9). Significant differences at $P < 0.05$ between leaves of control and treated plants are marked by an asterisk.

From the first day to the second day after fungicide application, significant difference was noticed between control and fdx treated plants (Fig. 1a). Pn was inhibited after fdx spraying and remains stable after fhd spraying. Significant differences in Pn were registered 7 days after fhd and fdx spraying. The reduction represented 9.4±3.4% and 14.3±4.0% in fdx and fhd treatments, respectively. Regarding g_s, disruptions were also noticed: one and two day after treatment, fdx induced a similar reduction of g_s by 30.8 ± 19.9% and 16.6 ± 7.7%, respectively (Fig. 1b). The value of g_s decreased by about 21.8 ± 12.1% 7 days after fhd treatments while it remained stable after fdx treatment. Ci was no affected by fungicide treatment (Fig. 1c). Two botryticides induced inhibition of Pn accompanied by decreased g_s while Ci was unaffected. Pn reduction was mostly caused by non-stomatal factors.

Chlorophyll a fluorescence

In order to further investigate the non-stomatal factors involved in Pn decrease, ΦPSII was followed in parallel. It was not significantly modified after treatment (Fig. 2).

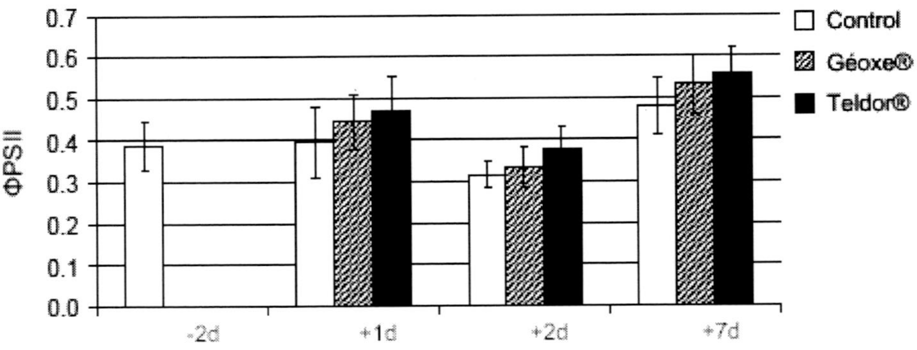

Figure 2. Changes in relative quantum yield of PSII (ΦPSII) in control and fdx-treated leaves of grapevine. Measurements were made 2 days before treatment (-2d) then 1 (+1d), 2 (+2d) and 7 (+7d) days after treatment Data are means ± standards errors (n=9). Significant differences at $P < 0.05$ between leaves of control and treated plants are marked by an asterisk.

Expression of photosynthesis-related genes

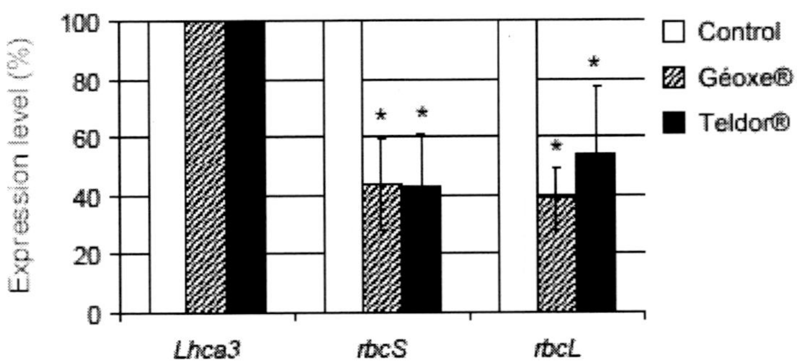

Figure 3. Expression level of Lhca3, rbcS and rbcL in control and fdx-treated leaves of grapevine, one day after treatment. Expression level for each analysed gene was fixed at 100%. Data are the mean ± standard error (n=3). Significant differences at $P < 0.05$ between leaves of control and treated plants are marked by an asterisk.

Real-time RT-PCR was used to determine changes in gene expression and thus to identify molecular mechanisms leading to photosynthesis modifications after botryticide treatments. A decline of expression was noticed for two photosynthesis-related genes whatever the fungicide sprayed (Fig. 3). Fdx and fdh treatment led to a 40-50% repression of rbcS and rbcL genes. Reduction of rbcS and rbcL expression were similar after fdx and fhd applications. Nevertheless, none fdx and fhd effect were noted with Lhca3 expression.

DISCUSSION

In leaves of Pinot Meunier, net photosynthesis was affected after both fdx and fhd treatments, as revealed by modifications of gas exchange. The time of inhibition was different with both fungicides. Photosynthesis disruption was detected after fdx spraying as early as one day after treatment and until the end of the experiment. After fhd spraying, Pn was reduced after 7 days. In grapevine plantlets grown in vitro, fdx has also been shown to cause a decrease of CO_2 fixation and photosynthetic pigment concentrations. On the contrary, fdx stimulated CO_2 assimilation and increased pigment concentration in vineyard (Saladin et al., 2003b). Nevertheless, most of other works reported an inhibitory effect of fungicides on plant photosynthesis. For example, photosynthesis decreased in cucumber plants after copper exposure (Alaoui-Sossé et al., 2004). Cu-stressed leaves accumulate carbohydrates which in turn induce a feedback inhibition of photosynthesis. Other study on cucumber plants treated with cuproxat or cyazofamid fungicides showed a reduction in net photosynthesis attributed to stomatal or non-stomatal factors, respectively (Xia et al., 2006). Indeed, decreased photosynthesis with curpoxat was accompanied by both decreased stomatal conductance and Ci, thus stomatal closure appeared to be the main factor to depressed photosynthesis. On the contrary, decreased photosynthesis with cyazofamid was associated with increased Ci, which means that cyazofamid-induced inhibition of photosynthesis was mostly attributed to non-stomatal factors. Benzimidazole fungicides also reduced net photosynthesis in bedding plants (Van Iersel and Bugbee, 1996). Similarly the application of the triazole fungicide, epoxiconazole, caused photosynthesis alteration of cleavers, partly due to the modification of phytosterol profiles and thylakoid function (Benton and Cobb, 1997). Gas exchange analysis indicates that photosynthesis disturbance was mostly generated by non-stomatal limitation with both fungicides as revealed by a g_s decrease associated without change in Ci. To further assess biochemical limitations to Pn, chlorophyll a fluorescence was also analysed after botryticide application. Results of this study show that both fdx and fhd not inhibit ΦPSII. Both fungicides did not modify PSII activity.

The expression of two essential photosynthesis-related genes coding proteins involved in rubisco small subunit (rbcS) and large submit (rbcL), was repressed one day after fdx and fhd treatments. These data suggest that photosynthesis was regulated transcriptionally in response to fungicide application. To our knowledge, it is the first time that the fungicide effects on photosynthesis-related gene expression were evaluated. Thus, fungicide treatments generated repression of genes encoding proteins, involved in different pathways of the photosynthetic process, which could lead to photosynthesis disruption. Rubisco is composed of 8 small and 8 large subunits that are respectively encoded by nuclear rbcS and chloroplastic rbcL genes. It has been shown that a reduction in the amount of rubisco may be caused by a decrease in the level of endogenous rbcS and rbcL transcripts (Jiang et al., 1994). In tobacco, a nuclear rbcS antisense mutant resulted in decreased photosynthetic activity (Zhang et al., 2002). Altogether, these results suggest that the repression of the rbcS and the rbcL genes after fdx or fhd treatments is directly involved in the reduction of the rubisco amount and the subsequent decrease of Pn.

Other environmental factors are known to disrupt photosynthesis and to decrease expression of rbcS and rbcL genes. For example, exposure to ozone (O_3), a stress known to speed up foliar senescence, reduces photosynthetic capacity and efficiency (Reich, 1983; Held et al., 1991; Nie et al., 1993; Guidi et al., 1997), accelerates loss of rubisco (Pell and Pearson, 1983; Nie et al., 1993; Pell et al., 1997), reduces transcript levels rbcS (Bahl and Kahl, 1995; Conklin and Last, 1995; Glick et al., 1995) and rbcL (Dizengremel, 2001; Pelloux et al., 2001). Similarly, ultraviolet-B radiation has been shown to impair the biochemical composition of the chloroplast (Bornman, 1989) and to reduce ΦPSII, mainly caused by disruptions in PSII function (Strid et al., 1994). The alteration of the photosynthetic apparatus following UV-B exposure is characterized by (i) a decrease in the content of rubisco which is a primary cause for the decline in photosynthetic rate (Allen et al., 1997) and (ii) a repression of rbcS gene expression (A-H-Mackerness et al., 1997). Finally, drought, cold or high-salinity stresses also downregulated rbcS gene (Seki et al., 2002). These various environmental stresses can affect the activities of antioxidative enzymes (Hernández et al., 1995; Zhang and Kirkham, 1995; Moran et al., 1994; Baier et al., 2005; Dai et al., 1997). Similarly, some fungicides stimulate the activity of antioxidative enzymes (Wu and von Tiedman, 2002; Garcia et al., 2003), suggesting that fdx and fhd may induce oxidative stress on grapevine leaves.

The Lhca3 protein is a member of the light-harvesting complex (LHC) with Lhca1, Lhca2 and Lhca4. These proteins are nuclear-encoded and related to light-harvesting cab binding proteins in the photosystem I (Haworth et al., 1983; Knoetzel et al., 1992; Green and Durnford 1996; Jansson, 1999). During photosynthesis, light energy is captured by pigments in the LHC and transferred to the reaction centres of the thylakoid membrane. The Lhca3 gene expression was not modified by fungicides, the light-harvesting in the photosystem I was intact and no leads to a decline of Pn.

Nevertheless, the downregulation of photosynthetic gene expression and the reduced rate of photosynthesis observed after fdx and fhd treatments may lead to a decrease in synthesized sugars. This may induce a perturbation in flower and berry development (Srinivasan and Mullins, 1981; Lebon et al., 2005).

Acknowledgments

This work was partly supported by Europol'Agro (Reims, France). We are grateful to CIVC for providing a vineyard for the experiments.

References

A-H-Mackerness, S., Thomas, B., and Jordan, B. R. (1997). The effect of supplementary ultraviolet-B radiation on mRNA transcripts, translation and stability of chloroplast proteins and pigment formation in Pisum sativum L. *Journal of Experimental Botany,* 48, 729–738.

Alaoui-Sossé, B., Genet, P., Vinit-Dunand, F., Toussaint, M., Epron, D., and Badot, P. M. (2004). Effect of copper on growth in cucumber plants (*Cucumis sativus*) and its relationships with carbohydrate accumulation and changes in ion contents. *Plant Science,* 166, 1213–1218.

Allen, D. J., McKee, I. F., Farage, P. K., Baker, N. R. (1997). Analysis of the limitation to CO_2 assimilation on the exposure of leaves of two *Brassica napus* cultivars to UV-B. *Plant Cell and Environment,* 20, 633–640.

Bahl, A., and Kahl, G. (1995). Air pollutant stress changes the steady-state transcript levels of three photosynthesis genes. *Environmental Pollution,* 88, 57-65.

Baier, M., Kandlbinder, A., Golldack, D., and Dietz, K. J. (2005). Oxidative stress and ozone: perception, signalling and response. *Plant, Cell and Environment,* 28, 1012-1020.

Baroffio, C. A., Siegfried, W., and Hilber, U. W. (2003). Long-term monitoring for resistance of *Botryotinia fuckeliana* to anilinopyrimidine, phenylpyrrole, and hydroxyanilide fungicides in Switzerland. *Plant Disease,* 87, 662-666.

Benton, J. M., and Cobb, A. H. (1997). The modification of phytosterol profiles and *in vitro* photosynthetic electron transport of *Galium aparine* L. (cleavers) treated with the fungicide, epoxiconazole. *Plant Growth Regulation,* 22, 93–100.

Bézier, A., Lambert, B., and Baillieul, F. (2002). Study of defense-related gene expression in grapevine leaves and berries infected with *Botrytis cinerea. European Journal of Plant Pathology,* 108, 111-120.

Bornman, J.F. (1989). Target sites of UV-B radiation in photosynthesis of higher plants. *Journal of Photochemistry and Photobiology,* 4, 145-158.

Bulit, J., and Dubos, B. (1988). Botrytis bunch rot and blight. In Pearson, R.C. and Goheen, A. C. (Eds.), *Compendium of grape diseases* (APS Press, pp. 13-14). Paul, Minnesota, USA.

Cabras, P., Angioni, A., Garau, V. L., Melis, M., Pirisi, F. M., Minelli, E. V., Cabitza, F., and Cubeddu, M. (1997). Fate of some new fungicides (cyprodinil, fludioxonil, pyrimethanil and tebuconazole) from vine to wine. *Journal of Agricultural and Food Chemistry,* 45, 2708-2710.

Cabras, P., Angioni, A., Garau, V. L., Pirisi, F. M., Cabitza, F., Pala, M., and Farris, G. A. (2001). Fenhexamid residues in grapes and wine. *Food additives and contaminants,* 18, 625-629.

Chapin, F. S. III., Autumn, K., and Pugnaire, F. (1993). Evolution of suites of traits in response to environmental stress. *American Naturalist,* 142, S78-S92.

Chaumont, M., Morot-Gaudry, J. F., and Foyer, C. (1994). Seasonal and diurnal changes in photosynthesis and carbon partitioning in *Vitis vinifera* leaves with and without fruit. *Journal of Experimental Botany,* 45, 1235-1243.

Conklin, P. L., and Last, R. L. (1995). Differential accumulation of antioxidant mRNAs in *Arabidopsis thaliana* exposed to ozone. *Plant Physiology,* 109, 203-212.

Dai, Q., Yan, B., Huang, S., Liu, X., Peng, S., Miranda, M. L. L., Chavez, A. Q., Vergara, B. S., and Olszyk, D. M. (1997). Response of oxidative stress defense systems in rice (*Oryza sativa*) leaves with supplemental UV-B radiation. *Physiologia Plantarum,* 101, 301-308.

Debieu, D., Bach, J., Hugon, M., Malosse, C., and Leroux, P. (2001). The hydroxyanilide fenhexamid, a new sterol biosynthesis inhibitor fungicide efficient against the plant pathogenic fungus *Botryotinia fuckeliana* (*Botrytis cinerea*). *Pest Management Science,* 57, 1060-1067.

Garcia, P. C., Rivero, R. M., Ruiz, J. M., and Romero, L. (2003). The role of fungicides in the physiology of higher plants: implications for defense responses. *The Botanical Review,* 69, 162–172.

Genty, B., Briantais, J. M., and Baker, N. R. (1989). The relationship between the quantum yield of photosynthetic electron transport and quenching of chlorophyll fluorescence. *Biochimica et Biophysica Acta,* 990, 87-92.

Glick, R. E., Schlagnhaufer, C. D., Arteca, R. N., and Pell, E. J. (1995). Ozone-induced ethylene emission accelerates the loss of ribulose-1,5-bisphosphate carboxylase/oxygenase and nuclear-encoded mRNAs in senescing potato leaves. *Plant Physiology,* 109, 891-898.

Green, B. R., and Durnford, D. G. (1996). The chlorophyll-carotenoid proteins of oxygenic photosynthesis. *Annual Review of Plant Physiology and Plant Molecular Biology,* 47, 685-714.

Guidi, L., Nali, C., Ciompi, S., Lorenzini, G., and Soldatini, G. F. (1997). The use of chlorophyll fluorescence and leaf gas exchange as methods for studying the different responses to ozone of two bean cultivars. *Journal of Experimental Botany,* 48, 173-179.

Hänβler, G., and Pontzen, R. (1999). Effect of fenhexamid on the developement of *Botrytis cinerea*. *Pflanzenschutz-Nachrichten Bayer,* 52, 158-176.

Haworth, P., Watson, J. L., and Arntzen, C. J. (1983). The detection, isolation and characterization of a light-harvesting complex which is specifically associated with photosystem I. *Biochimica et Biophysica Acta,* 724, 151–158.

Held, A. A., Mooney H. A., and Gorham J. N. (1991). Acclimation to ozone stress in radish: leaf demography and photosynthesis. *New Phytologist,* 118, 417-423.

Hernández, J. A., Olmos, E., Corpas, F. G., Sevilla, F., and del Rio, L. A. (1995). Salt-induced oxidative stress on chloroplasts of pea plants. *Plant Science,* 105, 151-167.

Jansson, S. (1999). A guide to the Lhc genes and their relatives in *Arabidopsis*. *Trends in Plant Science,* 4, 236–240.

Jiang, C. Z., Kliebenstein, D., Ke, N., and Rodermel, S. (1994). Destabilization of *rbcS* sense transcripts by antisense RNA. *Plant Molecular Biology,* 25, 569-576.

Knoetzel, J., Svendsen, I., and Simpson, D. J. (1992). Identification of the photosystem I antenna polypeptides in barley: isolation of three pigment-binding antenna complexes. *European Journal of Biochemistry,* 206, 209-215.

Krugh, B. W., and Miles, D. (1996). Monitoring the effects of five "nonherbicidal" pesticide chemicals of terrestrial plants using chlorophyll fluorescence. *Environmental Toxicology and Chemistry,* 15, 495–500.

Leroux, P., Chapeland, F., Desbrosses, D., and Gredt, M. (1999). Patterns of cross-resistance to fungicides in *Botryotinia fuckeliana* (*Botrytis cinerea*) isolates from French vineyards. *Crop Protection,* 18, 687-697.

Leroux, P., Fritz R., Debieu, D., Albertini, C., Lanen, C., Bach, J., Gredt, M., and Chapeland, F. (2002). Mechanisms of resistance to fungicides in field strains of *Botrytis cinerea*. *Pest Management Science,* 58, 876-888.

Lebon, G., Duchêne, E., Brun, O., and Clément, C. (2005). Phenology of flowering and starch accumulation in grape (*Vitis vinifera* L.) cuttings and vines. *Annals of Botany,* 95, 943-948.

Lidon, F. C., Ramalho, J. C., and Henriques, F. S. (1993). Copper inhibition of rice photosynthesis. *Journal of Plant Physiology,* 142, 12-17.

Likas, D. T., Tsiropoulos, N. G., and Miliadis, G. E. (2007). Rapid gas chromatographic method for the determination of famoxadone, trifloxystrobin and fenhexamid residues in tomato, grape and wine samples. *Journal of Chromatography A,* 1150, 208-214.

Liu, C., and Huang, B. (2000). Carbohydrate accumulation in relation to stress tolerance in two creeping bentgrass cultivars. *Journal of the American Society for Horticultural Science,* 125, 442-447.

Maxwell, K., and Johnson, GN. (2000). Chlorophyll fluorescence: a practical guide. *Journal of Experimental Botany,* 51, 659-668.

Moran, J. F., Becana, M., Iturbe-Ormaetxe, I., Frechilla, S., Klukas, R. V., and Aparicio-Tejo, P. (1994). Drought induces oxidative stress in pea plants. *Planta,* 194, 346-352.

Nie, G. Y., Tomasevic, M., and Baker, N. R. (1993). Effects of ozone on the photosynthetic apparatus and leaf proteins during leaf development in wheat. *Plant, Cell and Environment,* 16, 643-651.

Oxborough, K. (2004). Imaging of chlorophyll *a* fluorescence: theoretical and practical aspects of an emerging technique for the monitoring of photosynthetic performance. *Journal of Experimental Botany*, 55, 1195-1205.

Pell, E. J., and Pearson N. S. (1983). Ozone-induced reduction in quantity of ribulose-1,5-bisphosphate carboxylase in alfalfa foliage. *Plant Physiology,* 73, 185-187.

Pell, E.J., Schlagnhaufer, C. D., and Arteca, R. N. (1997). Ozone-induced oxidative stress: mechanisms of action and reaction. *Physiologia Plantarum,* 100, 264-273.

Petit, A. N., Vaillant, N., Boulay, M., Clément, C., Fontaine, F. (2006). Alteration of photosynthesis in grapevines affected by esca. *Phytopathology,* 96, 1060-1066.

Pillonel, C., and Meyer, T. (1997). Effect of phenylpyrroles on glycerol accumulation and protein kinase activity of *Neurospora crassa*. *Pesticide Science,* 49, 229-236.

Reich, P. B. (1983). Effects of low concentrations of O_3 on net photosynthesis, dark respiration, and chlorophyll contents in aging hybrid poplar leaves. *Plant Physiology,* 73, 291-296.

Roháček, K. (2002). Chlorophyll fluorescence parameters: the definitions, photosynthetic meaning, and mutual relationships. *Photosynthetica,* 40, 13-29.

Rosslenbroich, H. J., and Stuebler, D. (2000). *Botrytis cinerea* – history of chemical control and novel fungicides for its management. *Crop Protection,* 19, 557-561.

Saladin, G., Magné, C., and Clément, C. (2003)a. Physiological stress responses of *Vitis vinifera* L. to the fungicides fludioxonil and pyrimethanil. *Pesticide Biochemistry and Physiology,* 77, 125-137.

Saladin, G., Magné, C., and Clément, C. (2003)b. Effects of fludioxonil and pyrimethanil, two fungicides used against *Botrytis cinerea*, on carbohydrate physiology in *Vitis vinifera* L. *Pest Management Science,* 59, 1083-1092.

Saladin, G., and Clément, C. (2005). Physiological effects of pesticides on cultivated crops. In J. V. Livingston, *Agriculture and Soil Pollution: New Research* (Nova Science Publishers, pp. 53-86). USA.

Seki, M., Narusaka, M., Ishida, J., Nanjo, T., Fujita, M., Oono, Y., Kamiya, A., Nakajima, M., Enju, A., Sakurai, T., Satou, M, Akiyama, K., Taji, T., Yamaguchi-Shinozaki, K., Carninci, P., Kawai, J., Hayashizaki, Y., and Shinozaki, K. (2002). Monitoring the expression profiles of 7000 *arabidopsis* genes under drought, cold and high-salinity stresses using a full-length cdna microarray. *The plant journal,* 31, 279–292.

Srinivasan, C., and Mullins, M. G. (1981). Physiology of flowering in the grapevine: a review. *American Journal of Enology and Viticulture,* 32, 47-63.

Strid, A., Chow, W. S., and Anderson, J. M. (1994). UV-B damage and protection at the molecular level in plants. *Photosynthesis Research,* 39, 475–489.

Teixeira, M. J., Aguiar, A., Afonso, C. M. M., Alves, A., and Bastos, M. M. S. M. (2004). Comparison of pesticides levels in grape skin and in the whole grape by a new liquid chromatographic multiresidue methodology. *Analytica Chimica Acta,* 513, 333-340.

Terrier, N., Glissant, D., Grimplet, J., Barrieu, F., Abbal, P., Couture, C., Ageorges, A., Atanassova, R., Léon, C., Renaudin, J. P., Dédaldéchamp, F., Romieu, C., Delrot, S., and Hamdi, S. (2005). Isogene specific oligo arrays reveal multifaceted changes in gene expression during grape berry (*Vitis vinifera* L.) development. *Planta,* 222, 832–847.

Tort, N., and Türkyilmaz, B. (2003). Physiological effects of captan fungicide on pepper (*Capsicum annuum* L.). *Plant Pakistan Journal of Biological Sciences,* 6, 2026-2029.

Untiedt, R., and Blanke, M. M. (2004). Effects of fungicide and insecticide mixtures on apple tree canopy photosynthesis, dark respiration and carbon economy. *Crop Protection,* 23, 1001-1006.

Van Iersel, M. W., and Bugbee, B. (1996). Phytotoxic effects of benzimidazole fungicides on bedding plants. *Journal of the American Society for Horticultural Science,* 121, 1095-1102.

Wu, Y. X., and von Tiedman, A. (2002). Impact of fungicides on active oxygen species and antioxidant enzymes in spring barley (*Hordeum vulgare* L.) exposed to ozone. *Environmental Pollution,* 166, 37-47.

Xia, X. J., Huang, Y. Y., Wang, L., Huang, L. F., Yu, Y. L., Zhou, Y. H., and Yu, J. Q. (2006). Pesticides-induced depression of photosynthesis was alleviated by 24-epibrassinolide pretreatment in *Cucumis sativus* L. *Pesticide biochemistry and physiology,* 86, 42-48.

Zhang, J., and Kirkham, M. B. (1995). Enzymatic responses of the ascorbate-glutathione cycle to drought in the sorghum and sunflower plants. *Plant Science,* 113, 139-146.

Zhang, X. H., Ewy, R. G., Widholm, J. M., Portis, A. R. Jr. (2002). Complementation of the nuclear antisense *rbcS*-induced photosynthesis deficiency by introducing an *rbcS* gene into the tobacco plastid genome. *Plant and Cell Physiology,* 43, 1302-1313.

In: Photochemistry Research Progress
Editors: A. Sánchez, S. J. Gutierrez

ISBN: 978-1-60456-568-3

Chapter 2

THE INTERDISCIPLINARY NATURE OF CHEMICAL ENGINEERING RESEARCH

Donia Friedmann
Department of Physics and Astronomy, University of Aarhus, Denmark

Chemical engineering research is interdisciplinary by nature. This certainly holds true from the author's own experience in basic and applied research of photocatalytic processes, particularly TiO_2 photocatalysis for water treatment applications. Semiconductor photocatalysis involves the interaction between light (having sufficient energy) and a semiconductor photocatalyst, with the subsequent activation of the photocatalyst to generate highly active species including hydroxyl radicals and conduction band electrons. These active species can be targeted to destroy organic and inorganic pollutants in photocatalytic reactions. Titanium dioxide (TiO_2) is the most commonly used photocatalyst. This is a common material and is most encountered in its inactive form as a pigment.

The advances in photocatalytic technologies have been significant in recent years. Applications include environmental technologies for air and water purification and architectural applications such as self-cleaning surfaces and anti-bacterial coatings. Additionally, photocatalysis has been applied in the biomedical field, for example, in the use of self-decontaminating surfaces for medical devices. Photocatalytic systems are complex, with fundamental processes of photocatalysis encompassing a number of disciplines including catalysis, photochemistry, organic and inorganic chemistry, materials science, surface science and reaction engineering, to name a few. While many of its fundamental processes have been clarified, the complexity of photocatalysis is yet to allow for definite conclusions on some important aspects. For instance, further clarification is still needed on the mechanistic role of the different photogenerated active species and the influence of parameters such as the crystal phase of TiO_2, particle size, defects and impurities.

This chapter covers the basic principles of photocatalysis and gives an overview of developments of applications and trends. It also highlights the interdisciplinary nature of photocatalysis research by discussing contributions and progress made from the various research fields. Emphasis is placed on the author's research work on the design of photocatalysts of improved efficiencies and analysis of the intimate relationship between structure, electronic properties and reactivity of titanium dioxide, for application in photocatalytic water treatment processes. Conclusions are made on the present state of photocatalysis with recommendations for further key research into this very interesting and promising technology.

PHOTOCATALYSIS: BACKGROUND AND HISTORICAL DEVELOPMENTS

Over the past 30 years, photocatalysis has emerged as a valuable process to explore, with a wide range of applications. The commercialization of TiO_2 based products began in the mid 1990s with Japanese companies leading the way. Applications of photocatalysis can be classed according to one of three modes of action; photominerlization, photo-sterilization and photoinduced superhydrophylicity, and can be found in exterior construction materials, interior furnishing materials, road construction, purification facilities and household goods (Mills et al.., 2002). Specific examples include air conditioning units, defogging mirrors in cars, TiO_2 coatings on street lights and roads, and even a mosquito trap.

Heterogeneous photocatalysis is a complex process, and as such, it has drawn the interest of researchers from an array of different research disciplines. From the theoretical studies to the practical applications, key contributors to this research field include Akira Fujishima, Detlef Bahnemann, David Ollis, Pierre Pichat, Nick Serpone, Ralph Mathews, Michael Henderson and Ulrich Diebold, whose backgrounds span chemical engineering, organic chemistry, environmental sciences, semiconductor physics, photochemistry, and materials and surface sciences.

Research in photocatalysis to promote redox reactions was initiated in the field of semiconductor photoelectrochemistry of interest to solar conversion (Kalyanasundaram, 1983, Carey et al., 1976). Initial studies began on the photocurrent response of ZnO semiconductor electrodes. Around this time, Frank and Bard were to first to explore applying TiO_2 photocatalysis to problems of environmental interest by enhancing the solar driven water purification process specifically to treat water contaminated with cyanide and sulfite (Frank et al., 1977). The majority of the early works into semiconductor photocatalysis in fact focused on the removal of organic pollutants from water (Mills et al., 2002). The heterogeneous photocatalytic treatment approach has been shown to be particularly suited for the treatment of wastewaters contaminated with low levels of pollutants (typically less than 50ppm). Applications may be in the form of tertiary treatment at water treatment facilities and the remediation of polluted surface and ground waters. Up until now, photocatalytic water treatment applications have been limited. Other commercial potential lies in the designing and manufacturing of portable water purifiers for the medical and electronic industries. In more recent years, greater interest has been in the removal of organic pollutants from the gas phase (Mills et al., 2002), with commercial success. Target contaminants are typically volatile organic contaminants (VOCs) where the process is applied to destroy nuisance odours. Currently, there is also a strong research momentum in the photo-induced hydrophilicity of TiO_2. TiO_2 surface coatings for self-cleaning purposes particularly for architectural applications already have an established market with tremendous commercial potential.

Similarly with anti-bacterial applications which combine both the surface photoreactivity and photoinduced hydrophilicity of TiO_2.

THE PHOTOCATALYTIC PROCESS

Describing it simply, a photocatalytic process involves a semiconductor, typically TiO_2, being activated when it absorbs ultraviolet light. The result is the production of strongly oxidizing hydroxyl radicals (and other oxidizing species). OH radicals are the second best oxidizing agents after fluorine.These can decompose a wide range of organic compounds. At the same time, conduction band electrons are also produced, these can reduce polluting species such as some heavy metals. These reactions occur on or in close proximity to the surface of the photocatalyst. A heterogeneous photocatalyzed process is described by reaction 1 (Serpone et al., 2005):

$$A + Cat + hv\ B + Cat \quad (1)$$

Semiconductors such as TiO_2 have a filled valence band (VB) and an empty conduction band (CB), as shown in Figure 1. During the photocatalytic process, upon illumination, photons of the required energy (wavelength) that hit an electron in the occupied valence band of the semiconductor atom, can elevate that electron to the unoccupied conduction band leading to excited state conduction band electrons and positive valence band holes. These charge carriers can get trapped in the semiconductor, they can recombine or they can take part in various redox reactions.

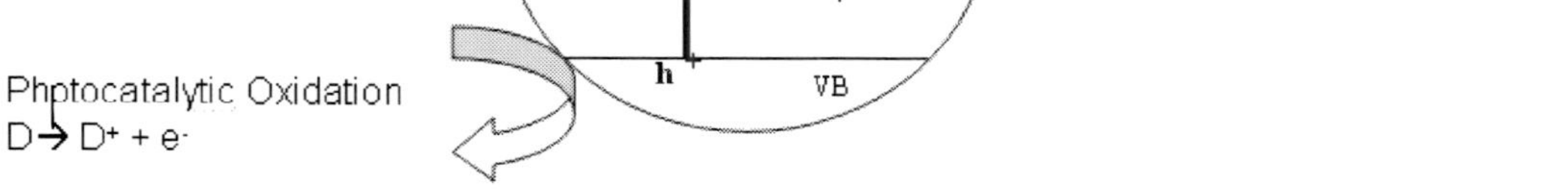

Figure 1. Charge transfer processes in a semiconductor particle upon bandgap excitation. hυ: energy of photons, CB: cband, VB: valence band, rec: recombination, D: electron donor; A: electron acceptor.

The heterogeneous photocatalytic process can also be expressed as a series of reactions. The main reactions are listed below, refer to Litter et al. (Litter et al., 1999) for a complete list of these reactions. Reaction (2) shows the formation of the charge carriers upon the illumination of the semiconductor. The hydroxyl radicals are formed by the reaction between the photogenerated holes and adsorbed water or surface hydroxyls (reactions 3 and 4). Superoxide ions are formed when the photoelectrons react with the dissolved oxygen (reaction 5). Electron donors (D) are oxidized by the hydroxyl radicals (reaction 6), they can also directly be oxidized by the holes (reaction 7). The electrons can directly react with electron acceptors (A) (reaction 8).

$TiO_2\ TiO_2\ (e^-_{cb}, h^+_{vb})$ (2)

$TiO_2\ (h^+_{vb}) + H_2O_{ads}\ TiO_2 + HO_{ads} + H^+$ (3)

$TiO_2\ (h^+_{vb}) + HO^-_{ads}\ TiO_2 + HO_{ads}$ (4)

$TiO_2\ (e^-_{cb}) + O_{2ads} + H^+\ TiO_2 + HO_2\ O^-_2 + H^+$ (5)

$HO_{ads} + D_{ads}\ D_{oxid}$ (6)

$TiO_2\ (h^+_{vb}) + D_{ads}\ TiO_2 + D^+_{ads}$ (7)

$TiO_2\ (e^-_{cb}) + A_{ads}\ TiO_2 + A^-_{ads}$ (8)

The overall efficiency of heterogeneous photocatalysis is determined by the population and lifetime of the photogenerated charge carriers and the time scale of interfacial charge transfer processes (Colombo et al., 1996). The efficiency can be increased by improvements in charge separation, inhibition of charge carrier recombination, and improvements in interfacial charge transfer (Bedja et al., 1995).

Photocatalysis for Water Treatment

One of the major advantages of the photocatalytic water treatment process over existing technologies is that there is no further requirement for secondary disposal methods. Photocatalysis competes well with adsorption by activated carbon and air stripping, which merely concentrate the chemicals present, by transferring them to the adsorbent or air, but they do not convert them into non-toxic wastes. Another advantage of photocatalytic water treatment is that when compared to other advanced oxidation technologies, especially those using oxidants such as hydrogen peroxide and ozone, expensive oxidising chemicals are not required as ambient oxygen is the oxidant (Matthews, 1993).

Applications of photocatalysis for water treatment are dependent on the quality and quantity of the water to be treated (Agrios et al., 2005). Some of the organic compounds which have been photocatalytically destroyed in laboratories and field studies include alcohols, aromatic hydrocarbons, chlorinated aromatic compounds, chlorinated alkanes, polychlorinated biphenyls, carboxylic acids, amines, and aldehydes. More recently, there has been interest in degrading chlorine disinfecting byproducts, endocrine disruptors (Auriol et al., 2006) and particularly biocides which cannot be treated by conventional biological treatment methods (Zalazar et al., 2005). Photocatalysis is well suited here due to the typical low concentrations of such pollutants. Another interesting application of photocatalysis is the recovery of precious metals such as platinum, palladium, gold, rhodium, and silver. Photocatalysis can also be used to eliminate toxic metals such as lead, mercury, and chromium from industrial waste effluents (Khalil et al., 1998, Herrmann, 2005)

Photocatalysis for Air Purification

For air purification, photocatalysis is again suited for low pollutant concentrations (parts per million levels). This is why this technology has seen commercial success in air conditioning units for the purification of indoor air containing volatile organic contaminants (VOCs) (Mills et al., 2002). For the purification of outdoor air and industrial air effluents, the technology is suitable only when the pollutant levels are low (Fujishima et al., 2006). Large air purification systems have been developed for the treatment of industrial sources of VOCs including the off-gas emissions from soil vapor extraction, groundwater air stripping and process air (Mills et al., 2002). TiO_2 photocatalysis can also be used to reduce nitrogen oxide levels in outdoor air by incorporating the photoactive TiO_2 into building materials.

Photocatalytic Self-Cleaning and Anti-Bacterial Properties

Research into photocatalysis for self-cleaning and anti-bacterial applications is another strong research area. Self-cleaning applications such as self-cleaning tiles, windows, and antifungal coatings rely on the photooxidation properties of TiO_2. Commercial success of such applications has continued since anti-bacterial tiles were first developed by the company TOTO in 1994 (Baba et al., 2005). Research into self-cleaning surfaces has shown that organics such as octadecane, glycerol trioleate and polyethylene glycol could be decomposed at significant rates under weak UV illumination (1 mW cm^{-2}) (Fujishima et al., 2006). Such surfaces have prolonged cleanliness since soot and dust particles are less likely to collect. The phenomenon of super-hydrophilicity of these surfaces makes them less wettable upon UV exposure and contributes to their self-cleaning properties. This will be discussed in the following section.

In the last decade semiconductor photocatalysis has also been widely utilized as a self-sterilizing material in biomedical applications since it has the ability to destroy bacteria, viruses and moulds (Mills et al., 2002). The microbial destruction achieved takes place through direct damage to cell walls caused by the hydroxyl radicals. Applications include coating clinical tools such as sanitary ware and items for use in hospitals. Currently, there is also interest in food packaging applications (Maneerat et al., 2006). The toxicity of nanoparticles is relevant for such applications and needs to be addressed. Other issues to consider when applying self-cleaning and anti-bacterial coatings are that abrasive cleaners cannot be used since they will damage the photocatalytically active coatings (Mills et al., 2002), additionally, the effectiveness of these coatings will depend on the amount of sunlight, rainfall, accumulation rate of soilage (Fujishima et al., 2006).

Another recent development are self-cleaning cotton textiles in which TiO_2 nano-clusters are deposited into the textile surface (Yuranova et al., 2006). This imparts anti-bacterial, self-cleaning and deodorizing properties aimed at prolonging the lifetime of the textile. The Lion corporation have a patent in the treatment of textile fabrics (Mills et al., 2002).

UV Induced TiO_2 Super-Hydrophilicity

Hydrophilicity has been of great interest to the TiO_2 research community since the early 1990s. In fact this process has been said to have had the biggest impact of semiconductor photochemistry due to its commercial success as anti-bacterial tiles and self-cleaning surfaces (Mills et al., 2002). When in contact with these surfaces, water forms a film rather than droplets. This was discovered by Fujishima and Hashimoto who were the first to coat TiO_2 photocatalysts on solids. TOTO developed self-cleaning tiles in 1997 (Baba et al., 2005). A consequence of super-hydrophilicity is that these materials do not fog. For the automotive industry, this process has been applied as antifogging glass and mirrors (Mills et al., 2002). At present, there exists a debate as to what causes the super-hydrophilicity of UV activated TiO_2 surfaces. According to the surface studies by Michael Henderson and his group, titania surfaces are rendered hydrophilic under UV illumination due to the degradation and removal of adsorbed hydrophobic organic contaminant molecules (White et al., 2003). Takeuchi et al. studied TiO_2 surfaces during UV light irradiation by near-infrared (NIR) absorption spectroscopy and their results showed that the time scale for the UV induced hydrophilicity was in good agreement with the decrease in the amount of H_2O molecules adsorbed on the TiO_2 surfaces but not the amount of the hydrocarbons eliminated by the photocatalytic oxidation reactions (Takeuchi et al., 2005). According to Fujishima and Zhang, hydrophilicity is related to the density of surface hydroxyl groups (Fujishima et al., 2006). It is clear that a complete understanding of the mechanism of hydrophilicity is significant for current applications and for the development of materials having a more stable hydrophilicity in the dark (Agrios et al., 2005).

The Photocatalytic Water Treatment Process

A typical photocatalytic water treatment process is highly complex with many factors affecting the photocatalytic reaction rates. A distinction can be made between extrinsic parameters and intrinsic parameters. Extrinsic parameters are external factors such as operating conditions. These include pH of the solution, initial concentration of the organic compound, light intensity (photonic flux), catalyst dosage, temperature, circulating flow rate, oxygen flow rate and presence of ions in solution (Litter, 1999, Basca et al., 1998, Surender et al., 1998, and Xu et al., 1999). Pre-treatment to remove inorganic ions, such as chemical precipitation, is usually required as these can hinder the photocatalytic reactions.

Intrinsic parameters are those related to the properties of the photocatalyst itself. Many studies have been carried out to investigate the link between the photocatalytic performance of titanium dioxide particles and their characteristics, both physical and chemical. Such studies have confirmed that the activity is highly influenced by a number of intrinsic features of the catalyst. As a heterogeneous catalyst reaction, a photoreaction has been described as being a structure sensitive reaction which is closely related to the crystallographic phase, the inter-atomic spacing and the crystallite size of the catalytic material (Xu et al., 1999, and Ying et al., 1997). Other reports have confirmed the dependence on the surface properties including the number of active sites, surface area, particle size, mesoporosity as well as the presence of surface acid or basic groups (Basca et al., 1998, Xu et al., 1999)

Effects Of Extrinsic Parameters on Photocatalytic Activity

Light Intensity

Light intensity (I) has been reported to be a very important reactor design parameter when considering light sources. This is due to the high cost of solar concentrators and artificial illumination (Prairie et al., 1993). Degradation rates have been found to have a non-linear dependence on the light intensity. It has also been shown that an excess of light promotes a faster electron-hole recombination (Litter, 1999) and it has long been accepted that degradation rates are proportional to $I^{0.5}$ at high intensity and $I^{1.0}$ at low intensity respectively (Chen et al., 1999). Recently, Serpone and Emeline (Serpone et al., 2005) discussed that the square root dependence of the reaction rate on photon flow arising from bimolecular recombination of the free charge carriers is a misconception often considered in heterogeneous photocatalysis since this pathway would require that the electrons and holes involved have the same momentum. The data presented by Emeline et al. showed the interdependence of reaction rates on photon flow and on reactant concentration (Emeline et al., 2005). According to this study, depending on the concentration of the reactant molecules, the dependence of the reaction rate on light intensity varied from zero-order kinetics at low concentrations to first order kinetics at high concentrations. The square root dependence of reaction rates on photon flow is simply a particular case lying within this range.

Reaction Temperature

The overall process of photocatalytic degradation has been reported as not being very temperature sensitive. The low activation energy of photocatalytic reactions (5-20 kJ/mol) compared with the ordinary thermal reactions was said to be the main reason for this (Chen et al., 1999). Hence, photocatalytic systems do not require heating and are operated at room temperature. The optimum temperature is generally between 20 and 80C (Herrmann, 2005).

Solution pH

The pH of the solution has been found to have a number of influences on the photocatalytic degradation rates. Theses can be interpreted in terms of electrostatic interactions between the charged TiO_2 particles and the contaminants to be degraded. This in turn affects adsorption processes, important since photocatalytic reactions are believed to be surface reactions (Litter, 1999). A variation in pH also affects surface speciation of TiO_2 due to its amphoteric nature (Kumar et al., 1997). Serpone et al. (Serpone et al., 2005) showed that the TiO_2 surface charge which is a function of pH is also a cause for the flow of carriers towards or away from the surface. The surface charge was said to be closely connected to the surface structure, defectiveness, disorder, adsorbed impurities and H^+/OH^- exchange with the surrounding aqueous phase, among others (Serpone et al., 2005) .

The pH also affects the position of the flatband potential of the semiconductor in solution. The flat band potential follows a Nernstian pH dependence and decreases by 59 mV per pH unit. Thus, this affects the ability of electrons and holes to take part in subsequent redox reactions (Litter, 1999). In slurry-type photoreactors the pH has also been reported to have an influence on the dispersion stability, and therefore the observed photodegradation rates (Dyk et al., 1998).

Catalyst Loading

The required mass of the photocatalyst is dependent on the geometry and working conditions of the photoreactor (Hermann, 2005). In slurry-type systems, catalyst loading is an important design parameter for effective use of the reactor space and photocatalyst (Chen et al., 1999). Typically, there is an optimum catalyst loading which differs depending on reactor configurations. The rate initially increases with an increase in the catalyst loading, as this translates into an increase in active sites. However, as the catalyst loading is increased beyond an optimum concentration, the increase in the number of active sites is offset by a decrease in the penetration depth of the UV radiation.

Effects of Intrinsic Parameters on Photocatalytic Activity

Crystalline Structure

It is generally accepted that the anatase crystal phase of TiO_2 is more active than the rutile phase. Although a debate still exists as to why this is the case. The rutile and anatase structure have many similarities. They both have a tetragonal structure although they in the number of terminal TiO_6 octahedrons. The mass density of the anatase phase is lower, and the bandgap of anatase is 3.2 eV while that of rutile is 3.0 eV. This could affect their photocatalytic activity.

Some reports have proposed the degree of hydroxylation as being one of the most important characteristics which gives anatase the edge in terms of the photocatalytic activity as compared to rutile (Surender et al., 1998). This study also pointed to the facilitated adsorption of oxygen as O^{2-} or O^{-} forms and hence a slower electron-hole recombination rate (interpreted from photo-conductance measurements) as possible explanations for the higher activity of anatase with respect to rutile. The typically higher surface area of anatase, due to a less intense heat treatment step during its synthesis, was also said to be a contributing factor for its higher activity when compared with rutile.

More recently, there have been discussions that a pure anatase sample might not necessarily lead to the best photocatalytic performance, with studies suggesting that a mixture of rutile and anatase might be the best combination. Basca et al. found that the presence of the rutile phase introduced mesoporosity and a wider pore size distribution into the sample (Basca et al., 1998). In this study, Basca et al. also suggested an appropriate crystal plane on which adsorption takes place, might be responsible for the enhancement effect on the photocatalytic activity observed upon combining both the anatase and rutile phases. Additionally, charge transfer at the interface between the anatase and rutile crystallites has been said to lead to improved charge carrier separation (Miyagi et al., 2004). Finally, the difference in photoactivity could be dependent on the type of reactants and reaction conditions, where anatase could be more active than rutile for O_2 photooxidation, but not necessarily for all photocatalytic processes (Diebold et al., 2003). Further investigations are required on the relationship between photocatalytic performance and phase composition.

Surface Area

From a catalysis point of view, a high surface area is an advantage in terms of a greater concentration of active sites per square metre, and hence generally leads to higher reactivity. However, it has often been reported that while the activity of the TiO_2 catalyst is affected by its external surface, photocatalytic activity is not a function of the surface area alone (Basca et al., 1998, Surender et al., 1998 and Kato, 1992). For example, while an increase in the extent of heat treatment applied leads to a reduction in the surface area, the impact of this on the photocatalytic activity was said not to be as significant as that of the phase transformation. The concomitant increase in crystallite size and changes in the microstructure with the decrease in surface area was said to greatly influence the photocatalytic properties of the catalyst (Kato, 1992). The effect of particle size on the photocatalytic activity can be interpreted in terms of its effect on the surface area. All things being equal, the smaller the particle size, the larger the surface area, and the higher the expected activity, however, if the particle size falls below a certain value (10 nm), other issues will need to be considered with the onset of quantum size effects (Bahnemann, 1993).

Reaction Engineering

Reactor Design

Photoreactors can take various forms. For water treatment, they can generally be classified as slurry or immobilized-catalyst photoreactors. In slurry photoreactors, the catalyst particles are suspended in the contaminated water to be treated. These can be magnetically stirred or mechanically agitated. In the immobilized-catalyst photoreactors, such as coated tubular photoreactors, annular and spiral photoreactors, where the catalyst particles are fixed onto the surface of various inert substrates. The inert substrates may be the walls of the photoreactor, membranes, glass beads, or glass-wool sleeves (Herrmann, 2005). While immobilized-catalyst photoreactors do provide a solution for separating the photocatalyst from the water after treatment, slurry photoreactors offer significant advantages in terms of the catalyst surface availability, and superior mass transfer properties (Zeltner et al., 1996, Litter, 1999). At lower solute concentrations, the photocatalytic rates in immobilized-catalyst photoreactors could be half of the rates obtained for slurry systems. Furthermore, the reactor design of immobilized systems may be limited by optical constraints (unless the substrate is UV transparent). Immobilized phase systems also have the disadvantage of the need for recoating the surface after prolonged use, due to the erosion of the TiO_2 from the surface of the support. Realistically, to treat a large amount of contaminated water, a slurry photoreactor is the preferred option.

There are a few studies concerning the separation of the TiO_2 photocatalyst from slurry photoreactors following water decontamination. One study investigated charge neutralization, coagulation and flocculation as a basis for the enhanced settling and recycling of the TiO_2 photocatalyst (Watts et al., 1995). For this approach, the additional cost for chemicals and the requirement for critical process control need to be considered. Others have looked at centrifugation to collect the photocatalyst particles. The additional cost for equipment and energy usage are drawbacks to this approach. Photoactive TiO_2 has also been attached onto larger support particles such as quartz, silica gel, pillared clay and zeolite to enhance its settling. This has a clear disadvantage given the loss in the specific surface area. The development of a novel magnetic photocatalyst with enhanced separation addressed this issue (Beydoun et al. 1998, 2000, 2001, 2002 (a, b), Watson et al., 2002, Amal et al., 2000 and 2003). The stable magnetic photocatalyst consisted of a magnetic core (Fe3O4) insulated by a layer of silica (SiO2) and coated with the photoactive titanium dioxide (TiO2). These magnetic photocatalysts provide the higher surface areas of submicron particles while allowing for easy separation from slurry photoreactors by an inexpensive magnetic separation process. The large scale production of these magnetic photocatalysts at reduced costs remains an issue.

It is clear that the design of effective photoreactors, whether slurry or immobilized, needs to take into account a number of parameters. For instance, the availability of the magnetic photocatalyst allows for a new generation of photocatalytic reactors to be developed. A small bench scale of such a reactor has been built and consisted of a spiral, re-circulating photoreactor with a near-UV illumination source (wavelength □350 nm). It included a magnetic separator (based on a series of permanent magnets) to separate and recover the catalyst from the treated water. A schematic diagram of a photo-reactor based on the magnetic photocatalyst with a magnetic separation unit is shown in Figure 2.

Other important components which need to be optimized for an integrated photocatalytic reactor system may include parabolic collectors, support materials and light sources. For the latter, the most common light sources are the black light type UV lamps, which emit in the UV-A region or the germicidal lamps emitting in the UV-C region. These lamps are relatively cheap. One issue is their size and geometry which inhibit the design of miniaturized photocatalytic devices. Recent developments here have been the introduction of UV LED devices. These are however still too expensive for practical applications (Fujishima et al., 2006). More research is needed in this area.

Reaction Kinetics

From basic chemical engineering principles, the design of any reactor relies on a mass balance in which the reaction term needs to be adequately described. When studying the kinetics of a photocatalytic reaction, the series of elementary steps usually considered are mass transport onto the catalyst surface, adsorption, electron-hole pair generation, charge transfer and desorption processes (Ilisz et al., 1999). In a well-mixed slurry photoreactor

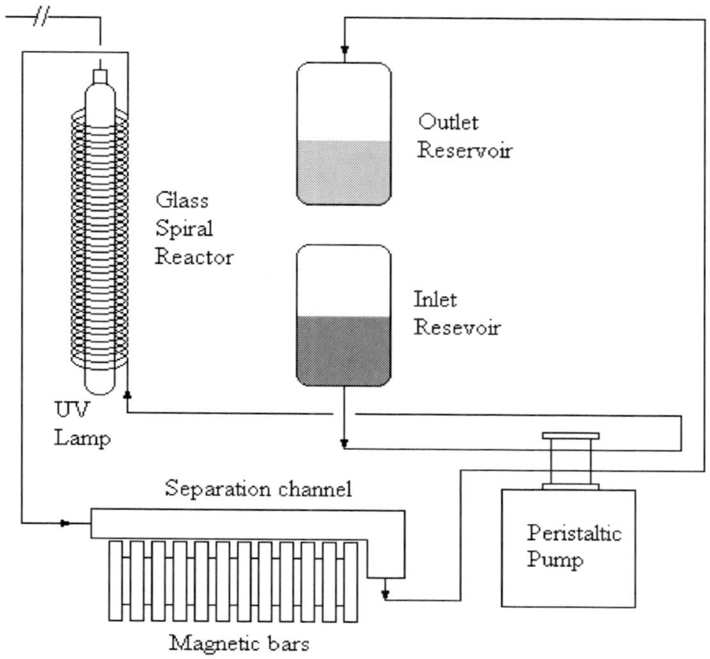

Figure 2. Schematic diagram of photocatalytic reactor with magnetic separation unit.

system, the mass transfer limitations have been found to be negligible, hence the proposed reaction mechanism is usually based on a reaction controlled by adsorption kinetics (Matthews, 1990, Basca et al., 1998).

The description of kinetic parameters by accurate kinetic models is necessary for the scaling up of laboratory scale experiments and the minimization of costly pilot plant experiments (Zalazar et al., 2005). According to Zalazar et al., missing information include the field distributions of concentrations, photons and velocities that cannot be extrapolated and directly used for scaling-up purposes. Adsorption models, with Langmuir-Hinshelwood (L-H) being the most commonly applied, are generally used to interpret the kinetic data obtained and to describe the degradation rate expressions in terms of the disappearance of the compounds and the appearance of CO_2 (Minero, 1992). The L-H based kinetic models relate the rate of surface-catalysed reactions to the surface covered by the substrate. According to the L-H model, the rate of a unimolecular surface reaction is proportional to the surface coverage.

Attention is needed when applying the L-H model to describe the kinetics of photocatalytic reactions. Emeline et al. (Emeline et al., 2005) in fact described the unquestioning application of the L-H model as one of the dogmas which continue to propagate in the heterogeneous photocatalysis. Limitations include that it describes a catalyst surface covered by a monolayer of adsorbed reactant molecules, with only one adsorbed species per surface site of equal energy. Other kinetic models have been proposed, namely the Langmuir-Rideal (L-R) model. This kinetics model describes those reactions occurring in the bulk solution as well as surface reactions (Minero, 1992).

The inadequacies of the L-H model have also been discussed to explain the spectral dependence of quantum yields of photocatalytic reactions (Serpone et al., 2005). From the general solution of the continuity equation, the absorption coefficient was identified as a key parameter, believed to be the cause of the spectral variation of the quantum yields. Emeline et al. also discussed that unlike thermal catalysis adsorption/desorption equilibria may not be established during a photocatalytic reaction (Emeline et al., 2005). Accordingly, the derived L-H constants may have very little, if any physical meaning. Hence a proposed mechanism must always be supported by other evidence outside the field of kinetics. A main conclusion from this work was that theoretical considerations of photocatalysis need to take into account bulk and surface processes in the photocatalyst as well as primary and secondary chemical surface reactions (Serpone et al., 2005).

Reaction mechanisms

It is clear from the above discussion that to optimise Existing knowledge includes that the primary step in the transformation of organic substrates is the attack by hydroxyl radicals or holes (Chiron et al., 2000). The reaction conditions can determine whether it is the holes or the hydroxyl radicals which are the main oxidizing agents. The reaction pH is important. For example, under acidic conditions, the direct oxidation of dichloroacetic acid by the hole is believed to be the prevailing mechanism (Bahnemann et al., 1997). This reaction pathway may also be dependent on the point of zero charge of the titanium dioxide that is being used due to its effect on the adsorption and desorption of reactants and intermediates (Zalazar et al., 2005). It is also known that dearomatization is rapid; and if an aliphatic chain is bound to the aromatic ring, the breaking of the bond is easy (Herrmann, 2005).

Charge Carrier Dynamics of TiO_2 Photocatalysis

The photogenerated holes and electrons in an activated TiO_2 photocatalyst are the key players to initiate surface chemical transformations (Serpone et al., 2005). Studying these charge carriers is very important as it can reveal crucial mechanistic information. The charge carriers in illuminated semiconductors exhibit strong optical absorption and hence can be studied spectroscopically. The use of semiconductor nanoparticles has enabled more definitive mechanistic studies of photocatalytic processes by spectroscopic methods. The dispersions of nano-sized semiconductor particles with diameters smaller than the wavelength of the incident light are optically transparent. The dimensions of these particles also render light scattering negligible in their suspensions.

The charge carrier dynamics are revealed by laser flash photolysis as follows. The trapped electrons exhibit strong optical absorption around 650 nm while the trapped holes absorb predominately at wavelengths around 430 nm (Colombo et al., 1995). The positions of the absorption maxima are strongly affected by the presence of suitable electron acceptors and donors in the surrounding aqueous phase. For example, in the presence of an electron acceptor such as methyl viologen, when a TiO2 colloidal solution is flashed with a 347 nm laser, an absorption band with □max = 394 nm is observed, and is attributed to the formed MV□+ radicals. In the presence of an electron donor, such as phenothiazine, a broad absorption band with □max = 520 nm is observed, and is attributed to the formed PMZ+ radicals. With nano-sized semiconductors it is therefore possible to directly monitor the charge injection into or from the semiconductor to reagents present in solution from the obtained absorption spectra. In laser flash photolysis experiments, the concentrations of transient species are determined by recording the transient absorbance spectra characteristic of each species, over the time domain of interest (for example, nanosecond or picosecond). From these absorption spectra results, both quantitative and qualitative information can be obtained on the photocatalytic reaction kinetics and mechanisms.

Laser flash photolysis experiments have been used to study a spectrum of photocatalytic reactions. From such studies Bahnemann et al. have contributed greatly to the basic understanding of elementary processes in illuminated semiconductors (Bahnemann et al., 1983, 1984 (a), 1984 (b), 1987 and 1997, Hilgendorff et al., 1996). For example Bahnemann et al. (1997) were the first to reveal that there are at least two different types of hole traps in excited TiO2 particles. Deeply trapped holes, which are long-lived and unreactive, and shallowly trapped holes, which exhibit a high oxidation potential. In the same study Bahnemann et al. also investigated the electron transfer kinetics to donors and acceptors. They found that the adsorption of the model compounds DCA and SCN on the TiO2 surface prior to the band gap excitation, was a prerequisite for efficient hole scavenging.

Laser flash photolysis has also been used to study photocatalytic reduction reactions on illuminated TiO2 (Navio et al., 1991 and 1999, Bahnemann et al., 1984 (a) and 1987). Navio et al. studied the reduction of methyl viologen (MV2+) as a probe photoreaction using an iron-titanium oxide photocatalyst (1991), and later in 1999 using ZrTiO4 and its parent oxides TiO2 and ZrO2. From the 1991 study, Navio et al. showed that the presence of a small amount of Fe ions (less than 0.5 weight %) in TiO2 matrices was beneficial for the photoreduction of methyl viologen. An increase in the amount of Fe doping in the TiO2 samples sharply reduced the MV+ yield to a value similar to that when pure TiO2 was used.

Particular attention has also been given to sensitized TiO2 systems (Argazzi et al., 1995 and 1997, Rabani et al., 1997, Wang et al., 1997, Salafsky et al. 1998, Zhou et al., 2002, Zhang et al., 2002 and Takeyama et al. 2003). More recently, Pan et al. (Pan et al., 2004) used laser flash photolysis to study the effectiveness of a self assembly of a carotenoid and pheophytin on the surface of TiO2 as a novel way to lengthen the lifetime of the charge carriers. This is seen as an approach for constructing dye sensitized solar cells and related systems to mimic the electron transfer chain in natural photosynthesis. Laser flash photolysis experiments have also been used to study the charge carrier dynamics in composite semiconductor systems, for example, Sant et al. (Sant et al., 2002) studied inter-particle electron transfer in CdS/TiO2 composite systems in which they observed an increase in quantum yield from 2.3% in bare TiO2 to 8.8% in CdS/TiO2.

Laser flash photolysis studies on the charge carrier dynamics of metal doped semiconductor systems are not frequently encountered in the literature. When these were carried out they have been directed mostly at Pt-TiO2 systems (Bahnemann et al. 1984 (a), 1984 (b) and 1997). Bahnemann et al. (1997) were able to clarify the role of the Pt deposits which helped extend the lifetime of the generated holes. In these experiments, by studying the reaction kinetics of bare and doped TiO2 under air saturated, O2 saturated and N2 saturated solutions, Bahnemann et al. were also able to provide a mechanistic scheme for the photo-initiated processes. A laser flash photolysis study on TiO2 suspensions containing silver ions revealed the nanosecond time scale photodepostion of metallic silver on the surface of the TiO2 particles and provided possible evidence for an antenna type effect during this photocatalytic reduction reaction (Friedmann et al., 2006).

Materials Engineering

An ideal photocatalyst should be stable, inexpensive, non-toxic and, of course, highly photoactive. The production of titanium dioxide as a photocatalyst is quite limited with the vast majority of TiO_2 production being for use in paints. While Degussa P25 TiO_2 is one of the most commonly recognized photocatalyst, its use as a photocatalyst is in fact limited to development work only, in laboratory investigations, process development and large-scale feasibility studies (Levy, 1997). It is in fact manufactured for use as a thermal stabilizer for heat resistant silicone rubber (Fujishima et al., 1999). Other commercially available photocatalysts include ST series, produced by Ishihara Sangyo Kaisha ltd, PC series, produced by Titan Kogyo, SSP series produced by Sakai Chemical Industries (Fujishima et al., 1999), Hombikat UV100, supplied by Sachtleben Chemie, PC-50 (produced by Millennium Performance Chemicals) and TKC-301 (produced by Tayca).

A bottleneck for future market expansions of photocatalysis is the inherently low quantum efficiency of TiO2 (Wang et al., 2003). The improvement of TiO2 as a photocatalyst is most important and is a main focus.

However, it remains that the photocatalytic process has the following inherent problem. Excited metal oxide semiconductor particles behave as short-circuited electrodes, with both oxidation and reduction processes occurring on their surfaces (Kamat, 1995). At least in colloidal sols of TiO2, it has been found that 90% of the charge carriers undergo recombination approximately 1 nanosecond after excitation (Serpone et al, 1996). Hence the quantum efficiency of any surface photoredox reaction cannot be greater than ~10%. This is still the current bottleneck of photocatalysis (Wang et al, 2003).

The improvement and the optimisation of TiO2 as a photocatalyst are most important for future technical applications of heterogeneous photocatalysis. Hence, aof main focus of current and future researchresearch. Increased efficiencies can be achieved by preparing inherently highly active TiO2 powders (examples are the studies by Bockelmann et al., 1996, Wang, 2000 and Wang et al., 2002). Magnetic photocatalyst particles allow for easy separation while providing the high surface areas (Beydoun et al. 1998, 2000, 2001, 2002 (a) and 2002 (b)), Watson et al., 2002, Amal et al., 2000 and 2003). Some success in enhancing the photocatalytic activity has also been achieved by coupling TiO2 with other semiconductor oxides (for example refer to the study by Bedja et al., 1995). Combinations with adsorbents such as activated carbon and zeolites to enable synergistic effects between adsorption and photocatalytic degradation is another approach (Herrmann et al., 1999). This relies on the diffusion of pollutants from adsorption sites to photocatalytic sites.

Another treatment method to improve the photocatalytic activity of TiO2 is surface modification, either with redox couples or by doping with noble metals (Choi et al., 1994, Aruna et al., 1996, Emilio et al., 2004, Hufschmidt et al., 2002 and Wang et al., 2003, Teoh et al., 2005). Doping with noble metals has in fact been identified as a promising means of improving semiconductor photocatalysts, electrodes, and selective electrocatalysts (Kulak et al., 2003). Figure 3 illustrates charge transfer processes in an illuminated metallized semiconductor particle. In the presence of the metal particles, additional electronic surface states are formed under bandgap excitation. The metal particles provide sites for more efficient utilization of the photogenerated electrons. These electrons can strongly interact with oxygen or react with other electron acceptors in the system. This improves the electron-hole separation hence the holes are available to take part in the oxidation reaction. The net effect is therefore an increase in quantum efficiency.

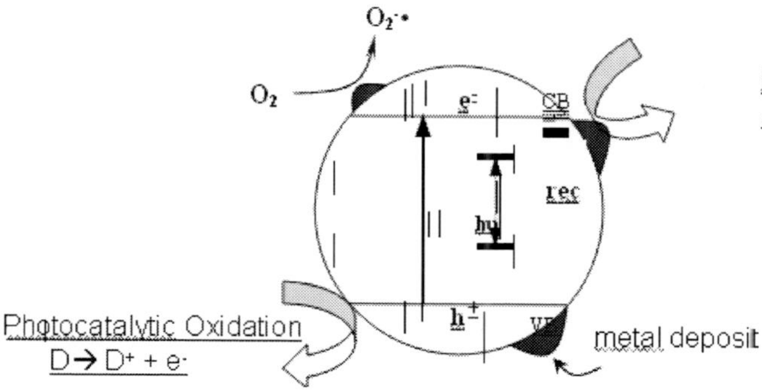

Figure 3. Charge transfer processes in a metallized photocatalyst particle upon bandgap excitation; hυ: energy of photons, CB: conduction band, VB: valence band, rec: recombination, D: electron donor; A: electron acceptor; O2: oxygen molecule; O2-•: superoxide radical (Adapted from Vamathevan, 2003).

The effect of metal doping on photoreactivity has been found to be dependent on the type of reaction, size and dispersion of the metal deposits. While doping TiO2 with an optimum amount of nano-Ag clusters greatly enhanced the mineralization of sucrose (Vamathevan et al., 2002 and 2004) and the photoreduction of selenate ions (Tan et al., 2003), the silver deposits had no significant influence on the mineralization of phenol or salicylic acid (Vamathevan, 2003). More recently, depositing an optimum amount of Pt on TiO2 was also found to lead to enhancements in the photocatalytic mineralization of sucrose (Teoh et al., 2005). A detailed understanding of the processes occurring at the metal/oxide interfaces and their relationship with the properties of the photocatalysts is important for improvements in efficiency and performance. Remaining unresolved issues are the role of the substrate, the size and shape of the metal deposits, the related changes in electronic structure as a function of these parameters and their effect upon the interaction with atoms and molecules (Canario et al., 2003). A better understanding of the charge carrier dynamics of these composite systems is important for tailoring their properties. Finally, an assessment of the stability and shelf-life of metal deposited TiO_2, particularly for their applications as photocatalysts, after prolonged UV irradiation is needed. This is critical for commercial applications. Surface deformations have been for example reported for Au particles (mean size 3 nm) supported on TiO_2 particles when exposed to irradiation (Kawahara et al., 2005). This occurred in aqueous solutions containing halogen ions. Preliminary studies on the reuse of Ag-TiO_2 photocatalysts for the degradation of dichloroacetic acid under acidic have shown little effect on activity.

VISIBLE LIGHT PHOTOCATALYSTS

Solar driven photocatalytic processes are highly desired as they offer potentially environmentally sound technologies at low cost. Pure titanium dioxide can be activated only by UV-A irradiation or stronger, that is, by illumination with light of wavelengths 390 nm or lower. This constitutes only 4% of the energy available within the solar spectrum. Additionally indoor illumination, by fluorescent lamps for instance, has mostly a visible light component. Hence there is a current strong research focus on developing efficient, visible light activated photocatalysts. One approach is to sensitize the TiO_2 with organic and organo-metallic dyes. The instability of these photocatalysts is a concern (Howe, 1998). Another approach has been to couple TiO_2 with a narrow-bandgap semiconductor which is activated by visible light. This then injects electrons into the conduction band of TiO_2. As an example, a schematic of the charge transfer under illumination of a coupled CdS/TiO_2 system is shown in Figure 4. A comprehensive review on these composite systems was undertaken by Beydoun et al. (Beydoun et al., 1999).

The introduction of ions into the TiO_2 matrix has been the other approach to decrease the bandgap of this semiconductor. Bahnemann et al. have prepared stable mixed Ti(IV)/Fe(III) oxide photocatalysts which have a new energy level below the conduction band and a photoresponse to 450 nm light (Bahnemann et al., 1993, Wang et al., 2002 and 2003). This photocatalyst had increased activity for the destruction of dichloroacetic acid (DCA). A recent and promising development in the search for visible light photocatalysts has been to introduce N, C or S ions into the TiO_2 matrix to give rise to an energy level above the valence band. For example with nitrogen doping, the N 2p orbitals mix with the oxygen O 2p orbitals. A present problem with these catalysts remains their lower activity under visible illumination compared to activation with UV light. More recently F ions have also been combined with nitrogen doping to enhance their activity (Li et al., 2005). The color of these photocatalysts is an issue (Agrios et al., 2005).

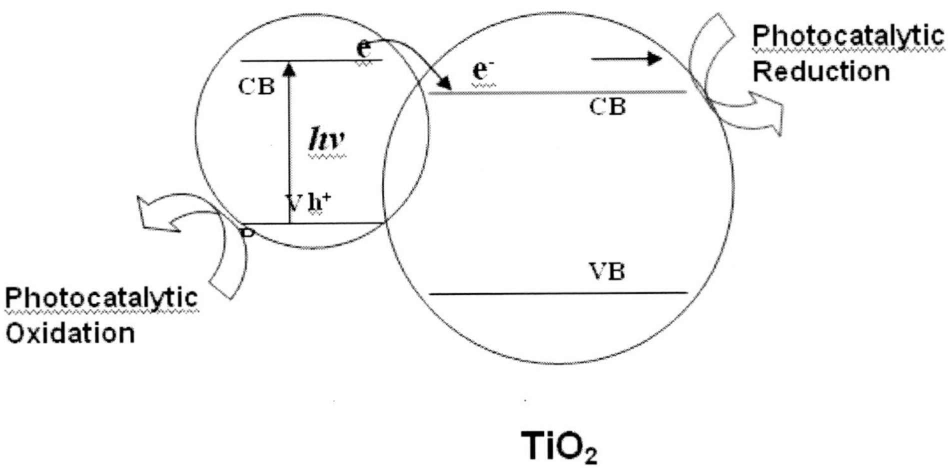

Figure 4. Schematic illustrating sensitization of TiO_2 by coupling it with the narrow bandgap semiconductor, CdS.

Particle Synthesis Technologies

Preparation methods of TiO_2 can be classified into two categories, the wet chemistry methods, such as the sol-gel technique, and the dry methods, such as the aerosol technique and the plasma arc process. The origin and the methods of preparation affect a photocatalyst's physicochemical properties such as its particle size, morphology, crystalline phase composition and the presence of dopant ions. High purity TiO_2 is typically made in a flame process from titanium tetrachloride. Sol -gel methods are also widely used. For current applications, photocatalysts are typically immobilized and produced using the sol-gel process or chemical vapor deposition (CVD) (Mills et al., 2002).

The introduction of new photocatalysts brings with it the need to modify existing particle manufacturing processes or develop new preparation methods which can produce particles with intricate properties at favourable capital and running costs. For metal modified TiO_2, the size and dispersion of metal deposits on TiO_2 are critical to its photoreactivity (Teoh et al., 2005). Teoh et al. discussed the development of Flame Spray Pyrolysis (FSP) as a preparation method for these photocatalyst which provides such close control over deposit size and dispersion. FSP is in fact seen as a technology which is starting to pave its way as an economical route to producing high precision nanoparticles. The production of coating solutions for transparent self-cleaning surfaces with a high solids content remains a challenge, as are coating methods compatible with organic substrates which allow for the fabrication of photocatalytically active flexible plastics or membranes (Agrios et al., 2005).

Standardization of the Photocatalytic Activity of TiO_2

Given that the chemistry of photocatalytic reactions is not fully understood, their behaviour is often difficult to predict. Additionally, the photocatalytic activity of TiO2 not only depends not only on intrinsic bulk and surface properties of the samples but also on the nature of the photocatalytic reaction (Siemon et al., 2002). At present, there is a lack of an appropriate protocol to compare the activities of different photocatalysts in different heterogeneous systems (Serpone et al., 2005). This presents a hindrance to the commercialization of heterogeneous photocatalysis as consumers need to be able to make comparisons between different photocatalytic products. The development of a reliable and simple photoactivity test method is needed to screen photocatalytically active samples. A standardization committee in Japan was started in 2002. Europe, China and South Korea have also started to develop standardization procedures. Harmonizing such testing methods is very important.

For heterogeneous catalysis, methods have been developed for testing and evaluating catalysts for various processes. Examples include the microactivity test (MAT) for Fluid Catalytic Cracking (FCC) catalyst screening (for example refer to ASTM, 2003 and Ng et al., 2001), selective oxidation catalysts (Paul et al., 2003) and dehydrogenation catalysts (Urschey et al., 2003 (a) and (b)). Research efforts continue with the combinatorial approach for catalyst screening. The advantage here is the vast number of materials which can be evaluated quickly. The main existing issue is the high costs associated with this method. Combinatorial methods are also often restricted to the screening of catalyst samples that are prepared by wet precipitation techniques, which allow compositional variations by robot pipetting whose properties can differ from conventional materials (Paul et al., 2003).

The development of reliable and quick methods for assessing the photocatalytic properties of TiO_2 have received much attention over the years. Important to this field are studies on the photoactivities of pigmentary TiO_2 in which the photoactivity needs to be suppressed. Early studies by Cundall et al. (1976, 1978) suggested that the photocatalytic oxidation of iso-PrOH can be used in the rapid determination of the photoactivity of TiO_2 pigments. Recent developments are the non-contact evaluation methods. Examples include the studies by Allen et al. (1991, 1993 and 2002) who utilized photoconductive and microwave techniques for the evaluation of the activity of TiO_2 pigments in polymeric systems. In their studies Allen et al. (1991) compared their results to the Carbonyl Index as a reference measure of photoactivity. The Carbonyl Index is used to monitor the degradation of pigmentary films when exposed to sunlight and weathering. It is a measure of the change in surface properties and is defined as the ratio of the maximum absorbance bands at 1710 (assigned to C=O) to absorbance at 1170 cm^{-1} as a reference. The chemiluminescence method has also been used to probe TiO_2 in pigments (Janes et al., 2001, Corrales et al., 2003).

Combinatorial synthesis and high-throughput evaluation of doped TiO_2 thin films for photocatalytic applications have also been investigated (Matsumoto et al., 2000, Nakayama, 2002 (a) and (b)). In the studies by Matsumoto et al. (2000), TiO_2 thin films doped with various transition metals were systematically fabricated by the combinatorial Laser-Molecular Beam Epitaxy (Laser-MBE) technique. The probe reaction was the photo-decomposition of water in the presence of a small amount of oxidizing or reducing agent which induced pH changes. A potentiometric pH field effect transistor was used for sensing such pH changes.

Other approaches have involved characterizing the physicohemical, optical and electronic properties of TiO_2 to predict its photoactivity. Examples include studying the TiO_2 photoresponse upon irradiation (Sumita et al., 2002 and Takesaki et al., 2004) and changes in TiO_2 surface potential as a function of light irradiation (Tadashi et al., 1999). Others have taken a whollistic approach by correlating physical properties of TiO_2 particles, such as XRD, BET, and surface area measurements, with their photoactivity from laboratory scale probe reactions and up to pilot scale investigations (Kominami et al., 2002 and Herrmann et al., 2002). While this approach provides valuable knowledge on photocatalytic system, and is an essential part of the catalyst development process, it is again expensive, time consuming and hence not suited to industrial laboratories.

Additional issues arise when results of different testing methods are presented for the same catalytic material. This can significantly affect the ranking of catalyst performance (Moorehead et al., 1989) and highlights that an understanding of test methodologies and implications of results are essential. Photonic efficiency ad quantum yield are the two most common ways of quantifying photocatalytic efficiency for comparison between different studies. (Serpone et al., 2005). Photonic efficiency describes how many molecules are transformed per photon incident on the system. Quantum yield is the ratio between the rate of molecules formed or degraded in the system and the rate of photons absorbed by the system. According to Serpone et al., a comparison between different photocatalytic systems based on these factors often ignores that the reaction rates must scale linearly with photon flow and the reaction rate must be independent of the concentration of reagent, and therefore such a comparison is often erroneous. The conclusion here is that screening tests need to be harmonized and need to provide a quick and cost effective tool while realistically simulating the environment in which the material will be applied.

Surface Science Studies of Photocatalytically Active TiO_2 Surfaces

As discussed earlier in this chapter, high surface area photocatalysts have been prepared using various techniques and their photocatalytic activity tested for the degradation of an extensive range of environmental contaminants. While great knowledge has been gained from these studies with regards to reaction mechanisms and the significance of various reaction parameters, they have also revealed the complexity of photocatalytic reactions. Currently, it is still not possible to design photocatalysts with optimum activity for the degradation of a specific contaminant, with technological application requiring extensive developmental studies and pilot scale testing for specific conditions. There is therefore a strong need for new fundamental insights on the nature of the photoactivated TiO_2 surface. The knowledge to be gained from these fundamental surface studies is highly relevant to research efforts in photocatalysis and for the design of improved photocatalysts. The extensive review by Diebold (Diebold, 2003) provides valuable information on TiO_2 surfaces. Much knowledge is still needed on the behaviour of key reactants, including the contaminant to be degraded and its intermediates, O_2 and H_2O, on the TiO_2 surface.

Scanning Tunnelling Microscopy (STM) with atomic resolution is an extremely useful characterizing tool for studying the surface of TiO_2 at the atomic-scale and under controlled conditions. STM studies have mostly focused on rutile TiO_2 (110) which is now well characterized. Initial studies by Fujishima were in fact also on a single crystal of rutile TiO_2 in which he demonstrated the photosplitting of water into oxygen. Fujishima continues to use single TiO_2 crystals in his research for example for fundamental wettability studies (Wang et al., 1999). Thus far, there have been minimal STM studies of photocatalytic reactions on the surface of illuminated TiO_2. STM studies by Henderson et al. (for example refer to Henderson, 2006 and Henderson et al., 2005) have been very useful in providing preliminary insights into the complex nature of photocatalytic reactions. It was not until recently that key features on a reduced TiO_2 surface (oxygen vacancies and hydroxyl groups) were correctly assigned (Wendt et al., 2005). Further STM studies on illuminated TiO_2 surfaces are needed, particularly on anatase TiO_2 single crystals. This may clarify why certain structures of TiO_2 have a better catalytic effect than others. Surface studies may also shed light on the selectivity of photocatalysts, an area which may prove useful for photocatalytic synthesis and in avoiding toxic intermediates.

Conclusion

The research field of photocatalysis has pulled together researchers with various expertise. The continued research efforts in the area of TiO_2 photocatalytic detoxification of waste waters have led to remarkable developments. While some aspects of the basic mechanisms and degradation pathways are still debated, present fundamental knowledge may be sufficient for application (Agrios et al., 2005). With the growth in global environmental pollution, and the increasingly stringent environmental regulations, photochemical water pollution remediation processes are being increasingly acknowledged as a real alternative. While a heterogeneous photocatalytic process for water purification is yet to be commercialized, commercial success has been achieved with air purification applications and self-cleaning and anti-bacterial coatings. Developments continue in terms of improved photoreactor design and process efficiencies. Progress in enhanced photocatalytic and optical properties of the material TiO_2 have been achieved through doping, coupling and sensitization. Research will continue into developing visible light activated photocatalysts, as will fundamental studies on reaction kinetics and mechanisms using spectroscopic and surface science techniques. The scaling up of synthesis methods to manufacture novel photocatalysts of intricate and controlled structures presents an engineering challenge, although, technologies such as flame spray pyrolysis already present a possible solution. Finally, it will be the consumer who will decide the success or failure of commercialized photocatalytic processes. These consumers need to be well informed and guided by accepted standards.

REFERENCES

Agrios, A., Pichat, P. (2005). *J. App. Electrochem*, 25, 655-663.

Allen, N., Chow, Y., Thompson, F., Jewitt, T. (1993). *Dyes Pigm*, 21, 135-156.

Allen, N., Chow, Y., Thompson, F., Jewitt, T., Hornby, M. (1991). *Polym. Degrad. Stab*, 32, 119-129.

Allen, N., Edge, M., Sandoval, G., Ortega, A., Liauw, C., Stratton, J., McIntyre, R. (2002). *Polym. Degrad. Stab*, 76, 305-319

Amal, R., Beydoun, D., Low, G., McEvoy, S. (2000). PCT WO0061290.

Amal, R., Beydoun, D., Low, G., McEvoy, S. (2003). United States Patent, No. US6558553B1, 6.5.

Argazzi, R., Bignozzi, C. (1995). *J. Am. Chem. Soc*, 117, 11815-11816.

Argazzi, R., Bignozzi, C., Heimer, T., Castellano, F., Meyer, G. (1997). *J. Chem. B*, 101, 2591-2597.

Aruna, S. T., Patil, K. C. (1996). *J. Mater. Synth. Process*, 4, 175-179. ASTM D 5154-03, ASTM International.

Auriol, M., Filali-Meknassi, Y., Tygayi, D., Adams, C., Surampalli, R. (2006).*Process Biochem*, 41, 525-539.

Baba, Y., Schichijo, N., Nagahara, Y., Yarime, M. (2005). In Proceedings of DRUID Tenth Anniversary Summer Conference, Dynamics of Industry and Innovation: Organizations, Networks and Systems. Danish Research Unit for Industrial Dynamics (DRUID): Copenhagen, Denmark, June 27-29, 2005.

Bacsa, R. R., Kiwi, J. (1998). *Appl. Catal. B*, 16, 19-29.

Bahnemann, D. W.. Dillert, R., Robertson, P. (2003). In *Chemical Physics of Nanostructured Semiconductors*. Kokorin, A., Bahnemann, D. Eds. VSP BV: Utrecht, Netherlands, 183-200.

Bahnemann, D. W., Fischer, C., Janata, E., Henglein, A. (1987). *Phys. Chem. Condens. Phases*, 83, 2559-2571.

Bahnemann, D. W., Henglein, A., Lilie, J., Spanhel, L. (1984). *J. Phys. Chem. (a)*, 88, 709-711.

Bahnemann, D. W., Henglein, A., Spanhel, L. (1984). Faraday Discuss. *Chem. Soc. (b)*, 78, 151-163.

Bahnemann, D. W., Hilgendorff, M., Memming, R. (1997). *J. Phys. Chem. B*, 101, 4265-4275.

Bahnemann, D. W. (1993). *Isr. J. Chem*, 33, 115-136.

Bahnemann, D. W., Henglein, A., Lilie, J., Spanhel, L. (1983). Ges. Dtsch. Chem, 8, 145-150.

Bedja, I., Kamat, P. V. (1995). *J. Phys. Chem*, 99, 9182-9818.

Beydoun, D., Amal, R. (2002). *J. Mater. Sci. Eng. B, (a)*, 94, 71-81.

Beydoun, D., Amal, R., Low G., McEvoy, S. (1999). *J. Nanopart. Res*, 1, 439-458.

Beydoun, D., Amal, R., Low, G., McEvoy, S. (2000). *J. Phys. Chem. B*, 104, 4387-4396.

Beydoun, D., Amal, R., Low, G., McEvoy, S. (1998). In Proceedings of the 3rd World Congress on Particle Technology in Brighton UK, 6-9 July 1998; AICHE: New York, MY, paper number 385.

Beydoun, D., Amal, R., Low, G., McEvoy, S. (2002). *J. Mol. Catal. A, (b)*, 180, 193-200.

Beydoun, D., Amal, R., Scott, J., Low, G., McEvoy, S. (2001). Chem. Eng. Technol, 24, 745-748.

Beydoun, D., Tse, H., Amal, R., Low, G., McEvoy, S. (2002). *J. Mol. Catal. A,* ©, 177, 265-272.

Bockelmann, D., Lindner, M., Bahnemann, D. W. (1996). In Fine Particles Science and Technology, Pelizzetti, E. Z., Ed. Inst. NATO ASI Series, Series 3: High Technology; Kluwer Academic Publishing: Netherlands, vol 12, 675-689.

Brezova, V., Blazkova, A., Karpinsky, L., Groskova, J., Havlinova, B., Jorik, V., Ceppan, M. *J. Photochem. Photobiol. A*, 109, 177-183.

Canario, A. R., Sanchez, E. A., Bandurin, Y., Esaulov, V. A. (2003). Surf. Sci. Lett, 547, L887-L894.

Carey, J.H., Lawrence, J., Tosine, H. M. (1976). Bull. Environ. Contam. Toxicol, 16, 697-702.

Chen, D., Ray, A. K. (1999). Appl. Catal. B, 23, 143-157.

Chiron, S., Fernandez-Alba, A., Rodriguez, A., Garcia-Calvo, E. (2000). Water Res, 34, 366-377.

Choi, W., Termin A., Hoffmann, M. R. (1994). *J. Phys. Chem*, 98, 13669-13679.

Colombo, A., Roussel, K., Saeh, J., Skinner, D., Cavaleri, J., Bowman, R. (1995). Chem. Phys. Lett, 232, 207-214.

Colombo, D. P. Jr., Bowman, R. M. (1996). *J. Phys. Chem*, 100, 18445-18449.

Corrales, T., Peinado, C., Allen, N., Edge, M., Sandoval, G., Catalina, F. (2003). *J. Photochem. Photobiol. A*, 156, 151-160.

Cundall, R., Hulme, B., Rudham, R., Salim, M. (1978). *J. Oil Color Chem. Assoc*, 61, 351-355.

Cundall, R., Rudham, R., Salim, M. (1976). *J. Chem. Soc., Faraday Trans*, 1, 72, 1642-1651.

Diebold, U. (2003). *Appl. Phys. A*, 76, 681-687.

Dyk, A., Heyns, A. (1998). *J. Colloid Interface Sci*, 206, 381-391.

Emeline, A. V., Ryabchuk, V. K., Serpone, N. (2005). *J. Phys. Chem. B*, 109, 18515-18521.

Emilio, C., Testa, J., Hufschmidt, D., Colon, G., Navio, J., Bahnemann, D. W., Litter, M. (2004). *J. Ind. Eng. Chem*, 10, 129-138.

Frank, S. N., Bard, A. (1977). *J. Phys. Chem*, 81, 1484-1486.

Friedmann, D., Hansing, H., Bahnemann, D. Z. (2006). *Phys. Chem, Submitted.*

Fujishima, A., Hashimoto, K., Watanabe, T. (1999). In *TiO_2 Photocatalysis Fundamentals and Applications*. BKC Inc, Tokyo, Japan.

Fujishima, A., Zhang X. C. R. (2006). *Chim*, 9, 750-760.

Henderson, M. (2005). *J. Phys. Chem. B*, 109, 12062-12070.

Henderson, M., White, J., Uetsuka, H., Onishi, H. (2006). *J. Catal*, 238, 153-164.

Herrmann, J. M., Matos, J., Disdier J., Guillard C., Laine J., Malato, S., Blanco, J. (1999). *Catal. Today*, 54, 255-265.

Herrmann, J. M. (2005). Top. Catal, 34, 49-65.

Herrmann, J., Guillard, C., Disdier, J., Lehaut, C., Malato, S., Blanco, J. (2002). *Appl. Catal. B*, 35, 281-294.

Hilgendorff, M., Bahnemann, D. W. (1996). In Book of Abstracts 211th ACS National Meeting New Orleans, LA, March 24-28, American Chemical Society: Washington, DC, vol 41, no. 1.

Howe, R. F. Dev. (1998). *Chem. Eng. Miner. Process*, 6, 55-84.

Hufschmidt, D., Bahnemann, D. W., Testa, J., Emilio, C., Litter, M. (2002). *J. Photochem. Photobiol. A*, 148, 223-231.

Ilisz, I., Laszlo, Z., Dombi, A. (1999). *Appl. Catal. A,* 180, 25-33.

Janes, R., Edge, M., Rigby, J., Mourelatou, D., Allen, N. (2001). *Dyes Pigm*, 48, 29-34.

Kalyanasundaram, K. (1983). In *Energy Resources Through Photochemistry and Catalysis*. Gratzel, M. Ed. Academic Press Inc. New York, MY, 217-260.

Kato, K. (1992). *Bull. Chem. Soc. Jpn*, 65, 34-38.

Kawahara, T., Soejima, T., Mitsui, T., Kiyonaga, T., Tada, H., Ito, S. (2005). *J. Colloidal Interface Sci*, 286, 816-819.

Khalil, L. B., Mourad, W. E., Rophael, M. W. (1998). *Appl. Catal. B*, 17, 267-273.

Kominami, H., Murakami, S., Kato, J., Kera, Y., Ohtani, B. (2002). *J. Phys. Chem. B,* 106[40], 10501-10507.

Kumar S., Davis A. P., (1997). *Environ. Res*, 69, 1238-1245.

Levy, B. (1997). *J. Electroceram*, 1, 239-272.

Li, D., Ohashi N., Hishita, S., Kolodiazhnyi, T., Haneda H. (2005). *J. Solid State Chem*, 178, 3293–3302.

Litter, M. (1999). *Appl. Catal. B*, 23, 89-114.

Maneerat, C., Hayata, Y. (2006). *Int. J. Food Microbiol*, 107, 99-103.

Matsumoto, Y., Murakami, M., Shono, T., Hasegawa, T., Fukumura, T., Kawasaki, M., Ahmet, P., Chikyow, T., Koshihara S., Koinuma, H. (2001). *Sci. Mag*, 291, 854-856.

Matthews, R. (1993). In *Photocatalytic Purification and Treatment of Water and Air, Ollis.* D.F., Al-Ekabi, H., Eds, Elsevier Science Publishers, New York, MY, 121-139.

Matthews, R. (1990). Water Res, 24, 653-660.

Mills, A., Lee, S. (2002). *J. Photochem. Photobiol. A*, 152, 133-247.

Minero, C., Catozzo, F., Pelizzetti, E. (1992). *Langmuir*, 8, 481-486.

Miyagi, T., Kamei M., Mitsuhashi T., Ishigaki T., Yamazaki A. (2004). *Chem. Phys. Lett*, 390, 399–402.

Moorehead, E., Margolis, M., McLean, J. (1989). ACS Symposium Series 411 (Charact. Catal. Dev.); American Chemical Society: New York, MY, 120-134.

Nakayama, C. (2002). *Koatsu Gasu (b),* 39, 917-921.

Nakayama, C. (2002). *Kogyo Zairyo (a),* 50, 22-25.

Navio, J. A., Hidalgo, M. C., Roncel, M., De la Rosa, M. A. (1999). *Mater. Lett*, 39, 370-373.
Navío, J. A., Marchena, F., Roncel M., De la Rosa, M. A. (1991). *J. Photochem. Photobiol. A*, 55, 319-322.
Ng, S., Yang, H., Wang, J., Zhu, Y., Fairbridge, C., Yui, S. (2001). *Energy Fuels*, 15, 783-785.
Pan, J., Xu, Y., Sun, L., Sundstroem, V., Polvka, T. (2004). *J. Am. Chem. Soc*, 126, 3066-3067.
Paul, J., Urschey, Jacobs, P., Maier, W., Verpoort, F. (2003). *J. Catal*, 220, 136-145.
Prairie, M. R., Stange, B. M., Evans, L. R. (1993). *Photocatalytic Purification and Treatment of Water and Air, Ollis*. D. F., Al-Ekabi, H, Eds. Elsevier Science Publishers. New York, MY, 353-363.
Rabani, J., Ushida, K., Yamashita, K., Johannes, G. S., Kira, A. (1997). *J. Phys. Chem. B*, 101, 3136-3146.
Rabani, J., Yamashita, K., Ushida, K., Stark, J., Kira, A. (1998). *J. Phys. Chem. B*, 102, 1689-1695.
Salafsky, J., Lubberhuizen, W., van Faassen E., Schropp, R. (1998). *J. Phys. Chem. B*, 102, 766-769.
Sant, P., Kamat, V. (2002). *Phys. Chem. Chem. Phys*, 4, 198-203.
Serpone, N., Emeline, A. V. (2005). *Res. Chem. Intermed*, 31, 391-432.
Siemon, U., Bahnemann, D. W., Testa, J., Rodriguez, D., Litter, M., Bruno, N. (2002). *J. Photochem. Photobiol. A*, 148, 247-255.
Sumita, Y., Yamamoto, H. (2002). *Jpn. Kokai Tokkyo Koho*, JP 2002031612.
Surender, G. D., Fotou, G. P., Pratsinis S. E. (1998). *Trends Chem. Eng*, 4, 145-159.
Takasaki, Y., Sagara, H., Shikawa, J. (2004). *Jpn. Kokai Tokkyo Koho*, JP 2004138387.
Takeuchi, M., Sakamoto, K., Martra, G., Coluccia, S., Anpo, M. (2005). *J. Phys. Chem. B*, 109, 15422-15428.
Takeyama, Y., Mamoru, F., Yasuyuki, A., Osamu, I. (2003). In Proceedings of Electrochemical Society (Fullerenes and Nanotubes) 2003-15; The Electrochemical Society, Pennington, NJ, vol 13, 128-139.
Tan, T., Yip C., Beydoun, D., Amal R., Zaw, M. (2003). *Chem. Eng. J*, 95, 179-186.
Teoh, W. Y., Mädler, L., Pratsinis, S. E., Beydoun, D., and Amal, R. (2005). *Chem. Eng. Sci*, 60, 5852-5861.
Urschey, J., Kuhnle, A., Maier, W. (2003). Appl. Catal. A, (b), 252, 91-106.
Urschey, J., Weiss, P., Scheidtmann, J., Richter, R., Maier, W. (2003). *Solid State Sci.* (a) , 5, 909-916.
Vamathevan, V. (2003). PhD Dissertation, UNSW, Sydney.
Vamathevan, V., Amal, R., Beydoun, D., Low G., McEvoy, S. (2004). *Chem. Eng. J.* 98, 127-139.
Vamathevan, V., Amal, R., Beydoun, D., Low, G., McEvoy, S. (2002). *J. Photochem. Photobiol. A*, 148, 233-245.
Wang, C. C., Zhang, Z., Ying, J. Y. (1997). *Nanostruct. Mater*, 9, 583-586.
Wang, C. Y., Bottcher, C., Bahnemann, D. W., Dohrmann, K. J. (2003). *Mater. Chem*, 13, 2322-2329.
Wang, C. Y., Rabani, J., Bahnemann, D. W., Dohrmann, J. (2002). *J. Photochem. Photobiol. A*, 148, 169-176

Wang, R., Sakai, N., Fujishima, A., Watanabe T., Hashimoto, K. (1999). *J. Phys. Chem. B*, 103, 2188-2194.

Wang, Y. (2000). *Water Res*, 34, 990-994.

Watson, S., Beydoun, D., Amal, R. (2002). *J. Photochem. Photobiol. A*, 148, 303-313.

Watson, S., Beydoun, D., Scott, J., Amal, R. (2003). *Chem. Eng. J.* 95, 179-186.

Watts, R., Kong, S., Lee, W. (1995). *J. Environ. Eng*, 730-735.

White, J. M., Szanyim J., Henderson, M. (2003). *J. Phys. Chem. B*, 107, 9029-9033.

Xu, N., Shi, Z., Fam, Y., Dong, J., Shi, J., Hu, M. Z. C. (1999). *Ind. Eng. Chem. Res*, 38, 373-379.

Ying, J., Sun, T. (1997). *J. Electroceram*, 1, 219-238.

Yuranova, T., Msteo, R., Bandara, J., Laub, D., Kiwi, J. (2006). *J. Mol. Catal. A*, 244, 160-167.

Zalazar, C., Martin, C., Cassano, A. (2005). *Chem. Eng. Sci*, 60, 4311-4322.

Zeltner, W. A., Anderson, M. A. (1996). In *Fine Particles Science and Technology*. Pelizzetti, E. Z., Ed. Inst. NATO ASI Series, Series 3, High Technology; Kluwer Academic Publishing: Netherlands, vol 12, 1996. 643-656.

Zhang, H., Yalin, Z., Zhang, M., Shen, T., Li, Y., Zhu, D. (2002). *J. Phys. Chem. B*, 106, 9597-9603.

Zhou, Z., Qian, S., Yao, S., Zhang, Z. (2002). *Phys. Chem*, 65, 241-248.

In: Photochemistry Research Progress
Editors: A. Sánchez, S. J. Gutierrez
ISBN: 978-1-60456-568-3

Chapter 3

PHYTOCHROME A: TWO NATIVE POOLS IN THE CELL WITH DIVERSE SPECTROSCOPIC, PHOTOCHEMICAL AND FUNCTIONAL PROPERTIES

V. A. Sineshchekov*

Dept. of Biology, M. V. Lomonosov Moscow State University,
Moscow, Russia 119992

ABSTRACT

Phytochrome (phy), a chromoprotein, is the major photoreceptor in plants, determining the strategy of their growth and development throughout their lifecycle. The chain of the regulatory steps leading to the photoresponse includes (i) light signal perception via the photoisomerization of the photoreceptor, (ii) transduction of the light signal from the activated photoreceptor through the complex network of mediators, and (iii) realizations of the light stimulus via modulation of expression of light-regulated genes. In recent years, we have been witnessing great progress in our understanding of the nature of this pigment and intimate mechanisms of its action. Among the fundamental discoveries are structural and functional phytochrome polymorphism with phyA and phyB as major ones and concrete mechanisms of their action including cytoplasm-to-nucleus translocation of the photoreceptor and interaction with transduction chain partners. It was shown that phyA is unique among other phys in many respects. In particular, it mediates photoresponse modes specific only for this pigment (the so-called very low fluence responses, VLFR, and high irradience response, HIR) and also low fluence responses, LFR, characteristic of phyB and other minor phys. The nature of this uniqueness remains, however, obscure. In the group of the author, the polymorphism of phyA itself was discovered with the use of an original method of fluorescence spectroscopy and photochemistry of phytochrome in the cell, which may explain, at least partially, the specific phenomenological and functional properties of phyA. In this review, the data on the structural and functional polymorphism are summarized and the literature on phyA is analyzed from this point of view.

* Chair of Physico-Chemical Biology, Biology Faculty. M. V. E-mail: vsineshchekov@umail.ru

1. Introduction

Plants possess a highly sophisticated apparatus of light perception comprised of a number of photoreceptor systems (phytochromes, cryptochromes, phototropins, receptors of green and UV-B light of unknown nature). With the aid of the phytochrome (phy) system, the best investigated, they adapt their growth and development to the ambient light conditions within the spectral interval from UV-B to near IR and intensities range of 7-8 orders of magnitude. During the last two decades, it was found that there exists a small family of phytochromes with each of its member possessing similar structure (biliproteins) and spectral and photochemical features but diverse phenomenological and functional properties. Their fundamental trait is a photorevesible photoconversion from the initial physiologically inactive form absorbing maximally in the red region (Pr) into the active one with maximum absorption in the far-red region (Pfr). The phytochromes are thus most active in the red (R) - far- red (FR) region of the spectrum and mediate a number of key photoregulation processes including induction of seed germination, transition from scotomorphogenesis to photomorphogenesis (de-etiolation), shade-avoidance, synchronization of circadian rhythms, modification and induction of phototropism and gravitropism (Batschauer, 2004; Briggs and Spudich, 2005; Schäfer and Nagy, 2006; Smith, 1997; Wada *et al.*, 2005; Wang and Deng, 2002).

The major phytochromes are phyA and phyB, they dominate in the cell both in content and functionally. In fact, they account for the majority of the photomorphogenic and photoregulation functions in plants. phyA, the subject of this review, is unique among all the other phytochromes in many respects. It is the most abundant phytochrome in dark-grown seedlings and is most active in mediating light-induced transition from scoto- to photomorphogenesis. It is the most sensitive phytochrome mediating the so-called very low fluence responses (VLFR) and its action spectrum spreads to the far-red region. It is the only phy species which performs the far-red – induced high irradience response (FR-HIR) (Casal *et al.*, 1997; Whitelam and Devlin, 1997). At the same time it is capable of the "classical" phytochrome R-induced / FR – reversible photoreactions in the intermediate light intensity range, the low fluence responses (LFR) characteristic of phyB and other minor phytochromes (see review in Sineshchekov, 2004). Distinctive feature of phyA is also its high light lability. Upon illumination its content steeply declines due to light-induced destruction and negative photoregulation of its biosynthesis. As a result, phyB, which is light-stable and does not undergo down-regulation of its synthesis, becomes a dominating species both in content and functionally under light growing conditions. However, in spite of the fact that the content of phyA is low in light-grown plants, its function is observed throughout the whole lifecycle of the plant (Franklin *et al.*, 2007).

These unique features of phyA is the result of phy evolution. According to Mathews *et al.* (2003), the high sensitivity of the pigment (phyB requires 4 orders of magnitude more light to trigger the R/FR reversible responses than phyA) and its ability to induce photomorphogenesis under FR is a result of positive selection at limited number of sites (a. a. residues) in the photosensory domain of phytochrome A. It is believed that the later became the key event in the evolution of higher plants and their decisive evolutionary advantage. This evolution have required the capacity to begin life in dimly lit environments and phyA is unique among flowering plants phytochromes in its capacity to serve a transient role under conditions where an extremely high sensitivity is required. The VLFR and HIR of phyA in the FR region became instrumental in providing the plant with the capacity to develop and grow under dense canopies giving the first angiosperms some adaptive advantage.

The structure-function relationships in the phyA molecule, which could account for the unique phenomenological and functional properties remain, however, unclear. The fact that phyA may perform photoregulation reactions of the different response modes can be explained, at least partially, by its polymorphism. It was found that phyA is represented in the cell by two native pools, major and light-labile phyA' and minor and relatively light-stable phyA'' (Sineshchekov 2004, 2005). The major objective of the present review is to consider the literature data related to phyA primarily from the viewpoint of the heterogeneity of the photoreceptor, its nature and possible functional implications.

2. Phytochrome A: Perception and Transduction of the Light Signal

2.1 Molecular structure and photocycle

2.1.1 The initial, red-light absorbing Pr form and its direct Pr→Pfr phototransformation

The pigment molecule is a homodimer whose monomers are chromoproteins with a molecular mass of approx. 125 kDa comprising about 1200 amino acid residues. Domination of the hydrophilic residues in the polypeptide chain of the pigment makes it water-soluble, cytosolic holoprotein. There are, however, amphiphilic regions at the terminal parts of the molecule which can participate in the protein-protein interactions. Each of the apoproteins is covalently connected by a thioether bond to a chromophore – phytochromobilin (PΦB) – via a cystein residue (Fig. 1, top). The chromophore is an open tetrapyrrole with a system of conjugated double bonds synthesized by reduction of biliverdin IX-α. It binds to the apoprotein autocatallytically in the hydrophobic pocket inaccessible to the external hydrophilic medium (Wu and Lagarias, 2000; Rockwell *et al.*, 2006).

The architecture of the phytochrome A monomer includes α-helices, random coils, β-turns and β-sheets as the secondary structure (Sommer and Song, 1990). Its structure-function relationships followed in the last decade with the use of molecular biological approaches can be briefly summarized as follows (Quail, 1997; Rockwell *et al.*, 2006). The molecule is comprised of two major domains (Fig. 1, bottom) – the N-terminal photosensory domain (approx. 65 kDa, 600-650 a. a.) and the regulatory C-terminal domain (approx. 55 kDa, 450-500 a. a.) connected by a hinge region. The N-terminal domain in its turn includes several subdomains: P1 – N-terminal extension, NTE, (65 a. a. residues, 10 kDa, serine-rich region with 8 residues of the first 12 being serines) is necessary for biological activity and is

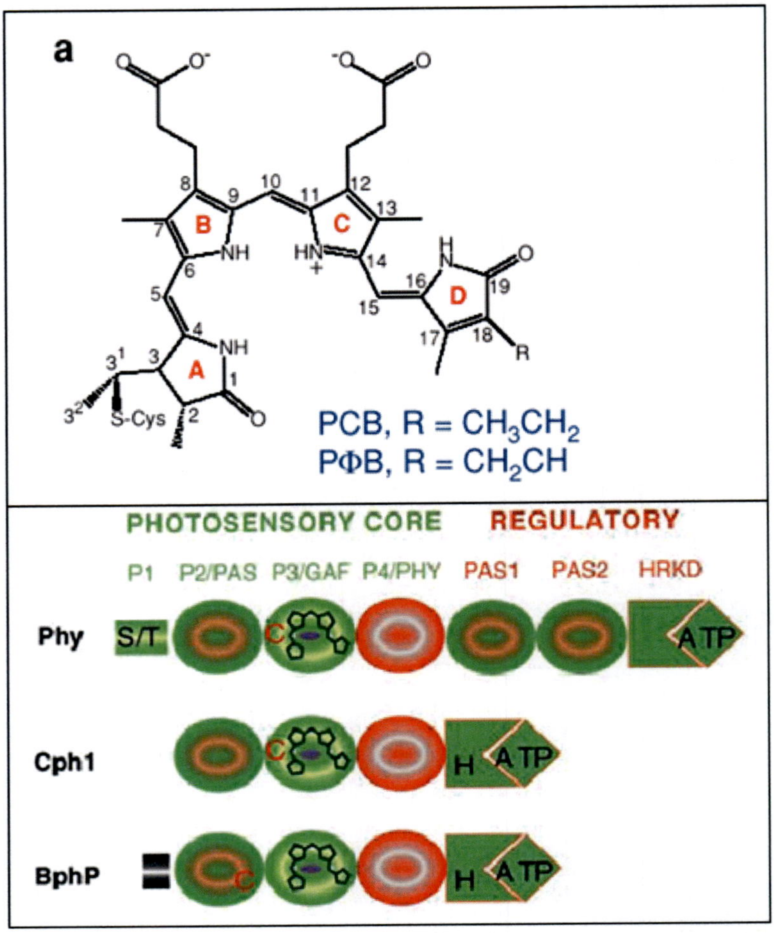

Figure 1. Molecular structure of phytochrome. Top panel: the structures of phytochromobilin (PΦB) and phycocyanobilin (PCB) chromophores differing only at the C18 side chain. These chromophores are utilized by plant phys and cyanobacterial Cph1s, respectively. The chromophore is shown in the C5–*Z,syn* C10–*Z,syn* C15–*Z,anti* configuration adopted in the Pr state. Bottom panel: domain architecture of the plant (Phy), cyanobacterial (Cph1) and bacterial (BphP) phytochromes. The phytochromes possess an N-terminal photosensory core region and typically share regulatory output domains related to those found on two-component histidine kinases (HKRD). The phytochromes share also P3/GAF domain associated with the bilin chromophore and P2/PAS and P4/PHY photosensory domains. Plant phytochromes (Phys) possess two additional PAS domains within the regulatory region. P1 in Phy is the N-terminal serine-rich extension. From Rockwell *et al.*, 2006 (modified).

involved in downstream signaling (Casal *et al.*, 2002). Boylan *et al.*, (1994) specify that the N-terminal 52 amino acids are necessary for completion of productive interactions only under FR-HIR conditions. The N-terminal extension is not exposed in the native conformation of phyA as probed by interactions with monoclonal antibodies (Natori *et al.*, 2007). The structures responsible for the difference in Pfr stability both between phyA and phyB and between the native pools of phyA (see below) are localized at the 6-kDa N-terminus; P3/GAF (domain acronyms connected with different types of protein folds, see in Rockwell *et al.*, 2006) is the bilin lyase subdomain; P2/PAS and P4/PHY domains are important for tuning the spectroscopic and photochemical properties of the bound bilin chromophore and P4/PHY is necessary for the intactness and stability of the pigment in the Pfr form (Oka *et al.*, 2004; Rockwell *et al.*, 2006). It is believed that the N-terminal domain in Pr is active in retaining phy in the cytoplasm (Nagatani, 2005).

The complex character and importance of functional interactions in the N-terminus can be illustrated by the effects of point-mutations in this region of the phyA molecule on its properties. Mateos *et al.* (2007) revealed a role of a distant residues of the GAF subdomain in regulation of phytochrome bilin-lyase activity. They detected a mutant allele of *Arabidopsis thaliana* (*phyA-303*) bearing an R384K substitution in the GAF subdomain, 61 a. a. residues away from the chromophore-bearing cystein residue (a distance, more than 13 Å), which showed poor incorporation of the chromophore. Other GAF-domain mutants have been reported in *Arabidopsis*: (i) the phyA (*eid4*) allele (substitution E229K) hypersensitive to FR and with normal PHYA protein levels and phyA spectral activity (Dieterle *et al.*, 2005); the *phyA-110* mutant (R279S), with normal WT protein levels but unable to form the Pr state from Pfr and the *phyA-109* mutant (G367S) with poor responses to FR and reduced levels of spectrally active phyA (Xu *et al.*, 1995). In pea, a mutation was found in an N-terminal region characterized by strong sequence conservation in phyA and high diversity from phyB (Weller *et al.* 2004; Sineshchekov and Weller 2004). The mutation is a substitution of alanine by valine at the position 194. It impaired phyA destruction and strongly enhanced its signaling ability, also suggesting a role of this locus in modulation of phyA activity.

The C-terminal segment is comprised of a PRD domain bearing a motif implicated in protein-protein interactions and interdomain communications and a histidine-kinase-related domain, HKRD (Yeh and Lagarias, 1998) (Fig. 1, bottom). The PRD motif contains two subdomains, PAS1 and PAS2, and also the regulatory core region (Quail box) which is responsible for the dimerization of the photoreceptor and protein-protein interactions between phytochrome and phytochrome-interacting proteins (PIPs). The HKRD segment, a histidine kinase paralog with Ser/Thr kinase specificity (Yeh and Lagarias, 1998), is involved in auto- and trans-phosphorylation activity of the pigment (see below). Finally, there is a stretch overlapping with HKRD which is associated with the nuclear localization activity (nuclear localization signal, NLS) (Kircher *et al.*, 1999).

The functional role of the C-terminal domain is far from being completely understood. Initially, it was believed that it initiates the signal in phytochrome-interacting transduction chain intermediates after its perception by the receptor N-terminal domain (Song *et al.* 1997). Later on it was shown (for phyB) that removal of the C-terminal domain was not critical to signaling but even enhanced the activity of the photoreceptor in inhibition of hypocotyl elongation provided the N-terminal half of the molecule was fused to dimerization and nuclear localization sequences (Matsushita *et al.*, 2003; Nagatani 2005). Quite recently, however, the same research group has found that root greening under red light induced by phyB was substantially reduced in the absence of the C-terminal domain (Usami *et al.*, 2007). They also found that overexpression of the isolated C-terminal domain of phyB enhanced cryptochrome signaling. The authors suggested that different domains of phyB can play various roles depending on light conditions and specific physiological response. The question of how this conclusion may relate to phyA remains, however, open. It is interesting to note in this connection that the PAS2 subdomain of the C-terminal domain in phyA is necessary to initiate signaling in HIR but not in VLFR (likely via interaction with a specific partner). This was shown by isolation of a. a. substitution mutants in *Arabidopsis* phyA PAS2 (*phyA-302*) retaining VLFR and lacking HIR (Yanovsky *et al.*, 2002).

The higher-order structures of phyA (tertiary and quaternary) was recently reevaluated with the use of an improved small-angle X-ray scattering technique and electron microscopy (Nakasako and Tokutomi 2005). They confirmed the dimeric association of Pr and determined the radius of gyration of 57.1±0.1 Å and maximum dimension of 183±5 Å. According to their low-resolution molecular model of phyA it is composed of two identical flat lobes stacked in an off-centered manner (Fig. 2) with a dimension of one lobe of 50 x 120 x 120 Å. A lobe is divided into two parts: one facing to the two-fold symmetry axis and the other located apart by ca. 60 Å from the axis. They were assigned to the N-terminal photosensory and C-terminal regulatory domains of the molecule, respectively.

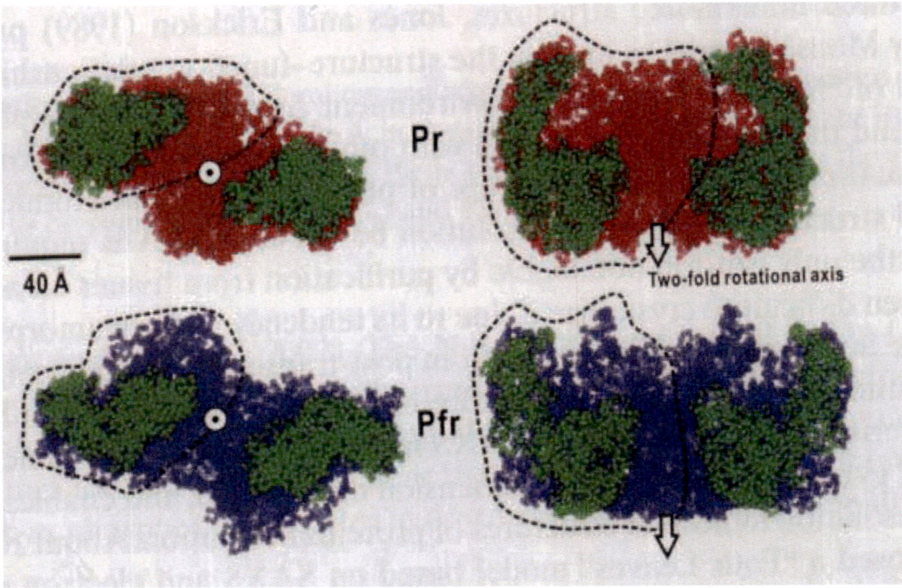

Figure 2. Molecular structures of the Pr and Pfr forms (red-colored models in the upper panel and blue-colored models in the lower panel, respectively) of pea large (118 kDa, lacking the N-terminal 6-kDa segment) phytochrome restored by *ab initio* simulation for small-angle X-ray scattering (SAXS) data. Two views of molecular models are displayed as assemblies of dummy scatters used in the simulation. The dashed lines indicate plausible borders of two subunits. The small circles in the left panel and arrows in the right indicate the twofold rotational symmetry axes. The green-colored model of pea small phyA (corresponding to the N-terminal domain) restored from the SAXS is superimposed on the models. (From Nakasako and Tokutomi, 2005.)

A great step forward in structural studies of phytochromes has been made recently with the unraveling of a crystal structure for the biliverdin (BV) containing P2/PAS and P3/GAF domains of the bacteriophytochrome from *Deinococcus radiodurans* (Wagner *et al.*, 2005). The authors confirmed the PAS and GAF folds for the P2 and P3 domains and showed that BV is covalently attached to Cys and is deeply buried within the conserved pocket in the P3/GAF domain. The two domains are connected by a trefoil knot – like structure which is highly conserved among other phytochromes. Rockwell *et al.* (2006) point out that the proper folding of this domain is critical for the phy functioning as is shown by loss-of-function and gain-of-function plant phytochrome mutations in the interface between the PAS domain and the trefoil knot and in the helices of the GAF domain itself. The authors proposed a role of these helices in signal transduction via light-mediated regulation of either intramolecular interactions or direct intermolecular interactions with downstream signaling components. They suggest also that the N-terminal P1 extension in plant phytochromes can affect the folding process in this region of the molecule and thus could play a potential regulatory role.

PhyA after its biosynthesis undergoes post-translational modifications. In the context of this review, of major importance is phosphorylation because it affects the stability of the pigment and interaction with putative phyA signaling proteins (phyA-signaling attenuation mechanism). It is believed also that phosphorylation could be involved in the differentiation of phyA into its two native populations, phyA' and phyA'' (see below). PhyA is a phosphoprotein *in vivo* (Quail *et al.* 1978; Hunt and Pratt, 1980) and it was shown to be phosphorylated after extraction (Biermann *et al.*,1994; McMichael and Lagarias, 1990; Wong *et al.*, 1986). A protein kinase activity was found (Wong *et al.*, 1986) to be associated with purified oat phy and red irradiation of coleoptiles or extracts reduced phosphorylation of phy whereas subsequent FR irradiation reversed the R light effect. Thus, a substrate for this activity was phy in the Pr form whereas Pfr was poorly phosphorylated by this kinase. Serine was the major phosphorylated residue on phy regardless of the phy form. According to Wong *et al.* (1986) the preferred sites of Pr phosphorylation was at the N-terminus (within the 10 kDa fragment) whereas in the Pfr form – at the C-terminus. Upon photoconversion into Pfr the site at the N-terminus could not be longer phosphorylated. McMichael and Lagarias (1990) confirmed and detailed these observations: phosphorylation of the Pr form occurred predominantly on Ser17 and of the Pfr form, on Ser598. This evidences for light-induced conformational alterations in the exposure of the region at Ser598 and at the N-terminus which seem to be important for phytochrome-mediated signal transduction. These data were in agreement with earlier observations that the region at the N-terminus is exposed in the Pr form and that photoconversion to Pfr shields this region and exposes the central hinge region (although this is in a seeming contrast with the recent finding that the N-terminal extreme is not exposed in the native conformation of phyA (Natori *et al.*, 2007)).

The importance of the N-terminus for preservation of the native spectral properties of phy was noted earlier by Vierstra and Quail (1986). Later on, Lapko *et al.* (1999) localized the sites of phosphorylation of oat phytochrome A *in vivo* at Ser7 and Ser598 and at Ser17 *in vitro*. Whereas the extent of phosphorylation at Ser7 was the same for phyA isolated from dark-grown and R- illuminated seedlings, the site at Ser598 was found only in phyA from the R-illuminated seedlings supporting earlier observations (McMichael and Lagarias 1990). This is consistent with the notion that the protein phosphorylation may play a functional role in the phytochrome-mediated light signal transduction. Thus, there are three sites of phosphorylation in phyA, at Ser7 and Ser17 at the N-terminus and at Ser598 in the hinge region between the two domains which are shown to be of the key functional importance, see below (Kim *et al.*, 2005).

Since phyA is the most abundant of all the phytochromes, the main body of the data on the physico-chemical properties of phytochrome and its photocycle obtained so far is related to it, more specifically, to the mixture of the two phyA species with different photophysical and photochemical properties (see for a review Sineshchekov 1995a). This heterogeneity may explain, at least partially, their complex character. In molecular terms, three major factors determine the spectroscopic and photochemical properties of phyA – stretching the chromophore, protonation of the Pr form and interaction with charged groups residing nearby the chromophore (Rüdiger, 1986).

Phytochrome absorbs in a diapason from ca. 300 nm to 800 nm, the most effective being red - far-red bands with absorption coefficient for Pr and Pfr of 70-130 and 40-50 mM^{-1} cm^{-1}, respectively. The absorption spectrum of the native full length 124 kDa phyA (spectrum *in vivo* cannot be obtained because of low phy content and high light scattering) in the Pr form have a maximum at 665-666 nm, a shoulder at 608 nm and a small maximum in the Soret band at 379 nm (see Fig. 3, left panel). Deep freezing of the samples causes a 5 -10 nm red shift and narrowing of the main bands. Oscillator strength ratio for the red and Soret band is approx. unity what is typical for extended conformation of the chromophore. And finally, the spectra of circular dichroism with a strong negative band for Pr suggest chromophore chirality due to the interaction of the chromophore with the protein.

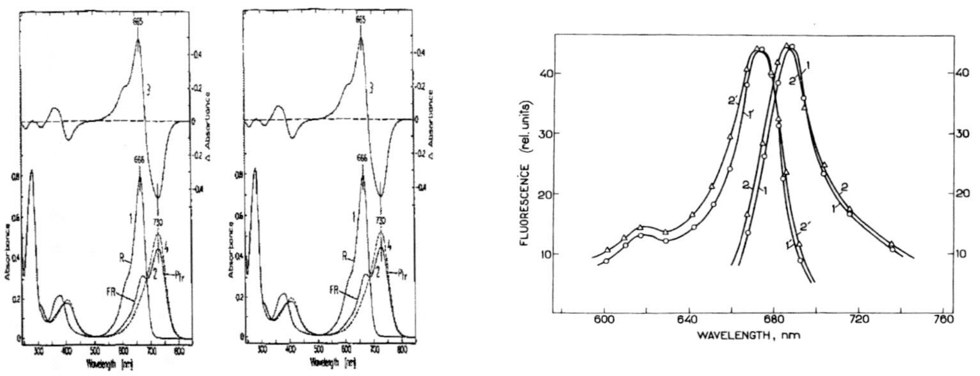

Figure 3. Absorption and fluorescence spectra of phytochrome. Left panel : Initial (1, 2), difference (3) and absolute (1, 4) absorption spectra of 124 kDa *Avena* phytochrome in the Pr (1) and Pfr (4) forms. Initial spectra were measured after far-red (FR) (1) and red (R) (2) illumination. Difference spectrum (3) is spectrum (1) minus spectrum (2). Initial spectrum (1) represents absolute spectrum of Pr because practically all phytochrome is in the Pr form after FR illumination; the absolute Pfr spectrum (4) was calculated from spectrum (2) after correction for incomplete photoconversion of Pr to Pfr assuming relative Pfr content of 0.86 at photoequilibium (From Rüdiger, 1986). Right panel: Low-temperature (85 K) emission (1,2, λ_e = 630 nm) and excitation (1',2', λ_m = 720 nm) fluorescence spectra of the red phytochrome form Pr in the cells of coleoptiles of etiolated oat seedlings (1, 1') and full-length 124 kDa phytochrome isolated from them (2, 2'). (From Sineshchekov *et al.*, 1990).

Fluorescence of phyA in its Pr form was detected both *in vitro* and *in vivo* with emission and excitation maxima at 680-687 nm and 666-673 nm depending on plant species and tissues used and experimental conditions (Fig. 3, right panel). The quantum yield (ϕ_f) is pretty low at ambient temperature (ϕ_f) is in the range ca. (1- 5) x 10^{-3} according to different sources) but it increases up to ca. 0.3 and more upon freezing at 77 K and below. The Pr fluorescence decay kinetics reveal a number of components with lifetimes (τ_f) ranging from several ps to 1 ns. The components do not depend on the degree of intactness of the phy molecule suggesting that the properties of the Pr excited states are governed primarily by the immediate protein surrounding of the chromophore. The time-resolved Pr fluorescence spectra of the kinetic components had different position of the maxima (λ_{max}) which varied within 10 nm.

Photoreaction, structural aspects. According to earlier data on plant phytochrome, (Rüdiger, 1986) the points of fixation of the chromophore in phyA are at the A-ring (the covalent bond) and between the arginine and propionic side chain of the ring B. The positive charge of the protonated phytochromobilin at ring B could be balanced by negatively charged aspartate or glutamate. Rings C and D have freedom for photoinduced conrotation about the C14-C15 and C15-C16 bonds. The model by Parker *et al.* (1994) suggest that Arg318, Ala319, the methylene of Ser322, Leu325 and Tyr327 form a pocket in which one side of the chromophore (bound to Cys323 in Avena) is anchored hydrophobically, covalently, and electrsotatically while the other side of the chromophore lacks hydrogen-bond donors and is involved only in van der Waals contacts.

The discovery of prokaryotic phytochromes, and, in particular, of the cyanobacterial phytochrome Cph1 (Hughes *et al.* 1997; Yeh *et al.* 1997; see Hughes and Lamparter 1999 for review) promises new insights into the structure-function relationships in the phyA molecule especially in view of a perspective of possible Cph1 crystallization. It may represent a model for the phytochrome family, although the PAS domain is missing from the pigment (Fig. 1, bottom). Cph1 autoassembles, with the reaction being a complex process of the chromophore (phycocyanobilin, PCB) incorporation (Borucki *et al.* 2003; Sineshchekov *et al.*, 2002). In general, the pigment possesses spectroscopic and photochemical characteristics close to those of phyA, more specifically, to its phyA'' form (the Pr'' type, see below) (Lamparter *et al.* 1997; Lamparter *et al.* 2001, Sineshchekov *et al.* 1998; Sineshchekov *et al.* 2002). Of particular interest is the fact that Cph1 photoconversion is associated with proton exchange (Remberg *et al.* 1997; van Thor *et al.* 2001) and that the Cph1 chromophore is fully protonated in both Pr and Pfr forms (Strauss *et al.* 2005). This suggests that the chromophore deprotonation in the Pr→Pfr photoconversion is only transient. To what extent this can be related to phyA remains, however, an open question because the proton exchange is seen in phyA only after a proteolytic attack (Tokutomi *et al.* 1982). It is of interest to note in this connection that the proteiolitic attack leads to the conversion of the phyA' form into phyA'' form (V. Sineshchekov, unpublished results). Thus, this is consistent with the fact that photochemically Cph1 is very close to phyA'' and could be considered as its model (see below).

At present, the state of the chromophore in the apoprotein pocket of phyA and of other plant phytochromes can be modeled proceeding from the situation in bacteriophytochrome (Wagner *et al.* 2005) based on the structural homologies of the P2/PAS and P3/GAF domains among the members of the phytochrome family (Rockwell *et al.*, 2006). The chromophore deeply buried in the GAF domain is on one side packed onto the highly conserved motif formed by Asp207, Ile208, and Pro209 and on the other closely apposed to His260. The A-ring is sandwiched between structure elements of the GAF domain and is covalently attached to Cys24. The B-ring carboxylate is associated with Arg254 while that of the C-ring with His260, Ser272 and Ser274. The D-ring is in a looser environment that would sterically permit more ready rotation (Fig. 4). The authors thus conclude that the conversion of Pr to Pfr proceeds with rotation of the D-ring and only the D-ring of the chromophore. The chromophore is in the ZZZ,ssa configuration in the Pr state. This is in general agreement with the earlier models according to which isomerization about C15 double bond between rings C and D is the most likely mechanism of the photoreaction (see above). Thus according to

Rockwell *et al.* (2006) the conversion of Pr to Pfr is best described by a single photochemical isomerization of the chromophore about C15—C16 double bond with lumi-R and Pfr adopting a C15—E,anti conformation. The authors believe that the Pfr chromophore does not have a more extended conformation than the initial in the Pr form and that the 60 nm red shift of the spectrum arises from greater electron density on the D–ring due to two proton transfers between the chromophore and the protein extending the effective length of the conjugated system. The Pfr structure could be stabilized through the donor/acceptor interaction within the chromophore, proton transfers and conformational changes of the protein matrix. Recent investigations with the use of solid-state nuclear magnetic resonans (NMR) of the 59-kDa N-terminal module of the cyanobacterial phytochrome Cph1 with ^{15}N-labeled phycocyanobilin cofactor (Rohmer *et al.*, 2006) revealed that photoisomerization at the C-15 methine bridge is linked to a loss of mobility of the entire chromophore in the Pfr state and its strong tension in contrast to a more relaxed Pr state. The outer rings (A and D) of the chromophore are significantly affected by the phototransformation, whereas the inner rings (B and C) remain in a conserved environment during phototransformation. The authors thus suggest that phototransformation include rotations between the A – B and C – D methine bridges and that the intramolecular signal transduction to the protein surface is achieved by the movement of a unit composed of the central rings B and C.

The energetics and kinetics of the initial photoprocesses. Picosecond kinetics measurements on large phyA reveal a decay of the Pr excited states (Pr*) within ca. 15-40 ps to prelumi-R (batho-R) which in its turn decays with a delay (up to 100 ps) to the photoproduct lumi-R (see reviews Braslavsky *et al.*, 1997; Gärtner *et al.*, 2004; Schaffner *et al.*, 1990; Sineshchekov, 1995a). The question arises, however, if the appearance of lumi-R proceeds via an intermediate prelumi-R or directly (Lippitsch *et al.* 1993). The quantum yield of the lumi-R formation in oat phyA was ca. 0.1 in agreement with the literature data for the total Pr→Pfr transformation. The upper value of the energy accumulated in lumi-R is 180 kJ mol^{-1}, and the sum of the constants of the initial photoprocesses of Pr is 0.3×10^{10} s^{-1} as estimated with the use of laser-induced optoacoustic spectroscopy.

At low temperatures (below 150-170 K), the photoprocess of the Pr→Pfr conversion can be stopped at the stage of lumi-R and the direct and reverse photoreactions (Pr↔lumi-R) can be investigated with the use of the conventional absorption and fluorescence technique (see Kendrick and Spruit., 1977; Rüdiger and Thümmler, 1991, 1994; Sineshchekov, 1995a for a review). Absorption (and fluorescence excitation) spectrum of lumi-R is characterized by the main band at 696-698 nm with a shoulder at 630 nm, by the better resolved structure, and higher oscillator strength than Pr. It has an emission maximum at 702-705 nm and a quantum yield of fluorescence close to that of the Pr form, 0.3 at 77 K (see discussion in Sineshchekov 1995a).

The kinetics of the light induced absorption and fluorescence intensity changes in the Pr↔lumi-R conversion at low temperatures is complex. It reveals several components whose rate and relative amplitudes change depending on temperature and direction of transformation, Pr→lumi-R or reverse. This was interpreted as the existence of several Pr ground state populations which differ by the activation barrier of the photoreaction (from several hundreds to several thousands J mol^{-1}) and thus by its quantum yields at low temperatures (see Sineshchekov and Sineshchekov, 1990; Sineshchekov and Akhobadze, 1992; Sineshchekov, 1995a and below).

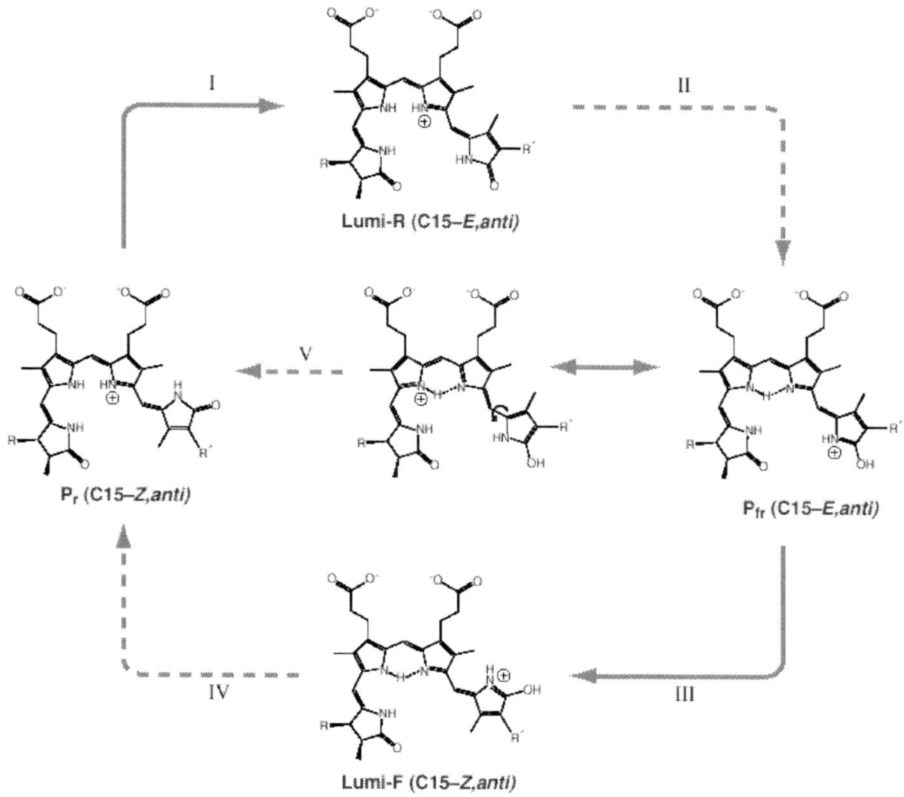

Figure 4. Top: Simplified cycle of the flower plant phytochrome conversion between the initial red-light – absorbing Pr form into the functionally active far-red-light – absorbing Pfr form. The initial photoreactions can be stopped at the stage of the formation of the photoproduct lumi-R for Pr (and lumi-F for Pfr) at low temperatures. Upon thawing the sample above indicated temperatures, the pigment undergoes further dark thermal relaxation to Pr (Pfr) via meta-R (meta-F). In contrast to the relatively stable Pr, Pfr is subject to (i) dark reversion into Pr, (ii) destruction and (iii) transport from the cytoplasm to the nucleus, where it interacts with the transduction chain partners. Bottom: Conformation changes of the chromophore in this cycle proposed in (Rockwell *et al.*, 2006). The scheme incorporates the crystal structure data on the bacterial phytochrome DrBphP (Wagner *et al.*, 2005). Red light (R) triggers photoisomerization about the C15-C16 double bond to give the C15 -*E,anti* primary photoproduct lumi-R, which is subsequently converted to Pfr. Far-red light (FR) triggers the reverse photoisomerization to yield the C15-*Z,anti* lumi-F primary photoproduct, which is subsequently converted to Pr. Dark Pfr →Pr reversion proceeds through the thermal rotation about the C15 -C16 bond which has a single-bond character in the Pfr resonance form. (From Rockwell *et al.*, 2006.)

Several schemes of the initial Pr photoreactions were proposed. The schemes by Holzwarth *et al.* (1992) and Lippitsch *et al.* (1993) had a common characteristic feature – the lack of the activation barrier for the Pr→lumi-R photoreaction in the excited Pr state and the scheme by Lippitsch *et al.* (1993) did not consider the reaction as an energy conserving process (see discussion in Sineshchekov, 1995a). An alternative scheme was proposed in (Sineshchekov and Sineshchekov, 1990) which explains the following aspects of the process

(Fig. 5). (1) Violation of the mirror symmetry rule between the absorption (excitation) spectrum and the emission spectrum of Pr. (2) The photoreaction is an activation photoprocess with an energy barrier in the excited state. (3) The photoreaction (completed or uncompleted) is the main (if not the only) temperature dependent process for the deactivation of the excited state. (4) A branching between the completed and uncompleted photoreactions takes place at the orthogonal 'hot' ground state between Pr and lumi-R (prelumi-R or batho-R state). (5) The photoreaction is the energy conserving process which is then used to drive the Pr→Pfr semi cycle. (6) The photoconversion lumi-R→Pr proceeds along a symmetrical scheme (but may not via the same intermediate) since lumi-R is pretty close by its photophysical and photochemical properties to Pr (Sineshchekov and Akhobadze, 1992).

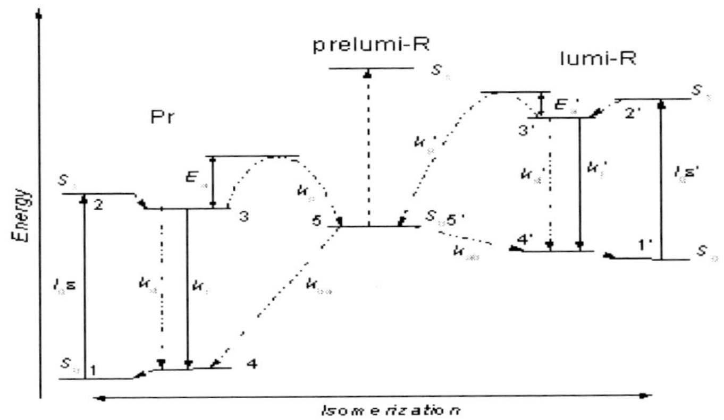

Figure 5. The energy level scheme of the photoreaction of the initial red-absorbing phytochrome form (Pr) into the first photoproduct (lumi-R) stable at low temperatures via a short-lived unstable orthogonal intermediate (prelumi-R); $I_0\varepsilon$, $I_0\varepsilon'$ - absorption coefficients; *k*, *k'* - constants of fluorescence, (k_f, k_f') temperature-independent degradation of excitation (k_d, k_d'), initial photoreaction (k_p, k_p'), phototransformation into photoproduct (k_{ab}), and return into the initial state (k_{ba}); E_a and E_a', activation energy barriers of the direct and reverse photoreactions. (From Sineshchekov and Sineshchekov, 1990.)

Dark transformations of the lumi-R photoproduct are shown in Fig. 6. In the Pr→Pfr phototransformation 4-5 kinetic intermediates were revealed with lifetimes ranging form several μs to hundreds ms (Zhang *et al.*, 1992; Scurlock *et al.*, 1993; Braslavsky *et al.*, 1997; Song *et al.*, 1997). The data are compatible with the models involving two parallel pathways or a sequential pathway with equilibrium at certain stages. The last step (formation of Pfr) involves an unbranched reaction. Low-temperature absorption investigations gave the picture of the Pr→Pfr phototransformation essentially similar to that of kinetics measurements (Kendrick and Spruit, 1977; Rüdiger and Thümmler, 1994): lumi-R→ meta-Ra→ meta-Rc → Pfr. They revealed a change in the sign of the circular dichroism (CD) bands upon formation of lumi-R and a decline in CD intensity upon generation of meta-Ra, meta-Rc and Pfr. This suggests a decrease in the interaction between the chromophore and the aromatic residue(s). FTIR studies also point to changes in the interaction of the carbonyl groups at rings A and D with the protein in the intermediates (Foerstendorf *et al.*, 1995; Siebert *et al.*, 1990). In general, the steps Pr→lumi-R→meta-Ra reflect changes in the chromophore while further steps involve protein rearrangement (Björling *et al.*, 1992; Eilfeld and Eilfeld, 1988).

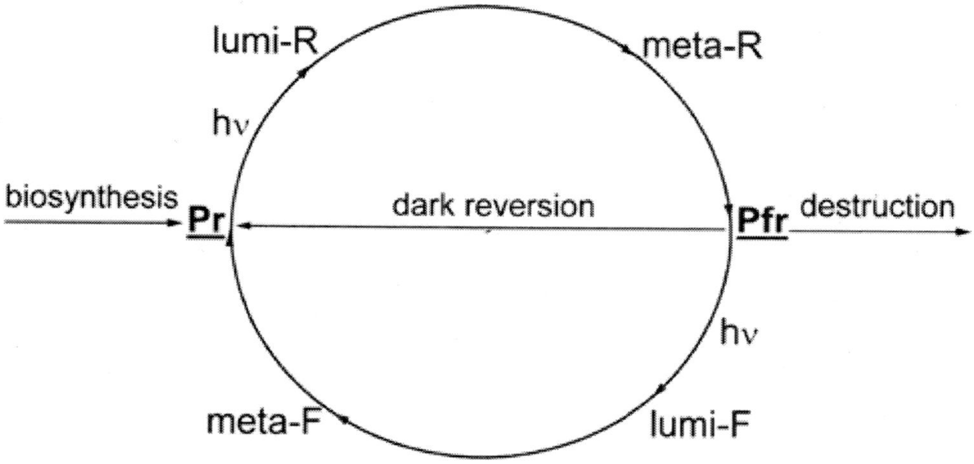

Figure 6. Simplified cycle of the phytochrome photoconversion and turnover.

2.1.2 The active, far-red-light absorbing Pfr form, and its reverse Pfr→Pr photoransformation

The chromophore in the Pfr form is different by its state of isomerization from that of Pr and possibly by the state of its protonation (see above). Besides, linear dichroism studies, energy transfer methods, experiments with oxidizing or reducing substances and molecular modeling revealed chromophore movement from its binding pocket during photoconversion (Song, 1985; Sommer and Song, 1990, Parker *et al.*, 1994). It should be mentioned that the results on higher plant phytochrome are in seeming contrast with the data obtained on the bacterial phytochromes, which predict relatively restricted movement of the chromophore in the apoprotein pocket. The state of protonation of the chromophore in phyA is also different form that in Cph1 (see above). This suggests that there should be specificity of the chromophore-protein interaction in phyA and that the results obtained on the model phytochromes (Cph1) cannot be readily applied to phyA. At the same time there is close similarity between the Cph1 photochemical properties and those of the Pr'' photochemical phytochrome type (see below). Obviously, this problem needs further detailed experimental and theoretical consideration.

Structural changes take place in the apoprotein as well, in particular, the α-helical folding increases as shown with the time-resolved CD spectroscopy in the far-UV region, differential sensitivity to proteolysis and binding studies with antibodies (Chen *et al.*, 1993; Deforce *et al.*, 1994; Song *et al.*, 1997). They are localized primarily within the 10 kDa N-terminus, at the junction of the N- and C-terminal domains, and around glutamate-354. (Nakazawa *et al.*, 1993). Also, the regions around four of the 10 tryptophanes may represent conformationally photoresponsive areas (Wells *et al.*, 1994). These alterations in the chromophore produce changes of its vibronic structure revealing themselves in modifications of spectroscopic properties. Firstly, the absorption spectrum shifts bathochromically giving the maxima at 730 and 400 nm. Secondly, the most pronounced changes are seen in the fluorescence yield of Pfr: its fluorescence was not detected neither *in vitro* nor *in vivo* even at cryogenic temperatures (Sineshchekov 1995a). The absence of the Pfr fluorescence was interpreted (Sineshchekov,

1995a, 1999) either as an indication to the reordering of the Pfr excited states so that the state with a prohibited transition becomes the lowest or that the photoreaction and internal conversion compete very efficiently with fluorescence even at low temperatures. This demands that there is no, or very little energy barrier for the Pfr→lumi-F photoreaction. The latter is at odds with the fact that it is an activation process with the potential barrier similar to that of Pr (ca. 5 kJ mol^{-1}) (see Sineshchekov, 1995a for discussion). Rockwell *et al.* (2006) explained the fact that the fluorescence yield of phyA in the Pfr form is extremely low by a proton transfer in the excited state via tunneling between the B- and C-rings or at the D-ring carbonyl. However, the intimate mechanism of this quenching still remains to be elucidated. Finally, modifications in Raman and IR spectra were also observed (Farrens *et al.*, 1989; Rospendovski *et al.*, 1989). This and the CD spectra of Pfr with a relatively weak positive signal for Pfr (Eilfeld and Eilfeld, 1988; Björling *et al.*, 1992; Chen *et al.*, 1993) (see above) suggest weaker interaction between the chromophore and the aromatic residue(s) in this form.

The photoreaction of Pfr is the reversion of the direct Pr photoreaction, that is a 15—E,anti – to – 15Z,anti photoisomerization around the C15—C16 double bond (Fig. 4). Three kinetic species were identified in the Pfr→Pr photoconversion and a sequential pathway model was proposed (Chen *et al.*, 1996). Low-temperature steady state measurements identify two intermediates (Fig. 6), lumi-F and meta-F, although the picture may be more complex (Kendrick and Spruit, 1977).

2.2.3 Pfr-to-Pr dark reversion and Pfr destruction

The above structural changes in the Pr→Pfr transition expose hydrophobic regions in the molecule which are considered to be responsible for the biological activity of the pigment. The photophysiological effect depends, in its turn, on the Pfr turnover in the cell determined by its degradation and dark reversion into the initial inactive form (see Clough and Vierstra, 1997 for a review). Phenomenologically, three patterns of Pfr decline were distinguished in plants depending on the prevalence of degradation or reversion (Furuya, 1989): 1. only dark reversion (light-grown cauliflower head, cauliflower type); 2. only decay (monocot coleoptile type); 3. both reversion and decay (dicot tissues type). This diverse picture can be explained by the existence of the light-labile phyA (and also of native pools within phyA differing in light-lability, see below) and of the light-stable phyB.

The Pfr→Pr dark reversion depends on plant species (characteristic of dicotyledons), phytochrome types (phyA and phyB) and their native state (Eichenberg *et al.*, 1999). As found on transgenic phytochromes from tobacco expressed in yeast and reconstituted with PCB dark reversion for phyB was faster than for phyA (Ruddat *et al.*, 1997) while according to (Kunkel *et al.*, 1995) phyA revealed no dark reversion in yeast. Kunkel *et al.* (1995) suggest that the molecular basis for the dark reversion is the formation of heterodimers between the Pr and Pfr monomers. Later on Eichenberg *et al.*, (1999) have found that dark reversion is possible both from Pr/Pfr and Pfr/Pfr dimers and hypothesized that there exists mechanism preventing it. From the kinetics point of view, the Pfr→Pr dark reversion should depend on the potential barrier for the process in the ground state which vary depending on phy species and their native state in the cell. In monocots, where this process is not observed (see above), phyA destruction becomes responsible for the attenuation of the signal transduction.

PhyA in the Pfr form is subject to rapid degradation (half-life, tens of minutes) whereas Pr is relatively stable (tens of hours). Considerable variability of degradation depending on fluence rate and wavelength among different species was followed. In monocots proteolytic degradation of phyA is equally fast in FRc and Rc whereas in dicots it is much slower in FRc than in Rc. Half-life of phyA in *Arabidopsis* in R is 30 min, in FR, 2 h, and in the dark, 35 h (see Hennig *et al.*, 1999 and the literature cited therein). PhyA degradation includes ubiquitination of the pigment, i.e. formation of conjugates between phytochrome and ubiquitin, a small (76 a. a.) highly conserved polypeptide, and its aggregation with the effect of formation of sequestered areas of phyA (SAPs) in the cytosol (Clough and Vierstra, 1997; Kim *et al.*, 2000). This and also light-independent SAP formation of phyA apoprotein (observed in experiments on transgenic yeast expressing phytochrome) is considered to be a molecular property of phyA connected with relative hydrophobicity of the molecule distinguishing it from light-stable phyB. Interestingly, the light-independent SAP formation correlates with the fact that phyA in yeast is represented primarily by the presumably membrane (protein)-associated phyA'' (Sineshchekov *et al.* 2001, see below). In transgenic plants, foreign phyA is also rapidly degraded whereas phyB levels appear unaffected by light. However, the steady-state levels of ectopic phyA reached after R illumination in overexpressors are considerably higher than that of the endogenous phyA as a result of (1) constitutive resynthesis of Pr and (2) slower rate of Pfr destruction, see (Clough and Vierstra, 1997) and the literature cited therein.

The first step in the phyA degradation process is its tagging by phyA phosphorylation (at the N-terminal and hinge Ser residues) whose level is higher for Pfr than for Pr and also for Pr cycled through Pfr, Pr*: in monocots and dicots, the Pfr-induced degradation of Pr was observed (Clough and Vierstra, 1997; Hennig *et al.*, 1999; Shinomura *et al.*, 2000). According to Song *et al.* (1997), the phosphorylation induces only a subtle local change in conformation and this serves as a surface label, rather than a transducer-activating trigger. In more general terms, the tagging could be the modification which mediates the light conditions and subcellular localization of the pigment, its stability and interaction with the partners (see Neff *et al.*, 2000 and below).

The decline of phyA in the Pfr form depends on the degradation machinery in the cell. This is pointed at by the fact (Kunkel *et al.*, 1995) that recombinant phyA and phyB in yeast cells, which do not posses the phy degradation systems, reveal similar light-stability. In plants, this machinery works via the ubiquitin/26S proteasome proteolytic pathway (with determinants for phyA degradation at the 6-kDa N-terminal domain, between residues 742 and 790 and also possibly near the chromophore, residues 323–360 (in oat phyA) (Clough and Vierstra, 1997; Clough *et al.*, 1999; Quail, 1997; Wagner *et al.*, 1996). The process includes (1) ATP-dependent activation of ubiquitin by ubiquitin-activating enzyme (E1) and production of an E1-ubiquitin intermediate; (2) ligation of activated ubiquitin to phyA with the help of ubiquitin-conjugating enzyme (E2) and ubiquitin-protein ligase (E3, more specifically, CONSTITUTIVELY PHOTOMORPHOGENIC 1, COP1); (3) degradation of the ubiquitin-phyA conjugates by the ubiquitin/26S proteasome (Clough and Vierstra, 1997; Lin and Wang, 2007; Seo *et al.*, 2004).

It is interesting in the context of the notion of the two phyAs, light-labile and relatively light stable (see below), that two pools of phyA-ubiquitin conjugates exist (Clough and Vierstra, 1997): one that is more quickly turned over and a second pool that is less efficient proteolytic substrate. The authors suggest that phyA could be modified and this reduces its recognition as a substrate for degradation. Supporting this, a high steady-state level of [P_{tot}] under FR and a significant fraction of stable phyA Pfr after a R pulse that undergoes neither destruction nor dark reversion were also observed in *Arabidopsis* (Hennig *et al.*, 2000).

2.2 Transduction of the light signal

The discovery in the recent years of the principles and mechanisms of the light signal transduction with participation of phytochrome is connected with the finding of the two types of photomorphogenic mutants. The first ones are defective in transition from scoto- to photomorphogenesis, that is the plant remains partially etiolated even under illumination. This defect is due to the lack of the photoreceptor – phytochrome or of the so-called positive regulators of photomorphogenesis In the other group of mutants, the light strategy of development, on the contrary, is achieved in darkness: the plant assumes shape and size characteristic of light-grown plants. These mutants are defective in the negative regulators of photomorphogenesis. These proteins belong to the class of ligases (see above) and carry out proteolysis of the positive regulators thus blocking the photomorphogenic route of development. Investigations of the two group of the mutants revealed ever-growing number of the key transduction chain intermediates (see for a review Wang and Deng, 2002; Whitelam and Devlin, 1997).

Another break-through in the phy research in the recent years is the discovery of the light-induced transport of the photoreceptor from the cytoplasm to the nucleus (Nagatani, 2004; Nagy and Schaefer, 2000, 2002; Nagy *et al.*, 2001). It was shown with the use of labeled phytochromes, phyA and phyB fused to green fluorescent protein (phyA-GFP and phyB-GFP), that upon illumination both of them are transferred to the nucleus, although with different kinetics – phyA within minutes and phyB within hours.

Based on these discoveries, the chain of events in the process of signal transduction downstream from phyA can be schematically presented as follows (Fig. 7). Initially, phyA in its Pr form is uniformly distributed in the cytosol (Pratt, 1994). The pigment is diffuse and light-labile. Here, in the cytoplasm the photoreceptor interacts with partner proteins and these interactions are of importance with regard to its native state. This especially relates to the problem of phyA polymorphism because its two native species are found already in the Pr form and these interactions are of potential effect on their formation (see below).

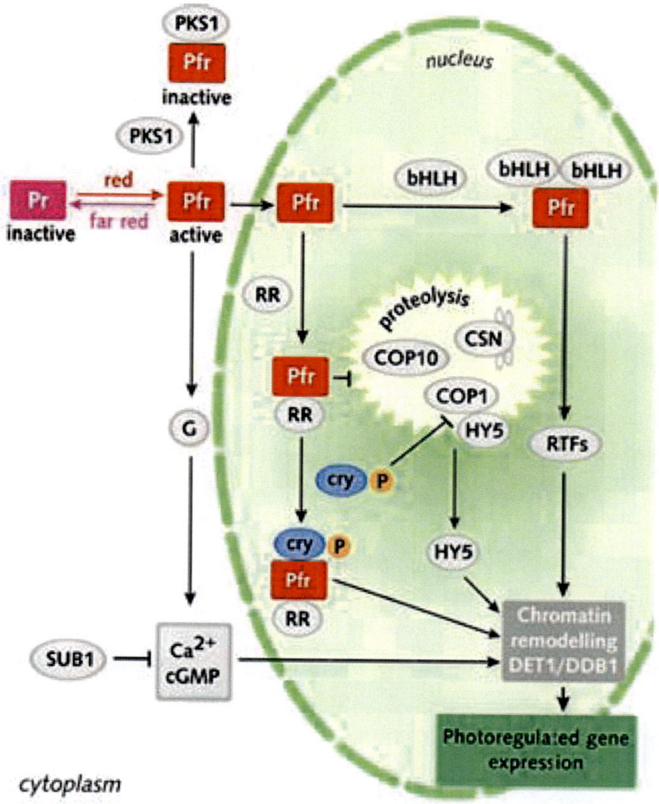

Figure 7. A model of phytochrome signal transduction. Activated phytochrome (Pfr) is proposed to regulate transcription through several parallel pathways. A rapid response involves Pfr translocation to the nucleus, where it binds transcription factors of the bHLH family (in particular PIF3). Key regulatory transcription factors (RTFs) that are responsible for inducing a range of light-regulated genes are subsequently activated. In a second nuclear-localized pathway, phytochromes are proposed to bind response regulators (RR), which stabilize them in the activated form and can induce light-regulated gene expression by inhibiting COP1-, COP10- and CSN-dependent proteolysis of the HY5 transcription factor and by binding to activated cryptochromes (cry). In all cases, regulation of the genes responsible for photomorphogenesis is predicted to require chromatin remodeling mediated by the DET1/DDB1 nucleosome-binding complex. In the cytoplasm, phytochrome may activate gene expression through G-proteins (G), calcium and cGMP-dependent pathways, which are regulated by SUB1. In addition, phytochromes may be sequestered away from the signaling-competent pool by PKS1. Elements involved in signaling from specific photoreceptors or controlling specific responses have not been included. From Schafer and Bowler, 2002.

These partner proteins are:

(i) phytochrome kinase substrates, PKS (Fankhauser *et al.*, 1999; Lariguet *et al.*, 2003). PKS1 and PKS2 locate in the cytoplasm and physically interact with phyA both in its Pr and Pfr forms. PKS1 was shown to be tightly associated with the plasma membrane. PKS1 and PKS2 are phosphorylated on serine and threonine residues by phyA which is more effective in the Pfr form. They act as negative regulators of phyA (and phyB) signaling this is believed to be achieved by (a) inhibition of postulated phytochrome kinase activity and/or (b) by retention of phyA in the cytoplasm and limiting the rate of its nuclear-cytoplasmic partitioning. *Arabidopsis* species modified in PKS content (PKS1OVX, *pks1* and *pks2*) are characterized by the increased VLFR whereas HIR revealed less sensitivity. A more complex fact for interpretation was, however, that, contrary to the expectations, the double mutant *pks1pks2* did not show hypersensitivity. To explain these data one has to employ an elaborated scheme of interaction of PKS1 and PKS2 with each other and with the other signal transduction components (see Lariguet *et al.*, 2003).

(ii) blue-light receptor cryptochrome 1, cry1; 2–4,114. Cry1 and cry2 interact with phytochrome A and photoactivation of phyA stimulates phosphorylation of cry1 *in vivo* enhancing its activity. (Ahmad *et al.*, 1998). Recombinant oat phyA *in vitro* phosphorylated recombinant preparations of cry1 and cry2 on serine residues with no difference between Pr and Pfr.

(iii) nucleoside diphosphate kinase 2 (NDPK2) – an enzyme implicated in many developmental processes. It binds to phyA with some preference for Pfr resulting in an increase in its enzymatic activity. It is a positive factor and may act as a transcriptional regulator (Choi *et al.*, 1999). Missense mutants of NDPK2 in *Arabidopsis* (*ndpk2* mutants) have reduced responsiveness to both R and FR suggesting that NDPK2 functionally interacts *in vivo* with both phyA and phyB probably acting as a transcriptional regulator. NDPK2 is localized also to the nucleus. Since R stimulates phosphorylation of NDPK2 *in vivo* it is likely to be a substrate for the kinase activity of phytochrome (Hamasa *et al.*, 1996, Tanaka *et al.*, 1998, Ogura *et al.*, 1999). PAS-related domains are implicated in phyA – NDPK2 interaction as shown by experiments with mutated pigment.

(iv) PAT1 - a phyA positive signaling component localized in the cytoplasm (Bolle *et al.*, 2000). This protein of 490 amino acids has sequence homology to the plant-specific GRAS regulatory protein family. It acts at an early step of phyA signal transduction.

Localization of phytochrome in the cell changes dramatically in response to light and this process exhibits a complex pattern. Upon photoconversion into Pfr, major part of the pigment rapidly aggregates in the cytoplasm forming speckles, the process followed by its destruction (see above). Nevertheless, the photoregulatory processes are likely to be initiated by Pfr in the cytoplasm (Bolle *et al.*, 2000; Rösler *et al.*, 2007; Schäfer and Bowler, 2002; Schäfer *et al.*, 1997). They involve the action of the "classical" biochemical mediators – G-proteins, Ca^{2+}/calmodulin, cGMP leading to the gene regulation (Fig. 7). On the other hand, they proceed at the biophysical level and include relatively rapid (seconds, minutes) changes in ion concentrations and electric potentials and osmotic phenomena, such as etioplast swelling, changes in the cytoplasmic motility and others (Long and Iino 2001; Roux, 1994; Takagi, 2005; Volotovski, 1998).

Part of the photoreceptor is transported into the nucleus forming there "speckles" which are relatively light stable and the photoregulation of intracellular partitioning of phyA differs significantly in different genetic backgrounds (Hisada *et al.*, 2000; Kircher *et al.*, 1999; Kim *et al.*, 2000). Initially the cytoplasm-to-nucleus partitioning was observed for phyB (Sakamoto and Nagatani, 1996; Kircher *et al.*, 1999; Yamaguchi *et al.*, 1999; Nagy *et al.*, 2001). Later on this process was also shown for phyA but the kinetics, light quality and quantity of the process were different from those of phyB (Kircher *et al.*, 1999; Kim *et al.*, 2000). The import was 10-fold faster (it takes minutes instead of hours for phyB) and induced by FR and R, although in R, the appearance of phyA in the nucleus was transient. The speckles formed by phyA in the nucleus were smaller than those of phyB and they were more numerous. PhyA in its Pr form cycled through the Pfr from (Pr*) was also capable of translocation in the nucleus. The light-induced intracellular location of phyA is wavelength dependent (different kinetics and pattern of the intracellular localization and destruction of phyA were observed under FRc vs. Rc) and this is believed to be important for its activity (Hisada *et al.*, 2000).

It is believed (Nagy and Schäfer, 2000) that the speckles represent loci for transcriptional activity. Very schematically the process can be represented as follows. In the nucleus, Pfr interacts with the negative regulators, which block photomorphogenesis. This interaction causes their proteolysis or transfer from the nucleus to the cytoplasm. As a result, the positive regulators are accumulated and the program of light development is switched on through activation of light-regulated genes.

Factors interacting with the far-red form of phyA in the nucleus are the following:

negative factors –

(i) PIF3 – phy interacting factor 3, (Bauer *et al.*, 2004; Kim *et al.* 2003; Ni *et al.*,1998, 1999; Quail, 2000; Zhy *et al.*, 2000). It is active in both phyA and phyB signal pathways although it has a greater effect on the phyB-regulated processes. It is necessary for VLFR and not for HIR of phyA. PIF3 is localized in the nucleus in the dark, physically interact with phytochrome and it undergoes rapid, light-induced degradation mediated by phyA, phyB and phyD. Although PIF3 functions negatively, rather than positively, as initially reported (Ni *et al.*, 1998) in the overall process of seedling de-etiolation, it nevertheless acts positively on a subset of rapidly responsive genes (Monte *et al.*, 2004; Quail, 2005).

(ii) SPA1 is a 114-kD protein with a motif (tryptophane-aspartic acid, WD, repeats) common to proteins involved in protein-protein interactions and with nuclear localization motifs (Hoecker *et al.*, 1998). An *Arabidopsis* mutant *spa1* is hypersensitive only to light signals perceived by phyA amplifying its signaling both in Rc and FRc. More specifically, SPA1 reduces the efficiency of phyA signaling with strong influence on VLFR and a weaker influence on HIR (Hoecker *et al.*, 1998, 1999; Zhou *et al.*, 2002). SPA1, representing a novel class of protein kinases, possibly post-translationally modifies the phyA molecule. It may down-regulate the signal activity of phyA by phosphorylation of serine residues at the extreme N-terminus (Hoecker *et al.*, 2003). Alternatively, it may degrade transcriptional factors involved in light signaling via COP1-SPA1 interaction (Saijo *et al.*, 2003). On the other hand, phyA was shown to induce increase in SPA1 mRNA levels under Rc what may provide negative feedback regulation of phyA signaling. Also, the regulatory steps with participation of phyA and SPA1 may involve specific phyA-induced modification of SPA1, thereby activating its repressor function.

Thus, SPA1 may play an important role in switching off phyA signaling in Rc and in conferring the FRc-specific sensitivity of phyA. According to Hoecker *et al.* (1998), at least three mechanisms exist that normally confine phyA responsiveness primarily to FRc with respect to deetiolation: first, transcription of the PHYA gene is downregulated by Rc (Somers and Quail, 1995); second, the Pfr conformation of phyA most abundant in Rc, is rapidly degraded (Clough and Vierstra, 1997) and third, signaling from phyA appears to be inhibited by an SPA1-dependent process. We may propose the forth mechanism that may regulate participation of phyA in R- or FR- initiated photoprocesses – changes in the equilibrium between the light-labile phyA' active under FR and relatively light-stable phyA'' which may function under R (see below).

(iii) EID1 is another negative regulator of phyA activity (Hoecker *et al.*, 2003; Zhou *et al.*, 2002) acting in HIR (with no effect on VLFR), possibly, by targeting activated components of the phyA signaling pathway to ubiquitin-dependent proteolysis (Dieterle *et al.*, 2001). It shifts the responsiveness of the phyA signaling from R to FR. EID1 is a novel nuclear F-box protein that contains a leucine zipper whose integrity is necessary for its biological function. Interestingly, *eid1* and *spa1* mutants had very similar phenotypes under weak-light conditions. However, SPA1 had a strong influence on the regulation of VLFR and a weaker influence on HIR whereas EID1 severely altered HIR and caused almost no change on VLFR. Measurements on *eid1spa1* double mutants indicate that EID1 and SPA1 are involved in different but interacting phyA-dependent signal transduction chains (Zhou *et al.*, 2002).

positive factors –

(i) HFR1 (Fairchild *et al.*, 2000). The *hfr1* mutant reveals reduction in responsiveness specifically to FRc and HFR1 mRNA is 30-fold more abundant in FRc than in Rc. It is a short-lived bHLH protein and is degraded through 26S proteasome-dependent pathway. Light enhances HFR1 accumulation by abrogating COP1-mediated its destabilization (Yang *et al.*, 2005). It is a transcription factor and has strong similarity to PIF3 and binds it. The HFR1/ PIF3 complex in turn binds phyA, preferentially to Pfr. This is likely to be the mechanism of HFR1 participation in the phyA signaling(Fairchild *et al.*, 2000). HFR1 does not act through the regulation of the phyA level or spectral activity. HFR1 does not bind phyA directly probably because such interaction requires a plant specific post-translational modification of HFR1 or phyA which is lacking in the system used (yeast or *in-vitro* coimmunoprecipitation assay). It can be pointed out in this connection that phyA obtained in yeast is different from native phyA expressed in plants (see Sineshchekov *et al.*, 2001a and below).

(iii) RSF1, an *Arabidopsis* bHLH transcription factor which is homologous to PIF3. It takes part in the FR-induced phyA mediated de-etiolation effects, although there is no evidence for direct interaction of this protein with phyA. (Fairchild *et al.*, 2000; Fankhauser and Chory, 2000; Soh *et al.*, 2000; Spiegelman *et al.*, 2000).

(iv) LAF1 - a nuclear protein specifically involved in phyA signal transduction under FR (Ballesteros *et al.*, 2001). LAF3 might regulate nucleo-cytoplasmic trafficking of an intermediate(s) involved in phyA signaling (Hare *et al.*, 2003)

(v) FHY1 and its homolog FHL (FHY1-like) – FHY1 is a 224 kDa protein - positive regulator of photomorphogenesis in FR participating directly in relaying the phyA signal. It is involved in the transport of phyA in the nucleus or in a regulatory complex formation of phyA, PIF3, and FAR1 (Hudson *et al.*, 1999; Zhou *et al.*, 2005). It regulates transcription by interacting with DNA-binding complexes as a coactivator. FHY1is most abundant in young seedlings grown in darkness and phyA is directly involved in its down-regulation (Shen *et al.*, 2005; Zeidler *et al.* 2004). FHL possesses functional nuclear localization and nuclear export signals and is essential for complete responsiveness to phyA-mediated signals. FHY1 and FHL are capable of homo- and heterodimerization and physical interaction with phyA (preferentially to the Pfr form). It is suggested that FHY1 might transmit phyA signals through direct interaction with the photoreceptor (Rösler *et al.*, 2007; Zhou *et al* 2005).

(vi) FAR1 - *Arabidopsis* mutant *far1* has reduced responsiveness to FRc but responds normally to other light wavelength what suggests a specific effect of the loss of FAR1 on phyA signaling (Hudson *et al.*, 1999). FAR1 must either be an integral component of the transduction pathway or a positive regulator of it. The FAR1 protein has no significant sequence similarity to any proteins of known function but it contains a nuclear localization signal (NLS) and is localized in the nucleus. FAR1 does not bind DNA itself but regulates transcription by interacting with DNA-binding complexes in the manner of a coactivator. FAR1 gene family may represent a new class of transcriptional regulators.

The sites in the phytochrome molecule involved in the recognition by a number of partners (NDPK2, PKS1, PKS2, PIF3 and others) are conserved between phyA and phyB. At the same time there is a specificity of interaction of these factors with phytochrome depending on the cell type. Neff *et al.* (2000) point out that mutants defective in PKS1, PIF3 and NDPK2 have different phenotypes affecting different tissues in the seedling. PKS1 and PIF3, for instance, are active in hypocotyls and NDPK2, in cells of the hook region.

To the problem of the membrane-associated phytochrome. In the context of the nuclear-cytoplasmic partitioning of phytochromes and their association with proteins retaining phytochrome in the cytoplasm (see above), the long-standing problem of possible association of phytochrome with cellular fractions again acquires importance and new dimension. Terry *et al.* (1992) have shown association of phytochrome with plasma membranes purified from etiolated wheat leaves. Plasma membranes contained about 0.1% of the total cellular phytochrome. Under R, the onset of the loss of phytochrome from plasma membranes was delayed for one hour as compared with the loss of P_{tot}. Similarly, the inhibition of synthesis by FRp was less for plasma membrane-associated than for P_{tot}. It is essential to note that the membrane binding was saturable. It was concluded that plasma membrane phytochrome represents a pool whose regulation is partially independent of the total type I phytochrome. Plasma membrane and total phy pools are affected differentially by treatments which lower P_{tot}. The association of phytochrome with plasma membrane is saturable when the total phytochrome content reaches about half the level in the dark controls. This indicates a limited number of available phytochrome-binding sites in the plasma membrane preparations which argues against the association being a non-specific, artifactual one. The results are, however, consistent with phytochrome being associated with a membrane-bound reaction partner. In dark-grown seedlings the number of plasma membrane-bound reaction partners control the amount of plasma membrane-associated phytochrome. These properties of the membrane-bound phytochrome are in remarkable parallels with the minor phyA'' pool found in our works (see below).

2.3 Regulation of phytochrome A activity

Two processes – phosphorylation/dephosphorylation of phyA and its ubiquitination in the Pfr form (see above) – play a central role in regulating its functions. They modulate (i) photoreceptor abundance, (ii) rate of the nuclear transport and (iii) its interactions with transduction chain partners. Attenuation of the photoreceptor abundance is connected with the light-induced degradation of phyA in the Pfr form and with the Pfr→Pr dark reversion (see above). PhyA is also subject to negative regulation of biosynthesis both at the protein and the mRNA level by a light-induced reduction in PHYA transcript abundance. PhyA and phyB are required for the negative regulation induced by FR and R, respectively, with monocots appearing to respond more strongly than dicots in this process (Quail, 1994). The effect depends on plant's organ and tissue and their physiological conditions: for instance, the regulation occurs predominantly in shoots, whereas in roots near constitutive expression is observed. PhyA level is thus regulated by multiple signals, including environmental, organ-specific and developmental signals (Cantón and Quail, 1999).

PhyA transport to the nucleus and its interaction with partner proteins are modulated by its phosphorylation as shown by the experiments on mutated phyA (Fankhauser *et al.*, 1999; Jordan *et al.* 1997; Kim *et al.* 2002, 2004; Stockhaus *et al.* 1992). The site critical for light regulation of phytochrome autophosphorylation and kinase activity is at Ser598. PhyA phosphorylates PKS1- PKS4 and thus modifies the process of its nuclear translocation, possibly, by inhibition of phyA kinase activity and/or by retention of phyA in the cytoplasm and limiting the rate of its nuclear–cytoplasmic partitioning (Fankhauser 2001). Phytochrome phosphorylation at Ser598 (see above) is involved in inhibition of its action also via hampering interaction of the pigment with its putative signal transducers (NDPK2 and PIF3).

It was followed in line with this that a type 5 protein phosphatase (PAPP5) specifically dephosphorylates phytochrome in its biologically active Pfr form and enhances phytochrome-mediated photoresponses (Ryu *et al.* 2005). Dephosphorylation of phyA by another phosphatase (FyPP) (Park *et al.*, 2000), which occurs in the cytoplasm, would modify its light-induced nuclear localization. Kim *et al.* (2002) point out that unphosphorylated Pfr has a reduced affinity for PKS1 (Fankhauser *et al.*, 1999) but exhibits an increased affinity for PIF3 (Ni *et al.*, 1998). Thus, the light-activated phytochrome is desensitized by phosphorylation at Ser598 but sensitized by FyPP-mediated dephosphorylation. This is a general sensitization / desensitization mechanism of phy action (Kim *et al.*, 2004; Park *et al.*, 2000).

2.4 Sensory modes and phenomenology of phytochrome A photoresponses

2.4.1 Modes of phyA action

Changes in the activity of the photoregulated genes modulate protein biosynthesis resulting in photophysiological reactions of the plant. These reactions are divided into several types (modes) based on the relationship between a response and the relative content of the physiologically active Pfr form, $[P_{fr}]/[P_{tot}]$: (i) very-low-fluence responses (VLFR), ($[P_{fr}]/[P_{tot}] \leq 0.1\%$ at total light fluence of 0.1 – 1 μmol m^{-2}); (ii) low-fluence responses (LFR) ($[P_{fr}]/[P_{tot}] \geq 1\%$, 1-1000 μmol m^{-2}) and (iii) far-red induced high-irradience responses (FR-HIR) ($\geq$1000 μmol m^{-2}). The first two modes are inductive responses to light pulses while for the last one fluence rate is critical. The three types of the responses were also defined with regard to the spectral range of activity and to the R /FR reversibility, the LFR being the classical red-induced /far-red –reversible photoresponses obeying the Bunsen-Rosco law of reciprocity (Casal *et al.*, 1997). On the contrary, VLFR do not show FR-reversibility and manifest themselves in the whole spectral range, including the far-red region, the effect being achieved by short pulses of R or FR. Also, kinetics of VLFR and LFR is different: the VLFR are transitionally while LFR cause a more persistent effect (see Casal *et al.*, 2002 and the literature cited therein). Experimentally the VLFR are measured as an effect of hourly light pulses whereas the FR-HIR mode is determined as the difference between the effect of continuous FR light (FRc) and hourly light pulses (FRp) at equal spectral composition and total fluence (Casal *et al.*, 1997).

The high-irradience responses induced by FR were considered to be irreversible by R and to disobey the reciprocity law. In elegant experiments with brief far-red pulses (FRp) of high frequency, Shinomura *et al.* (2000) showed, however, that this light treatment could produce the same response as continuous FR light (FRc) and in this case FR-HIR followed the reciprocity law and were R-reversible. The authors suggested that in these FR – induced / R-reversible HIR the physiologically active form is Pr*, that is Pr cycled through Pfr. Alternatively, it is believed that the FR-HIR require a certain concentration of a short-lived intermediate generated in the Pfr→Pr conversion which act as functional photoreceptor species (Cerdán *et al.*, 1997). Still the FR-HIR can be explained exclusively based on the classical old-standing notion that the active form of the light-labile phyA is Pfr and that its highest content for a prolonged period needed for a reaction can be reached upon actinic illumination in the FR region (see Hennig *et al.*, 1999 and the literature cited therein). The above definitions of the three photoresponse modes concern primarily seeds and etiolated seedlings, however, according to Casal (2000), in light grown seedlings both LFR and HIR are likely to operate and some VLFR cannot be ruled out. Interestingly, all the three types of the responses can converge to control transcriptional activity of a single gene (Cerdán *et al.*, 1997). Other types of reactions of the two photoreceptors, phyA and phyB, are also distinguished – end-of-day far-red responses, i.e. effect of enrichment of the light by the far-red components at dusk mediated by phyA, and R/FR ratio perception (reaction of vegetative shade or proximity detection) which are sensitized by phyB (Smith, 1997; Quail, 1997).

With the use of null mutants in the PHY genes it was found that responses to FR are eliminated in *phyA* null mutants while responses to R are strongly reduced in *phyB* mutants implying that phyA alone mediates the FR responses and that phyB functions specifically in R perception (see for review Whitelam and Devlin 1997; Casal, 2000). Thus, there was a general notion that VLFR and HIR are exclusively mediated by phyA while LFR, by phyB and by other minor phytochromes. Evidence is being accumulated, however, that phyA is capable of mediating all the types of the response modes, including the LFR (see below).

The differences between VLFR and HIR are traced back to the domain functions in phyA: this is the N-terminal serine-rich stretch between residues 6 and 12 in the phyA molecule. It is believed that the regulation of the two modes could involve cellular context-dependent phosphorylation of this subdomain (see Casal *et al.*, 2002 and above). Also, missense mutation in the PAS2 sequence in the C-terminal domain (Glu777Lys) of phyA eliminates HIR but not VLFR of phyA implicating this region of the molecule in functional specificity of phyA (Yanovsky *et al.* 1997, 2000,).

The differences between the photoresponse modes go beyond the photoreceptor molecules. Signaling downstream of phyA and phyB splits into at least three branches (Deng and Quail., 1999; Neff *et al.*, 2000, Fankhauser and Chory, 2000 and the literature cited therein): a phyA-specific branch, a branch implicated in both phyA and phyB signaling and a phyB-specific branch. Casal *et al.* (1998) found also that the kinetics of phototransduction with participation of phyB for a number of responses in de-etiolation in *Arabidopsis* was faster than that of phyA suggesting divergence of signal transduction pathways downstream from phyA and phyB. More to that, phyA action in FR-HIR proceeds via a special branch for signal transduction with unique components for FR/R signal which are not shared by other phytochromes (see above). PhyA is thus able to initiate two different pathways that converge to control the morphology of de-etiolating *Arabidopsis* seedlings (Yanovsky *et al.*, 1997, 2000; Casal *et al.*, 2000, Cerdán *et al.*, 2000).

2.4.2 Phenomenology of phyA-mediated responses

Dark-grown plants, in general, respond to light of very low fluence rates while light-grown plants, to high fluence rates. Photophysiological activity of phyA was clearly shown in dark-grown plants in the early stages of development and the most sensitive phyA signaling system is, thus, was shown to be the most important in early seedling establishment and survival. Nevertheless, the activity of the light-labile phyA was also followed in green plants (Quail *et al.*, 1995; Casal et, 1998). Of particular interest in this respect are the observations of the Whitelam group (Franklin *et al.* 2007) that phyA is capable of mediating R-induced photoresponses in light-grown *Arabidopsis* seedlings as revealed by the use of quadruple *phyBphyCphyEphyD* mutants containing only functional phyA. The authors have shown that phyA is an irradience-dependent red light sensor and that its functional activity under R light is observed at high (>160 μmol m^{-2} s^{-1}) photon irradience. The authors point out that the enhanced activity of phyA in continuous R (Rc) of high photon irradience correlates with retarded degradation of the endogenous phyA protein in wild-type plants, the effect called by the authors irradience-dependent 'photoprotection'. In their view, the fact that phyA was shown to be inactive under Rc in the majority of photomorphogenesis experiments, can be explained by lower intensity of R light (< 50 μmol m^{-2} s^{-1}) used in these experiments.

PhyA specifically participates in the perception of:

(a) brief light pulses that can promote seed germination in the VLFR mode. In *Arabidopsis* seeds, phyA appears in the dark ca. 2 days after the onset of imbibition and participates in light-induced germination (Botto 1996; Shinomura *et al.*, 1996.) PhyA detects light in buried seeds and during soil cultivation and in seeds germinating under dense canopies (VLFR) (Casal *et al.*, 1997). The mode of action of phyA in inducing germination varies, however, depending on plants used. In some species, FRp promotes germination but FRc inhibits germination (McCormac *et al.*, 1993a, Casal and Sanchez, 1998).

(b) difference between darkness and the FR-rich environment that initiates de-etiolation beneath dense canopies (Quail, 1998); phyA participates also in R-induced enhancement of seedling de-etiolation (Mazzella, 1997).

(c) phyA is engaged in early neighbor detection responding to FR reflected in green tissues of neighboring vegetation. It affects the sensitivity and/or responsivity to changes in R/FR ratio by phyB and other minor phytochromes (Casal *et al.*, 1997, Smith and Whitelam, 1997). Because of the very high sensitivity of the phyA action, which is saturated by extremely low levels of Pfr, it cannot participate directly in the perception of the proportion of R/FR light and thus in the shade avoidance syndrome (Casal *et al.*, 1997, Smith and Whitelam, 1997).

(d) phyA is involved in the measurement of photoperiod (Casal, 2000; Chory *et al.*, 1996) promoting flowering in response to a FR extension of a short day in long-day plants (LDP). The resetting of the circadian oscillator is a VLFR (Casal *et al.*, 1997; Jackson and Thomas, 1997). PhyA shortens the clock period in response to low fluence red light (Chory *et al.*,

1996; Neff *et al*., 2000; Somers *et al*., 1998) whereas phyB shortens the period of the clock in response to high fluence red light (Somers *et al*. 1998; Neff *et al*., 2000.)

(e) It is also active in the modification of gravitropism and phototropism: gravitropism is controlled by phyA in a VLFR mode, under FR it induces agravitropism of hypocotyls (Hangarter, 1997). In phototropism, phyA enhances first positive curvature under R (VLFR) (Janoudii, 1997; Parks *et al*., (1996) and reduces the time threshold in second positive curvature (Hangarter, 1997). Gravitropism is controlled by phyA in a VLFR (Franklin and Whitelam, 2004; Hangarter, 1997) It is believed that the phytochrome-mediated randomization of *Arabidopsis* hypocotyl orientation provides fitness advantage to seedlings developing in low-light environments (Allen *et al*. 2006).

Adam *et al*. (1997) suggested that phyA may have additional functions, based on the characteristic expression pattern of PHYA genes in tobacco phyA. They are (i) regulation of root developments; (ii) control of transition from scoto- to photomorphogenesis in specialized phloem cells (mobilization of reserves and their translocation), flowering and bud dormancy; (iii) control of flower morphogenesis. High concentration of phyA was detected in stem and petiole vascular tissue implicating vascular tissue as a potential site of phyA action (Jordan *et al*., 1995)

Although there is a general notion that responses to Rc are mediated by phyB and other photostable phytochromes (Botto *et al*.,1996, Casal, 2000) there is a number of indications that phyA can participate in the R/FR reversible LFR characteristic of phyB. Barnes *et al*. (1996) have shown using an *Arabidopsis* mutant deficient in FHY1, a component of the phyA signaling pathway (see above), that R-induction of chalcone synthase (CHS) is mediated almost entirely by phyA. Jackson and Thomas (1997) pointed out that that the action spectra for the night-break promotion of flowering in the long-day plants (LDP) obtained in (Downs 1956 cited in Jackson and Thomas, 1997), which is mediated by phyA, shows maximum effectiveness of red light and reversibility by far-red light. The red induced / far-red reversible reactions (LFR) were observed by Long and Iino (2001). The authors have shown with the use of phytochrome-deficient mutants that phyA participated in light-dependent osmoregulation in pea stem protoplasts in a totally R/FR reversible manner as did phyB. Takano *et al*. (2005) observed the R/FR photoreversibility of the induction of chlorophyll a/b binding protein (CAB) gene expression in WT and *phyB* and *phyBphyC* mutants of rice. This suggests that phyA may mediate the R/FR reversible LFR and its mode of action may depend on the experimental conditions and the type of photoresponse under investigation.

The attribution of a particular mode of action to phyA or phyB (VLFR vs. LFR) may depend on the experimental setup (Parks and Spalding, 1999). In kinetic studies with high time resolution the authors have shown that phyA initiated hypocotyl growth inhibition upon the onset of Rc. This inhibition persisted until phyA disappeared (≈3 h) and only after that an exclusively phyB-dependent phase of inhibition began. In FRc, the inhibition proceeded through exclusively phyA-mediated pathway. The timing of the phyA and phyB action under Rc correlated with the time of their light-induced transfer in the nucleus.

Phototropism is another effect where phyA acts under R enhancing curvature. Parks *et al*. (1996) have shown that WT and *phyB Arabidopsis* show a marked enhancement of first positive curvature whereas this response is absent in *phyA* seedlings. The enhancement was found in response to very low-to low fluence R; phyA (and also phyB) plays essential role during second positive curvature. The time threshold for second positive curvature is also reduced by exposure to red light (Briggs and Huala, 1999; Hangarter, 1997). Photomorphogenic mutants of *Arabidopsis* that had been selected for their inability to respond to constant illumination were tested for their ability to carry out R-induced enhancement of phototropism, which is an inductive phy response. phyA is the primary receptor regulating this response and it has a common regulatory domain in the phyA polypeptide which functions for both high-irradience and inductive phytochrome responses (Parks *et al.*, 1996). Of interest in this context is mediation of phototropism of roots by phyA together with phyB under red light (Kiss *et al.* 2003; Ruppel *et al.* 2001).

2.4.3 Functional interactions of phyA with phyB and redundancy between them

PhyA plays a role in perception of R in *Arabidopsis* seedlings as does phyB. According to (Reed *et al.* 1994) *phyAphyB* double mutants exhibit reduced Rc-induced de-etiolation and induction of chlorophyll a/b binding protein (CAB) gene expression compared with either single mutant. Also, transgenic *Arabidopsis* seedlings that overexpress phyA show enhanced deetiolation not only in FRc but also in Rc (Boylan and Quail 1991). According to Hoecker *et al* (1998), *spa1* mutation in the phyA-specific pathway increases both Rc and FRc sensitivity. This implies a certain degree of interaction between phyA and phyB in mediating the light signal.

As outlined by Casal (2000), interactions between photoreceptors can be positive, negative, additive, or synergistic and they can be different for one pair of receptors for different responses . For phyA and phyB, two types of interaction were established using different regimes of R and FR treatment and *phyA*, *phyB* and double *phyAphyB Arabidopsis* mutants: (1) antagonism – de-etiolation under red light (continuous or hourly pulses) is mediated by phyA and phyA antagonizes phyB-mediated de-etiolation; in general, under red the effects are primarily mediated by phyB and secondarily by phyA. (2) synergism – under FR light there is no evidence for phyA-phyB interaction, however, if a red light pulse is preceded by a short exposure to FR de-etiolation is significant suggesting synergism between phyA and phyB.

Thus, the positive or negative signal of the interactions between phyA and phyB appears to depend on the mode of action of phyA: VLFR exerts negative effect on phyB action and HIR causes amplification of the phyB action. Interaction between phyA and phyB occurs at subcellular level downstream from the photoreceptor. Antagonistic and synergistic effects could involve cross-talk between players of each signaling cascade, modulation of phytochromes transport to the nucleus, transcriptional synergy, etc. (Casal, 2000).

Overlapping action of phyA and phyB substantially promote de-etiolation in sparse vegetation (Smith *et al.*, 1997). At low Pfr/P (under dense vegetation), inhibition of hypocotyls extension was strong because of the action of phyA under FR while at high Pfr/P (open day light) it was strong because of phyB. At intermediate equilibria the inhibition was less strong. Kinetic measurements (Parks and Spalding, 1999, see above) show that the picture of the phyA and phyB functional interaction is even more complex.

PhyA is conditionally redundant to action of R or white light (W) and thus to phyB since the response to R is only seen in the phyB background (Reed *et al.*, 1994, Whitelam and Devlin, 1997). This can be illustrated by the fact that *phyA* or *phyB* mutant grown under 10 h light/14 h dark photoperiod has the same mass as WT, while *phyAphyB*, only 30-40% (Devlin *et al.*, 1996). Casal (2000) points out, however, that the redundancy between phyA and phyB is an example of generic redundancy rather than true redundancy because these photoreceptors have different photoperception functions (see above).

2.4.4 Structural aspects of phyA functional specificity

Functional specificity of phyA is determined at different levels of its action and turnover. Firstly, at the molecular level, phyA differs from phyB and other phytochromes by determinants within the N-terminal chromophore domain. Secondly, it possesses specific patterns of expression in cells and tissues, turnover characteristics, light stability and abundance in Rc and FRc, thermal stability of the Pfr form and dark Pfr→Pr reversion. Finally, the kinetics of the cytoplasm-to-nucleus partitioning and transduction chains of phyA are also different from that of phyB (Fairchild and Quail 1998; Fairchild *et al*., 2000; Hennig *et al* 1999; Quail, 1997; Ruddat *et al*., 1997).

Considering the molecular aspects of the specificity of the photoperception by phyA there are data that they may relate both to the chromophore and the apoprotein. Although the chromophore is identical and the mechanisms of its attachment to the apoprotein are probably very similar for phyA and phyB (Kunkel *et al*., 1993) and the photophysical and photochemical characteristics of phyB (potato) appear in general to be qualitatively similar to those of phyA (oat) (Braslavsky *et al*., 1997), there are differences in the requirements for the chromophore structure of the two pigments. According to Hanzawa *et al*. (2002), the sensitivity of phyA depends on the structure of the linear tetrapyrrole (bilin) prosthetic group: PΦB (phytochromobilin with the double bond in the vinyl side chain of D-ring) is required for phyA responsiveness under FRc although photochemically and spectroscopically functional holophytochrome was obtained for phyA with phycocyanobilin (PCB with saturated D-ring substituents). In contrast, phyB appears to be more flexible accepting both PΦB and PCB as a functional chromophore. It is suggested that the double bond of PΦB is crucial for interaction with the α-helix-forming N-terminal 6-kDa fragment of phyA which is important for physiological activity of phyA.

Differences in the apoprotein between phyA and phyB are associated primarily with different determinants within the N-terminal chromophore domain of the molecule (Quail, 1997). The extra 37 amino acids at the N-terminus participate in the phyB apoprotein-chromophore interaction and results in the phyB-specific property in *Arabidopsis* (Quail 1997). In general, however, the amino acid residues chain can be either shorter for phyA (*Arabidopsis*) or longer (Pisum) than that for phyB and thus the size of the molecule does not seem to be the cause for their functional specificity.

Different regions within the phyA holoprotein are believed to participate in signaling via the VLFR or HIR pathways. In experiments with transgenic *Arabidopsis* overexpressing truncated or mutated phyA (see Casal *et al*. 2002 and the literature cited therein) it was found that one important domain is a serine-rich Δ6-12 stretch whose deletion (or substitution) produces hyperactive phyA in VLFR (with regard to hypocotyl growth inhibition) but with a dominant negative effect for HIR. According to Casal *et al*. (2002) this was consistent with a differential role of the N-terminus on VLFR and HIR whose deletion enhanced the VLFR. Recent observations by the same group (Trupkin *et al*., 2007), however, allow a quite different interpretation of the role of this part of the N-terminus in the phyA-mediated responses. They re-evaluated the above conclusion that the serine-rich region is involved in the negative regulation of phyA signaling (VLFR) using transgenic *Arabidopsis* seedlings of the phyA-null background expressing *Arabidopsis* phyA with deletion at 6-12 a. a. under the control of the native phyA gene promoter. These plants showed normal responses to pulses of FR (FRp) (VLFR) and impaired responses to FRc (HIR) and, importantly, deleted phyA revealed reduced stability under R or FR. These observations are consistent, on the other hand, with the data that phyA turnover is regulated by phosphorylation if one keeps in mind that phosphorylation takes place at the N-terminal serine residues. The data by Trupkin *et al.* (2007) suggesting that the Δ6-12 phyA is more light-labile are also in agreement with our observation (Sineshchekov 2004, 2005) that it is represented almost entirely by the light-labile phyA' form whereas in wild-type *Arabidopsis* approx. one half of total phyA is relatively light-stable phyA''. Thus, the N-terminal serine-rich part of phyA is necessary for its stability. The above discrepancies can be explained by the fact that phyA with the deletions in the N-terminal part of the molecule expressed ectopically and under strong promoter reveals higher concentration under illumination than the endogenous normal phyA and this may explain its observed higher physiological activity.

3. Polymorphism of Phytochrome A

Since the phytochrome discovery, there was a controversy between the existence of the different modes of the pigment action and the notion that there exists only one its molecular type. Later on, however, polymorphism of the pigment was discovered by the detection of (1) its light-labile and light-stable pools (Brockmann and Schäfer, 1982); (2) immunologically distinguishable phytochromes (Pratt, 1994; Tokuhisa *et al.*, 1985); and (3) native phytochrome species with distinct fluorescence and photochemical properties (Sineshchekov and Sineshchekov and Sineshchekov, 1987; 1989). With the seminal discovery of different phytochrome genes (Sharrock and Quail, 1989) these observations could be most naturally interpreted as a manifestation of different PHY gene products with the light-labile phyA and light-stable phyB as major ones (see above). However, the phytochrome system proved to be more complex with further levels of complexity – at least for phyA there exist (i) post-translationally modified phy states and (ii) conformers of the chromophore within one phyA molecular type (see for a review Sineshchekov 1995a, 1998, 1999, 2004, 2005).

3.1 Different phenomenological types of phytochrome

Circumstantial data on different phytochrome A pools. Initially there was a notion of soluble and membrane (protein)-associated phy species and this problem is not yet solved. Terry *et al.*, (1992) point, in particular, to the possibility that there is a separately transcribel population of type I-like phy (that is phyA-like) which can associate with the plasma membrane (see above). Two pools of phyA in the Pfr form in *Arabidopsis*, one of which forms speckles in the nucleus and the other is homogeneously distributed in the cytoplasm, were observed (Nagy *et al.*, 2000). Cantón and Quail (1999) have found two phyA pools in *Arabidopsis* differing in their response to the light regulation of its synthesis – the basal and the light-regulated pools. Boylan and Quail (1991) have reported that oat phyA overexpressed in *Arabidopsis* possessed functional properties differing it from the endogenous phyA: it induced inhibition of the hypocotyls extension growth by R and this effect was reversed by FR (which is not observed in the WT). Further, phyA population in the cell may differ by the state of phosphorylation of the molecule (see above). Circumstantial evidence suggesting that there is phyA heterogeneity is the detection of phyA populations with relatively high light-stability, what is not characteristic of the bulk phyA. A high steady-state level of phyA in far-red-light (25%) and the significant fraction of stable Pfr after a red-light pulse (30%) that undergoes neither destruction nor dark reversion was observed by Hennig *et al.* (1999). Clough and Vierstra (1997) reported on relatively light-stable phyA in green plants which was characterized by a half-life time of several hours (vs. tens of min in etiolated seedlings). In this connection it is of interest to mention the observation by Franklin *et al.* (2007) that the light stability of phyA depends on the irradience of the actinic light (see above): under high intensity illumination phyA becomes relatively light stable. Thus, there are indications that phyA population may differ by light stability, localization in the cell and light regulation of biosynthesis.

In our group, the problem of phytochrome heterogeneity is tackled by investigating the chromophore component of the pigment with the use of *in-vivo* low-temperature fluorescence spectroscopy (for a review, see Sineshchekov, 1988, 1995a, 1998, 1999). In etiolated plants, phytochrome fluorescence was detected and, based on that, a highly sensitive, informative and relatively simple method of phytochrome assay was developed (Fig. 8) (Sineshchekov and Sineshchekov, 1986, 1987a-d; 1989a,b). It allowed *in-planta* characterization of phytochrome by a number of parameters: (1) total content ($[P_{tot}]$ proportional to the fluorescence intensity of phytochrome in its Pr form); (2) fluorescence emission and excitation (absorption) spectra (position of the maximum, λ_{max} , and half-band width, $\Delta\lambda$); (3,4) extent of the photoconversion of the initial form Pr into the first photoproduct stable at low temperature, Pr→lumi-R (85 K), and into the far-red form, Pr→Pfr (273 K) at ambient temperature (γ_1 and γ_2, respectively). The pigment was also characterized by the kinetic and energetic parameters, including activation barrier, E_a, of the Pr fluorescence decay and of the Pr→lumi-R photoreaction. It is essential to mention that the invented method of the low-temperature fluorescence spectroscopy of phytochrome *in-planta* is much more sensitive (by approx. two orders of magnitude) and more informative than the traditional "classical" method of difference absorption spectroscopy. In fact, initial (not difference) absorption spectra of phytochrome in plant tissues cannot be obtained because of high light scattering of the plant tissues and very low concentration of the pigment hampering, in particular, direct evaluations of the key γ_1 and γ_2 photochemical phytochrome parameters.

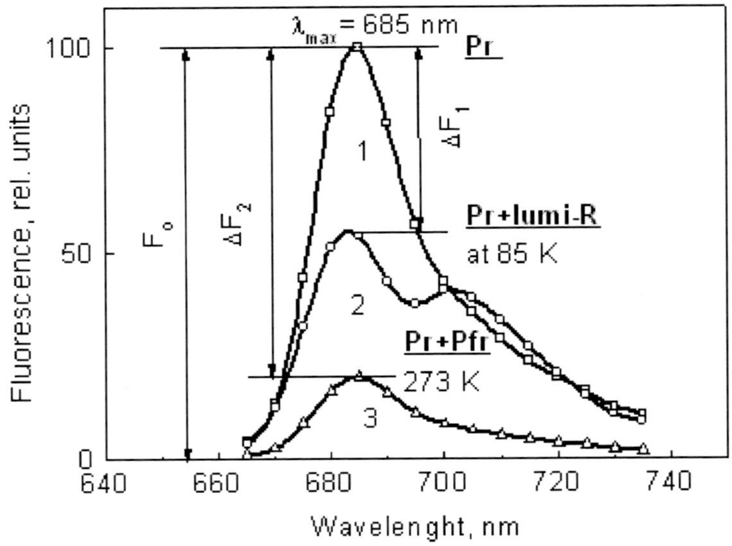

Figure 8. The low-temperature method of phytochrome *in-situ* assay. To characterize the pigment, three fluorescence spectra of phytochrome were measured at 85 K: 1- in etiolated tissues (of wheat in this figure) when all phytochrome is in its Pr form; 2 – in the same sample after saturating red illumination at 85 K partially converting Pr into lumi-R, the first stable photoproduct, and 3 – in the same sample after thawing at 273 K, saturating red illumination converting Pr into Pfr and freezing again at 85 K. Four major parameters were obtained from these spectra: position of the emission spectrum (λ_{max}); total phytochrome content proportional to the fluorescence intensity ($[P_{tot}] \approx F_0$); extent of the Pr→lumi-R photoconversion, equal to the relative fluorescence decline ($\gamma_1 = \Delta F_1/F_0$) and characterizing the initial photoreaction; and the extent of the Pr→Pfr photoconversion ($\gamma_2 = \Delta F_2/F_0$), characterizing the whole phytochrome cycle. (From Sineshchekov, 2004, 2005)

These and other phytochrome parameters were found to depend on plant species and their phytochrome mutants and over- and underexpressors used, on localization of phytochrome in organs and tissues, plant age, on light conditions and other physiological factors (see, for instance, Fig. 9). In general, it was observed that, with the decrease of $[P_{tot}]$, γ_1 dropped down from ≈ 0.5 to 0 (Fig. 10), γ_2, from 0.85 to ≤ 0.7 and λ_{max} of emission from 685-687 to 680-682 nm. These variations were interpreted as a manifestation of the existence of at least two phenomenological Pr populations in the cell: Pr' – the longer-wavelength, major, variable by its content in tissues and light-labile with λ_{max}=687/673 nm (emission/absorption) and γ_1=0.5 and Pr" – the shorter-wavelength, minor, more evenly distributed in the plant and relatively light-stable with λ_{max}=682/668 nm and γ_1=0 (Sineshchekov, 1994). The photochemical difference between Pr' and Pr'' can be explained in terms of the energy level scheme of the Pr→lumi-R photoreaction (see Fig. 5 and Sineshchekov, 1995a; Sineshchekov and Sineshchekov, 1990)) by the fact that the activation barrier, E_a, in the excited state for the Pr→lumi-R photoreaction is much lower in the case of Pr' (Pr' type) than Pr'' (Pr'' type) - several hundreds J/mol versus tens kJ/mol). This may affect the quantum yield of the direct Pr→lumi-R photoreaction at low temperatures and as a consequence determine the Pr↔lumi-R photoequilibrium. In Pr'', the quantum yield of the direct photoreaction is considerably lower than that of the reverse so that the photoequilibrium is shifted to Pr (γ_1''→0) while in Pr', they are comparable (Sineshchekov and Akhobadze, 1992). At ambient temperature the activation barrier, E_a, does not restrict the Pr→lumi-R conversion so that the yield of the Pr→Pfr conversion (γ_2) are practically equal for both Pr types. These photochemical distinctions between the two Pr pools are evidently connected with deep changes of the chromophore state in the protein pocket of the molecule. The extreme photochemical differences between Pr' and Pr'' at low temperatures (with the individual γ_1 values of 0.5 and 0, respectively) allowed simple determination of the content of the two phy types in plant tissues based on the experimental γ_1 value of phytochrome in the sample and its [Ptot] (see Sineshchekov, 1994 and Fig. 11).

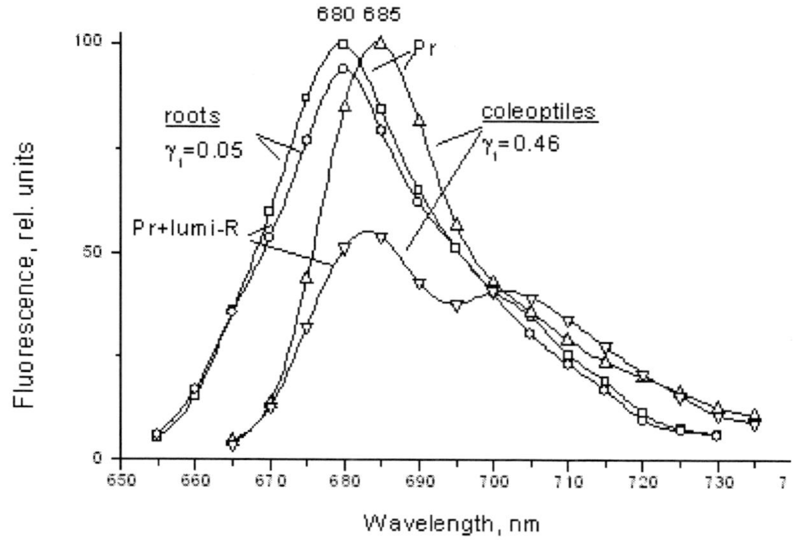

Figure 9. Fluorescence emission spectra (85 K, λ_{ex}=633 nm) of phytochrome in coleoptiles (right) and roots (left) of etiolated wheat taken immediately after darkness, when all the pigment is in the Pr form, and after saturating red (λ_a=633 nm) illumination at 85 K partially converting Pr into photoproduct lumi-R stable at low temperatures (photoequilibirum Pr + lumi-R). Note the differences in the position of the bands and in the extent of the Pr→lumi-R photoconversion, γ_1, in the case of roots and coleoptiles.

3.2 Two phyA native pools with the properties of Pr' and Pr'' – phyA' and phyA''

Proceeding from the fact that there exist several phytochromes, the major ones being phyA and phyB, we tried to correlate the detected Pr' and Pr" species with the different phytochromes by using phytochrome mutants and transgenic plants with modified phyA and

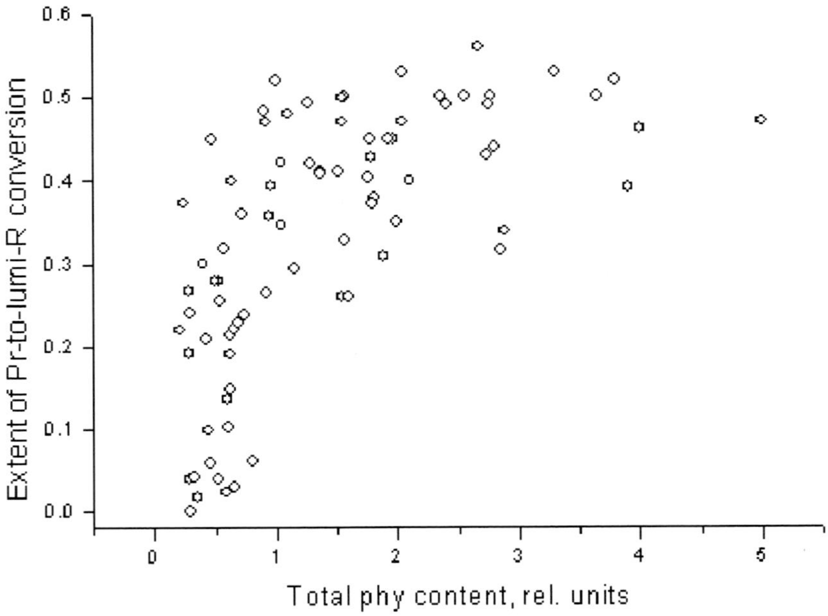

Figure 10. Dependence of the extent of the Pr→lumi-R conversion at 85 K (parameter γ_1) on the total phytochrome content in different etiolated wheat seedlings. (From Sineshchekov *et al.*, 2001b.)

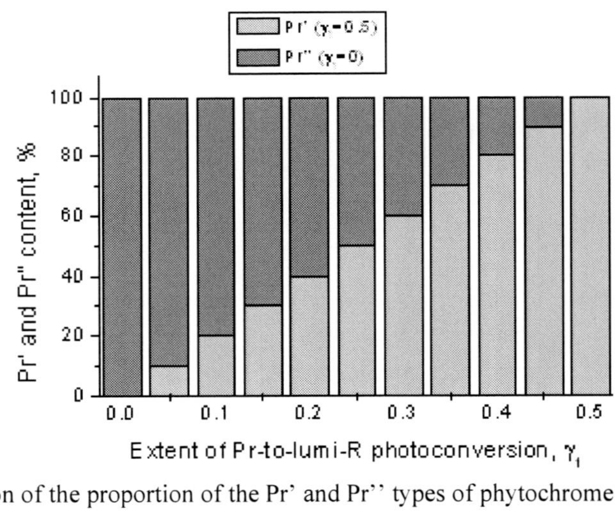

Figure 11. Evaluation of the proportion of the Pr' and Pr'' types of phytochrome in plant tissues from the individual γ_1 values (0.5 and 0 for Pr' and Pr'', respectively) and experimental γ_1 values of the sample according to (Sineshchekov, 1994).

phyB content. First of all, the use of *Arabidopsis* (Sineshchekov *et al.* 1998a) and pea (Sineshchekov *et al.* 1999a) mutants deficient both in phyA and phyB (*phyAphyB* double mutants) have shown that the phytochrome fluorescence belongs to phyA and phyB and the input of minor phytochromes (C-E) into the emission is negligible. Pr' and Pr" are thus can be associated with phyA and phyB. Further, phyB was attributed to Pr" because the phyA-deficient mutants of these species (*Arabidopsis,* pea) contained only Pr" and overexpression of phyB (native phyB and rice phyB in *Arabidopsis* (Sineshchekov *et al.* 1998a) and

Arabidopsis phyB in potato (Sineshchekov *et al.* 2000a)) brought about an increase of only [Pr”]. In these experiments, phyB was for the first time characterized *in vivo* to have a fluorescence maximum, λ_{max} , at 682 nm, $\gamma_1 = 0$ and relatively low $\gamma_2 = 0.3$-0.5 and the latter parameter revealed an early red drop (upon actinic illumination at $\lambda_a = 680$ nm $\gamma_2 \rightarrow 0$). phyB made up less than 10% of [P_{tot}] in *Arabidopsis* and pea, it was light-stable and was evenly distributed in the plants (Sineshchekov *et al.*, 1998a, 1999a, 2000a).

In the case of phyA attribution, the picture was, however, more complex. Overexpression of phyA (of oat in *Arabidopsis* (Sineshchekov *et al.* 1998a)) and suppression of phyA synthesis (in anti-sense potato (Sineshchekov *et al.* 1996)) led to the increase and decrease of [Pr’], respectively, while [Pr”] remained practically unchanged suggesting that phyA is identical with Pr’. However, in phyB-deficient mutants of cucumber (Sineshchekov, 1995b), *Arabidopsis* (Sineshchekov *et al.*, 1998a) and pea (Sineshchekov *et al.*, 1999a), both Pr’ and Pr” were observed and their content was almost the same as in the respective wild-type plants (Fig. 12). This implies that there are two subpopulations of phyA (major phyA’ and minor phyA”) with characteristics of Pr’ and Pr”, respectively (see above). In contrast to phyB, both phyA’ and phyA” have high γ_2 (0.7-0.85) and this parameter drops down less steeply with the increase of the wavelength of actinic light.

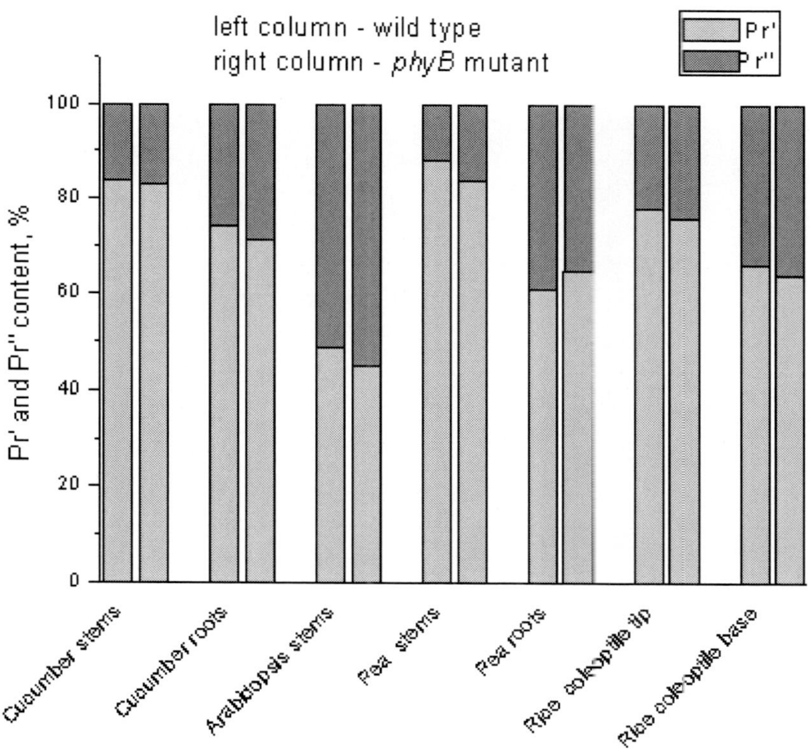

Figure 12. Relative content of the two phytochrome phenomenological types, Pr’ and Pr’’, in etiolated dicotyledonous and monocotyledonous plants and their *phyB*-deficient mutants. Close similarity in the content of Pr’ and Pr’’ between the wild type plants and the mutants strongly suggest that there exist two phyA species, phyA’ and phyA’’, with the properties of Pr’ and Pr’’, respectively. (From Sineshchekov 1995b; Sineshchekov *et al.*, 1998a, 1999a, 2006).

The fact that phyA is represented *in vivo* by the two isoforms with the properties of the Pr' and Pr'' phenomenological pools was thus rigorously shown for the dicotyledonous plants because of the availability of their phy mutants. The two Pr types were also found in monocots [oat, wheat, barley, rice, maize (Sineshchekov, 1994; Sineshchekov *et al.*, 2001b, 2004)] which could be attributed to phyA' and phyA'' similarly to the situation with the dicots based on their spectroscopic and photochemical properties and also abundance and distribution in plant species and tissues under different physiological conditions. However, the lack of phyA and phyB mutants of monocots did not allow us to present a rigorous proof to the fact that phyA in these plants also exists in the two isoforms. The isolation of the phy mutants of rice (Takano *et al.*, 2001, 2005a,b) opened this perspective and we have shown (Sineshchekov *et al.*, 2006) that this holds true for the monocotyledonous plants as well. In fluorescence investigations of phytochrome (phy) in rice (*Oryza sativa* L. cv. Nipponbare) mutants deficient in phyA, phyB and phyAphyB the total content of the pigment, $[P_{tot}]$, and its spectroscopic and photochemical characteristics were for the first time determined in situ. As expected, the input of phyC was negligible whereas phyA made up to 95% of $[P_{tot}]$ in the coleoptiles of etiolated wild-type (WT) rice seedlings and virtually 100%, in the *phyB* mutant, most of the pigment being located in the upper and middle parts of the coleoptile. The extent of the Pr→lumi-R conversion at 85 K was shown to vary in the same way both in the WT and *phyB* mutant depending on the part of the coleoptile used (Fig. 12). This proved the fact that phyA is heterogeneous in the monocot rice and comprises two phyA populations, phyA' and phyA'' with the phyA'/ phyA'' proportion varying in different parts of coleoptiles from ≈80/20% at the tip of the coleoptile to ≈60/40% at the middle. Thus, we can firmly conclude that in flower plants two phyA pools (phyA' and phyA'') exist with the properties of the two phenomenological Pr' and Pr'' phytochrome types (Table 1).

Table 1. Two phenomenological phytochrome A types with the characteristics of Pr' and Pr'' in mono- and dicotyledonous plants

Phy type / Parameter	Pr' = phyA'	Pr'' = phyA'' + phyB
Position of spectra, λ_{max}	longer-wavelength (685/672 nm)	shorter-wavelength (680/667 nm)
Half-band width, $\Delta\lambda$	22-24 nm	30 nm
Pr→lumi-R conversion, 85 K, γ_1	0.49±0.03	≤0.05
Activation energy, E_a	≤ 1 kJ mol^{-1}	≥ 10 kJ mol^{-1}
Pr→Pfr conversion, ≥273 K, γ_2	0.80-0.85	0.75
Content in tissues	major, variable	minor, conserved, saturable
Light lability	light-labile	relatively light-stable

Finally, there are subpopulations within the main phyA' pool, which is active at cryogenic temperatures, which differ by the fluorescence and photochemical parameters – position of the emission/absorption maximum, kinetics of the fluorescence decay, activation energy of the photoreaction (Sineshchekov and Akhobadze, 1992). They are likely to be due to the existence of different structural chromophore states, conformers of the chromophore, in the pigment molecule characterized by different degree of interaction with the protein environment hindering the photoisomerization and, thus, possessing different activation barrier for the photoreaction. Using different experimental approach (thermochromy – temperature-induced absorption spectra changes) Schmidt *et al.* (1998) have also detected Pr isoforms in native and recombinant (65 kDa) oat phyA. Although these states within phyA' may initiate parallel photocycles of the pigment and be potential participants of different photoregulation reactions, their functional role remains unclear. What is clear is that investigations of these phyA states are of importance for understanding the nature of the chromophore-protein interactions determining activation parameters of the Pr photoreaction (see above and Fig. 5) (scheme). Interestingly, investigations (Sineshchekov *et al.*, 1998b, 2002) of the recombinant phytochrome from cyanobacterium *Synechocystis*, Cph1, which shows extensive similarities to plant phytochromes, suggest that intrinsic heterogeneity is characteristic for this phy too as is the case of phyA'. Cph1 heterogeneity may be due to the chromophore photoproduct formation under excitation light. So, this heterogeneity is probably a universal property of phytochromes and it is to be taken into account in investigations on the dynamics of the initial processes in phy and its crystallization.

3.3 Similarity of phyA'' to phyB and cryptogam and cyanobacterial phytochromes

Comparative fluorescence and photochemical investigations of fern *Adiantum capillus-veneris* phy1 overexpressed in *Arabidopsis* (Koppel *et al.*, 1999), and of recombinant phytochromes of the moss *Ceratodon purpureus* CP2 (Sineshchekov *et al.*, 2000b) and of the cyanobacterium *Synechocystis* Cph1 (Sineshchekov *et al.*, 1998b, 2002) *in vitro* allowed characterization of the holoproteins in the Pr form by their fluorescence emission and excitation spectra, temperature dependence of the fluorescence and photoreaction, and kinetic and energetic parameters of the initial photoprocesses. Cph1 was also compared with its deletion mutant Cph1Δ2 (with PCB, PΦB and PEB as a chromophore). The process of autoassembly of the Cph1 holoprotein was followed by fluorescence spectroscopy in which two stages were distinguished. Cph1 and Cph1Δ2 were shown to be closely similar in their spectroscopic and photochemical characteristics, suggesting that the sensory module is photochemically autonomous. All the phytochromes examined were shown to belong to the Pr'' photochemical type, that is they were incapable of performing the photoisomerization at cryogenic temperatures ($\gamma_1 = 0$). Thus, phyA'', phyB and the cryptogam and cyanobactrial phytochromes turned out to be of the Pr'' type, whereas phyA' is a unique species, the only found so far to belong to the Pr' type. As in the case of phyA'', the photochemical properties of Cph1 and the cryptogam phytochromes were interpreted in terms of the relatively high activation barrier for the Z-E isomerization which cannot be overcome at cryogenic temperatures. Although the activation barrier for the photoreaction was not measured for phyB because of its low content in plant tissues, it is believed that its photochemical properties are also governed by it. At the same time, these phytochromes revealed photochemical properties at ambient temperatures very similar to those of higher plant phytochromes. This similarity of Cph1 to plant Pr"-type phytochromes suggests Cph1 as a model for this phytochrome type, what is of particular importance given that its 3D structural data is more likely to be obtained in the future than those of the plant phys.

3.4 Nature of phyA' and phyA''

The nature of the two phyA species, phyA' and phyA", remains unclear. In general, we may consider three possible major causes of the phyA' and phyA'' structural differences: (i) post-translational modification of the phyA molecule, (ii) its interaction with a reaction partner(s) and (iii) specific intracellular localization. With regard to the first possibility we may conclude that phyA' and phyA'' are post-translationaly modified products of one PHYA gene and the two phyA species differ by the small 6-kDa part of the N-terminal chromophore domain. This is suggested by the facts (i) that N-terminally (Δ7-69) truncated oat phyA overexpressed in transgenic tobacco is represented primarily, if not only, by the phyA", while full-length and C - terminally (Δ786-1129) truncated phyA overexpressed in the same system comprised both phyA' and phyA" (Sineshchekov *et al.*, 1999b). Our experiments on Δ6-12 truncated oat phyA (fused to GFP) expressed in *Arabidopsis* in the phyA-null background (V. Sineshchekov, A. Sudnitsin, F. Nagy and E. Schäfer, unpublished results; Sudnitsin *et al.*, 2003) showed that the Δ6-12 phyA mutant was represented primarily by phyA' (the phyA'/phyA'' ≈86/14 % vs. ≈50/50 % in wild-type *Arabidopsis* (Sineshchekov *et al.* 1998a).

Recent investigations of recombinant *Arabidopsis* and rice (*Oryza*) phyA expressed in yeast cells and reconstituted with the chromophore in situ (Sineshchekov *et al.*, 2001b) have shown that it is close to phyA", although not identical, and differ considerably from phyA'. PhyA overexpressed in *E. coli* was represented, on the contrary, almost entirely by the phyA' fraction (V. Sineshchekov, L. Koppel, I. Schepens, C. Fankhauser, unpublished results). This suggests that phyA *in vivo* should undergo plant-specific post-translational modification and the site of this modification is in the small region near the N-terminus. The fact that phyA reconstituted in yeast does not form phyA' may explain the lack of interaction of phyA in expression systems with some of potential transduction chain partners , for instance, HFR1 which acts in phyA signaling (Fairchild *et al.*, 2000). The authors hypothesize that interaction of HFR1 with phyA with complex formation in transgenic yeast requires a plant-specific post-translational modification of HFR1 or phyA. The expression system, thus, affects the properties of photoactive chromoprotein implying that a proper conformation of recombinant phy is important for its photochemical and spectroscopic properties. In support of this, Hill *et al.* (1994) used coexpression of chaperones to improve the folding of the pigment. Gärtner *et al.* (1996) based on selectivity of chromophore incorporation and spectral properties conclude that interactions between protein domains of phytochrome control protein folding and Pr/Pfr absorption characteristics. Proceeding from this it would be natural to expect that modification(s) of phyA yielding either phyA' or phyA'' affects the overall conformation of the pigment which tells, in its turn, on the spectroscopic and photochemical properties of the phyA pools.

The presumed modification could be phosphorylation because dephosphorylation of the native (oat) phyA brings about a ≈2.5-fold decrease of the phyA'/phyA" proportion suggesting that phyA' is a phosphorylated species (see Sineshchekov, 1995a). In order to further explore this possibility, we have been looking for a correlation between changes in the phosphorylation state of phyA in the cell and in the balance between its pools. Experiments with phosphatase inhibitors on the phyA'/phyA'' equilibrium *in situ* support the notion on the involvement of phosphorylation / dephosphorylation of the photoreceptor in the phyA'/ phyA'' formation and/or maintenance of their equilibirum. We have observed that inhibition of phosphatases in etiolated tissues of maize brought about a considerable changes in the phyA'/ phyA'' proportion and the sign and extent of these changes depended on the state of phyA in the cell (in particular, on the phyA'/phyA'' content), on the nature of the inhibitor (specific and unspecific) and other experimental conditions. Firstly, an unspecific protease and phosphatase inhibitor phenylmethylsulfonyl fluoride (PMSF) had a strong effect on the phyA' and phyA'' abundance in roots of *Zea mais* and its sign and extent were developmentally determined. A 10 min incubation of young (length $\leq$ 2 mm) etiolated roots in a 0.5 mM PMSF solution shifted the phyA'/ phyA" ratio from the domination of phyA' (phyA'/phyA' $\approx$ 92/8%) to approx. balance of the species (55/45%). A reverse picture was in the case of older roots (length $\approx$ 7-10 mm): the same treatment resulted in a shift from the phyA'/phyA'' proportion of 50/50 % to 94/6% (Koppel' and Sineshchekov 2003). Secondly, treatment of etiolated maize shoots with the specific inhibitor of the Ser/Thr phosphatases PP1 and PP2A, okadaic acid produced a considerable decline of [phyA'] and a concomitant rise of [phyA''] in coleoptile and mesocotyl tissues (V. Sineshchekov, L. Koppel, E. Orekhova and P. Galland, unpublished data). Although precise interpretation of the obtained data is difficult at present, they implicate phosphorylation / dephosphorylation in maintaining a proper phyA'/phyA'' balance in the cell.

Still another indication of involvement of phosphorylation / dephosphorylation in the regulation of the phyA'/ phyA'' equilibrium are the effects on it of the phyA interaction with substrates of its phosphorylation activity (PKS1-PKS4) and cryptochrome (cry1). It should be noted that phyA interaction with nuclear partners are to be excluded from consideration in this context because phyA' and phyA'' are observed in the dark-grown plants when the photoreceptor is localized in the cytoplasm. Besides, short term (minutes) red illumination and more prolonged far-red illumination (hours) do not produce noticeable shift in the phyA'/phyA'' equilibrium suggesting that the events connected with translocation of phyA into the nucleus are not involved in the process of the phyA differentiation into phyA' and phyA''. In order to test the possibility that the interaction with the proteins of the PKS family (Fankhauser, 2000, see above) could be the cause of the differences between the two phyA pools we investigated *Arabidopsis* mutants deficient in PKS1, PKS2, PKS1 and PKS2 and also *Arabidopsis* PKS1 overexpressor (Sineshchekov and Fankhauser, 2004). It was found that the share of phyA' and phyA'' does not differ much in all the lines except the double mutant *pks1pks2* where phyA' drops down more than 2-fold and phyA'' rises at its expense by $\approx$1.5-fold (Fig. 13). Thus, there is an indication that PKS1 together with PKS2 can affect the state of phyA in the cell by changing the relative content of its two native species, phyA' and phyA". Since the PKS proteins are considered to be the negative regulators of phyA activity, possibly by their retention of phyA in the cytoplasm via physical interaction, our data suggest that such an interaction could influence the state of phyA' and phyA'' in the cell such that the interaction of phyA with both PKS1 and PKS2 favors formation of phyA'. It should be noted that there seem to be a redundancy between PKS1 and PKS2 because the single *pks1* and *pks2* mutants do not practically differ from the WT with regard to the ratio between phyA' and phyA''.

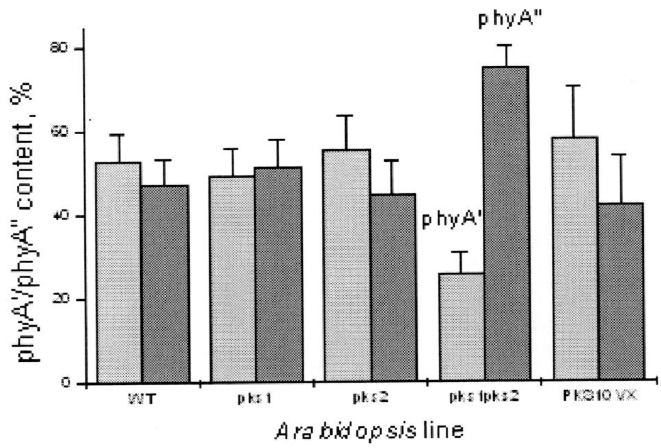

Figure 13. PhyA interaction with Phytochrome Kinase Substrates (PKSs) affects the phyA'/phyA'' equilibrium as shown with mutant *Arabidopsis* strains deficient in *pks1* and/or *pks2* and PKS1 overexpressor. Effect is observed in the case of the double *pks1pks2* mutant suggesting redundancy between *pks1* and *pks2*. Light grey columns – phyA', grey columns – phyA''. (From Sineshchekov and Fankhauser, 2004).

Cry1 can also affect the properties of phyA and the equilibrium between phyA' and phyA'' (L. Koppel' and V. Sineshchekov, unpublished results). In etiolated seedlings of the *Columbia hy4* mutant lacking cry1, we observed very close content of phyA to that in the wild type (0.93±0.07 and 1.02±0.08 rel. un., respectively). However, the experimental γ_1 values were different, 0.31±0.02 and 0.24±0.04 suggesting that in the mutant the proportion between phyA' and phyA'' is shifted towards phyA', from approx. 48/52 % in WT to 62/38 % in the mutant. These data indicate that the interaction of phyA with cry1 is thus in favored of the phyA'' formation. Interpretation of these facts renders difficult at present, nevertheless they suggest that, directly or indirectly, the substrates of the phyA kinase activity could in turn have an effect on the state of phyA in the cell and, in particular, on the equilibrium between the two phyA components.

3.5 To the localization of phyA' and phyA'' in the cell

The problem of the nature of phyA' and phyA'' and their functions, see below, is obviously closely related to the problem of localization of the pigment in the cell (cytosolic and membrane-associated fractions) and its light-induced translocation from the cytosol to the nucleus. In this connection, we tried to correlate the found phenomenological types of phyA with its water-soluble and membrane-associated pigment fraction (Sineshchekov *et al.*, 1994) extracted in (Lamparter *et al.*, 1992). It was found that the water-soluble fraction of phyA was characterized by the parameters close to those of phyA' *in situ*, while the minor fraction of the pigment connected with the membrane was close to phyA". An argument for the existence of the membrane-associated fraction is also the fact that the proportion of phyA' and phyA" in a number of plants *in vivo* correlates with that of the two fractions extracted from them (in the order of the increase of relative content of Pr" and membrane-associated fraction they are oat, pea, maize) (Sineshchekov, 1994). Thus, one may assume that phyA'' is the membrane (protein)-associated fraction of the pigment.

In this connection, the question arises which of the two phyA species (or both of them) is transferred to the nucleus. We were trying to approach this problem by investigating the phyA' and phyA'' content of the gene engineered phyA fused to the green fluorescent protein, phyA-GFP, for which the process of the nuclear translocation was investigated in detail (Kircher *et al.*, 1999). Experiments were carried out on transgenic tobacco overexpressing rice phyA-GFP and *Arabidopsis* phyA-deficient mutant overexpressing authentic phyA-GFP. Preliminary data suggest that the rice phyA-GFP in tobacco contains both phyA' and phyA'' in quantities almost identical to those of endogenous phyA', 70/30%. In *Arabidopsis* phyA-GFP, we observed a shift of the phyA'/phyA'' content as compared with that for the endogenous phyA in the WT, 65/35 % and 48/52 % respectively. The fact that the phyA fusion proteins comprise both forms of phyA makes both phyA forms potential participants in the process of the nuclear translocation. It is of interest to mention in this connection that Nagy *et al.* (2001) distinguish two pools of phyA in the Pfr form one of which forms speckles in the nucleus and the other is homogeneously distributed in the cytoplasm. To what extent this may be connected with the two phyA populations remains to be solved.

It should be emphasized that the above differences between the two phyA populations are not connected with the effect of its sequestering - light-induced localization of the initially homogeneously distributed phy in the cytoplasm in cell loci (Nagy *et al.*, 2001). It is believed that it is connected with the destruction of the pigment in the Pfr form. In our experiments, the light adaptation of the etiolated samples at room temperature to red and then to far-red light (condition under which the sequestering is observed) did not bring about changes of the phyA fluorescence and photochemical parameters and of the concentration of its two pools (Sineshchekov 1994, Sineshchekov *et al.*, 1998a).

Finally, we have investigated the effect of dehydration of plant tissues on the phyA state in the cell (Sineshchekov, 2006). On etiolated seedling of barley and maize, it was found that even a considerable loss of water (up to 75–85% of the initial fresh weight) by coleoptiles does not bring about noticeable alterations of the spectroscopic and photochemical parameters of phytochrome pointing to a relative stability of the phyA'/phyA'' system in this regard. However, extreme dehydration (loss of weight $\geq$ 90%) of plant tissues (air- or freeze-drying) caused dramatic changes of the phytochrome properties – blue shift of the emission maximum and its widening and reduction in the extent of the Pr photoconversion into lumi-R at 85 K and into Pfr at 273 K. Rehydration of the dried tissues did not reverse the spectroscopic changes and did not recover the Pr $\rightarrow$lumi-R phototransformation at 85 K but restored the ability of Pr to photoconvert into Pfr at ambient temperatures. At the same time, the total phytochrome content was not affected by these treatments. These effects were interpreted as an irreversible transformation of phyA' into phyA'' upon extreme loss of water by plant tissues suggesting that water may play a role in stabilizing the conformation of the major and soluble phyA' species. The data also imply that phyA in dry and imbibing seeds is likely represented primarily by its phyA'' isoform. These data support the notion that phyA' is water-soluble while phyA'' is an amphiphilic, membrane (protein)-associated fraction in agreement with our earlier observations (see above).

Thus, the native phyA' and phyA'' species are likely to differ with regard to all the three factors mentioned above – by the structure of the molecule, interaction with partner proteins and also, probably, by localization in the cell. This implies that the photochemical properties of phyA in plants are not only the consequence of its inherent encoded polypeptide sequence, but are largely due to post-translational modifications of the pigment.

4. Functional Implications of the PhyA Polymorphism

4.1 Different modes of phyA' and phyA'' action

The discovery of the two native phyA pools suggests that the complex phenomenology of the phyA physiological action in plants (see above) might be due, at least partially, to its polymorphism. It is likely that phyA' and phyA'' could mediate different types of photoresponses that are currently attributed to the homogeneous phyA. We may hypothesize also that the different phenomenological and functional properties of phyA observed under different physiological conditions may be the result of changes in the phyA'/ phyA'' equilibrium. A straightforward way to elucidate the functions of phyA' and phyA'' could be to investigate plants deficient in phyA' or phyA''. However, because of the lack of them, we followed another approach – we have tried to establish a correlation between changes in the content of phyA' and phyA'' and in the phenotype of phyA of transgenic plants expressing normal and mutant phyA.

Initially, we turned to transgenic potato with altered content of phyA and modified phenotype (Heyer *et al*., 1995) According to the authors, the overexpressor (*BIN PS2*) with increased (2-fold) phyA levels had a more pronounced FR-HIR and a decreased shade avoidance response. The underexpressors (*AP15/9*, *AP15/11* and *AP15/13*), with a 10-fold decline in the phyA content, on the contrary, showed a delayed FR-HIR and increased shade avoidance response. They were less sensitive to red light with regard to stem extension and expression of the ribulose bisphosphate carboxilase (RBCS) gene. Our fluorescence and photochemical investigations have shown (Sineshchekov *et al*., 1996) that phyA' decreased 30-60 - fold in the sequence *BIN PS2* - wild type - *AP15/9* or *AP15/11* while phyA'' remained relatively constant (changes of its content amounted to 4-6 – fold in the same sequence). Because the most dramatic variations were observed in the concentration of phyA', one may suggest that the changes in the phenotype are primarily connected with this pool and to a lesser extent with the minor phyA''. It should be noted, however, that strong enhancement of phyA responses requires only a 2- to 3-fold increase in the level of the pigment and further increase in phyA level has little additional effect (Boylan and Qauil, 1991; Jordan *et al*., 1995). Since we observed changes in the phyA'' content of this magnitude, one cannot exclude entirely its participation in the FR-induced effects in the transgenic potato.

More decisive were the results (Sineshchekov *et al*., 2001b) with transgenic wheat obtained by Shlumukov and co-authors (2001). This wheat line overexpresses an oat *PHYA* gene, leading to the appearance of FR-HIR (inhibition of coleoptile extension, synthesis of anthocyanin and leaf unrolling) that were not characteristic of the wild type. At the same time, in the transgenic line, we observed an increased content of phyA' while phyA" remained essentially the same as in the WT. This indicates, in support the results on the transgenic potato, that the phyA' pool is primarily responsible for the observed modification of the phenotype and suggests that even in WT plants phyA' may mediate the FR-HIR.

The mutant phyA with the deletions at the N-terminus was shown to be unable to form phyA' or phyA'' (see above) and this provided an opportunity to look more closely into the individual properties of the two phyA species. On transgenic tobacco overexpressing full-length and truncated oat phyA (Cherri *et al.*, 1992, 1993), it was found (Sineshchekov *et al.*, 1999b) that the pigment missing the 6-kDa N-terminal domain (NA with Δ7-69) fails to form phyA' (phyA'' overexpressor) while the full-length (FL) species produced both phyA' and phyA'' with the domination of phyA' (phyA'+phyA'' overexpressor). Thus, differences of the NA phenotype from that of WT and FL could be associated with phyA''. As a measure of the phenotype alterations, we followed changes in the content of the inactive and active protochlorophyllide emitting at 633 and 655 nm, $Pchlide^{633}$ and $Pchlide^{655}$ (Sineshchekov *et al.*, 2004a) and growth responses in dark-grown and FRc- and Rc-grown plants (V. Sineshchekov, unpublished results). Constant FR illumination brought about an increase of both Pchlides in all the transgenic lines (in contrast to the literature data on the FRc inhibition of $Pchlide^{655}$ accumulation, see below) (Fig. 14). However, in NA the effect was considerably lower than in the WT while it was the highest in FL suggesting that phyA'' interfered with the endogenous phyA responses in seedlings grown under FR. The shape and length of the NA plants were the same under FRc as in the dark. However, under Rc ($\lambda \geq 660$ nm), NA revealed a higher effect of cotyledon unfolding with the angle between cotyledons $\alpha° = 64 \pm 2$ than WT ($\alpha° = 39 \pm 3$), although not as high as in the overexpressor of both phyA' and phyA'' (FL), $\alpha° = 136 \pm 4$. In the case of FRc illumination, $\alpha°$ was 0 for NA, 8 ± 2 for WT and 10 ± 2 for FL. Thus, the phyA activity under FRc ($\lambda \geq 720$ nm) with regard to the Pchlide formation and growth effects could be associated with phyA'. On the other hand, these observations indicate that under FRc phyA'' is inactive and even suppresses the action of endogenous phyA' although it may function under Rc. This phenomenology of the NA action is close to the de-etiolation effects on oat phyA missing the 52 N-terminal amino acids, ΔN52 phyA, overexpressed in *Arabidopsis* and tobacco (Boylan *et al.* 1994), and on rice phyA, missing the 80 N-terminal amino acids in transgenic tobacco (Emmler *et al.* 1995). It is also of interest to note in this connection that overexpressed phyB behaves itself similarly to phyA'' causing dominant negative interference on phyA-mediated inhibition of hypocotyl growth under FRc (Casal, 2000). Finally, the effect of phyA'' overexpression was observed in the dark: the level of the active $Pchlide^{655}$ and inactive $Pchlide^{633}$ in stems was much higher in the NA tobacco line overexpressing phyA'' than in the WT (Sineshchekov *et al.*, 2004a) (Fig. 14). It should be noted that phyB was also shown to have some activity in the absence of red light activation (Reed *et al.*, 1993).

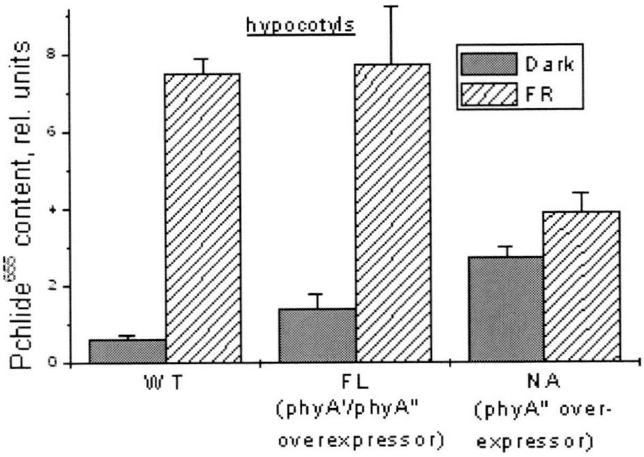

Figure 14. PhyA'' is inactive under far-red light and inhibits phyA' action. This is revealed by induction of Pchlide^{655} biosynthesis in hypocotyls of transgenic tobacco overexpressing full-length oat phyA (FL, represented by both phyA species) and Δ7-69 deleted phyA (NA, consisting of only phyA''). The seedlings were grown in darkness (left column) and in far-red ($\lambda_a \geq 720$ nm, I = 0.1 W m^{-2}) light (right column). (From Sineshchekov *et al.*, 2004).

The truncated *Arabidopsis* phyA with the deletion of the 6-12 a. a. stretch provides an opportunity to examine the functional properties of phyA' more directly because this mutant phyA is represented primarily by this species (see above). This truncated phyA revealed either exaggerated (Casal et al., 2002) or normal (Trupkin *et al.*, 2007) VLFR , depending on the character of the promoter, presence of native phyA in the transgenic plant and other experimental conditions, and reduced HIR activity. Thus, this supports the above conclusion that phyA' is active in VLFR. The fact that the activity of the mutant phyA' is hampered in HIR might suggest, however, that it is incapable of this type of the photoresponses and that HIR is mediated instead by phyA''. This is, nevertheless, at odds with the above data on the phyA' and phyA'' functions. Also, our experiments with point-mutated phyA defective in HIR but with normal VLFR have shown that this deficiency is not connected with violation in the phyA'/phyA'' equilibrium. Mateos *et al.* (2007) identified the phyA-303 mutant allele of *Arabidopsis thaliana* bearing an R384K substitution in the GAF subdomain of the N-terminal half of phyA. Mutated phyA showed reduced phyA content, almost normal VLFR, and severely impaired HIR. Similarly, Yanovsky *et al.* (2002) have isolated phyA-302 *Arabidopsis* mutant retaining VLFR and lacking HIR, in which Glu 777 changed to Lys in the PAS2 motif of the carboxy terminal domain. The *phyA-302* mutant showed a 50% reduction in phytochrome A levels compensated for by a higher stability under continuous far-red light. In our experiments (V. Sineshchekov, O. Ogorodnikova and J. Casal, unpublished results) both phyA-302 and phyA-303 had a higher proportion of the phyA'/ phyA'' pools than that in the wild type. This implies that the impairment in HIR in the two mutants is not connected with the disappearance of phyA' and suggests, together with the other data (see above), that phyA' is likely to mediate both VLFR and HIR. We cannot exclude, however, a possibility that the violation in the HIR responses could be due to the lowering in the phyA'' content. Finally, overexpressed oat phyA in tobacco and rice mediates VLFR, but not LFR in the de-etiolation process (Casal *et al.*, 1997). Because overexpressed oat phyA in tobacco according to our data is primarily of the phyA' type (see above), this supports the notion that phyA' is employed in VLFR.

The regulation of the active $Pchlide^{655}$ accumulation by phyA under FRc merits more detailed discussion. It is well known that phyA mediates de-etiolation under FRc with the effect similar to those observed under Rc (including inhibition of stem growth and promotion of hook opening, cotyledon expansion, and anthocyanin synthesis) and it is believed that in natural conditions phyA thus helps seedlings to survive under dense canopies enriched with FR light (Yanofsky *et al.*, 1995; Smith *et al.*, 1997). However, Apel and co-workers (Apel, 1981, Apel *et al.*, 1980) and Barnes *et al.* (1996) have documented a decline in the protochlorophyllide oxidoreductase (POR) formation and active protochlorophyllide accumulation and even the destruction of the prolamellar body under FRc. According to Yanovsky *et al.* (2000) it is predominantly a HIR and shows little or no VLFR and the inhibition of white light-induced subsequent greening, brought about by prolonged exposures to FRc in *Arabidopsis* (the FR-preconditioned block of greening), becomes the test for the phyA activity under FR (Yanovsky *et al.*, 2000; Hoecker *et al.*, 1998; Casal *et al.*, 2002). Our observations suggest that the observed down-regulation of the $Pchlide^{655}$ synthesis is not a universal phenomenon – the sign and the magnitude of the FRc effect on $Pchlide^{655}$ depend on the plant species and, within the plant, on the organ or tissue used (Sineshchekov *et al.*, 2004a). In cotyledons of tomato and *Arabidopsis,* we observed a decline of the $Pchlide^{655}$ production in line with the data in the literature (Runge *et al.*, 1996; Barnes *et al.*, 1996). Quite on the contrary, in cotyledons of tobacco and leaves of pea and also in upper stems of all the plants investigated, i.e. tobacco, pea, tomato and *Arabidopsis*, a positive effect of FRc on $Pchlide^{655}$ of different magnitude was observed.

The fact that the sign and the magnitude of the effect on $Pchlide^{655}$ change depending on the tissue (or organ) resembles the effect of the light regulation of phyA mRNA (Cantón and Quail, 1999), which together with the POR gene belongs to the small group of FR - down-regulated genes. The authors have found three *PHYA* transcripts that were differentially regulated by light depending on their localization in the plant. This was interpreted as a manifestation of the fact that *PHYA* gene is subject to regulation by multiple signals, including environmental, developmental, and organ-specific signals. In line with this, we may assume the existence of similar properties of the POR gene. The diverse regulation effects of FRc on the $Pchlide^{655}$ biosynthesis in the different etiolated plants and their tissues are likely to be connected with POR A and POR B which are detected in etiolated seedling and young plants (Frick *et al.*, 2003).

It should be noted that the different signs of the FRc effect on $Pchlide^{655}$ is not connected with the availability of the chromophore. It is obviously not a limiting factor in the $Pchlide^{655}$ formation in the case of the negative effect, because in all the plants used and their organs the free chromophore, $Pchlide^{633}$, experienced a marked increase under FRc. Also, it was shown earlier (Schendel *et al.*, 1996) that the limiting factor for the $Pchlide^{655}$ formation is the apoprotein because feeding of the etiolated seedlings with the Pchlide precursor, δ-aminolevulinic acid, increased the $Pchlide^{633}$ content by 10-100- fold while the content of $Pchlide^{655}$ remained practically unchanged. Thus, we may conclude that the sign and the extent of the phyA action on the POR gene expression is modulated by the plant- and organ – specific signals.

Experiments with monocotyledounous rice, wild type and phyA mutants, supports the above conclusions on phyA functions based primarily on dicots (Sineshchekov *et al.*, 2006).

In these experiments, the existence of phyA' and phyA'' was rigorously proved for monocots (see above) and functional properties of phyA and its pools were determined in regard to its autoregulation, inactive and active protochlorophyllide (Pchlide^{633} and Pchlide^{655}) biosynthesis and growth responses under constant (FRc) and pulsed (FRp) far-red illumination at equal fluences. Constant FR drastically reduced the content of Pchlide^{633} and Pchlide^{655} in the WT and *phyB* mutant (in agreements with the data on barley, *Arabidopsis* and tomato, see above) and brought about coleoptile shortening and appearance of the first leaf, whereas FRp was of low effectiveness. This allowed attribution of the responses primarily to the HIR type (in agreement with Yanovsky *et al.*, 2000), which are likely to be mediated by the major species phyA'. The effects on protochlorophyllide biosynthesis and growth responses were more pronounced in the *phyB* mutant than in WT suggesting that phyB may interfere with the action of phyA', see above. On the other hand, the fact that the *phyB* mutant has higher content of phyA and protochlorophyllides than the WT and its coleoptile is longer in the dark-grown seedlings points to a possibility that phyB accumulated in seeds may affect the development of the WT seedlings in the dark.

An interesting picture was observed on the *phyA* rice mutant. The content of the inactive and active protochlorophyllide, Pchlide^{633} and Pchlide^{655} in the mutant in the dark was similar to that in WT and did not change under FRp. However under FRc, the content of the active Pchlide^{655} increased by 100 %. This points to a possible participation of phyC in this process (with the sign opposite to that of phyA) in agreement with the observation by Takano *et al.* (2001, 2005) of the positive effect of FRc on the light-dependent CAB genes in the *phyA* mutant and lack of it in the double *phyAphyC* mutant.

The phenomenon of the reversion of the sign of the FRc effects mediated by phyA was also observed (Sineshchekov *et al.*, 2004a) on an another type of rice (*Oryza sativa* L. *Japonica* cv Nihonmasari) mutants – the *hebiba* mutant deficient in the hormone jasmonic acid (Riemann *et al.*, 2003). In the dark, the mutant had a long mesocotyl and a short coleoptile, whereas the situation was reverse under FRc ($\lambda_a \geq 720$ nm) – short mesocotyl and long coleoptile – suggesting that the effect is mediated by phyA. In contrast, under these light conditions the WT displayed a short coleoptile and emergence of the first leaf. In the dark, the spectroscopic and photochemical properties of phyA, its content and the proportion of its two pools, phyA' and phyA'', were virtually identical between WT and *hebiba*. However, the total content of protochlorophyllides was higher in the mutant (Fig. 15). Upon illumination with FRc, [phyA] declined in the WT and the ratio between phyA' and phyA'' shifted towards phyA''. In *hebiba*, the light-induced decline of [phyA] was less pronounced and the ratio between phyA' and phyA'' did not shift. Moreover, in the WT, pulsed FR (FRp) stimulated the biosynthesis of Pchlide^{655}, whereas FRc of equal fluence was inhibiting. In contrast, in the mutant, both FRp and FRc stimulated the synthesis of Pchlide^{655}. This means that FRc caused the opposite effect in *hebiba*. This difference correlates with a slower photodestruction of primarily the light-labile phyA' pool in *hebiba* which is believed to be responsible for the mediation of the FR effects.

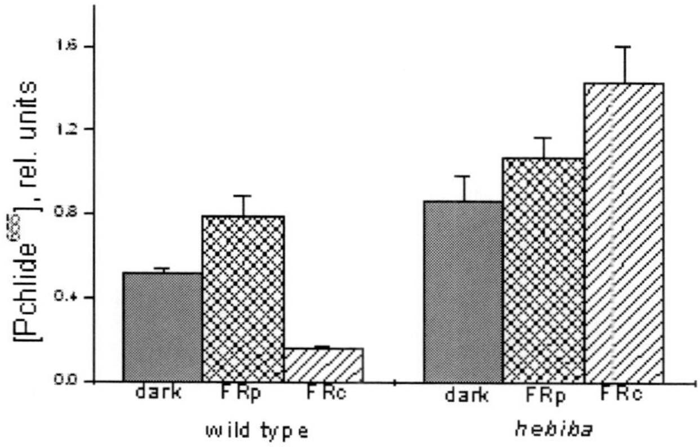

Figure 15. Content of the active protochlorophyllide (Pchlide^{655}) in coleoptiles of wild-type rice and its mutant *hebiba* deficient if the hormon jasmonic acid grown in darkness and under pulsed (FRp) and constant (FRc) far-red light of equal fluence. Note the opposite effects of FRp and FRc in wild type and of FRc in wild type and *hebiba*. (From Sineshchekov *et al*., 2004b).

Thus, because phyA' is a prevailing species and labile while phyA'' is minor and conserved, one may hypothesize that these properties are closely associated with their functions. Indeed, the light-labile phyA' is responsible for de-etiolation while the relatively light-stable phyA" may function under constant illumination throughout the plant life cycle.

4.2 Light regulation of the phyA'/phyA'' equilibrium as a means of fine-tuning of the phyA action

Of particular interest in the context of the phyA' and phyA'' functioning and interaction is the fact that their content in the plant is differentially regulated by Rc and FRc. Both short-term illumination with Rc (2 - 4 h) and long-term illumination with FRc (FR-grown seedlings) bring about considerable lowering of the phy content in epicotyls. The effect under Rc is due primarily to the destruction of phyA' (Sineshchekov *et al*., 1999a). Interestingly, in the WT, the rate of the [phyA'] decline was twice as high as that in the *phyB* mutant. This might be accounted for by the fact that phyB exerts negative regulation on the phyA synthesis in the WT in agreement with (Furuya *et al*., 1991, Cantón and Quail, 1999). Under FRc, on the contrary, the decline of the phyA content in epicotyls was not followed by the enrichment of the phyA pool by phyA'', even a decline of phyA'' was observed and phyB did not affect the extent of the phyA decline (Sineshchekov and Weller, 2004) (Fig. 16). It should be noted that this effect is likely to have a universal character because a very similar situation was observed on dark- and FRc-grown seedlings of maize (V. Sineshchekov, unpublished data). Interestingly, an effect similar to the FRc-induced lowering of [phyA] without changes in the phyA'/phyA'' ratio was observed in the dark in stems of the de-etiolated (*lip*) mutant of pea as compared to the wild-type (Sineshchekov *et al*., 1995) (Fig. 17). An approx. 3-fold lowering was folowed both for the WT pea grown under FRc (Fig. 16) and for the *lip* mutant in the dark (Fig. 17) as compared to etiolated WT pea.

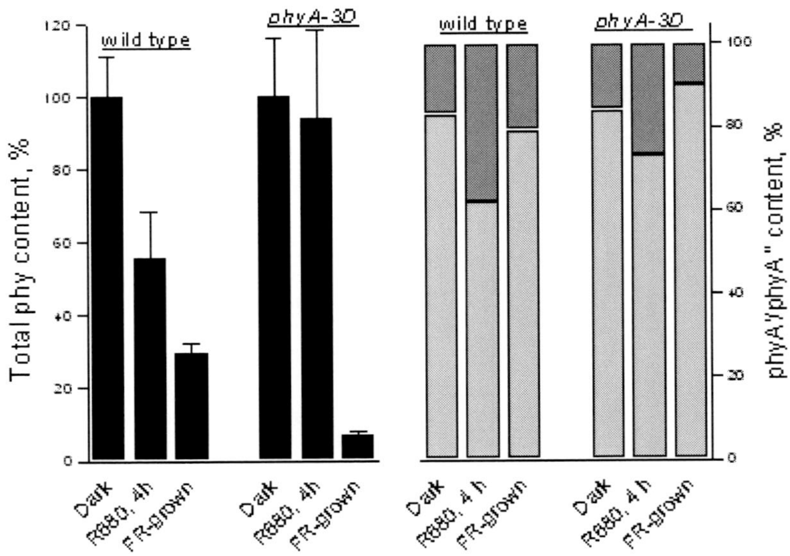

Figure 16. Two modes of light-induced phyA decline – with and without changes in phyA'/phyA'' equilibrium. Epicotyls of wild type pea and its point mutated *phyA-3D* species with alanine to valine substitution at position 194, A194V. Note that the decrease in the total phyA content in epicotyls of etiolated seedlings after saturating red illumination (λ_a = 680 nm, 4 h) is followed by a shift in the phyA'/phyA'' equilibirum towards the less light-labile phyA'' form, whereas even much more pronounced decline of [P_{tot}] in seedlings grown under far-red light ($\lambda a \geq 720$ nm, I = 0.1 W m^{-2}) is practically not accompanied by violation in the equilibrium. Note also a much lower effect of R and a much higher effect of FR on the total phytochrome content illumination in the case of the pea mutant suggesting that this mutation makes the photoreceptor less light-labile and more physiologically active. (From Sineshchekov and Weller, 2004; Weller *et al*., 2004).

Experiments on a point-mutated phyA in pea (mutation *phyA-3D* with a substitution of alanine by valine at the position 194 in its N-terminal domain producing a more stable and active phyA (Weller *et al.*, 2004)) have shown that the effect of the lowering of the total phyA content under Rc was more pronounced in WT than in *phyA-3D* (Sineshchekov and Weller, 2004)(Fig. 16). On the contrary, under FRc the decline was more pronounced in the mutant. Also, in contrast to the effects under Rc, contribution of phyB in the FRc effects is low, if any as revealed by the experiments on *phyA-3DphyB* and *phyB* vs. *phyA-3D* and WT. Thus, phyA in the mutant was more light stable under Rc but more labile under FRc. Such a phenomenology is likely to be due to the down-regulation of the phyA synthesis under FRc: in the case of the phyA destruction one would have expected to observe a decline of the more light labile phyA' whereas in the experiment there is no changes in the phyA'/phyA'' ratio or even more affected fraction is phyA''. More pronounced effects under FRc in the *AF05* and *AF05phyB* mutants suggests hyperfunction of the mutant phyA'. This interpretation of the FRc effects in the *phyA-3D* pea mutant is in agreement with the observation of an opposite light effects on [phyA] as a general consequence of reduced phyA signal-transduction feedback activity. Higher levels of phyA under FRc and increased Rc sensitivity have been reported for the FRc hyposensitive *fhy3* (Fairchild and Quail 1998; Whitelam 1993) and *far1* (Hudson *et al.*, 1999) mutants.

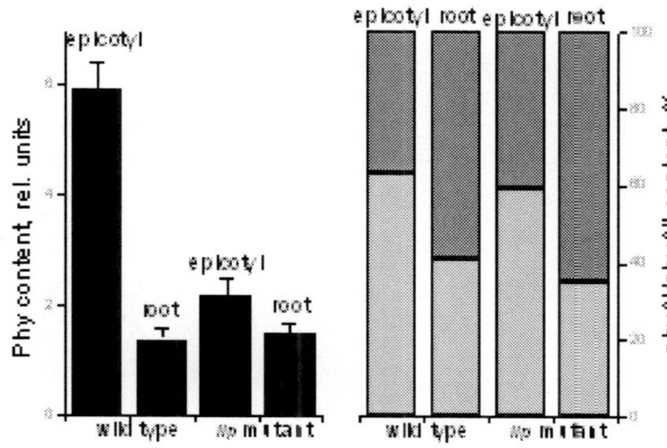

Figure 17. PhyA decline without violation of the phyA'/phyA'' equilibrium in de-etiolated (*lip*) mutant of pea. The total phytochrome content and the phyA'/phyA'' ratio were determined in etiolated epicotyls of dark-grown wild type pea and its mutant (*l*ight *i*ndependent *p*hotomorphogenesis, *lip*), deficient in negative regulators of photomorphogenesis resulting in "photomorphogenic" pattern of development in darkness. (From Sineshchekov *et al.*, 1995).

In roots the patterns of the changes in the total phy content and the [phyA']/ [phyA"] ratio under Rc and FRc were essentially the same as those described above for the epicotyls of the seedlings of the respective lines, namely, (i) a decline of the total phytochrome content under R and considerable shift in the phyA'/phyA'' proportion towards phyA'', especially pronounced in the case of the wild type, and (ii) a decline of the total phyA content without changes in the phyA'/phyA'' proportion in seedlings grown under FR. It should be noted, however, that the differences between the *phyA-3D* and the wild-type lines and also *phyA-3DphyB* and *phyA* lines were much more pronounced in the case of epicotyls than roots suggesting the involvement in these effects of those light-regulated phyA genes which are also subject to the control by tissues specific signals (Cantón and Quail, 1999).

Thus, we may conclude that under Rc the decline in [phyA] in epicotyls is primarily due to the phyA' destruction, whereas under FRc, to the inhibition of the phyA biosynthesis. This supports the notion that phyA' is dispensable under Rc and is active under FRc, while, on the contrary, phyA'' is active under Rc and dispensable under FRc.

5. Conclusion

Recent progress in our understanding of the mechanism of phytochrome action is associated with seminal discoveries related to all stages of the light-signal processing. These findings are (i) phytochrome gene family and their structurally and functionally different products; (ii) (cyano)bacterial phytochrome, crystallization and determination of the 3D structure of the chromophore domain at atomic resolution, determination of the functional role of different domains in the phytochrome molecule and interdomain signal transduction; (iii) detection of enzymatic activity of phytochrome A (capacity of light-induced auto- and transphosphorylation); (iv) light-induced nuclear-cytoplasmic partitioning of the pigment; (v) understanding of the principles of the transduction of the signal downstream of phytochromes and determination of a number of master regulators (negative and positive) common or specific for phytochromes A and B.

One of unanswered and intriguing questions is, however, the nature of the uniqueness of phyA. It is distinguished from phyB and other minor phytochromes by high content in etiolated plants and efficient mechanisms of its light-induced decline modulating its content and activity in a very wide range under different light conditions. This enables it to mediate all the photoresponse types (modes), namely, VLFR, LFR and FR-HIR. PhyA is involved in photoregulation throughout the whole life cycle of angiosperm plants and its functions show little overlapping with that of other phytochromes. These versatile phyA functions are believed to provide adaptive advantages to the higher plants. Chemical evolution of phyA has deviated from that of other phytochromes towards different photoperseptive functions, in particular, those initiated by FR, and this is one of the key factors in the evolution of flower plants.

Now it is becoming more and more likely that the unique features of phyA are connected to a certain degree with its polymorphism – existence in the two spectroscopically, photochemically, phenomenologically and, possibly, functionally different isoforms, major and light labile phyA' and minor and relatively light-stable phyA'' – the products of one and the same *PHYA* gene differing by post-translational modification, possibly, by phosphorylation / dephosphorylation of serine residue(s) at N-terminus of the pigment molecule. Their existence was rigorously shown with the use of phytochrome mutants of dicotyledounous and monocotyledonous plants and the complex phenomenon of phyA action is likely to be associated with this fact: phyA' is responsible for HIR and VLFR whereas phyA'' may account for LFR, be active in de-etiolated plants and suppress the action of phyA'. Thus, the balance between the two phyA isoforms may be part of the mechanism of fine tuning of its activity.

The most intriguing questions with regard to the phyA pools, however, cannot be considered as finally solved – these are (i) the mechanism of the post-translational modification of phyA and structural differences between its two pools and (ii) exact functional distinctions between them. One can envisage several further steps for their solution. Firstly, elucidation of the site(s) (a. a. residue(s)) of the presumed phosphorylation of the phyA molecule can be done by searching for point mutated phyA at the N-terminal serine cluster, thought to be involved in the modification process, which is represented only by one of the phyA species, phyA' or phyA''. Secondly, understanding of the mechanism of the modification can be achieved by analyzing the causes of the fact that recombinant phyA in different expression systems (as revealed on yeast and *E. coli*) is represented either by phyA' or phyA''. Phosphorylation / dephosphorylation experiments on these recombinant phyA species *in vitro* would prove (or disprove) involvement of this process in their formation. Finally, phyA' and phyA'' functions can be strictly followed by changes in the phenotype of transgenic plants deficient in their native phyA and expressing point-mutated phyA represented either by phyA' or phyA''. Recent developments in genetics, molecular biology, photophysiology and physico-chemical biology of the phytochrome system promise successful solution of the problem of the heterogeneity of phyA in the near future.

Acknowledgments

The author is very grateful to his co-workers and co-authors of his publications. This work was partially supported by the Russian Foundation for the Fundamental Investigations (grant 05-04-49549 to V. S.).

References

Adam, E., Kozma-Bognar, L., Schäfer E., and Nagy, F. (1997). Tcbacco phytochromes: genes, structure and expression. *Plant Cell Environ*, 20, 678-684.

Ahmad, M., Jarillo, J. A., Smirnova, O., and Cashmore, A. R. (1998). The CRY1 blue light photoreceptor of Arabidopsis interacts with phytochrome A in vitro. *Mol. Cell*, 1, 939-948.

Allen, T., Ingles, P. J., Praekelt, U., Smith, H., Whitelam, G. C. (2006). Phytochrome-mediated agravitoprism in Arabidopsis hypocotyls requires GIL1 and confers a fitness advantage. *Plant J*, 46, 641-648.

Apel, K. (1981). The protochlorophillide holochrome of barley (Hordeum vulgare, L.). Phytochrome induced decrease of transletable mRNA coding for the NADPH: protochlorophyllide oxidoreductase. *Eur. J. Biochem*, 120, 89-93.

Apel, K., Santel, H.-J., Redlinger, T.E., and Falk, H. (1980). The protochlorophyllide holochrome of barley. Isolation and characterization of the NADPH:protochlorophylide oxidoreductase. *Eur. J. Biochem*, 111, 251-258.

Ballesteros, M. L., Bolle, C., Lois, L. M., Moore, J. M., Velle-Calzada, J.-P., Grossniklaus, U., and Chua, N.-H. (2001). LAF1, a MYB transcription activatcr for phytochrome A signaling. *Genes Dev*, 15, 2613-2625.

Barnes, S.A., Nishizawa, N.K., Quaggio, R.B., Whitelam, G.C., and Chua, N.-H. (1996) Far-red light blocks greening of Arabidopsis seedlings via phytochrome A –mediated change in plastid development. *The Plant Cell*, 8, 601-615.

Batschauer, A. (ed.) (2004) *Photoreceptors and light signaling. Comprehensive series in photochemical and photobiological sciences*. Häder, D.P., and Jori, G. (series eds.). Cambridge, UK: Royal Soc. Chemistry.

Bauer, D., Viczian, A., Kircher, S., Nobis, T., Nitschke, R., Kunkel, T., Panigrahi K. C. N., Adam, E., Fejes, E, Schafer, E., and Nagy, F. (2004). Photomorphogenesis 1 and multiple photoreceptors control degradation of Phytochrome Interacting Factor 3, a transcription factor required for light signaling in Arabidopsis. *Plant Cell*, 16, 1433-1445.

Biermann, B. J., Pao, L. I., and Feldman, L. J. (1994). Pr-specific phytochrome phosphorylation in vitro by a protein kinase present in anti-phytochrome maize immunoprecipitates. *Plant Physiol*, 105, 243-251.

Björling, S. C., Song, P.-S. and Kliger, D. S. (1992). Time-resolved absorption studies of native etiolated oat phytochrome. *J. Am. Chem. Soc*, 114, 4569-4580.

Bolle, C., Koncz, C., and Chua, N.-H. (2000). PAT1, a new member of the GRAS family, is involved in phytochrome A signal transduction. *Genes Dev*, 14, 1269-1278.

Borucki, B., Otto, H., Rottwinkel, G., Hughes, J., Heyn, M., and Lamparter, T. (2003). Mechanism of Cph1 phytochrome assembly from stopped-flow kinetics and circular dichroism. *Biochemistry*, 42, 13684-13697.

Botto, J.F., Sanchez, R.A., Whitelam, G.C., and Casal, J.J. (1996). Phytochrome A mediates the promotion of seed germination by very low fluences of light and canopy shade light in Arabidopsis. *Plant Physiol*, 110, 439-444.

Boylan, M.T. and Quail, P.H. (1991). Phytochrome A overexpression inhibits hypocotyl elongation in transgenic Arabidopsis. *Proc. Natl. Acad. Sci*, USA, 88, 10806-10810.

Boylan, M., Douglas, N., and Quail, P.H. (1994). Dominant negative suppression of Arabidopsis photoresponses by mutant phytochrome A sequences identifies spatially discrete regulatory domains in the Photoreceptor. *Plant Cell*, 6, 449-460.

Braslavsky, S. E., Gärtner, W. and Schaffner, K. (1997). Phytochrome photoconversion. *Plant Cell Environ*, 20, 700-707.

Briggs, W. R., and Huala, E. (1999). Blue-light photoreceptors in higher plants. *Annu. Rev. Cell Dev. Biol*, 15, 33-62.

Briggs, W. R., and J. L. Spudich (Eds.) (2005) *Handbook of photosensory receptors*. Weinheim, FRG: WILEY-VCH Verlag Gmbh and Co. KgaA.

Brockman, J., and Schäfer, E. (1982). Analysis of Pfr destruction in Amarantus caudatus L. – evidence for two pools of phytochrome. *Photochem. Photobiol*, 35, 555-558.

Cantón, F. R., and Quail, P. H. (1999). Both phyA and phyB mediate light-imposed repression of PHYA gene expression in Arabidopsis. *Plant Physiol*, 121, 1207-1216.

Casal, J. J. (2000). Phytochromes, cryptochromes, phototropin: Photoreceptor interactions in plants. *Photochem. Photobiol*, 71, 1-11.

Casal, J. J., and Sánchez, R. A. (1998). Phytochromes and seed germination. *Seed Sci. Res*, 8, 317-329.

Casal, J. J., Sánchez, R. A., and Yanovsky, M. J. (1997). The function of phytochrome A. *Plant Cell Environ*, 20, 813-819.

Casal, J. J., Sánchez, R. A., and Botto, J. F., (1998). Modes of action of phytochromes. *J. Exp. Botany*, 49, 127-138.

Casal, J. J., Davis, S. J., Kirchenbauer, D., Viczian, A., Yanovsky, M. J., Clough, R. C., Kircher, S., Jordan-Beebe, E. T., Schäfer, E., Nagy, F., and Vierstra, R. D. (2002). The serine-rich N-terminal domain of oat phytochrome A helps regulate light responses and subnuclear localization of the photoreceptor. *Plant Physiol*, 129, 1127-1137.

Cerdán, P. D., Staneloni, R. J., Casal J. J., and Sánchez, R. A. (1997). A 146 bp fragmant of the tobacco Lhcb 1*2 promoter confers very-low-fluence, low-fluence and high-irradience responses of phytochrome to a minimal CaMV 35S promoter. *Plant Mol. Biol*, 33, 245-255.

Chen, E. F., Parker, W., Lewis, J. W., Song, P. S. and Kliger, D. S. (1993). Time-resolved UV circular dichroism of phytochrome A: Folding of the N-terminal region. *J. Amer. Chem. Soc*, 115, 9854-9855.

Chen, E. F., Lapko, V. N., Lewis, J. W., Song, P. S., and Kliger, D. S. (1996). Mechanism of native oat phytochrome photoreversion - a time-resolved absorption investigation. *Biochemistry*, 35, 843-850.

Cherri, J.R., Hondred, D., Walker, J.M. and Vierstra, R.D. (1992). Phytochrome requires the 6-kDa N-terminal domain for full biological activity. *Proc. Natl. Acad. Sci*, USA 89, 5039-5043.

Cherri, J. R., Hondred, D., Walker, J. M., Keller, J. M., Hershey, H. P., and Vierstra, R. D. (1993). Carboxy-terminal deletion analysis of oat phytochrome-A reveals the presence of separate domains required for strucutral and biological activity. *Plant Cell*, 5, 565-575.

Choi, G., Yi, H., Lee, J., Kwon, Y.-K., Soh,, M.-S., Shin, B., Luka, Z., Hahn T.-R., and Song, P.-S. (1999). Phytochrome signaling is mediated through nucleoside diphosphate kinase 2. *Nature*, 401, 610-613.

Chory, J., Chatterjee, M., Cook, R. K., Elich, T., Fanhauser, C., Li, J., Nagpal, P., Neff, M., Pepper, A., Poole, D., Reed J. and Vitart, V. (1996). From seed germination to flowering, light controls plant development via the pigment phytochrome. *Proc. Natl. Acad. Sci.* USA, 93, 12066-12071.

Clough, R. C., and Vierstra, R. D. (1997). Phytochrome degradation. *Plant Cell Environ*, 20, 713-721.

Clough, R. C., Jordan-Beebe, E. T., Lohman, K. N., Marita, J. M., Walker, J. M., Gatz C., and Vierstra, R. D. (1999). Sequences within both the N- and C-terminal domains of phyA are required for Pfr ubiquitination and degradation. *Plant J*, 17, 155–67.

Deng, X.-W., and Quail, P. H. (1999). Signaling in light-controlled development. *Semin Cell Dev Biol*, 10, 121-129.

Devlin, P.F., Halliday, K. J., Harber, N. P. and Whitelam, G. C. (1996). The rosette habit of Arabidopsis thaliana is dependent upon phytochrome action: novel phytochromes control internode ellongation and flowering time. *Plant J*, 10, 1127-1134.

Dieterle, M., Zhou, Y.-C., Schäfer, E., Funk M., and Kretsch, T. (2001) EID1, an F-box protein involved in phytochrome A-specific light signaling. *Genes Dev*, 15, 939 –944.

Dieterle, M., Bauer, D., Buche, C., Krenz, M. Schäfer, E., and Kretsch, T. (2005). *Plant J*, 41, 146-161.

Eichenberg, K., Kunkel, T., Kretsch, T., Speth V., and Schäfer, E. (1999). In vivo characterization of chimeric phytochromes in yeast. *J. Biol. Chem*, 274, 354-359.

Eilfeld, P.H. and Eilfeld P.G. (1988). Circular dichroism of phytochrome intermedates. Physiol. *Plant*, 74, 169-175.

Emmler, K., Stockhaus, J., Chua N.-H., and Schäfer, E. (1995). An aminoterminal deletion of rice phytochrome A results in a dominant negative suppression of tobacco phytochrome A activity in transgenic tobacco seedlings. *Planta*, 197, 103–110.

Fairchild C. D. and Quail, P. H. (1998). *The phytochromes: Photosensory perception and signal transduction*. In A. J. Greenland, E. M. Meyerowitz, and M. W. Steer (Eds.) Control of plant development: Genes and signals (pp. 85-92). Cambridge: Company of Biologists.

Fairchild, C. D., Schumaker, M. A., and Quail, P. H. (2000). HFR1 encodes an atypical bHLH protein that acts in phytochrome A signal transduction. *Genes Dev*, 14, 2377-2391.

Fankhauser, C. (2000). Phytochromes as light-modulated protein kinases. *Semin. Cell Devel. Biol*, 11, 467–473.

Fankhauser, C. (2001). The phytochromes, a family of red/far-red absorbing photoreceptors. *J. Biol. Chem*, 276, 11453-11456.

Fankhauser C. and Chory, J. (2000). RSF1, an Arabidopsis locus implicated in phytochrome A signaling. *Plant Physiol*, 124, 39-45.

Fankhauser, C., Yeh, K.-C., Lagarias, J., C. Zhang, H., Elich T. D., and Chory J. (1999). PKS1, a substrate phosphorylated by phytochrome that modulates light signaling in Arabidopsis. *Science*, 284, 1539–1541.

Farrens, D. L., Holt, R. E., Rospendowskii, B. N., Song, P.-S., and Cotton, T. M. (1989). Surface-Enhanced Resonance Raman Scattering (SERRS) spectroscopy applied to phytochrome and its model compounds. II. Phytochrome and phycocyanin chromophores. *J. Am. Chem. Soc*, 111, 9162-9169.

Foerstendorf, H., Scheer, H., Rudiger, W., Schäfer, E., and Siebert, F. (1995). *FT-IR difference spectroscopic studies of intermediates of the photoreaction of phytochrome*. In: Tetrapyrrole Photoreceptors, Freising, Germany, Book of Abstracts, P4.1.

Folta, K. M., and Spalding, E. P. (2001). Opposing roles of phytochrome A and phytochrome B in early cryptochrome-mediated growth inhibition. *Plant J*, 28, 333-340.

Franklin, K. A., and Whitelam, G. C. (2004). Light signals, phytochromes and cross-talk with other environmental cues. *J. Experim. Botany*, 55, 271-276.

Franklin, K. A., Allen, T, and Whitelam, G. C. (2007). Phytochrome A is an irradience-dependent red light sensor. *Plant Journal*, 50, 108-117.

Frick, G., Su, Q., Apel, K., and Armstrong, G.A. (2003). An Arabidopsis porB porC double mutant lacking light-dependent NADPH: protochlorophyllide oxidoreductase B and C is highly chlorophyll-deficient and developmentally arreste. *Plant J*, 35, 141-153.

Furuya, M. (1989). Molecular properties and biogenesis of phytochrome I and II. *Adv. Biophys*, 25, 133-167.

Furuya, M. (1993). Phytochromes: their molecular species, gene families, and functions. *Annu. Rev. Plant Physiol. Plant Mol. Biol*, 44, 617-645

Furuya, M., and Song, P.S. (1994). *Assembly and properties of holophytochrome*. In, R. E. Kendrick and, G. H. M. Kronenberg (eds.) Photomorphogenesis in plants (2nd edition, pp. 105-140), Dordrecht/Boston/London: Kluver Academic Publishers.

Furuya, M., Ito, N., Tomizava, K, and Schäfer, E. (1991). A stable pool phytochrome pool regulates the expression of the phytochrome I gene in pea seedlings. *Planta*, 183, 218-221.

Gärtner, W., and Braslavsky, S. E. (2004). *The phytochromes: Spectroscopy and function*. In A. Batschauer (Ed.) Photoreceptors and light signaling. Comprehensive series in photochemical and photobiological sciences (vol. 3, pp. 136-180). D.P. Häder and G. Jori (series eds.). Cambridge, UK: Royal Soc. Chemistry.

Gärtner, W., Hill, C., Worm, K., Braslavsky, S. E., and Schaffner, K. (1996). Influence of expression system on chromophore binding and preservation of spectral properties in recombinant phytochrome. *A. Eur. J Biochem*, 236, 978-983.

Hamasa, T., Tanaka, N., Noguchi, T., Kimura N., and Hasunuma, K. (1996). Phytochrome regulates phosphorylation of a protein with characteristics of a nucleoside diphosphate kinase in the crude membrane fraction from stem sections of etiolated pea seedlings. *J. Photochem. Photobiol. B: Biol*, 33, 143–151.

Hangarter, R. P. (1997). Gravity, light and plant form. *Plant Cell Environ*, 20, 796-800.

Hanzawa, H., Shinomura, T., Inomata, K., Kakiuchi, T., Kinoshita, H., Wada, K. and Furuya, M. (2002). Structural requirement of bilin chromophore for the photosensory specificity of phytochromes A and B. *Proc. Natl. Acad. Sci*, USA, 99, 4725-4729.

Hare, P. D., Moller, S. G., Huang, L.-F., and Chua, N.-H. (2003). LAF3, a novel factor required for normal phytochrome A signaling. *Plant Physiol*, 133, 1592-1604.

Hennig, L. Büche, C. Eichenberg, K. and Schäfer, E. (1999). Dynamic properties of endogenous phytochrome A in Arabidopsis seedlings. *Plant Physiol*, 121, 571–577.

Hennig, L. Büche C., and Schäfer, E. (2000). Degradation of phytochrome A and the hight irradience response in Arabidopsis: a kinetic analysis. *Plant Cell Environ*, 23, 727–734.

Heyer, A., G., Mozley, D., Landschutze, V., Thomas, B., and Gatz, C. (1995). Function of phytochrome A in potato plants as revealed through the study of transgenic plants. *Plant Physiol,* 109, 53-61.

Hill, C., Gärtner, W., Towner, P., Braslavsky, S. E., and Schaffner, K. (1994). Expression of phytochrome apoprotein from Avena sativa in Escherihia coli and formation of photoactive chromoproteins by assembly with phycocyanobilin. *Eur. J Biochem*, 223, 69-77.

Hisada, A., Hanzava, H., Weller, J. L., Nagatani, A., Reid J. B., and Furuya M. (2000). Light-induced nuclear translocation of endogenous pea phytochrome A visualized by immunocytochemical procedures. *Plant Cell*, 12, 1063-1078.

Hoecker, U., Xu, Y., and Quail, P.H. (1998). SPA1: a new genetic locus involved in phytochrome A-specific signal transduction. *Plant Cell*, 10, 19-33.

Hoecker , U. Tepperman, J. M. Quail, P. H. (1999). SPA, a WD-repeat protein specific to phytochrome A signal transduction. *Science*, 284, 496-499.

Hoecker, U., Baumgardt L., and Laubinder, S. (2003). *Investigation of the functions of the SPA gene family*. Plant Photobiology Meeting, Marburg 2003, Abstract 6.4.

Holzwarth, A. R., Venuti, E., Braslavsky, S. E., and Schaffner, K. (1992). The phototransformation process in phytochrome. 1. Ultrafast fluorescence component and models for the intital Pr->Pfr transformation steps in native phytochrome. Biochim. Biophys. *Acta*, 1140, 59-68.

Hudson, M., Ringli, C., Boylan M. T., and Quail, P. H. (1999). The FAR1 locus encodes a novel nuclear protein specific to phytochrome. *A signaling. Genes Dev*, 13, 2017-2027.

Hughes, J. and Lamparter, T. (1999). Prokaryotes and phytochrome: the connection to chromophores and signaling. *Plant Physiol*, 121, 1059-1068.

Hughes, J., Lamparter, T., Mittmann, F., Hartmann, E., Gärtner, W., Wilde, A., and Börner, T. (1997). A prokaryotic phytochrome. *Nature*, 386, 663.

Hunt, R. E. and Pratt, L. H. (1980). Partial characterization of undegradedoat phytochrome. *Biochemistry*, 19, 390-394.

Jackson, S. and Thomas, B. (1997). Photoreceptors and signals in the photoperiodic control of developments. *Plant Cell Environ*, 20, 790-795.

Janoudi, A. K., Gordon, W. R., Wagner, D., Quail P., and Poff, K. L. (1997). Multiple phytochromes are involved in red-light-induced enhancement of first positive phototropism in Arabidopsis thaliana. *Plant Physiol*, 113, 975-979.

Jordan, E. T., Hatfield, P. M., Hondred, D., Talon, M., Zeevaart, J. A., and Vierstra, R.D. (1995). Phytochrome A overexpression in transgenic tobacco. Correlation of dwarf phenotype with high concentrations of phytochrome in vascular tissue and attenuated gibberellin levels. *Plant Physiol*, 107, 797-805.

Jordan, E. T., Marita, J. M., Clough R. C., and Vierstra, R. D. (1997). Characterization of regions within the N-terminal 6-kilodalton domain of phytochrome A that modulate its biological activity. *Plant Physiol*, 115, 693-704.

Kendrick, R., and Nagatani, A. (1991). Phytochrome mutants. *Plant J,* 1, 133-139.

Kendrick, R. E., and Spruit, C. J. P. (1977). Phototransformations of phytochrome. Photochem. *Photobiol,* 26, 201-214.

Kim, L. Kircher, S. Toth, R. Adam, E. Schäfer E., and Nagy, F. (2000). Light-induced nuclear import of phytochrome-A:GFP fusion proteins is differentially regulated in transgenic tobacco and Arabidopsis. *Plant J*, 22, 125–134.

Kim, J., Yi, H., Choi, G., Shin, B., Song, P. S., and Choi, G. (2003). Fucntional characterization of phytochrome interaccting factor 3 in phytochrome-meditated light signal transduction. *Plant Cel*l, 15, 2399-2407.

Kim, J.I., Shen, Y., Han, Y.J., Park, J.E., Kirchenbauer, D., Soh, M.S., Nagy, F., Schäfer, E., and Song, P.S. (2004). Phytochrome phosphorylation modulates light signalling by influencing the protein-protein interaction. *Plant Cell*, 16, 2629-2640.

Kim, J.I., Park, J.E., Zarate, X. and Song, P.S. (2005) Phytochrome phosphorylation in plant light signalling. *Photochem. Photobiol. Sci*, 4, 681- 687.

Kircher, S., Kozma-Bognar, L. Kim,L. Adam, E. Harter, K., Schäfer, E, and Nagy, F. (1999). Light quality-dependent nuclear import of the plant photoreceptors phytochrome-A and B. *Plant Cell*, 11, 1445–1456.

Kiss, J. Z., Mullen, J. L., Correll, M. J., and Hangarter, R. P. (2003). Phytochrome A and B mediate red-light-induced positive phototropism in roots. *Plant Physiol*, 131, 1411-1417.

Koppel, L., and Sineshchekov, V. (2003). *Effect of a non-specific phosphatase inhibitor, PMSF, on the two phyA populations in roots of maize*.International Plant Photobiology Meeting, Marburg, Abstract P10.

Koppel, L., Okamoto, H., Wada, M., and Sineshchekov, V. (1999). *Phytochrome in transgenic Arabidopsis overexpressing the fern Adiantum capillus-veneris PHY1: fluorescence spectroscopy and photochemistry*. In Program and abstracts of European Symposium on Photomorphogenesis (P-25). Berlin, March 21-26.

Kunkel, T. Speth, V. Büche C. and Schäfer, E. (1995). In vivo characterization of phytochrome-phycocyanin adducts in yeast. *J. Biol. Chem*, 270, 20193-20200.

Kunkel, T., Tomizawa, K., Kern, R., Furuya, M., and Schäfer, E. (1993). In vitro formation of a photoreversible adduct of phycocyanobilin and tobacco apophytochrome B. *Eur. J. Biochem*, 215, 587-594.

Lamparter, T., Lutterbuse, P., Schneider-Poetsch, H.A.W. and Hertel, R. (1992). A study of membrane-associated phytochrome: hydrophobicity test and native size determination. Photochem. *Photobiol*, 56, 697-707.

Lamparter, T., Mittmann, F., Gärtner, W., Börner T., Hartmann, E., Hughes, J. (1997). Characterization of recombinant phytochrome from the cyanobacterium Synechocystis. *Proc. Natl. Acad. Sci*, USA., 94, 11792-11797.

Lamparter, T., Esteban, B., Hughes, J. (2001). Phytochrome Cph1 from the cyanobacterium Synechocystis PCC6803: purification, assembly, and quarternary structure. *Eur. J. Biochem*, 268, 4720-4730.

Lapko, V. N. Jiang, X. Y. Smith D. L. and Song, P. S. (1999). Mass spectrometric characterization of oat phytochrome A: isoforms and posttranslational modifications. *Protein Sci* , 8, 1032-1044.

Lariguet, P., Boccalandro, H. E., Alonso, J. M., Ecker, J. R., Chory,J., Casal, J. J., and Fankhauser, C. (2003). A growth regulatory loop that provides homeostasis to phytochrome a signaling. *Plant Cell*, 12, 2966-2978.

Lin, R, and Wang, H., (2007). Targeting proteins for degradation by Arabidopsis COP1: Teamwork is what matters. *J. Integr. Plant Biol*, 49, 35-42.

Lippitsch, M. E., Hermann, G., Brunner, H., Muller, E., and Aussenegg, F. R. (1993). Picosecond events in the photoransformation of phytochrome - a time-resolved absorption study. *J. Photochem. Photobiol. B: Biol*, 18, 17-25.

Long, C., and Iino, M. (2001). Light-dependent osmoregulation in pea protoplasts. Photoreceptors, tissues specificity, ion regulation, and physiological implications. *Plant Physiol*, 125, 1854-1869.

Lorrain, S., Genoud, T., and Fankhauser, C. (2006). Let there be light in the nucleus. *Curr. Opin. Plant Biol*, 9, 509-514.

Mateos, J. L., Luppi, J. P., Ogorodnikova, O. B., Sineshchekov, V. A., Yanovsky, M. J., Braslavsky, S. E., Gärtner, W., and Casal J. J. (2006). Functional and biochemical analysis of the N-terminal domain of phytochrome A. *J. Biol. Chem*, 281, 34421–34429.

Mathews, S., and Sharrock, R. A. (1997). Phytochrome gene diversity. *Plant Cell Environ*, 20, 666-671.

Mathews, S., Burleigh, J. G., and Donoghue, M.J. (2003) Adaptive evolution in the photosensory domain of phytochrome A in early angiosperms. *Mol. Biol. Evol*, 20, 1087-1097.

Matsushita, T., Mochizuki, M., and Nagatani, A. (2003). Dimers of the N-terminal domain of phytochrome B are functional in the nucleus. *Nature,* 424, 571–574.

Mazzella, M. A., Alconada Magliano, T. M., and Casal, J. J. (1997). Dual effect of phytochrome A on hypocotyl growth under continuous red light. *Plant Cell Environ*, 20, 261-267.

McCormac, A. C., Smith, H., and Whitelam, G. C. (1993). Photoregulation of germination in seed of transgenic lines of tobacco and Arabidopsis which express an introduced cDNA encoding phytochrome A or phytochrome B. *Planta*, 191, 386-393.

McMichael, R. W., and Lagarias, J. C. (1990). Phosphopeptide mapping of Avena phytochrome phosphorylated by protein kinases in vitro. *Biochemistry*, 29, 3872-3878.

Monte, E., Tepperman, J. M., Al-Sady, B., Kaczorowski, K. A , Alonso, J. M., Ecker, J. R., Li, X., Zang, Y., and Quail, P. H. (2004). The phytochrome-interacting transcription factor, PIF3, acts early, selectively, and positively in light-induced chloroplast devlopment. *Proc. Natl. Acad. Sci. USA*, 101, 16091-16098.

Nakazawa, M., Hayashi, H., Yoshida, Y. and Manabe, K. (1993). Identification of surface-exposed parts of red-light- and far-red-light-absorbing forms of native pea phytochrome by limited proteolysis. *Plant Cell Physiol*, 34, 83-91.

Nagatani, A. (2004). Light-regulated nuclear localization of phytochromes. *Curr. Opin. Plant Biol*, 7, 708-711.

Nagatani, A. (2005). *Functions of different domains of phytochrome*. In Wada, M., Shimazaki, K., and Iino, M. (Eds.) (2005). Light sensing in plant (pp. 59-68). Tokyo, Japan: Springer-Verlag.

Nagy, F., and Schäfer, E. (2000). Nuclear and cytosolic events of light-induced, phytochrome-regulated signaling in higher plants. *Embo J*, 19, 157-163.

Nagy, F., and Schäfer, E., (2002). Phytochromes control photomorphogenesis by differentially regulated, interacting signaling pathways in higher plants. *Annu. Rev. Plant Biol*, 53, 329-355.

Nagy, F., Kircher, S., and Schäfer, E. (2001). Intracellular trafficking of photoreceptors during light-induced signal transduction in plants. *J. Cell Sci*, 114, 475-480.

Natori, C., Kim, J.-I., Bhoo, S. H., Han, Y.-J., Hanzava, H., Furuya, M., and Song, P.-S. (2007). Differential interactions of phytochrome A (Pr vs. Pfr) with monoclonal antibodies probed by a surface plasmon resonance technique. Photochem. *Photobiol. Sci*, 6, 83-89.

Neff, M. M., Fankhauser, C., and Chory, J. (2000). Light: an indicator of time and place. *Genes Dev*, 14, 257-271.

Ni, M., Tepperman, J. M., and Quail, P. H. (1998). PIF3, a phytochrome-interacting factor necessary for normal photoinduced signal transduction, is a novel helix-loop-helix protein. *Cell*, 95, 657-667.

Ni, M. Tepperman J. M. and Quail, P. H. (1999). Binding of phytochrome B to its nuclear signaling partner PIF3 is reversibly induced by light. *Nature*, 400, 781-784.

Ogura, T., Tanaka, N., Yabe, N., Komatsu, S., and Hasunuma, K. (1999). Characterization of protein complexes containing nucleoside diphosphate kinase with characteristics of light signal transduction through phytochrome in etiolated pea seedlings. *Photochem. Photobiol*, 69, 397-403.

Oka, Y., Matsushita, T., Mochizuki, N., Tokutomi, S., and Nagatani, A. (2004). Functional analysis of a 450- amino acid N terminal fragment of phytochrome B in Arabidopsis. *Plant Cell*, 16, 2104-2126.

Park, C.-M., Bhoo, S.-H., and Song, P.-S. (2000). Interdomain crosstalk in the phytochrome molecules. *Semin. Cell Dev. Biol*, 11, 449-456.

Parker, W., Goebel, P., Ross, II, C. R., Song, P.-S., and Stezowski, J. J. (1994). Molecular modeling of phytochrome using constitutive C-phycocyanin from Fremyella diplosiphon as a structural template. *Bioconjugaate Chem*, 5, 21-30.

Parks, B. M., and Spalding, E. P. (1999). Sequential and coordinated action of phytochromes A and B during Arabidopsis stem growth revealed by kinetic analysis. *Proc. Natl. Acad. Sci*, USA, 96, 14142-14146.

Parks, B. M., Quail, P. H., and Hangarter, R. P. (1996). Phytochrome A regulates red-light induction of phototropic enhancement in Arabidopsis. *Plant Physiology*, 110, 155-162.

Pratt, L.H. (1994). *Distribution and localization of phytochrome within the plant.* In R. E. Kendrick, and G. H. M. Kronenberg (Eds.), Photomorphogenesis in Plants (2nd edition, pp. 163-186). Dordrecht/Boston/London: Kluver Academic Publishers.

Quail, P.H. (1994) *Phytochrome genes and their expression.* In R. E. Kendrick, and G. H. M. Kronenberg (Eds.), Photomorphogenesis in Plants (2nd edition, pp. 71-104). Dordrecht/Boston/London: Kluver Academic Publishers.

Quail, P. H. (1997). An emerging molecular map of the phytochromes. *Plant Cell Environ*, 20, 657-665.

Quail, P. H. (1998). The phytochrome family: Dissection of functional roles and signaling pathways among family members. *Phil. Trans. R. Soc*, London B, 14, 257-271.

Quail P. H., (2000). Phytochrome-interacting factors, Semin. *Cell Dev. Biol*, 11, 457-466.

Quail, P. H. (2002). Photosensory perception and signaling in plant cells: new paradigms? *Curr. Opin. Cell Biol*, 14, 180-188.

Quail, P. H. (2005). *Phytochrome overview*. In M. Wada, K. Shimazaki, and M. Iino (Eds.), Light sensing in plants (pp. 21-35). Tokyo, Japan: Springer-Verlag.

Quail, P. H., Briggs, W. R., and Pratt, L. H. (1978). *In vivo phosphorylation of phytochrome*. Carnegie Insititution Yearbook 77, 342-344.

Quail, P. H., Boylan, M. T., Parks, B. M., Short, T. W., Xu, Y., and Wagner, D. (1995). Phytochromes: photosensory perception and signal transduction. *Science*, 268, 675-680.

Reed, J. W., Nagpal, P., Poole, D. S., Furuya, M., and Chory, J. (1993). Mutations in the gene for the red/far-red light receptor phytochrome B alter cell elongation and physiological responses throughout Arabidopsis development. *Plant Cell*, 5, 147–157.

Reed, J. W., Nagatani, A., Elich, T. D., Fagan, M., and Chory, J. (1994). Phytochrome A and phytochrome B have overlapping but distict functions in Arabidopsis development. *Plant Cell*, 5, 147-157.

Remberg A., Lindner I., Lamparter T., Hughes J., Kneip K., Hildebrandt P., Braslavsky S.E., Gärtner W., Schaffner, K. (1997). Raman spectroscopic and light-induced kinetic characterization of a recombinant phytochrome of the cyanobacterium Synechocystis. *Biochemistry*, 36, 13389-13395.

Riemann, M., Müller, A., Korte, A., Furuya, M., Weiler, E.W., and Nick, P. (2003). Impaired induction of the jasmonaate pathway in the rice mutant hebiba. *Plant Physiol*, 133, 1820-1830.

Rohmer, T., Strauss, H., Hughes, J., de Groot, H., Gärtner, Schmieder, P., and Matysik, J. (2006). ^{15}N MAS NMR studies of Cph1 phytochrome: Chromphore dynamics and intramolecular signal transduction. *J. Phys. Chem*, 110, 20580-20585.

Rockwell, N. C., Su, Y.-S., and Lagarias, C. (2006). Phytochrome structure and signaling mechanism. *Annu. Rev. Plant. Biol*, 57, 837-858.

Rösler, J., Klein, I., and Zeidler, M. (2007). Arabidopsis fhl/fhy1 double mutant reveals a distinct cytoplasmic action of phytochrome A. *Proc. Natl. Acad. Sci*, U S A. 104, 10737-10742.

Rospendovski, B. N., Farrens, D. L., Cotton, T. M., and Song, P.-S. (1989). Surface Enhanced Resonance Raman Scattering as a probe of the structural differences between the Pr and Pfr forms of phytochrome. *Febs Letters*, 258, 1-4.

Roux, S. J. (1994). *Signal transduction in phytochrome responses*. In R. E. Kendrick and, G. H. M. Kronenberg (eds.) Photomorphogenesis in plants (2nd edition, pp. 187-209), Dordrecht /Boston /London: Kluver Academic Publishers.

Ruddat, A. Schmidt, P. Gatz, C. Braslavsky, S. E. Gärtner W. and Schaffner, K. (1997). Recombinant type A and B phytochromes from potato. Transient absorption spectroscopy. *Biochemistry*, 36, 103-111.

Rüdiger, W. (1986). *The Chromophore*. In R. E. Kendrick and G. H. M. Kronenberg (Eds.), Photomorphogenesis in plants (pp. 17-33). Dordrecht, The Netherlands: Martinus Nijhoff.

Rüdiger, W., and Thümmler, F. (1991). Phytochrome, the visual pigment of plants. *Angew. Chem. Int. Ed. Eng*, 30, 1216-1228.

Rüdiger, W., and Thümmler, F (1994). The phytochrome chromophore. In R. E. Kendrick and, G. H. M. Kronenberg (Eds.) Photomorphogenesis in plants (2nd edition, pp. 51-70), Dordrecht/Boston/London: Kluver Academic Publishers.

Runge, S., Sperling, U., Frick, G., Apel, K., and Armstrong, G.A. (1996). Distinct roles for light-dependent NADPH:protochlorophillide oxidoreductases A and B during greening in higher plants. *Plant J*, 9, 513-523.

Ruppel, N. J., R. P. Hangarter and J. Z. Kiss (2001) Red-light-induced positive phototropism in Arabidopsis roots. *Planta*, 212, 424-430.

Ryu, J. S., Kim, J. I., Kunkel, T., Kim, B. C., Cho, D. S., Hong, S. H., Kim, S. H., Fernandez, A. P., Kim, Y., Alonso, J. M., Ecker, J. R., Nagy, F., Lim, P.O., Song, P.-S., Schäfer, E., and Nam, H. G. (2005). Phytochrome-specific type 5 phosphatase controls light signal flux by enhancing phytochrome stability and affinity for a signal transducer. *Cell*, 120, 290-292.

Saijo, Y., Wang, H., Yang, J., Shen, Y., Rubio, V., Ma, L., Hoecker, U., and Deng X. W. (2003). The COP1-SPA1 interaction defines a critical step in phytochrome A- mediated regulation of HY5 activity. *Gene Dev*, 17, 2642-2647.

Sakamoto, K. and Nagatani, A. (1997). Nuclear localization activity of phytochrome *B. Plant J*, 10, 859-868.

Schäfer, E., Kunkel, T, and Frohnmeyer, H. (1997). Signal transduction in the photocontrol of chalcone synthase gene expression. *Plant Cell Environ*, 20, 722-727.

Schäfer, E., and Bowler , C. (2002). Phytochrome-mediated photoperception and signal transduction in higher plants. *Embo reports*, 3, 1042-1048.

Schäfer, E, and Nagy, F. (Eds.) (2006). *Photomorphogenesis in Plants and Bacteria* (3rd edition). Dordrecht, Netherlands: Springer.

Schaffner, K., Braslavsky, S. E., Holzwarth, A. R. (1990). Photophysics and photochemistry of phytochrome. *Adv. Photochem*, 15, 229-277.

Schendel, R., Dörnemann, L. E. D., Rüdiger, W., and Sineshchekov, V. (1996). Comparative investigations of the effect of 5-aminolevulinate feeding on phytochrome and protochlorphyll(ide) content in dark-grown seedlings of barley, cucumber and cress. *J. Photochem. Photobiol. B: Biology*, 36, 245-253.

Schmidt, P., Gensch, T., Remberg, A., Gärtner, W., Braslavsky S. E., and Schaffner, K. (1998). The complexity of the Pr to Pfr phototransformation kinetics is an intrinsic property of native phytochrome. *Photochem. Photobiol*, 68, 754-761.

Scurlock, R. D., Braslavsky, S. E., and Schaffner, K. (1993). A phytochrome study using two-laser/two-color flash photolysis: I700 is a mandatory intermediate in the Pr→Pfr phototransformation. *Photochem. Photobiol*, 57, 690-695.

Seo, H.S., Watanabe, E., Tokutomi, S., Nagatani, A., Chua, N. H. (2004). Photoreceptor ubiquitination by COP1 E3 ligase desensitizes phytochrome A signaling. *Genes Dev*, 18, 617-622.

Sharrock, R. A., and Quail, P. H. (1989). Novel phytochrome sequences in Arabiddopsis thaliana: structure, evolution, and differential expression of a plant regulatory photoreceptor family. *Genes Dev*, 3, 1745-1757.

Sharrock, R. A., and Clack, T. (2002). Patterns of expression and normalized levels of the five Arabidopsis phytochromes. *Plant Physiol*, 130, 442–456.

Sharrock, R. A., and Quail, P. H. (1989). Novel phytochrome sequences in Arabidopsis thaliana: structure, evolution, and differential expression of a plant regulatory photoreceptor family. *Genes Dev*, 3, 1745-1757.

Shen, Y., Feng, S., Ma, L., Lin, R., Qu, L.-J., Chen, Z., Wang, H., and Deng, X. W. (2005). Arabidopsis FHY1 proein stability is regulated by light via phytochrome A and 26S proteosome. *Plant Physiol*, 139, 1234-1243.

Shinomura, T. Nagatani, A. Hanzawa, H. Kubota, M. Watanabe, M. and Furuya, M. (1996). Action spectra for phytochrome A- and B-specific photoinduction of seed germination in Arabidopsis thaliana. *Proc. Natl. Acad. Sci*, USA, 93, 8129-8133.

Shinomura, T., Uchida, K., and Furuya, M. (2000). Elementary responses of photoperception by phytochrome A for high irradience response of hypocotyl elongation in Arabidopsis thaliana. *Plant Physiol*, 122, 147-156.

Shlumukov, L., Barro, F., Barcelo, P., Lazzeri P., and Smith, H. (2001). Establishment of far-red high irradiance responses in wheat through transgenic expression of an oat phytochrome A gene. *Plant Cell Environ*, 24, 703–712.

Smith, H. (ed.) (1997). Photomorphogenesis. Special issue. Plant Cell Environ., 20, 657-844.

Smith, H., and Whitelam, G. C. (1997). The shade avoidance syndrome: multiple responses mediated by multiple phytochromes. *Plant Cell Environ*, 20, 840-844.

Smith, H., Xu, Y., and Quail, P.H. (1997). Antagonistic but complementary actions of phytochromes A and B allow optimum seedling de-etiolation. *Plant Physiol*, 114, 637-641.

Sineshchekov, V. A. (1988). *Comparative fluorescence investigations of bacteriorhodopsin, phycobilins and related pigments - rhodopsin and phytochrome.* In I. N. Gogotov (Ed.) Phototrophic microorganisms (pp. 82-94). Pushchino, USSR: Acad. Sci. USSR.

Sineshchekov, V. A. (1994). Two spectroscopically and photochemically distinguishable phytochromes in etiolated seedlings of monocots and dicots. *Photochem. Photobiol*, 59, 77-86.

Sineshchekov, V. A. (1995a). Photobiophysics and photobiochemistry of the heterogeneous phytochrome system. *Biochim. Biophys. Acta*, 1228, 125-164.

Sineshchekov, V. A. (1995b). Evidence for the existence of two phytochrome A populations. *J. Photochem. Photobiol, B: Biology*, 28, 53-55.

Sineshchekov, V. A. (1997). *Phytochromes in the plant cell: fluorescence and photochemical approach* (pp. 82-114). In Godnev's Lectures IV. Minsk, Belarus: National Academy of Sciences.

Sineshchekov, V. A. (1998). The system of phytochromes: photobiophysics and photobiochemistry in vivo. *Membr. Cell Biol*, 12, 691-720.

Sineshchekov, V. A. (1999). *Phytochromes: molecular structure, photoreceptor process and physiological function.* In G.S. Singhal, G. Renger, S.K. Sopory, K.D. Irrgang and Govindjee (Eds.), Concepts in photobiology: Photosynthesis and photomorphogenesis (pp. 755-795). New Delhi, India: Narosa Publ. House.

Sineshchekov, V. (2004) Phytochrome A: functional diversity and polymorphism. *Photochem. Photobiol. Sci*, 3, 596-607.

Sineshchekov, V. A. (2005). *Polymorphism of phytochrome A and its functional implications.* In M. Wada, K. Shimazaki, and M. Iino (Eds.), Light sensing in plants (pp. 95-102). Tokyo, Japan: Springer-Verlag.

Sineshchekov, V. A. (2006). Extreme dehydration of plant tissues irreversibly converts the major and variable phyA′ into the minor and conserved phyA″. *J. Photochem. Photobiol. B: Biol*, 85, 85-91.

Sineshchekov, V. A. and Akhobadze, V.V. (1992). Phytochrome states in etiolated pea seedlings: fluorescence and primary photoreactions at low temperatures.*Photochem. Photobiol*, 56, 743-749.

Sineshchekov, V. and C. Fankhauser (2004). PKS1 and PKS2 affect phyA state in etiolated Arabidopsis seedlings. *Photochem. Photobiol. Sci*, 3, 608-611.

Sineshchekov, V.A., and Rüdiger, W. (1992). Fluorescence spectroscopy of stable and labile phytochrome in stems and roots of etiolated cress seedlings. *Photochem. Photobiol*, 56, 735-742.

Sineshchekov, V. A., and Sineshchekov, A. V. (1986). *Method of determination of the state of phytochrome in the plant cell.* Patent No.1450788, priority of February 2, 1986, USSR Committee of Inventions and Discoveries.

Sineshchekov, V. A., and Sineshchekov, A. V. (1987a). *Fluorescence of phytochrome in the cell.* Doklady AN USSR (Russ.), 292, 477-481.

Sineshchekov, V. A., and Sineshchekov, A. V. (1987b). *Fluorescence of phytochrome in the cells of etiolated pea seedlings.* Biofizika (Russ.) 32, 110-115.

Sineshchekov, V. A., and Sineshchekov, A. V. (1987c). Fluorescence and absorption investigations of phytochrome in the cells of etiolated seedlings. *Plant Physiol* (Russ.), 34, 730-741.

Sineshchekov, V. A., and Sineshchekov, A. V. (1987d). Procedure for combined fluorescence-absorption investigation of phytochrome in plant tissues. *Plant Physiol* (Russ.), 34, 832-836.

Sineshchekov, V. A., and Sineshchekov, A. V. (1989a). Fluorescence of phytochrome in the cells of dark-grown plants and its connection with the phototransformations of the pigment. *Photochem. Photobiol*, 49, 325-330.

Sineshchekov, V. A., and Sineshchekov, A. V. (1989b). Fluorescence and photoconversions of the native states of the red phytochrome form in the cells of etiolated pea seedlings. *Biofizika* (Russ.) 34, 611-617.

Sineshchekov, V. A., and Sineshchekov, A. V. (1990). Different photoactive states of the red phytochrome form in the cells of etiolated pea and oat seedlings. *J. Photochem. Photobiol. B: Biol*, 5, 197-217.

Sineshchekov, V. A., and Weller, J. L., (2004). Two modes of the light-induced phytochrome A decline – with and without changes in the relative content of its native pools (phyA' and phyA''): evidence from in-situ fluorescence investigations of wild-type and mutant phyA-3D pea. *J. Photochem. Photobiol. B: Biology*, 75, 127-35.

Sineshchekov, V. A., Lapko, V. N., Sineshchekov, A. V., Kozhukh, G. V., and Udal'tsov, A. V. (1990). Fluorescence studies of phytochrome in the cells of etiolated oat seedlings and 124 kDa phytochrome isolated from them. *J. Photochem. Photobiol*, 5, 219-229.

Sineshchekov, V., Lamparter, T., and Hartmann, E. (1994). Evidence for the existence of membrane-associated phytochrome in the cell. *Photochem. Photobiol*, 60, 516-520.

Sineshchekov, V. A., Frances, S., and White, M. J. (1995) Fluorescence and photochemical characterization of phytochrome in de-etiolated pea mutant lip. *J. Photochem. Photobiol*, B: Biol., 28, 47-51.

Sineshchekov, V.A., Heyer, A.G., and Gatz, C. (1996) Phytochrome states in transgenic potato plants with altered phyA levels. *J. Photochem. Photobiol, B: Biol*, 34, 137-142.

Sineshchekov, V. A., Ogorodnikova, O. B., Devlin, P. F., and Whitelam G. C. (1998a) Fluorescence spectroscopy and photochemistry of phytochromes A and B in wild-type, mutant and transgenic strains of Arabidopsis thaliana. *J. Photochem. Photobiol, B: Biol*, 42, 133-142.

Sineshchekov, V., Hughes, J., Hartmann, E., and Lamparter, T. (1998b). Fluorescence and photochemistry of recombinant phytochrome from the cyanobacterium Synechocystis. *Photochem. Photobiol*, 67, 263-267.

Sineshchekov, V. A., Ogorodnikova, O. B., and J. L. Weller (1999a). Fluorescence and photochemical properties of phytochromes A and B in etiolated pea seedlings. *J. Photochem. Photobiol. B: Biol*, 49, 204-211.

Sineshchekov, V. A., Clough, R. C., Jordan-Beebe, E. T., Vierstra, R. D. (1999b). Fluorescence analysis of oat phyA deletion mutants expressed in tobacco suggests that the N-terminal domain determines the photochemical and spectroscopic distinctions between phyA' and phyA". *Photochem. Photobiol*, 69, 728-732.

Sineshchekov, V., Ogorodnikova, O., Thiele, A., and Gatz, C. (2000a). Fluorescence and photochemical characterization of phytochromes A and B in transgenic potato expressing Arabidopsis phytochrome B. *J.Photochem. Photobiol. B: Biol*, 59, 139-146.

Sineshchekov, V., Koppel', L., Hughes, J., Lamparter, T., and Zeidler, M. (2000b). Recombinant phytochrome of the moss Ceratodon purpureus (CP2): fluorescence spectroscopy and photochemistry. *J.Photochem. Photobiol, B. Biol*, 56, 145-153.

Sineshchekov, V., Hennig, L.,. Lamparter, T., Hughes, J., Gärtner, W., and Schäfer, E. (2001a). Recombinant phytochrome A in yeast differs by its spectroscopic and photochemical properties from the major phyA' and is close to the minor phyA": evidence for post-translational modification of the pigment in plants. *Photochem. Photobiol*, 73, 692-696.

Sineshchekov, V., Koppel, L., Shlumukov, L., Barro, F., Barcelo, P., Lazzeri, P., and Smith, H. (2001b). Fluorescence and photochemical properties of phytochromes in wild-type wheat and a transgenic line over-expressing an oat phytochrome A (PHYA) gene: functional implications. *Plant Cell Environ*, 24, 1289-1297.

Sineshchekov, V., Koppel', L. B. Esteban, Hughes, J., and Lamparter, T. (2002). Fluorescence investigation of the recombinant cyanobacterial phytochrome (Cph1) and its C-terminally truncated monomeric species (Cph1Δ2): implication for holoprotein assembly, chromophore-apoprotein interaction and photochemistry. *J. Photochem. Photobiol. B: Biol*, 67, 39-50.

Sineshchekov, V., Belyaeva, O., and Sudnitsin, A. (2004a). Phytochrome A positively regulates biosynthesis of the active protochlorophyllide in dicots under far-red light. *J. Photochem. Photobiol. B: Biol*, 74, 47-54.

Sineshchekov, V. A., Loskovich, A. V. Riemann, M., and Nick, P. (2004b). The jasmonate-free rice mutant hebiba is affected in the response of phyA'/phyA'' pools and protochlorophyllide biosynthesis to far-red light. *Photochem. Photobiol. Sci*, 3, 1058 – 1062.

Sineshchekov, V., Loskovich, A., Inagaki, N., and Takano, M. (2006). Two native pools of phytochrome A in monocots: evidence from fluorescence investigations of phytochrome mutants of rice. *Photochem. Photobiol*, 82, 1116 –1122.

Soh, M. S., Kim, Y. M., Han, S. J., and Song, P. S. (2000). REP1, a basic helix-loop-helix protein, is required for a brach pathway of phytochrome a signaling in Arabidopsis. *Plant Cell*, 12, 2061-2073.

Somers, D. E. and Quail, P. H. (1995). Temporal and spatial expression patterns of PHYA and PHYB genes in Arabidopsis. *Plant J*, 7,, 413-427.

Somers, D. E., Devlin, P. F., and Kay, S. A. (1998). Phytochromes and cyptochromes in the entrainment of the Arabidopsis circadian clock. *Science*, 282, 1488-1490.

Sommer, D., and Song, P.S. (1990). The molecular topography and secondary structure fo 124 kDa Avena phytochrome probed by Zn2+-induced chromophore modification. *Biochemistry*, 29, 1943-1948.

Sommer, D., and Song, P.S. (1995). *Phytochromes: structure and function*. In: Tetrapyrrole Photoreceptors, Frising, Germany, Book of Abstracts, O 22.

Song, P.-S. (1985). *Primary molecular events in aneural cell photoreception*. In: G. Colombetti, F. Lenci, and P.-S. Song (eds.) Sensory Perception and Transduction in Aneural Organisms (pp. 47-59). London: Plenum Press.

Song, P.-S. (1999). Inter-domain signal transduction within the phytochromes. *J. Biochem. Molecular Biol*, 32, 215-225.

Song, P.S., Park, M.H., and Furuya, M. (1997). Chromophore:apoprotein interactions in phytochrome A. *Plant Cell Environ*, 20, 707-712.

Spalding, E. P., and Folta, K. M. (2005). Illuminating topics in plant photobiology. *Plant Cell Environ*, 28. 39-53.

Spruit, C. J. P., and Kendrick, R. E. (1972). Phytochrome decay in seedlings under continuous incandescent light. *Planta* (Berl.), 103, 319-326.

Stockhaus, J., Nagatani, A., Halter, U., Kay, S., Furuya, M., and Chua, N.H. (1992). Serine-to-alanine substitutions at the amino-terminal region of phytochrome A results in an increase in biological activity. *Genes Dev*, 6, 2364-2372.

Strauss, H. M., Hughes, J., Schmieder, P. (2005). Heteronuclear solution-state NMR studies of the chromophore in cyanobacterial phytochrome Cph1. *Biochemistry*, 44, 8244-8250.

Sudnitsin, A., Adam, E., Nagy, F., Schäfer E., and Sineshchekov, V. (2003). *phyA-GFP is spectroscopically and photochemically close to phyA and comprises both phyA' and phyA''*. Plant Photobiology Meeting, Marburg, Abstract P9.

Takagi, S. (2005). *Photoregulation of cytoplasmic motility*. In M. Wada, K. Shimazaki, and M. Iino (Eds.), Light sensing in plants (pp. 87-94). Tokyo, Japan: Springer-Verlag.

Takano, M., Kanegae, H., Shinomura, T., Miyao, A., Hirochika, H., and Furuya, M. (2001). Isolation and characterization of rice phytochrome A mutants. *Plant Cell*, 13, 521–534.

Takano, M., Xie, X., Inagaki, N., and Shinomura, T. (2005a). *Distinct functions of phytochromes on the photomorphogenesis in rice*. In M. Wada, K. Shimazaki, and M. Iino (Eds.), Light sensing in plants (pp. 111-117). Tokyo, Japan: Springer-Verlag.

Takano, M., Inagaki, N., Xie, X., Yuzurihara, N., Hihara, F., Ishizuka, T., Yano, M., Nishimura, M., Miyao, A., Hirochika, H., and Shinomura, T. (2005b). Distinct and cooperative functions of phytochromes A, B, and C in the control of deetiolation and flowering in rice. *Plant Cell*, 17, 3311-3325.

Tanaka, N., Ogura, T., Noguchi, T., Hirano, H., Yabe, N., and Hasunuma, K. (1998). Phytochrome-mediated light signals are transduced to nucleoside diphoshate kinase in Pisum sativum L. cv. Alaska. *J. Photochem. Photobiol*, B, 45, 113-121.

Tepperman, J. M., Hwang, Y.S., and Quail, P. H. (2006). phyA dominates in transduction of red-light signals to rapidly responding genes at the initiation of Arabidopsis seedling de-etiolation. *Plant J*, 48, 728-742.

Terry, M. J., Hall, J. L., and Thomas, B. (1992). The association of type I phytochrome with wheat leaf plasma membranes. *J. Plant Physiol*, 140, 691-698.

Tokuhisa, J. G., Daniels, S. M., and Quail, P. H. (1985). Phytochrome in green tissues: spectral and immunochemical evidence for two distinct molecular species of phytochrome in light-grown Avena sativa L. *Planta*, 164, 321-332.

Tokutomi, S., Yamamoto, K. T., Miyoshi, Y., and Furuya. M. (1982). Photoreversible changes in pH of pea phytochrome solutions. *Photochem Photobiol*, 35, 431-433.

Trupkin, S. A., Debrieux, D., Hiltbrunner, A., Fankhauser C., and Casal J. J. (2007). The serine-rich N-terminal region of Arabidopsis phytochrome A is required for protein stability. *Plant Mol. Biol*, 63, 669-678.

Tu S.L., and Lagarias, J.C. (2005). *The phytochromes*. In W. R. Briggs, and J. L. Spudich (Eds.), Handbook of photosensory receptors (pp. 122-139). Weinheim, FRG: WILEY-VCH Verlag Gmbh and Co. KgaA.

Usami, T., Matsushita, T., Oka, Y., Mochhizuki, N., and Nagatani, A. (2007). Roles for the N- and C-terminal domains of phytochrome B in interacitons between phytochrome B and cryptochrome signaling cascades. *Plant Cell Physiol*, 48, 423-433.

van Thor, J., Borucki, B., Lamparter, T., Hughes, J., Hellingwerf, K., and Heyn, M. (2001). Light-induced proton release and uptake in the cyanobacterial phytochrome, Cph1. *Biochemistry*, 40, 11460-11471.

Vierstra, R. D., and Quail, P. H. (1986). *Phytochrome: the protein*. In R. E. Kendrick and G. H. M. Kronenberg (Eds.), Photomorphogenesis in plants (pp. 35-60). Dordrecht, The Netherlands: Martinus Nijhoff.

Volotovski, I. D. (1998). Ca^{2+} and intracellular signaling in plant cells: a role on phytochrome transduction. Membr. *Cell Biol*, 12, 721-742.

Wada, M., Shimazaki, K., and Iino, M. (Eds.) (2005). *Light sensing in plants*. Tokyo, Japan: Springer-Verlag.

Wagner, D., Koloszvari, M., and Quail, P. H. (1996). Two small spatially distinct regions of phytochrome B are required for efficient signaling rates. Plant Cell, 8, 859–871.

Wagner, J. R., Brunzelle, J. S., Forest, K. T., and Vierstra, R. (2005). A light-sensing knot revealed by the structure of the chromophore-binding domain of phytochrome. Nature, 43, 325-331.

Wang, H., and Deng, X. W. (2002). Phytochrome signaling mechanism, The Arabidopsis Book, 1–30.

Wang, H., and Deng, X. W. (2003). Dissecting the phytochrome A-dependent signaling network in higher plants, Trends Plant Sci, 8, 172-178. Weller, J. L., Batge, S. L., Smith, J. J., Kerckhoffs, L. H. J., Sineshchekov, V. A., Murfet, I. C., and Reid, J. B. (2004). A dominant mutation in the pea PHYA gene confers enhanced responses to light and impairs the light-dependent degradation of phytochrome A. *Plant Physiol*, 135, 2186-2195.

Wells, T. A., Nakazawa, M., Manabe K., and Song P. S. (1994). A conformational change associated with the phototransformation of Pisum phytochrome A as probed by fluorescence quenching. *Biochemistry*, 33, 708-712.

Whitelam, G. C., and Devlin, P. F. (1997). Roles of different phytochromes in Arabidopsis photomorphogenesis. *Plant Cell Environ*, 20, 752-759.

Wong, Y. S., Cheng, H. C. and Lagarias, J. C. (1986). Phosphorylation of Avena phytochrome in witro as a probe of light-induced conformational changes. *J. Biol. Chem*, 261, 12089-12097.

Xu, Y., Parks, B. M., Short, T. W., and Quail, P. H. (1995). Missense mutations define a restricted segment in the C-terminal domain of phytochrome A critical to its regulatory activity. *Plant Cell*, 7, 1433-1443.

Yamaguchi, R., Nakamura, M., Mochizuki, N., Kay S. A., and Nagatani A. (1999). Light-dependent translocation of a phytochrome B-GFP fusion protein to the nucleus in transgenic Arabidopsis. *J. Cell Biol*, 145, 437-445.

Yanofsky, M.J., Casal, J.J., and Whitelam, G.C. (1995). Phytochrome A, phytochrome B and HY4 are involved in hypocotyl growth-responses to natural radiation in Arabidopsis weak de-etiolation in the phyA mutant under dense canopies. *Plant Cell Environ*, 18, 788-794.

Yanovsky, M. J., Casal, J. J., and Luppi, J. P. (1997). The VLF loci, polymorphic between ecotypes Landsberg erecta and Columbia, dissected two branches of phytochrome A signaling pathways that correspond to the very-low fluence and high-irradience responses of phytochrome. *Plant J*, 12, 659-667

Yanovsky, M. J., Whitelam, G. C., and Casal, J. J. (2000). fhy3-1 retains inductive responses of phytochrome A. *Plant Physiol*, 123, 235-242.

Yanovsky, M. J., Luppi, J. P., Kirchbauer, D., Ogorodnikova, O. B., Sineshchekov, V. A., Adam, E., Kircher, S., Staneloni, R. J., Schafer, E., Nagy, F., and Casal, J. J. (2002). Missense mutation in the PAS2 domain of phytochrome A impairs subnuclear localization and a subset of responses. *Plant Cell*, 14, 1591-1603.

Yang, J., Lin, R., Sullivan, J., Hoecker, U., Liu, B., Deng, X. W., and Wang, H. (2005). Light regulates COP1-mediated degradation of HFR1, a transcription factor essential for light signaling in Arabidopsis. *Plant Cell*, 17, 804-821.

Yeh, K.-C., and Lagarias J. C. (1998). Eukaryotic phytochromes: Light-regulated serine/threonine protein kinases with histidine kinase ancestry. *Proc. Natl. Acad. Sci*, USA, 95, 13976-13981.

Yeh, K. C., Wu, S., Murphy, J., and Lagarias, J. C. (1997). A cyanobacterial phytochrome two-component light sensory system. *Science*, 277, 1505-1508.

Zeidler, M., Zhou, Q., Sarda, X., Yau, C. P., Chua, N. H. (2004). The nuclear licalization signal and the C-terminal region of Fhy1 are required for transmission of phytochrome A signals. *Plant J*, 40, 355-365.

Zhang, C.F., Farrens, D.L., Björling, S.C., Song, P.S., and Kliger, D.S. (1992) Time-resolved absorption studies of native etiolated oat phytochrome. *J. Am. Chem. Soc*, 114, 4569-4580.

Zhou, Q., Hare, P. D., Yang, S.W., Zaidler, M., Huang, L.F., Chua, N.H. (2005). FHL is required for full phytochrome A signaling and shared overlaping functions with FHY1. *Plant J*, 42, 356-370.

Zhou, Y. C. , Dieterle, M., Buche, C., and Kretsch, T. (2002). The negatively acting factors EID1 and SPA1 have distinct functions in phytochrome A-specific light signaling. *Plant Physiol*, 128,1098-108.

Zhu, Y., Tepperman, J. M., Fairchild, C. D., and Quail, P. H. (2000). Phytochrome B binds with greater apparent affinity than phytochrome A to the basic helix-loop-helix factor PIF3 in a reaction requiring the PAS domain of PIF3. *Proc. Natl. Acad. Sci*, USA, 97, 13419-13424.

In: Photochemistry Research Progress
Editors: A. Sánchez, S. J. Gutierrez

ISBN: 978-1-60456-568-3

Chapter 4

Photochemical Reactions of Iron(III) Complexes: Classification, Mechanisms and Application

Jozef Šima*
Department of Inorganic Chemistry, Slovak Technical University, Radlinského 9, 812 37 Bratislava, Slovak Republic

Abstract

In this chapter, the primary photochemical deactivation processes and secondary thermal reactions of iron(III) compounds are reviewed. Along with the theoretical aspects and deactivation mechanisms, particular stress is paid to application importance of the processes occurring in irradiated systems of iron(III) compounds. Both the high-spin and low-spin iron(III) complexes are dealt with here.

A dominant share of the accumulated knowledge concerns photoredox reactivity involving both the photoreduction of central atom Fe(III) to Fe(II), and photooxidation of Fe(III), mainly to Fe(V). The chapter covers intramolecular photoredox decompositions as well as outer-sphere electron-transfer processes. While the majority of the results have been obtained investigating mononuclear complexes, a deep insight into the mechanisms of photochemical and photophysical deactivation processes has been reached by also studying polynuclear complexes and solid polymeric compounds.

The complexes are classified according to the bonding properties of their ligands into groups with monodentate, chelating and macrocyclic ligands. Due to their peculiar properties, native proteins containing the iron(III) central atom are discussed in a separate section. In addition to solution photochemistry, the photoredox processes occurring in microheterogeneous systems, polymer matrices and solid complexes are evaluated. In each type of complex, the current knowledge is assessed and the main directions for future photochemical research are tentatively offered.

One of the driving forces in focusins long-lasting and systematic attention to the photochemical properties of iron(III) compounds is their practical importance. Due to this

* Email jozef.sima@stuba.sk

fact, application aspects of the photochemistry of such complexes are documented as well, with particular attention paid to their contribution to environmental protection, the role played in biological systems, therapeutical significance, photomagnetic materials, understanding the processes occurring in the marine environment, and technological applications.

1. INTRODUCTION

The first stage of modern inorganic photochemistry concluded with the publishing of the excellent monograph, *Photochemistry of Coordination Compounds*, compiled by Balzani and Carassiti [1]. Most of the research activity was focused on photochemical properties of classical Werner-type complexes of first-raw transition metals in solution. The chapter of the mentioned monograph devoted to iron compounds ended with the highest number of references (317). The other chapters were devoted to other metal compounds. Photochemical investigation of iron(III) compounds has a very rich long history. The first scientific paper devoted to the photochemical properties of inorganic compounds can be traced back to 1790, when the photochemical preparation of $FeCl_2$ through irradiation of solution of $FeCl_3$ in ether was published. Several original papers appeared thereafter, and some reviews and monographs also deal with various aspects of the photochemistry of iron compounds [2-14].

Later on, mainly due to the first energy crisis arising at the beginning of the 1970s and owing to the necessity to search for alternative energy sources, the effort of photochemists has expanded significantly to photophysical investigation of light-harvesting systems represented mainly by the complex $[Ru(bpy)_3]^{2+}$ and its analogues. In spite of this tendency, iron compounds have still remained a significant object of photochemical research. The permanent interest in the photochemistry of iron compounds lies in their important role played in industrial, environmental, and biological processes. There is a huge number of original papers and conference presentations dealing with individual, isolated aspects of photochemistry of iron compounds. It seems to be obvious that the time is ripe to offer a comprehensive review of the field stimulated by the following three main objectives:

1. to evaluate the state of the art in the field and offer (photo)chemists a general view on the topic;
2. to enable (photo)chemists to utilize the results obtained in the area of iron compounds photochemistry for general understanding of photochemical behaviour of other compounds via knowledge-transfer procedure;
3. to stimulate a wider application of the theoretical knowledge into practice.

Most papers quoted in this chapter were published in the last 15 years, but some important older sources, containing the results which are necessary to formulate more general conclusions, are also referred to. The complexes are classified according to the bonding properties of their ligands into groups with monodentate, chelating and macrocyclic ligands. Due to their peculiar properties, native proteins containing the iron(III) central atom are discussed in a separate section. In addition to solution photochemistry, the photoredox processes occurring in microheterogeneous systems, polymer matrices and solid complexes are evaluated as well. In each type of complexes, the current knowledge is assessed and the main directions for future photochemical research are tentatively offerred.

This chapter provides an overview of the pure and applied photochemistry of iron(III) compounds. At the beginning of any discussion and rationalization of the results obtained within the investigation of photochemical behaviour of iron(III) complexes, some particular features, differing these complexes from complexes of many other transition metals (mainly Cr(III), Pd(II), Pt(II), Pt(IV), Ru(II), Os(II), Ir(I), Ir(III)), must be taken into account.

One of the differences is the almost total absence of radiation deactivation processes involving the central atom orbitals. This lack of any luminescence creates a serious problem in interpreting the intramolecular transitions between individual excited states and between excited states and the ground state.

Other difference, connected with the above mentioned, lies in the lack of kinetic data related to intramolecular changes in multiplicity, energy, orbital occupation. It has been tacitly supposed that intramolecular electronic transitions are too fast to be measured. Up to now, femtosecond techniques have been exploited in a monitoring of intramolecular transitions and determination of corresponding kinetic parameters for some iron(II) polypyridyl complexes [15], such data are not, however, available for iron(III) complexes.

A further issue concerns the kinetic lability of high-spin iron(III) complexes and equilibria established in their solution, mainly in polar solvents. Except for low-spin complexes with strong ligand-field ligands, such as CN^-, bpy, phen or macrocyclic ligands, iron(III) forms mostly high-spin, substitutionally labile complexes [1,14]. This lability has a very trouble-creating consequence for investigation of solutions of such complexes. Particularly in cases of the complexes with monodentate and/or chelating ligands, complicated equilibria may be established in solutions and it is not so simple to attribute a specific value (e.g. quantum yield value) to specific individual species. The situation can be exemplified [16] on the distribution of nitrilotriacetato iron(III) complexes in aqueous solutions with $c(Fe^{III}) = c(H_3NTA) = 2.0 \times 10^{-3}$ M, and $c(Fe^{III}) = 2.0 \times 10^{-3}$ M and $c(H_3NTA) = 6.0 \times 10^{-3}$ M at pH = 7. The system with the molar ratio of $c(Fe^{III}) : c(H_3NTA) = 1 : 1$ contains 16% of the total iron(III) in the form of $[Fe(OH)_2(NTA)]^{2-}$, 68% as $[Fe(OH)(NTA)]^-$, and 16% as $[Fe_2O(NTA)_2]^{2-}$, while the system with $c(Fe^{III}) : c(H_3NTA) = 1 : 3$ contains 13% of the total iron(III) as $[Fe(OH)_2(NTA)]^{2-}$, 44% as $[Fe(OH)(NTA)]^-$, 3% as $[Fe_2O(NA)_2]^{2-}$, and 38% as $[Fe(NTA)_2]$. Moreover, any change in the total iron(III) concentration, $c(Fe^{III}) : c(H_3NTA)$ molar ratio, pH, or ionic strength has a profound effect on the distribution of iron(III) containing species as well as those not containing iron(III), such as free protonated and deprotonated ligand. It is obvious that in most cases with fast equilibria of species, a substantial barrier must be overcome in order to ensure the correct interpretation of photochemical data.

This chapter is devided into 8 sections. Some general remarks and a brief description of evolution of iron(III) photochemistry are given in introductory section 1. Information on the structure and electronic spectra of iron(III) complexes are briefly provided in section 2. Photochemical processes involving the iron(III) central atom are reviewed in sections 3 – 6. Photochemical reactions virtually not involving the iron(III) central atom are described in section 7. The most important applications of photochemical properties of iron(III) complexes are indicated in section 8.

2. Structure and Electronic Spectra of Iron(III) Complexes

Detailed investigation of molecular and crystal structures of hundreds of crystalline iron(III) mononuclear and polynuclear complexes has led to a conclusion that in the vast majority of cases the iron(III) central atom forms the hexacoordinated centre of octahedral or pseudo-octahedral coordination polyhedra [17,18]. Hexacoordination has been found also in solutions of iron(III) complexes [19,20].

In their electronic absorption spectra, the spin-forbidden ligand-field (LF) bands are generally very weak and hidden in the tails of intense spin-allowed ligand-to-metal charge transfer (LMCT) and/or intraligand (IL) bands [21]. In polynuclear mixed-valence complexes, metal-to-metal charge transfer (MMCT) can be identified; in some systems, the ion-pair charge transfer (IPCT) nature has been attributed to the bands which are missing in the spectra of individual ions.

The overwhelming majority of iron(III) complexes provide no luminescence at either room or lower temperatures, which is tacitly related to the extremely short lifetime of their excited states. A very rare case of a fluorescence of high-spin iron(III) complexes $[Fe^{III}(L)_2]Cl$ complexes, where L are acyclic fully or partly deprotonated tridentate salicylaldehyde 2-pyridyl hydrazone and its methylated derivative, has recently been announced [22,23]. On excitation at 390 nm, the ligand and the complexes display ligand-localized ($\pi^* \rightarrow \pi$) fluorescence in the blue region (λ_{em} = 440–450 nm) of the spectrum. The corresponding luminescent excited states do not involve the central atom.

The tetrahedral anions $[FeCl_4]^-$ and $[FeBr_4]^-$ luminesce when irradiated into their LF or LMCT bands at 85 K. The maximum of the broad featureless phosphorescence band is localized in the near-IR region (945-1030 nm for $[FeCl_4]^-$ and 1005-1050 nm for $[FeBr_4]^-$) and varies slightly with the excitation wavelength [24].

3. Complexes with Monodentate Ligands

This section covers photochemistry of iron(III) complexes containing only monodentate ligands bonded in the primary coordination sphere. Frequently, also when exciting complexes with both monodentate and chelating/macrocyclic ligands, it is a monodentate ligand which undergoes a photochemical (usually photoredox) change. Due to the possibilities offered to tune the photoreactivity of a monodentate ligand through modifying electronic and /or steric properties of polydentate ligand(s), these complexes are dealt with in the following sections 4 and 5 devoted to chelates and macrocyclic iron(III) complexes.

In studying the photochemical properties of Fe(III) complexes with monodentate ligands, attention has been focused on four main topics:

1. photoassisted Fenton process and its exploitation in degradation of organic pollutants;
2. the identification of the primary photoredox step in Fe(III) photoreduction;
3. the energy transfer processes and their chemical consequences;
4. the wavelength dependence of the quantum yield of Fe(II) formation;
5. the electron transfer processes in dinuclear-bridged complexes and ion pairs.

From the viewpoint of applied photochemistry, the most important photoinduced redox reaction of Fe(III) compounds is Fe(III) photoreduction associated with the formation of $^{\bullet}OH$ radicals. This photoreaction

$$Fe(OH)^{2+} \xrightarrow{h\nu} Fe^{2+} + {}^{\bullet}OH \quad (1)$$

occurring under UV-visible irradiation of aqueous solutions of iron(III) salts containing aquated Fe^{3+} and $Fe(OH)^{2+}$ ions, enhances the concentration of Fe^{2+} ions and $^{\bullet}OH$ radicals formed simultaneously in the spontaneous process, called Fenton process [25]

$$Fe^{2+} + H_2O_2 \rightarrow Fe^{3+} + OH^- + {}^{\bullet}OH \quad (2)$$
$$Fe^{3+} + H_2O_2 + H_2O \rightarrow Fe^{2+} + H_3O^+ + {}^{\bullet}O_2H \quad (3)$$

The combination of all three reactions (1) – (3) represents photoassisted Fenton process (shortly also photo-Fenton process). The practical importance of reaction (1) lies in increasing the concentration of $^{\bullet}OH$ radicals acting as a photooxidant to many organic substances, e.g. pollutants (for details, see Section 8). The optimum pH for this reaction is about 3.0, it occurs, however, at neutral media as well. The main primary photoactive species at this pH is aquated $Fe(OH)^{2+}$ absorbing radiation in the range 290-410 nm of the spectrum, other mono-, di- or polynuclear hydroxo iron(III) species can, however, be involved in the $^{\bullet}OH$ radicals generation at higher pH too.

The difficulties emerging in the field of primary photoredox step identification have their origin in the mentioned kinetic lability of Fe(III) complexes. Attempts to rationalize the photochemical reactions occurring in systems of complexes with monodentate ligands will illustrate the matter. Of the originally proposed four modes of the primary electron transfer step with each of them involving the absorption of a photon by the Fe(III) complex as the primary photophysical step, only two are nowadays generally accepted, namely an inner-sphere electron transfer from a coordinated ligand to the central atom, forming the radical and an Fe(II) complex, e.g. in octahedral $[Fe(CH_3OH)_6]^{3+}$

$$[Fe(CH_3OH)_6]^{3+} \xrightarrow{h\nu} [Fe(CH_3OH)_5]^{2+} + H^+ + CH_2OH^{\bullet} \quad (4)$$

or an outer-sphere electron transfer from an organic molecule RH to iron(III) complex, e.g. to $FeCl_3$

$$FeCl_3 + RH \xrightarrow{h\nu} FeCl_2 + HCl + R^{\bullet} \quad (5)$$

The former mode (4) is similar to the generally accepted release of the oxidized ligand $X^{\bullet}$ from the LMCT excited states of the kinetically inert halogeno complexes of Co(III), Ir(IV), etc. [1,8]. The formation of electron paramagnetic resonance (EPR) detectable radicals $R^{\bullet}$ was supposed to occur in solutions of $FeCl_3$ in non-polar uncoordinating solvents, in which $FeCl_3$ does not undergo any observable ionization (e.g. in aromatics or in dissolved $FeCl_3$ at photochemical synthesis of $FeCl_2$ in ether) [1,8].

The two other modes, an outer-sphere electron transfer from an anionic oxidizable ligand X^- (such as Cl^-, Br^-, NCS^-) to the central atom Fe(III) within an ion pair [1,6], e.g. in a red thiocyanato ion pair

$$Fe^{3+}(solv).NCS^- \xrightarrow{h\nu} Fe^{2+}(solv) + NCS^{\bullet} \quad (6)$$

or an outer-sphere electron transfer from an anion X^- located in the secondary coordination sphere to the excited Fe(III) complex [12,26]

$$[FeCl_4]^- + Cl^- \xrightarrow{h\nu} FeCl_2 + Cl^- + Cl_2^{\bullet-} \quad (7)$$

do not appear in currently published mechanisms.

Based on the study of the composition and structure of Fe(III) complexes in solution [20], which proves the presence of well-coordinated X^- ligands in the primary coordination sphere of Fe(III) complexes, possibility of ion-pair involvement (6) can be excluded as a common mode of photoreaction. However, neither the identification of various radicals, nor the dependence of the quantum yield of Fe(II) on the composition of the system [26-29] allows an unambiguous answer to be given to the question concerning the primary photoredox step or the discrimination between the mentioned modes. A very detailed study of the processes occurring in irradiated systems containing $[FeCl_4]^-$, free Cl^- and ethanol [26] has led to the conclusion that, in these and analogous systems, the primary photoredox step may involve also outer-sphere extraction of an electron from an anion Cl^- by the excited complex $*[FeCl_4]^-$ leading to the radical $Cl_2^{\bullet-}$ (7).

The photochemistry of complexes was used as a tool to clarify the mechanism of decarboxylation of halocarboxylic acids [30] and carboxylic acids [31-33]. An inner-sphere decomposition was postulated to be the primary photoredox step in solvated carboxylato complexes

$$Fe^{III}(RCO_2)^{2+} \xrightarrow{h\nu} Fe^{2+} + R^{\bullet} + CO_2 \quad (8)$$

However, no details were given. The irradiation of $FeCl_3$ in water-ethanol mixtures gives rise to the formation of gaseous H_2 [34]; the primary photoredox step is believed to involve homolytic splitting of an Fe^{III}-Cl^- bond.

In the photochemistry of Fe(III) complexes there are (at least) two perennial problems, namely the mechanisms of the photoreactions of $[Fe(C_2O_4)_3]^{3-}$ and of $[Fe(CN)_5L]^{q-}$ complexes. The former case will be discussed in Section 4, while the following lines are devoted to the cyano complexes.

The principal problems arising in the study of Fe(III) cyano complexes are that their photochemistry is strongly influenced by the experimental conditions (pH value, presence of other species in the investigated systems, etc.) and complicated photochemical and dark reactions occur simultaneously. Even if we limit ourselves to the processes involving an excited complex, the question concerning the primary photochemical steps is still difficult to answer. Based on the data gathered in [1,6,35,36], it seems to have been proven definitely that one of the primary steps is Fe(III) photoreduction, the other being photosubstitution of one of the ligands. The photoreduction of Fe(III) can be accompanied by an inner-sphere photooxidation of the neutral ligand L or CN^- or by an outer-sphere oxidation of a molecule located in the secondary coordination sphere (e.g. an alcohol molecule). Not infrequently, hydrated Fe_2O_3 or Prussian blue is formed as the final product of the reactions occurring in systems of irradiated iron(III) cyano complexes.

Of the pentacyano complexes, the complex $[Fe(CN)_5(NO)]^{2-}$ attracts virtually continuous attention. One of the reasons lies in its application in cardiology [36], where the compound $Na_2[Fe(CN)_5(NO)]\cdot 2H_2O$ is used in clinical practice as an effective hypotensive drug under the commercial name Nipride.

Before starting any discussion on the photochemistry of cyano-nitrosyl complexes, the question of the oxidation state of the central atom and of the NO-ligand is to be answered. This necessity stems from a fact that some results of theoretical approaches and experimental methods used to describe bonding properties of nitrosyl complexes, e.g. $[Fe(CN)_5(NO)]^{2-}$, contradict each other. Results obtained by such methods, as IR spectroscopy, single-crystal X-ray diffraction, and quantum-chemical calculations, lead to a frequently accepted opinion that the mentioned complex should be characterized as an Fe(II) complex with the NO^+ ligand. However, the results obtained by methods directly determining the charge localized on the atoms, such as electron spectroscopy for chemical analysis (ESCA), revealed that the actual positive charge located on the iron central atom in the complex $[Fe(CN)_5(NO)]^{2-}$ is higher than that in $[Fe(CN)_6]^{3-}$. It clearly follows from the binding energy E_b of electrons occupying the atomic core orbitals. For a free, uncharged NO molecule, $E_b(N1s) = 410.3$ eV while for the NO in the complex $[Fe(CN)_5(NO)]^{2-}$ it is $E_b(N1s) = 403.4$ eV. As for the iron central atom, the situation is even more convincing, $E_b(Fe2p_{3/2})$ are 708.3 eV, 709.7 eV, and 710.4 eV for $[Fe^{II}(CN)_6]^{4-}$, $[Fe^{III}(CN)_6]^{3-}$, and $[Fe^{?}(CN)_5(NO)]^{2-}$, respectively. The same trend was found for $E_b(Fe1s)$ with the following experimental binding energies 7111.8 eV, 7112.0 eV and 7112.6 eV for $[Fe^{II}(CN)_6]^{4-}$, $[Fe^{III}(CN)_6]^{3-}$, and $[Fe^{?}(CN)_5(NO)]^{2-}$, respectively [37]. It means that the iron central atom bears in fact a more positive charge in the complex $[Fe(CN)_5(NO)]^{2-}$ than in $[Fe^{III}(CN)_6]^{3-}$. Based on this charge distribution, and on the general scientific principle requiring the compliance of artificial, man-introduced quantities (such as oxidation number) with objective, man-independent experimental data, the complex $[Fe(CN)_5(NO)]^{2-}$ may be (more correctly) formulated also as Fe(IV) complex with coordinated anionic NO^-, i.e. as $[Fe^{IV}(CN)_5(NO^-)]^{2-}$. In accordance with „the golden mean" principle, the complex will be characterized as Fe(III) containing a neutral NO molecule. In reality, owing to the strong covalency of the bonds in the moiety {FeNO}, the most adequate approach should be to consider the whole moiety {FeNO} as an integral part of the complex. Anyway, excited-state chemistry of the nitrosyl iron complexes is so rich that it deserves some lines.

Up-to-now, several kinds of primary photoinduced reaction steps have been identified and characterized in the complex $[Fe(CN)_5(NO)]^{2-}$. Its excitation can lead to elimination of CN^- anion or NO molecule from the primary coordination sphere, the abstraction of an electron from a neighbouring (solvent) molecule, electron ejection to the complex environment, addition reaction localized on the NO ligand [38]. Photoelimination of molecular NO was found to be the primary photochemical step in many cases and it can occur via photooxidation-substitution, photoreduction, and ligand rearrangement or decomposition as a result of photoreaction [36,39], keeping still in mind that the applicability of the terms "photooxidation" and "photoreduction" depend on the oxidation state ascribed to the iron central atom.

Charge transfer photolysis of the complex $[Fe(CN)_5(NO)]^{2-}$ in methanol produced $[Fe^{III}(CN)_5(CH_3OH)]^{2-}$ (in the presence of KOH the complex $[Fe^{II}(CN)_5(CH_3OH)]^{3-}$ was formed) and the quantum yield of these photosubstitution products did not depend on the presence of dioxygen in the irradiated systems [38]. The complex $[Fe(CN)_5(NO)]^{2-}$ reacts with a large variety of nucleophiles yielding complexes of the $[Fe(CN)_5N(O)L]^{q-}$ type, where L depicts C-, N-, O-, S-, or Se-nucleophiles. The thermally stable $[Fe(CN)_5N(O)SR]^{3-}$ complexes undergo photodecomposition via the unstable $[Fe^{III}(CN)_5\{N(O)\text{-}SR^{\bullet-}\}]^{3-}$ charge-separated state, which in turn decomposes via heterolytic cleavage of the Fe-N and N-S bonds. The general reaction mode is thus similar to that of $[Fe(CN)_5(NO)]^{2-}$, but in the case of $[Fe(CN)_5N(O)SR]^{3-}$, the nitrosothiyl anion radical is generated instead of NO. Photogenerated nitrosothiyl anion radical may decompose to the thiolate anion and nitric oxide or undergo oxidation yielding nitrosothiol. The starting complexes and products of the primary photochemical changes underwent further dark and photochemical substitution and/or redox processes yielding a final product, composition of which depended on the experimental conditions [36,38,39].

The photoreduction mode is important in iron(III) sulfur nitrosyl monomers and clusters such as Roussin's red salt, $[Fe_2S_2(NO)_4]^{2-}$, and Roussin's black salt, $[Fe_4S_3(NO)_7]^-$. These compounds are thermally stable, but irradiation of these species results in photodegradation of the cluster accompanied with almost quantitative nitric oxide release. DFT calculations made for model monomeric compounds indicated that these complexes consist of Fe(III) centers and NO ligands which are strongly antiferromagnetically coupled. Easy interconversion between $[Fe_4S_3(NO)_7]^-$, $[Fe_2S_2(NO)_4]^{2-}$, $[Fe_2(SR)_2(NO)_4]$, and $[Fe(SR)_2(NO)_2]^-$ makes the system susceptible to ligand-controlled NO photogeneration with the quantum yield depending strongly on oxygen concentration. All these may lead to the tissue-selective NO photogeneration, which is important from a photomedical point of view [36,39].

Electrochemically formed pentacoordinated $[Fe(CN)_4(NO)]^{2-}$ undergoes a photochemical oxidation in the presence of oxidizing agents (such as $KMnO_4$ or O_2); however, the evaluation of the mechanism calls for further investigation [40].

Possibilities offered by new experimental techniques have been applied in investigating photoinduced processes occurring in nitrosyl complexes at low temperature. It has been documented that the nitrosyl ligand changes its coordination mode from common ground-state M-NO to the metastable isonitrosyl M-ON and side-on coordination in the form of π-complex [41]. These modes have been observed in nitrosyl complexes of iron(III) as well as other central atoms. As for iron complexes, those with the iron central atom in a lower oxidation state were investigated too.

Energy transfer and its consequences for redox processes occurring in irradiated systems of Fe(III) complexes were followed in systems containing chloro complexes or bromo complexes of both Fe(III) and Fe(II) in methanol-ketone (acetone, propiophenone or acetophenone) mixtures [5,29,42,43]. The systems were irradiated with light absorbed only by the Fe(III) complexes ($\lambda_{irr} > 365$ nm). Detailed investigation of the redox properties of such systems led to the conclusion that, in addition to the spontaneous oxidation of Fe(II) by the ketone due to the mutual influence of halogeno and ketone ligands via the central atom [44] and the photoreduction of excited Fe(III)* accompanied by the oxidation of methanol, oxidation of Fe(II) sensitized by the excited Fe(III)* complexes also occurred. The limiting quantum yield of photosensitized Fe(III) formation was calculated for the systems. It was shown that the higher the ketone concentration, the higher the quantum yield [43].

The transfer of energy from *Fe(III) to Fe(II), followed by subsequent oxidation of *Fe(II), was also observed in acidic aqueous solutions containing both $[Fe(H_2O)_6]^{3+}$ and $[Fe(H_2O)_6]^{2+}$ ions and $(C_4H_9)_4P^+$ ions [45]. The mechanism proposed is analogous to that evidenced for halogeno iron complexes in methanol-ketone mixtures.

The wavelength dependence of the photoformation of $OH^{\bullet}$ and $SO_4^{\bullet-}$ radicals was followed in aqueous solutions of $[Fe(H_2O)_5(OH)]^{2+}$ and $[Fe(H_2O)_5(SO_4)]^+$ complexes [46]. The photoreduction of Fe(III) to Fe(II) was not followed directly although its occurrence was declared. Using the technique of radical scavenging, it was found that the quantum yields of both radicals formation decreased with increasing wavelength of incident radiation. The absolute $SO_4^{\bullet-}$ quantum yield increased from 1.6×10^{-3} at 350 nm to 7.9×10^{-3} at 280 nm (the corresponding values for the $OH^{\bullet}$ radical were 0.07 at 370 nm and 0.31 at 280 nm). It is highly probable that the quantum yield of Fe(II) formation exhibits the same trends (this phenomenon has been known for Fe(III) complexes for many decades) [1]. The decrease in the quantum yield of photosubstitution of the NO ligand for CH_3OH or CH_3O^- ligands in $[Fe(CN)_5(NO)]^{2-}$ with increasing irradiation wavelength [38] is also in accordance with the mentioned general trend.

As for ion pairs and dinuclear-bridged mixed-valence complexes, the attention has been focused mainly on two phenomena: the energetics of optical electron transfer (the energy of IPCT and MMCT transitions) and photoinduced electron-transfer processes in such systems. The spectral properties are correlated with the redox potential values and optical electronegativities of the redox centres involved, the electric properties of the media and other parameters. The experimental data and their theoretical evaluation have been reviewed in many articles [47-49].

A representative selection of some recent results on the photoredox chemistry of ion pairs and polynuclear complexes are given in [6,47,48]. Due to the kinetic lability of the majority of Fe(III) complexes, polynuclear complexes containing an Fe(III) moiety with unidentate ligands are relatively rare.

Irradiation of the complex $[(PPh_3)_3Cu^{I}\text{-}Cl\text{-}Fe^{III}Cl_3]$ in CH_2Cl_2 into the MMCT band ($\lambda_{irr} =$ 546 nm) leads to decomposition to Fe(II) and Cu(II) mononuclear species with a quantum yield Φ of 0.02 [50]. The small quantum yield was explained as being a consequence of rather effective thermodynamically driven back electron transfer within the photochemically formed dinuclear $[(PPh_3)_3Cu^{II}\text{-}Cl\text{-}Fe^{II}Cl_3]$ intermediate. The electronic absorption spectra, namely the energy of the absorption maximum of the MMCT band, of the symmetrical complex $[(CN)_5Fe^{III}\text{-}L\text{-}Fe^{II}(CN)_5]^{5-}$, where L = bis(pyridyl)ethylene, are influenced by the presence of cations (Na^+, Ca^{2+}, La^{3+}) which, at low concentrations, preferably bind to the more negative $\{Fe^{II}(CN)_5\}$ moiety. The absorption of a photon in the MMCT region leads to an electron transfer from Fe(II) to Fe(III), i.e. from the aspects of composition and symmetry, the reactant and product are the same species [51]. Femtosecond and picosecond photolysis of Prussian blue was exploited to study the dynamic processes associated with the wavepacket motion on the ground and MMCT potential energy surfaces [52].

In contrast with the above-mentioned systems in which the Fe(III) complexes themselves absorbed photons and underwent monomolecular or outer-sphere photoredox processes, such complexes can also act as electron-transfer quenchers of long-lived excited species, producing Fe(II) and an oxidized product of the electron donor. The complexes $[Fe(H_2O)_6]^{3+}$ and $[Fe(CN)_6]^{3-}$ were found to play the role of quenchers, whereas excited triplet zinc-substituted myoglobine complex Zn(Mb) [53], $[Ru(bpy)_3]^{2+}$ [54], $[Zn(TMpyPor]^{4+}$ [55] and $[Ru(Pc)L_2]$ [56] (where L = pyridine, DMSO, DMF or CO; Pc^{2-} = dianion of phthalocyanine; $TMpyPor^{2+}$ = dication of tetrakis(*N*-methyl-4-pyridinio)porphyrin) acted as electron donors. As an example [53], the reaction

$$^3[Zn^{II}(Mb)]^* + [Fe^{III}(CN)_6]^{3-} \rightarrow {}^1[Zn^{II}(Mb^+)] + [Fe^{II}(CN)_6]^{4-} \quad (9)$$

can be shown. Poly-L-lysine inhibits the reaction due to the formation of a complex with $[Fe(CN)_6]^{3-}$.

There are several papers devoted to practical application of photochemical processes involving iron(III) complexes. In many cases such papers are rather focused on discussing pros and cons of the exploitation of complexes than on detailed theoretical aspects and mechanisms of the processes. Morever, it is not so simple to declare unambiguously the composition of the complexes playing a demanded role in the systems. To such systems, Section 8 will be devoted.

4. Complexes with Chelating Ligands

Two types of iron(III) complexes with chelating ligands have been examined from the viewpoint of their photochemical behaviour: those containing in their primary coordination sphere both chelating and monodentate ligands, and those containing only chelating ligands. Such classification is not naught since the mentioned types differ significantly in their photochemical properties. A characteristic feature of the complexes containing also monodentate ligand(s) is that when undergoing intramolecular decomposition, it is the monodentate ligand which is redox changed and chelating ligands behave virtually as innocent part of the complex.

The primary photoredox step in iron(III) complexes containing only chelating ligands is usually intramolecular electron transfer from a ligand to the central atom. The course of subsequent spontaneous processes depends on several factors, such as the ability of surrounding molecules to participate in redox processes, redox reversibility of the oxidized chelating ligands, etc.

Investigation concerning the photochemistry of iron(III) complexes with chelating ligands has dealt with the following main problems:

1. the tuning and optimization of the photoredox reactivity, i.e. searching out correlations between the photoredox reactivity parameters of excited complexes and the ground state parameters of the complexes, their central atom and ligands;
2. the mechanism of detachment of a monodentate or chelating ligand from the coordination sphere of excited iron(III) complexes;
3. the photodecomposition of chelating ligands induced by their complexation with iron(III);
4. the outer-sphere photoinduced electron-transfer reactions and optical electron-transfer processes in electronically weakly coupled metal complexes (one of them being an Fe(III) complex) or in binuclear-bridged complexes.

In the chemistry of ground state compounds, the existing knowledge and methods allow the formulation of a number of relationships between the structure, physico-chemical properties and chemical behaviour of chemical compounds. It may be said that in fact all laws and rules of chemistry stem from the information on ground-state structure, chemical and physical properties of the chemical elements and compounds. Searching out the structure-photoreactivity relationships is still in its infancy except for some more or less successful attempts to empirically find and theoretically rationalize correlations between the excited state (ES) reactivity parameters (quenching rate constant k_q, quantum yield Φ, and threshold energy E_{th}) and the ground state (GS) properties of complexes and their constituents. The state-of-the-art in the issues mentioned in point (1) has been thoroughly considered in our previous reviews [3,11,43,57]. The term "optimization" means approaching or reaching required photochemical behaviour based on a targeted selection of the relevant ground state parameters of irradiated compounds. In the field of photochemistry of coordination compounds it means the capacity to control the course and photoreactivity parameters stemming from available ground state characteristics of complexes, their central atoms and ligands. Within the aim to assess the possibilities and limitations in this area, the photoredox reactivity of Fe(III) complexes with bidentate and/or open-chain tetradentate ligands was followed. The problems concerning the relationships between the GS and ES parameters of the complexes can be expressed by three questions:

1. Does any correlation exist between the parameters characterizing the electronic properties of polydentate ligands and their ability to undergo photooxidation (expressed via the ES parameter $\Phi_{Fe(II)}$?)
2. Do the peripheral groups of one ligand influence the tendency of another ligand to undergo photo-oxidation?
3. Is it justifiable and possible to optimize the efficiency of photoredox processes stemming from the GS parameters of the complexes or their constituents?

To answer the first question, the complexes $[Fe^{III}(R\text{-}K)_3]$ were investigated [58-60], where H(R-K) denotes kojic acid (5-hydroxo-2-(hydroxymethyl)-4H-pyran-4-one) or its derivatives (5-hydroxo-2-(R-methyl)-4H-pyran-4-ones). The distance between the peripheral R groups of the ligands and the central Fe(III) atom is five σ-bonds, and the electronic absorption spectra (positions of the LMCT bands) are nearly identical for all the complexes. The overall quantum yield of Fe(II) formation ($\Phi_{Fe(II)}$) is sensitive to the R groups, but there is no correlation between the Hammett constants of the R groups, the potentials of anodic oxidation of the R-K$^-$ anions to radicals R-K$^•$, (E_{pa}(R-K)$^{•/-}$) or the half-wave potentials of the complexes ($E_{1/2}$(Fe$^{III/II}$)) (for some selected representative data, see Table 1). The molar ratio of $c(Fe^{II})$: $c(CH_2O)$ is approximately 2 : 1 in irradiated methanolic solutions of [Fe(R-K)$_3$], and the formation of the R-K$^•$ and $^•CH_2OH$ radicals during irradiation was indicated by spin-trapping EPR. The total concentration of the kojic acid derivative H(R-K) is, however, constant, and no photooxidation product other than CH_2O was found after irradiation. Given the above observations, the proposed mechanism for the occuring redox processes is as follows (for the sake of simplicity, coordinated solvent molecules and unobservable non-redox processes are omitted)

$$_{LMCT}[Fe^{III}(R\text{-}K)_3]^* \rightarrow [Fe^{II}(R\text{-}K)_2] + R\text{-}K^• \quad (10)$$
$$[Fe^{II}(R\text{-}K)_2] + R\text{-}K^• \rightarrow [Fe^{III}(R\text{-}K)_3] \quad (11)$$
$$R\text{-}K^• + CH_3OH \rightarrow H(R\text{-}K) + CH_2OH^• \quad (12)$$
$$CH_2OH^• + [Fe^{III}(R\text{-}K)_3] \rightarrow CH_2O + H^+ + [Fe^{II}(R\text{-}K)_3] \quad (13)$$

Table 1. Integral quantum yields of Fe(II) formation, $\Phi(\lambda_{irr})$ for the iron(III) kojato complexes [Fe(R-K)$_3$] irradiated in methanol with λ_{irr} of 254 nm, 313 nm, 365 nm or 436 nm [62-64], the sum of Hammett constants $\Sigma\sigma$ of the peripheral groups R, electrode potentials E^0(Fe$^{III/II}$) for the complexes and anodic potentials E_{pc}(R-K)$^{-/•}$) for the ligands (both in Volts *vs* a ferricenium/ferrocene couple) [43,57,58,61]

R	$\Sigma\sigma$	E^0(Fe$^{III/II}$)	E_{pc}(R-K)$^{-/•}$)	Φ(254 nm)	Φ(313 nm)	Φ(365 nm)	Φ(436 nm)
OH	0.70	-0.70	0.36	0.16	0.10	0.029	0.018
SC_6H_{11}	0.25	-0.70	0.35	0.16	0.15	0.049	< 0.0001
SC_8H_{17}	0.23	-0,71	0.395	0.30	0.086	0.047	< 0.0001
Cl	0.19	-0.66	0.38	0.17	0.14	0.061	0.038
H	0.00	-0.79	0.37	0.011	0.0025	0.0010	0.00075

The first scientific investigation of the influence of one ligand on the ability to other ligand(s) to undego chemical change can be traced back to the beginning of the 20th century and it is associated with the terms of *trans*-effect and *cis*-effect [44,65]. Later on, this influence was generalized, named *mutual influence of ligands*, rationalized by the quantum-chemical methods and exploited for explanation and prediction of such particular phenomena as the plasticity of coordination polyhedra and distortion isomerism [44,65].

In the field of photochemistry of iron(III) complexes, mutual influence of ligands was studied [62-64,66-69] using the complexes *cis*-[Fe(N_2O_2)(C_2O_4)] and *trans*-[Fe(N_2O_2)(CH_3OH)L_{ax}]^q, where axial ligand L_{ax} = CH_3OH (q = 1) or F^-, Cl^-, Br^-, I^-, NCS^-, N_3^-, CH_3COO^- (q = 0), and N_2O_2 are open-chain, tetradentate, Schiff-base, dianionic ligands of salen and benacen type (R_2-salen^{2-} = *N,N'*-ethylenebis(R-salicylideniminato); R_2-benacen^{2-} = *N,N'*-ethylenebis(4-R-benzoylacetoneiminato) dianions. Experiments on photoredox behaviour of the complexes led to the following general conclusions.

1. The primary photoredox step involves the photoreduction of Fe(III) to Fe(II), accompanied by inner-sphere oxidation of the ligand L_{ax} or $C_2O_4^{2-}$. The fate of the radical products and the mechanisms of the reactions are similar to those expressed by Eqs. (10)-(13). Tetradentate N_2O_2 ligands obviously participate neither in the primary photoredox nor in subsequent dark redox processes.
2. The quantum yield of Fe(II) formation and, therefore, of ligand L_{ax} or $C_2O_4^{2-}$ photooxidation, depends on the periphery of the N_2O_2 ligands (some illustrative data see in Table 2). In other words, the electronic properties of the remote parts of one ligand influence the ability of another coordinated ligand to undergo photooxidation. This influence is manifested both in *trans*-complexes, where the electronic effects of the peripheral groups can be transmitted to the central Fe(III) atom and to its bond with the axial L_{ax} ligands via σ- and π-bonding systems, and in *cis*-complexes, where due to a strong distortion of the N_2O_2 ligands the effect of π-conjugation is strongly reduced (if any). The effect of the peripheral groups R of the N_2O_2 ligands to the ligands L_{ax} or $C_2O_4^{2-}$ is transmitted through at least seven σ-bonds.
3. Analogously as in case of the mentioned kojato complexes, there is no correlation between any relevant GS property of the complexes, ligands or their peripheral groups (redox potential, ionization energy, optical electronegativity, Hammett constants, etc.) and the efficiency of photoreduction of Fe(III) to Fe(II) expressed by means of $\Phi_{Fe(II)}$ (ES reactivity parameter of the complexes).
4. Except for azido complexes (explanation see below), the observed wavelength dependence of $\Phi_{Fe(II)}$ meets expectations, i.e. the longer the wavelength, the lower the quantum yield. However, in the complexes studied, an increase in the photon energy also means an increase in the distance between the site of photon absorption (in the UV region, the absorption of a photon is connected with N_2O_2 intraligand ES formation) and the central atom. Communication between the populated intraligand and photoredox-reactive spin-allowed sextet or spin-forbidden quartet LMCT states must be very effective.

Detailed study of photodeactivation processes in the complexes revealed, however, also differences in their mechanism which are worth mentioning.

The complex [Fe(benacen)(CH_3OH)F] (as well as its [Fe(R_2-benacen)(CH_3OH)F] analogues) exhibits a high redox stability not only in the ground state (which is a typical feature of iron(III) fluoro complexes) but also in excited states. When irradiating the complex with $\lambda_{irr} \geq$ 366 nm, the photoreduction of Fe(III) to Fe(II) in almost undetectable and even in case of applying energy-higher UV radiation, the efficiency of Fe(II) formation is still very low. The molar ratio of $c(Fe^{II}) : c(CH_2O) \cong 2 : 1$ and the presence of $^{\bullet}CH_2OH$ radicals identified by EPR spin trapping documents that the redox processes obey a general mechanism consisting of the following processes [79] (for the sake of simplicity, spin-states

Table 2. Quantum yields of Fe(II) formation, $\Phi_{Fe(II)}$ in irradiated methanolic solutions of [Fe(R$_2$-benacen)X(CH$_3$OH)] and [Fe(benacen)(C$_2$O$_4$)] [3,66-80]

Complex / λ_{irr}	254 nm	313 nm	366 nm	436 nm
	$\Phi_{Fe(II)}$			
[Fe(benacen)(CH$_3$OH)F]	0.00049	0.00006	< 0.00001	<0.00001
[Fe(benacen)(CH$_3$OH)Cl]	0.11	0.041	0.0059	0.0027
[Fe(benacen)(CH$_3$OH)Br]	0.032	0.016	0.0012	0.00095
[Fe(benacen)(CH$_3$OH)I]	0.027	0.010	0.0055	0.0032
[Fe(benacen)(CH$_3$OH)N$_3$]	0.0018	0.018	0.00090	0.00047
[Fe(benacen)(CH$_3$OH)(NCS)]	0.091	0.039	0.0098	0.0050
[Fe(benacen)(CH$_3$OH)$_2$]$^+$	0.065	0.035	0.0089	0.0043
[Fe(benacen)(C$_2$O$_4$)]$^-$	0.079	0.047	0.037	0.033
[Fe(F$_2$-benacen)(CH$_3$OH)I]	0.18	0.13	0.0085	0.0065
[Fe(Cl$_2$-benacen)(CH$_3$OH)I]	0.35	0.022	0.0090	0.0070
[Fe(Br$_2$-benacen)(CH$_3$OH)I]	0.20	0.17	0.065	0.040
[Fe(CH$_3$O)$_2$-benacen)(CH$_3$OH)I]	0.48	0.40	0.25	0.13

of the complexes and non-redox chemical deactivations are not included; *ic* and *isc* mean internal conversion and intersystem crossing, respectively):

$$_{GS}[Fe^{III}(benacen)(CH_3OH)F] \xrightarrow{h\nu(UV)} {}_{IL}[Fe^{III}(benacen)(CH_3OH)F] \quad (14)$$

$$_{IL}[Fe^{III}(benacen)(CH_3OH)F] \xrightarrow{ic\ /\ isc} {}_{LMCT}[Fe^{III}(benacen)(CH_3OH)F] \quad (15)$$

$$_{GS}[Fe^{III}(benacen)(CH_3OH)F] \xrightarrow{h\nu(VIS)} {}_{LMCT}[Fe^{III}(benacen)(CH_3OH)F] \quad (16)$$

$$_{LMCT}[Fe^{III}(benacen)(CH_3OH)F] \rightarrow [Fe^{II}(benacen)F]^- + H^+ + {}^{\bullet}CH_2OH \quad (17)$$

$$[Fe^{III}(benacen)(CH_3OH)F] + {}^{\bullet}CH_2OH \rightarrow [Fe^{II}(benacen)(CH_3OH)F]^- + H^+ + CH_2O \quad (18)$$

Owing to the high electrode potential $E^{o}(F_2/F^-)$, a possibility of fluorine atoms formation can be ruled out. From the viewpoint of electron density distribution, the only excited states suitable for inner-complex redox decomposition (14) are LMCT states (both spin-allowed sextets and spin-forbidden quartets which are, however, due to the absence of any luminescence, indistinguishable).

The reaction (15) is supported by thermodynamic parameters and up-to-now obtained experience [14]. As documented by the value of $E^0({}^{\bullet}CH_2OH/CH_2O) = -1.180$ V [80], the radical ${}^{\bullet}CH_2OH$ is a strong reducing agent able to reduce further Fe(III) in subsequent secondary thermal redox steps (15) forming thus the final products Fe(II) and CH_2O.

"Redox innocent" behaviour of tetradentate ligands is supported by dark reoxidation of Fe(II) in the system expressed by stoichiometry

$$4\ [Fe^{II}(benacen)F]^- + O_2 + 4\ H^+ + 4\ CH_3OH \rightarrow 4\ [Fe^{III}(benacen)(CH_3OH)F] + 2\ H_2O \quad (19)$$

When compared to its fluoro analogue, deactivation processes of chloro, bromo and tiocyanato complexes obey an analogous mechanism (a difference is in quantum yield $\Phi_{Fe(II)}$ values).

A different mechanism is, however, applied in systems of the iodo complex $[Fe^{III}(benacen)(CH_3OH)I]$ [74-76]. A new feature lies in identification of solvated electrons in addition to $^{\bullet}CH_2OH$ radicals. The formation of the solvated electrons and their reactions can be expressed as follows

$$CH_2OH + I^- \rightarrow ICH_2OH + e^- \rightarrow HI + CH_2O + e^- \qquad (20)$$

$$_{GS}[Fe^{III}(N_2O_2)(CH_3OH)I] + e^- \rightarrow [Fe^{II}(N_2O_2)(CH_3OH)I]^- \qquad (21)$$

No sign of accumulation of I_3^- anions during irradiation was observed in electronic absorption spectra, i.e. iodide ligands were not oxidized to iodine atoms within the redox processes.

The $\Phi_{Fe(II)}/\lambda_{irr}$ dependence for the complex $[Fe(benacen)(CH_3OH)N_3]$ differs from that found for the other investigated complexes. It reaches the highest value at 313 nm and decreases to a lower values both when increasing (to 254 nm) and decreasing (to 366 nm and 436 nm) the energy of incident photons [77,78]. Such a photochemical dependence was named "bell-shaped" [81]. Along with a commonly found decrease in $\Phi_{Fe(II)}$ with increasing radiation wavelength, an unusual drop in $\Phi_{Fe(II)}$ with decreasing radiation wavelength was observed. This is caused by the absorption of part of the 254 nm photons by anion N_3^- populating ($n \rightarrow \pi^*$) excited $^*N_3^-$ and its decomposition producing the nitrene intermediate anion N^-.

$$^*N_3^- \rightarrow N_2 + N^- \qquad (22)$$

As rationalized previously [82], this electronically unsaturated six-electron nitrene anion is extremely reactive and abstracts electrons and/or atoms from surrounding molecules, in our case from methanol molecules producing additional formaldehyde

$$N^- + 2\,CH_3OH \rightarrow CH_2O + CH_3O^- + NH_3 \qquad (23)$$

The sequence initiating by azide excitation and terminating by formaldehyde formation (Eq. 23) was proved independently by 254 nm irradiation of a methanolic solution of 2.0×10^{-3} mol dm^{-3} NaN_3 in the absence of iron(III) complex.

A decrease in $\Phi_{Fe(II)}$ with decreasing incident radiation wavelength is thus a consequence of opening a new chemical deactivation channel (decomposition of azido group) not involving the central atom competing with a common Fe(III) photoreduction.

IR spectra of solid residues obtained by vacuum room-temperature evaporation of methanol from solutions of $[Fe(benacen)(CH_3OH)N_3]$ irradiated at 254 nm or 313 nm showed no characteristic peak attributable to a triple Fe(V)≡N bond [83], i.e. Fe(V) nitrido complexes were not formed. This is in agreement with observation [83,84] that such complexes are stable only at low temperature and in case of complexes with macrocyclic (cyclam, porphyrin) ligands.

The complex $[Fe(benacen)(CH_3OH)_2]^+$ is present in solutions in the absence of "well-coordinating" acido ligands, i.e. when the counter anions are, e.g. ClO_4^- or NO_3^-. The complex behaves similarly to fluoro complex under the influence of radiation, the difference lies in a significantly higher quantum yield of Fe(II) formation. The identification of $^{\bullet}CH_2OH$ radical (and the absence of $CH_3O^{\bullet}$ signal) may be taken as evidence that instead of free (non-coordinated) CH_3OH or CH_3O^- anions, rather coordinated CH_3OH molecules take part in the primary electron transfer process, and, consequently, the primary redox process is of inner-sphere nature [85,86]. The mechanism expressed by equations (14) – (18) may be applied to $[Fe(benacen)(CH_3OH)_2]^+$ too.

The complex $[Fe(benacen)(C_2O_4)]^-$ belongs to the group of complexes having only chelating ligands in the primary coordination sphere. It differs from the above mentioned complexes at least in three features: it contains a "bent" benacen ligand with the donor oxygen atoms of the oxalato ligand coordinated in the *cis*-position [17]; oxalato ligand is a typical redox irreversible ligand (once being oxidized to oxalyl radical anion $C_2O_4^{-\bullet}$ and decomposed to CO_2 molecule and $CO_2^{-\bullet}$ anion radical, it cannot be reverted to the parent $C_2O_4^{2-}$ anion); for some of oxalato-containing complexes the quantum yield of the central atom reduction does not depend significantly on the incident radiation wavelength (at least over a certain spectral region, a typical example is the ferrioxalate actinometer) and also "bell-shaped" dependence was observed in some cases as exemplified [87] by $[Fe(phen)_2(C_2O_4)]^+$. A relatively small wavelength effect on the quantum yield can be rationalized as a consequence of different excited states forming the manifold of redox reactive LMCT states participating in the primary photoredox deactivation steps [87,88].

A substantially different mechanism was proposed for Fe(III) dialkyldithiocarbamates, $[Fe(R_2\text{-}dtc)_3]$ irradiated in the absence or in the presence of halogenated hydrocarbons RX [89,90]. In the absence of RX, the quantum yield of the complex disappearance is of the order 10^{-3}. The photolysis of $Fe^{III}(Et_2dtc)_3$ (Et_2dtc^- = diethyldithiocarbamate) in inert solvents (e.g., CH_3CN) and in the presence of 1,2-bis(diphenylphosphino)ethane (dppe) leads to the formation of $Fe^{II}(Et_2dtc)_2(dppe)$ and $(Et_2dtc)_2$ [90]. This photoreaction, which is initiated by LMCT excitation, generates $Fe^{II}(Et_2dtc)_2$ and a $Et_2dtc^{\bullet}$ radical in the primary photochemical step. Product formation occurs by the interception of $Fe^{II}(Et_2dtc)_2$ with dppe.

In the presence of RX, the higher the content of RX, the higher quantum yield (in pure RX solvent reaching the values of the order 10^{-1}). As for the primary process, two redox steps were suggested. The former one can be expressed by two processes (for the sake of clearness, the charges of ligands and oxidation state of the central iron atom are given):

$$[Fe^{III}(Et_2dtc^-)_3]^* \rightarrow [Fe^{II}(Et_2dtc^-)_2(Et_2dtc^{\bullet})] \qquad (24)$$

followed by

$$[Fe^{II}(Et_2dtc^-)_2(Et_2dtc^{\bullet})] + RX \rightarrow [Fe^{III}(Et_2dtc^-)_2X] + Et_2dtc\text{-}R \qquad (25)$$

Based on the knowledge of electron accepting capacity of RX molecules, the former mechanism was revised and the following set of processes, initiating by an intermolecular electron transfer, was suggested

$$[Fe^{III}(Et_2dtc^-)_3]^* + RX \rightarrow [Fe^{IV}(Et_2dtc^-)_3^+..RX^-] \quad (26)$$
$$[Fe^{IV}(Et_2dtc^-)_3^+..RX^-] \rightarrow [\{Fe^{III}(Et_2dtc^-)_2X^-\}... Et_2dtc^{\bullet}...R^{\bullet}] \quad (27)$$
$$[\{Fe^{III}(Et_2dtc^-)_2X^-\}... Et_2dtc^{\bullet}...R^{\bullet}] \rightarrow [Fe^{III}(Et_2dtc^-)_2X^-] + Et_2dtc\text{-}R \quad (28)$$

The final complex $[Fe^{III}(Et_2dtc^-)_2X^-]$ was formed also at irradiation of $[Fe^{IV}(Et_2dtc^-)_3][BF_4]$. The involvement of Fe(IV) species in the mechanism should not be considered too exotic since dithiocarbamates are known to form complexes with the central atoms in unusual high oxidation states.

The mechanisms for successive attachment of a polydentate ligand to the central metal atom in the evolution of complex formation have been known for a long time [91,92]. On the other hand, photochemical investigations aimed at recognizing individual steps in the mechanism of detachment of a chelating ligand have been rather rare and the majority of the results are linked to photosubstitutions and photoisomerizations [1,5-8,93,94]. Evolution of the knowledge in the field of photoredox chemistry can be demonstrated with the anion $[Fe(ox)_3]^{3-}$, the photosensitivity of which has been known for a long time [1,7]. The primary photoredox step occurring from a LMCT state is the homolytic splitting of one Fe(III)-O bond, leading to the formation of the coordinated oxalyl radical $C_2O_4^{-\bullet}$. The radical can be released from the coordination sphere and either itself, or in the form of $CO_2^{-\bullet}$ being a product of its decomposition. Both radicals can react with another Fe(III), forming the final products Fe(II) and CO_2. The overall quantum yield of Fe(II) is close to 1.2. The presence of Cu(II) ions causes an increase in the quantum yield of Fe(II) via an increasing rate of decomposition of the radical $C_2O_4^{-\bullet}$ [95]. The process can be schematized as follows [6,95]

$$\begin{array}{lll}
 & Fe(II) + 2\,CO_2 + Cu(I) \xrightarrow{Fe(III)} Fe(II) + Cu(II) & \\
 & \uparrow + Cu(II) & \\
[Fe^{III}(C_2O_4)_3]^{3-} \xrightarrow{h\nu} & [(C_2O_4)_2Fe^{II}\text{-}OCO\text{-}COO^{\bullet}]^{3-} & \xrightarrow{Fe(III)} 2Fe(II) + 2CO_2 \\
 & \downarrow & \\
 & Fe(II) + CO_2 + CO_2^{-\bullet} \xrightarrow{Fe(III)} Fe(II) + CO_2 & \\
\end{array} \quad (29)$$

In connection with the photochemistry of $[Fe(ox)_3]^{3-}$, two observations which are not common in photochemistry are worth mentioning. The first is the formation of non-equilibrium spatial dissipative structures during irradiation of thin layers of solutions of the complex in the presence of $[Fe(CN)_6]^{3-}$. The pattern is formed by Turnbull's blue on a yellow background [96]. The second observation concerns the photodecomposition of $[Fe(ox)_3]^{3-}$ occurring in $NaMg[Al(ox)_3].9H_2O$ or $K_3[Al(ox)_3].3H_2O$ single crystal hosts studied by EPR technique [97]. Interpretation of a very detailed study led to the conclusion that the primary redox products of the photolysis are $[Fe^{II}(ox)_2]$ and $C_2O_4^{-\bullet}$ coupled by an antiferromagnetic interaction. No evidence for the formation of the radical $CO_2^{-\bullet}$ was found; in the single crystals, no protonation of $C_2O_4^{-\bullet}$ occurred [96].

A systematic study of the mechanism of a chelate ligand photoinduced detachment linked to partial photodegradation of the ligand was carried out using Fe(III) carboxylate complexes [93,98-100]. The quantum yields for the photoreduction of these complexes vary with the nature of the carboxylate ligand and the solution pH. The photoreduction of Fe(III) to Fe(II) is accompanied by the oxidative decarboxylation of the carboxylate ligand, and can be observed even in the solid state. The efficiency of the photodecarboxylation in solution depends mainly on two factors: the pH and the initial ligand-to-metal ratio. While the inclusion of a free carboxylate radical is a virtual necessity in the mechanisms of Fe(III) oxalate systems, where mononuclear complexes are the sole light absorbers, the pH and speciation dependence presented in the Fe(III) hydroxypolycarboxylates [93] justifies a different approach. The simplest mechanism consistent with the data in [93] involves initial LMCT excitation, followed by the formation of a radical anion carboxylate ligand and Fe(II). At very low pH, no net photoreaction is observed, probably due to non-radiative decay in the chelated mononuclear complex. At moderate pH, the light is absorbed by the dinuclear species, and the carboxylate radical anion is formed in the immediate presence of another Fe(III) atom, which acts as the other half of the dimer ($R = CH_2COO^-$)

$$\{(OH)C(R)_2(COO)Fe^{III}\}_2(\mu\text{-}O) \xrightarrow{h\nu} (OH)C(R)_2(COO^{\bullet})Fe^{II}\text{-}(\mu\text{-}O)\text{-}Fe^{III}(COO)(R)_2C(OH) \quad (30)$$

This allows very rapid completion of the ligand oxidation in a thermal step

$$(OH)C(R)_2(COO^{\bullet})Fe^{II}\text{-}(\mu\text{-}O)\text{-}Fe^{III}(COO)(R)_2C(OH) \xrightarrow[-H^+,-CO_2]{} O{=}C(R)_2Fe^{II}\text{-}(\mu\text{-}O)\text{-}Fe^{III}(COO)(R)_2C(OH) \quad (31)$$

The two Fe(II) ions produced will be released from the carboxylate coordination owing to the lower affinity of Fe(II) for oxygen coordinating sites. The acetonedicarboxylic acid can further undergo two subsequent non-oxidative decarboxylations to form acetone as the ultimate organic product. However, the suggested mechanism needs definitive proof using flash photolysis.

The photodegradation of radiation-transparent organic pollutants present in natural water, induced by complexation with Fe(III), is of great environmental importance (for a more detailed discussion, see Section 8). Investigation of the mechanism of photodegradation processes has been focused on aminopolycarboxylic and aminopolyphosphonic acids [101-103], which are strong chelating agents widely used for metal cation removal both in industry and in the house and are subsequently dispersed in wastewater. The phototransformation of nitrilotriacetato(3–) (NTA) iron(III) complex, $Fe^{III}(R_2NCH_2COO)$, where $R = CH_2COO^-$, has been studied at 20°C on monochromatic excitation at different pH [102]. The photochemical behaviour can be adequately described by the following simplified mechanism, involving a redox step between Fe(III) and a carboxylate group

$$^{*}Fe^{III}(R_2NCH_2COO) \rightarrow Fe^{2+} + R_2NCH_2COO^{\bullet} \rightarrow Fe^{2+} + R_2NCH_2^{\bullet} + CO_2 \rightarrow Fe^{2+} + CH_2O + NHR_2 \quad (32)$$

Similar formation of formaldehyde accompanied by the central atom photoreduction was observed at degradation of iminodiacetic acid due to complexation with Fe(III) and subsequent excitation at 365 nm in aqueous solution (the other final products were not determined) [103]. Contrary to the above-mentioned oxalato complexes, the addition of Cu(II) ions to Fe(III) iminodiacetate complex solution makes the process less efficient. The photodegradation of ethylenediaminetetra(methylenephosphonic) acid in the presence of Fe(III) was followed by measuring the concentration of phosphate groups formed together with Fe(II) [101]. The final photoproduct is *N*-methylaminomethylenephosphonic acid, which is photostable under the experimental conditions. Glyphosate, (*N*-phosphonomethyl)glycine, coordinated to Fe(III), undergoes an analogous photodecomposition yielding Fe(II), CO_2 and NH_3 [104]. Photochemistry of iron(III) polycarboxylate complexes were investigated as part of study of the processes occurring in atmospheric and surface waters [105]. The determined quantum yields of Fe(II) formation reached 0.28 for irradiation of $[Fe(OH)(citrate)]^-$ at 436 nm; 0.027 for $[Fe(malonate)]^{2-}$ at 366 nm and 1.0 for $[Fe(C_2O_4)_3]^{3-}$ at 436 nm.

With regard to the photoredox processes, optical charge-transfer transitions between a reducing and an oxidizing metal centre have been studied, either with both metals connected via bridging ligands in a dinuclear complex (inner-sphere process) [106] or in close contact in separate metal complexes (outer-sphere process) [107]. Rare outer-sphere transitions of LMCT nature, where L is a ligand of one complex and M is the central atom of another complex, can be observed in ion pairs and in pairs consisting of an ion and a neutral molecular complex. A mixture of neutral bis(*N*-ethylsalicylaldiminato)nickel(II), $[Ni(Et\text{-}sal)_2]$ and anionic $[Fe(ox)_3]^{3-}$ displays absorption at λ_{max} = 472 nm. This new absorption band, which is not present in the electronic absorption spectra of the individual complexes, was attributed to an outer-sphere LMCT transition from the coordinated ligand Et-sal$^-$ of $[Ni(Et\text{-}sal)_2]$ to Fe(III) of $[Fe(ox)_3]^{3-}$. Mixing free Et-sal$^-$ or $[M(Et\text{-}sal)_2]$, where M = Co(II), Cu(II) or Zn(II), and $[Fe(ox)_3]^{3-}$ shows the same phenomenon. The maximum of the LMCT band is independent of the nature of the metal. A comparison of the energies of the Et-sal$^-$ to Fe(III) inner-sphere LMCT transition of $[Fe^{III}(Et\text{-}sal)_3]$ and the outer-sphere LMCT transition of $[Ni(Et\text{-}sal)_2]\text{-}[Fe(ox)_3]^{3-}$ may help to answer questions concerning the character of the LMCT transition. It should be pointed out that in spite of the labile nature of the involved complexes, no (at least partial) dissociation of the ligands was taken into account.

The unit Fe(III)-O-Fe(III) plays a very important role in biology and chemistry. The majority of results has come from investigation of dinuclear bridged macrocyclic tetrapyrrole complexes (see Section 5). Here we will limit ourselves to complexes with chelating ligands. In this context, the photoredox behaviour of the complex anion $[\{Fe^{III}(edta)\}_2O]^-$ was reported [106]. The reversible photolysis of aqueous $[\{Fe(edta)\}_2O]^-$ leads to the evolution of oxygen and the formation of $[Fe^{II}(edta)]^{2-}$. It has been concluded that MMCT excitation leads to disproportionation yielding Fe(II) and Fe(IV)=O in the primary photochemical step. Two oxoferryl fragments combine to generate Fe(II) and an O_2 molecule. The quantum yield of Fe(II) formation decreases with increasing wavelength of the incident radiation. This system represents a simple model for the photochemical oxidation of water or oxides. The results presented in [106] are of considerable interest with regard to the photochemistry of mixed--valence compounds, although in this case the mixed-valence system does not exist in the ground state but is generated by MMCT excitation.

Outer-sphere photoinduced electron-transfer reactions, in which the composition and symmetry of the redox-reactive complex are preserved, were found [108] to occur in some homogeneous and micellar systems of the complex $[Fe(bpy)_2(CN)_2]^+$. The quantum yield of photoinduced reduction of Fe(III) to Fe(II) was of the order of 10^{-2} in both aqueous homogeneous and micellar systems. The participation of detergent dodecylsulphate anions DS$^-$ in the electron transfer

$*[Fe^{III}(bpy)_2(CN)_2]^+ + DS^- \rightarrow [Fe^{II}(bpy)_2(CN)_2] + DS^{\bullet}$ (33)

was suggested, however, not proved. In homogeneous systems, the formation of hydroxyl radicals $^{\bullet}OH$ is indicated. However, no radical was experimentally identified.

In the majority of cases, photoredox processes of iron(III) chelates have been initiated by their photoexcitation, i.e. by population of an excited state of the iron(III) complex under question by the absorption of a photon of incident radiation. A different type of mechanism was proposed for the photoreduction of tris(salicylato)iron(III) complex, $[Fe(sal)_3]^{3-}$ in the presence of methylene blue (MB^+). Irradiation of the systems containing both $[Fe(sal)_3]^{3-}$ and MB^+ by radiation absorbed solely by MB^+ populating its excited singlet state $^1(MB^+)^*$ led to the photoreduction of Fe(III) to Fe(II) via the static quenching in an ion pair composed from excited $^1(MB^+)^*$ and $[Fe(sal)_3]^{3-}$ in its ground state [109]. The processes were described as follows:

$_{GS}MB^+ \xrightarrow{h\nu} {}^1(MB^+)^*$ (34)

${}^1(MB^+)^* + {}_{GS}[Fe^{III}(sal)_3]^{3-} \rightarrow *(MB^+...[Fe^{III}(sal)_3]^{3-})$ (35)

$*(MB^+...[Fe^{III}(sal)_3]^{3-}) \rightarrow (MB^{2+}...[Fe^{II}(sal)_3]^{4-}) \rightarrow MB^{2+} + [Fe^{II}(sal)_3]^{4-}$ (36)

In addition, other processes, not involving iron complexes and not presented here (singlet triplet intersystems crossing in $^1(MB^+)^*$, protonation of MB^+, etc.) form part of the mechanism.

5. Complexes with Macrocyclic Ligands

The basic structural features of iron(III) macrocyclic complexes are similar to those of other central atoms. Macrocyclic, usually tetradentate N_4-ligands form the equatorial plane of the complexes, and one or two ligands are coordinated in the axial position(s). In binuclear-bridged complexes, the macrocyclic ligands are in a mostly parallel face-to-face position. The iron(III) central atom can be in the low-spin (S = 1/2), intermediate-spin (S = 3/2) or high-spin (S = 5/2) state. The energy difference between the spin states is small and some complexes (e.g. [Fe(TPP)(NCS)(py)] and $[Fe(OEP)(3\text{-}Clpy)_2]ClO_4$) show spin crossovers in the solid state [110]; in solutions of cytochrome *c* the presence of all spin states was found [111]. Luminescence originating from a triplet-sextet intraligand-localized excited state was observed for Fe(III) complexes with tetraphenylporphyrin (luminescence maximum) λ_{max} = 718 nm, $\Phi_{lum} = 7 \times 10^{-5}$, complex irradiated in methylcyclohexane glass at 77 K [112]) and with mesoporphyrin IX dimethyl ester (λ_{max} = 681 nm, complex irradiated in EPA glass (glass consisting of ethanol, ether and isopentane) at 77 K [113]).

Iron macrocyclic (generally MC) complexes, especially those with tetrapyrrole ligands (porphyrin, Por, and phthalocyanine, Pc, derivatives), have attracted the interest of a large number of researchers for the following three reasons.

1) Iron macrocycles act as structural and functional models for the *in vivo* behaviour of some iron redox enzymes, even though their function is not a consequence of their photochemical reaction. However, fast photochemical methods (operating in pico- or femtosecond time domain) can contribute substantially to

the understanding and rationalization of the biological processes involving these enzymes.

2) Iron macrocycles, through reversible changes in the oxidation state of the central iron atom, exhibit very promising photocatalytic properties which can find application in the synthesis of organic compounds and the degradation of some pollutants.
3) In spite of their ostensible complicated structures, the macrocyclic ligands themselves do not undergo chemical changes within the photoredox reactions of iron(III) macrocyclic complexes; this is a very advantageous property when investigating and interpreting the mechanisms of the reactions.

The known photoredox reactions of iron(III) macrocyclic complexes can be classified as follows:

1) reactions in which the photoreduction of Fe(III) is associated with the inner-sphere photooxidation and subsequent detachment of an axial ligand;
2) reactions in which an axial ligand is decomposed and part of it remains coordinated to the redox changed central iron atom;
3) photoredox decomposition of binuclear complexes, forming usually mononuclear Fe(II) and Fe(IV) macrocyclic complexes;
4) formation of a binuclear-bridged mixed-valence complex induced by irradiation of a mononuclear complex;
5) outer-sphere photoinduced electron abstraction from a surrounding redox-reactive species (e.g. solvent molecule, triplet excited closed-shell metalloporphyrin) by an Fe(III) macrocyclic complex.

The results obtained are summarized in Table 3.

Table 3. Photoredox properties of excited macrocyclic Fe(III) complexes

Starting complex*	Products; efficiency; conditions; notes	Ref.
[Fe(TPP)Cl]	$[Fe^{II}(TPP)]$; $\lambda_{irr} > 418$ nm, $\Phi = 8 \times 10^{-6}$; MeTHF	[114,115]
$[Fe(TPP)N_3]$	$[Fe^{II}(TPP)]$; $N_3^{\bullet}$ identified by spin-trapping EPR; $\lambda_{irr} > 400$ nm; benzene, toluene, CH_2Cl_2	[116]
$[Fe(TPP)(NO_3)]$	$[\{Fe^{III}(TPP)\}_2O]$; $[Fe^{IV}O(TPP^+)$ formed as an intermediate; $\lambda_{irr} = 350$-450 nm; benzene	[117,118]
$[Fe(OEP)(MeIm)(CH_3OH)]$	$[Fe^{II}(OEP)(MeIm)]$; visible light; CH_2Cl_2; both axial ligands necessary for Fe(III) photoreduction	[119,120]
[Fe(OEP)L], L = F^-, Cl^-, Br^-, I, ClO_4^-, Im, 1-MeIm, 2-MeIm, 1,2-Me_2Im	$[Fe^{II}(OEP)]$; visible light; Fe(III) reduced only for L = 2-MeIm or 1,2-Me_2Im; CH_2Cl_2	[121]

Table 3. (Continued)

Starting complex*	Products; efficiency; conditions; notes	Ref.
$[Fe(Por)N_3]$, Por = TPP, OEP	$[(Por)Fe^{III}\text{-N-}Fe^{IV}(Por)]$; λ_{irr} = 413.1 nm for solid starting complex or radiation from a high pressure Hg lamp for THF solution of complex	[122,123]
[Fe(Por)Cl], Por = TPP, TFPP, TTP, OEP, TMP, TDCPP, ProtoIX	$[Fe^{II}(Por)]$; $\lambda_{irr} > 350$ nm, $\Phi \sim 10^{-2}\text{-}10^{-3}$; water + pyridine + alcohols or cyclohexene or polystyrene matrices; complexes used as catalysts	[124-131]
$[Fe(Por)N_3]$, Por = TPP, OEP, TMP	$[Fe^{V}N(Por)]$; λ_{irr} = 514.5 or 488 nm (laser); T = 30 K; solid complexes; Fe≡N bond evidenced by Raman spectroscopy; $[Fe(OEP)]_2N$ formed by local heating	[83,122,123]
$[Fe(Me_3\text{-cyclam-ac})(N_3)](PF_6)$	$[Fe^{IV}(Me_3\text{-cyclam-ac})(N_3)](PF_6)_2$ as a product of electrochemical oxidation of the reactant gives $[Fe^{VI}(Me_3\text{-cyclam-ac})(N)](PF_6)_2$ at λ_{irr} = 650 nm, T = 77K	[132]
[Fe(TMP)X], X^- = Cl^-, OH^-	Four-coordinated $[Fe^{II}(TMP)]$; λ_{irr} = 300 nm; aromatic hydrocarbons	[133]
$[Fe(tim)(CH_3OH)(CH_3O)]$	$[Fe^{II}(tim)(CH_3OH)_2]^{2+} + CH_2O$; UV or visible radiation; $\Phi_{Fe(II)} \sim 2$, $\Phi_{CH2O} \sim 10^{-2}$; methanol	[134,135]
$[Fe(MC)X_2]^+$, X^- = Cl^-, Br^-, N_3, NCS^-; MC = [15]pydieneN_5, [15]pyaneN_5	$[Fe^{II}(MC)X]^+$, radicals $X^{\bullet}$ scavenged by alcohols; ACN; threshold energies of Fe(III) photoreduction increase with increase in ionization energy of X^-	[136]
$[\{Fe(Por)]\}_2O]$, P = TPP, TTPC, TTPS	$[Fe^{II}(Por)] + [Fe^{IV}O(Por)]$; visible light; $\Phi \sim 10^{-4}$; water, pyridine, hydrocarbons, CH_2Cl_2; FeO(Por) behaves as a strong oxidant	[137-143]
$[\{Fe(TdPc)\}]_2O]$	$[Fe^{II}(TdPc)]$; products of oxidation unidentified; visible light; $\Phi \sim 10^{-2}$; DMF or THF	[144]
$[\{Fe(TSPc)\}]_2(\mu\text{-}O_2^{2-})]$	$[Fe^{II}(TSPc)] + {}^1O_2$; λ_{irr} = 333 nm; $\Phi = 5 \times 10^{-4}$, water/methanol (99:1), reactive LMCT state reached from populated IL state	[145,146]
[Fe(TPP-lipid)Cl], spherical micelles or fibres	Unreactive at LMCT irradiation, reduced to high-spin N_5-coordinated Fe(II) complex in the presence of saccharides or hyaluronic acid at λ_{irr} = 362 nm;	[147,148]

*Abbrevation of macrocyclic ligands: TPP = tetraphenylporphyrinato(2–); OEP = octaethylporphyrinato(2–); TFPP = tetrakis(pentafluorophenyl)porphyrinato(2–); TTP = tetrakis(4-tolyl)porphyrinato(2–); TMP = tetrakis(mesityl)porphyrinato(2–); TDCPP = tetrakis(2,6-dichlorophenyl)porphyrinato(2–); ProtoIX = protoporphyrinato IX(2–); tim = 2,3,9,10-tetramethyl-1,4,8,11-tetraazacyclotetradeca-1,3,8,9-tetraene; TTPC = tetrakis(4-carboxyphenyl)porphyrinato(6–); TTPS = tetrakis(4-sulfonatophenyl)porphyrinato(6–); TSPc = tetrasulfonatophthalocyaninato(2–); cyclam = 1,4,8,11-tetraazacyclotetradecane; Me_3-cyclam-ac = 4,8,11trimethyl1,4,8,11tetraazacyclotetradecane1acetato(1);TdPc=tetrakis(dodecylsulfonamido)phthalocyaninato(2–).

In general, reactions in which the photoreduction of Fe(III) is associated with the inner-sphere photooxidation and subsequent detachment of an axial ligand can be expressed as

$$*[L\text{-}Fe^{III}(MC)\text{-}(X^{q-})] \rightarrow [L\text{-}Fe^{II}(MC)] + X^{1-q} \qquad (37)$$

where L denotes the redox unchanged axial ligand (if any), MC is the macrocyclic ligand and X^{q-} is the axial ligand undergoing photooxidation (it can also be a neutral molecule). This type of primary photoredox step was postulated for Fe(III) complexes with various tetraazamacrocycles [134-136] and porphyrins [114-131,149-152]. The quantum yields of Fe(II) formation strongly depend on the composition of the primary coordination sphere of the complexes, solvent properties and experimental conditions, and reach values of the order of $10^{-2} - 10^{-7}$ (Table 3). It is assumed that the LMCT states (where L denotes an axial ligand) are responsible for the primary homolytic axial ligand-Fe(III) bond splitting even in instances when LMCT bands are hidden under an envelope of very intense IL bands or when complexes are irradiated into IL bands. For the low-spin complexes $[Fe(tim)(CH_3O)(CH_3OH)]$ and $[Fe(dmg)(CH_3O)(CH_3OH)]$, where dgm = dimethylglyoximato(1−) ligand, the threshold energy for the photoreduction of Fe(III), associated with the oxidation of the CH_3O^- ligand, was measured to be larger than or equal to 2.65 eV [135], which is more than the threshold for LMCT absorption (the complexes absorb photons at energies lower than 1.80 eV). In the absence of oxygen, irradiation of methanolic solutions of the complexes leads to reactions which can, in a simplified form, be expressed by two steps

$$*[Fe^{III}(MC)(CH_3O)(CH_3OH)] \rightarrow [Fe^{II}(MC)(CH_3OH)] + CH_3O^{\bullet} \qquad (38)$$
$$[Fe^{III}(MC)(CH_3O)(CH_3OH)] + CH_3O^{\bullet} \rightarrow [Fe^{II}(MC)(CH_3OH)_2] - CH_2O \qquad (39)$$

The molar ratio $c(Fe^{II}) : c(CH_2O) = 2 : 1$ determined in these systems [134,135] was taken as a clear evidence that other reactions of the methoxy radical $CH_3O^{\bullet}$ (its disproportionation to CH_2O and CH_3OH or the formation of $HOCH_2\text{-}CH_2OH$) have negligible (if any) efficiency.

The threshold energy for the high-spin complexes $[Fe([15]pydieneN_5)X_2]^+$ and $[Fe([15]pyaneN_5)X_2]^+$ increases with increasing ionization energy of the acido ligand X^- (X = Cl, Br, N_3, NCS). Also, in this case, the threshold energies of the photoredox reactions are 0.2 – 1.6 eV larger than those of the LMCT bands [136].

Usually, only the photoreduction of Fe(III) is followed directly, and it is *a priori* supposed that it is the ligand X^{q-} which is oxidized in the primary photoredox step. In many cases, research has been directed to practical applications (phototransformation of organic substrates catalysed by a macrocyclic complex) and the mechanisms of the reactions have not been investigated in detail.

Contrary to the reactions of previous type, in which Fe(III) is photoreduced to Fe(II), in the atom transfer reactions the central atom Fe(III) is, as a rule, oxidized to Fe(IV) or Fe(V). Such reactions have been observed only for porphyrin complexes of Cr(III), Mn(III) and Fe(III) [4].

Photoredox generation of nitrido complexes by denitrogenation of corresponding azido complexes can be, in general, expressed as

$$[(L)_xM^n{-}N_3] \xrightarrow{h\nu\ (IL)} [(L)_xM^{n+2}{\equiv}N] + N_2 \tag{40}$$

Since the nitride anion N^{3-} is one of the strongest π-donor ligands, a necessary prerequisite for the formation of stable nitride species is the availability of suitable empty d-orbitals of the central atom. Thus, the vanadium(V), chromium(V), manganese(V) and even iron(V) central atoms should be ideally suited for the existence of nitrido complexes. Due to the enormous *trans*-influence of the nitrido ligand, a (potential) ligand bonded in a trans-position is labilized to such an extent that it is not able to coordinate and the majority of nitrido complexes are thus five-coordinate, square-pyramidal with the nitrido ligand in the apical position.

Up to now obtained results have shown that the photochemically produced nitrido complexes can be grouped into four main families, namely:

1. mononuclear complexes with unusually high oxidation number of the central atom,
2. dinuclear complexes with the different oxidation number and/or spin state of the central atoms linked via a bridging nitrido ligand,
3. dinuclear complexes with the identical oxidation number of the central atoms connected by a bridging nitrido ligand,
4. polymeric complexes with bridging nitrido ligands.

As documented by several experiments, a mode of photoredox reactivity of some complexes may be tuned by variation of the experimental conditions and the structure of investigated complexes. Not infrequently the irradiation of one complex may lead to, along with nitrido complexes, other products too. As an instructive example, iron(III) azido complexes with chelating or macrocyclic ligands can serve. Photolysis of macrocyclic complexes with porphyrinato(2−) (Por), phthalocyaninato(2−) (Pc) or 1,4,8,11-tetraazacyclotetradecane (cyclam) ligands at room temperature usually leads to the photoreduction of the iron(III) central atom to iron(II) accompanied by oxidation of azido ligands to its radical.

At irradiation of thin solid films of $[Fe(Por)(N_3)]$ at 30 K [152], *trans*-$[Fe(cyclam)(N_3)_2]$ [153] or $[Fe(cyclam\text{-}ac)(N_3)](PF_6)$ [154] in frozen acetonitrile solutions at 77 K, mononuclear nitrido iron(V) complexes with an Fe≡N triple-bond, i.e. $[Fe(Por)(N)]$, $[Fe(cyclam)(N_3)(N)]$ and $[Fe(cyclam\text{-}ac)(N)](PF_6)$, respectively, are formed in the primary processes. The products remain stable only at the given low temperatures.

Photochemical formation of the catalytically active nitrido iron(V) species from the potent precursor *trans*-$[Fe(cyclam)(N_3)_2]^+$@NaY encapsulated in zeolite-Y was envisaged in paper [155]. The authors believed that the encapsulation of the azido complexes within the supercages of zeolites could restore the highly reactive mononuclear nitrido iron(V) species preventing the formation of dinuclear molecules even at ambient temperature.

The photochemical behaviour of cyclam-containing complexes depends, along with the temperature, also on their stereochemistry [153]. Photolysis of the low-spin *trans*-$[Fe(cyclam)(N_3)_2]^+$ in acetonitrile at the temperatures of −35°C and 20°C leads to the dinuclar mixed-valence complex $[\{trans\text{-}(cyclam)Fe^{III}(N_3)\}(\mu\text{-}N)\{trans\text{-}Fe^{IV}(cyclam)(N_3)\}]^{2+}$ with the localized antiferromagnetically coupled intermediate-spin iron(III) (S = 3/2) and low-spin iron(IV) (S = 1). On the other hand, photolysis of the high-spin *cis*-$[Fe(cyclam)(N_3)_2]^+$ in acetonitrile at −35°C gives rise to a [{*cis*-(cyclam)$Fe^{III}(N_3)$}(μ-N){*trans*-Fe^{IV}(cyclam)(N_3)}$]^{2+}$ with the localized antiferromagnetically coupled high-spin iron(III) (S = 5/2) and low-spin iron(IV) (S = 1).

Along with complexes containing at least one axial ligand bonded to the central atom iron(III) and exhibiting photoredox reactivity, there are a few examples of tetracoordinated planar iron(III) complexes, e.g. $[Fe(TPPS)]^{3-}$, which is photochemically almost non-reactive [156-158] ($\Phi \sim 10^{-6}$ both in the presence and in the absence of oxygen). It differs thus from its iron(II) analogue, $[Fe(TPPS)]^{4-}$, which is not planar, but adopt sitting-atop structure [156] with distorted porphyrin ligand (similar structural feature is known in the case of deoxyhemoglobin). This distorsion allows a photoinduced transfer of charge from the macrocyclic ligand to Fe(II) giving finally rise to the macrocyclic ligand decomposition. The formed Fe(I) intermediate is subsequently reoxidized.

Analogous difference in both iron central atoms was documented also for the dinuclear [{L(Cl_4-cat)Fe^{III}}(μ-N){Fe^{IV}(Cl_4cat)L}], [{L(Ph_2acac)Fe^{III}}(μ-N){Fe^{IV}(Cl_4cat)L}], and [{L(nadiol)Fe^{III}}(μ-N){Fe^{IV}(nadiol)L}], formed at irradiation of corresponding mononuclear azido iron(III) complexes, where L represents tridentate 1,4,7-trimethyl-1,4,7-triazacyclononane, Cl_4cat is the tetrachlorocatecholato(2−), Ph_2acac is the 1,3-diphenylpropane-1,3-dionato(1−), nadiol is naphtaline-2,3-dionato(2−) ligands [159].

The results obtained within the investigation of the complexes with the mentioned flexible coordination sphere differ from those found under the photolysis of structurally more rigid porphinato or phthalocyaninato azido iron(III) complexes. In the latter case both iron atoms in dinuclear complexes are equivalent with the valence delocalization in the linear {$Fe^{3.5}$=N=$Fe^{3.5}$} core [160]. Such delocalization seems to be a general phenomenon for porphyrinato and phthalocyaninato complexes and was found even in cases with two different macrocyclic ligands, such as in [(TPP)$Fe^{3.5}$(μ-N)$Fe^{3.5}$(Pc)] [161].

Detail investigation indicated that the (μ-nitrido)diiron products were formed by the combinantion of iron(II) and iron(V) generated concurrently via photoreduction and photooxidation of the parent iron(III) azido complexes, respectively (charges of the complexes are omitted):

$$[(L_x)Fe^{II}] + [(L_y)Fe^{V}{\equiv}N] \rightarrow [(L_x)Fe^{III}(\mu\text{-}N)Fe^{IV}(L_y)] \text{ or } [(L_x)Fe^{3.5}(\mu\text{-}N)Fe^{3.5}(L_y)] \quad (41)$$

The formation of mononuclear porphyrinatoiron(V) nitrido intermediate was proved by Nakamoto [152], production of porphyrinatoiron(II) associated with radical $N_3^{\bullet}$ releasing from the coordination sphere was evidenced by spin trapping EPR [116].

Besides reaction (38), alternative mechanisms, summarized in a review [162], were proposed for the formation of dinuclear nitrido-bridged complexes. One of them described iron(II) complexes generation as the following process

$$[Fe^{V}(Por)(N)] + [Fe^{III}(Por)(N_3)] \rightarrow 2\,[Fe^{II}(Por)] + 2\,N_2 \quad (42)$$

another suggested that iron(V) species were products of a back reaction of the radical $N_3^{\bullet}$, and $[Fe^{II}(Por)]$ produced by the following reaction

$$N_3^{\bullet} + [Fe^{II}(Por)] \rightarrow [Fe^{V}(Por)(N)] + N_2 \quad (43)$$

A priori, none of the mechanisms may be excluded and it might happen that simultaneusly more of them are realized in irradiated systems with the proportion depending on experimental conditions and the complex composition.

Along with investigation of azido complexes, photoinduced oxidation of Fe(III) was observed within the study of porphyrinato iron(III) complexes having other monodentate ligands, such as O-donor (e.g. NO_3^-) or N-donor (NR) ligands coordinated in the apical position. The reactions can be indicated as follows

$$(Por)Fe^{III}\text{-}OR \xrightarrow{h\nu} OFe^{IV}(Por+) + R \quad (44)$$

$$(Por)Fe^{III}\text{-}NR \xrightarrow{h\nu} NFe^{V}(Por) + R \quad (45)$$

In the former case (Eq. (44)), the absorption of a photon induces heterolytic bond cleavage within the ligand OR yielding the formation of an oxo-ferryl π-cation radical, a strong oxidant capable of initiating C-H hydroxylation processes [117] and other photooxidations [163,164] (see Section 8). Of the processes discussed in this section, it is the only one involving a redox change localized on both the central atom and the macrocyclic ligand. In the latter case (Eq. (45)), nitridoiron(V) porphyrins with a triple Fe≡N bond are formed when thin films of the corresponding azido complexes [$(Por)Fe^{III}N_3$] are irradiated by visible light at approximately 30 K. In this case, it was demonstrated that the formation of π-cation radicals can be ruled out [83,122,152]. At higher temperatures, the nitridoiron(V) porphyrins convert to the μ-nitrido mixed-valence dimers [$(Por)Fe^{III}\text{-}N\text{-}Fe^{IV}(Por)$]. Very detailed studies involving isotopically labelled complexes prove that the remaining (O, N) atoms have their origin in the formerly anionic axial ligands.

The third class of photoredox processes of Fe(III) macrocycles involves the photoinduced decomposition of porphyrinato and phthalocyaninato binuclear-bridged complexes. So far, the attention of various research groups has been focused on the oxo-bridged complexes [$(MC)Fe^{III}\text{-}O\text{-}Fe^{III}(MC)$] due to their highly promising photocatalytic activity (MC = porphyrinato TPP, TPPC and TPPS [137-143] and phthalocyaninato TdPc ligands [144,165]). The quantum yield of photodisproportionation is rather low (10^{-3}-10^{-5}) and strongly dependent on the wavelength of the incident radiation. The photoredox-reactive excited state(s) is not specified; however, a contribution from a highly distorted MMCT state mixed with an LMCT state can be expected [4,6].

The primary photoredox step is believed to involve iron-oxygen bond splitting

$$[(MC)Fe^{III}(\mu\text{-}O)Fe^{III}(MC)] \xrightarrow{h\nu} [Fe^{II}(MC)] + [O{=}Fe^{IV}(MC)] \quad (46)$$

The strongly oxidizing oxo-ferryl complex behaves as a two-electron oxidant or as an oxygen atom donor towards amines, hydrocarbons and phosphines (see Section 8). The fact that, in some systems [143,144], the formation of an oxo-ferryl species was not observed does not exclude Eq. (46) as the primary photoredox step. Due to their very fast reactions, the characterization of oxo-ferryl complexes is a difficult task and their formulation as π-cation radicals [$O{=}Fe^{III}(MC^+)$] with strong antiferromagnetic coupling of both paramagnetic centres cannot be excluded either [117,118].

Very interesting photochemical reaction represents inner-sphere reduction of Fe(III) in the peroxo-bridged dinuclear tetrasulfonatophthalocyaninato iron(III) complex

$[\{Fe(TSPc)\}]_2(\mu\text{-}O_2^{2-})]$. Population of LMCT state (L means peroxo-ligand) from directly populated macrocycle-localized IL state(s) leads to formation of mononuclear Fe(II) complex [Fe(TSPc)]. Peroxo ligand behaves, therefore, as a reductant towards Fe(III) being itself converted to singlet oxygen [145,146].

There is no direct evidence to indicate that the formation of a binuclear complex on irradiation of a mononuclear complex is a true photochemical reaction involving an excited species with enhanced Lewis acidity or basicity, i.e. a phenomenon found on exciplex and excimer formation [4,5]. The formation of a polynuclear species is usually understood to be a secondary thermal reaction of the reactive species formed in the primary photochemical step. The insertion of such reactions in this section is conditioned by two reasons. Firstly, bearing in mind the mechanisms of exciplex and excimer formation [4,5], there is no reason to exclude an analogous mode of excited species involvement from the reaction of polynuclear complex formation. Secondly, in some cases, dinuclear complexes were prepared for the first time by a photochemical synthetic route. The first binuclear iron macrocyclic complex prepared by photochemical synthesis [123] was the μ-nitrido tetraphenylporphyrinato (TPP) mixed-valence complex $[(TPP)Fe^{III}(\mu\text{-}N)Fe^{IV}(TPP)]$. Later on an analogous complex with octaethylporphyrinato ligand was prepared [122]. In general, it is meaningless to try to find a relationship between the nature of the excited states involved and the ability of mononuclear complexes to undergo polymerization, since the ground state molecules participate in addition reactions leading to polynuclear complex formation.

Outer-sphere electron transfer from a surrounding molecule to an energy-rich excited Fe(III) complex can hardly be omitted in systems containing easily oxidizable molecules. However, this does not apply only to complexes with macrocyclic ligands. In systems with kinetically labile complexes, it is currently generally impossible to distinguish whether the photoreduction of the central atom is associated with the oxidation of a molecule (e.g. alcohol molecule) representing a ligand in the primary coordination sphere or of a molecule from the solvent cage of the excited complex. In all the systems discussed so far containing oxidizable solvent molecules, reactions of this kind should be taken into account at least as an alternative pathway. The situation is clearer in the systems in which an outer-sphere electron transfer from an excited inert molecule to the ground state Fe(III) complex occurs. As an example, the photoinduced electron transfer

$$^3[Zn^{II}(cyt)]^* + [Fe^{III}(UroP)(CN)_2] \rightarrow [Zn^{II}(cyt^+)] + [Fe^{II}(UroP)(CN)_2] \quad (47)$$

can be used, in which a long-range electron transfer occurs with a rate constant of $9.1 \times 10^4\ s^{-1}$ within a preformed electrostatically associated ion pair, consisting of zinc-substituted cytochrome *c* in its highly reducing triplet state and ferric uroporphyrin [166]. The constituents of the 1 : 1 aduct (zinc-substituted cytochrome and ferric porphyrin) are bound via the electrostatic interaction between the positively charged lysine residues of the cytochrome and the negatively charged carboxylate residues on the porphyrin periphery. The actual charges of the species depend on the pH. In a series of analogous electron transfer processes, a clear observation of the Marcus inverted region was revealed for the thermal processes, but no such region (over the same range of driving force) was found for photoinduced electron transfer. A system consisting of cofacial covalently bound Zn(II) and Fe(III) porphyrinato complexes behaves similarly, producing $[Zn^{II}(OEP^{-}]$ radical and $[Fe^{II}(OEP)]$ species [167] In this case, however, the light-driven electron transfer rate constant is higher ($k > 10^{11}$ s^{-1}). The investigation of several cofacial diporphyrinato systems by picosecond photochemistry led to a conclusion that electron transfer reactions occur from singlet excited states in such dimers within less than 6 ps. Analogous reactions involving Fe(III) proteins acting as an electron-transfer quencher are mentioned in Section 6.

The extensive search for new purposefully exploitable materials has resulted in a shift from the investigation of pure compounds and/or their solution to more complex systems containing porphyrin complexes in inorganic or organic (polymeric) host materials such as clays, layered semiconductors, nanotubes, mesoporous materials [168,169] or polymeric membranes [170]. The structure, chemical, photochemical and photophysical properties of such complexes accommodated in host materials differ from those characteristic for „free“ complexes. For example [170], photoexcitation of various cationic porphyrinato iron(III) complexes, such as $[Fe(TPP]^+$, $[Fe(TDCPP]^+$, and $[Fe(TFPP]^+$ (where TPP, TDCPP, and TFPP are *meso*-tetraphenylporphyrin, *meso*-tetrakis(2,6-dichlorophenyl)porphyrin, and *meso*-tetrakis(pentafluorophenyl)porphyrin, respectively) in homogeneous alcoholic (ethanol, 2-propanol) solution and caged into inert Nafion membrane (Nafion is sulfonated polytetrafluoroethylene) gives rise to significantly different behaviour. In solutions containing cyclohexene or cyclohexane, photooxidation processes are slower, no ether is produced, and in addition to allylic oxidation products, cyclohexanol and cyclohexanone are formed. Photoproduced Fe(II) is oxidized by oxygen and the complexes can be characterized as photocatalysts.

An amphiphilic Fe^{III}(TPP) derivative with four alkylphosphocholine groups and a proximal imidazole (lipidporphyrinato-iron(III) chloride), self-assembled in the form of spherical micelles with a diameter of 10 nm is photochemically stable when irradiated into LMCT bands under argon atmosphere [147]. An addition of glucose led to complete Fe(III) photoreduction, giving a five-N-coordinated high-spin iron(II) complex [147]. In the presence of hyaluronic acid, the morphology of the derivative assembly is changed from the micelle to a thin fiber in which Fe(III) is also easily reduced to Fe(II) [148]. The photoreduced fibers can reversibly bind and release O_2 similar to the same fibers which were prepared by chemical reduction using ascorbic acid.

6. Iron(III) Proteins

Photochemistry of iron(III) complexes has significantly penetrated also into the field of biochemistry, biology and medicine. In this area the photochemical investigation has contributed to understanding not only typical photoinduced processes but also the background of processes, which do not occur photochemically. In many cases photobiological processes are investigated using rather structurally and functionally artificial model compounds than native proteins.

From the viewpoint of the inorganic chemist in general and the bioinorganic chemist in particular, artificial iron porphyrinato complexes are suitable models of native biologically active enzymes (mainly of cytochromes), since the structure of the site governing their physiological function is identical for both kinds of compounds. However, the actual biological function of the enzymes is tuned by their "organic" protein part. In addition, the protein part contains several sites which can be used to bind certain species, and such a modified enzyme offers new opportunities to investigate the chemical basis of its biological action. These new horizons are one of the main reasons for very intensive study of the relationships between the structure, chemical properties and biological activity of cytochromes. Cytochromes provide photochemical reactions which are not common in the photochemistry of artificial porphyrinato complexes and this is why their photochemistry is discussed separately.

In the study of photoreactions involving cytochromes, two main issues can be identified:

1) the *in vivo* and *in vitro* mechanisms of the chemical reactions of cytochromes using photochemical methods as a tool for investigation;
2) the theoretical aspects of the dependence of the kinetic parameters of long-range electron transfer reactions on the edge-to-edge distance between the donor and acceptor centres, the difference between their redox potentials, the geometric dispositions for the electron transfer step and the solvent/structural reorganization accompanying the transfer.

Both problems mutually overlap and will be discussed as one topic. In all photochemical reactions of cytochromes, one of the reaction sites is the central iron atom. The photochemical processes involving cytochromes can be grouped into four categories according to the localization of the redox counterpartner and the mode of its interaction with cytochrome:

1) the photoreduction of Fe(III) associated with the inner-sphere oxidation of an axial ligand coordinated to the central iron(III) atom;
2) the photoreduction of Fe(III) accompanied by the outer-sphere oxidation of an anion forming an ion pair with cytochrome by electrostatic attraction to a positively charged site of the protein part of cytochrome;
3) photoinduced long-range inner-sphere electron transfer between the iron central atom of cytochrome and its redox-reactive partner covalently bound to the protein;
4) long-range outer-sphere electron-transfer quenching of excited molecules by ground state cytochrome.

The photoredox reactions were investigated in homogeneous and microheterogeneous systems (cytochrome embedded in micelles or bilayers). These investigations can be understood as essential part of photobioinorganic chemistry.

In the absence of small anionic ligands, which can penetrate to the central Fe(III) atom and coordinate to it, both ferric and ferrous cytochromes *c* (denoted Fe(III)cyt*c* and Fe(II)cyt*c*, respectively) contain histidine-18 and methionine-80 as axial ligands at neutral pH. Irradiation of Fe(III)cyt*c* leads to the reduction of Fe(III) associated with the oxidation of the sulphur donor atom of the methionine ligand [131,171,172]. Thermal reoxidation of Fe(II)cyt*c* to Fe(III)cyt*c* does not lead to the parent Fe(III)cyt*c*. The hexacoordination of the central Fe(III) atom is preserved, but in the axial position tyrosine-67 with an oxygen donor atom becomes coordinated instead of methionine. The photoreduction quantum yield was observed to depend on the pH reaching a maximum value (about 6×10^{-4}) at pH 6. The fate of the oxidized methionine and the potential participation of the protein part in the redox changes are topics still open to investigation. The presence of the biological reductant NADH in irradiated systems of Fe(III)cyt*c* has a strong impact on both the efficiency of photoreduction (the corresponding quantum yield is five times higher than in the absence of NADH) and the mechanism of the thermal redox processes (due to the participation of NADH in these reactions, reoxidation of Fe(II)cyt*c* yields the parent Fe(III)cyt*c* having histidine-18 and methionine-80 in the axial positions [173]).

Some small ligands can penetrate to the vicinity of the central Fe(III) atom to displace the methionine ligand from the primary coordination sphere and to occupy the axial position. Irradiation of cytochrome *c* in the presence of N_3^- anions induces the formation of the radical $N_3^{\bullet}$ formed by an inner-sphere electron transfer from the coordinated anion N_3^- to the central iron(III) atom. After its detachment from the Fe(II)cyt*c* formed, the radical $N_3^{\bullet}$ can be identified by spin-trapping EPR [171].

Fe(III)cyt*c* is a cation with a charge as high as 7+ [174] which can attract anionic species on to the surface of its protein part forming ion pairs (in accordance with common practice, hereafter the charge of cytochromes will not be indicated). The investigation of systems containing Fe(III)cyt*c* and strongly redox irreversible anions $C_2O_4^{2-}$ [175] or reversible anions R-K$^-$ of kojic acid and its derivatives [176] has shown that these anions do not penetrate to the central Fe(III) atom; however, their presence increases the quantum yields of photoreduction of Fe(III)cyt*c* to Fe(II)cyt*c*. It was concluded that the anions are attracted to the positively charged protein part of Fe(III)cyt*c* and, together with the intramolecular transfer of an electron from the methionine axial ligand to the central Fe(III) atom (discussed above), electron transfer from anionic $C_2O_4^{2-}$ or R-K$^-$ to the central atom occurs. The higher the number of attracted anions, the higher the probability of the electron transfer within such ion pairs and, therefore, the higher the quantum yield of the reaction. For example, using 313 nm excitation, $\Phi_{Fe(II)cytc}$ increases from 2.3×10^{-4} (for free hydrated Fe(III)cyt*c*) to 5.6×10^{-3} (for ion pairs Fe(III)cyt*c*,(OH-K$^-$)$_n$) or 7.0×10^{-3} (for ion pairs Fe(III)cyt*c*,$(C_2O_4^{2-})_n$).

The validity of the proposed reaction model was checked using liposome (in the case of $C_2O_4^{2-}$ anions) and sodium dodecylsulphate (SDS) (in the case of R-K$^-$ anions), which can cage both free cytochromes and their ion pairs with the anions in liposome bilayers or SDS micelles. The results obtained can be summarized as follows. Due to the rigidity of the SDS cages containing Fe(III)cyt*c* with on-surface-bound R-K$^-$, the molecules of the geminate pair Fe(II)cyt*c* and R-K$^{\bullet}$ cannot separate; thermodynamically driven back electron transfer occurs and the parent reactants Fe(III)cyt*c* and R-K$^-$ reappear. In systems containing redox pairs Fe(III)cyt*c*,R-K$^-$ embedded in SDS micelles, no net photoredox change is observed on continuous photolysis. However, irradiation of the ion pairs Fe(III)cyt*c*,$C_2O_4^{2-}$ caged in liposome bilayers gives rise to the photoreduction of Fe(III)cyt*c*, since CO_2 molecules, formed from irreversibly oxidized $C_2O_4^{2-}$, do not back react to yield $C_2O_4^{2-}$. In systems containing Fe(III)cyt*c* caged in SDS micelles and $C_2O_4^{2-}$ or R-K$^-$ located outside the micelles, no net photochemical change is observed, probably due to the large distance between the reaction centres Fe(III) and the anions (inner diameter of the SDS micelles, Stern layer included, is estimated to be $r > 1500$ pm [177]). The effect of incorporation of one of the reaction partners into micelles was monitored following the electron transfer between Fe(III)cyt*c* and N-dodecyl-N'-(2-phosphonoethyl)-1,4,5,8-naphthalenetetracarbolylic diimide (DNDI). Having both reactants „free“ in aqueous solutions, UV irradiation resulted in photoreduction of Fe(III)cyt*c* to Fe(II)cyt*c*. When DNDI was kept incorporated in cetyl trimethyl ammonium bromide micelles, electron transfer process was inhibited [178].

In the systems discussed above, the interaction between Fe(III) and the anions involves the attraction between ions within ion pairs. Several surface-modified cytochromes with redox partners covalently bound to a specific site of the protein (usually at the histidine residue) were prepared, and electron transfer between the redox sites, induced by pulse radiolysis and flash photolysis, was studied [179-183]. Owing to their favourable properties (long-lived emission, substitutional stability, possibility to vary the redox potential value in a relatively large range), ruthenium complexes covalently bound to cytochromes were mainly used as redox partners.

In systems containing histidine-modified cytochromes with a ruthenium redox centre, the kinetic parameters for both spontaneous and photoinduced electron transfer were followed. In photochemical investigations, the excited centre can be, in principle, either an iron or ruthenium centre; however, the excitation of the ruthenium centre prevails. As an example, the electron transfer of ruthenium(II) polypyridine cytochrome *c* labelled at lysine-13 is introduced [179]. The redox changes in the irradiated systems of the labelled ferric cytochromes *c* can be, in their simplest way, shown as follows

$$\text{Ru(II)-Fe(III)} \xrightarrow{h\nu} {}^{*}\text{Ru(II)-Fe(III)} \tag{48}$$

$$^{*}\text{Ru(II)-Fe(III)} \xrightarrow{k_d} \text{Ru(II)-Fe(III)} \tag{49}$$

$$^{*}\text{Ru(II)-Fe(III)} \xrightarrow{k_f} \text{Ru(III)-Fe(II)} \xrightarrow{k_b} \text{Ru(II)-Fe(III)} \tag{50}$$

where k_d, k_f and k_b are the rate constants of deactivation, forward and back electron transfer, respectively. In the case of $[(bpy)_2Ru^{II}(dcbpy)\text{-}Fe^{III}cytc]$, the corresponding rate constants are $k_d = 8 \times 10^6\ s^{-1}$, $k_f = 1.6 \times 10^7\ s^{-1}$ and $k_b = 2.6 \times 10^7\ s^{-1}$. Modification of cytochrome at another lysine group leads to slightly different values of the rate constants [179]. Analogous electron transfer was observed in the systems containing a negatively charged metallopeptide, $[Ru^{II}(bpy)_2(phen\text{-}am)\text{-}Cys\text{-}(Glu)_5\text{-}Gly]^{3-}$ attached to the cystein side chain of positively charged iron(III)cyt*c* via acetamido or sulfido linker [184]. An electron passes from triplet Ru(II) localized state to the central iron(III) atom of Fe(III)cyt*c* by both the static and dynamic quenching.

Electron transfer (ET) reactions play essential role in the important biological processes such as photosynthesis or respiration and typically involve the transport of electrons through a series of distinct protein-protein complexes. Photochemistry, mainly its very fast flash techniques, is able to contribute to understanding the mechanisms of the ET processes on the molecular level and its potential has been successfully utilized. In several research groups, ET processes have been investigated using a zinc-substituted protein (Zn-myoglobin, Zn-cytochrome *c*, and its derivatives) and an iron(III) protein such as Fe(III)cyt*b*, Fe(III)cyt*c* [185-191]. Irradiation of such systems leads first to the population of the lowest singlet excited state of the zinc-substituted protein followed by intersystem crossing giving rise to its triplet excited state and continues by the abstraction of an electron from the organic part of zinc-substituted protein to the iron(III) central atom of the counter partner. The process is formally similar to that expressed by equation (42), however, in this Section ET reactions involving Fe(III) proteins are discussed. Electron transfer from various conformation forms (folded, molten-globule or unfolded forms) of zinc-substituted cytochrome *c* in its triplet excited state to several inorganic and biocomplexes reductants, comprising $[Fe(CN)_6]^{3-}$ and Fe(III)cyt*c*, were followed [192] and it was found that the form of the zinc-substituted cytochrome *c* exhibits a significant effect on the ET efficiency.

Outer-sphere electron-transfer quenching was observed to occur in some systems containing GS ferric cytochromes together with long-lived excited molecules, such as zinc porphyrins, metal-free porphyrins [193], acridine [194] and Mg(II)-substituted cytochrome peroxidase [195,196]. The rate of dipolar energy transfer, like that of electron transfer, varies strongly with the distance r and is proportional to r^{-6} [195]. The redox processes occurring in systems containing ground-state Fe(III)cyt*c* and the excited intraligand triplet of metal-free porphyrins or Zn(II) porphyrins consist of two steps. The first one involves an electron transfer from the excited porphyrin to Fe(III)cyt*c* and the second step involves back electron transfer from Fe(II)cyt*c* to the GS porphyrin π-cation [194]. The mechanism is believed to be a simple electron transfer. Ferric cytochrome b_5 and cytochrome *c* quench the excited state of acridine with rate constants of 8.7×10^8 M^{-1} s^{-1} and 1.8×10^9 M^{-1} s^{-1} respectively [194]. In systems containing ferrous cytochrome b_5 and ferric cytochrome *c*, ground-state electron transfer occurs between the iron central atoms. The interfacial regulation of long-range electron-transfer processes between many closed-shell excited metalloporphyrins, magnesium- or zinc-modified proteins and ground-state iron(III) proteins was investigated and the results contributed to the understanding of the relationships governing electron transfer in biology [197]. Outer-sphere electron transfer is also responsible for the photodecomposition of the hydroperoxide derivative of biphenyl in the presence of Fe(III)cyt*c* [198]. The reduction of Fe(III)cyt*c* is mediated by the electron transfer from a light-induced product of the biphenyl derivative. Details concerning the mechanism of the reduction are, however, not given.

Given the results in [196,197], it is found that in *in vivo* systems, an interaction between Fe(III)cyt*c* and cytochrome *c* peroxidase is a critical prerequisite for electron transfer between these two enzymes. An essential element of biological specificity is the position of the amino acids on the surface of peroxidase which controls enzyme-enzyme recognition.

In addition to cytochrome *c* investigation, other cytochromes were studied as well. The redox behaviour of ferric cytochrome c_{551} (Pseudomonas aeruginosa) exhibits some differences compared with that of cytochrome *c* (horse heart, yeast or tuna). Irradiation of aqueous ferric cytochrome c_{551} in the same conditions as Fe(III)cyt*c* gave no observable photoreduction of Fe(III) to Fe(II); instead the formation of a high-spin iron(III) compound was observed [172]. A possible reason is that ferric cytochrome c_{551} does not contain a tyrosine residue in the vicinity of the heme group and ferrous cytochrome c_{551} can undergo fast autooxidation. In the presence of anions N_3^-, the photochemical behaviour of ferric cytochrome c_{551} is similar to that of Fe(III)cyt*c*.

Investigation of photochemical properties of cytochromes P-450 and their enzymes forms part of their comprehensive study stimulated by their key role played at a huge range of catalytic reactions, using oxygen to attack thousands of different molecules [198,199] such as acetaminophen, caffeine, nicotine, diazepam, aniline and benzene.

All cytochromes P-450 enzymes contain an iron atom held in a heme group. This iron atom takes electrons and uses them to charge an oxygen atom, making it highly reactive. Safranine T is a basic compound which mediates electron transfer at illumination. It was found with flash photolysis that an electron from photoreduced safranine T reaches the heme iron of cytochrome P-450 with the electron-transfer rate constant 2.1 x 10^6 $M^{-1}s^{-1}$. Both the inter- and intramolecular electron-transfer mechanism of P-450 was proposed.

Intramolecular long-range electron transfer was also studied in hemoglobin derivatives [200]. In the human hemoglobin chain, the single sulphydryl group (cysteine-140) was first converted to disulphide and subsequently to its anion radical by the pulse radiolytically generated radical $CO_2^{-\bullet}$. Disulphide radical anion acts as an electron donor and hemoglobin Fe(III) as an electron acceptor.

UV and visible irradiation of ferritin or apoferritin induces the release of Fe(II) into the medium [201,202]. The Fe(II) release depends strongly on the pH, buffer and oxygen concentrations. The quantum yield of Fe(II) formation is of the order of 10^{-3} and markedly increases with the frequency of the incident radiation. The importance of ferritin to skin photobiology has been indicated [198].

Methemoglobin, metmyoglobin and other ferric protoheme complexes can be excited via their chemical reaction with H_2O_2 and, consequently, provide chemiluminescence. An excited state was formed as a result of electron transfer from an axial ligand to heme iron and its creation was found only for compounds with pentacoordinated heme iron [203].

Sunlight was found to induce the formation of *N*-(carboxymethyl)lysine (CML) from glycated polylysine in the presence of Fe(III) ion as catalyst [204]. Photoinduced oxidative C–C bond cleavage of the Amadori compound (*N*-(1-deoxy-D-fructos-1-yl)glycine) was assumed to be initiated by photochemical electron transfer from the ligand to Fe(III) in the Fe(III)–Amadori compound complex affording the Fe(II)–Amadori compound radical intermediate, which eventually yields either CML or active oxygen species. CML is thus a useful oxidative stress marker. The mechanism of the processes was proposed and a high accumulation of CML in lens protein and skin actinic elastosis explained.

For years, significant attention has been devoted to understanding relationships between cutaneous pathologies such as skin cancer, photosensitization, photoaging, necrotic cell death and other adversed consequences induced in the human body and its exposition to UV radiation [205]. It has been evidenced that the generation and action of reactive oxygen species (lipid peroxidation products, hydroxyl radicals), frequently catalyzed by iron(III)-containing species, play the key role in the above detrimental effects [206-208]. The iron-storage protein ferritin is present in the basal layer of skin [209] and can participate, together with "free" iron ions, in cyclic (photo)redox processes as catalyst of the generation of reactive oxygen species. Topical application of iron chelators, e.g. deferoxamine, H_4edta, 1,2-diethyl-3-hydroxypyridin-4-one, or *meso*-2,3-dimercaptosuccinic acid, reduces erythema, epidermal and dermal hypertrophy, wrinkle formation, tumour appearance both at short- and long-term UV exposure due to binding iron ions into chelates and supressing thus iron-driven generation of lipid peroxides and hydroxyl radicals. The influence of chelators lies, therefore, in moderating the amount of labile iron promoting oxidative damage and necrotic cell death [209,210]. It should be pointed out that the actual photochemical processes include rather organic biomolecules than iron ions or their chelates.

Investigation of photochemical properties of carcinogenic nitriloacetate complexes Fe^{III}(NTA) revealed that Fe(III) is reduced by visible light to Fe(II), the quantum yield of Fe(II) photoformation depends on pH. Formed Fe^{II}(NTA) transfer an electron to molecular oxygen forming reactive radical O_2^- and regenerating Fe^{III}(NTA) complex. The generation of reactive oxygen species and its relation to Fe(III) photoreduction are rationalized [211,212].

7. Photoprocesses Not Involving Iron(III) Central Atom

Photochemical reactions discussed in the previous Sections 3 – 7 involve the central iron(III) atom localized changes, usually of redox nature. Due to the kinetic substitutional lability of high-spin iron(III) complexes, other type chemical changes (non-redox substitution, elimination, isomerization, etc.) can take place in the mechanism, however, not being observed in final products. Low-spin inert complexes, such as cyano or macrocyclic complexes, may undergo also non-redox photochemical changes, usually involving the central atom.

Along with dominant photoredox processes induced by excitation of iron(III) complexes, also non-redox photochemical reactions were identified. In a recent paper [22], photoinduced azo-imine tautomerism of the free ligand acyclic tridentate salicyl-2-pyridyl hydrazone and the ligand coordinated to the iron(III) central atom was observed. The reaction is ligand-localized and involves neither the iron(III) central atom nor the ligand donor atoms.

There are a few examples of processes which obviously occur on a ligand without any direct participation of the iron(III) central atom. The coordination of the ligand involved may be a precondition of such reactions and they may, therefore, be understood as mediated or facilitated by the central atom. As a classical example, photoisomerization of a coordinated ligand can be introduced [213,214]. Within investigation photochemical properties of spin-crossover iron(III) complexes it was fould that 315 nm irradiation of $[Fe_2(salten)_2(az)](BPh_4)_2$, $[Fe_2(3\text{-}MeO\text{-}salten)_2(pyet)](BPh_4)_2$ and $[Fe(acacen)(stpy)](BPh_4)$ [213] dissolved in acetonitrile at 298 K gave rise to the *trans*→*cis* isomerization of coordinated ligands *trans*-4,4'-azopyridine(az), *trans*-4-styrylpyridine(pyet), and *trans*-1,2-bis(4-pyridyl)ethylene(stpy), but neither the oxidation nor the spin-state of iron(III) changed. *Cis*→*trans* isomerization of coordinated azobenzene was observed at irradiation of a supramolecular system containing magnetic ionic liquid (iron(III) chloride with 1-butyl-3-methylimidazolium chloride) and azobenzene [214]. Dependence of magnetic moment on magnetic field exhibited neither iron(III) photoreduction nor change in the iron(III) spin state or its paramagnetic behaviour.

8. Applications

Theoretical and practical aspects of the photochemical behaviour of chemical compounds and materials strongly overlap. A brief, but comprehensive, evaluation of the perspectives in applied inorganic photochemistry is given in paper [215]. Moreover, there are a number of review papers dealing with the preparation and characterization of photocatalytically active species [7,215,216]. Use of solar energy to drive chemical processes is still an investigation area of great importance [217]. In nearly all branches of applied photochemistry iron compounds are involved.

Major potential and/or realized applications of the photochemistry of Fe(III) complexes can be grouped into three main categories:

1. photocatalytic processes aimed at the preparation of required chemical (namely organic) compounds using Fe(III) complexes as photocatalysts;

2. the photocatalytic destruction of harmful or otherwise undesirable compounds in the presence of Fe(III) complexes playing an active part in decomposition reactions through photo-assisted Fenton process and advanced oxidation processes;
3. the use of Fe(III) complexes as photoinitiators, photostabilizers and photodestructors in polymer chemistry;
4. photoinduced magnetic phenomena, including photochemical processes intended to prepare materials with required magnetic properties and photoinduced changes in crossover compounds,
5. rationalization and exploitation of the knowledge on Fe(III) photochemistry to solve environmental and health protection issues.

In addition, Fe(III) compounds are applied on a narrower scale as photoconductors, active substances of non-conventional photographic materials, etc.

The ability of Fe(III) compounds to act as photocatalysts in the oxidation of alcohols to aldehydes or ketones has been well known for a long time [1]. In recent years, the investigation has been extended to diols which are oxidized in the presence of gaseous oxygen and an iron(III)-based catalyst ($FeCl_3$ or *meso*-porphyrin derivatives) to aldehydes, ketones and acids [10,170,219-222]. Depending on the system composition and experimental conditions, product formation in 99% yield and turnover numbers of the order of 10^3 were reached in some cases. A characteristic feature of the photolytic reactions of 1,2-diols is the regioselective oxidative cleavage of the 1,2-C-C bond.

Unsaturated hydrocarbons can be photocatalytically oxidized selectively to alcohols, ketones, aldehydes, epoxides and acids. Iron(III) catalysts (usually *meso*-porphyrin derivatives) can be entrapped in a polymer matrix [126] or dissolved in a non-aqueous solvent [9,141,163,222,223]. Turnover numbers as high as 10^3 and excellent selectivity in product formation were found in some cases (e.g. 100% epoxide formation from 1-methylcyclooctene in photocatalytic oxygenation and 100% alcohol formation from the same alkene in photoinduced auto-oxidation [141]). The photocatalytic oxygenation of α-pinene gives rise to six major products [9,163,222], the ratio of which can be purposefully modified by the catalyst used. Of the products, pinene oxide, *trans*-verbenol and verbenone are of practical interest. The oxygenation can involve the radical $^{\bullet}O_2H$ or ferryl $[O{=}Fe^{IV}(TPP)]$ as oxygen atom donors. Based on product analysis, it was concluded that alkenes with strained carbon double bonds give epoxides preferentially, whereas unstrained alkenes undergo mainly allylic oxygenation. Moreover, it was suggested that artificial porphyrinatoiron(III) complexes and microsomal cytochrome PA50 act in a very similar mode [163]. Comparison of the results, expressed in turnover number or (stereo)selectivity of required products formation, obtained using „free“ complexes with those embedded in heterogeneous systems, documented that in many cases the incorporation of an iron(III)-based photocatalyst significantly improved the course of the followed process [170]. As an example, results obtained within oxidation of cyclohexene by oxygen photocatalyzed by $[Fe(TDCCP)]^+$ dissolved in ethanol and incorporated into Nafion membranes in ethanol can be given. Irradiation was conducted in both cases at the same conditions and the product formation is expressed in mol of product formed per mol of consumed complex. The photocatalyzed oxidation resulted in the formation of (homogeneous solution/heterogeneous system) cyclohex-2-en-1-one (110/380), cyclohex-2-en-1-ol (79/1700), and *trans*-cyclohexane-1,2-diol monoethyl ether (0/520). It is obvious that heterogenization increased significantly the stability of the complex and the efficiency of the oxidation process [170].

Using $FeCl_3$ instead of iron porphyrins leads, in addition to carbon-carbon bond cleavage and oxygenation, to the formation of chloro derivatives [223]. However, such formations of α-chloro or ω-dichloroketones from cycloalkenes are not catalytic but stoichiometric reactions.

The photoaddition of alcohols to epoxides is catalysed by solid Fe_2O_3 or Fe^{3+} ions, and this reaction can be performed in the absence of oxygen [223].

Fe(II) formed by the photoreduction of the acetylacetonato complex $[Fe(acac)_3]$ participates in the addition of O_2 to benzaldehyde to form benzoic peracids and benzoic acid [224]. Irradiation of systems containing salicylic acid and Fe^{3+} ions induces the formation of the hydroxylation products 2,3- and 2,5-dihydroxybenzoic acid. The mechanism of the reaction involves the excited singlet state of methylene blue [109].

Nanostructured iron titanate films prepared from iron(III) chloride and titanium tetraisopropylate and containing at least traces of $FeCl_3$ are able to photocatalyze dinitrogen fixation and reduction to ammonia in the presence of ethanol [225]. Hydrazine was identified as intermediate.

From a very critical point of view, the situation in the area of purposeful photochemical transformations of organic substrates may be, when assessed frankly, insufficiently mature to formulate general conclusions. However, it is definitely stimulating, with many potential prospects in chemistry.

Efficient exploitation of the combination of solar radiation and water to convert the solar energy into thermal energy, and catalytically into chemical and/or electrical energies is still one of the greatest challenges for science and technology. Application of Fe(III) compounds in this field is focused mainly on water splitting to produce hydrogen as fuel for energy utilization using mixed-oxide semiconductor photocatalysts [226]. Among the several materials, Pt/Bi_2FeNbO_7 was investigated and it was found that its efficiency in producing hydrogen and oxygen from water-methanol mixtures was even higher than that of using Pt/TiO_2 photocatalyst. However, the efficiency is too low and price too high to be considered for practical application.

Due to the large-scale development of agrochemical industry and world-wide application of its products, pollution of continental and marine natural waters, soil and air has reached dramatical level. Most pesticides, but also other organic pollutants, are resistant to chemical, photochemical and biological degradation under common environmental conditions. Such compounds may cause chronic effects as carcinogenesis, neurotoxicity, effects on reproduction, etc. One of the most promising remediation modes of elimination of organic pollutants from the environment lies in application of advanced oxidation processes (AOPs). Three classes of AOPs, all of them involving the use of UV and/or visible radiation, have been investigated at the laboratory scale and offered for industrial application: direct photolysis of H_2O_2 in aqueous homogeneous systems by UV radiation shorter than 300 nm; photoactivation of TiO_2 by UV radiation of wavelength shorter than 387 nm; photo-assisted Fenton process involving redox active transition metal compounds acting as (photo)catalysts, iron(III) compounds including. The advances, obstacles and results obtained applying AOPs processes in destruction of organic pollutants has recently been evaluated in some review papers [25,227-229].

When performing the pollutant destruction in aqueous media, it is generally accepted that the key role is played by attacking radicals $HO^{\bullet}$ formed within Fenton (photo-assisted Fenton) process (see Eqs. 5 – 7). The disadvantage of applying photo-assisted Fenton processes to oxidize organic matter is frequent incomplete mineralization and the necessity of using expensive chemicals (e.g. H_2O_2).

When acting in oxidizing processes relating to the environment, iron(III) is usually present in the form of aquated Fe^{3+} cations and its hydrolytic products, iron(III) complexes, Fe^{3+} cations or iron(III) complexes immobilized on a membrane, zeolite or other solid component of the system, insoluble iron(III) oxides, dopant of TiO_2 or other semiconductor. Simultaneously with photoinduced degradation of undesirable compounds present in the environment, attention is devoted to the use of ecologically friendly catalysts. As an example, preparation, characterization and evaluation of an efficient heterogeneous Fe(III)-HY catalyst for photo-assisted Fenton reaction can be offered [230]. Fe^{3+} ions are immobilized on HY zeolite and the activity of the catalyst was evaluated by degradation of phenol, examining the effect of Fe(III) loadings, H_2O_2 concentration, pH and quenching on phenol degradation. The results obtained showed clearly that Fe(III)-HY was more efficient than free Fe^{3+} ions. In addition, heterogeneous Fe(III)-HY allows a wide range of pH for reaction against the narrow pH range in homogeneous system. As for photoinduced transformation of organic pollutants in the aqueous environment, good performance was found when using transition metal ions, iron(III) including, incorporated together with TiO_2 into mesoporous molecular sieves MCM-41 [231]. An alternative way lies in the use of immobilized photo-assisted Fenton reactions via membranes, fabrics (e.g. Nafion), mats (Nafion-glass mats) and modified copolymers (polyethylene block copolymers with maleine anhydride anchoring groups) [232] resistant to the leaching of Fe^{3+} ions into solution and to corrosion.

Studies on the photolysis of herbicides and pesticides in solution in the presence of Fe^{3+} ions and corresponding hydroxo iron(III) species, resulting in simple inorganic species, e.g. CO_2, Cl^- or NH_3, have demonstrated the major role of Fe(III) species in both the transformation and mineralization of these pollutants [102,232,234]. Considerable attention has been focused on Fe^{3+} ion-assisted photooxidation of haloacetates in connection with the use of haloacetates as possible CFC replacement compounds [30]. At highly acidic aqueous media, Fe^{3+} ions are dominant species, in less acidic media, however, mononuclear $Fe(OH)^{2+}$ and polynuclear iron(III) hydroxo complexes prevail. These hydroxo species undergo very efficient photoreduction of Fe(III) associated with production of oxidizing $^{\bullet}OH$ radicals. Hydroxyl radicals are able to decompose organic compounds, however, in some cases the level of degradation is rather low. For example, only about 5% of 3-chlorophenol was degraded using 365 nm radiation [235]. The active role of $^{\bullet}OH$ radicals in the photooxidation of organic substances was declared in the complete mineralization of the herbicide asulam (4-amino-benzosulfonyl-methylcarbamate) [236] or phenol [237], degradation of benzoic acid [238], degradation of 2,4,6-trinitrotoluene (TNT) producing 1,3,5-trinitrobenzene, 1-methyl-2,4-dinitrobenzene, 2,5-dinitrobenzoic acid and 1,3-dinitrobenzene [239]. Photo-assisted Fenton process was successfully used in degradation of the pesticides monuron [240], isoproturon [241], carbofuran [242], 2,4-dichlorophenol [243], herbicide imazapyr [244], 4-chloroaniline where $Fe(ClO_4)_3$ was applied as a source of Fe^{3+} ions [245], herbicide chlortoluron [246], pesticide amitrole (3-amino-1,2,4-triazole) [247]. The complete degradation of triphenyltin was achieved by the attack of $^{\bullet}OH$ radicals produced at photolysis of aquated $[Fe(OH)]^{2+}$ [248]. 4,4'-diaminostilbene-2,2'-disulfonic acid is an important intermediate in the synthesis of a great number of fluorescent whitening agents and frequently is scattered in the aquatic environment. It was degraded via photo-assisted Fenton process, with aquated $[Fe(OH)]^{2+}$ acting as the most photoactive species [248]. The final products of total mineralization using $Fe(ClO_4)_3$ as a source of Fe^{3+} ions, reached within 2 days exposure, are CO_2 molecules, SO_4^{2-}, and NH_4^+ ions. Photo-assisted Fenton process is responsible for oxidation of benzoic acid in atmospheric liquids to 2-, 3- or 4-hydroxybenzoic acids also under solar radiation [249]. The ability of Fe^{3+} ions and/or their hydrolytic products to function as photocatalyst were documented within the study of degradation of supramolecular host-guest complexes, namely naphthalene with α-cyclodextrine and 4-chloroaniline with *p*-sulfonic-calix[4]arene in aqueous solutions [250]. The influence of the host on the reactivity of the guest was observed and rationalized. In the majority of laboratory experiments, acidic aqueous solutions of organic compounds were irradiated by 365 nm UV radiation in the presence of Fe(III) species and H_2O_2. Mechanisms of occurring processes were elucidated based on kinetic measurements, product analysis and intermediate identification. As a characteristic feature, level of photoinduced degradation was monitored.

Along with aquated Fe^{3+} cations and its hydrolytic products, some iron(III) complexes were applied in the photodestruction of undesirable organic substances. Several Fe(III) porphyrin complexes can act as effective catalysts in the degradation processes of CCl_4 by alcohols in wastes [124,125,251]. The catalytic properties of $[Fe(OEP)]^+$ and [Fe(TDCPP)] complexes have been compared and found that the catalytic efficiency of [Fe(TDCPP)] is higher when expressed in turnover numbers. Investigation of several porphyrinato and phthalocyaninato complexes in photocatalytic degradation of atrazine documented that some of them, e.g. iron(III) phthalocyanines degraded atrazine more efficiently that TiO_2 [252]. Iron(III) *meso*-tetrakis(2,2-dichloro-3-sulfophenyl)porphyrin was used to oxidize 4-chlorophenol in aerated aqueous solution to *p*-benzoquinone and *p*-hydroquinone [253]. In the absence of oxygen, no *p*-benzoquinone was formed. It was evidenced that the oxidation occurred through two mechanisms, by the reaction between excited porphyrin complex and 4-chlorophenol, and by the action of singlet oxygen. Iron(III) porphyrinato derivatives were applied in photooxidation catalysis not only as dissolved compounds but also in the form of solid, microporous material. Iron(III) carboxylate complexes having metalloporphyrins, Fe[M(TCPP)], where M = Fe, Cu, Pd, Mn, Co and protonated H_2TCPP = 4,4',4'',4'''-(21H,23H-porphine-5,10,15,20-tetrayl)tetrakis(benzoic acid) successfully catalyzed the photooxidation of hydrocarbons with 1 atm oxygen using visible light. Within 3 hours, the complex showed 37 turnover numbers [254]. Irradiation of aqueous systems containing Fe^{3+} ions by visible light ($\lambda > 420$ nm) led to the photodegradation of the present Acid Orange 7. Photodegradation was preceeded by coordination of Acid Orange 7 to the iron(III) central atom [255]. The detailed study revealed that the primary photoredox step was the light-induced electron transfer from the azo chromophoric group to the iron(III) central atom. The particulal feature of the degradation are as follows: only Acid Orange 7 molecules coordinated to the iron(III) central atom undergo the degradation; $^{\bullet}$OH radicals do not participate in the degradation; the process does not require the addition of hydrogen peroxide and might be thus developed into an economically viable method to pretreat or decolorize azo-dye wastewaters using sunlight.

Ferrioxalato anions $[Fe(C_2O_4)_3]^{3-}$ was successfully exploited also in total degradation of tetracycline in several different kinds of water (pure water, surface water, sewage treatment plant effluent) [256], analogous results were obtained using $Fe(NO_3)_3$ as a source of iron(III).

Photocatalytic degradation of the azo dye Acid Red 1 occurs at irradiation of the dye in TiO_2 suspension in the presence of Fe^{3+} ions [257], the mechanism differs from that proposed for degradation of Acid Orange 7 [255]. The main difference lies in the observation that Acid Orange 7 photodegradation is conditioned by its coordination to Fe^{3+} ions in solution, whereas in case of Acid Red 1, adsorption of the dye on the iron(III) modified semiconductor surface is the pre-degradation step. Other dye, malachite green is photo-assisted Fenton degraded in the presence of aromatic compounds (such as salicylic acid, hydroquinone, quinone, carboxylic and amino aromatic derivatives) when exposed to visible light [258]. It is possible that iron(III) complexes with the mentioned derivatives participate in the degradation.

Photo-assisted Fenton process might be successfully applied in treatment of wastewaters originating from various large-scale processes when the present Fe(III) is stable in solution at neutral media. Phoroassisted Fenton process improved biodegradability of wastewater generated in textile industry [259]. In addition to degradation of organic pollutants, ferrioxalato ions are able to inactivate also biological microorganism in the wastewater treatment. A simple photo-assisted Fenton process, exploiting aquated Fe^{3+} ions or their hydrolytic products, frequently cannot be applied since it is effective only at acidic media (pH < 3). When using ferrioxalato ions, stable also in neutral aqueous media, modified Fenton process exploiting $[Fe(C_2O_4)_3]^{3-}$ anions instead of Fe^{3+} cations allow to inactivate *Escherichia coli* [260]. The process is more effective than disinfection by ozone, it is, however, slower than convenient chemical disinfectants. Dark brown coloured coffee effuent was decolorized by photo-assisted Fenton process using UV radiation [261] achieving the complete decolorization. A comparative study for decolorization documented that photo-assisted Fenton process was more effective than processes exploiting TiO_2 or ZnO as photocatalysts. When expressing in the terms of total organic carbon, 95% purification level was achieved in the treatment of winery wastewaters by photo-assisted Fenton process [262]. This advanced oxidation process offers a particularly useful alternative to other procedures applied in wine and grape wastewaters treatment containing alcohols, phenols, acids, sugars and unpleasantly odouring substances as principal species. Advanced oxidation technology was applied in EDTA destruction using $[Fe(C_2O_4)_3]^{3-}$ and H_2O_2, the total organic carbon removal reached almost 100% at solar exposure under the best conditions [263].

Iron oxides, widespread in nature, take part in the photocatalytic processes of pollutant destruction occurring in ecosystems. α-Fe_2O_3 has been found to be stable in the transformations of sulfites, oxalates, formates, haloacetates, peroxosulfates and phenols [30,33,264-266]. The degradation of the organic compounds in aerated iron oxide suspensions could be due to the following: *i*) LMCT within the Fe(III) surface, with concomitant oxidation of the organic compound and formation of Fe(II) on the iron surface, which is subsequently re-oxidized in the presence of air; *ii*) photo-Kolbe-type reactions in which the semiconductor holes are scavenged by pollutant molecules and electrons by the surface iron atoms; *iii*) $OH^{\bullet}$ radicals resulting from the oxidation of surface OH-groups. While in many photodegradation processes, hydrogen peroxide acts as oxidant of organic pollutants, its photoinduced decomposition by iron(III) compounds was investigated too. It was observed that iron(III)-bearing clays of montmorillonite type efficiently catalyzed the decomposition of H_2O_2 under UV irradiation, the reduction of Fe(III) present in clay to Fe(II) being the essential step in the catalytic process [267]. Iron(III) phthalocyanines supported on silica or encapsulated inside zeolite Y as solid photocatalysts are able to degrade phenols and sulfur heterocycles using hydrogen peroxide as oxidant [268]. While silica-based catalyst cannot be used in organic solvents due to self-degradation and/or desorption of the complex from silica surface to the solution, zeolite-based catalyst can be applied for oxidation both in aqueous and organic media.

The use of Fe^{3+} ions as dopants to enhance the photocatalytic activity of TiO_2 semiconductors is well documented in several papers [269-273]. In recent years, there has been increasing interest in the use of doped semiconductors in water purification processes [270]. The semiconductor-sensitized photomineralization of organic pollutants, e.g. alkanes, haloalkanes, aliphatic alcohols, aliphatic and aromatic carboxylic acids, alkenes, haloalkenes, aromatics, haloaromatics, phenols, polymers, surfactants, herbicides, pesticides and dyes, has been successfully studied. The addition of Fe_2O_3 to TiO_2 powder remarkably increases the catalytic activity for NO_x removal from polluted air [274]. TiO_2 showed to act as an efficient catalyst of photocatalytic oxidation of various deicing agents (2-ethoxyethanol; diethylene glycol monoethyl ether; ethylene glycol) in aqueous systems containing Fe^{3+} ions [275]. Unlike photo-assisted Fenton process, photocatalytic oxidation is based on the action of positively charged holes on the semiconductor surface, i.e. on electron abstraction by the holes. The effect of Fe^{3+} ions lies is supressing the hole-electron recombination by scavenging conduction band electrons from the TiO_2 surface. Iron(III)-doped titania nanoparticles absorb light up to 475 nm. It was shown that these nanoparticles are able to photocatalyze the degradation of dye pollutant sulforhodamine B in aqueous media [276] when irradiated with visible light.

Knowledge of the photochemical and photophysical properties of Fe(III) complexes is applied in polymer chemistry in the area of photoinitiation of polymerization reactions, photostabilization and photodestruction of polymers. The research activity in this field, however, declines. This trend is probably caused by the fact that in large-scale industrial processes, thermal and photochemical polymerization of monomeric compounds is usually initiated by organic initiators (e.g. peroxo compounds).

The radical polymerization of methyl methacrylate, styrene, acrylonitrile, vinylacetate and acrylamide has been initiated by $[Fe(SCN)_3(py)_3]$ and glucosato complex $[Fe^{III}\text{-HGl}]$ [5]. Complex initiator is applied in the first step of initiation. The primary radical product is an oxidized ligand or solvent molecule. In aqueous media, $^{\bullet}OH$ radicals generated in photo-assisted Fenton process are believed to act as initiators. The radical further reacts with the monomer molecule in a thermal step forming the monomer radical. Other reactions of the monomer radical result in polymerization. Various iron arene organometallics act as cationic and free radical photoinitiators of epoxide and acrylate polymerization. In the catalytic cycles of polymerization, Fe(III) arenes are formed and involved [277-280]. Recently, iron(III) complexes were applied mainly in polymerization of alkyl methacrylate monomers. Methyl methacrylate underwent photopolymerization in DMSO under UV radiation of 254 nm using a novel iron(III) oxalato photoinitiator $Fe[Fe(C_2O_4)_3].4H_2O$ [281]. Based on kinetic study, a mechanism of the photopolymerization was proposed.

The photoinitiated polymerization of methyl methacrylate (MMA) by Fe(III) complexes was monitored in aqueous and organic solvents using 254 nm radiation [282,283]. $[Fe(bpy)_3]^{3+}$ and $[Fe(DMSO)_6]^{3+}$ initiate the polymerization both in the presence and absence of co-initiators (e.g. triethylamine). The complexes $[Fe(CH_3OH)_6]^{3+}$ and $[Fe(MMA)_6]^{3+}$ initiate the polymerization when irradiated in the visible LMCT band. The photoinitiation by both complexes proceeds by different mechanisms due to the different binding of MMA and methanol to the central atom in the complexes. It was found that the methanol-containing complex is more effective in producing initiating radicals than the MMA-containing one.

Nanoparticles of iron oxides (as well as of ZnO or CdS) are efficient photoinitiators of butyl methacrylate polymerization [284]. On the basis of the kinetic parameters of the process a free radical mechanism of the butyl methacrylate photopolymerization induced by investigated nanosized semiconductors was inferred. A relation between the specific rate of the butyl methacrylate photopolymerization and nanoparticles diameter was explained and schemes of the mechanism of the photopolymerization was proposed.

Along with research on photoinitiation leading to polymer formation, attention has been paid to the preparation of polymers exhibiting particularly designed properties. The film-forming properties of divinyl triethylene glycol bis-*o*-phthalate in air in the presence of a redox system and β-diketonates of transition metals were studied [285]. A novel magnetic polymer microsphere displaying photoconductivity was prepared by covalently bonding iron(III) phthalocyanine onto a polystyrene microsphere with hydroxyl groups (PSA) [286]. PSA was synthesized by dispersion polymerization in the presence of Fe_3O_4 magnetic fluid. The structure and properties of the magnetic polymer microsphere were identified by infrared spectroscopy, elemental analysis and scanning electron microscopy. The results showed that this kind of microsphere combines the merits of magnetic responsiveness and photoconductivity. The application of magnetic polymer microspheres for absorption of bovine serum albumin (BSA) was characterized at different temperatures, different incubation times, different pH values and different initial BSA concentrations.

The photostabilization of light-sensitive polymers involves retardation or elimination of various undesirable photoprocesses occurring during polymer decomposition. A dialkyl-dithiocarbamate complex of Fe(III) acts as an efficient antioxidant reacting with the radical products of polymer destruction [287]. Tris(dibenzoylmethanato) chelates of Fe(III) are quenchers of potentially active excited singlet and triplet states of both aliphatic and aromatic carbonyl compounds and polynuclear aromatic hydrocarbons [288].

Growing interest has been devoted to the problems of photochemical degradation of polymers. Depending on the concentration, $[Fe(HAOX)_3]$ in polyethylene acts as a catalyst of oxidation and hence of polymer destruction, where as at higher concentrations it functions as an inhibitor of the destruction reaction. Dialkyldithiocarbamate radicals $R_2dtc^{\bullet}$, produced in the irradiation of $[Fe(R_2dtc)_3]$ complex, were studied as efficient initiators of polymer oxidation [5,287], as were the $OH^{\bullet}$ radicals generated during the photolysis of $FeCl_3$ [289]. Dehydrochlorination of polyvinylchloride generated by UV irradiation is associated with C—Cl bond cleavage. Iron pigments have been used in PVC laminates and, as might be expected, this affects the degradation rate [290].

Investigation of relationship between photochemical and magnetic behaviour of iron(III) compounds has been driven mainly by an effort to reveal theoretical principles governing behaviour of materials with required optical memory and to design them for practical purposes [291,292]. Contrary to the previously discussed area of application (polymer chemistry), the field of photo-magnetic effects manifests a clear sign of expansion. Up to now performed studies involving iron(III) compounds have been focused on three main areas: photoinduced aggregation processes, light and/or ligand-induced spin change, and development of phototunable molecular complexes.

Research on photoinduced aggregation processes has been focused mainly on depositing magnetic iron oxide films and on formation supramolecular systems. Magnetic iron oxide films deposited on heated fused-quartz substrates via the plume generated by the excimer laser ablation of polymeric iron(III) methacrylate in an oxygen atmosphere [293]. The investigation by various methods documented that these films consist of iron-rich Fe_3O_4 and α-Fe_2O_3 phases that are made up of ball-like nanograins. Magnetic properties (hysteresis loop, saturation and remanence magnetization, and coercivity) of the deposited films was measured and rationalized. The first photocontrollable magnetic vesicles were prepared by intercalating Prussian blue into photoresponsive azobenzene-containing multibilayer vesicles. The control of magnetic properties was achieved, however, only at very low temperature (2 K). A new type of photofunctional magnetic vesicles, exhibiting super-paramagnetic properties at room temperature, designated to control the magnetic behaviour by photoillumination, were prepared from the commercially available dye SP 1822 (1'-octadecyl-3',3'-dimethyl-6-nitro-8-[docosanoyl-oxymethyl]-spiro-[2H-1-benzopyran-2,2'-indoline]) and Fe_3O_4 particles in aqueous solution of poly(vinyl alcohol) acting as a matrix [294]. On UV irradiation, photoinduced isomerization (SP 1822 to photo-merocyanine isomer) and aggregation of the vesicles occurred, resulting in changes of both optical and magnetic properties.

A new attractive domain in which photochemistry and magnetochemistry overlap, is photoinduced aggregation behaviour of supramolecular systems consisting of iron(III)-based paramagnetic ionic liquids and a diamagnetic organic compound [214]. Mixtures of magnetic ionic liquid formed from iron(III) chloride and 1-butyl-3-methylimidazolium chloride with *cis*-azobenzene exhibit paramagnetic properties similar to pure ionic liquid at 298 K. Irradiation of such mixtures resulted in a gradual increase of magnetic moment which was rationalized in the terms of irreversible aggregation of paramagnetic entities forming a short-range order in the originated supramolecular system. The aggregation is conditioned by reversible *trans* ↔ *cis* isomerization of azobenzene accompanying conformation changes in the organic cation. Dependence of magnetic moment on magnetic field exhibited neither iron(III) photoreduction nor change in the iron(III) spin state or its paramagnatic behaviour.

High-spin ↔ low-spin transitions in spin-crossover iron(II) complexes caused by the absorption of radiation belong to well-known phenomena [295,296]. Photochemical properties of the spin-crossover iron(III) complexes are investigated in a lesser extent. A comparison of properties of Fe(II) and Fe(III) shows that the high-spin → low-spin relaxation is generally faster in iron(III) complexes owing to their smaller bond lengths [297]. Optically switchable molecular solids based on iron(III) spin-crossover complexes were prepared and their photochromism and photoinduced magnetism investigated in [298]. The results led to suggestions that such solids should exhibit an abrupt phase transition or hysterese loop and must possess at least two bistable states. The achievement of optical switching effects in molecular compounds is facilitated through intermolecular interactions of π-π, hydrogen bonding of ion-dipole nature [298-301]. In the paper [298], the first observation of LIESST (Light-Induced Excited Spin State Trapping) effect for iron(III) complex, namely in $[Fe(pap)_2]ClO_4.H_2O$ (where pap = N-2-pyridyl-methylidene2-hydroxy-phenylaminato) is declared. It is pointed out that to design practically usable photomagnets, attention should be paid not only to the optical switching properties at a single-molecule level but to the magnetic interactions between spin sources. In the field of photo-switching systems a shift from molecular system through polymeric compounds to composite materials is obvious [302]. It is worth mentioning that along with inorganic systems possessing paramagnetic central atom localized centres, much attention is paid to inorganic-organic hybrid composite materials, e.g. spiropyran vesicles containing iron oxides [302] behaving as photochromic magnetic materials. An actually acting photomagnet can be exemplified [298,303,304] by FeCo Prussian Blue having two proximal electronic states, namely the low-spin diamagnetic Fe^{II}-CN-Co^{III} in dinuclear complex and high-spin Fe^{III}-CN-Co^{II} in polymeric form of the compound. Magnetic interactions of the paramagnetic centres mediated by CN-bridges allow the spin to align both ferromagnetically and antiferromagnetically producing thus ferro and ferrimagnets below critical temperature. Depending on the chemical environment of Fe-CN-Co moieties, a reversible photoenhancement of magnetization based on photoinduced electron transfer between diamagnetic state and valence isomers exhibiting ferromagnetism or ferrimagnetism was observed. Photoinduced paramagnet ↔ ferrimagnet transitions in thin films of iron-cobalt polycyanide, $K_{0.4}Co_{1.3}[Fe(CN)_6].5H_2O$ containing part of cobalt atoms in the form of paramagnetic Co(II) was characterized in [305]. The photoinduced transition to ferrimagnet was achieved by red light induced MMCT transition (Fe^{II}-CN-Co^{III} → Fe^{III}-CN-Co^{II}) increasing the magnetization value. The change persists for several days at 5 K, however, magnetic properties relaxed to the initial state quickly at 150 K. It was shown that irradiation of ferromagnetic samples at 5 K gave rise to reversible electron transfer process and the hysteresis loop disappeared too. In the photoinduced changes, Fe(II) and Fe(III) are in their low-spin states, however, Co(II) is both in low-spin doublet and high-spin quartet states. Photoinduced magnetization and temperature-dependent magnetic phenomena were investigated and the mechanisms of processes involving metal-to-metal charge transfer transitions and spin-state changes for several other Prussian blue analogues were proposed. The results on heterobimetallic (where metal central atoms include along with transition metals also lantanoids) are summarized in a review paper [306]. The obtained results seem to be promising in developing new types of magneto-optical devices, e.g. opto-magnetic memory devices, by selecting specific laser wavelengths to induce required transitions.

As mentioned in Section 7, different effect of light absorption was identified at irradiation of acetonitrile solutions of $[Fe_2(salten)_2(az)](BPh_4)_2$, $[Fe_2(3\text{-}MeO\text{-}salten)_2(pyet)](BPh_4)_2$ and $[Fe(acacen)(stpy)](BPh_4)$ [213] (H_2salten = 4-azaheptamethylene-1,7-bis(salicylideneiminate) where the *trans→cis* isomerization of coordinated ligands *trans*-4,4'-azopyridine(az), *trans*-4-styrylpyridine(pyet), and *trans*-1,2-bis(4-pyridyl)ethylene(stpy) was observed, but the spin-state transition of iron(III) by radiation absorption was not confirmed. It should be pointed out, however, that in their original paper [307], for the iron(III) complex $[Fe(salten)(Mepepy)]BPh_4$ containing one photoisomerizable ligand (Mepepy = 1-(pyridin-4-yl)-2-(*N*-methylpyrrol-2-yl)ethene), both *trans-to-cis* photoisomerization of the ligand and thermally-induced spin crossover in the solid state and in solution were evidenced. The ligand isomerization is sufficient to cause a partial spin change of the iron(III) central atom. This photoinduced spin change is seen for the first time from a high-spin state to a low-spin state.

Ligand-driven light-induced spin change (LD LISC) is a photomagnetic effect, in which a change in ligand field strength due to ligand isomerization leads to the change in the central atom localized spin state [308]. The experiments performed on several iron(II) and iron(III) compounds in solution or polymeric matrices showed that the spin conversion can be triggered at room temperature too.

To the category of practical application, examination and exploitation of photoprocesses with the participation of iron(III) complexes both in seawater and fresh water can be included. In aquatic ecosystems, iron plays a central role in numerous spontaneous and photoinduced chemical and biological processes, such as decomposition of polluants, growth of phytoplankton, etc. The investigation is devoted mainly to the photocatalytic oxidation of organic pollutants and to general understanding of the mechanisms of (photo)redox processes occurring in waters under the influence of light. The former direction was discussed above, in this part the later processes are reviewed.

It has been shown by a number of researchers that more than 99% of iron in the open ocean is complexed with organic ligands [309]. It is generally believed that strong ligands are produced by plankton while the weaker ligands through decomposing the strong ligands, one of the possible mechanism of the decomposition being the photochemical degradation of the ligands complexed to the iron(III) central atom in the surface (upper 1 – 5 m) layer.

Iron is a limiting nutrient for primary production in large areas of the oceans [13,310]. Dissolved iron(III) in the upper oceans occurs almost entirely in the form of complexes with strong organic ligands presumed to be of biological origin. Although the importance of organic ligands to aquatic Fe(II) ↔ Fe(III) cycling is becoming clear, the mechanism by which they are involved in this process remains still open. The questions of enhancement or inhibition of Fe(III) photoreduction by various ligands in open ocean seawater [310-313] were studied in laboratory conditions and the effect of five individual iron(III) binding ligands: phaeophytin, ferrichrome, desferrioxamine B (DFOB), inositol hexaphosphate (phytic acid), and protoporphyrin IX (PPIX) on the Fe(II) photoproduction using seawater of the open Southern Ocean was investigated. As a medium, open Southern Ocean seawater was used. In the absence of the obove mentioned ligands, amorphous Fe(III) hydroxides were formed. These amorphous Fe(III) hydroxides are the major source of iron(III) for the photoproduction of iron(II). The addition of the model ligands changed the Fe(II) photoproduction considerably and in various ways. Phaeophytin, ferrichrome, and phytic acid increased the concentration of photoproduced Fe(II), contrary to DFOB and PPIX which strongly decreased the photoproduced Fe(II) concentration. The results were rationalized on the basis of complex formation and aggregate effects. It was suggested [310] that Fe(III)-binding ligands can enhance the photolytic production of reactive iron species in the euphotic zone and so influence the iron availability in aquatic systems. The research shows that translation to natural conditions is not possible without further characterization of the natural occurring ligands.

Acquisition of iron bound to strong organic complexes, with different iron-binding groups and photochemical reactivities, by plankton communities in iron-limited subantarctic waters were examined in [314]. The obtained results demonstrate that iron is available for uptake and growth by indigenous plankton, but that photolability of these complexes does not determine iron bioavailability. In contrast, light significantly enhances iron acquisition from the *in situ* ligands, suggesting that the *in situ* iron ligands are photolabile, and that photochemistry in surface waters may play a significant role in iron uptake from the dissolved organic iron pool by oceanic plankton.

Mutual relationships of marine inorganic photochemistry, as one of the important branches of marine chemistry, with marine biology, marine ecology, marine geochemistry and marine environment are underlined in [315]. It is emphasized that investigation in the field should cover along with iron complexes also complexes of other photoredox reactive metals (manganese, copper) to approach a more comprehensive knowledge.

Thermodynamic, kinetic and day/night aspects of Fe(II) ↔ Fe(III) light-induced redox cycling were investigated also in fresh waters. As an example [316], results obtained within the study of the cycle in eutrophic, natural water body (Greifensee Lake) and oligotrophic, artificial mountain lake (Malchsee, both in Switzerland) may be presented. It was suggested that superoxide may be a key parameter for the light-induced redox cycling in both lakes. At the surface of both lakes, Fe(II) is produced mainly via photochemical reduction of Fe(III) and its concentration during day is about 6 times higher than that at night.

Fe^{3+} ions can participate in the production of hazardous substances in water, as exemplified in [317] on Fe(III)-induced photochemical production of dissolved gaseous mercury in lake and pond freshwater. It was proposed that the generation of dissolved gaseous mercury is part of a chain of (photo)redox processes, initiated by the photoreduction of Fe(III) to Fe(II) associated with formation of highly reducing organic free radicals through photolysis of Fe(III)-organic coordination compounds and terminated by subsequent reduction of Hg(II) to Hg(0) by the organic radicals. Hg(0) can be dark oxidized by photochemically originated oxidants (e.g. $^{\bullet}OH$ radicals).

Concluding this short review on the applications of Fe(III) complex photochemistry, the photoelectric properties of axially polymerized phthalocyaninato macrocyclic Fe(III) complexes, e.g. $[PcFe(tz)]_n$, $[PcFe(pyz)]_n$, $[Cl_{16}PcFe(pyz)]_n$ $[Me_8PcFe(dib)]_n$ and $[Me_8PcFe(Cl_4dib)]_n$, should be commented upon. The electronic influences of the substituents of the macrocycles and of the bridging ligands result in different photoelectric sensitivities [318].

The utilization of the photochemistry of Fe(III) complexes in analytical chemistry is also of interest. Photochemical titration in systems containing Fe(III) complexes has been used for the quantitative determination of various species in solution [319,320], and the rapid photochemical method for the determination of the Fe(II) and Fe(III) contents in minerals using 1,10-phenanthroline as chelating agent has been applied [321].

Photochemically active Fe^{3+} cations and some Fe(III) complexes have been applied in the field of photographic processes. Fe(III) cyano complexes (Turnbull's and Prussian blue) and oxalato complexes are worth mentioning in connection with unconventional photography [322]. Fe(III) complexes with polydentate ligands, such as 1,2-ethylenediamine and *N,N'*-ethylenediaminetetraacetic acid, have been evaluated as silver halide development accelerators [323,324]. A series of experiments, in which exposed emulsion coatings have been treated with redox buffer solutions composed of citrate complexes and sulfates of Fe(III), have been carried out to investigate the redox stability of the latent-image and latent-subimage centres [324,325]. Furthermore, the role of Fe(III) photochemically active cations, i.e. impurity ions in commercial photographic emulsions, has been discussed within the framework of the elementary process of latent image formation [326]. In addition, a role of iron(III) polycarboxylates with a low light sensitivity in „non-silver“ printing processes might be mentioned [327]. The essence of non-silver processes lies in a substitution of silver (formed by decomposition of silver halide at illumination of a negative) by platinum or palladium which are much more stable. Formation of platinum at illumination can be described by two subsequent redox processes

$$2\,[Fe^{III}(C_2O_4)_3]^{3-} \xrightarrow{h\nu} 2\,[Fe^{II}(C_2O_4)_2]^{2-} + 2\,CO_2 + C_2O_4^{\,2-} \qquad (51)$$

$$[Pt^{II}Cl_4]^{2-} + 2\,[Fe^{II}(C_2O_4)_2]^{2-} + 2\,C_2O_4^{\,2-} \rightarrow Pt^0 + 2\,[Fe^{III}(C_2O_4)_3]^{3-} + 4\,Cl^- \qquad (52)$$

Thus a "print-out" process results, in which the final image is generated during the light exposure and no "development bath" is required, just a bath to remove the excess soluble chemicals.

UV-radiation induced Fe(III) photoreduction to Fe(II) was exploited in electroplating or electrodepositing iron onto a target substrate where the electroplating solution included chelating compounds, e.g. hydroxycarboxylic acids and their lactones. Specifically, sugar acids, such as saccharic or glucaric acid, are especially preferred [328]. Fe(III) in the complexes is easily reduced to Fe(II), i.e. to that form of iron which is suitable for electrodepositing.

9. Conclusion

In each of the previous sections, a specific group of iron(III) compounds is classified by type of complexes and/or tasks solved, and their photochemistry evaluated. This section is aimed at presenting the author's opinion on what is missing, predicting future trends in this field and formulating some conclusions of a more general nature.

There is still a wide gap in possibilities of optimizing the course, selectivity and efficiency of photochemical processes and in applicability of the attained knowledge and experience in practice. Taking into account such mutually overlapped and influenced issues of vital importance as energy conversion, storage and utilization processes, preparation of new compounds and materials possessing designed and desired performance, preservation and improvement of the environment and human life quality, it is clear that there is no thread of unemployment for the photochemists – challenges are too inspiring and necessities too urgent. Considering these challenges and needs, and comparing the state-of-the-art in the field of iron(III) photochemistry with the level of pure and applied photochemistry and photophysics of inorganic and organic compounds in general, some imbalances are obvious at first sight. We have almost no information on exciplex and excimer formation processes involving iron(III) compounds. The primary cause may lie in the simple fact that nobody has tried to find them. Another reason may be the extremely short lifetime of excited iron(III) molecules. Investigation focused on the lifetime, rate constant of internal conversion and intersystem crossing in iron(III) complexes, and the results obtained could significantly improve and deepen the interpretation of photochemical behaviour of this class of compounds. No systematic research has been performed in the field of the exploitation of photochemistry as a tool of synthesis of new iron compounds. In our previous paper [5] we documented the benefits offered by photochemistry in this area and collected data available at that time. As a curiosity, it is worth mentioning that iron(III) iodide, FeI_3, (the existence of which is still hidden for students in many university textbooks on inorganic chemistry) was prepared by photochemical synthesis [329]. Photochemistry enabled preparation of an iron(VI) compound other than FeO_4^{2-} [132] and, on the other hand, several complexes with the iron central atom in zero or negative oxidation state were prepared by means photochemical synthesis [5,330,331].

In the introductory section of this chapter, several reviews dealing with various aspects of iron(III) photochemistry are mentioned. It is the author's hope that the process of publishing fundamental papers on photochemistry and photophysics of transition metal compounds has been triggered, but not terminated, by the two monographs edited by Balzani and Campagna [332], and that there would be a space for a chapter devoted to iron complexes, too.

ACKNOWLEDGMENT

Financial support of this work by the VEGA grant (1/0353/08) is acknowledged.

REFERENCES

[1] Balzani, V., Carassiti, V. (1990). *Photochemistry of Coordination Compounds*. Academic Press: New York, USA, pp 145-192.

[2] Kalyanasundaram, K. (1992). *Photochemistry of Polypyridine and Porphyrin Complexes*. Academic Press: New York,USA.

[3] Šima, J. (1992).*Comments Inorg. Chem*, 13, 277-291.

[4] Šima, J. (1995). *Str. Bond*, 84, 135-193.

[5] Sýkora, J., Šima, J. (1990). *Coord. Chem. Rev*, 107, 1-225.

[6] Horváth, O., Stevenson, K.L. (1993). *Charge Transfer Photochemistry of Coordination Compounds*. Verlag Chemie: New York, USA, pp. 207-266.

[7] Hennig, H., Rehorek, D. (1987). *Photochemische und Photokatalytische Reaktionen von Koordinationsverbindungen*. Acad. Verlag: Berlin,Germany.

[8] Kryukov, A.I., Kutschmij, S.J. (1989). *Fotochimiya Kompleksov Perechodnych Metallov*, Naukova Dumka: Kiev, pp. 128-154.

[9] Hennig, H., Weber, L., Rehorek, D. (1993). *Adv. Chem. Ser*, 238, 351-357.

[10] Salomon, R.G., Ghosh, S., Raychaudhuri, S. (1993). *Adv. Chem. Ser*, 238, 315-333.

[11] Šima, J. (1999). *Roum. Chem. Quart. Rev*, 7, 159 – 171.

[12] Khmelnickij, I.V., Amosov, K.A., Pliusnin, V.F., Grivin, V.P. (1990). *Koord. Khim*, 16, 1373-1380.

[13] Barbeau, K. (2006). *Photochem Photobiol*, 82, 1505-1516.

[14] Šima, J., Makáňová, J. (1997). *Coord. Chem. Rev*, 160, 161-189.

[15] Juban, E.A., Smeigh, A.L., Monat, J.E., McCusker, J.K. (2006). *Coord. Chem. Rev*, 250, 1783-1791.

[16] Sanchiz, J., Esparza, P., Domínguez, S., Brito, F., Mederos, A. (1999). *Inorg. Chim. Acta,* 291, 158-165.

[17] Melník, M., Ondrejkovičová, I., Vančová, V. (1997). *Rev. Inorg. Chem*, 17, 55-286.

[18] Melník, M., Vančová, V., Ondrejkovičová, I., Holloway, C.E. (1998). *Rev. Inorg. Chem*,18, 1-181.

[19] Funahashi, S., Inada, Y. (2002). *Bull. Chem. Soc, Japan,* 75, 1901-1925.

[20] Ohtaki, H., Radnai, T. (1993). *Chem. Rev*, 93, 1157-1204.

[21] Solomon, E.I. (2006). *Inorganic Electronic Structure and Spectroscopy*, Wiley: New York,

[22] Basu, C., Chowdhury, S., Banerjee, R., Evans, H.S., Mukherjee, S. (2007).*Polyhedron,* 26, 3617-3624.

[23] Thomas, F., Serratrice, G., Beguin, C., Aman, E.S., Pierre, J.L., Fontecave, M., Laulhere, J.P. (1999). *J. Biol. Chem,* 274, 13375-13383.

[24] Flint, C.D., Greenough, P. (1977). *J. Chem. Phys*, 56, 5771-5774.

[25] Teixeira, A.C.S.C., Mendes, L., Stollar, G., Guardani, R., Nascimento, C.A.O. (2005). *Braz. Arch. Biol. Technol*, 48, 207-218.

[26] Khmelnickij, I.V., Pliusnin, V.F., Grivin, V.P. (1989). *Russ. J. Phys. Chem,*63, 1494-1497.

[27] Rehorek, D., Grikos G., Hennig, H. *Z.* (1978). *Chem*, 27, 372-373.

[28] Šima, J. (1992). *Pol. J. Chem*, 66, 936-938.

[29] Šima, J. (1994). *Acta Chim. Hung*,131, 273-279.

[30] Maruthamuthu, P., Huie, R.E. (1995). *Chemosphere*, 30, 2199-2207.

[31] Vincze, L., Papp, S.(1987). *J. Photochem*, 36, 279-298.

[32] Bideau, M., Claudel, B., Faure, L., Rachimoellah, M. (1987). *J. Photochem*, 39, 107-128.

[33] Bideau, M., Faure, L., Assemian Y., Claudel, B.(1988). *J. Mol. Catal*, 43, 267-279.

[34] Tennakone, K., Ketipearachchi, U.S., Ileperuma, O.A., Punchihewa, S. (1991). *J. Mol. Catal*, 64, 155-162.

[35] De Oliveira, M.G., Langley L., Rest, A.J. (1995). *J. Chem. Soc, Dalton Trans*, 2013-2019.

[36] Szacilowski, K., Macyk, W., Drzewiecka-Matuszek, A., Brindell, M., Stochel, G. (2005). *Chem. Rev*, 105, 2647-2694.

[37] Johansson, L.Y., Blomquist, J. (1975). *Chem. Phys. Lett*, 34, 115-118.

[38] Szacilowski, K., Strasicka, Z. (2002). *Coord. Chem. Rev*, 229, 17-26.

[39] Szacilowski, K., Macyk, W., Stochel, G., Strasicka, Z., Sostero, S., Traverso, O. (2000). *Coord. Chem. Rev*, 208, 277-297.

[40] Fiedler, J. (1993). *Coll. Czech. Chem. Commun*. 58, 461-473.

[41] Bitterwolf, T.E. (2006). Coord. Chem. Rev. 250, 1196-1207.

[42] Zliechovcová, H., Šima, J., Valigura D., Sýkora, J. (1982). *Chem. Papers* 36, 59-63.

[43] Šima, J. (2008). In *Coordination Chemistry Research Progress*, Cartere, T.W., Verley, S.K. Eds. Nova Science Publishers: New York,USA.

[44] Gažo, J., Boča, R., Jóna, E., Kabešová, M., Macášková L., Šima, J. (1983). *Pure Appl. Chem*. 55, 65-76.

[45] Vincze, L., Papp, S. (1986). *J. Photochem*. 34, 175-181.

[46] Benkelberg, H.J., Wameck, P. (1995). *J. Phys. Chem*. 99, 5214-5221.

[47] Billing, R. (1997). *Coord. Chem. Rev.* 159, 257-270.

[48] Endicott, J.F., Song, X., Watzky, M.A., Buranda, T. (1994). *J. Photochem. Photobiol. A:Chem*. 82, 181-190.

[49] Vogler, A., Kunkely, H.(1990). *Top. Curr. Chem*, 158, l-30.

[50] Kunkely, H., Vogler, A. (1995). *Inorg. Chem*, 34, 2468-2469.

[51] Blackboum, R.L., Dong, Y., Lyon, L.A., Hupp, J.T. (1994).*Inorg. Chem*, 33, 4446-4452.

[52] Amett, D.C., Vähringer, P., Scherer, N.F. (1995). *J. Am. Chem. Soc*, 117, 12262-12272.

[53] Tsukahara, K., Asami, S. (1991). *Chem. Lett*, 1337-1338.

[54] Takashima, H., Tara, C., Namikawa, S., Kato, T., Araki, Y., Ito, O., Tsukahara, K. J. (2006). *Phys.Chem,* B, 110, 26413-26423.

[55] Joselevich, E., Willner, I. (1995). *J. Phys. Chem*, 99, 6903-6912.

[56] Gerasimov, O.V., Limar, S.V., Pannon, V.N. (1991). *J. Photochem. Photobiol. A:Chem*, 56, 275-285.

[57] Šima, J. (1997). *Coord. Chem. Rev*, 159, 195-204.
[58] Šima, J., Havašová, K., Ducárová, T. *Z.* (1990). *Chem*, 30, 446-447.
[59] Šima, J.,Chochulová, B.,Veverka,M., Makáňová, J.,Hajšelová, M.,Bradiaková, A. (1993). *Pol. J. Chem*, 67, 1369-1374.
[60] Kotočová, A., Šima, J.(1994). *Chem, Papers*, 48, 175-176.
[61] Šima, J.(2007). *J. Chil. Chem. Soc*, 52, 1123-1125.
[62] Šima, J., Ducárová, T., Šramko, T. (1992). *Pol. J. Chem*, 66, 53-58.
[63] Šima, J., Makáňová J., Štibrányi, L. (1994). J. *Photochem. Photobiol. A:Chem*, 84, 173-176.
[64] Šima, J., Makáňová, J., Veverka, M. (1995). *Monatsh. Chem*, 126, 149-154.
[65] Gažo, J., Boča, R., Jóna, E., Kabešová, M., Macášková, Ľ., Šima, J., Pelikán, P., Valach, F. (1982). *Coord. Chem. Rev*, 43, 87-131.
[66] Šima, J., Izakovič, M., Šípoš, R., Švorec, J. (2007). *Acta Chim. Slo*, 54, 268-272.
[67] Šramko, T., Šima J., Fodran, P. (1992). *Acta Chim. Hung*,129, 215-223.
[68] Šima, J., Ducárová, T., Šramko, T., Kotočová, A. (1991). *Bull. Soc.Chim. Belg*,100, 193-194.
[69] Šima, J., Mašlejová-Vojtašová, M., Ducárová, T., Kotočová, A. *Z.* (1990). *Anorg. Allg. Chem*, 589, 207-213.
[70] Izakovič, M., Šima, J., Brezová, V. (2004). *J. Photochem Photobiol. A:Chem*, 167, 81-86.
[71] Šima, J., Izakovič, M. (2005).*Croat. Chem. Acta*, 78, 29-35.
[72] Šima, J., Izakovič, M., Žitňanský, M. (2006). *Inter. J. Photoenergy*, 8, 1-5.
[73] Šima, J. (2006). *Coord. Chem. Rev*, 250, 2325-2334.
[74] Šima, J., Lauková, D., Brezová, V. (2001).*Coll. Czech. Chem. Commun*, 66, 109-118.
[75] Šima, J., Brezová, V. (2001). *Monatsh. Chem*, 132, 1492-1500.
[76] Šima, J., Brezová, V. (2002). *Coord. Chem. Rev*, 229, 27-35.
[77] Šima, J., Horváthová, L. (2003). *Pol. J. Chem*, 77, 1857-1866.
[78] Šima, J., Horváthová, L., Izakovič, M. (2004). *Monatsh. Chem*, 135, 5-12.
[79] Šima, J., Mrázová, J., Kotočová, A. (1999). *J. Chin. Chem. Soc*, 46, 993-997.
[80] Schwarz, H.A., Dodson, R.W. (1989). *J. Phys. Chem*, 93, 409-414.
[81] Bartocci, C., Maldotti, A., Carassiti, V., Traverso, O., Ferri, A. (1985). *Inorg. Chim. Acta*,107, 5-12.
[82] Reed, J.L., Gafney, H.D., Basolo, F. (1974). *J. Am. Chem. Soc*, 96, 1363-1369.
[83] Wagner, W.D., Nakamoto, K. (1988). *J. Am. Chem.Soc*, 110, 4044-4045.
[84] Eikey, R.A., Abu-Omar, M.M. (2003). *Coord. Chem. Rev*, 243, 83-124.
[85] Castellanos, M.M., Reyman, D., Sieiro, C., Calle, P. (2001), *Ultrasonics Sonochem*, 8, 17-22.
[86] Chen, F., Xie, Y., He J., Zhao, J. (2001). *J. Photochem. Photobiol. A:Chem*, 138, 139-146.
[87] Šima, J. (2003). In: *Progress in Coordination and Bioinorganic Chemistry*, Melník, M., Sirota, A., Eds. Slovak Technical University Press: Bratislava, pp. 291-296.
[88] Jakúbek, V., Lees, A. (2000). J. *Inorg. Chem*, 39, 5779-5784.
[89] Plyusnin, V.F., Grivin, V.P., Larionov, S.V. (1997). *Coord. Chem. Rev*, 159, 121-133.
[90] Kunkely, H., Vogler, A. (2002). *Inorg. Chem. Commun*. 5, 730-732.
[91] Knoch, R., Elias H., Paulus, H. (1995). *Inorg. Chem*, 34, 4032-4040.

[92] Wilkins, R.G. (1991). *Kinetics and Mechanism of Reactions of Transition Metal Complexes*, 2nd ed. VCH: Weinheim, p. 199.
[93] Abrahamson, H.B., Rezvani, A.B., Brushmiller, J.G. (1994). *Inorg. Chim. Acta,* 226, 117-127.
[94] Scandola, F. (1980). *Rearrangements in Ground and Excited States*, Vol. 3, Academic Press: New York, USA, p. 549.
[95] Rehorek, D., Grikos, H., Billing, R. *Z.* (1990). *Chem*, 30, 378-379.
[96] Avnir, D., Kagan, M.L., Ross, W. (1987). *Chem. Phys. Lett*, 135, 177-181.
[97] Doetschman, D.C., Dwyer, D.W., Trojan, K.L. (1989). *Chem. Phys*, 129, 285-294.
[98] Duka, G.G., Skurlatov, J.I., Tchub, L.S., Sytchev, A.J. (1987). *Koord. Khim*, 13, 1074-1077.
[99] Duka, G.G., Batyr, D.G., Romantchuk, L.S., Sytchev, A.J. (1990). *Koord. Khim*, 16, 93-106.
[100] Loginov, A.V., Katenin, S.B., Vojakin, I.V., Shagisultanova, G.A. (1986). *Koord. Khim*, 12, 1621-1626.
[101] Bolte, M., Adrianirinaharivelo, S., Mailhot, G. (1995). *Sol. Energy Mater. Sol. Cells*, 38, 459-474.
[102] Adrianirinaharivelo, S., Pilichowski, J.F., Bolte, M. (1993).*Transition Met. Chem*,18, 37-41.
[103] Mailhot, G., Bordes, A.L., and Bolte, M. (1995). Chemosphere, 30, 1729-1737.
[104] Makáňová, J., Šima, J., Daučík, P., Horváth, A. (1997). *Chem. Papers*, 51, 258-262.
[105] Faust, B., Zepp, R. (1993). *Environ. Sci. Technol*, 27, 2517-2522.
[106] Kunkely, H., Vogler, A. (1994). *J. Chem. Soc. Chem. Commun*, 2671-2672.
[107] Osman, A.H. (1990). *Inorg. Chim. Acta*,176, 283-285.
[108] Papula, L., Horváth O., Papp, S. (1990). *J. Photochem. Photobiol. A:Chem*, 54, 205-212.
[109] Lang, K., Luňák, S., Sedlák, K., Kubát P., Brodilová, J. (1995). *Z. Phys. Chem*, 190, 203-210.
[110] Konig, E. (1991). *Str. Bond*,76, 51-152.
[111] Yoshimura, T. (1988). *Arch. Biochem. Biophys*, 264, 450-461.
[112] Harriman, A. (1981). *J. Chem. Soc. Faraday Trans, I.* 77, 369-377.
[113] Becker, R.S., Allison, J.A. (1963). *J. Phys. Chem*, 67, 2662-2669.
[114] Hendrickson, D.N., Konnaird, M.G., Suslik, K.S. (1987). *J. Am. Chem. Soc*, 109, 1243-1244.
[115] Imamura, T., Jin, T., Suzuki, T., Fujimoto, M. (1985).*Chem. Lett*, 847-848.
[116] Rehorek, D., Berthold, T., Hennig, H., Kemp, T.J. (1988). *Z. Chem*, 28, 72-73.
[117] Suslick, K.S., Watson, R.A. (1991). *Inorg. Chem*, 30, 912-919.
[118] Guest, C.R., Straub, K.D., Rentzepis, P.M. (1989). *Res. Chem. Intermed*, 12, 203-216.
[119] Ogura, T., Fidler, V., Ozaki, Y., Kitagawa, T. (1990). *Chem. Phys. Lett*, 169, 457-459.
[120] Fidler, V., Ogura, T., Dato, S.I., Aoyagi, K., Kitagawa, T. (1991). *Bull. Chem. Soc. Jpn*, 64, 2315-2322.
[121] Ozaki, Y., Iriyama, K., Ogoshi, K., Kitagawa, T. (1987). *J. Am. Chem. Soc*, 109, 5583-5586.
[122] Wagner, W.D., Nakamoto, K. (1989). *J. Am. Chem. Soc*, 111, 1590-1598.
[123] Buchler, J.W., Dreher, C. *Z.* (1984). *Naturforsch, B.* 39, 222-230.

[124] Bartocci, C., Maldotti, A., Varani, G., Battistoni, P., Carassiti, V., Mansuy, D. (1991). *Inorg. Chem*, 30, 1255-1259.
[125] Maldotti, A., Bartoci, C., Amadelli, R., Carassiti, V. (1989). *J. Chem. Soc. Dalton Trans*, 1197-1201.
[126] Polo, E., Amadelli, R., Carassiti, V., Maldotti, A. (1992). *Inorg. Chim. Acta,* 192, 1-3.
[127] Bartocci, C., Scandola, F., Ferri, A., Carassiti, V. (1980). *J. Am. Chem. Soc*, 102, 7067-7072.
[128] Maldotti, A., Bartocci, C., Chiorboli, C., Ferri, A., Carassiti, V. (1985). *J. Chem. Soc. Chem. Commun*, 881-882.
[129] Maldotti, A., Bartocci, C., Amadelli, R., Carassiti, V. (1983). *Inorg. Chim. Acta*,74, 275-278.
[130] Bartocci, C., Maldotti, A., Traverso, O., Bignozzi, C.A., Carassiti, V. (1983). *Polyhedron,* 2, 97-102.
[131] Maldotti, A., Amadelli, R., Bartocci, C., Carassiti, V., Polo. E., Varani, G. (1993). *Coord. Chem. Rev*, 125, 143-154.
[132] Berry, J.F., Bill, E., Bothe, E., DeBeer George, S., Mienert, B., Neese, F., Weighardt, K. (2006). *Science,* 312, 1937-1941.
[133] Tohara, A., Sato, M. (1989). *Chem. Lett*, 153-154.
[134] Reichgott, D.W., Rose, N.J. (1977). *J. Am. Chem. Soc*, 99, 1813-1818.
[135] Ferraudi, G., Carrasco, C. (1980). *Inorg. Chem*, 19, 3466-3469.
[136] Ferraudi, G. (1980). *Inorg. Chem*, 19, 438-444.
[137] Richman, R.M., Peterson, M.W. (1982). *J. Am. Chem. Soc*, 104, 5795-5796.
[138] Peterson, M.W., Rivers, D.S., Richman, R.M. (1985). *J. Am. Chem. Soc*, 107, 2907-2915.
[139] Berthold, T., Rehorek, D., Hennig, H. (1986). *Z. Chem*, 26, 183-184.
[140] Guest, C.R., Straub, K.D., Hutchinson, J.A., Rentzepis, P.M. (1988). *J. Am. Chem. Soc*,110, 5276-5280.
[141] Weber, L., Haufe, G., Rehorek, D., Hennig, H. (1991). *J. Chem. Soc. Chem. Commun*, 502-503.
[142] Peterson, M.W., Richman, R.M. (1985). *Inorg. Chem*, 24, 722-725.
[143] Hatano, K., Ishida, Y. (1982). *Bull. Chem. Soc. Jpn*, 55, 3333-3334.
[144] Lever, A.B.P., Licoccia, S., Ramaswamy, B.S. (1982). *Inorg. Chim. Acta,* 64, L87-L90.
[145] Vogler, A., Kunkely, H. (2006). *Coord. Chem. Rev*, 250, 1622-1626.
[146] Kunkely, H., Vogler, A. (2005). *Inorg. Chim. Acta,* 358, 4086-4088.
[147] Yanagimoto, T., Nakagawa, A., Komatsu, T., Tsuchida, E. (2001). *Bull. Chem. Soc. Jpn,*74, 2123-2128.
[148] Komatsu, T., Yanagimoto, T., Nagakawa, A., Tsuchida, E. (2000).*Chem. Lett*. 29, 84-85.
[149] Sato, S., Kamogawa, K., Aoyagi, K., Kitagawa, T. (1992). *J. Phys. Chem*. 96, 10676-10681.
[150] Sato, S., Kamogawa, K., Aoyagi, K., Kitagawa, T. (1992).In *Time Resolved Vibrational Spectroscopy V.*, Takahashi, H.; Ed. Springer: Berlin, p. 109.
[151] Gilbert, B.C., Lindsay Smith, J.R., MacFaul, P., Taylor, P. (1993). *J. Chem. Soc. Perkin Trans, 2*, 2033-2037.
[152] Nakamoto, K. (2002). *Coord. Chem. Rev,* 226, 153-165.

[153] Meyer, K., Bill, E., Mienert, B., Weyhermüller, T., Wieghardt, J. (1999). *J. Am. Chem. Soc*, 121, 4859-4876.

[154] Grapperhaus, C.A., Mienert, B., Bill, E., Weyhermüller, T., Wieghardt, K. (2000). *Inorg. Chem*, 39, 5306-5317.

[155] Hu, X., Meyer, K. (2003). *Inorg Chim Acta* 337, 53-58.

[156] Horváth, O., Huszánk, R., Valicsek, Z., Lendvay, G. (2006). *Coord. Chem. Rev*, 250, 1792-1803.

[157] Huszánk, R., Horváth, O. (2005).*Chem. Commun*, 224-226

[158]Huszánk, R., Lendvay, G., Horváth, O. (2007). *J. Biol. Inorg. Chem,*12, 681-690.

[159] Jüstel, T., Müller, M., Weyhermüller, T., Kressl, C., Bill, E., Hildebrandt, P., Lengen, M., Grodzicki, M., Trautwein, A.X., Nuber, B., Weighardt, K. (1999). *Chem. Eur. J*, 5, 793-810.

[160] Ercolani, C., Jubb, J., Pennesi, G., Russo, U., Trigiante, G. (1995). *Inorg. Chem*, 34, 2535-2541.

[161] Donzello, M.P., Ercolani, C., Kadish, K.M., Ou, Z., Russo, U. (1998). *Inorg. Chem*, 37, 3682-3688.

[162] Suslick, K.S., Watson, R.A. (1992). *New J. Chem*, 16, 633-642.

[163] Weber, L., Hommel, R., Behling, J., Haufe, G., Hennig, H. (1994). *J. Am. Chem. Soc*, 116, 2400-2408.

[164] Machii, K., Watanabe, Y., Morishima, I. (1995). *J. Am. Chem. Soc*, 117, 6691-6697.

[165] Shen, S., Lin, S., Shou, H., Vu, Q., Xu, H., Chen, D. (1989). *Chin. Sci. Bull*, 34, 259-264.

[166] Zhou, S., Rodgers, M.A.J.(1991). *J. Am. Chem. Soc*, 113, 7728–7734.

[167] Fujita, I., Netzel, T.L., Chang, C.K., Wang, C.B. (1982). *Proc. Natl. Acad. Sci*, 79, 413-417.

[168] Takagi, S., Eguchi, M., Tryk, D.A., Inoue, H. (2006). *J. Photochem. Photobiol. C:Rev*, 7, 104-126.

[169] Hibino, T., Jones, W. (2001). *J. Mater. Chem,* 11, 1321-1323.

[170] Maldotti, A., Molinari, A., Andreotti, L., Fogagnolo, M., Amadelli, R. (1998). *Chem. Commun*, 507-509.

[171] Bartocci, C., Maldotti, A., Carassiti, V., and Traverso, O. (1985). *Inorg. Chim. Acta*, 107 5-12.

[172] Maldotti, A., Bartocci, C., Locatelli, C., Carassiti, V., Ferri, A., and Bortolotti, F. (1986). *Inorg. Chim. Acta*, 125 129-133.

[173] Ferri, A., Patti, D., Chiozzi, P., Cattozzo, M., Bartocci, C., Maldotti, A. (1988). *J. Photochem. Photobiol. B:Biol*, 2, 341-353.

[174] Tiede, D.M., Vashishta, A.C. (1991). J. *Mol. Cryst. Liq. Cryst*, 194, 191-200.

[175] Šima, J., Ducárová, T., Havašová, K., Antalík, M. (1990). *Inorg. Chim. Acta,* 176, 15-18.

[176] Šima, J., Bradiaková, A., Antalík, M. (1996). *J. Photochem. Photobiol. A:Chem*, 94, 163-166.

[177] Horváth, O., Papula, L., Kraut, B., Papp, S. (1989). *J. Photochem. Photobiol. A:Chem*, 47, 91-95.

[178] Figueiredo, K.M., Marcon, R.O., Campos, I.B., Nante, I.L., Brochsztain, S. (2005). *J. Photochem. Photobiol. B:Biol***,** 79, 1-9.

[179] Durham, B., Pan, L.P., Hahn, S., Long, J., Millett, F. (1990). *Adv. Chem. Ser*, 226, 181-193.

[180] Osvath, P., Salmon, G.A., Sykes, G. (1988). *J. Am. Chem. Soc*, 110, 7114-7118.

[181] Sun, J., Wishart, J.F., Gardineer, M.B., Cho, M.P., Isied, S.S. (1995). *Inorg. Chem*, 34, 3301-3309.

[182] Casimiro, D.R., Richards, J.H., Winkler, J.R., Gray, H.B. (1993). *J. Phys. Chem*, 97, 13073-13077.

[183] Bellelli, A., Brunori, M., Brzezinski, P., Wilson, M.T. (2001). *Methods* 24, 139-152.

[184] Liu, L., Hong, J., Ogawa, M.Y. (2004). *J. Am. Chem. Soc*,126, 50-51.

[185] Liang, Z.X., Nocek, J.M., Huang, K., Hayes, R.T., Kurnikov, I.V., Beratan, D.N., Hoffman, B.M. (2002). *J. Am. Chem. Soc*, 124, 6849-6859.

[186] Furukawa, Y., Matsuda, F., Ishimori, K., Morishima, I.(2002). *J. Am. Chem. Soc*, 124, 4008-4019.

[187] Liang, Z.X., Nocek, J.M., Kurnikov, I.V., Beratan, D.N., Hoffman, B.M. (2000). *J. Am. Chem. Soc*, 122, 3552-3553.

[188] Naito, N.R., Huang, H., Sturgess, A.W., Nocek, J.M., Hoffman, B.M. (1998). *J. Am. Chem. Soc*, 120, 11256-11262.

[189] Hayashi, T., Hitomi, Y., Ogoshi, H. (1998). *J. Am. Chem. Soc*, 120, 4910-4915.

[190] Nocek, J.M., Sishta, B.P., Cameron, J.C., Mauk, A.G., Hoffman, B.M. (1997). *J. Am. Chem. Soc*, 119, 2146-2155.

[191] Furukawa, Y., Ishimori, K., Morishima, I. (2002). *Biochem*, 41, 9824-9832.

[192] Tremain, S.M., Kostic, N.M. (2002). *Inorg. Chem*, 41, 3291-3301.

[193] Cho, K.C., Che, C.M., Ng, K.M., Choy, C.L.(1986). *J. Am. Chem. Soc*, 108, 2814-2818.

[194] Qin, L., Rodgers, K.K., Sligar, S.G. (1991). *Mol. Cryst. Liq. Cryst*, 194, 311-316.

[195] Hake, R., Zhang Q., Harohn, G.J., McLendon, G., Corin, A. (1991). *Mol. Cryst. Liq. Cryst*, 194, 351-356.

[196] Zhang, Q., Hake, R., Billstone, V., Simmons, J., Falvo, J., Holzscher, D., Lu, K., McLendon, G., Corin, A. (1991). *Mol. Cryst. Liq. Cryst*, 194, 343-349.

[197] Hoffman, B.M., Natan, M.J., Nocek, J.M., Wallin, S.A. (1991). *Str. Bond*, 75, 85-108.

[198] Matsugo, S., Kayamori-Sato, N., Konishi, T. (1994). *Photochem. Photobiol*, 60, 415-420.

[199] Nakano, R., Konami, H., Sato, H., Ito, O., Shimizu, T. (1995). *Biochim. Biophys. Acta*, 1252, 245-250.

[200] Faraggi, M., Klapper, M.H. (1988). *J. Am. Chem. Soc*, 110, 5753-5756.

[201] Aubailly, M., Santus, R.,Salmon, S. (1991). *Photochem. Photobiol*, 54, 769-773.

[202] Lauhére, J.P., Laboué, A.M., Briat, J.F. (1990). *Biochem. J*, 269, 79-84.

[203] Liu, Y., Nohl, H. (1995). *Photochem. Photobiol*, 62, 433-438.

[204] Sakurai, T., Fujimori, K., Ueda, T., Shindo, H., Shibusawa, Y., Nakano, M. (2001). *Photochem. Photobiol*, 74, 407-411.

[205] M. Kitazawa, K. Iwasaki, K. Sakamoto, J. (2006). Cosmet. Dermatol. 5, 210-217.

[206] Giordani, A., Martin, M.-E., Beaumont, C., Santus, R., Morliere, A. (2000). *Photochem. Photobiol*,72, 746-752.

[207] Pourzand, Ch., Watkin, R.D., Brown, J.E., Tyrrel, R.M. (1999). *Proc. Nat. Acad. Sci*, 96, 6751-6756.

[208] Brenneisen, P., Wenk, J., Klotz, O., Wlaschek, M., Briviba, K., Krieg, T., Sies, H., Scharffetter-Kochanek, K. (1998). *J. Biol. Chem*, 273, 5279-5287.

[209] Seité, S., Popovic, E., Verdier, M. P., Roguet, R., Portes, P., Cohen, C., Fourtanier, A., Galey, J. B. (2004). *Photodermatol. Photoimmunol. Photomed*, 20, 47-52.

[210] Yiakouvaki, A., Savovič, J., Al-Qenaei, A., Dowden, J., Pourzand, C. (2006). *J. Investig. Dermatol*, 126, 2287-2295.

[211] Akai, K., Tsuchiya, K., Tokumura, A., Kogure, K., Ueno, S., Shibata, A., Tamaki, T., Fukuzawa, K. (2004). *Free Radical Res*, 38, 951-962.

[212] Tsuchiya, K., Akai, K., Tokumura, A., Abe, S., Tamaki, T., Takiguchi, Y., Fukuzawa, K. (2005). *Biochim. Biophys. Acta*, 1725, 111-119.

[213] Yonezo, M., Shigeyuki, H., Shinya, H. (1998). *Haii Kagobutsu No Kokagaku, Toronkai Koen Yoshishu*,11, 14-15.

[214] Akitsu, T., Einaga, Y. (2006). *Inorg. Chem. Commun*, 9, 1108-1110.

[215] Carassiti, V. (1993). *Coord. Chem. Rev*, 125, 351-360.

[216] Bickley, R.I., Gonzales-Carreno, T., Gonzales-Elipé, A., Manuera, G., Palmisano, L. (1994). *J. Chem.Soc. Faraday Trans*, 90, 2257-2264.

[217] Nezeeruddin, M.K., Grätzel, M. (2007). *Str. Bond*, 123, 113-175.

[218] Wilkins, F.W., Blake, D.M. (1994). *Chem. Eng. Progr*, 90, 41-49.

[219] Ito, I., Kunimoto, K., Miyachi, S., Kato, T. (1991). *Tetrahedron Lett*, 32, 4007-4010.

[220] Ito, Y. (1991). *J. Chem. Soc. Chem. Commun*, 622-624.

[221] Tokamoto, K., Sasaki, K., Oka, S. (1988). *J. Am. Chem. Soc*, 110, 1187-1196.

[222] Hennig, H., Rehorek, D., Stich, R., Weber, L. (1990). *Pure Appl. Chem*, 62, 1489-1494.

[223] Kolle, U. (1993).In *Photosensitization and Photocatalysis Using Inorganic and Organometallic Compounds*, Kalyanasundaram, K., Grätzel, M. Eds. Kluwer: Amsterdam, p. 331.

[224] Luňák, S., Lederer, P., Stopka, D., Veprek-Šiška, J. (1981).*Coll.Czech.Chem. Commun*. 46, 2455-2465.

[225] Linnik, O., Kisch, H. (2006).*Photochem. Photobiol. Sci*, 5, 938-942.

[226] Zou, Z., Arakawa, H. (2003). *J. Photochem. Photobiol. A:Chem*, 158, 145-162.

[227] Feng, W., Nasheng, D. (2000). *Chemosphere* 41, 1137-1147.

[228] Burrows, H.D., Canle L, M., Santaballa, J.A., Steenken, S. (2002). *J.Photochem. Photobiol. B:Biol*. 67, 71-108.

[229]T eixeira, A.C.S.C., Guardani, R., Nascimento, C.A.O. (2003). *Ind. Eng. Chem. Res*. 42, 5751-5761.

[230] Machiraju, S., Noorjahan, D.K.M. (2006).US Patent 20060138057, issued June 29,

[231] Reddy, E.P., Sun, B., Smirniotis, P.G. *J. Phys. Chem. B*, (2004). 108, 17198-17205.

[232] Fernandez, J., Nadtochenko, V., Enea, O., Bozzi, A., Yuranova, T., Kiwi, J. (2003). *Int. J. Photoenergy* 5, 107-114.

[233] Pignatello, J.J., Sun, Y. (1993). *ACS Symp. Ser*. 518, 77-105.

[234] Parlar, H. (1990). In *Environmental Fate of Pesticides*, Hutson, D.H.; Roberts, T.R.; Eds. Wiley: New York,USA, p. 245.

[235] Mazellier, P., Bolte, M. (2001). *Chemosphere*, 42, 361-366.

[236] Catastini, C., Sarakha, M., Mailhot, G., Bolte, M. (2002). *Sci. Total Environ*, 298, 219-228.

[237] Feng, H.E., Le-cheng, L.E.I. (2004). *J. Zhejiang Univ. Sci*, 5, 198-205.

[238] Andreozzi, R., Marotta, R. (2004). *Water Res*, 38, 1225-1236.

[239] Liou, M.J., Lu, M.C., Chen, J.N. (2004). *Chemosphere,* 57, 1107-1114.

[240] Chu, W., Chan, K.H. (2007). *Ind. Eng. Chem. Res,* 46, 1505-1510.

[241] Galichet, F., Mailhot, G., Bonnemoy, F., Bohatier, J., Bolte, M. (2006).*Pest Manag. Sci*, 58, 707-712.

[242] Keisuke, M., Hideyuki, K., Satoshi, K., Tooru, S., Kiyohisa, O. (2004). *Nippon Kagakki Koen Yokoshu,* 84, 311-312.

[243] Al-Momani, F., Gonzalez, O., Sans, C., Esplugas, S. (2004). *Water Sci. Technol*, 49, 293-298.

[244] Kaichouh, G., Oturan, N., Oturan, M.A., El Kacemi, K., El Hourch, A. (2004). *Environ. Chem. Lett,*2, 31-33.

[245] Mailhot, G., Hykrdová, L., Jirkovský, J., Lemr, K., Grabner, G., Bolte, M. (2004). *Appl. Catal. B: Environ*, 50, 25-35.

[246] Poulain, L., Mailhot, G., Chung, P.W.W., Bolte, M. (2003). *J. Photochem. Photobiol. A:Chem*, 159, 81-88.

[247] Catastili, C., Rafqah, S., Mailhot, G., Sarakha, M. (2004). *J. Photochem. Photobiol. A:Chem*, 162, 97-103.

[248] Chung, P.W.W., Mailhot, G., Bolte, M. (2003). *Int. J. Photoenergy,* 5, 37-42.

[249] Deng, Y., Zhang, K., Chen, H., Wu, T., Krzyaniak, M., Wellons, A., Bolla, D., Douglas, K., Zuo, Y. (2006)/ *Atm. Environ*, 40, 3665-3676.

[250] Hykrdová, L., Jirkovský, J., Grabner, G., Mailhot, G., Bolte, M. (2003). *Photochem. Photobiol. Sci*, 2, 163-170.

[251] Bartocci, C., Maldotti, A., Varani, G., Carassiti, V., Battioni, P., Mansuy, D. (1989). *J. Chem. Soc. Chem. Commun,* 964-965.

[252] Héquet, V., Cloirec, P.L., Gonzáles, C., Meunier, B. (2000). *Chemosphere,* 41, 379-386.

[253] Silva, E., Pereira, M.M., Burrows, H.D., Azenha, M.E., Sarakha, M., Bolte, M. (2004). *Photochem. Photobiol. Sci*, 3, 200-204.

[254] Wasuke, M., Nozaki, K.C. (2006). *Sci. J. Kanawaga Univ*, 17, 89-90.

[255] Park, H., Choi, W. (2003). *J. Photochem. Photobiol. A:Chem*,159, 241-247.

[256] Bautitz, I.R., Pupo Nogueira, R.F. (2007). *J. Photochem. Photobiol. A:Chem*,187, 33-39.

[257] Mrowetz, M., Selli, E. (2004). *J. Photochem. Photobiol. A:Chem*,162, 89-95.

[258] Chen, F., He, J., Zhao, J., Yu, J.C. (2002). *New J. Chem,* 26, 336-341.

[259] Rodriguez, M., Sarria, V., Esplugas, S., Pulgarin, C. (2002). *J. Photochem. Photobiol. A:Chem,* 151, 129-135.

[260] Cho, M., Lee, Y., Chung, H., Yoon, J. (2004). *Appl. Environ. Microbiol*,70, 1129-1134.

[261] Tokumura, M., Ohta, A., Znad, H.T., Kawase, Y. (2006). *Water Res,* 450, 3775-3784.

[262]Ormad, M.P., Mosteo, R., Ibarz, C., Ovelleiro, J.L. (2006). *Appl. Catal. B:Environ*, 66, 58-63.

[263] Emilio, C.A., Jardin, W.F., Litter, M.I., Mansilla, H.D. (2002). *J. Photochem. Photobiol. A:Chem*, 151, 121-127.

[264] Pulgarin, C., Kiwi, J. (1995). *Langmuir, II.* 519-526.

[265] Ansari, A., Peral, J., Domenech, X., Rodrigues-Clemente, R., Roig, A., Molins, E. (1995). *J. Photochem.Photobiol. A:Chem*, 87, 121-125.

[266] Chatterjee, S., Sakar, S., Bhattacharyya, S.N. (1994). *J.Photochem. Photobiol. A:Chem*, 81, 199-203.

[267] Song, W., Chen, M., Ma, J., Ma, W., Chen, Ch., Zhao, J. (2006). *Environ. Sci. Technol*, 40, 4782-4787.

[268] Alvaro, M., Carbonell, E., Esplá, M., Garcia, H. (2005). *Appl. Catal. B: Environ*,57, 37-42.

[269] Choi, W., Termin, A., Hoffmann, M.R. (1994). *Angew. Chem. Int. Ed. Engl*, 10, 1091-1092.

[270] Mills, A., Davies, R.H., Worsley, D. (1993). *Chem. Rev*, 93, 417-425.

[271] Serpone, N., Maruthamuthu, P., Pichat, P., Pelizzetti, E., Hidaka, H. (1995). *J. Photochem. Photobiol. A:Chem*, 85, 247-255.

[272] Bideau, M.,Claudel, B., Faure, L., Kazouan, H. (1994). *J.Photochem. Photobiol. A:Chem*, 84, 57-67.

[273] Litter, M., Navio, J.A. (1994). *Photochem. Photobiol. A:Chem*, 84, 183-193.

[274] Ibusuki, T., Takeuchi, K. (1994). *J. Mol. Catal*, 88, 93-102.

[275] Klauson, D., Preis, S. (2007). *Int. J. Photoenergy,* Article ID 89359, 7 pp.

[276] Kumbhar, A., Chumanov, G. (2005). *J. Nanopart. Res*, 7, 489-498.

[277] Bowser, R., Davidson, R.S. (1994). *J. Photochem. Photobiol. A:Chem*, 77, 269-276.

[278] Roloff, A., Meier, K., Riediker, M. (1986). *Pure Appl. Chem*, 58, 1267-1272.

[279] Meier, M., Buhler, N., Zweifel, H., Berner G., Lohse, F. (1984). European Patent. 0094915.

[280] Klingert, B., Riediker, M., Roloff, A. (1988). *Comments Inorg. Chem*, 7, 109-138.

[281] Goswami, A., Baruah, S.D., Dass, N.N. (1995). *Polym. Bull,* 35, 561-566.

[282] Neumann, M.G., Schmitt, C.C., Rigoli, I.C.(2003).*J.Photochem.Photobiol. A:Chem*,159, 145-150.

[283] Goswami, A. Baruah, S.D.(1997). *Polym. Inter, 43, 22-26.*

[284] Stroyuk, A.L., Granchak, V.M., Korzhakand, A.V., Kuchmii, S.Y. (2004).*J. Photochem. Photobiol. A:Chem,*162, 339-351.

[285] Mironovich, L.M., Nikozyat, Y.B., Ivashchenko, E.I. (2005). *Zh. Prikl. Khim*, 78, 658-661.

[286] Zhang, J., Ding, X., Peng, Y., Mang, W. (2002). *Polym. Inter,*51, 617-621.

[287] G. Scott. (1979). *Developments in Polymer Stabilization, Part l,* Applied Science Publishers: London,UK.

[288] Allen, N.S. (1986). *Chem. Soc. Rev*, 15, 373-404.

[289] Lindén, L.A., Rabek, J.F., Kaczmarek, H., Karninska, A., Scoponi, M.(1993).*Coord. Chem. Rev*, 125, 195-217.

[290] Allen, N.S. (2002). In *Photochemistry Vol. 33*; Gilbert, A.,Ed. Roy. Soc. Chem.: London, 337-404.

[291] Feringa, B.L. (2001). *Molecular Switches*, Wiley-VCH: Weinheim.

[292] Gutlich, P. van Koningsbruggen, P.J., Renz, F. (2004). *Str. Bond*, 107, 27-75.

[293] Ouyang, M., Hiraoka, H. (1995). *Appl. Phys. Lett,* 67, 1781-1782.

[294] Einaga, Y., Taguchi, M., Li, G., Akitsu, T., Gu, Z., Sugai, T., Sato, O. (2003). *Chem. Mater*, 15, 8-10.

[295] Decurtins, C., Gutlich, P., Kohler, C.P., Spiering, H., Hauser, A. (1984). *Chem. Phys. Lett*,105, 1-4.

[296] van Koningsbruggen, P.J., Maeda, Y., Oshio, H. (2004). *Top. Curr. Chem*, 233, 259-324.

[297] Enachescu, C., Hauser, A., Girerd, J.J., Boillot, M.L. (2006). *Chem. Phys. Chem*, 7, 1127-1135.

[298] Sato, O. (2003). *Acc. Chem. Res*, 36, 692-700.

[299] Hayami, S., Hashiguchi, K., Inoue, K., Maeda, Y. (2004). *J. Nucl. Radiochem. Sci*, 5, N1-N3.

[300] Juhasz, G., Hayami, S., Sato, O., Maeda, Y. (2002). *Chem. Phys. Lett*, 364, 164-170.

[301] Hayami, S., Gu, Z.Z., Shiro, M., Einaga, Y., Fujishima, A., Sato, O. (2000). *J. Am. Chem. Soc*, 122, 7126-7127.

[302] Einaga, Y. (2006). *J. Photochem. Photobiol. C:Rev*, 7, 69-88.

[303] Sato, O., Iyoda, T., Fujishima, A., Hashimoto, K. (1996). *Science*, 272, 704-705.

[304] Shimamoto, N., Okhoshi, S., Sato, O., Hashimoto, K. (2002). *Inorg. Chem*, 41, 678-684.

[305] Ohkoshi, S.I., Hashimoto, K. (2001). *J. Photochem. Photobiol. C.Rev*, 2, 71-88.

[306] Sato, O. (2004). *J. Photochem. Photobiol. C:Rev*, 5, 203-223.

[307] Sour, A., Boillot, M.L., Riviere, E., Iwsot, P. (1999). *Eur. J. Inorg. Chem*, 12, 2117-2119.

[308] Boillot, M.L., Zarembowitch, J., Sour, A. (2004).*Top. Curr. Chem*, 234, 261-276.

[309] Powell, R.P., Wilson-Finelli, A. (2003). *Aquatic Sciences – Research Across Boundaries*, 65, 367-374.

[310] Barbeau, K., Rue, E.L., Bruland, K.W., Butler, A. (2001). *Nature*, 413, 409-413.

[311] Rijkenberg, M.J.A., Gerringa, L.J.A., Carolus, V.E., Velzeboer, I., de Baar, H.J.W. (2006). *Geochim. Cosmochim. Acta,* 70, 2790-2805.

[312] Rijkenberg, M.J.A., Fischer, A.C., Kroon, J.J., Gerringa, L.J.A., Timmermans, K.J., de Baar, H.J.W. (2005). *Marine Chem*, 93, 119-129.

[313] Barbeau, K., Rue, E.L., Trick, C.G., Bruland, K.W., Butler, A. (2003). *Limnol. Oceanogr*, 48, 1069-1078.

[314] Maldonado, M.T., Strzepek, R.F., Sander, S., Boyd, P.W. (2005). *Global Biogeochem. Cycl*, 19, GB4S23.

[315] Yang, G., Zhao, X., Qi, J. (2003). *J. Ocean Univ. Quigdao,* 2, 79-84.

[316] Emmenegger, L., Schonenberg, R., Sigg, L., Salzberger, B. (2001). *Limnol. Oceanogr*, 46, 49-61.

[317] Zhang, H., Lindberg, S. E. (2001). *Environ. Sci. Technol,* 35, 928-935.

[318] Meier, H., Albrecht, W., Hanack, M. (1991). *Mol. Cryst. Liq. Cryst*, 194, 75-83.

[319] Almnássy, G., Dezso, I. (1985). *Magyar Kemn. Foly*, 61, 300-309.

[320] Péter, A., Csányi, L.J. (1979). *Acta Chim. Acad. Sci. Hung*, 100, 163-172.

[321] Komadel, P., Stucki, J.W. (1988). *Clays Clay Miner*, 36, 379-381.

[322] Ware, M. (2006). *Gold Bull*, 39, 124-131.

[323] Mikhailov, O.V., Polovnyak, V.K. (1991). *J. Imag. Sci*, 35, 125-128.

[324] Foster, D.G., Marrese, C.A., Gray, A.M. (2001).*US Patent.* 6197483, issued March

[325] Matsubara, T., Levy, M. (1988). *J. Imag. Sci*, 32, 178-182.

[326] Mitchell, J.W. (1987). *J. Imag. Sci*, 31, 239-242.

[327] Arentz, D. (2005). *Platinum and Palladium Printing*. Elsevier: Focal Press, Boston.

[328] Filas, R.W. (2006). *US Patent* 7144489, issued December 5.

[329] Yoon, K.B., Kochi, J.K. (1988). *Z. Anorg. Allg. Chem,* 561, 174-184.

[330] Geoffroy, G.L., Wrighton, M.S. (1979).*Organometallic Photochemistry*. Academic Press: New York,USA.

[331] King, R.B., Bitterwolf, T.E. (2000). *Coord. Chem. Rev,* 2006-2007, 563-579.
[332] *Photochemistry and Photophysics of Transition Metal Complexes, I. and II.* Balzani, V., Campagna, S. Eds. *Top. Cur. Chem*, Vols. 280 and 281, Springer: Heidelberg (2007).

In: Photochemistry Research Progress
Editors: A. Sánchez, S. J. Gutierrez

ISBN: 978-1-60456-568-3

Chapter 5

HYDROGEN BONDING EFFECTS ON THE PHOTOCHEMISTRY OF CHROMOPHORES IN SOLUTION

Guang-Jiu Zhao [*] ***and Ke-Li Han*** [**]
State Key Laboratory of Molecular Reaction Dynamics,
Dalian Institute of Chemical Physics, Chinese Academy of Sciences,
Dalian 116023, P. R. China

ABSTRACT

Intermolecular hydrogen bonding, as a site-specific interaction between hydrogen donor and acceptor molecules, is a very important type of solute-solvent interactions. It is central to the understanding of the microscopic structure and function in many molecular systems. It also plays an important role in the photochemistry of the hydrogen-bonded systems in their electronically excited states. In this chapter, we have reviewed our recent investigations about the hydrogen bonding effects on the photochemistry of chromophores in solution. The electronically excited states of chromophores were theoretically studied using the time-dependent density functional theory (TDDFT) method. At the same time, the infrared spectra in both the ground state and electronically excited states of chromophores were also calculated. By monitoring the spectral shifts of some characterized vibrational modes involved in the formation of hydrogen bonds in different electronic states, it has been demonstrated for the first time that the intermolecular hydrogen bonds between chromophores and hydrogen bonding solvents are significantly strengthened upon photoexcitation. This excited-state hydrogen bonding dynamic occurs on the ultrafast timescale mainly set by vibrational motions of the hydrogen donor and acceptor groups, and it controls the excited-state dynamics of hydrogen-bonded complexes in a significant way. We have demonstrated that the radiationless deactivation via internal conversion (IC) from the fluorescent state to

---

[*] Corresponding Author.. Tel: +86-411-84379293; Fax: +86-411-84675584;E-mail Address:
E-mail: gjzhao@dicp.ac.cn; Homepage: http://beam.dicp.ac.cn/members/perscnalinfo/GuangjiuZhao.htm
[**] klhan@dicp.ac.cn; Group Homepage: http://beam.dicp.ac.cn

ground state can be enhanced due to the hydrogen bond strengthening in the fluorescent state. The ultrafast solute-solvent intermolecular photoinduced electron transfer (PET) reaction, which is facilitated by the hydrogen bond strengthening, has been proposed to account for the drastic fluorescence quenching behaviors of chromophores in alcoholic solvents. Moreover, the site-specific solvation dynamics can be initiated by the intermolecular hydrogen bond strengthening in the electronically excited state.

Keywords: Hydrogen bonding dynamics, photochemistry, fluorescence quenching, internal conversion, photoinduced electron transfer, site-specific solvation

1. INTRODUCTION

Photochemistry is the study of chemical reactions and physical processes of electronically excited molecules [1]. The electronically excited molecules, which are prepared by absorbing UV/vis light, can be deactivated to the ground state through radiative and/or nonradiative channels [2]. The molecules in the singlet excited state can directly emit fluorescence [3,4], or emit phosphorescence from the triplet state after intersystem crossing (ISC) [5-7]. At the same time, the radiative deactivations are completed with several important nonradiative proceses, such as internal conversion (IC) [8-12], vibrational relaxation (VR) [13-16], photoisomerization [17-20], excited-state electron transfer (ET) [21-29] or proton transfer (PT) [30-33], twisted intramolecular charge transfer (TICT) [34,35], etc. All of these nonradiative processes significantly affect the fluorescence or phosphorescence emission and are given more and more attention in recent photochemistry research.

Photochemistry of organic and biological chromophores in solution has been widely studied using modern experimental and theoretical techniques. The absorption and fluorescence or phosphorescence emission properties can be measured by the steady-state spectroscopy [1]. On the other hand, the nonradiative deactivation processes of the electronically excited state always take place in the ultrafast timescale (from tens of femtoseconds to hundreds of picoseconds) [8-35]. Therefore, the femtosecond and picosecond time-resolved spectroscopies become good candidates to investigate the ultrafast excited-state dynamic processes. Until now, transient absorption [36], time-resolved fluorescence [37], fluorescence up-conversion [9,38], time-resolved infrared (IR) [39] and Raman [11,40] spectra, and time-resolved photon echos [41,42] have been the conventional spectroscopic methods to study ultrafast excited-state dynamics in solution. Moreover, some other techniques, such as time-resolved multidimensional spectroscopy [43-45], time-resolved X-ray diffraction [46,47], ultrafast electron diffraction, crystallography, and microscopy [48] have been developed and applied widely to study the structural dynamics in real time.

The solute-solvent interactions play a fundamental role in the photochemistry of the organic and biological chromophores in solution. As we know, solute-solvent interactions are mainly due to the dielectric properties of the solvents [49]. Since the range of the Coulomb interaction is large compared to the structure in the radial distribution function of the solvent, such interactions are often termed “non-specific”, implying that no specific chemical bonds are formed or broken between the solute and the solvent [50,51]. Intermolecular hydrogen bonding, as a site-specific interaction between hydrogen donor and acceptor molecules, is another important type of solute-solvent interactions. It is central to the understanding of microscopic structure and function in many molecular systems, such as hydrogen-bonded water or alcohol networks, organic chromophores, proteins, and DNA building blocks of the life [52-57]. Furthermore, the dynamic behaviour of hydrogen bonds in electronically excited states plays an important role in determining the rates of many chemical, physical and biochemical processes that occur in hydrogen bonding surroundings [58-67]. It has been found that the fluorescence emission behaviour of the organic and biological chromophores can be strongly influenced by the intermolecular hydrogen bonding interactions between chromophores and hydrogen bonding solvents. For example, one can always find that the organic and biological chromophores in the polar aprotic solvents can emit strong fluorescence, while the fluorescence of these chromophores in the polar protic solvents is usually quenched (see Figure 1) [68-76].

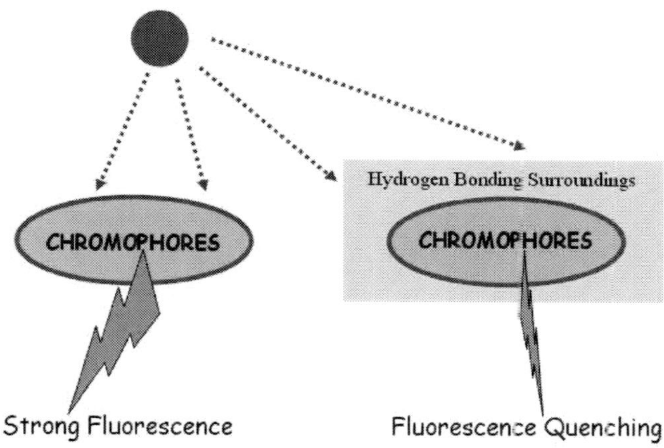

Figure 1. The schematic view of the different fluorescence emission behaviours of chromophores in nonpolar and polar protic solvents.

The difference in the fluorescence behaviours in aprotic and protic solvents suggests that the intermolecular hydrogen bonding interactions between chromophores and protic solvents play important roles in the influence of the fluorescence emission. A question about how the intermolecular hydrogen bonds influence the fluorescence emission will quickly emerge. Consequently, the hydrogen bonding dynamics in the electronically excited states of the chromophores is ineluctably considered [77]. For extensively understanding how the fluorescence emission is influenced by the intermolecular hydrogen bonding, we have studied the electronically excited states of some organic and biological chromophores in hydrogen bonding surroundings. It has been demonstrated that the intermolecular hydrogen bonding can significantly affect the nonradiative deactivation of the electronically excited chromophores, and then strongly influence the fluorescence emission [78]. In this chapter, we will review our recent investigations about the hydrogen bonding effects on the ultrafast nonradiative deactivation processes for electronically excited chromophores in solution.

The outline of the present review is as follows: In Section 2, we present the experimental and theoretical methods employed in our studies. In Section 3, some recent results about the hydrogen bonding effects on the photochemistry of chromophores in solution are discussed. The main conclusions are summarized in Section 4.

2. Experimental and Theoretical Methods

2.1 Steady-State and Femtosecond Time-Resolved Spectroscopy

Ultrafast excited-state dynamics was investigated using the femtosecond time-resolved stimulated emission pumping fluorescence depletion spectroscopy (FS TR SEP FD) technique [12,13,29,79,80]. Two femtosecond laser pulses are used in the experiment. One femtosecond laser pulse excites the sample to the electronically excited state in order to produce fluorescence. With a varying delay time, the other femtosecond laser pulse initiates stimulated emission pumping (SEP) from the electronically excited state to the ground state [12,13]. The variation in the decrease of the fluorescence induced by the SEP probe pulse as a function of delay time can reflect the ultrafast excited-state dynamics of the samples [29]. The spectra-Physics Hurricane system is used as the laser source. This system comprises a seeded laser (Mai Tai, cw diode-pumped laser and a mode-locked Ti sapphire pulse laser), a pump laser (Evolution, diode-pumped Q-switched Nd-YLF laser), a stretcher, a Ti sapphire regenerative amplifier, and a compressor. The output of the system consists of pulses with central wavelength of 800 nm, energy of 1mJ, FWHM (Full Width at Half Maximum) of 100 fs, and a repetition rate of 1 KHz. The pulses from the amplifier are used to pump an optical parametric amplifier (OPA). The delay time at the sample position between the pump and probe laser pulses can be controlled by a motorized translational stage (Unidex-100, Aerotech Inc.). In the experiment, the laser pulse of 620 nm was used to excite the sample to generate fluorescence and the 800 nm laser pulse as the probe pulse. Fluorescence was collected using a lens and focused into a monochromator (Jarrel-Ash, Division of Fisher Scientific Company) and then detected by a PMT (R456, semiconductor cooling, HANANATSH Corp.). The pmt signal was then sent to a Boxcar (SR250, Standford Corp.). All the fluorescence depletion signals were measured at room temperature.

UV/Vis spectra were recorded on a Perkin–Elmer Lambda 35 UV/Vis Spectrophotometer and HP 8453 spectrophotometer, with a quartz cuvette (path length: 1cm). Fluorescence spectra were recorded with a JASCO FP-6500 spectrometer. All the chromophores in our experimental studies were purchased from Exciton and used as received without further purification.

2.2 Theoretical and Computational Details

The ground-state geometric optimization for all the chromophores and their hydrogen-bonded complexes was performed by the density-functional theory (DFT). Both the generalized gradient approximation (GGA) functional BP86 and the Becke's three-parameter hybrid exchange function with Lee-Yang-Parr gradient-corrected correlation functional B3-LYP were carried out in our theoretical investigations [81,82]. Moreover, the resolution-of-the-identity (RI) approximation was also used in the BP86 method to improve the efficiency without sacrificing the accuracy of the results [83-85]. The triple-ζ valence quality with one set of polarization functions (TZVP) was chosen as basis sets throughout and the corresponding auxiliary basis sets for the RI approximation [86,87]. Furthermore, the fine quadrature grid with 4 was employed.

The electronic excitation and the excited-state geometric optimization were performed using the time-dependent density functional theory (TDDFT) method with the same functionals (RI-BP86 or B3-LYP) and basis set (TZVP) as the ground state [88-93]. The self-consistent field (SCF) convergency thresholds of the energy for both the ground-state and excited-state geometric optimization were set to be 10-8 (default settings are 10-6). Harmonic vibrational frequencies in the ground state and the electronically excited state were determined by diagonalization of the Hessian [94-96]. The excited state Hessian was obtained by numerical differentiation of analytical gradients using central differences and default displacements of 0.02 Bohr [94,95]. The infrared (IR) intensities were determined from the gradients of the dipole moment [95,96]. It should be noted that the calculation of the IR spectra in the electronically excited state is difficult and very time-consuming. In addition, the configuration interaction singles (CIS) [90-93] and coupled-cluster with resolution-of-the-identity approximation (RI-CC2) [97,98] methods were also introduced in our studies to confirm the reliability of the TDDFT results. All the electronic structure calculations were carried out using the TURBOMOLE program suite.

The potential energy surfaces (PES) were obtained by calculating the electronic excitation at the partially optimized geometric conformation along the related coordinates. Moreover, Conductor-like Screening Model (Cosmo) [99], as a continuum solvation model, was used to treat the solvation effect in the outer solvation shell. The basis set superposition error (BSSE) [100] was also considered in the theoretical calculations of the intermolecular hydrogen bond binding energies. In our studies, the calculated BSSE value for the intermolecular hydrogen bond binding energy using the above methods is at the range of 1-3 kJ/mol. It is distinct that the calculated BSSE value is much smaller than the corresponding intermolecular hydrogen bond binding energies.

3. Results and Discussions

3.1 Hydrogen Bonding Dynamics in Electronically Excited States

The phenomenon of hydrogen bonding has been recognized for its importance in physics, chemistry, and biology. The nature of the hydrogen bonding in solution is of particular interest and has been investigated extensively by diverse experimental and theoretical methods, since solute-solvent interactions play a fundamental role in molecular non-equilibrium processes in solution [52-78]. Upon photoexcitation, the solute and solvent molecules that are involved in the formation of intermolecular hydrogen bonds need to reorganize themselves due to the significant difference in charge distribution for the different electronic states [77,78]. This process is defined as hydrogen bonding dynamics.

The excited-state hydrogen bonding dynamics occurs on ultrafast timescales mainly set by vibrational motions of the hydrogen donor and acceptor groups, which controls the excited-state dynamics of hydrogen-bonded complexes in a significant way [29,78]. In addition, changes in the site-specific hydrogen bonding interactions can induce spectral shifts of some characteristic vibrational modes involved in the formation of hydrogen bonds [77]. Thus, femtosecond time-resolved vibrational spectroscopy has shown the potential to give a good insight into the microscopic dynamics and provide information on local structures

[61,64,66]. On the other hand, time-dependent density functional theory (TDDFT) method has been demonstrated as a reliable tool to calculate the vibrational absorption spectra in electronically excited states [101-109]. Sobolewski et al. have investigated the equilibrium geometries and vibrational spectra of anthranilic acid (AA) and salicylic acid (SA) in the S1($\pi\pi$*) excited state using the TDDFT method [101]. Their theoretical results were in good agreement with the recent UV-IR double-resonance experiments performed by Southern et al. [110]. This suggests that the TDDFT method is a reliable tool for the calculation of the infrared spectra in the electronically excited state. So the TDDFT method has become very useful for theoretically studying the hydrogen bonding dynamics by monitoring the spectral shifts of some characterized vibrational modes involved in the formation of hydrogen bonds in different electronic states [29,77,78].

Coumarin 102 (C102) is used as an ideal probe molecule to study the hydrogen bonding dynamics, since the C=O group is the only site that is responsible for the hydrogen bond formation in hydrogen donating environments [77,111,112]. So in the studies on the ultrafast response of C102 in hydrogen donating solvents to the photoexcitation using the femtosecond vibrational spectroscopy performed by Chudoba et al., insights into local changes in site-specific hydrogen bonding between C102 and the hydrogen donating solvents have been provided by monitoring the C=O stretching mode [61]. Upon electronic excitation, the vibrational absorption of the C=O stretching band for the C102 chromophore in CHCl3 or phenol solvents, which is drastically redshifted due to the formation of the intermolecular hydrogen bond C=O···H in ground state, shows an ultrafast spectral blueshift within a 200 fs timescale [61,63]. Furthermore, the femtosecond photon echo experiments have demonstrated that the electronic state hopping from ground state to the initially excited state of C102 completes around 200 fs, i.e., the C102 molecule can get to its equilibrium positions of the initially excited state within 200 fs timescale after electronic excitation [63]. Since the hydrogen bonding dynamics and the electronic state hopping take place in the same timescale, it has been demonstrated that they are coupled each other [77]. Due to the limited spectral-resolution for the laser pulses in the femtosecond IR spectroscopy, the hydrogen bonding dynamics occurred in the early time of photoexcitation is difficult to be directly monitored in the time-resolved spectroscopic experiments [77]. Therefore, to delineate the detailed aspects of the early-time hydrogen bonding dynamics, we have employed the TDDFT method for the calculations of the optimized geometries, energetics, vibrational frequencies, and IR intensities of the isolated C102 and hydrogen-bonded C102-phenol complex in the electronically excited state.

We focused our attention on the transient changes of intermolecular hydrogen bonds in the early time of the electronic excitation. Since the hydrogen bonding dynamics takes place within the ultrafast timescale mainly set by vibrational motions of the hydrogen donor and acceptor groups, only the solvent molecules in the inner solvation shell of the photoexcited chromophores can be attributed to the early-time hydrogen bonding dynamics occurring in the ultrafast timescale [77,78]. Herein, we have been motivated to theoretically investigate the hydrogen-bonded C102-phenol complex including only the phenol moleules in the first solvation shell. As is well-known, the vibrational frequencies of the stretching vibrations of C=O and O–H groups involved in the formation of hydrogen bonds can provide a clear-cut signature of the hydrogen bonding dynamics [77,78]. Thus, the IR spectra of the isolated molecules and the hydrogen-bonded complexes in the spectral regions of both the C=O and O–H stretching bands are presented here.

Figure 2 shows the calculated IR spectra of isolated C102 and its hydrogen-bonded C102-phenol complex in different electronic states in the spectral region of C=O stretching band. The corresponding experimental values measured by the femtosecond vibrational spectroscopy are also shown as green lines. One can find that the calculated C=O stretching vibrational frequency in the ground state (1740 cm^{-1}) is in good agreement with the experimental value (1735 cm^{-1}) [61]. It should be noted that upon electronic excitation, the C=O stretching band is blueshifted from 1740 cm–1 in the ground state to 1761 cm^{-1} in the optimized excited state. In addition, the electronic state hopping from ground state to the electronically excited state, which is followed by the intramolecular charge redistribution in the excited state of C102, takes place in the same timescale as the spectral blueshift of the C=O stretching mode [63]. The spectral shift of the C=O stretching band in the electronically excited state can be driven by the intramolecular charge redistribution [77]. Therefore, we have demonstrated that the ultrafast blueshift of the C=O stretching band observed by the femtosecond vibarational spectroscopy should be attributed to the electronic state hopping from ground state to the electronically excited state upon photoexcitation of the C102 chromophore.

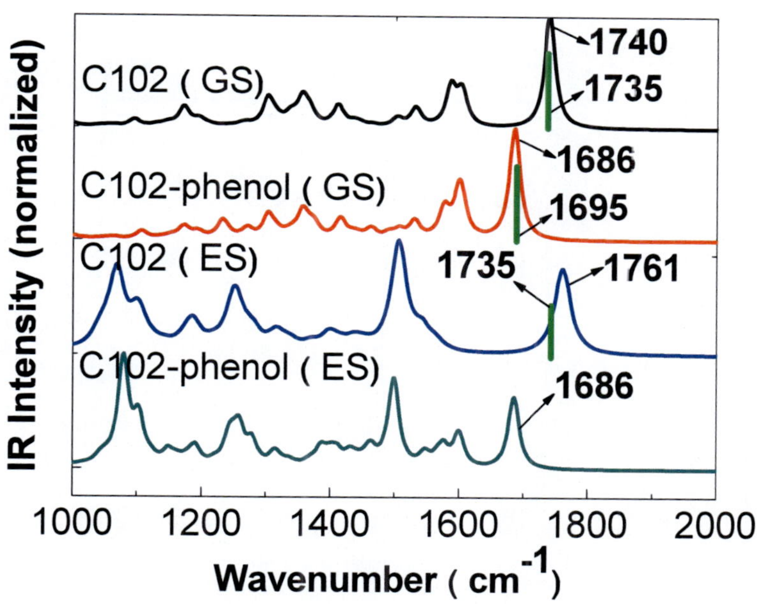

Figure 2. Calculated C=O stretching bands of isolated C102 and hydrogen-bonded C102-phenol complex in both the ground state (GS) and the electronically excited state (ES). The green lines show the corresponding C=O stretching vibrational frequencies observed in experiments. Taken from Ref. [77].

From Figure 2, one can also find that the C=O stretching vibrational frequency in the ground state is strongly redshifted by 55 cm^{-1} from 1741 to 1686 cm^{-1} due to the formation of intermolecular hydrogen bond C=O···H–O, which is in accordance with the experimental result which is redshifted from 1735 to 1695 cm^{-1} [61]. Moreover, the C=O stretching vibrational frequency of the hydrogen-bonded C102-phenol complex in the excited state is downshifted by 75 cm^{-1} from 1761 to 1686 cm^{-1} by comparison with that of the isolated C102 in the excited state. That is to say, the formation of the intermolecular hydrogen bond C=O···H–O between C102 and phenol molecules can induce a larger redshift of the C=O stretching mode in the electronically excited state (75 cm^{-1}) than that in the ground state (55 cm^{-1}).

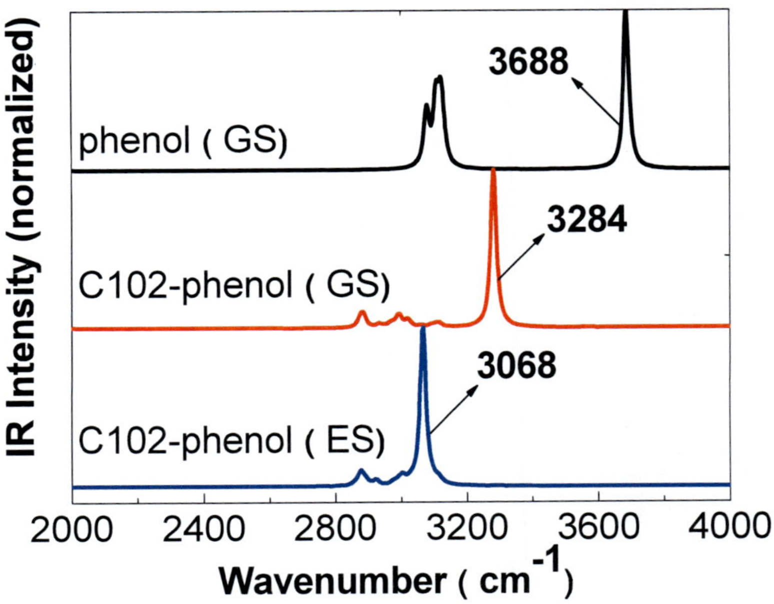

Figure 3. Calculated IR spectra of the hydrogen-bonded C102-phenol complex in both the ground state (GS) and electronically excited state (ES) in the spectral region of the O–H stretching band. Taken from Ref. [77].

Figure 3 shows the IR spectra of hydrogen-bonded C102-phenol complex in different electronic states in the spectral region of the O–H stretching band. It is distinct that the O–H stretching band in the ground state is strongly redshifted by 404 cm^{-1} from 3688 cm^{-1} for the free phenol to 3284 cm^{-1} for the hydrogen-bonded complex due to the formation of intermolecular hydrogen bond C=O···H–O between C102 and phenol. Upon photoexcitation to the electronically excited state of the hydrogen-bonded C102-phenol dimer, the O–H stretching band continues to redshift from 3284 to 3068 cm^{-1}. This evidently indicates that the intermolecular hydrogen bonding interactions can induce a larger spectral redshift for the O–H stretching band in the electronically excited state than that in the ground state.

As discussed above, both the C=O and O–H stretching bands in the ground state are strongly redshifted due to the formation of intermolecular hydrogen bond C=O···H–O between C102 and phenol. Moreover, in the electronically excited state, the intermolecular hydrogen bonding interactions can induce larger spectral redshifts for both the C=O and O–H stretching bands in the hydrogen-bonded C102-phenol complex. As we know, both the C=O and O–H stretching modes can be induced to undergo a shift to lower frequency by the intermolecular hydrogen bonding interactions [77,78]. Therefore, the larger redshift of the

C=O and O–H stretching bands in the electronically excited states than that in the ground state distinctly suggests that the intermolecular hydrogen bond C=O···H–O should be strengthened in the electronically excited state upon photoexcitation of the C102 chromophore.

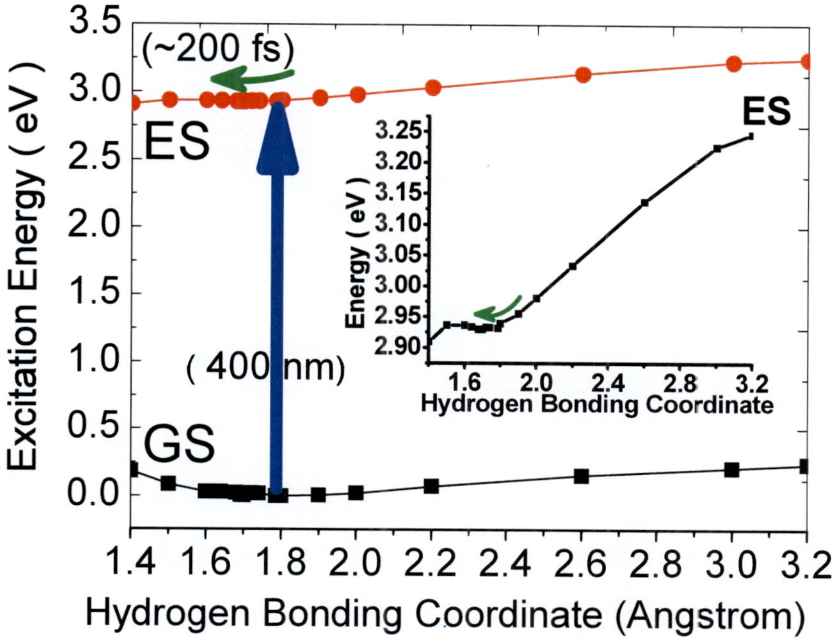

Figure 4. Potential energy curves for the ground state (GS) and the electronically excited state (ES) of the hydrogen-bonded C102-phenol complex with the hydrogen bonding coordinate (distance between O and H atoms in hydrogen bond C=O···H–O, Å). The inset shows the detailed potential energy curve of the excited state. Taken from Ref. [77].

In Figure 4, the calculated potential energy curves are shown for the ground state and the electronically excited state of the hydrogen-bonded C102-phenol complex with the hydrogen bonding coordinate. This will be helpful for understanding the mechanism of the hydrogen bonding dynamics in the electronically excited state.

As shown in Figure 5, the hydrogen-bonded C102-phenol complex is vertically excited from the equilibrium position of the ground state to the electronically excited state by a laser pulse with the wavelength at around 400 nm, which is in very good agreement with the experimental pulse [61]. Then the hydrogen-bonded C102-phenol complex in the Franck-Condon (FC) region of the excited state will be relaxed to the equilibrium position of the same excited state within 200 fs timescale [63]. At the same time, the excited-state hydrogen bonding dynamics takes place along the hydrogen bonding coordinate within the same timescale [77]. From our optimized geometric conformations of the hydrogen-bonded C102-phenol complex in both the equilibrium positions of ground state and electronically excited state, we know that the bond length between H and O atoms for the intermolecular hydrogen bond C=O···H–O is markedly shortened from 1.786 to 1.691 Å during the excited-state hydrogen bonding dynamics process.

The calculated hydrogen bond binding energies and the corresponding hydrogen bond lengths in different electronic states are listed in Table 1.

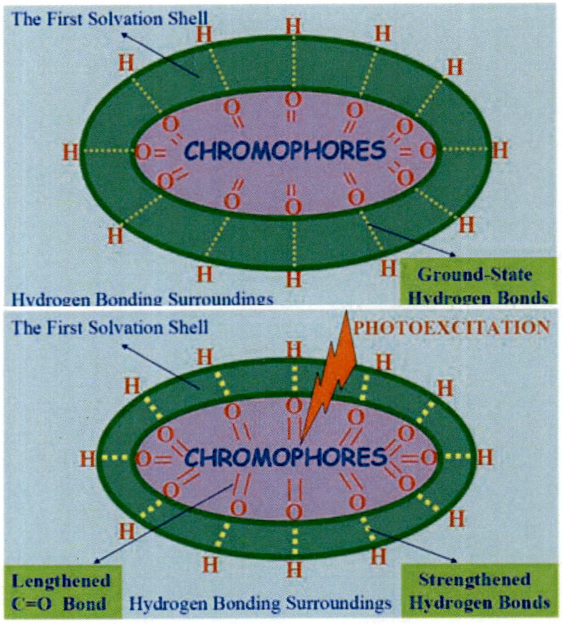

Figure 5. The schematic view of the intermolecular hydrogen bond strengthening in the electronically excited state of chromophores in hydrogen bonding surroundings.

Table 1. Calculated hydrogen bond binding energies E_{HB} (kJ/mol) and the corresponding hydrogen bond length $L_{O\cdots H}$ (Å) as well as the lengths of the hydrogen-bonded groups $L_{C=O}$ (Å) and L_{H-O} (Å)

	phenol	C102	C102-phenol			
	L_{H-O}	$L_{C=O}$	E_{HB}	$L_{C=O}$	$L_{O\cdots H}$	L_{H-O}
GS	0.974	1.215	36.8	1.231	1.786	0.995
ES	---	1.220	51.5	1.246	1.691	1.007

The binding energy of the intermolecular hydrogen bond C=O···H–O between C102 and phenol is calculated to be 36.8 kJ/mol in the ground state of the hydrogen-bond C102-phenol complex. However, the binding energy of hydrogen bond C=O···H–O in the electronically excited state is significantly increased to 51.5 kJ/mol. Thus, the intermolecular hydrogen bond C=O···H–O between C102 and phenol is evidently strengthened in the electronically excited state. One can also find that both the hydrogen-bonded C=O and O–H groups become lengthened in the ground state of the hydrogen-bonded C102-phenol complex due to the formation of the intermolecular hydrogen bond C=O···H–O between C102 and phenol. Moreover, it should be noted that both the hydrogen-bonded C=O and O–H groups continue to be lengthened in the electronically excited state of the hydrogen-bonded C102-phenol complex, which is followed by the strengthening of the intermolecular hydrogen bond C=O···H–O. Figure 5 gives a schematic view of the intermolecular hydrogen bond strengthening in the electronically excited state of chromophores in hydrogen bonding surroundings. Upon photoexcitation of chromophores in the hydrogen bonding surroundings, the intermolecular hydrogen bonding interactions in the first solvation shell are significantly strengthened within the ultrafast timescale. At the same time, the intermolecular hydrogen bond lengths are shortened in the electronically excited state, while the hydrogen-bonded C=O groups of chromophores are lengthened.

In this section, we have distinctly demonstrated for the first time that the intermolecular hydrogen bonds between chromophores and the hydrogen bonding solvents can be significantly strengthened in the early time of the photoexcitation. This novel intermolecular hydrogen bond strengthening behavior in the electronically excited state of chromophores in solution upon photoexcitation may facilitate the ultrafast radiationless deactivation processes of the fluorescent state, and then induce the fluorescence quenching [77].

3.2 Hydrogen Bonding Facilitates Internal Conversion (IC) of Chromophores

The primary photophysics of Fluorenone (FN) has been well studied because Fluorenone exhibits some rather unique spectroscopic and photophysical properties that have made it the subject of many investigations [113-119]. In general, Fluorenone is very weakly fluorescent in solution [113]. In aprotic solvents, processes following excitation to the S1 state are governed essentially by the dependence of intersystem crossing (ISC) to the triplet state on the solvent polarity [114]. However, in protic solvents, the triplet state yield falls and the internal conversion (IC) from the fluorescent state to ground state becomes the most important dissipative process [114]. It has been demonstrated that deactivation of excited state via IC can be strongly enhanced by the intermolecular hydrogen bonding interactions [114,115]. Herein, we proposed that the hydrogen bonding dynamics in the fluorescent state of Fluorenone chromophore in hydrogen donating solvents will be tightly associated with this enhanced IC process.

To delineate the detailed aspects of the hydrogen bonding dynamics for the Fluorenone chromophore in protic methanol (MeOH) solvent, we have theoretically studied the S_1 state of the hydrogen-bonded FN-MeOH complex using the TDDFT method [78]. We have demonstrated that all the calculated absorption and fluorescence spectra are in good agreement with the spectral results recorded in experiments [115]. Thus, it was confirmed that the hydrogen-bonded FN-MeOH complex presented in our work can be viewed as a reliable model to study the intermolecular hydrogen bonding interactions between Fluorenone chromophore and the methanol solvent without consideration of the bulk effect of the methanol molecules in the outer solvation shells [78].

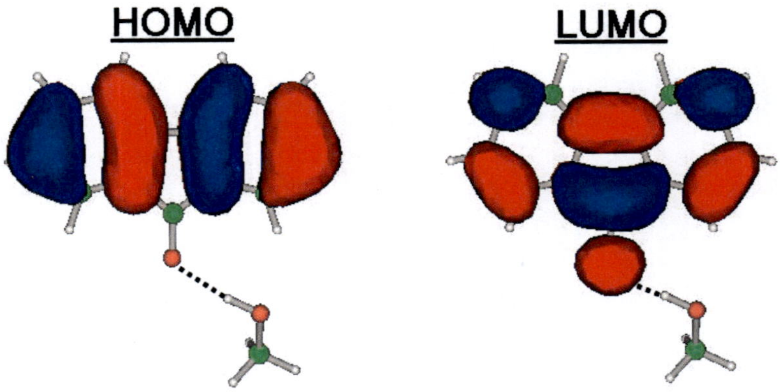

Figure 6. Frontier molecular orbitals (MOs) of the intermolecular hydrogen-bonded FN-MeOH complex. Taken from Ref. [78].

Figure 6 shows the frontier molecular orbitals (MOs) of the hydrogen-bonded FN-MeOH complex. Since the S_1 state of the hydrogen-bonded FN-MeOH complex is dominatively corresponding to the orbital transition from HOMO to LUMO [78], only the HOMO and LUMO orbitals are presented here. The π character for the HOMO orbital and the π^* character for the LUMO orbital can be clearly seen. So it is confirmed that the S1 state of the Fluorenone chromophore is of distinct $\pi\pi^*$ feature [114]. Moreover, one can find that the electron densities of both the HOMO and LUMO orbitals for the hydrogen-bonded FN-MeOH complex are strictly localized on the fluorenone moiety. As a result, the S1 state of the hydrogen-bonded FN-MeOH complex can be assigned as a locally excited (LE) state on Fluorenone molecule [120]. It is noted that the electron density of the C=O group is increased after transition from HOMO to LUMO orbital. Thus, the electronegativity of the C=O group in the S_1 state of the hydrogen-bonded FN-MeOH complex will be increased. So it can be predicted that the intermolecular hydrogen bond C=O···H–O between Fluorenone and methanol is strengthened in the S_1 state.

Since the S_1 state of the hydrogen-bonded FN-MeOH complex is a LE state localized on the Fluorenone moiety, the methanol molecule in the S_1 state of the hydrogen-bonded FN-MeOH complex always remains in the ground state. So the O–H stretching vibrational mode cannot be strongly influenced by the electronic excitation [78]. Thus, the hydrogen bonding dynamics can be distinctly monitored by changes of the O–H stretching mode in different electronic states.

From the calculated IR spectra of the hydrogen-bonded FN-MeOH complex in different electronic states shown in Figure 7, one can find that the O–H stretching mode is redshifted by 176 cm^{-1} due to the formation of intermolecular hydrogen bond C=O···H–O in the ground state of the hydrogen-bonded complex. In the S_1 state of the hydrogen-bonded FN-MeOH complex, a larger redshift of 335 cm^{-1} for the O–H stretching mode can be observed. The additional redshift of 159 cm-1 should be attributed to the hydrogen bond strengthening in the S1 state of the hydrogen-bonded FN-MeOH complex [78].

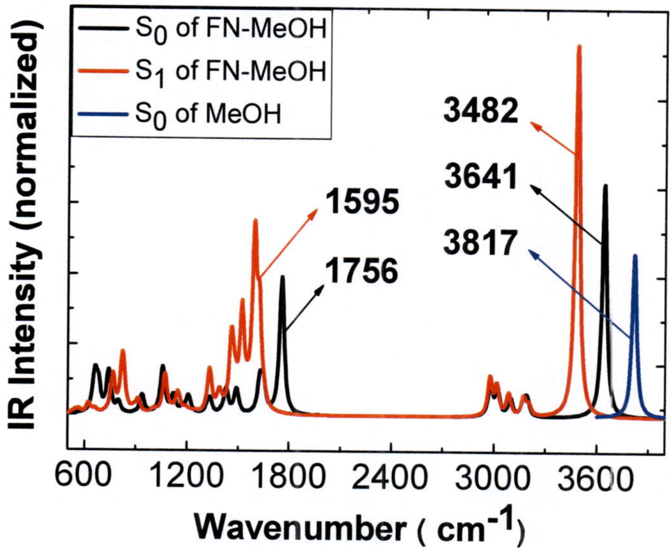

Figure 7. Calculated IR spectra of the hydrogen-bonded FN-MeOH complex in different electronic states. The free O–H stretching band in the ground state of methanol is also shown. Taken from Ref. [78].

The calculated hydrogen bond binding energies and the corresponding hydrogen bond lengths, as well as bond lengths of the hydrogen-bonded groups in both the ground state and the S_1 state of the hydrogen-bonded FN-MeOH complex are listed in Table 2.

Table 2. Calculated hydrogen bond binding energies E_b (kJ/mol) and corresponding hydrogen bond lengths $L_{O\cdots H}$ (Å) of the hydrogen-bonded FN-MeOH complex in different electronic states

	MeOH	FN	FN-MeOH			
	$L_{O\text{–}H}$	$L_{C=O}$	E_b	$L_{C=O}$	$L_{O\cdots H}$	$L_{O\text{–}H}$
S_0	0.963	1.212	27.85	1.219	1.906	0.972
S_1	---	1.250	42.62	1.259	1.802	0.981

It is noted that the binding energy of the intermolecular hydrogen bond C=O···H–O is markedly increased from 27.85 kJ/mol in the ground state to 42.62 kJ/mol in the S1 state of the hydrogen-bonded FN-MeOH complex. At the same time, the hydrogen bond length between oxygen and hydrogen atoms is drastically decreased from 1.906 Å in the ground state to 1.802 Å in the S_1 state. Meoreover, the bond lengths of hydrogen-bonded C=O and O–H groups are correspondingly increased in the S_1 state. As a result, it is evidently demonstrated that the intermolecular hydrogen bond C=O···H–O between Fluorenone and methanol is significantly strengthened in the S_1 state of the hydrogen-bonded FN-MeOH complex.

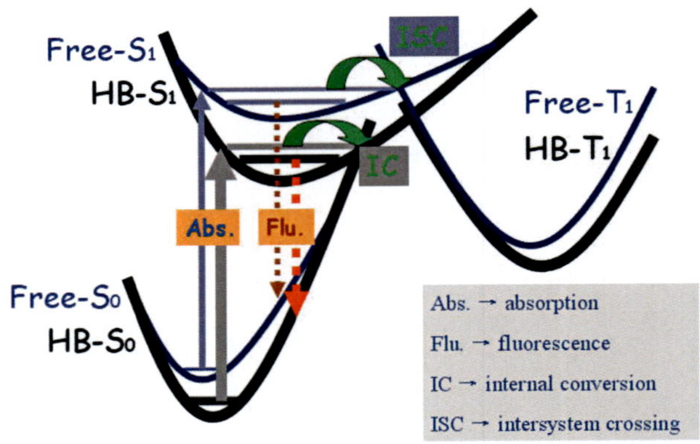

Figure 8. Schematic view of the intersystem crossing (ISC) and internal conversion (IC) processes in the different electronic states for the free and hydrogen-bonded Fluorenone.

Figure 8 presents a schematic view of the intersystem crossing (ISC) and internal conversion (IC) processes between the different electronic states for both the free and hydrogen-bonded Fluorenone. It indicates that there has a strong electronic coupling between the S_1 and T_1 states for the free Fluorenone. So the deactivation of the S_1 state for the Fluorenone in aprotic solvents is dominateively controlled by the ISC process from the S_1 state to the T1 state. For Fluorenone in the hydrogen bonding solvents, the energy levels of all the electronic states are lowered due to the formation of the intermolecular hydrogen bonds between Fluorenone and the hydrogen bonding solvents. Since we have demonstrated that the intermolecular hydrogen bond C=O···H–O between fluorenone and methanol is significantly strengthened in the S_1 state of the hydrogen-bonded FN-MeOH complex, the energy level of the S_1 state will be lowered much more than that of the ground state. Thus, the energy gap between the S_1 and S_0 states of the hydrogen-bonded Fluorenone is smaller than that of the free Fluorenone. As shown in Figure 8, the electronic coupling between the S_1 and S_0 states becomes stronger than the electronic coupling between the S_1 and T_1 states. So the deactivation of the S_1 state via IC is strongly facilitated by the intermolecular hydrogen bond strengthening in the electronically excited state [78]. As a result, it can be explained that the ISC yield markedly falls and the IC becomes the most important dissipative process for the S1 state of the Fluorenone in the protic solvents [114].

The steady-state spectral features in the absorption and fluorescence spectra of Fluorenone in the alcoholic solution can also be explained by the intermolecular hydrogen bond strengthening in the electronically excited state [78]. In the absorption spectra of Fluorenone in methanol, there are two absorption peaks at around 379 and 403 nm for the S1 state [115]. By comparison with the calculated S_1 absorption wavelength of 392 and 411 nm for the isolated Fluorenone and the hydrogen-bonded FN-MeOH complex, respectively, the two S1 absorption peaks at around 379 and 403 nm should be assigned as the free and hydrogen-bonded Fluorenone [78]. Moreover, it can be observed that the absorption peak at 379 nm is stronger than the peak at 403 nm in experiments [115]. So the free form is much more than the hydrogen-bonded form in the ground state. On the other hand, the calculated fluorescence wavelengths are located at 521 and 557 nm for the isolated fluorenone and the hydrogen-bonded FN-MeOH complex, which is in good agreement with the dual fluorescence bands with maxima at 514 and 565 nm [115]. Consequently, it can be confirmed that the dual fluorescence bands are originated from the free and hydrogen-bonded Fluorenone, respectively [78]. Interestingly, the fluorescence peak at 514 nm is weaker than that at 565 nm [115]. That is to say that the free form is much less than that of the hydrogen-bonded form.

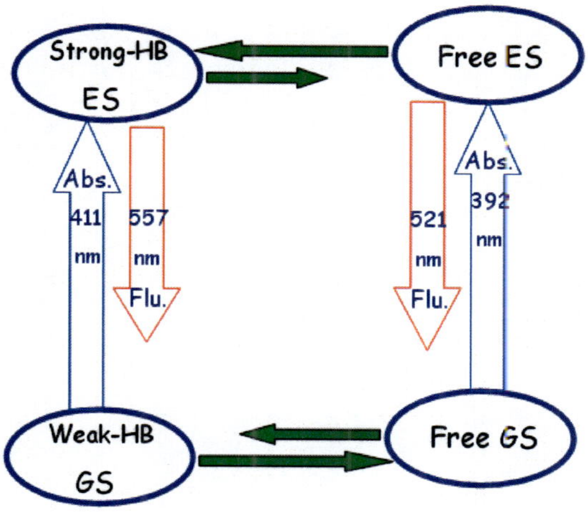

Figure 9. A schematic view of the dynamic equilibrium between the free and hydrogen-bonded forms in both the ground state (GS) and the electronically excited state (ES). The green arrows denote that the equilibrium remains in favor of the free form in ground state and becomes in favor of the hydrogen-bonded form in the electronically excited state due to the intermolecular hydrogen bond strengthening. Taken from Ref. [78].

Figure 9 shows a dynamic equilibrium view in both ground state and the electronically excited state of Fluorenone chromophore in hydrogen bonding solvents to explain the steady-state spectral features. Fluorenone chromophore in alcoholic solution is located in the equilibrium between the free and the hydrogen-bonded forms in the ground state [78]. Since the intermolecular hydrogen bonding interactions are weak in the ground state, the equilibrium is strongly in favor of the free molecules in ground state [78]. So the absorption band of the non-hydrogen-bonded fluorenone is stronger than that of the hydrogen-bonded form in ground state. Upon photoexcitation, the relatively weak hydrogen bonds between fluorenone and methanol solvents in ground state can be significantly strengthened in the electronically excited states. Therefore, the equilibrium in excited states will become markedly in favor of the hydrogen-bonded forms [78]. The fluorescence band of the hydrogen-bonded form is stronger than that of the non-hydrogen-bonded form upon electronic excitation to electronically excited states [78].

3.3 Hydrogen Bonding Facilitates Intermolecular Photoinduced Electron Transfer (PET)

We have proposed a new mechanism to explain the drastic fluorescence quenching behavior of the Oxazine 750 (OX750) chromophore in the polar protic solvents shown in Figure 10.

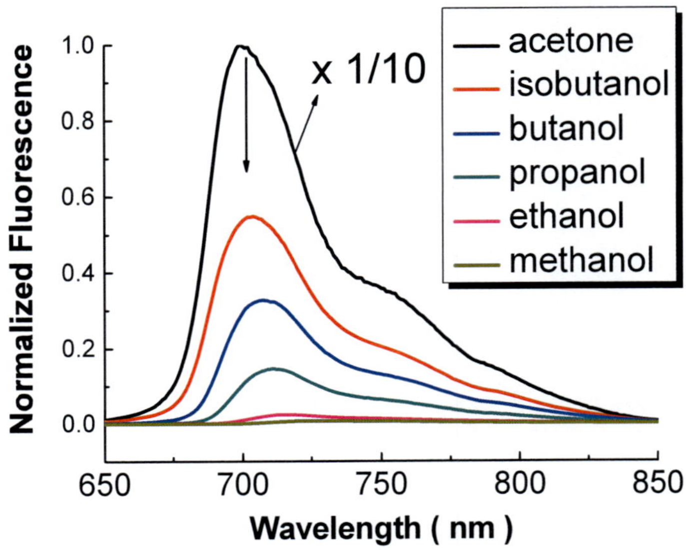

Figure 10. Steady-state fluorescence spectra of OX750 in different alcoholic solvents as well as in aprotic acetone. Taken from Ref. [29].

Compared with the strong fluorescence of OX750 in acetone, one can notice that the fluorescence of OX750 in the alcoholic solvents is drastically quenched [29].The difference of the fluorescence behavior of OX750 in aprotic and protic solvents suggests that intermolecular hydrogen bonding plays an important role on the fluorescence quenching behavior. In our studies, the solute-solvent intermolecular photoinduced electron transfer (PET), which is facilitated by the intermolecular hydrogen bonding, was proposed for the first time to account for the drastic fluorescence quenching of OX750 chromophore in alcoholic solvents [29].

Intermolecular hydrogen bonding as a site-specific interaction between hydrogen donor and acceptor molecules is strong enough to determine the structures and reactivities of many molecular systems [52-67]. Thus, the intermolecular hydrogen bonds in solutions can significantly influence the solute-solvent intermolecular PET reactions [121-127]. Investigations on the intermolecular ET reactions in some biological systems and their analogues have clearly established that as the photoexcitation initiates the transfer of an electron from the donor to the acceptor, the intermolecular hydrogen bond can act as an effective conduit to facilitate a biological ET process [128-131]. Intermolecular site-specific interactions via hydrogen bonding can remarkably increase intermolecular electronic coupling, and may eventually promote the intermolecular PET reaction [29,128-131]. Hence, the ultrafast intermolecular PET should be accompanied by and possibly be coupled with the hydrogen bonding dynamics [29].

To delineate extensively the intermolecular hydrogen bonds between the OX750 and alcohols, we have been motivated to study theoretically the hydrogen-bonded OX750-(alcohol)$_n$ complexes [29]. From our calculations, we have found that the nitrogen atom in the central ring of OX750 is the most negatively polarized and so is one of the preferred sites for the formation of the intermolecular hydrogen bond O–H···N [29]. In addition, another intermolecular hydrogen bond N–H···O at the protonation site of OX750 can also be formed

[29]. Both the hydrogen bonds O–H···N and N–H···O remain in the plane of the OX750 molecule. Moreover, we have theoretically demonstrated that the intermolecular hydrogen bond N–H···O is stronger than hydrogen bond O–H···N in the ground state [29].

In order to understand the nature of the electronically excited states for the hydrogen-bonded OX750-(alcohol)$_2$ complex, the electronic excitation energies and their corresponding oscillator strengths of the low-lying excited states of the hydrogen-bonded OX750-(EtOH)$_2$ trimer were calculated using the TDDFT method [29]. For comparison, the OX750 monomer and the hydrogen-bonded dimer which is bonded by the hydrogen bonds O–H···N and N–H···O, respectively, are also investigated.It has been noticed that in addition to the electronically excited states of the OX750 monomer, a further energetically lower state in the OX750-(EtOH)$_2$ trimer is located 0.5 eV below the lowest excited state of OX750 [29]. Since the S$_1$ state of the OX750-(EtOH)$_2$ trimer with weak oscillator strengths is lower in energy than any other intramolecular excited states of the OX750 monomer, it might be originated from interactions between the OX750 and ethanol molecules [132].Thus, this additional electronic state may exist as an energetically lowest intermolecular charge transfer (ICT) state [120].In this ICT state, the solute-solvent intermolecular PET will take place upon photoexcitation. Furthermore, the reliability of our TDDFT results has been confirmed by the configuration interaction singles (CIS) method [90-93].

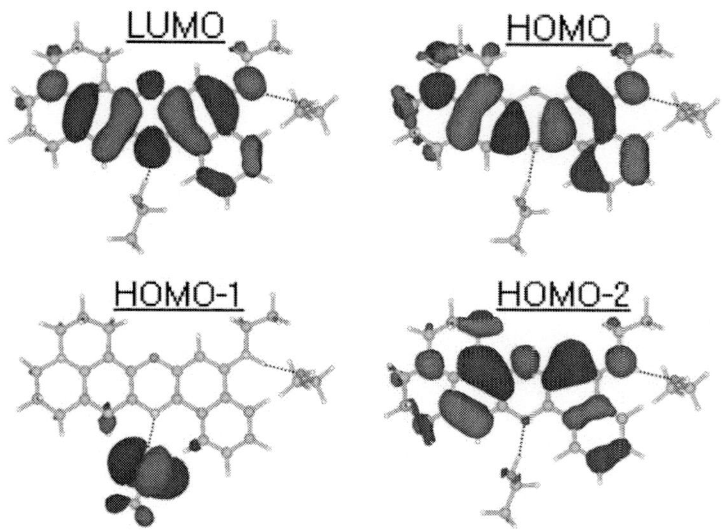

Figure 11. Frontier molecular orbitals (MOs) of the intermolecular hydrogen-bonded OX750-(EtOH)$_2$ complex (dotted lines denote the intermolecular hydrogen bonds). Taken from Ref. [29].

Figure 11 shows the frontier molecular orbitals (MOs) of the hydrogen-bonded OX750-(EtOH)$_2$ complex. One can find that the electron density of the HOMO-1 orbital is localized on the ethanol molecule that is bonded by the intermolecular hydrogen bond O–H···N, while electron densities of the other orbitals are always localized on the OX750 moiety. From our TDDFT results, the S$_1$ state of the hydrogen-bonded OX750-(EtOH)$_2$ trimer corresponds to the orbital transition from HOMO-1 to LUMO [29]. So the ICT character of S$_1$ state for the hydrogen-bonded OX750-(EtOH)$_2$ complex is distinctly confirmed. Moreover, it is also indicated that the intermolecular PET reaction takes place from the alcoholic solvent to the OX750 chromophore by passing through the relatively weak intermolecular hydrogen bond

O–H···N, while not the stronger intermolecular hydrogen bond N–H···O. As a result, the solute-solvent intermolecular PET passing through the site-specific intermolecular hydrogen bonds exhibits an unambiguous site-selectivity [29]. At the same time, it can also be demonstrated that the S_2 state of the hydrogen-bonded OX750-$(EtOH)_2$ complex with the largest oscillator strengths is a locally excited (LE) state on the OX750, since the S_2 state is corresponding to orbital transitions from both HOMO and HOMO-2 to LUMO [29].

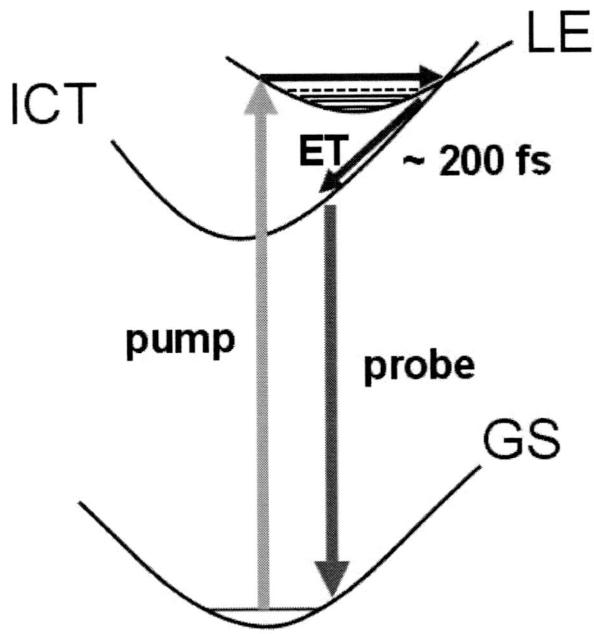

Figure 12. Schematic view of the ultrafast photoinduced electron transfer (PET) in the excited-state potential energy curves. Taken from Ref. [29].

From the our theoretical studies, the OX750 chromophore in the alcoholic solvents is initially photoexcited to the LE state of the hydrogen-bonded OX750-$(alcohol)_2$ complexes [29]. This initially-ready state can be assigned as the 'donor state' for an ultrafast intermolecular PET reaction occurred in the ICT state (see Figure 12). Accordingly, inspired by our calculated electronic excitation energies of the LE and ICT states [29], a 620 nm laser pulse was chosen to pump the systems and a laser pulse of 800 nm was used to probe the ultrafast intermolecular PET from the alcoholic solvents to the OX750 chromophore in our femtosecond time-resolved stimulated emission pumping fluorescence depletion spectroscopy (FS TR SEP FD) experiment.

The time constants of the ultrafast dynamical processes and their corresponding coefficients obtained by the time-resolved fluorescence depletion spectroscopy are listed in Table 3.

Table 3. Time constants for the ultrafast dynamical processes and their corresponding coefficients obtained from the time-resolved fluorescence depletion spectroscopy

Solvents	C_1	τ_1 (fs)	C_2	τ_2 (ps)
Methanol	0.79	185±15	0.21	2.49±0.27
Ethanol	0.80	184±13	0.20	3.49±0.27
Prapanol	0.81	178±11	0.19	7.21±0.47
Butanol	0.86	229±15	0.14	9.74±0.40
Isobutanol	0.81	265±17	0.19	12.9±0.54

A rapid decay and a much slower relaxation process can be found. The ultrafast solute-solvent intermolecular PET should account for the faster decay and can be found to take place as fast as 200 fs [29]. Thus, this ultrafast intermolecular PET from the alcoholic solvents to the chromophore should be coupled with the hydrogen bonding dynamics in the first solvation shell which occurrs in the same timescale [29]. Since the time constant of the slower process shows a very good linear relationship with the viscosity of the alcoholic solvents, this process can be described as a diffusive solvation which is mainly attributed to the bulk solvent effects in the outer solvation shell [133,134]. It is evident that the intermolecular PET is much faster than the diffusive salvation. Thus, the subpicosecond site-specific hydrogen bonding dynamics in the inner solvation shell and the picosecond diffusive dynamics in the outer solvation shell are distinctly separated [29].

In our previous studies on the ultrafast vibrational relaxation dynamics of the electronically excited OX750 chromophore in aprotic acetone solvent, the intramolecular vibrational redistribution (IVR) process was found to take place within 350 fs timescale [13,79]. Furthermore, we have found that the IVR process of the photoinduced OX750 is almost independent of the solvents, since it is controlled mainly by the coupling of intramolecular vibrational modes of OX750 chromophore [13]. Therefore, it is demonstrated that the ultrafast intermolecular PET process from alcoholic solvents to OX750 chromophore is even significantly faster than the IVR process of the OX750 in the aprotic solvents due to the site-specific intermolecular hydrogen bonding interactions. Thus, the ultrafast intermolecular PET takes place from the vibrationally excited electronic state [29]. By considering the microscopic mechanism responsible for the ultrafast intermolecular PET reactions much faster than the diffusive solvation process, it is reasonably to assume that the initially-ready excited state in the hydrogen bonding coordinate is close to the region where the effectively reacting vibrational channels are located, as shown in Figure 12 [69,135]. Thus, the ultrafast intermolecular PET reaction takes place quickly along the hydrogen bonding coordinate and without diffusive solvation [29].

To understand the important role of intermolecular hydrogen bonding interactions in the initially-ready excited state for the ultrafast intermolecular PET reaction, the LE states of all the hydrogen-bonded OX750-(alcohol)$_2$ complexes were fully optimized using the TDDFT method [29].

Table 4. Calculated intermolecular hydrogen bond binding energies (kJ/mol) and the the corresponding hydrogen bond lengths (Å, in the parenthesis) in different electronic states

Solvents	GS		LE	
	O–H···N	N–H···O	O–H···N	N–H···O
Methanol	17.67(1.987)	35.29(2.029)	21.05(1.914)	34.82(2.042)
Ethanol	16.61(2.000)	33.50(2.028)	20.03(1.931)	33.12(2.048)
Propanol	16.54(2.002)	33.66(2.029)	19.96(1.929)	33.30(2.048)
Butanol	16.49(1.995)	33.78(2.035)	19.85(1.930)	33.60(2.043)
Isobutanol	12.32(2.113)	34.63(2.023)	15.32(2.024)	34.10(2.040)

Table 4 lists the calculated hydrogen bond binding energies and the corresponding hydrogen bond lengths of O–H···N and N–H···O in both the ground state and the LE state. It is noted that the stronger intermolecular hydrogen bond N–H···O in the ground state remains nearly unchanged upon photoexcitation to the LE state, showing that the alcoholic molecules which are bonded by intermolecular hydrogen bond N–H···O are not influenced by the electronic excitation [29]. So the intermolecular hydrogen bond N–H···O cannot be attributed to the ultrafast intermolecular PET. However, the intermolecular hydrogen bond O–H···N can be significantly strengthened in the LE state, indicating that the electronic coupling between the solute and the solvent in this hydrogen bonding site is strongly increased [29]. So the effective reacting channels can be formed only in the hydrogen bonding site that is bonded by the intermolecular hydrogen bond O–H···N. Consequently, the intermolecular hydrogen bond O–H···N is selected as a conduit to facilitate the ultrafast intermolecular PET from the alcoholic solvents to the chromophore, which has been proposed to account for the fluorescence quenching of OX750 chromophore in the alcoholic solvents [29]. In our work, we have theoretically demonstrated for the first time that the selected intermolecular hydrogen bonds for facilitating the ultrafast solute-solvent intermolecular PET reaction are transiently strengthened in the electronically excited states.

3.4 Hydrogen Bonding Initiates Site-Specific Solvation Dynamics

Protochlorophyllide a (Pchlide a) is a precursor in the biosynthesis of chlorophyll, which is a sort of ubiquitous pigment of photosynthesis in plants, green algae and cyanobacteria [136,137]. Determination of the spectral properties for Pchlide a in different solvents would be helpful to understand the properties and function of Pchlide a *in vivo* [138-141].

The steady-state spectral properties of Pchlide a in various solvents have been extensively investigated in the previous work [142-148]. In all the studied solvents, the absorption spectra have a shape characteristic of chlorophyll and its derivatives in the monomeric form, consisting of Soret and Q_y bands that are accompanied with Satellite bands of lower intensity on the high-energy side [142].The observed absorption maxima are within the range of 432.5-451 nm and 624-633 nm, for Soret and Q_y bands, respectively [143]. For Pchlide a in non-polar benzene solvent, the absorption maxima for Soret and Q_y bands are 442 and 632.5 nm, respectively [144]. For Pchlide a in polar protic methanol solvent, the absorption maxima for

Soret and Q_y bands are 434 and 629 nm, respectively. Moreover, the fluorescence spectrum of Pchlide a in methanol exhibits a strong $S_1 \rightarrow S_0$ band centered at 641 nm [145,146]. The Stokes shift of Pchlide a in the investigated solvents shows a slight increase for increasing solvent orientation polarizability. This indicates that the spectral properties of Pchlide a have low sensitivity to the non-specific solvation [147,148]. Thus, the differences for fluorescence emission spectra of Pchlide a in various solvents are mainly due to the site-specific solvation between solutes and solvents [148].

It should be noted that there has a novel shoulder at around 690 nm in the fluorescence spectra of Pchlide a in polar solvents [149-151]. It has been excluded that the shoulder is formed due to the Pchlide a aggregations [148]. As a result, the novel fluorescence shoulder at 690 nm may be correlated with the site-specific solvation of the photoexcited Pchlide a in polar protic solvents due to the site-specific coordination and hydrogen bonding interactions [152]. So the coordinated and hydrogen-bonded Pchlide a-$(MeOH)_n$ complexes were investigated theoretically in both the ground state and electronically excited states, especially focusing on the intermolecular coordination and hydrogen bonding interactions between Pchlide a and methanol molecules in the electronically excited state [152].

The optimized geometric structures of the isolated Pchlide a and the complex formed by Pchlide a and methanol molecules are shown in Figure 13.

In addition to a strong intermolecular coordination bond (**CB**) between Mg atom in Pchlide a and O atom in methanol, six types of intermolecular hydrogen bonds (from **HB-I** to **HB-VI**) are formed in the coordinated and hydrogen-bonded Pchlide a-$(MeOH)_4$ complex. For comparison, some other hydrogen-bonded Pchlide a-$(MeOH)_n$ ($n<4$) complexes have also been studied [152]. Herein, we denote the complex formed by Pchlide a and the coordinated methanol molecule as Pchlide a-MeOH. The hydrogen-bonded trimer formed by the Pchlide a-MeOH dimer and the methanol molecule bonded by the intermolecular hydrogen bonds **HB-I** and **HB-II** is denoted as Pchlide a-$(MeOH)_2$. Moreover, the hydrogen-bonded complex formed by Pchlide a-$(MeOH)_2$ and the methanol molecule bonded by hydrogen bonds **HB-V** and **HB-VI** is labelled as Pchlide a-$(MeOH)_3^a$. Similarly, we use Pchlide a-$(MeOH)_3^b$ refers to the hydrogen-bonded complex formed by Pchlide a-$(MeOH)_2$ and the methanol molecule which is bonded by the intermolecular hydrogen bonds **HB-III** and **HB-IV**.

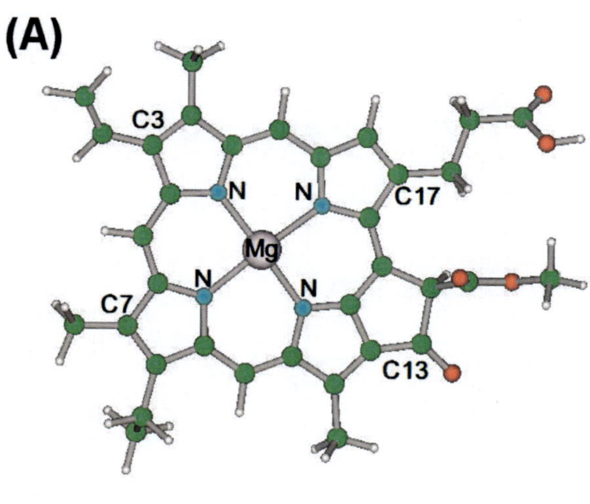

Figure 13 (Continued on next page).

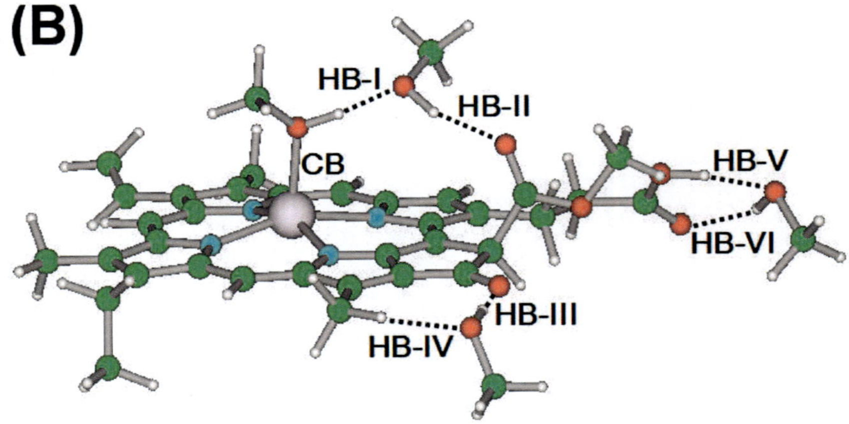

Figure 13. The optimized geometric structures of protochlorophyllide a (A); and its hydrogen-bonded complex with methanol (B). Nitrogen (N) and Magnesium (Mg) and some Carbon (C) atoms are labeled. Dotted lines denote the intermolecular hydrogen bonds. CB: coordination bond; HB: hydrogen bond. Taken from Ref. [152].

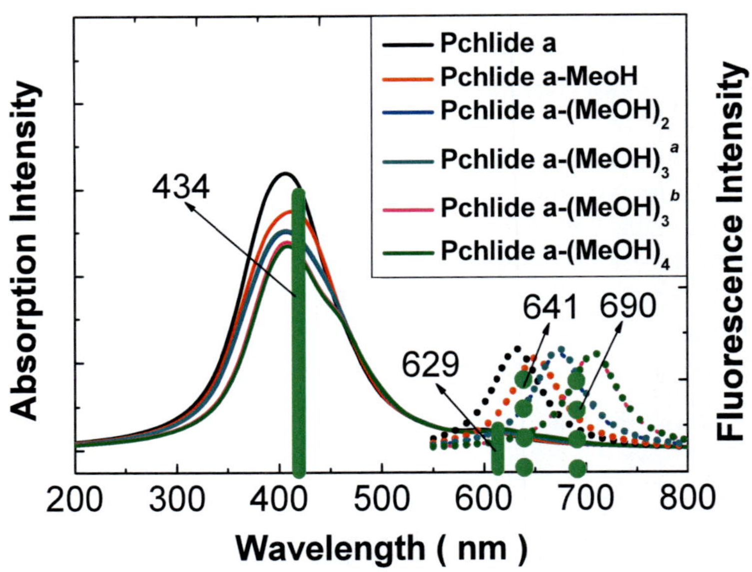

Figure 14. Calculated absorption (solid lines) and fluorescence spectra (dotted lines) of isolated Pchlide a and its various hydrogen-bonded Pchlide a-$(MeOH)_{1,2,3,4}$ complexes. Bold arrows indicate the order with increasing the number of MeOH molecules. Taken from Ref. [152].

Figure 14 shows the the calculated absorption and fluorescence spectra of the isolated Pchlide a and all the coordinated and hydrogen-bonded Pchlide a-$(MeOH)_n$ ($n \leq 4$) complexes. One can find that the calculated Soret band for the isolated Pchlide a and all the complexes is at around 434 nm, which is in good agreement with the experimental result [148]. Moreover, the calculated Q_y (S_1) band of the isolated Pchlide a is at 615 nm, while this band is redshifted to 668 nm for the coordinated and hydrogen-bonded Pchlide a-$(MeOH)_4$ complexes. So it is suggested that the only one peak located at 629 nm in the experimental Q_y (S_1) absorption band should be dominatively ascribed to the absorption of the isolated Pchlide a [152]. While no evident absorption peak of the coordinated and hydrogen-bonded Pchlide a-$(MeOH)_n$ complexes can be found in the experimental Q_y (S_1) band [151].

It is noted that both the absorption and fluorescence spectra of Pchlide a-$(MeOH)_2$ and Pchlide a-$(MeOH)_3{}^a$ as well as Pchlide a-$(MeOH)_3{}^b$ and Pchlide a-$(MeOH)_4$ are nearly the same. This indicates that the strong intermolecular hydrogen bonds **HB-V** and **HB-VI** cannot significantly influence the electronic spectra of the Pchlide a-$(MeOH)_n$ complexes [152]. This may be due to that the hydrogen bonds **HB-V** and **HB-VI** are far away from the conjugated moiety of Pchlide a molecule [152].

The calculated fluorescence of isolated Pchlide a is at 632 nm, while the fluorescence of the coordinated and hydrogen-bonded complexes is redshifted to be in the range from 651 to 708 nm. Thus, it can be demonstrated that the fluorescence maximum at 641 nm observed in experiments is originated from the isolated Pchlide a; while the novel fluorescence shoulder at around 690 nm of the Phclide a in methanol solvent should be ascribed to the coordinated and hydrogen-bonded Pchlide a-$(MeOH)_4$ complex [152]. On the other hand, the formation of this novel fluorescence shoulder can also indicate that the site-specific solvation plays a more important role on the influence of the spectral feature of Pchlide a in methanol solvent in the electronically excited state than that in the ground state. Consequently, the site-specific solvation which is attributed to the novel fluorescence shoulder at around 690 nm may be tightly associated with the hydrogen bonding dynamics in the electronically excited state [152].

The calculated binding energies of the intermolecular coordination and hydrogen bonds in both the ground state and S_1 state of the coordinated and hydrogen-bonded Pchlide a-$(MeOH)_4$ complex are listed in Table 5.

Table 5. Calculated binding energies (kJ/mol) for the intermolecular coordination bond (CB) and hydrogen bonds (HB) in different electronic states of the Pchlide a-$(MeOH)_4$ complex

	CB	HB-I + HB-II	HB-III + HB-IV	HB-V + HB-VI
S_0	56.42	68.83	25.41	49.76
S_1	61.76	73.91	32.09	50.28

One can find that the binding energy of the coordination bond **CB** is increased from the 56.42 kJ/mol in the ground state to the 61.76 kJ/mol in the S_1 state. The total hydrogen binding energy of **HB-I** and **HB-II** is 68.83 kJ/mol in ground state, while is increased to 73.91 kJ/mol in the S_1 state. In addition, the total hydrogen bond binding energy of **HB-III** and **HB-IV** is also increased to 32.09 kJ/mol in the S_1 state from 25.41 kJ/mol in ground state. It can be noted that the strong hydrogen bonds **HB-V** and **HB-VI** are nearly unchanged upon photoexcitation to the S_1 state. This coincides with that these strong intermolecular hydrogen bonds **HB-V** and **HB-VI** cannot affect the electronic spectra of the Pchlide a-$(MeOH)_n$ complexes as discussed above [152].

Table 6. Calculated bond lengths (Å) for the intermolecular coordination bond (CB) and hydrogen bonds (HB) in different electronic states of the Pchlide a-$(MeOH)_4$ complex

	CB	HB-I	HB-II	HB-III	HB-IV	HB-V	HB-VI
S_0	2.105	1.637	1.833	1.860	2.337	1.670	1.914
S_1	2.092	1.607	1.775	1.789	2.314	1.669	1.915

In Table 6, the bond lengths of all the intermolecular coordination and hydrogen bonds in the ground and electronically excited states are listed. One can find that the hydrogen bonds lengths of both **HB-V** and **HB-VI** remain unchanged upon photoexcitation to the S_1 state. While all the other intermolecular hydrogen bonds and coordination bond are shortened in the S_1 state. So the calculated bond lengths of all the intermolecular coordination and hydrogen bonds in the ground and electronically excited states are in accordance with the calculated binding energies [152]. At the same time, it has been confirmed that all the intermolecular coordination and hydrogen bonds except **HB-V** and **HB-VI** are significantly strengthened in the electronically excited state [152].

The intermolecular hydrogen bonding dynamics in the electronically excited states can be followed by the rearrangement of the intermolecular hydrogen-bonded solvent molecules [152]. As a consequence, the site-specific solvation could be initiated by the intermolecular coordination and hydrogen bonds strengthening upon photoexcitation to the S_1 state of the Pchlide a in methanol solvent [153]. Moreover, this site-specific solvation in the electronically excited state can induce a coordinated and hydrogen-bonded intermediate state, which is attributed to the formation of the novel fluorescence shoulder [152].

The ultrafast excited-state dynamics of Pchlide a was demonstrated to be strongly dependent on the solvent polarity in the time-resolved absorption and fluorescence spectroscopic experiments [149-151]. In the polar protic methanol solvent, the excited-state relaxation dynamics of Pchlide a is multiexponential with three distinguishable timescales of 4.0-4.5 ps for vibrational relaxation (VR) and vibrational energy redistribution (VER) of the initially excited S_1 state; 22-27 ps for the formation of an intermediate state; and 200 ps for the decay of this intermediate state back to the ground state [150]. While in nonpolar cyclohexane solvent, only the 4.5 ps relaxation process can be observed [151]. In addition to polarity, the viscosity of solvent can also affect the formation rate of the intermediate state. Upon increasing the viscosity by adding glycerol to a methanolic solution of Pchlide a, the formation process of the intermediate state is found to be decelerated [149-151]. Consequently, the formation of the intermediate state should be related to the dynamic solvation process of Pchlide a in the S_1 state [152,153]. Therefore, we assigned the intermediate state as the formation of a coordinated and hydrogen-bonded Pchlide a-$(MeOH)_n$ complex after site-specific solvation in the electronically excited state, which takes place within 22-27 ps timescale [152].

4. Summary and Conclusion

In this review, we have discussed our latest results of the hydrogen bonding effects on the photochemistry of chromophores in solution, especially focusing on the hydrogen bonding dynamics in the electronically excited states of chromophores and its influence on the ultrafast nonradiative deactivation processes of the fluorescent state. We have demonstrated for the first time that the intermolecular hydrogen bonds between chromophores and the hydrogen bonding solvents can be significantly strengthened upon photoexcitation to electronically excited states of chromophores. Furthermore, the ultrafast radiationless deactivation processes of the fluorescent state, such as internal conversion (IC), intermolecular photoinduced electron transfer (PET), etc. are strongly facilitated by the intermolecular hydrogen bonding due to the hydrogen bond strengthening in electronically excited states of chromophores. As a consequence, the fluorescence emission of chromophores in the hydrogen bonding solvents can be drastically quenched.

The hydrogen bonding dynamics of the intermolecular hydrogen bond C=O···H–O between C102 chromophore and the hydrogen donating solvents was theoretically investigated by monitoring the C=O and O–H stretching modes in different electronic states. We found that the C=O stretching band of the C102 chromophore can be induced a blueshift by the electronic excitation. On the contrary, both the C=O and O–H stretching bands can be induced by the intermolecular hydrogen bonding to redshift in the ground state. Moreover, the intermolecular hydrogen bonding induces a larger redshift for both the C=O and O–H stretching modes in the electronically excited state. Therefore, it was demonstrated that the intermolecular hydrogen bond C=O···H–O is significantly strengthened upon photoexcitation. It was also observed that the hydrogen bond length between O and H atoms is markedly decreased in the electronically excited state. At the same time, the bond lengths of hydrogen-bonded C=O and O–H groups are correspondingly increased.

The intermolecular hydrogen bonding can lower the energy levels of both the ground state and electronically excited states. Since the energy level of the electronically excited state will be lowered more than that of the ground state due to the hydrogen bond strengthening upon photoexcitation, the energy gap between the fluorescent state and ground state becomes smaller, so the electronic coupling between the two electronic states will be strongly increased. Thus, the IC process from the fluorescent state to the ground state for Fluorenone in alcoholic solvents is enhanced.

The intermolecular hydrogen bond strengthening can also enhance the electronic coupling between chromophores and hydrogen bonding solvents in the electronically excited state. Thus, the intermolecular hydrogen bonds could be regarded as a conduit for facilitating the ultrafast solute-solvent intermolecular PET, which we proposed to account for the fluorescence quenching of OX750 chromophore in alcoholic solvents. Moreover, the intermolecular PET was measured to take place within the same timescale as the hydrogen bonding dynamics by the FS TR SEP FD spectroscopy. Hence, the ultrafast solute-solvent intermolecular PET is tightly coupled with the intermolecular hydrogen bond strengthening in the electronically excited state. Consequently, the intermolecular PET can only pass through the strengthened intermolecular hydrogen bonds in the electronically excited state.

We also demonstrated that the intermolecular coordination and hydrogen bonds between Pchlide a and methanol molecules are strengthened in the electronically excited state. The excited-state site-specific solvation can be initiated by the strengthened intermolecular coordination and hydrogen bonding interactions. Moreover, the site-specific solvation of photoexcited Pchlide a in methanol induces the formation of an intermolecular coordinated and hydrogen-bonded complex in the electronically excited state. In our study, the excited-state intermolecular coordinated and hydrogen-bonded complex was used to account for the novel fluorescence shoulder at around 690 nm of Pchlide a in methanol.

ACKNOWLEDGMENT

This work was supported by NKBRSF (No: 2007CB815202) and NSFC (Nos: 20333050, and 20573110).

REFERENCES

[1] Turro, N. J. (1991). *Modern Molecular Photochemistry*. University Science Books: Sausalito, CA, USA.
[2] Han, K.L., He, G.Z. (2007). *J. Photochem. Photobiol. C: Photochem. Rev*, 8, 55.
[3] Bernardi, F., Olivucci, M., Robb, M. A. (1990). *Chem. Soc. Rev*, 25, 321.
[4] Mack, J., Bolton, J. R. (1999). *J. Photochem. Photobiol. A: Chem*, 128,1.
[5] Bitterwolf, T. E. (2004). *J. Organomet. Chem*, 689, 3939.
[6] Vaida, V. (2005). *Int. J. Photoenergy*, 7, 61.
[7] Garavelli, M. (2006).*Theo. Chem. Acc*, 116, 87.
[8] Macpherson, A. N., Gillbro, T. (1998). *J. Phys. Chem*, A 102, 5049.
[9] Zewail, A. H. (2000). *J. Phys. Chem*, A 104, 5660.
[10] Pecourt, J. M. L., Peon, J., Kohler, B. (2001). *J. Am. Chem. Soc*, 123, 10370.
[11] Rondonuwu, F. S., Watanabe, Y., Zhang, J. P., Furuichi, K., Koyama, Y. (2002). *Chem. Phys. Lett*, 357, 376.
[12] Shi, Y., Liu, J.Y., Han, K.L. (2005). *Chem. Phys. Lett*, 410, 260.
[13] Liu, J.Y., Fan, W.H., Han, K.L.,Deng, W.Q., Xu, D.L., Lou, N.Q. (2003). *J. Phys. Chem*, A 107, 10857.
[14] Owrutsky, J. C., Raftery, D., Hochstrasser, R. M. (1994). *Annu. Rev. Phys. Chem*, 45, 519.
[15] Nesbitt, D. J., Field, R. W. (1996). *J. Phys. Chem*, 100, 12735.
[16] Pal, S. K., Peon, J., Bagchi, B., Zewail, A. H. (2002). *J. Phys. Chem*, B 106, 12376.
[17] Waldeck, D. H. (1991).*Chem. Rev*, 91, 415.
[18] Dugave, C., Demange, L.(2003). *Chem. Rev*, 103, 2475.
[19] Zhao, G.J., Han, K.L., Lei, Y.B., Dou, Y. (2007). *J. Chem. Phys*, 127, 094307.
[20] Zhao, G.J., Liu, Y.H., Han, K.L., Dou, Y. (2008). *Chem. Phys. Lett*, 453, 29.
[21] Wasielewski,M. R. (1992). *Chem. Rev*, 92, 435.
[22] Weaver, M. J. (1992). *Chem. Rev*, 92, 463.
[23] Marcus, R. A. (1993). *Rev. Mod. Phys*, 65, 599.
[24] Bissell, R. A., de Silva, A. P., Gunaratne, H. Q. N., Lynch, P. L. M., Maguire, G. E. M., McCoy, C. P., Sundanayake, K. R. A. S. (1993). *Top. Curr. Chem*, 168, 223.
[25] Barbara, P. F., Meyer, T. J., Ratner, M. A. (1996). *J. Phys. Chem*, 100, 13148.
[26] Gray, H. B., Winkler, J. R. (1996). *Annu. Rev. Biochem*, 65, 537.
[27] Piotrowiak, P. (1999). *Chem. Soc. Rev*, 28, 143.
[28] Chu, T.S., Zhang, Y., Han, K.L. (2006). *Int. Rev. Phys. Chem*, 25, 201.

[29] Zhao, G.J., Liu, J.Y., Zhou, L.C., Han, K.L. (2007). *J. Phys. Chem*, B 111, 8940.
[30] Legon, A. C. (1993).*Chem. Soc. Rev*, 22, 153.
[31] Cukier, R. I., Nocera, D. G. (1998). *Annu. Rev. Biochem*, 49, 337.
[32] Tolbert, L. M., Solntsev, K. M. (2002). *Acc. Chem. Rev*, 35, 19.
[33] Bhattacharyya, K. (2003). *Acc. Chem. Rev*, 36, 95.
[34] Grabowski, Z. R., Rotkiewicz, K., Rettig, W. (2003). *Chem. Rev*, 103, 3899.
[35] Chen, R.K., Zhao, G.J., Yang, X.C., Jiang, X., Liu, J.F., Tian, H.N., Gao, Y., Han, K.L., Sun, M.T., Sun, L.C. (2008). *J. Mol. Struct*, 876, 102.
[36] Zhao, G.J., Han, K.L. (2008). *J. Comput. Chem,* DOI: 10.1002/jcc.20957.
[37] Zhao, G.J., Han, K.L. (2007). *J. Chem. Phys*, 127, 024306.
[38] Mataga, N., Shibata, Y., Chosrowjan, H., Yoshida, N., Osuka, A. (2000). *J. Phys. Chem*, B 104, 4001.
[39] Nibbering, E. T. J., Fidder, H., Pines, E. (2005). *Annu. Rev. Phys. Chem*, 56, 337.
[40] Kukura, P., McCamant, D. W., Mathies, R. A. (2007). *Annu. Rev. Phys. Chem*, 58, 461.
[41] Pshenichnikov, M. S., Duppen, K., Wiersma, D. A. (1995). *Phys. Rev. Lett*, 74, 674.
[42] Jordanides, X. J., Lang, M. J., Song, X. Y., Fleming, G. R. (1999). *J. Phys. Chem*, B 103, 7995.
[43] Kirkwood, J. C., Albrecht, A. C. (2000). *J. Raman Spectrosc*, 31, 107.
[44] Bredenbeck, J., Helbing, J., Behrendt, R., Renner, C., Moroder, L., Wachtveitl, J., Hamm, P. (2003). *J. Phys. Chem*, B 107,8654.
[45] Hochstrasser, R. M. (2007). *Proc. Natl. Acad. Sci*, USA, 104, 14189.
[46] Rischel, C., Rousse, A., Uschmann, I., Albouy, P. A., Geindre, J. P., Audebert, P., Gauthier, J. C., Forster, E., Martin, J. L., Antonetti, A. (1997). *Nature*, 390, 490.
[47] Schotte, F., Lim, M. H., Jackson, T. A., Smirnov, A. V., Scman, J., Olson, J. S., Phillips, G. N., Wulff, M., Anfinrud, P. A. (2003). *Science*, 300, 1944.
[48] Zewail, A. H. (2006). *Annu. Rev. Phys. Chem*, 57, 65.
[49] Jimenez, R., Fleming, G. R., Kumar, P. V., Maroncelli, M. (1994). *Nature*, 369, 471.
[50] Fleming, G. R., Cho, M. H. (1996). *Annu. Rev. Phys. Chem*, 47, 109.
[51] Tomasi, J. (2004) *Theo. Chem. Acc*, 112, 184.
[52] Abraham, M. H. (1993) *Chem. Soc. Rev*, 22, 73.
[53] Aakeroy, C. B., Seddon, K. R. (1993). *Chem. Soc. Rev*, 22, 397.
[54] Mcdonald, I. K., Thornton, J. M. (1994). *J. Mol. Biol*, 238, 777.
[55] Gordon, M. S., Jensen, J. H. (1996). *Acc. Chem. Res*, 29, 536.
[56] Meot-Ner, M. (2005). *Chem. Rev,* 105, 213.
[57] Steiner, T. (2002). *Angew. Chem. Int*, Ed. 41, 48.
[58] Kearley, G. J., Fillaux, F., Baron, M. H.,Bennington, S., Tomkinson, J. A. (1994). *Science,* 264, 1285.
[59] Luzar, A., Chandler, D. (1996). *Nature*, 379, 55.
[60] Woutersen, S., Emmerichs, U., Bakker, H. J. (1997). *Science*, 278, 658.
[61] Chudoba, C., Nibbering, E. T. J., Elsaesser, T. (1998). *Phys Rev. Lett*, 81, 3010.
[62] Fecko, C. J., Eaves, J. D.,Loparo, J. J.,Tokmakoff, A.,Geissler, P. L. (2003). *Science*, 301,1698.
[63] Nibbering, E. T. J., Tschirschwitz, F., Chudoba, C., Elsaesser, T. (2000). *J. Phys. Chem*, A 104, 4236.
[64] Palit, D. K., Zhang, T. Q., Kumazaki, S.,Yoshihara, K. (2003). *J. Phys. Chem*, A 107, 10798.

[65] Asbury, J. B., Steinel, T., Stromberg, C., Gaffney, K. J., Piletic, I. R.,Goun, A., Fayer, M. D. (2003). *Phys. Rev. Lett*, 91, 237402.
[66] Van Tassle, A. J., Prantil, M. A., Fleming, G. R. (2006). *J. Phys. Chem*, B 110, 18989.
[67] Pines, E., Pines, D., Ma, Y. Z., Fleming, G. R. (2004). *ChemPhysChem*, 5, 1315.
[68] Biczók, L., Bérces, T., Linschitz, H. (1997). *J. Am. Chem. Soc*, 119, 11071.
[69] Yoshihara, K. (1999). *Adv. Chem. Phys*, 107, 371.
[70] Chen, L. H., McBranch, D. W., Wang, H. L., Helgeson, R., Wudl, F., Whitten, D. G. (1999). *Proc. Natl. Acad. Sci*, USA, 96, 12287.
[71] Kyrychenko, A., Herbich, J., Wu, F., Thummel, R. P., Waluk, J. (2000). *J. Am. Chem. Soc*, 122, 2818.
[72] Zhuang, X., Ha, T., Kim, H. D.,Centner, T., Labeit, S., Chu, S. (2000). *Proc. Natl. Acad. Sci*, USA, 97, 14241.
[73] Sessler, J. L., Sationsatham, M., Brown, C. T., Rhodes, T. A., Wiederrecht, G. (2001). *J. Am. Chem. Soc*, 123, 3655.
[74] Klapstein, D., Pischel, U., Nau, W. M. (2002). *J. Am. Chem. Soc*, 124, 11349.
[75] Voityuk, A. A., Siriwong, K., Rösch, N. (2004). *Angew. Chem. Int*, Ed. 43, 624.
[76] Mataga, N., Chosrowjan, H., Taniguchi, S.(2005). *J.Photochem.Photobiol.C: Photochem. Rev*, 6, 37.
[77] Zhao, G.J., Han, K.L. (2007). *J. Phys. Chem*, A 111, 2469.
[78] Zhao, G.J., Han, K.L. (2007). *J. Phys. Chem*, A 111, 9218.
[79] Liu, J.Y., Fan, W.H., Han, K.L., Xu, D.L., Lou, N.Q. (2003). *J. Phys. Chem*, A 107,1914.
[80] Zhou, L.C., Liu, J.F., Zhao, G.J., Shi, Y., Peng, X.J., Han, K.L. (2007).*Chem. Phys*, 333, 179.
[81] Becke, A. D. (1993) *J. Chem. Phys*, 98, 5648.
[82] Ahlrichs, R., Bär, M., Horn, H., Kölmel, C. (1989). *Chem. Phys. Lett*, 162, 165.
[83] Weigend, F. (2002). *Phys. Chem. Chem. Phys*, 4, 4285.
[84] Ahlrichs, R. (2004). *Phys. Chem. Chem. Phys*, 6, 5119.
[85] Arnim, M. V., Ahlrichs, R. (1998). *J. Comput. Chem*, 19, 1746.
[86] Schäfer, A., Huber, C., Ahlrichs, R. (1994). *J. Chem. Phys*, 100, 5829.
[87] Eichkorn, K., Weigend, F., Treutler, O., Ahlrichs, R. (1997). *Theo. Chem. Acc*, 97, 119.
[88] Furche, F., Ahlrichs, R. (2002). *J. Chem. Phys*, 117, 7433.
[89] Furche, F., Ahlrichs, R. (2004). *J. Chem. Phys*, 121, 12772.
[90] Bauernschmitt, R., Ahlrichs, R. (1996). *Chem. Phys. Lett*, 256, 454.
[91] Bauernschmitt, R., Häser, M., Treutler, O., Ahlrichs, R. (1997). *Chem. Phys. Lett*, 264, 573.
[92] Grimme, S., Furche, F., Ahlrichs, R. (2002). *Chem. Phys. Lett*, 361, 321.
[93] Rappoport, D., Furche, F. (2005). *J. Chem. Phys*, 122, 064105.
[94] Treutler, O., Ahlrichs, R. (1995). *J. Chem. Phys*, 102, 346.
[95] Deglmann, P., Furche, F., Ahlrichs, R. (2002).*Chem. Phys. Lett*, 362, 511.
[96] Deglmann, P., May, K., Furche, F., Ahlrichs, R. (2004). *Chem. Phys. Lett*, 384, 103.
[97] Hättig, C. (2003). *J. Chem. Phys*, 118, 7751.
[98] Köhn, A., Hättig, C. (2003). *J. Chem. Phys*, 119, 5021.
[99] Klamt, A., Schüürmann, G. (1993). *J. Chem. Soc. Perkin Trans*, 2, 799.
[100] Boys, S. F., Bernardi, F. (1970). *Mol. Phys*, 19, 553.
[101] Sobolewski, A. L., Domcke, W. (2004). *J. Phys. Chem*, A 108, 10917.

[102] Sobolewski, A. L., Domcke, W., Hattig, C. (2006). *J. Phys. Chem*, A 110, 6301.
[103] Xie, D.Q., Zeng, J. (2004). *J. Comput. Chem*, 25, 1487.
[104] Dreuw, A., Weisman, J. L., Head-Gordon, M. (2003). *J. Chem. Phys*, 119, 2943.
[105] Wanko, M., Garavelli, M., Bernardi, F., Niehaus, T. A., Frauenheim, T., Elstner, M. (2004). *J. Chem. Phys*, 120, 1674.
[106] Hieringer, W., Görling, A. (2006). *Chem. Phys. Lett*, 419, 557.
[107] Zhou, L.C., Zhao, G.J., Liu, J.F., Han, K.L., Wu, Y.K., Peng, X.J., Sun, M.T. (2007). *J. Photochem. Photobiol, A: Chem*, 187, 305.
[108] Wan, S.B., Liu, S.S., Zhao, G.J., Chen, M.D., Han, K.L., Sun, M.T. (2007). *Biophys. Chem*, 129, 218.
[109] Barone, V., Polimeno, A. (2007). *Chem. Soc. Rev*, 36, 1724.
[110] Southern, C. A., Levy, D. H., Florio, G. M., Longarte, A., Zwier, T. S. (2003). *J. Phys. Chem*, A 107, 4032.
[111] Glasbeek, M., Zhang, H. (2004).*Chem. Rev*, 104, 1929.
[112] Das, K., Jain, B., Patel, H. S. (2006). *J. Phys. Chem*, A 110, 1698.
[113] Fujii, T., Sano, M., Mishima, S., Hiratsuka, H. (1996). *Bull. Chem. Soc. Jpn*, 69, 1833.
[114] Biczók, L., Bérces, T., Linschitz, H. (1997). *J. Am. Chem. Soc*, 119, 11071.
[115] Samant, V., Singh, A. K., Ramakrishna, G., Ghosh, H. N., Ghanty, T. K., Palit, D. K. (2005). *J. Phys. Chem*, A 109, 8693.
[116] Jamison, J. L., Davenport, L., Williams, B. W. (2006). *Chem. Phys. Lett*, 422, 30.
[117] Petkova, I., Mudadu, M. S., Singh, A., Thummel, R. P., Van Stokkum, I. H. M., Buma, W. J., Waluk, J. (2007). *J. Phys. Chem*, A 111, 11400.
[118] Zhang, J., Gembicky, M., Messerschmidt, M., Coppens, P. (2007). *Chem. Commun*, 23, 2399.
[119] Odoi, M. Y., Hammer, N. I., Rathnayake, H. P., Lahti, P. M., Barnes, M. D.(2007). *Chem Phys Chem*, 8, 1481.
[120] Halls, J. J. M., Cornil, J., Dos Santos, D. A. R., Silbey, D. H., Hwang, A. B., Holmes, J. L., Brédas, R., Friend, H. (1998). *Phys. Rev*, B 60, 5721.
[121] Prasad, E., Gopidas, K. R. (2000). *J. Am. Chem. Soc*, 122, 3191.
[122] Arimura, T., Ide, S., Suga, Y., Nishioka, T., Murata, S., Tachiya, M., Nagamura, T., Inoue, H. (2001). *J. Am. Chem. Soc*, 123,10744.
[123] Fukuzumi, S., Yoshida, Y., Okamoto, K., Imahori, H., Araki, Y., Ito, O. (2002). *J. Am. Chem. Soc*, 124, 6794.
[124] Cotlet, M., Masuo, S., Lor, M., Fron, E., Van der Auweraer, M., Műllen, K., Hofkens, J., De Schryver, F. (2004). *Angew.Chem. Int*, Ed. 43, 6116.
[125] Kwok, W. M., George, M. W., Grills, D. C., Ma, C. S., Matousek, P., Parker, A. W., Phillips, D., Toner, W. T., Towrie, M. (2003). *Angew. Chem. Int*, Ed. 42, 1826.
[126] Voityuk, A. A., Siriwong, K., Rösch, N. (2004). *Angew. Chem. Int*, Ed. 43, 624.
[127] Hirose, Y., Yui, H., Sawada, T. (2004). *J. Phys. Chem*, B 108, 9070.
[128] Ward, M. D. (1997). *Chem. Soc. Rev*, 26, 365.
[129] Jiménez, M. C., Miranda, M. A., Tormos, R. (2005). *Chem. Soc. Rev*, 34,783.
[130] Benniston, A. C., Harriman, A.(2006). *Chem. Soc. Rev*, 35,169.
[131] Lawrence, D. S., Jiang, T., Levett, M. (1995). *Chem. Rev*, 95, 2229.
[132] Lemaur, V., Steel, M., Beljonne, D., Brédas, J. L., Cornil, J. (2005). *J. Am. Chem. Soc*, 127, 6077.
[133] Benigo, A. J., Ahmed, E., Berg, M. (1996). *J. Chem. Phys*, 104, 7382.

[134]Sheu, S. Y., Yang, D. Y., Selzel, H. L., Schlag, E. W. (2003). *Proc. Natl. Acad. Sci*, USA, 100, 12683.

[135] Yoshihara, K., Tominaga, K., Nagasawa, Y. (1995). *Bull. Chem. Soc. Jpn*, 68, 696.

[136] Bollivar, D. W. (2006). *Photosynth. Res*, 90, 173.

[137] Zhong, D. P. (2007).*Curr. Opin. Chem. Biol*, 11,174.

[138] El Hamouri, B., Brouers, M., Sironval, C. (1981). *Plant Sci. Lett*, 21,375.

[139] Ryderg, M., Sundqvist, C. (1982). *Physiol. Plant*, 56,133.

[140] Böddi, B., Lindsten, A., Ryderg, M., Sundqvist, C. (1989).*Physiol. Plant*, 76,135.

[141] Franck, F., Bereza, B., Böddi, B. (1999). *Photosynth. Res*, 59, 53.

[142] Mysliwa-Kurdziel, B., Franck, F., Strzalka, K. (1999). *Photochem. Photobiol*, 70, 616.

[143] Shipman, L. I., Cotton, T. M., Norris, J. R., Katz, J. J. (1976). *J. Am. Chem. Soc*, 98, 8222.

[144] Renge, I., Avarmaa, R. (1985). *Photochem, Photobiol*, 42, 253.

[145] Krawczyk, S. (1989). *Biochim, Biophys. Acta*, 976,140.

[146]Limantara, L., Sakamoto, S., Koyama, Y., Nagae, H. (1997). *Photochem.Photobiol*, 65, 330.

[147] Vladkova, R. (2000). *Photochem. Photobiol*, 71,71.

[148] Mysliwa-Kurdziel, B., Kruk, J., Strzalka, K. (2004). *Photochem. Photobiol*, 79,62.

[149] Dietzek, B., Maksimenka, R., Siebert, T., Birckner, E., Kiefer, W., Popp, J., Hermann, G., Schmitt, M. (2004).*Chem. Phys. Lett*, 397,110.

[150] Dietzek, B., Kiefer, W., Popp, J., Hermann, G., Schmitt, M. (2006). *J. Phys. Chem*, B 110, 4399.

[151] Dietzek, B., Kiefer, W., Yartsev, A., Sundström, V., Schellenberg, P., Grigaravicius, P., Hermann, G., Popp, J., Schmitt, M. (2006). *ChemPhysChem*, 7, 1727.

[152] Zhao, G.J., Han, K.L. (2008). *Biophys. J*, 94, 38.

[153] Molotsky, T., Huppert, D. (2003). *J. Phys. Chem*, A 107, 2769.

In: Photochemistry Research Progress
Editors: A. Sánchez, S. J. Gutierrez

ISBN: 978-1-60456-568-3

Chapter 6

PHOTOCHEMICAL GENERATION OF SILICON-CENTERED RADICALS AND THEIR REACTIONS

Al Postigo*
Faculty of Science University of Belgrano
Villanueva 1324 CP 1426 Buenos Aires Argentina

ABSTRACT

This chapter is organized according to classes of silicon-containing compounds that give rise to silyl-centered radicals through photolytic methods, and then reviews their reactivity, their spectroscopic data, and their possible applications. Such diverse sources of silicon compounds include simple hydrosilanes, alkyl-, alkyl-aryl, and arylsilanes, disilanes, polysilanes, acylsilanes, silacycloalkanes, and silyl ethers.

Photochemically-induced homolytic cleavage of Si-X bonds (where X can be B, Ge, or other elements) will also be described as available sources of silyl radicals. The study of the photochemistry of different organic compounds, which bear common silicon-protective groups, their photolytic de-protection reactions, and the silyl radicals generated thereafter, will be presented.

It is worthwhile to emphasize that this chapter covers only silicon-centered radicals that are directly or indirectly generated by the interaction of silicon-containing compounds with light, thus excluding silyl radicals that are obtained through thermal methods. Thermal-decomposition methods for the production of silicon-centered radicals are widely distributed, and constitute the subject of future review articles. This chapter also circumvents other silicon-based intermediates, such as silylenes, silenes, and radical ions of silicon, which are also the subject of future review studies. Throughout the chapter it will become apparent that the photochemical generation of silicon-centered radicals is often accompanied by the production of other kinds of reactive silicon intermediates (silicon–centered biradicals, silylenes, silenes, etc.).

* Tel./Fax.: +54 115411-4700 e-mail: apostigo@ub.edu.ar

Introduction

Silicon-centered radicals can either be produced through thermal or photolytic reactions. Thermal initiation entails a radical precursor (obtained by a primary thermal decomposition reaction) and a silicon-containing compound, whereas direct or indirect irradiation can give rise to silyl radicals arising from different compounds containing Si atoms (ranging from simple organosilanes to silicon surfaces) and light.

Silyl radical reactivity has been widely studied and surveyed in the last decade, and a great deal of information regarding rate constants of reactions involving silicon-centered radicals has been presented in important review articles and book chapters. Chatgilialoglu has presented one of the most complete recent reviews on silyl radical reactivity in the book Organosilanes in Radical Chemistry.[1] Organosilane photochemistry is also regularly surveyed.[2]

Since a hydrogen atom of a organosilyl hydride is, however, resistant to abstraction by a carbon-centered radical, organosilyl radicals are more difficult to be produced from the corresponding silicon hydrides than the tin-centered radicals are from their respective tin hydrides. Polarity-reversal catalysis is reported to promote an efficient radical chain process, using silanes and catalytic amounts of a thiol. Thus, the carbon-centered radical (from initiation) abstracts a hydrogen atom from the thiol. The resulting thiyl radical (sulfur-centered radical) abstracts a hydrogen from the silane (silicon hydride), generating a silyl radical (silicon-centered radical). The thiol is regenerated, thus propagating the chain cycle.[3]

The photochemical production of silyl radicals is a very economical and convenient route for producing them. For example, photochemically produced t-BuO• radicals have been mainly used for the generation of silyl radicals to be studied by spectroscopic techniques. There exist several other methods for generation of silyl radicals using direct interaction of hydrosilanes with light. However, none of them is of general usefulness, being limited to some specific examples.[4] A second way involves the interaction with a ketone triplet state such as that of benzophenone that leads to hydrogen abstraction, which generates ketyl and silyl radicals.[5] Another indirect way of producing silyl radicals involves an Energy Transfer (ET) process in which a disilane in the presence of an acceptor such as a dicyanoanthracene, DCA, generates the radical cation of the disilane, which upon reaction with singlet oxygen, yields silyl radicals and silyl-substituted peroxyl radicals. The Si-Si bond homolysis of the disilane radical cation can also ensue after attack by other nucleophiles, such as acetonitrile, and water.[6] However, this latter approach has not found much synthetic application. Indirect methods for silyl radical production, such as photosensitizations, will be discussed in detail in this chapter.

The usefulness of organosilyl radicals in organic synthesis is very well documented. The replacement of organotin radicals with organosilyl radicals has driven an exhaustive study on these latter intermediates. Silicon-centered radicals are well known to react with C-C and C-O multiple bonds, with bromo- and iodocompounds as dehalogenation reagents, with azides as an ammination method, and with numerous other substrates, both in organic solvents, and more recently, in water.[7] Some synthetically-useful reactions of silyl radicals have been reported. Photooxidative Si-Si bond cleavage can result in silylation of electron-deficient arenes, although chemical yields are low. The major products of cyano-substituted benzenes are believed to derive from silyl radical addition to the arene radical anion followed by expulsion of cyanide. Ring silylation of trifluoromethyl-substituted benzenes, on the other hand, is a minor process of homolytic aromatic substitution by silyl radicals. Higher yields are obtained by phenanthrene cosensitized photolysis, and this technique has been applied successfully to the photosilylation of electron-deficient dicyano-substituted alkenes by hexamethyldisilane. Phenanthrene cosensitized photolysis of 1,1-dicyano-2-arylethenes with hexamethyldisilane in acetonitrile affords regioselectively more than 75 % yield of 1,1-dicyano-2-aryl-2-(trimethylsilyl)ethanes. Silylation involves addition of trimehylsilyl radicals to the arylethene anion radical to give a carbanion that is protonated by the solvent.

1.-Silyl Radicals From Hydrosilanes, Peralkylsilanes, Silyl Ethers, And Benzylsilanes:

Silicon-centered radicals have been known for quite some time to arise from Hg sensitization of simple hydrosilanes, such as SiH_4, $MeSiH_3$, Me_2SiH_2, and Me_3SiH, involving exclusive H atom loss from Si-H bond. The fate of these Si- centered radicals is Si-Si recombination and disproportionation.[8]

The laser flash photolysis of SiH_4 - O_2 mixture has been reported as early as 1996. The reaction conditions generate silyl radicals $H_3Si\bullet$ which react with oxygen to give an excited acyclic H_3SiOO species. It has been proposed that acyclic H_3SiOO cyclizes to the unsubstituted dioxasilirane H_2SiO_2 radical 1, which then rearranges to silaformic acid [HSi(=O)OH], (Figure 1)[9].

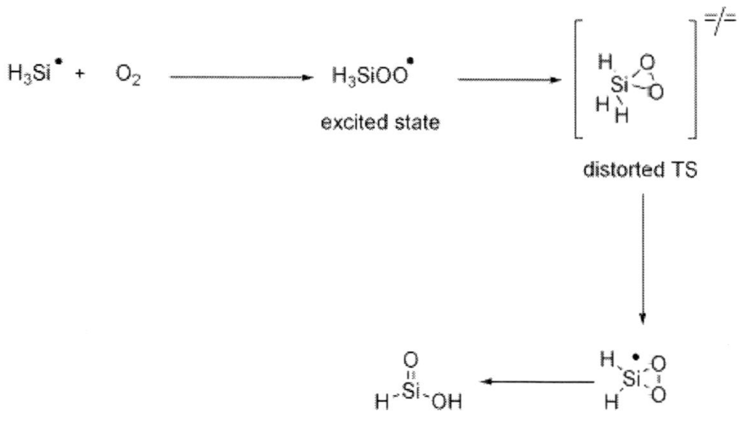

Figure 1. Reaction of silane-derived silyl radicals with dioxygen.

When ring-fused cyclopropylsilyl ethers undergo oxidative photoinduced electron transfer (PET), the cyclopropane ring opens and forms a reactive β-keto radical (and a silyl radical) which undergoes intramolecular cyclization.[10] This reaction is synthetically useful, since products derived from a radical-cascade are obtained. Thus, when compounds 2 and 3 (eq 1) are irradiated in acetonitrile in the presence of DCA, compounds 4 and 5 are obtained with a high degree of stereoselectivity.

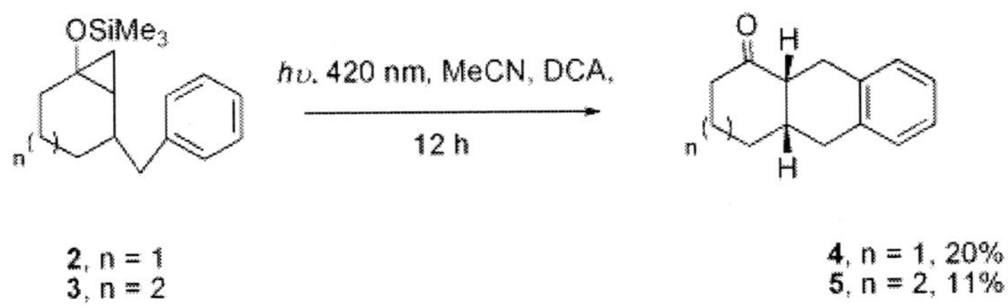

(1)

In this latter reaction, the sensitizer DCA oxidizes substrates 2 and 3 to their respective radical cations. Exocyclic cyclopropane ring opening leads to the formation of a β-keto radical, which further cyclizes to the tricyclic compounds 4 and 5. The last step is H atom elimination to regain aromaticity. Thus, in this way, the PET-induced ring opening of cyclopropylsilyl ethers is quite suitable for the production of polycyclic compounds with high degree of stereoselectivity.[11]

A new class of photochemically removable protecting groups for alcohols has been reported: the (hydroxystyryl)diisopropylsilyl (HDSIS) and the ((2-hydroxy-3-naphthyl)vinyl)-diisopropylsilyl groups (HNVDS).[12] The HSDIS group is removed with 254 nm irradiation, and the HNVDS group is removed with 350 nm irradiation. A variety of primary and secondary alcohols are protected with these groups, and can be deprotected cleanly and in high yields. In the photodeprotection reactions, however, no silyl radical intermediates are proposed. In these processes, two major classes of photochemical pathways are identified: *trans→cis* isomerization leading to an intramolecular nucleophilic substitution at silicon, and 1,5-silyl shift, leading to an unsymmetrical dialkoxysilane.

Benzylsilanes show a remarkable solvent effect. For the laser flash photolysis of benzyltrimethylsilane in methanol, absorption bands due to the benzyl type radicals and the triplet are observed, while in cyclohexane, only T-T absorption is observed.[13]

2. Silyl Radicals from Aromatic Silanes

Irradiation of certain aromatic silanes results in homolysis of the Si-C bond to produce silyl radicals. One of the few examples displaying this type of reactivity is tetraphenylsilane. The triphenylsilyl radicals are trapped by CCl_4 to give chlorotriphenylsilane, reportedly in high yield. When generated by laser flash photolysis in an emulsion of hexadecyl sulfate sodium salt, pentanol, and hexadecane, the triphenylsilyl radical exhibits a transient absorption at 315 nm. Under these heterogeneous conditions, the radical decay profile can be resolved into two components, a fast decay corresponding to geminate radical pairs, and a slow decay attributable to free silyl radicals derived from cage escape.

Transient absorption studies and magnetic field effects suggest that Si-C bond homolysis occurs in the triplet excited state of the arene to give, initially, the triplet geminate pair $Ph_3Si\bullet$ $\bullet Ph$. The rise time of the absorption of the transient correlates with the triplet-triplet absorption of the reagent. Transient absorption spectrum of triphenylsilyl radical at 328 nm is especially prominent in acetonitrile and can be monitored in THF with various concentrations of *trans*-piperylene (a triplet quencher). Stern-Volmer analysis of the radical yield at various diene concentrations provides an estimate of 6 ± 3 ns for the lifetime of the triplet excited state of the precursor aromatic silane.

It has been observed that the decay of the geminate pair is subject to field effects. At 1 to 1.35 T, the recombination of the geminate pair is inhibited due to the retardation of the triplet converting to the singlet. The result of this is an increase in the yield of cage-escape of the triphenylsilyl radicals. Triphenylsilyl radicals and methyldiphenylsilyl radicals (generated from alkoxy radical abstraction from the corresponding hydrosilanes) exhibit similar UV-transient absorption spectra, with maxima in the range 300-330 nm and a weak absorption at 380-450 nm.[14]

Photochemical desilylation reactions of aromatic silanes take place after excited state protonation of the arene, followed by Si-C bond cleavage in the intermediate para-silyl carbocation. The mechanism is not a simple homolytic cleavage to radicals, but more likely involves certain excited state interactions with the protic solvent methanol.

Certain aryl- substituted silacycloalkanes, such as phenyl-substituted silacyclobutanes, undergo Si-C bond cleavage to generate silicon-centered radicals that suffer internal rearrangement reactions.[15] As depicted in Figure 2, photolysis of compound 6 gives rise to a series of rearranged proposed intermediates that are trapped with methanol. In this account, an array of other intermediates such as silenes (intermediate bearing Si = C double bond) are proposed to be involved in the reaction.

The photochemistry of 4-diphenyl(trimethylsilyl)methyl-*N,N*-dimethylaniline (7) has been studied in various solvents.[16] In polar media like MeCN, 7 undergoes photoionization to the radical cation $7^{\bullet+}$ in a two-photon process, and in part dissociates to radical 8 via the triplet state (*vide infra*). The bulkiness of the triphenylmethyl group prevents the dissociation of $7^{\bullet+}$ via a nucleophile assisted C-Si bond cleavage to radical 8. In non-polar solvents such as hexane or cyclohexane, only the second process takes place, leading to radical 8 in high quantum yields. The complete sequence, $7 \rightarrow S1 \rightarrow T1 \rightarrow 8$ has been elucidated using nanosecond and picosecond laser flash photolysis.

In Compound 7, the dimethylanilino chromophore, as a result of its low ionization potential provokes the photoionization process in polar solvents, giving the corresponding radical cation which undergoes very efficient fragmentation in solution, generating radical 8, and the trimethylsilyl cation. This contrasts with the photobehaviour of compound 9, which upon photolysis in polar solvents generates trimethylsilyl radicals and p-(benzoyl)triphenylmethyl radicals in good yields. Compound 9 has been proposed as a good photoinitiator in radical polymerization reactions.

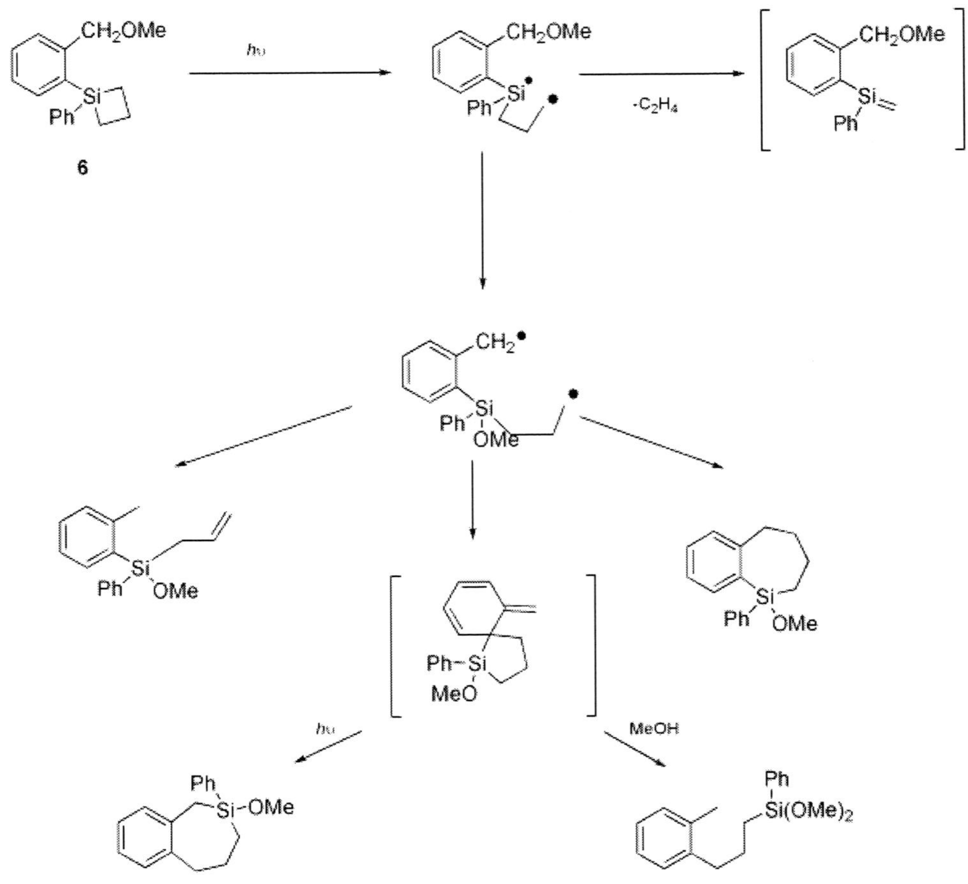

Figure 2. Silyl radicals from photochemistry of phenyl-substituted silacyclobutanes.

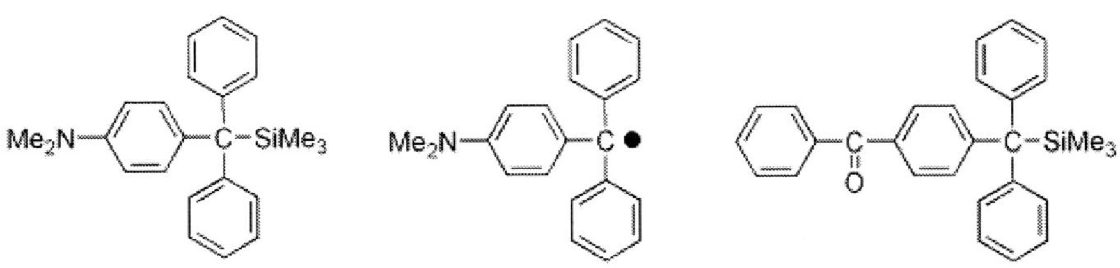

7 8 9

3. Silyl Radicals from Disilanes

Photolysis of aryldisilanes results in the production of three types of intermediates: free silyl radicals from homolysis of Si-Si bonds (path a); silatrienes arising from [1,3]-silyl migration into one of the aromatic rings (path b); and silenes, by a concerted dehydrosilylation reaction (path c) (Figure 3).

Direct photolysis of 1,2-ditertbutyl-1,1,2,2-tetraphenyldisilane in chloroform gives *tert*-butyldiphenylsilyl radicals, which are trapped by the solvent to afford chloro-*tert*-butyldiphenylsilane, in 95 % yield.[17] In this work, nanosecond laser flash photolysis studies have been used to demonstrate the formation of *tert*-butyldiphenyl silyl radicals in the photolysis of the disilane. Also, in this work, a series of bimolecular rate constants for the reactions of *tert*-butyldiphenyl silyl radicals with a variety of alkyl halides, alkenes and carbonyl compounds are informed.

Silenic products are derived from the lowest excited singlet state and are the major products in nonpolar solvents, while silyl radicals are derived from the lowest excited triplet state of the disilane and are the major products in polar solvents such as acetonitrile.[18] By laser flash photolysis experiments, it has been found that, saturation of MeCN solutions of triphenyltrimethyldisilane with *trans*-piperylene (a triplet quencher) provokes a reduction of the silyl radical transient absorption compared to that of the silatriene, verifying that the triplet excited manifold of the disilane is responsible for the silyl radical formation.

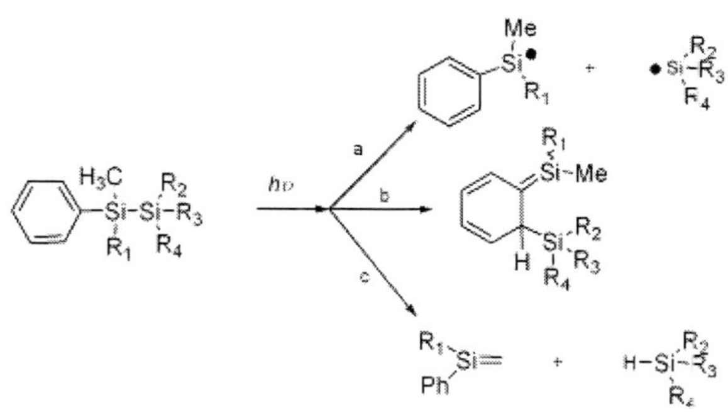

Figure 3. Different reaction pathways from photolysis of aryldisilanes.

It should be possible to design an arylalkyldisilane, which would give only silyl radicals upon photolysis. Such a compound could be extremely useful from diverse points of view, as a synthetic target, and as a potential photoinitiator. Organosilyl radicals undergo several well-known reactions such as halogen atom abstraction from alkyl halides and addition to unsaturated species such as alkenes and carbonyl compounds. Many of these reactions have been applied fruitfully in synthetic organic chemistry. While rate constants for many of these reactions have been reported, most of these data are qualitative and have been obtained by competition kinetics using product analysis or EPR spectroscopy. A few studies involving kinetic EPR spectroscopy or laser flash photolysis techniques to determine absolute rate constants have been reported but are mostly limited to reactions of trialkylsilyl radicals. Few absolute rate constants for the reactions of arylsilyl radicals with these substrates are available. The bulk of the kinetic data on silyl radical reactivity that does exist has been obtained by methods in which the radical is produced "indirectly", as for example by reaction of photochemically produced alkoxy radicals with a hydrosilane.

Measurement of bimolecular rate constants for reaction of silyl radicals with various substrates is then carried out by monitoring the decay of the silyl radical directly or by monitoring the growth of the product of reaction of the radical with either the substrate or a secondary probe (*e.g.*, benzyl) as a function of the concentration of added substrate. These methods, while extremely powerful, typically require high (>1 M) concentrations of silane and peroxide in order to be successful. This is because Si-H abstraction by alkoxy radicals is relatively slow ($\leq 10^7$ $M^{-1}s^{-1}$), and the absorptivity of dialkyl peroxides is fairly low above 300 nm. A system which yields arylsilyl radicals in a single chemical step, in any solvent, and using low precursor concentrations would clearly be interesting to study. While most aryldisilanes yield silyl radicals to some extent upon photolysis, their usefulness in spectroscopic studies of the radicals is limited because the competing silene (path c, Figure 3) absorptions obscure those due to the radicals of interest. Thus, in order for disilanes to be useful as silyl radical precursors, the alternative rearrangement pathways must be effectively blocked. The di*tert*butyltetraphenyldisilane 10 is a good source of silyl radicals from a disilane, as shown in Figure 4 below.

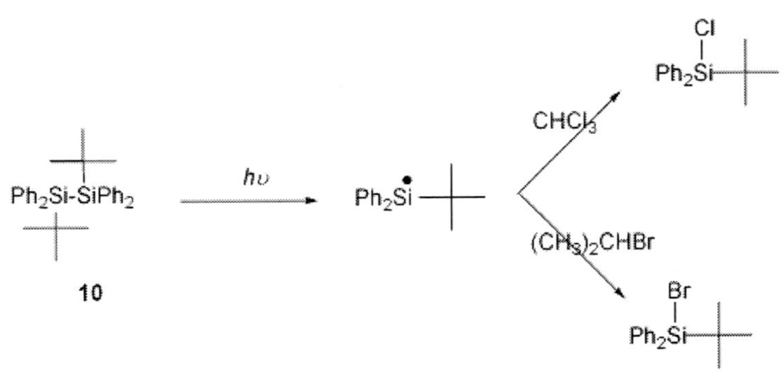

10

Figure 4. Silyl radicals from di*tert*butyltetraphenyldisilane.

Photolysis of hexaphenyldisilane leads to triphenylsilyl radicals, which can be studied and characterized by microsecond laser flash photolysis [19]. By nanosecond laser flash photolysis, the optical transient spectrum of triphenylsilyl radicals has been observed with an absorption maximum at 328 nm in *n*-hexane, *vide supra*.[20] In carefully-dried de-oxygenated MeCN, the triphenylsilyl radical absorption maximum extends to 450 nm. Oxygen, 2,3-dimethyl-1,3-butadiene (DMB), and CCl_4 affect the lifetime of the triphenylsilyl radical absorption at 330 nm as well as that of the longer wavelength transient. However, $CHCl_3$ at concentrations < 0.02 M does not affect the lifetime of the long wavelength transient absorption, but quenches that of the radical efficiently.

Photolysis of 1,1,1-trimethyl-2,2,2-triphenyldisilane in hydrocarbon solvent yields the transient 1,1-diphenyl-1-(sila)hexatriene. In contrast, photolysis in more polar solvents leads to high yields of the corresponding silyl free radicals. The high sensitivity of this compound to give one or other kind of intermediate according to solvent polarity threw light to the fact that silyl free radicals are formed from the triplet excited state of the disilane, and are not precursor to silene products. However, when 1,1,1-triphenylsila-2,2,2-trimethylgermane is photolyzed in the presence of DMB, radical-derived products are still observed, thus indicating that radical formation is probably both singlet- and triplet-derived. While photolysis of 1,1,1-trimethyl-2,2,2-triphenyldisilane leads to Si-Si bond homolysis only in polar solvents via reaction of the lowest triplet excited state, the authors could not rule out that in the case of 1,1,1-triphenylsila-2,2,2-trimethylgermane some portion of the reaction may be derived from the lowest singlet excited state.

It has been shown that 1,1,1-triisopropyl-2,2,2-triphenyldisilane (11, not shown), undergoes Si-Si bond homolysis upon direct irradiation (254 nm) to yield silyl radicals (triisopropylsilyl radicals) in competition with rearrangement to silatriene. The yield of silyl radicals is substantially higher in acetonitrile than in cyclohexane solution.[21] Steady-state photolysis of 11, indicates that the radical-derived products account for *ca.* 40 % of the photolyzed disilane in hexane solution at room temperature. In acetonitrile, the radical-derived products account for 90 % of the primary transient product mixture.

Irradiation of methylpentaphenyldisilane in cyclohexane at 254 nm, affords products derived from triphenylsilyl radicals and diphenylmethylsilyl radicals (plus 1,1-diphenylsilene, triphenylsilane, and silatriene). Utilizing both acetone as a silene scavenger together with chloroform to trap the silyl radicals, the yields of 1,1-diphenylsilene and silyl radicals are found to be *ca.* 35 % and 20 % respectively (eq 2). In addition, siloxatane 12 (eq 2) is detected in the crude photolysate by NMR spectroscopy.

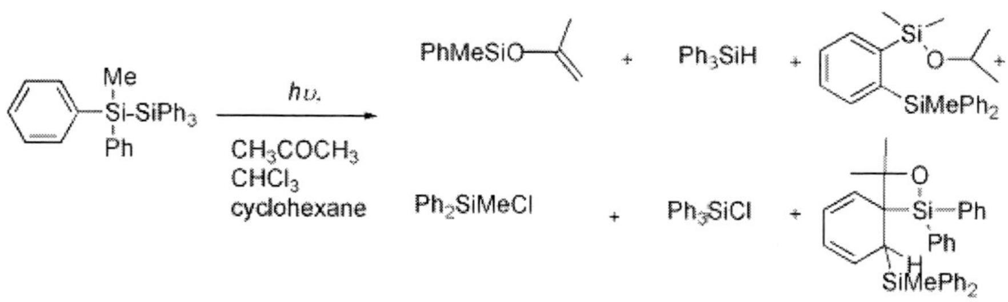

12

(2)

1,1-diphenylsilene and silyl radical intermediates can be distinguished from each other through selective quenching experiments. By quenching the silyl radicals through reaction with added chloroform, 1,1-diphenylsilene is observed alone at 325 nm (utilizing 248 nm laser pulses from KrF excimer laser), whereas scavenging silenes with acetone allows observation of the silyl radicals, which are detected in the same UV region. Further distinctions have been made on the basis of kinetics for trapping of these two species.

The photolyses of 1,1-*bis*(1-naphthyl)tetramethyldisilane (13) and 1,2-*bis*(1-naphthyl)tetramethyldisilane (14) afford photoisomers via homolytic scission of a Si-Si bond (Figure 5). In these photolyses, no silenes are produced.[22] 1,1-*Bis*(1-naphthyl)tetramethyldisilane 13 affords upon photolysis in the presence or absence of trapping agents 1-[methyl(1-naphthyl)silyl]-8-(trimethylsilyl)naphthalene (15), while irradiation of 1,2-*bis*(1-naphthyl)tetramethyldisilane 14, yields 2,3-dihydro-1,1,3,3-tetramethyl-1,3-disilaphenalene2-spiro-1′-2′-dihydronaphthalene (16) as the main product, according to Figure 5.

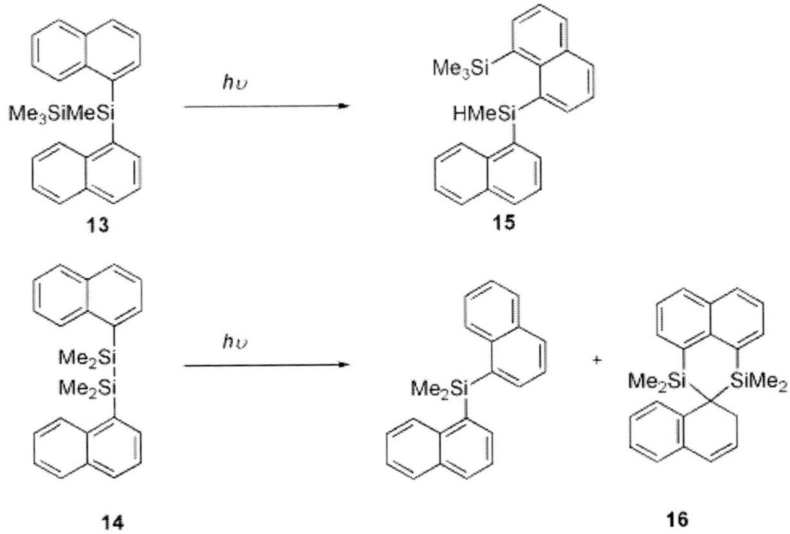

Figure 5. Photolysis of *Bis*-naphthyl-substituted Disilanes.

The photolysis of diasteromerically pure *meso*-isomer 17, Figure 6, in acetone gives a mixture of 18a, and 18b in a ratio 18a/18b = 93/7 in 64 % overall yield, together with 19 and 20, and unreactive 17 in 14 %, 14 %, and 1 % yields respectively, as depicted in Figure 6 below.[23] The authors had demonstrated previously that disilane 17 rearranged via a concerted suprafacial 1,3-silyl shift under the photochemical conditions to produce the respective silene diastereospecifically. The photolysis of 17 in toluene, renders *p-*, *m-*, and *o-*tolylethylmethylphenylsilanes, obtained in 14 %, 8 %, and 6 % yields respectively. These tolylsilanes may be derived from the reactions of the ethylmethylphenylsilyl radical with toluene. It seems likely that energy transfer from toluene in the triplet excited state to the disilane, leading to disilane triplets which cleave to give the silyl radicals, is involved in this reaction. In fact, when a similar reaction is carried out in the presence of *trans*-piperylene as a triplet quenching reagent, the formation of ethylmethylphenyltolylsilanes is completely suppressed.

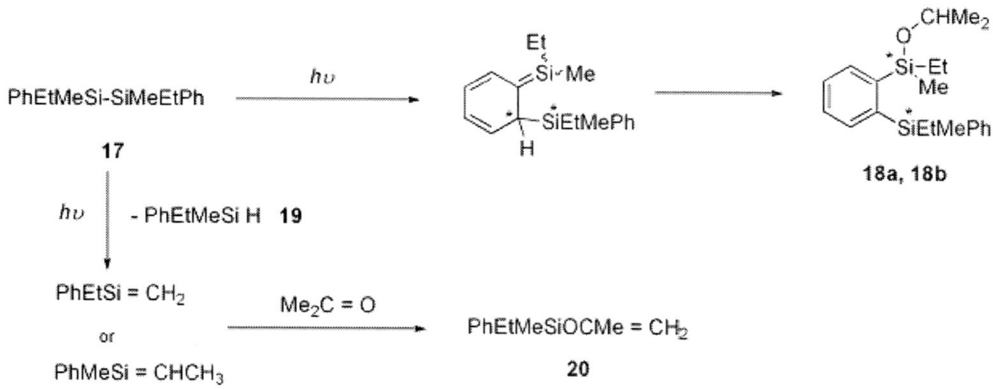

Figure 6. Photolysis of *meso*-1,2-diethyl-1,2-dimethyldiphenyldisilane 17.

Steric hindrance at silicon in aryldisilanes can have a major influence on the relative yields of silatriene (path b, Figure 3) and silyl radical intermediates (path a, Figure 3). Introduction of sterically bulky substituents and also radical stabilizing substituents such as phenyl groups on the silicon atoms and substitution to the *ortho* position in the aryl ring seem to facilitate the formation of silyl radicals. High yields of silyl radicals are produced upon irradiation of 1,2-dimesityltetramethyl disilane in cyclohexane, due to the combined effects of blocking 1,3-Si migration by *ortho* disubstitution and steric crowding of bulky vicinal substituents. The extent of excited state Si-Si bond scission has been estimated to be as high as 60 % on the basis of trapping of the silyl radicals with butadiene; dehydrosilylation is not completely quenched. The fraction that is not quenched corresponds to *ca.* 40 % of the photochemistry, and is due to either molecular elimination, directly in the singlet excited state, or to residual caged radical pair disproportionation. At relatively low concentrations, *n*-Bu_3SnH scavenges free silyl radicals that are believed to derive from cage escape.

Film surfaces containing a disilane moiety do also produce silyl radicals. Irradiation of solid films of disilanylphenylene polymers results in the appearance of Si-OH and Si-O-Si bands in the IR, possibly due to scavenging of silyl radicals by oxygen.[24]

More recently, the photochemistry of (2-allyloxyphenyl)pentamethyldisilane 21 (Figure 7) and [2-(3,3-dimethyl-2-propenyloxy)phenyl]pentamethyldisilane has been investigated.[25] In these molecules, novel photoproduts from the intramolecular reactions of the *ortho*-substituted group with the silatriene intermediate (22, Figure 7), which is formed from the radical scission of a silicon-silicon bond followed by migration of the resulting trimethylsilyl radical to the C6 position of the benzene ring are observed. The formation of product 23, Figure 7, can be rationalized in terms of initial formation of the silatriene intermediate (22, Figure 7), arising from 1,3-migration of trimethylsilyl radical which is formed via homolytic cleavage of Si-Si bond in the photoexcited state of 21 to the C6 position of the benzene ring. Compound 25 is in turn formed from recombination of radical 24, which is produced by the homolysis of oxygen-carbon bond in allyloxy group in the photoexcited state of 21 and pentamethyldisilanyl radical which is generated in turn by the homolysis of the silicon-carbon bond in the photoexcited state of 21. Compound 26 is likely to arise from 1,3-migration of the allyl radical, produced via homolytic cleavage of the oxygen-carbon bond in the allyloxy moiety to the C3 position of benzene ring

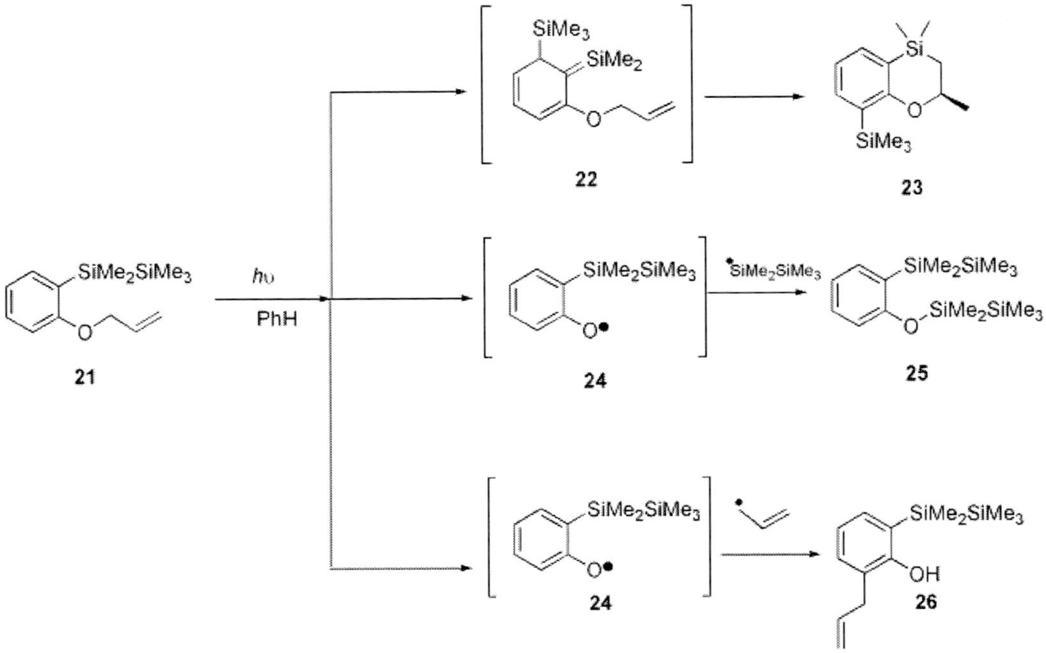

Figure 7. Irradiation of (2-Allyloxyphenyl)pentamethyldisilane.

Irradiation of 1-(p-cyanophenyl)-2-(pentamethyldisilanyl)cyclopentene generates a tricyclic compound, presumably by a mechanism involving Si-Si bond cleavage, radical cyclization, and disproportionation of the silyl radical pair.[26] Both locally-excited and charge transfer complex (CTC) have been invoked as the reactive states. As discussed in a previous section of this chapter, laser flash photolysis studies indicate that the homolysis of Si-Si bond is a triplet excited state photoprocess and that intersystem crossing is promoted by polar solvents. This contrasts with the direct photolysis of (E)-4-cyano-4′-(pentamethyldisilanyl)stilbene (27, not shown), which E,Z photoisomerization is quenched by methanol and tert-amyl alcohol. This substrate shows linear Stern-Volmer quenching plots, consistent with a singlet excited state as the reactive state. The fluorescence of 27 is also quenched by methanol, suggesting that the same state that E,Z isomerizes is the one that fluoresces.[27]

A partially monodeuterated Si (1,1,1) surface subsequently exposed to disilane at low temperatures is a suitable model for examining a mechanism on a fully monodeuterated surface dominated by photogenerated silyl radicals.[28] However, it has been suggested that silyl radicals are not the dominant photoinduced reactive intermediates leading to Si deposition. Similar observations were made from the Si (0,0,0) surface. These and other studies suggest that the photoinduced reaction pathway for disilanes is distinct from the thermal reaction pathway.

DCA-sensitized photoreaction of phenyltetramethyldisilanylbutanol (28, eq 3) gives silyl ether 29 in high yields together with concomitant formation of dimethylphenylsilyl radicals (eq 3). Chemical evidence for the intervention of the silyl radical is provided by the photolysis of 28 in the presence of DCA; namely, the formation of dimethyldiphenylsilane in benzene is readily rationalized by the reaction of dimethylphenylsilyl radical with the solvent benzene.[29] In CCl_4, the same reaction yields chlorodimethylphenylsilane, which arises from dimethylphenylsilyl radicals abstracting a chlorine atom from the solvent.

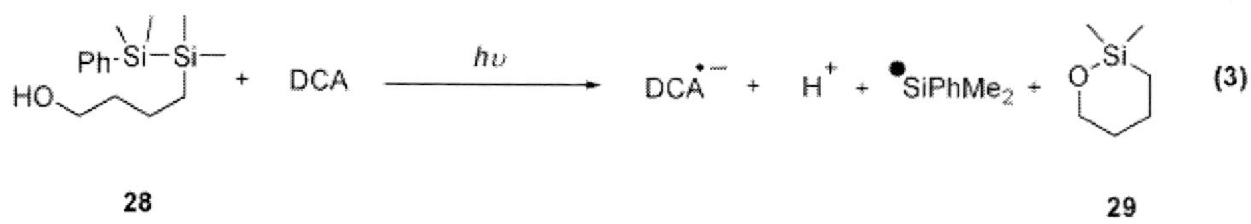

The photoreaction of 2-(pentamethyldisilanyl)naphthalene with *iso*butene affords the ene-type adduct 30 (not shown) consistent with the trapping of a conjugated silene resulting from 1,3-Si migration to the C1 position, although a silyl radical pair recombination mechanism has alternatively been proposed for this photoprocess.[30]

1,1,2,2-Tetramethyl-1,2-di-2′-thienyldisilane (31, eq 4) represents one of the few examples in which nucleophilic addition of methanol across the Si-Si bond is a major photoprocess. Deuterium labeling indicates that 65-77% of the hydrosilane and methoxysilane photoproducts arise from this mechanism, whereas 23-27% of the products are formed by dehydrosilylation, possibly via disproportionation of a silyl radical pair.[31] Unlike the aryldisilane described earlier in this section (Figures 3 and 7), 1,3-Si migration into the thiophene ring is not observed, eq 4.

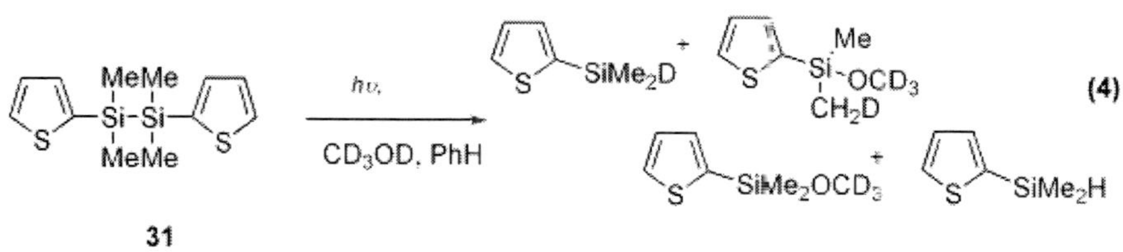

Recently, interesting disilanes have been proposed as photoinitiators in radical polymerization reactions.[32] In this study, the photoinitiator ability of compounds 32 and 33, Figure 8, have been investigated.

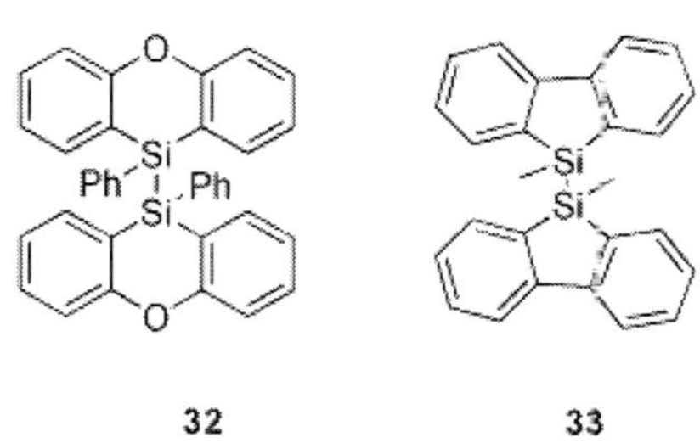

Figure 8. Novel disilanes used as photoinitiators.

These two latter compounds show a photochemical reactivity three times higher than that of dimethoxyphenyl acetophenone (DMPA), a well-known photoinitiator. Thus compound 32 is a remarkable photoinitiator. By laser flash photolysis experiments, new transients are observed. The spectra are relatively similar to that of $Ph_3Si\bullet$ (maxima at about 330 nm, *vide supra*). The transients are strongly quenched by oxygen and TEMPO, which are excellent radical trapping agents. The authors ascribe the transients observed to those of the silyl radicals generated through the homolytic cleavage of the Si-Si bond. It has been found that the dissociation rate constants are higher than 10^8 s^{-1}. The dissociation quantum yields are close to 1, demonstrating the high efficiency of the cleavage process. The calculated Si-Si BDEs are quite weak (ca. 60 Kcal/mol). The very high reactivity of the silyl radicals (generated upon photodecomposition of 22) for addition to methyl acrylate is remarkable, and higher than that reported for the well known phosphinoyl, hydroxyl isopropyl, or benzoyl initiating radicals.

Photoexcited anthryldisilanes exhibit some unique interactions between the σ(Si-Si) and π(anthryl)* groups.[33] These interactions are absent in related pyrenyldisilanes. These interactions cause anthryldisilanes to show novel photophysical properties, *e.g.* the rigidity of the Si-Si bond in these compounds, causes a decrease in the threshold of cleavage of Si-Si linkage from < 300 nm in silanes to 400 nm in anthryldisilanes.

Recently, Goodman et al. investigated the mechanism of fragmentation of Si-Si bonds through Electron Transfer (ET) processes.[34] They conclude that dissociative return electron transfer (DRET) can produce silyl radicals via bond fragmentation, and that this same mechanism also applies to oligo- and polysilanes (vide infra), using a variety of sensitizers with different spectral properties. Thus, steady-state excitation of tetrachlorophthalic anhydride (TCPA) in dry CHCl3 (λ > 350 nm) in the presence of disilanes 34 produces chlorosilane (not shown) in high chemical and quantum yields (Figure 9). Nanosecond excitation of TCPA (λ = 308 nm) in dry CHCl3 produces a triplet state absorption spectrum (transient at λ = 390 nm) within the laser pulse (20 ns). Quenching of this state 3TCPA by disilane occurs at diffusion rate. The authors attribute the quenching of 3TCPA by disilane to yield a triplet ion-radical pair 3(TCPA•–/34•+), based on energetic considerations, Figure 9.

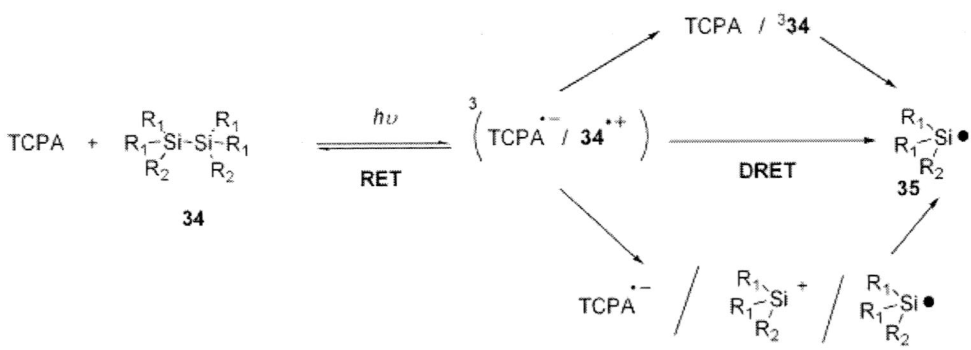

Figure 9. ET fragmentation pathways of disilanes.

Other triplet sensitizers with higher triplet energies such as benzophenone, are not quenched by 34. The concerted DRET mechanism would yield radical 35 directly. However, electron transfer and Si-Si bond dissociation could occur stepwise by either (i) initial Si-Si fragmentation of $34^{+\bullet}$ followed by electron transfer or (ii) initial electron transfer to yield $^{3}34$ followed by fragmentation. To rule out the initial fragmentation of $34^{+\bullet}$, this species was generated independently without the strong reducing $TCPA^{\bullet-}$ counterion. Under these conditions, the lifetime of $34^{+\bullet}$ is considerably longer than that of $^{3}(TCPA^{\bullet-}/34^{\bullet+})$, and consequently excludes unimolecular or solvent-assisted bimolecular fragmentation as a viable mechanism for radical 35 formation. The short lifetime of the ion-pair also excludes initial electron transfer to yield the localized $^{3}34$. Consequently, DRET is the only mechanism that is consistent with the experimental observations.

4. Silyl Radicals from Polysilanes

Polysilanes have attracted considerable attention as mid- and deep-UV photoresists because of their high sensitivity and photodegradation. Their unique electronic properties make them useful in various electronic applications such as non-linear optics, electric conduction, and photoconduction, electroluminescence, light emitting diodes, photoinitiation of polymerization, microlithography and sensors.[35] During the photolysis of linear polysilanes, silylene and silyl radicals are produced as intermediates. Three possible reactions for the photodegradation of polysilanes have been proposed on the basis of photodegradation product analysis [36]: first the simultaneous extrusion of a silylene fragment and the formation of two silyl radicals; second, the Si-Si bond homolysis and the formation of two silyl radicals; and third, the extrusion of a silylene fragment with simultaneous formation of Si-Si bond. It has also been pointed out that the photobehaviour of phenyl-substituted polysilanes is considerable more complex than that of alkyl-substituted polysilanes. The wavelength dependence of photodegradation products from poly(dialkyl)silanes has been noted previously by Michl and Miller.[37]

UV irradiation of high molecular weight poly(dialkylsilane)s dissolved in any of a variety of aliphatic hydrocarbon solvents has been monitored by EPR spectroscopy and results in the formation of a persistent radical.[38] The photodegradation of cyclic and linear permethylpolysilanes has been studied by Ishikawa and Kumada [39]; during the photolysis of cyclic compounds silylene is the main intermediate. However, in the photolysis of poly(cyclohexylmethylsilane), silyl radicals are the main intermediates, *vide infra.*

The *tris*(trimethylsilyl)silane (TTMSS)-derived radical has recently been studied by laser flash photolysis.[40] In this work, an interesting comparison between the TTMSS-derived radical and simple silyl radicals is established. Being the Si-H bond in TTMSS weaker than in most other silanes, a very high yield of silyl radicals production is obtained from triplet benzophenone energy transfer and from reaction of TTMSS with *t*-BuO• (quantum yields in the order of 0.95). The reactivity of this silyl radical is compared, by *ab-initio* calculation methods, with that of carbon-centered radicals, being the latter more selective toward addition to C-C double bonds and hydrogen abstraction from phenols. Both electron-poor and electron-rich alkenes react efficiently with the (TTMSS)-derived radical, exhibiting a very low selectivity toward the addition process. The high reactivity of TTMSS-derived radicals toward C-C double bonds is explained in terms of antagonists polar and enthalpy effects, being this behavior in full agreement with quantum mechanical calculations. The good hydrogen abstracting property of the TTMSS-derived radical can be exploited as an efficient initiator toward polymerization of an acrylate monomer. TTMSS has recently been used in water as a silicon-centered radical mediator in numerous reactions. The superb ability of TTMSS-derived radical towards halogen atom abstraction has been fruitfully used in the reduction of an array of both water-soluble and water-insoluble organic halides. These same authors also employ TTMSS as a hydrosilylating reagent of C-C and C-O multiple bonds in water.

Work on the photolysis of alkoxypolysilanes is limited. The sisyl ether (*tris*(trimethylsilyl)silyl ether) of methanol has been shown to extrude silylene by two different regiochemical pathways in a 2:1 ratio, as shown by trapping experiments with 1,3-butadiene. Sisyl ethers can be cleaved using photolysis at 254 nm during 30 minutes to yield the starting alcohols in yields ranging from 60 to 95%, generating the TTMSS-derived radical in the process or silylene intermediates. These sisyl ethers are shown to be stable towards selected fluoride reagents; KF + 18-crown-6, cesium fluoride, and tetrabutylammonium fluoride. Brook *et al.* have investigated the mechanism of sisyl ether deprotection using in the photolysis mixture a silylene trap, such as 2,3-dimethyl-1,3-butadiene (DMB). The photoproduct distribution shows the presence of products derived from a single silylene.[41]

Si-Si bond cleavage to silyl radicals appears to be especially prevalent in polysiloxane polymers incorporating phenyl and TMS substituents at silicon. Irradiation of thin films results in extensive cross-linking due to addition of silyl radicals to the pendant phenyl groups of the polymer chain. Similar reactivity is observed for the monomeric $(Me_3SiO)_2Si(Ph)SiMe_3$, and 254-nm irradiation in hexane containing *iso*propylbenzene results in homolytic substitution of the aryl ring. Hydrogen abstraction by silyl radicals from *iso*propylbenzene to give hydrosilanes is only a minor process and no silene intermediate is trapped by methanol.

Chatgilialoglu and coworkers have prepared a wide variety of silyl radicals by proton abstraction of *tert*-butoxy radicals from tri-organosilanes. All the silyl radicals they used show a very strong absorption band at 300 nm and a weaker band shoulder at longer wavelength (*vide infra*).

It has been established that photochemically-produced silyl radicals abstract hydrogen originally attached to the α-position of alkyl chains in polysilane materials, rather than from the solvent or from silylene traps such as Et_3SiH.[42]

UV-irradiation of $(Me_3Si)_3Si\text{-}SiMe_2\text{-}SiMe_2\text{-}Si(SiMe_3)_3$ (36, not shown) in CCl_4 affords products derived from cleavage of Si-Si bonds *i.e.:* $(Me_3Si)_3SiCl$ and $(Me_3Si)_3SiSiMe_2SiMe_2Cl$. Under the photolysis conditions used, the central Si-Si bond in 36 is the most susceptible to cleave. Photochemical irradiation of 36 in hexane leads to a more complex array of products, being $(Me_3Si)_3SiH$, $(Me_3Si)_3SiSiMe_2SiMe_2H$, and $(Me_3Si)_3SiSiMe_3H$ the most prevalent in the photoproduct mixture. These Si-Si bond homolysis products account for 35 % of the total product formation.[43] In contrast, photolysis of $(Me_3Si)_2SiMe\text{-}SiMe_2\text{-}SiMe_3$ does not result in Si-Si bond homolysis, whereas photolysis of $(Me_3Si)_3Si\text{-}SiMe_2\text{-}SiMe_3$ exhibits such cleavage to produce $(Me_3Si)_3Si\bullet$

radicals. The authors conclude that as the degree of steric congestion augments, so does the extent of Si-Si bond scission.

Matrix-isolated irradiation of 1,1,1,3,3,3-hexamethyl-2,2-*bis*(diazo)trisilane (37, eq 5) results in a complex photoreaction mixture. Besides product 38 depicted in eq 5, products arising from cleavage of Si-Si bond (forming silicon-centered radicals) are believed to participate.[44] By changing the wavelength of the light source, from visible to UV, a different photoproduct distribution is observed. However, when the photoreaction ($\lambda \geq 400$ nm) is carried out in ethanol at 77 K, two main products are identified, diazo compound 39 , and ethoxysilane (eq 5).

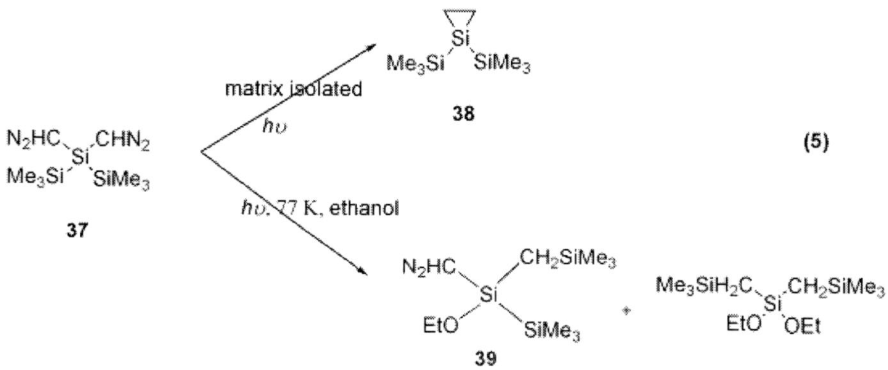

(5)

It is known that ketone triplets cleave the Si-Si bond of oligosilanes via S_H2 type radical displacement at the silicon atom. The photolysis of cyclic organosilane 40 (eq 6) and phenantroquinone (PQ) in the presence of CCl_4 affords $Cl(Me_2Si)_nCl$ (n = 4-6). This photoproduct distribution would support a mechanism involving silyl radical intermediates such as 41. Under the reaction conditions, the formation of $Cl(Me_2Si)_6Cl$ may indicate the electron transfer from 40 to $^3PQ^*$, and this affords the corresponding radical cation $40^{\bullet+}$ which readily reacts with CCl_4 to give $Cl(Me_2Si)_6Cl$.[45] Quenching experiments by anthracene reveal that PQ should act as triplet $^3PQ^*$ to undergo radical displacement at the silicon atom of 40. In polar solvents, ET from 40 to $^3PQ^*$ would generate PQ radical anion and the coupling of the radical ion pair formed would cause the Si-Si bond cleavage followed by intramolecular O-Si bond formation to give silylene transfer products.

When 2-phenyl-2-thienylhexamethyltrisilane is photolyzed in the presence of a cyclohexene and methanol mixture, an apparent radical reaction occurred.[46] The authors suggest that the sulfur atom of the thienyl group strongly stabilizes the silyl radical. This result is supported by both identifying its typical radical reaction products and examining the ESR spectrum of its adduct with a radical trap. An analogous behavior takes place when di(-thienyl)hexamethyltrisilane is photolyzed under the same reaction conditions.

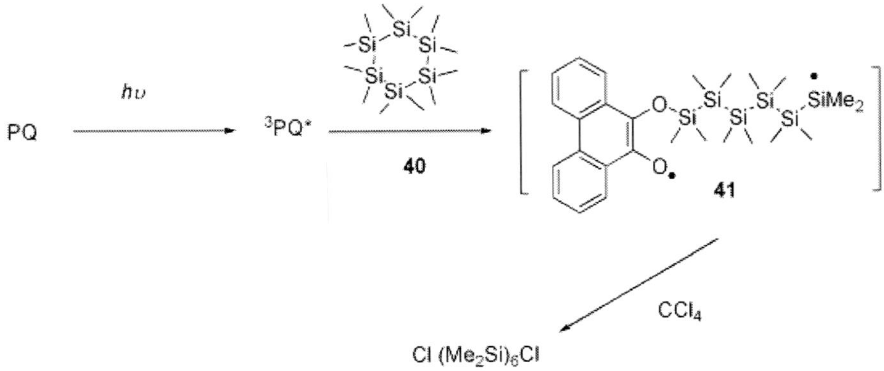

(6)

Photolysis of (*E*)-4-cyano-4′-(2-heptamethyltrisilanyl)stilbene in CH_2Cl_2 results in efficient intramolecular ET, which converts the initial π,π* excited state (^{1}LE) to the charge transfer (CT) excited state. The CT excited state fluoresces, undergoes *E,Z* photoisomerization, and reacts with methanol to produce hydrosilane via Si-Si bond cleavage and protonation of the central silicon, as shown by deuterium labeling experiments.[47] Azulene quenching experiments show that hydrosilane and the *E,Z* photoisomerization derive from the same excited state, which is assigned to the CT excited state.

Poly(cyclohexylmethylsilane) and poly(phenylmethylsilane) have been investigated by laser flash photolysis experiments. The time-resolved absorption spectra of poly(cyclohexylmethylsilane) show a sharp transient absorption peak at 370 nm with an absorption shoulder at 430 nm. These transient absorption bands are assigned to silyl radicals by quenching experiments using CH_2Cl_2 as radical quencher. For poly(phenylmethylsilane) a wavelength dependence is observed for the intermediate during the photodegradation. When poly(phenylmethyl)silane is excited at longer wavelengths (σ,σ* band), silyl radicals are produced. Silylene and silyl radicals are produced by shorter wavelength excitation (π,π* band). One explanation for such wavelength dependence is to consider the charge transfer excited state between the aromatic group and the disilyl group.[48] In these studies, a silylene trap such as triethylsilane has been used in order to assign the silicon centered radicals unambiguously. The charge transfer ability of polysilanes has been uncovered particularly for a disilane, the (phenylethyl)pentamethyldisilane, which is reported to exhibit intramolecular charge transfer character at 77 K, involving σ(SiSi)—π* charge transfer state.[49]

Branched polysilanes show upon irradiation in solution photoproducts derived mainly from silylene intermediates, which can be selectively trapped with 2,3-dimethyl-1,3-butadiene (DMB). However, irradiation (low pressure Hg lamp) of 2,2-*bis*(trimethylsilyl)octamethyltetrasilane (42) and 2,2,3,3-*tetrakis*(trimethylsilyl) hexamethyltetrasilane (43) in pentane yields hydrosilanes arising from homolytic scission of an Si-Si bond followed by hydrogen abstraction of the resulting silyl radicals. In these experiments, a silylene trap such as DMB is also used. Among the photoproducts derived from silyl radicals, TTMSS (*tris*(trimethylsilyl)silane) is observed from photolysis of 42 which is claimed to be formed from homolytic scission of an Si(2)-Si(3) bond and followed by hydrogen abstraction of the resulting silyl radical.[50] Pentamethyldisilane, which would be produced from the pentamethyldisilanyl radical, however, could not be detected by glc analysis.

When compound 43 is photolyzed in the presence of DMB in pentane, $(Me_3Si)_3SiMe$, TTMSS and $(Me_3Si)_3Si(SiMe_3)_2SiH$ are obtained in 4 %, 14 %, and 15 % yields, respectively. The formation of TTMSS and $(Me_3Si)_3Si(SiMe_3)_2SiH$ can be best understood by assuming the *tris*(trimethylsilyl)silyl radical and *pentakis*(trimethylsilyl)disilanyl radical arising from homolytic cleavage of the Si(2)-Si(3) bond and Si(1)-Si(2) respectively.

It has been reported that electron transfer from photoexcited polysilanes to electron acceptors such as *p*-dicyanobenzene, is responsible for the increase in the photolysis rates of polysilanes. The resultant polysilane radical cations cause cleavage of Si-Si bonds, which should be much more efficient than homolytic cleavage of Si-Si bonds in the absence of electron acceptors, because radical-radical recombination has to be considered in the latter process. It has been observed that the increase in the photolysis rates of poly(phenylhydrosilane) in the presence of 3,3′,4,4′-*tetra(tert*-butylperoxycarbonyl)benzophenone (44) can be explained by efficient PET reactions from photoexcited polysilane to 44, followed by decomposition of the peroxide moieties. This prevents the electrons from getting back to the polysilane radical cations.[51] Products formed from 44 by photoinduced electron transfers are aromatic compounds bearing carboxylic acids which promote higher dissolution rates in a basic aqueous solution. The comparison of 44 with benzophenone and C-60 implies that electron acceptors bearing labile bonds accelerate photodegradations of polysilanes and that the presence of peroxides is crucial.

KrF-laser irradiation of thin films of poly(4-trimethylsilylmethylstyrene) (PTMSMS) under atmospheric conditions induces oxygenation at the film surface and cross-linking inside the film. The photooxygenation of PTMSMS would possibly proceed via the C-Si bond cleavage as a major pathway. Generation of the radical pair (benzyl-type radical/trimethylsilyl radical) in the surface layer followed by an attack of molecular oxygen in air toward the benzyl-type radical, would lead to the formation of the carboxylic acid group.[52] It is expected that the quantum yield for the C-Si bond cleavage to give the benzyl-silyl radical pair is much larger than that for the C-H bond cleavages of the 4-methyl group in poly(4-methylstyrene); however, even at the film surface, most of the benzyl-silyl radical pair could reproduce the C-Si bond via recombination without its trap by oxygen. However, in the benzyl-hydrogen radical pair generated from irradiation of poly(4-methylstyrene) part of the hydrogen radical could diffuse, escaping out from the cage, giving the free benzyl-type radical that leads to the carboxylic acid formation. The diffusion of the trimethylsilyl radical would be much slower than that of the hydrogen radical owing to its bulkiness. The authors suppose that in the photooxygenation of the PTMSMS surface, the trimethylsilyl group induces selective C-Si bond cleavage and prevents the C-H bond cleavage of the polymer main chain but, on the other hand, decreases the efficiency of free radical formation.

Upon irradiation of (*E*)-poly(1,2-dimethyl-1,2-diphenyldisilanylene)ethenylene, photodegradation products arising from Si-Si bond cleavage in the polymer backbone are fromed, producing silyl radicals. The silyl radicals thus formed may add to ethenylene bonds and phenyl groups leading to the formation of polysilyl-substituted ethylene units and allylic units, probably in the cyclohexadienyl ring respectively.[53] The formation of Si-H bonds can be explained by disproportionation of the silyl radicals giving hydrosilanes and silenes, and also by hydrogen atom abstraction from other available sources.

More recently, two-laser beam photolysis experiments on polysilanes in the gas phase have thrown light into the reactive intermediates.[54] One-photon photolysis in the gas phase of non-cyclic polysilanes $PhMeSi(SiMe_3)_2$ and $PhSi(SiMe_3)_3$ gives radical products derived from the homolytic scission of a single Si-Si bond with little or no evidence of silylene being generated directly, whereas dimethylsilylene (Me_2Si:) is directly observed as the major one-photon photoproduct from the cyclic precursors $(Me_2Si)_6$ and 1,3-diphenyl-1,2,2,3-tetramethyl-1,2,3-trisilacycloheptane. The authors suggest a common mechanism for polysilane photolysis in the gas phase in which the primary photochemical event is the homolytic cleavage of a single Si-Si bond. For non-cyclic polysilanes, the resulting silyl radicals recoil from each other and become the ultimate detectable intermediates. Homolytic Si-Si cleavage in cyclic polysilanes, on the other hand, yields α,ω-diradicals which are capable of subsequent intramolecular reactions (Figure 10). Thus, silylene formation becomes possible by an S_H2 mechanism by which a Si-Si bond is attacked by the distal silyl radical. S_H2 displacements of Si-Si bonds are well known reactions.

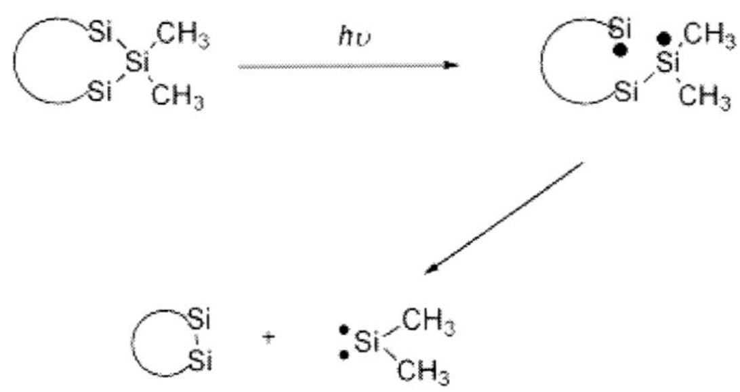

Figure 10. Silylene and silyl radical intermediates from common diradical precursor.

Although linear and branched polysilanes give only silyl radicals under molecular beam conditions, they are highly efficient silylene precursors in solution and cold matrices. The authors (*vide supra*) also suggest that this discrepancy is still reconcilable by a primary photochemical step. In the condensed phases, the radical pair generated is initially bound in either a solvent or matrix cage where further reaction may take place. The S_H2 displacement of the Si-Si bond of one radical by the other may effectively generate the silylene and disilane products in a mechanistic step similar to that proposed for the cyclic precursors in the gas phase. A second bimolecular step for the production of silylene must be very fast and efficient in order to be competitive with radical separation from the cage. However, theoretical studies suggest a concerted extrusion under these conditions.

5. SILYL RADICALS FROM ACYL-SILANES, SILYLIMINES, SILYLKETENES, AND ACYLOXY(METHYL)-SILANES

In addition to α-cleavage, acylsilanes 45 have been reported to lead to [1,2]-silyl shift to siloxycarbenes 46, which are extremely labile and rapidly rearrange back to the acylsilane 45 (eq 7), in

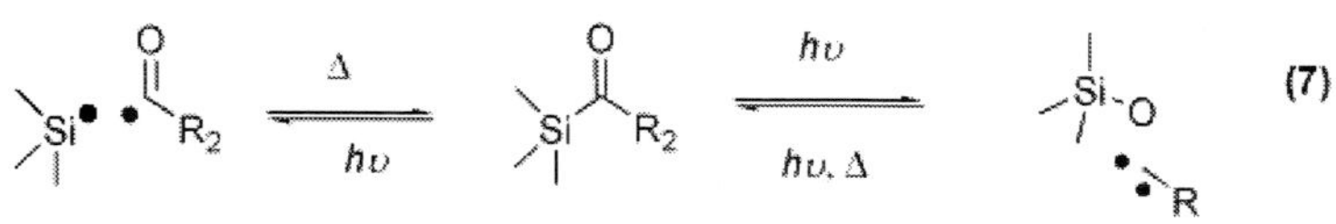

45 46

argon, and oxygen-doped argon matrices.[55] Thus UV-irradiation of formyltrimethylsilane, and acetyltrimethylsilane in Ar results in decarbonylation while UV-irradiation of benzoyltrimethylsilane does not lead to decarbonylation, and the substrate remains stable under the same photoreaction conditions. The major photoproduct from irradiation of acyltrimethylsilanes is a trimethyl(vinyloxy)silane. In O_2-doped argon matrices, irradiation leads to insertion of the oxygen molecule into the Si-CO bond.

Photolysis of acylsilanes in CCl_4 was suspected to proceed by Norrish type I cleavage to yield an acyl-silyl radical pair. By contrast, photolysis of triphenylacylsilane in CCl_4 is reported to yield triphenylsilylchloride, acetyl chloride, trichloroacetone, and hexachloroethane. Photolysis in alcohol solvents, however, affords carbenoid, rather than free radical intermediates. CIDNP experiments suggest that photolysis in CCl_4 proceeds through initial formation of acylsilane-CCl_4 exciplex, followed by selective collapse to acetyl-trichoromethyl radical pair and the silyl chloride.[56] An alternative to the exciplex mechanism is the Norrish-type I cleavage of the acetyldimethylphenylsilane from an excited singlet state followed by a rapid reaction with CCl_4 (pair substitution).

Photolysis of *bis*(organosilyl)ketones is different from that of acylsilanes. *Bis*(triphenylsilyl)ketone (a red-violet solid with a UV maximum absorption at 550 nm) rapidly undergoes photolysis in solution to yield hexaphenyldisilane and carbon monoxide.[57]

Upon photolysis, epoxy-substituted silyl ketones undergo n,π^*-excitation. In general, epoxy ketones render upon photolysis products arising from initial cleavage of the C(O)-bond and subsequent C(1)C(2) migration of the silyl moiety.[58] The photolysis of cyclopropyl silyl ketones instead, proceeds through a Norrish-type-II reaction involving -H abstraction, cyclopropyl ring cleavage followed by retro-enolization to the acylsilanes. As a common product, a dihydrofuran is formed via the alternative C(1)-C(2) cleavage of the cyclopropyl moiety.[59]

N-substituted *bis*(organosilyl)imines, on the other hand, exhibit red-shifted UV absorption around 400 nm, with relatively large extinction coefficients (150-200) ascribed to the $n\rightarrow\pi^*$ transition. UV-irradiation of bis(organosilylimines) 47 (eq 8) affords (in C_6D_6 containing CCl_4) organosilyl chlorides (48) and *iso*cyanides (49).[60]

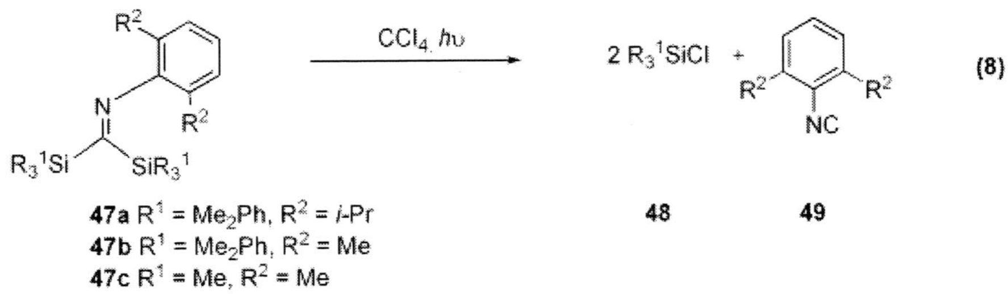

These results could suggest that the photolysis of *N*-substituted *bis*(organosilyl)imines involves a stepwise elimination of the two organosilyl radicals. This assumption, however, was ruled out by carrying out the photoreaction in the absence of CCl_4. Under these reaction conditions, no adducts are found. In contrast, by performing the photoreaction in the presence of alkyl-iodides in C_6D_6, organosilyliodide is obtained, which is readily hydrolyzed to the corresponding siloxane, along with the aromatic isocyanide 49. The photolysis of 47b with ethyl iodide in the presence of TEMPO, affords dimethylphenylsilyl TEMPO-adduct 50 (Figure 11) together with a small amount of dimethylphenylsilyl iodide and ethyl-TEMPO adduct (Figure 11). This photoproduct distribution may suggest that TEMPO is operating as an electron acceptor in the PET mechanism. The imidoyl radical 51 (Figure 11) can give the isocyanide 49 and the organosilyl radical, which affords another equivalent of 50 by trapping the silyl radical with TEMPO, as depicted in Figure 11.

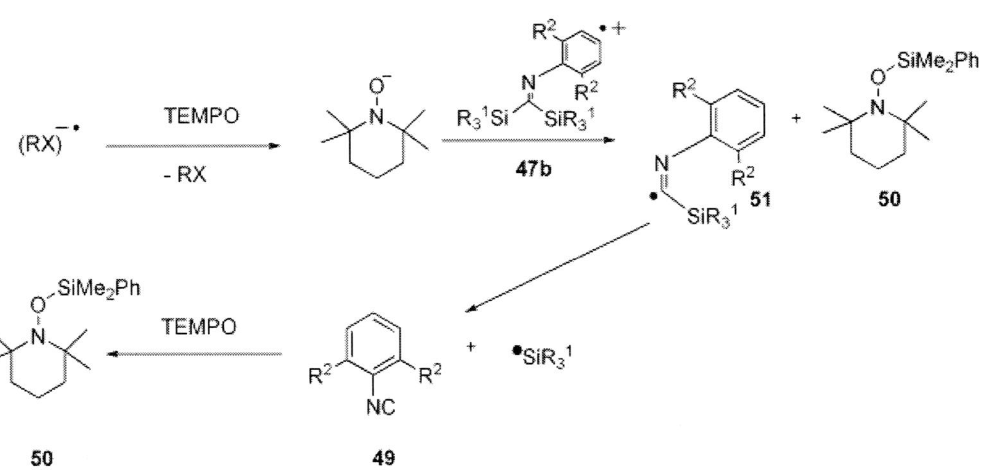

Figure 11. PET reaction of 46b in the presence of TEMPO

Irradiation of a mixture of 47b (eq 8) and iodobenzene in methylmethacrylate (MMA) and benzene produces poly(methylmethacrylate) (PMMA). This finding supports the intervention of organosilyl radicals in the polymerization reaction, since irradiation of MMA in the presence of iodobenzene and absence of 47b does not lead to a polymerization reaction. Intervention of organosilyl radicals as well as phenyl radical (this latter works as an initiating species), has been confirmed by ^{1}H NMR analysis of poly(methylmethacrylate) (PMMA) indicating that both of the dimethylphenylsilyl groups in 47b are included in the ratio 60:40, respectively, as terminal groups of the PMMA polymer chains.

The photocatalyzed cleavage of methyl trimethylsilyl ketene acetals in MeCN by using a N^{21},N^{22}-bridged porphyrin as catalyst, produces trimethylsilyl radicals that add to aldehydes and ketones. Coupling of the resulting α-silyloxyalkyl radicals with those resulting from the acetal leads to a photoinduced Mukayama reaction (aldehyde or ketone and a silylenol ether giving the silylated β-hydroxycarbonyl). The process shows an induction period before the formation of the products, but does not proceed in the dark, thus being considered as a photocatalytic reaction.[61]

Photochemistry of [(acyloxy)methyl]benzyldimethylsilanes 52 (eq 9) as well as benzyltrimethylsilanes affords benzyl radical-silyl radical pairs in the primary photochemical process and also 1,2-(C→Si) acyloxy migration, as shown in eq 9.

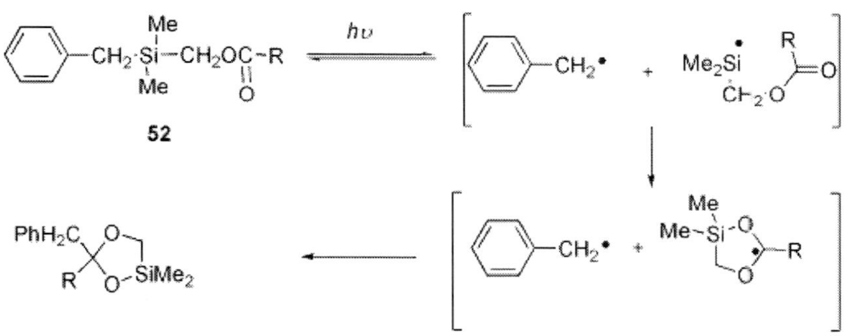

(9)

6. Silyl Radicals from Cleavage of Si-B Bonds

Organosiliconboranes having bulky substituents on the boron, *e g.* $R_3SiB[N(CHMe)_2]_2$, exhibit UV absorption at wavelengths longer than 300 nm.[62] Photolysis of this band affords a pair of silyl and boryl radicals that can be trapped quantitatively by nitroxide (TEMPO) as shown in reaction 10 below:

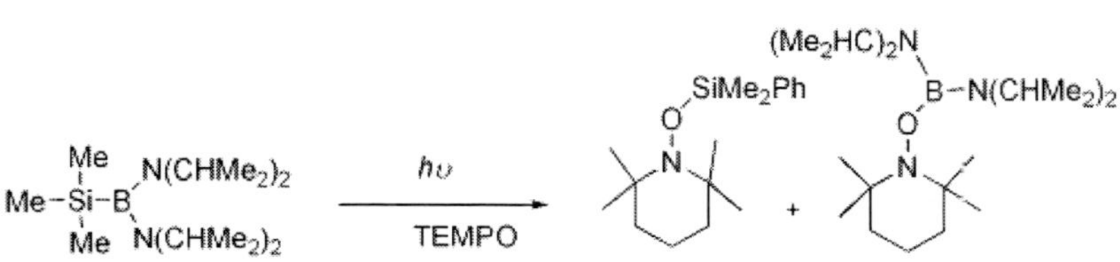

(10)

The absorption spectra of phenyl-substituted silylborates in THF have an absorption maximum at 270 nm, which can be explained in terms of a σ→σ* transition from the bonding orbital of the Si-B bond to its antibonding orbital. Irradiation of THF solutions of phenyl-substituted silylborates with a Xe lamp affords hydrides which are derived from silyl radicals generated from photocleavage of the Si-B bond. In the presence of chloroform, the photolysis affords the corresponding chlorosilanes.[63] Silylborates of the type Li[$Ph_nMe_{3-n}SiBPh_3$], with n = 1-3, undergo photoreaction to generate silyl radicals, presumably from cleavage of the Si-B bond in their triplet excited state. The primary process of these silylborates has been investigated by chemical trapping experiments and CIDEP (Chemical Induced Dynamic Electron Polarization) methods.

UV-irradiation of 53 (eq 11) in the presence of 1-octene and cyclohexene in *n*-hexane affords dimethyl-n-octylphenylsilane 54 (R = C_6H_{13}) and dimethyl(cyclohexyl)phenyl silane 55 (not shown), respectively.

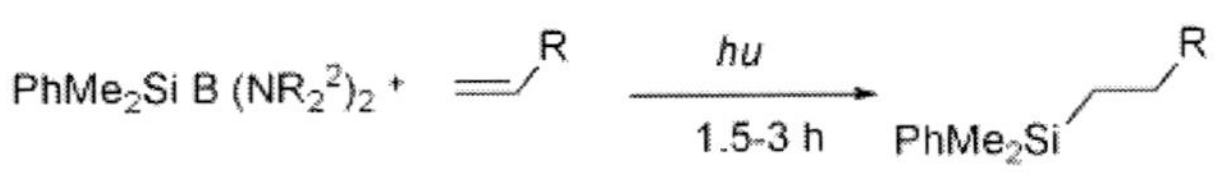

(11)

53 **54**

Irradiation of 1,6-dienes in the presence of **53** affords 5-*exo*-cyclized products **56** exclusively in a stereoselective manner without using conventional radical initiators (eq 12).

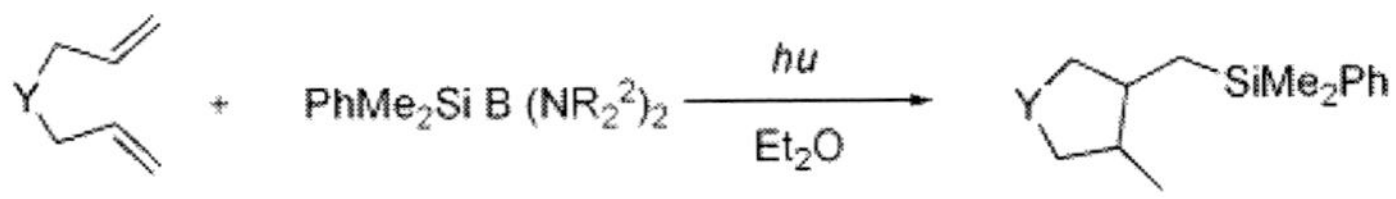

(12)

Y =		
O		67%
C(CO_2Et)$_2$		67%
CH_2		53%
NAc		58%

53 **56**

Organosilylborane 53 can also promote a radical polymerization of methylmetacrylate, acrylate, and vinyl acetate.

In contrast to the reactive organosilyl radical, the organoboryl radical does not add to olefins, probably due to steric hindrance. As is well known, silyl radicals are highly reactive towards halogen atom abstraction. The organosilylborane 53 promotes a reductive cyclization of 1-bromo-5-hexene to give methylcyclopentane. Halogen abstraction by the silyl radical and subsequent cyclization is followed by hydrogen abstraction to give the product. This finding has pointed out that organosilylborane 53 may function as an alternative to organotin hydride-radical initiation.

7. Silyl Radicals from Mesolysis of Si-Se and Si-Hg Bonds

Silyl radicals have been obtained by a complex but efficient method using $PhSeSiR_3$ as the reagent. The strategy is based on the mesolysis of $PhSeSiR_3^{\bullet -}$ to give $R_3Si^{\bullet}$ radical and $PhSe^-$.[64] Indeed, the selective formation of $PhSeSiR_3^{\bullet -}$ is accomplished by visible-light irradiation (410 nm) of solutions containing $PhSeSiR_3$, 9,10-dimethoxyanthracene (DMA) as the electron donor, and ascorbic acid (H_2A) as the co-oxidant. Figure 12 below shows the photoinduced electron transfer (PET) with the formation of $PhSeSiR_3$ radical anion and $DMA^{\bullet +}$, together with the regeneration of DMA at the expense of ascorbic acid. The choice of the substituents is limited by their stability. Trialkyl-substituted derivatives are highly sensitive to air and prone to hydrolysis, whereas the *tert*-$BuPh_2Si$ derivative has been found to be the most stable.

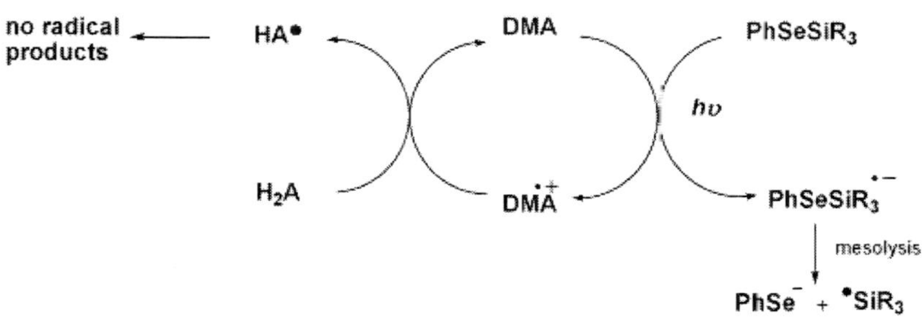

Figure 12. Mesolysis of $PhSeSiR_3$.

Subsequent studies have amply demonstrated that $PhSe\text{-}SiR_3$ could be utilized as an *in-situ* source of alkyl silyl radicals more halophilic than alkyl tin radicals could. In this fashion, the initiation of bimolecular group transfer radical reactions can take place as shown in Figure 13.[65]

It should be pointed out that among other carbocyclizations, bimolecular group transfer radical reactions have been applied successfully to the synthesis of phenylselenomethyl-substituted tetrahydrofurane rings.

Another case of the use of an auxiliary radical involves the generation of a silyl radical by photocatalytic reductive activation from 1,4-dimethoxynaphthalene (DMN) of a selenylsilane. The strong affinity for halogens makes this radical abstract a bromine atom from an α-bromoester. The resulting C-radical adds to a terminal olefin, and the adduct, in turn, reacts with the selenyl anion, finally giving a γ-selenyl ester.[66]

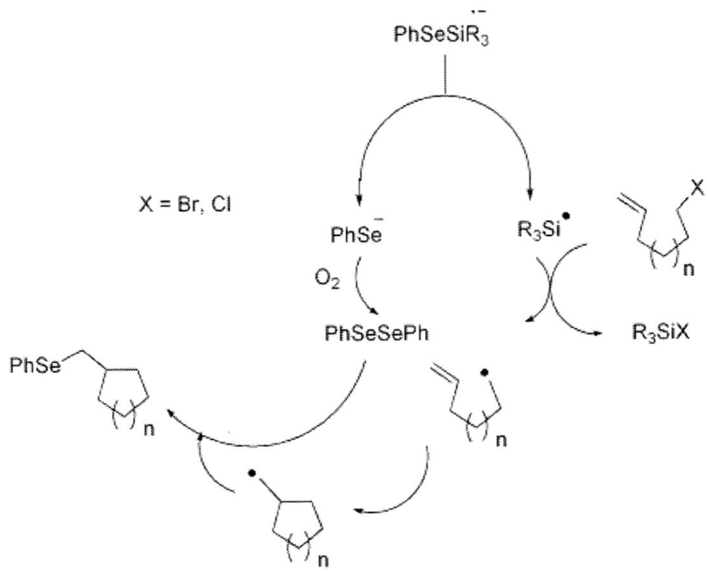

Figure 13. Silyl Radicals from PhSe-SiR$_3$,

Silylmercury compounds are excellent precursors for the selective generation of silyl radicals, and unlike silanes, there is no need to use initiators. In silylmercury precursors, only the Si-Hg bond is fragmented to produce the silyl radical, unlike polysilanes which show several fragmentation modes (*vide supra*).[67] Disilylmercury compounds $(R_3Si)_2Hg$ are usually colored, with colors ranging from green to yellow. Geminal dimercury silicon compounds of the type R_1-(Hg-SiR_2-Hg)-R_1 are also known to generate silyl radicals. Trimercurysilyl compound 57 (Figure 14), exhibits absorption maxima at 340 nm and 380 nm. Photolysis of 57 in C_6D_6 (Figure 14) results first in the formation of the silyl mercury radical 58 and later of silyl mercury radical 59, both as intermediates. This entails a reaction of the external Si-Hg bond. The photolytic cleavage of Si-Hg bonds is a convenient method for the generation of silyl radicals. Highly toxic $Hg(SiMe_2)_2$ has been used in the past as a source of $Me_3Si\bullet$ radicals for the photopolymerization of styrene.[68]

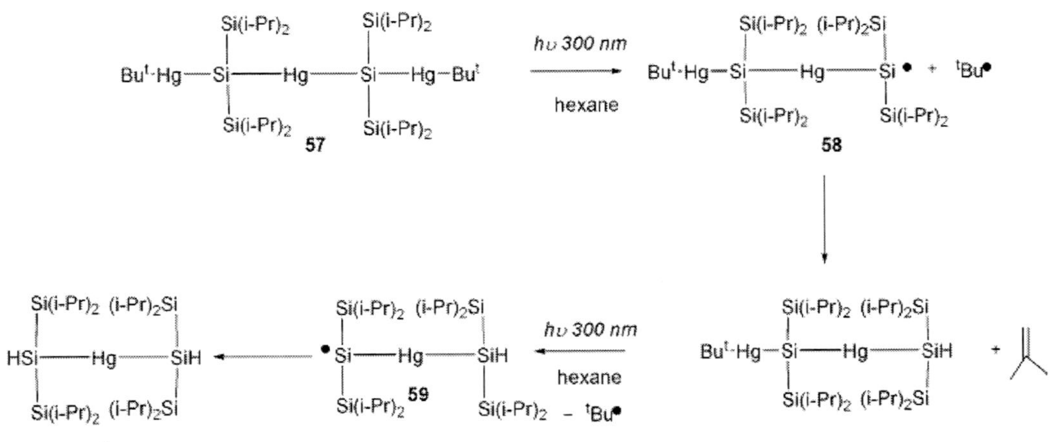

Figure 14. Silyl radicals from Si-Hg bond homolysis.

The reaction is monitored by EPR spectroscopy, assigning the EPR signals to mercurysilyl radicals. The R_1-Hg-SiR_2• radical is characterized by a downfield shifted g factor, by interaction of the unpaired electron with both ^{29}Si and ^{199}Hg nuclei, as indicated by the measured coupling constants, $\alpha(\alpha\text{-}^{29}Si)$ of 56G and $\alpha(^{199}Hg)$ of 63-105G (satellites due to ^{201}Hg nucleus were not observed); chemical trapping experiments by TEMPO are also used in this work to identify the radicals proposed.

8. Silicon-Centered Biradicals

Several silicon-centered biradical intermediates have been postulated in numerous reactions, among them, the photolysis of cyclopolysilanes. It is well known that photolysis of cyclosilanes (and cyclogermanes) brings about ring contraction products, giving the corresponding smaller rings, with extrusion of silylene (R_2Si:).[69] For instance, in the photolysis of 60a-c (254 nm), Figure 15, 1,3- and 1,4-silicon-centered biradicals have been postulated to generate the products observed. In these experiments, the 1,4-biradicals cyclize to the cyclotetrasilanes (path a, Figure 15), the 1,3-silicon-centered biradicals (path c, Figure 15) cyclize to the cyclotrisilane, respectively. The extruded silylene is trapped by 2,3-dimethylbutadiene (DMB), to render the cyclopentenylsilane (path b), and the extruded germylene (R_2´Ge:) is also trapped by DMB to render cyclopentenylgermane (path a). Parallel experiments have been also carried out by these same authors to trap the 1,3-silicon-centered biradicals with halogenated solvents and hydrocarbons. It is worthwhile noting that in these kinds of substrates (cyclosilanes), an array of reactive intermediates, *i.e.*, biradicals, silylenes, and silenes, must be invoked in order to rationalize the product distribution observed.

It has recently been observed by *in-computero* methods in 1,3-diradicals of the type $XSi(GeY_2)_2SiX$ (where X = H, CH_3, SiH_3, $C(CH_3)_3$, NH_2 for X = F; Y = H, CH_3, OH, NH_2, SiH_3 for X = H) that silicon-centered singlet diradicals are more stable than the lowest triplets and than the bicylic π-bonded isomers if the isomers exist. The most stable singlet diradicals are not the π-type diradicals, but the σ-type diradicals where the radicals interact with each other through the Si-Ge bonds in the four-membered rings.[70]

Dialkylsilicon porphyrins, upon excitation of the porphyrin ring with visible light, undergo homolytic cleavage of the axial Si-C bonds to generate a highly stable silicon diradical at the ground state, which can be activated by visible light to react with nitroxy compounds.[71] The axial Si-O bonds in the resulting dinitroxysilicon porphyrins are cleaved homolytically to afford free radicals upon irradiation with visible light. The authors suggest the possibility that the silicon diradical within a porphyrin π cloud could serve as a photoswitchable radical , whose reactivity can be controlled by visible light.

In the absence of nitroxy compounds, photolysis of dialkylsilicon porphyrins results in the formation of a long-lived (> 50 days at 25 °C) EPR –active silicon diradical with a g value of 2.0026, which does not react with nitroxy compounds in the dark.

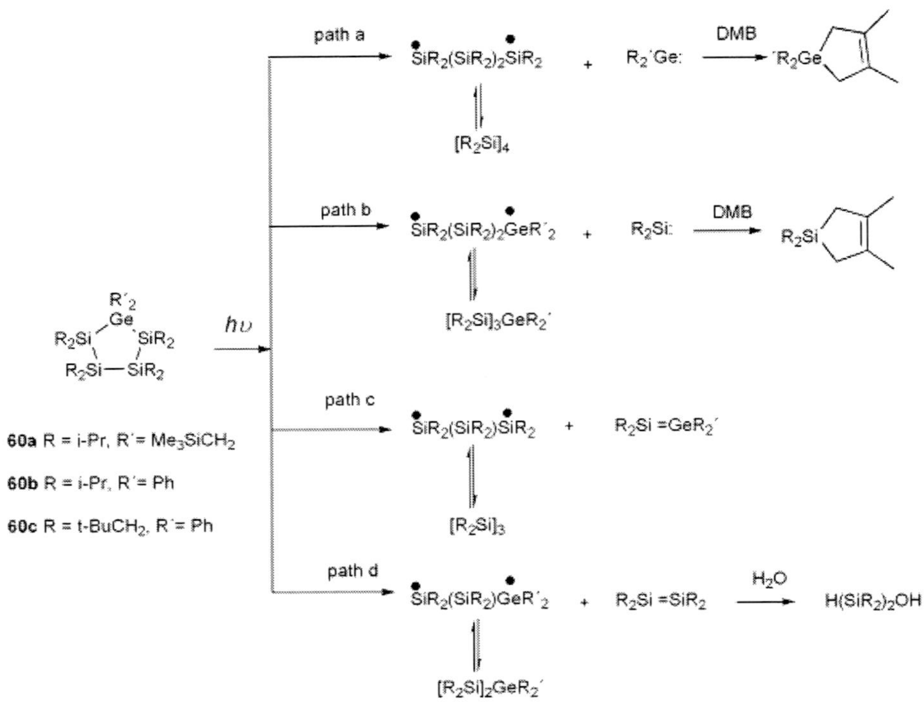

Figure 15. Silicon-centered biradicals.

The photoreaction between dodecamethylcyclohexasilane $(Me_2Si)_6$ and 9,10 phenanthroquinone (PQ) is shown to proceed via triplet radical ion pair formed by phenanthroquinone radical anion and cyclohexasilane radical cation, which leads to a product resulting from addition of linear 1,6-silicon-centered biradical $\bullet Si(Me)_2Si_4(Me_2)_4$-$(Me)_2Si\bullet$ to C=O bonds of quinone.[72] CIDNP investigations reveal that the first step is a single ET resulting in a triplet radical ion pair composed of phenanthroquinone radical anion and cyclohexasilane radical cation. The CIDNP results point out at two possible pathways for the radical cation of $(Me_2Si)_6$. After the back ET in the radical ion pair, it is transformed into biradical $\bullet Si(Me)_2Si_4(Me_2)_4$-$(Me)_2Si\bullet$. Also the radical anion of PQ is capable of abstracting the proton from the radical cation of $(Me_2Si)_6$, with the formation of semiquinone radical and a silene derivative.

Laser pulse photolysis and CIDNP experiments of 7-silanorbornadiene suggest that a silicon-carbon-centered biradical intermediate is generated. The authors could also point out that irradiation in the presence of tetraphenylnaphthalene (TPN), three biradical precursors can be identified (by a PET reaction) ; *i.e.:* a singlet biradical (generated through Si-C bond cleavage) and biradicals from disilabicyclooctadiene (both in singlet and triplet initial states).[73]

9.-Spectroscopic Characterization of Silyl Radicals

The pyramidal structure of silicon-centered radicals has first been noted from chirality studies on silicon compounds containing an asymmetric silicon atom. Thus, silyl radicals generated from optically active silanes, react with radical scavengers such as CCl_4 to give optically-active chlorosilanes mostly with retention of configuration. These early studies indicated that the silyl radical is chiral and adopts a pyramidal form (abstracting a chlorine atom faster than inverting its center). However, this observation is contrasted with that of the chiral $Ph_3SiSi(Ph)(Me)$ radical reacting with CCl_4, in which a racemic mixture of the respective chlorides is observed. This would indicate that the inversion rate of the disilyl radical is much faster than its rate of reaction with CCl_4, or that the radical center is planar.

Electronic absorption spectra of trialkylsilyl radicals are known both in the condensed and gas phases. Radicals $R_3Si\bullet$ (where R = Me, Et, *n*-Pr) exhibit absorption maxima *ca.* 260 nm. In isooctane, a continuously increasing absorption below 340 nm, with no maximum beyond 280 nm, and a symmetric band with maximum at 390 nm are representative of their UV spectra. By progressively substituting Me by Ph in $Me_3Si\bullet$, an absorption band between 360 and 550 nm shows up.

The spectral characterization of low molecular weight alkyl- and aryl-substituted silyl radicals has been carried out by Chatgilialoglu *et al.* by laser flash photolysis studies. Simple silyl radicals absorb strongly near 300 nm and weakly near 400 nm.[74] Assignments of silyl radicals has been confirmed by quenching experiments using CH_2Cl_2. Acyloxymethyl-substituted benzylsilanes have been designed as intramolecular silyl radical traps, in order to obtain evidence for the formation of the radical pair intermediate in the photoreaction. Intramolecular trapping of a silyl radical by an acyloxy carbonyl group, *i.e.*free radical 1,2-(C$\rightarrow$Si) acyloxy migration, may intercept the radical pair formed initially.[75]

Organosilyl radicals generated from hydrogen abstraction of organosilanes with photochemically generated tertbutoxy radicals show well-resolved ESR spectra which permit positive assignment of the radical species.[76] The reported high retention of configuration of silyl radicals is consistent with a sizeable barrier to inversion of these pyramidal radicals compared to the carbon analogs, as commented initially.[77] These ESR spectra of organosilyl radicals generated from silane, methylsilane, dimethylsilane, trimethylsilane, and triethylsilane have been studied in detail by Kochi *et al.*

Also by ESR spectroscopy, Nakamura et al. have identified the SiH_3 radical as produced in Kr and Xe matrices by reaction of H atoms with silane.[78]

The dynamic behavior of silyl radicals formed upon decomposition of the triplet states has been observed by laser flash photolysis studies, and the yields of cage escape of silyl radicals are shown to increase with magnetic field strength from 0 to 1.35 T. These studies have been conducted on micellar solutions in order to monitor the fate of the triplet radical pairs.[79] The lifetimes of the triplet radical pairs are also shown to increase with magnetic field strength. The authors explain the observations through a combination of the Hyperfine Coupling and Relaxation Mechanisms (HPCM and RM).

More recently, Sekiguchi et al. have succeeded in the isolation and complete characterization of the silyl radical 61, in Figure 16.[80]

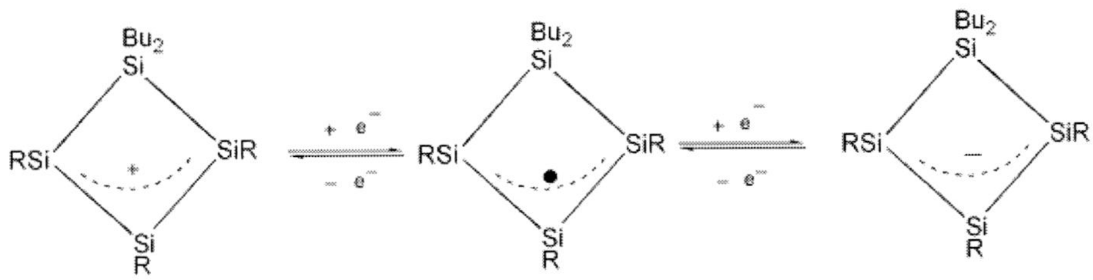

Figure 16. Isolation of persistent silyl radical 61.

Silyl radicals in the gas phase have also been reported to react with a fair number of stable molecules (O_2, Cl_2, S_2Cl_2, ClNO, NO_2, and C_3H_6) and labile intermediates (SiH_3 and H).[81]

EPR spectra of silyl radicals show a central set of lines due to ^{1}H hfs constants and weaker satellites due to the coupling with ^{29}Si (I=1/2, 4.7 %). From these types of experiments it has been concluded that increasing the electronegativity of the α-substituent on the silicon atom, the silyl radical becomes more structurally pyramidal (a higher content of 3s character in the single occupied molecular orbital).

EPR spectroscopy measurements on solutions of the sodium silanides Na*t*Bu_3Si in THF that were exposed to dry air show the presence of the supersilyl radical *t*Bu_3Si·.[82] EPR spectra of polysilanes P-(R_2Si) whose R is longer than methyl, and not branched at the α-position of the side chain, display a four-line pattern. As the length of the substituent is decreased from tetradecyl to butyl, the members of the central pair became broader, and their separation decreased.

When $MeNO_2$, *t*-$BuNO_2$, nitrobenzene, *m*-dinitro, *p*-dinitrobenzene and 1,3,5-trinitrobenzene are made react with Me_3SiH and TTMSS, strong EPR spectra due to silyloxy nitroxide RN(O•)OSiMe_3 and RN(O•)OSi(SiMe_3)$_3$ are formed from addition of silyl radicals onto the nitro group.[83]

More recently, new insight into the manifold of silyl radicals generated photochemically has been reported. Thus C-Si bond dissociation in highly-excited triplet states (T_n) of *p*-phenylbenzylphenylsilanes (PBPS) has been documented using stepwise two-color laser photolysis techniques.[84] The observation indicates that triplet PBPS undergoes C-Si bond scission in a highly excited triplet state to produce p-phenylbenzyl radical (PBR) along with the corresponding phenylsilyl radicals (PSR) whose absorption spectra are difficult to distinguish due to the relatively small molar extinction coefficients at 320 nm. When homolytic cleavage proceeds in the triplet excited state of PBPS, a triplet radical pair of PSR and PBR in the solvent cage, $^3(PSR + PBR)_{cage}$, is initially formed according to spin-conservation rules. The triplet radical pair is free of geminate recombination to the parent molecule. The rationalization of the photophysical process has been argued in detail. Mainly, when the T_1 (π,π*) state is excited with 355- nm laser pulse, the T_n (π,π*) state where triplet energy is localized on the biphenyl moiety of PBPS is produced at an energy level of 145 Kcal mol^{-1} . The T_n state will deactivate by internal conversion to a dissociative state T_R (π,π*). It is originally suggested that the bond dissociation in excited states proceeds by avoided crossings between the reactive excited state and dissociative potential energy surfaces of the same overall symmetry.[85] In the PBPS case, the electronic configuration of the dissociative potential for the C-Si bond rupture is of $\sigma,\sigma*$ which is allowed to interact with the T_R $(\pi,\pi*)$ state, leading to free radical formation. When the C-Si bond dissociates in the T_R state, a triplet σ-radical pair of PBR and PSR may be changed to that of π-radical due to stabilization accrued from π-delocalization in PBR and PSR.

In the same fashion, C-Si bond cleavage of *p*-trimethylsilylmethylacetophenone occurrs in a higher excited triplet state (T_n), giving mainly *p*-acetylbenzyl radical with a transient absorption in the region 295-360 nm, using the two-color laser flash photolysis techniques.[86] In contrast, the C-Si bond cleavage of p-trimethylsilylmethylbenzophenone, is absent in the T_n state, whose energy is larger than the C-Si BDE. Although the quantum yield of the C-Si bond cleavage from $^1(T_n)$ is calculated to be 0.046±0.008, and the non-reactive process is the main rout for relaxation of $^1(T_n)$, the disappeared $^1(T_1)$ changes mainly to *p*-acetylbenzyl and trimethylsilyl radicals through the C-Si bond cleavage. Thus the authors point out that the occurrence of the C-Si bond cleavage depends on both excitation energy and crossing of potential surfaces of the T_n and the C-Si bond dissociation energy of *p*-trimethylsilylmethylacetophenone.

REFERENCES

[1] *Organosilanes in Radical Chemistry. Principles, Methods and Applications*, by Chryssostomos Chatgilialoglu. John Wiley and Sons, New York, 2004. Chatgilialoglu, C. *Acc.Chem.Res*, 1992, 25, 188-194.

[2] Steinmetz, M.G. (1995). *Chem.Rev*, 95, 1527-1588.

[3] Jin, J., Newcomb, M. (2007). *J.Org.Chem*, 72, 5098-5103.

[4] Sharp, K.G., Sutor, P. (1977). *J.Am.Chem.Soc*, 99, 8046-8047.

[5] Lalevee, J., Allonas, X., Fouassier, J.P. (2002). *J.Am.Chem.Soc*, 124, 9613-9621.

[6] Mizuno, K., Tamai, T., Hashida, I., Otsuji, Y. (1995). *J.Org.Chem*, 60, 2935-2937.

[7] Postigo, A., Ferreri, C., Navacchia, M.L., Chatgilialoglu, C. (2005). *Synlett*, 2854-2856. Postigo, A.; Kopsov, S.; Ferreri, C.; Chatgilialoglu, C. *Org.Lett*. 2007, 9, 25, 5159-5162. DOI, 10.1021/ol7020803

[8] Nay, M.A., Woodall, G.N.C., Strausz, O.P., Gunning, H.E. (1965). *J.Am.Chem.Soc*, 87, 179-187.

[9] Sawwan, N., Greer, A. (2007). *Chem.Rev*, 107, 3247-3285.

[10] Waske, P.A., Mattay, J. (2005). *Tetrahedron*, 61, 10321-10330.

[11] Hasegawa, E., Yamaguchi, N., Muraoka, H., Tsuchida, H. (2007). *Org.Lett*, 9, 2811-2814.

[12] Pirrung, M., Fallon, L., Zhu, J., Lee, R.Y. (2001). *J.Am.Chem.Soc*, 123, 3638-3643.

[13] Sluggett, G.W., Leigh, W.J. (1992). *J.Am.Chem.Soc*, 114, 1195-1201.

[14] Leigh, W.J., Xiaojing, L. (2003). *J.Am.Chem.Soc*, 125, 8096-8097.

[15] Tasis, D.A., Siskos, M.G., Zarkadis, A.K., Steeken, S., Pistolis, G. (2000). *J.Org.Chem*, 65, 4274-4280.

[16] Sluggett, G.W., Leigh, W.J. (1994). *Organomet.* 1992, *11*, 3731-3736. *Organomet*, 13, 1005-1013.

[17] Leigh, W.J., Sluggett, G.W. (1993). *J.Am.Chem.Soc*, 115, 7531-7532.

[18] Ito, O., Hoteysha, K., Watanabe, A., Matauda, M. (1991). *Bull.Chem.Soc.Jpn,* 64, 962.
[19] Leigh, W.J., Toltl, N., Apodaca, P., Castruita, M., Pannell, K.H. (2000). *Organomet*, 19, 3232-3241.
[20] Toltl, N., Leigh, W.J. (1996). *Organomet*, 15, 2554-2561.
[21] Ohshita, J., Ohsaki, H., Ishikawa, M. (1991). *Organomet*, 110, 2695-2700.
[22] Ohshita, J., Niwa, H., Ishikawa, M. (1996). *Organomet*, 15, 4632-4638.
[23] Nate, K., Ishikawa, M., Ni, H., Watanabe, H., Saheki, Y. (1987). *Organomet.*, 6, 1673-1679.
[24] Park, S.K. (2005). *J.Photochem.Photobiol.A: Chem*, 173, 29-35.
[25] Steinmetz, M.G., Yu, C., Li, L. (1994). *J.Am.Chem.Soc*, 116, 932-943.
[26] Liu, G., Heisler, L., Li, L., Steinmetz, M.G. (1996). *J.Am.Chem.Soc*, 118, 11412-11422.
[27] Batinica, G.J., Crowell, J.E. (1998). *J.Phys.Chem. B*, 102, 4135-4142.
[28] Kako, M., Nakadaira, Y. (1998). *Coord.Chem.Rev*, 176, 87-112.
[29] Ishikawa, M., Oda, M., Miyoshi, N., Fabry, L., Kumada, M., Yamabe, T., Akagi, K., Fukui, K. (1979). *J.Am.Chem.Soc*, 101, 4612-4618.
[30] Hu, S.S., Weber, W.P. (1989). *J.Organomet.Chem*, 369, 155-163.
[31] Lalevee, J., El-Roz, M., Morlet-Savary, F., Graff, B., Allonas, X., Fouassier, J.P. (2007). *Macromol*, 40, 8527-8530.
[32] Yang, D.D.H., Yang, N.C.C., Steele, I.M., Li, H., Ma, Y.Z., Fleming, G.R. (2003). *J.Am.Chem.Soc*, 125, 5107-5110.
[33] Al-Kaysi, R.O., Goodman, J.L. (2005). *J.Am.Chem.Soc*, 127, 1620-1621.
[34] Kosa, M., Karni, M., Apeloig, Y. (2007). *Organomet*, 26, 2806-2814.
[35] Trefonas, P., West, R., Miller, R.D. (1985). *J.Am.Chem.Soc*, 107, 2737-2742.
[36] Miller, R.D., Michl, J. (1989). *J.Chem.Rev*, 89, 1359.
[37] McKinley, A., Karatsu, K., Wallraff, G.M., Thompson, D.P., Miller, R.D., Michl, J. (1991). *J.Am.Chem.Soc*, 113, 2003-2010.
[38] Ishikawa, M., Kumada, M. (1981). *Ad.Organomet.Chem*, 19, 51.
[39] Lalevee, J., Allonas, X., Fouassier, J.P. (2007). *J.Org.Chem*, 72, 6434-6439.
[40] Brook, M.A., Balduzzi, S., Mustafa, M., Gottardo, C. (1999). *Tetrahedron*, 55, 10027-10040.
[41] Davidson, I.M.T., Michl, J., Simpson, T. (1991). *Organomet*, 10, 842-844.
[42] Whittaker, S.M., Brun, M.C., Cervantes-Lee, F., Pannell, K.H. (1995). *J.Organomet.Chem*, 499, 247-252.
[43] Zielinski, M., Trommer, M., Sander, W. (1999). *Organomet*, 18, 2791-2800.
[44] Kato, M., Ninomiya, M., Gu, M., Nakadaira, Y., Wakasa, M., Hayasi, H., Akasaka, T., Wakahara, T. (2001). *J.Organomet.Chem*, 636, 63-68.
[45] Wu, S.H., Qian, H., Wu, G., Jiang, N. (2004). *Heteroat.Chem*, 3, 589-597.
[46] Steinmetz, M.G., Luo, C., Liu, G. (1999). *J.Org.Chem*, 64, 2057-2065.
[47] Watana, A., Matsuda, M. (1992). *Macromolecules*, 25, 484-488.
[48] Horn, K.A., Whitenack, A.A. (1988). *J.Phys.Chem*, 92, 3875-3880.
[49] Ohshita, J., Yoshitomi, T., Ishikawa, M. (1994). *Organomet*, 13, 3227-3232.
[50] Kani, R., Nakano, Y., Yoshida, H., Hayase, S. (1998). *Macromolecules*, 31, 8794.8801.
[51] Tamai, T., Ichinose, N., Kawanishi, S., Mizuno, K. (2000). *Macromolecules*, 33, 2881-2886.

[52] Ohshita, J., Kanaya, D., Watanabe, T., Ishikawa, M. (1995). *J.Organomet.Chem*, 489, 165-173.
[53] Borthwick, I., Baldwin, L.C., Sulkes, M., Fink, M. (2000). *Organomet*, 19, 139-146.
[54] Trommer, M., Sander, W. (1996). *Organomet*, 15, 189-193.
[55] Porter, N.A., Iloff Jr., P.M. (1974). *J.Am.Chem.Soc*, 96, 6200-6202.
[56] Brook, A.G., Peddlel, G.J.D. (1966). *J.Organomet.Chem*, 5, 106-107.
[57] Scheller, M.E., Frei, B. (1992). *Helv.Chim.Acta*, 75, 69-78.
[58] Scheller, M.E., Frei, B. (1990). *Helv.Chim.Acta*, 73, 922-931.
[59] Matsumoto, A., Ito, Y. (2000). *J.Org.Chem*, 65, 624-626.
[60] Fagnoni, M., Dondi, D., Ravelli, D., Albini, A. (2007). *Chem.Rev*, 107, 2725-2756.
[61] Matsumoto, A., Ito, Y. (2000). *J.Org.Chem*, 65, 5707-5711.
[62] Nanjo, M., Matsudo, K., Kurihara, M., Nakamura, S., Sakaguchi, S., Hayashi, H., Mochida, K. (2006). *Organomet*, 25, 832-838.
[63] Pandey, G., Rao, K.S.S.P. (1995). *Angew.Chem.Int.Ed.Engl*, 34, 2669-2671.
[64] Pandey, G., Gadre, S.R. (2004). *Acc.Chem.Res*, 37, 201-210.
[65] Pandey, G., Rao, K.S.S.P., Rao, K.V.N. (2000). *J.Org.Chem*, 65, 4309-4314.
[66] Bravo-Zhivotovskii: Ruderfer, I., Yuzefovich, M., Kosa, M., Botoshansky, M., Tumanskii, B., Apeloig, Y. (2005). *Organomet*, 24, 2698-2704.
[67] Hisayoshi, L., Yozo, M., Masayoshi, K. (1976). *Macromol.Chem*, 177, 2647.
[68] Suzuki, H., Tanaka, K., Yoshizoe, B., Yamamoto, T., Kenmotzu, N., Matuura, S., Akabane, T., Watanabe, H., Goto, M. (1998). *Organomet*, 17, 5091-5101.
[69] Wang, Y., Ma, J., Inagaki, S. (2007). *J.Phys.Org.Chem*, 20, 649-655.
[70] Zheng, J,Y., Konishi, K., Aida, T. (1998). *J.Am.Chem.Soc*, 120, 9838-9843.
[71] Taraban, M.B., Polyakov, N.E., Volkova, O.S., Kuibida, L.V., Leshina, T.V., Egorov, M.P., Nefedov, O.M. (2006). *J.Organomet.Chem*, 691, 1411-1418.
[72] Taraban, M.B., Volkova, O.S., Kruppa, A.I., Plyusnin, V.F., Grivin, V.P., Ivanov, Y.V., Leshina, T.V., Egorov, M.P., Nefedov, O.M. (1998). *J.Organomet.Chem*, 566, 73-83.
[73] Chatgilialoglu, C., Ingold, K.U., Lusztyk, J., Nazran, A.S., Scaiano, J.C. (1983). *Organomet*, 2, 1332-1335.
[74] Kira, M., Yoshida, H., Sakurai, H. (1985). *J.Am.Chem.Soc*, 107, 7767-7768.
[75] Krusic, P.J., Kochi, J.K. (1969). *J.Am.Chem.Soc*, 91, 3938-3940.
[76] Brook, A.G., Duff, J.H. (1969). *J.Am.Chem.Soc*, 91, 2118-2119.
[77] Nakamura, K., Masaki, N., Sato, S., Shimokoshi, K. (1985). *J.Chem.Phys*, 83, 4504-4510.
[78] Wakasa, M., Sakaguchi, Y., Hayashi, H. (1992). *J.Am.Chem.Soc*, 114, 8171-8176.
[79] Matsumo, T., Itchinohe, M., Sekiguchi, A. (2003). *Angew.Chem.Int.Ed.Engl*, 2000, 41, 1559. Weindenbruch, M.*Organomet*, 22, 4348-4360.
[80] Niiramen, J.T., Gutman, D. (1993). *J.Am.Chem.Soc*, 97, 4106-4109.
[81] Lerner, H.W. (2005). *Coord.Chem.Rev*, 249, 781-798.
[82] Alberti, A., Dellonte, S., Paradisi, C., Roffia, S., Pedulli, G.F. (1990). *J.Am.Chem.Soc*, 112, 1123-1129.
[83] Yamaji, M., Mikoshiba, T., Masuda, S. (2007). *Chem.Phys.Lett*, 438, 229-233.
[84] Dauben, W.G., Salem, L., Turro, N.J. (1975). *Acc.Chem.Res*, 8, 41-54.
[85] Cai, X., Sakamoto, M., Hara, M., Inomata, S., Yamaji, M., Tojo, S., Kawai, K., Endo, M., Fujitsuka, M., Majima, T. (2005). *Chem.Phys.Lett*, 407, 402-406.

In: Photochemistry Research Progress
Editors: A. Sánchez, S. J. Gutierrez

ISBN: 978-1-60456-568-3

Chapter 7

DEVELOPMENT AND APPLICATIONS OF NOVEL UV AND VUV EXCILAMPS IN PHOTOCHEMISTRY

Edward A. Sosnin [1]***, Irina V. Sokolova*** [2] ***and Victor F. Tarasenko*** [1]

[1]Optical Radiation Laboratory, High Current Electronics Institute, Siberian Branch of Russian Academy of Science, 634055, Russia, Tomsk, 2/3 Akademicheski Ave

[2]Laboratory of Photochemistry and Photophysics of Molecules, Tomsk State University, Novosobrnaya Sqr., Tomsk 34050, Russia

ABSTRACT

The chapter is concerned with ultraviolet (UV) and vacuum ultraviolet (VUV) excilamps and excilamp-based photoreactors and reviews our efforts in UV/VUV electrodeless barrier-discharge excilamps and their applications in photochemistry, namely, for oxidation and mineralization of organic substrates, VUV-initiated photodimerization and removal of water from hydrocarbon gas mixtures, and in actinometry.

Keywords: Actinometry, advanced oxidation processes (AOPs), aqueous photochemistry, carbamide, combined UV and biodergadation, dielectric barrier discharge, capacitive discharge, excimer, exciplex, excilamp, hydrocarbon gases, incoherent UV/VUV light source, photomineralization, photodimerization, phenolic compounds

1.Tel.: (8)3822-491685
Fax: (8)3822-492410 E-mail: badik@loi.hcei.tsc.ru
2 Tel.: (8)3822-533426 Fax: (8)3822-530043 E-mail: sokolova@phys. tsu.ru

1. Introduction

The invention and development of excilamps is a significant breakthrough in designing novel light sources. Excimer lamps are sources of spontaneous radiation in the ultraviolet (UV) or vacuum ultraviolet (VUV) spectral regions. They produce radiation due to dissociation of excimer molecules (so-called excimers – excited dimers, for molecules consist of the same atoms, e.g. Kr_2*) or exciplex molecules (so called exciplex – excited complex, for molecule consisting of hetero-atomic, e.g. XeBr*). Excimers and exciplexes are produced by various mechanisms, depending on the type of the working gas and conditions of its excitation, and the characteristic lifetime of the these molecules is 10^{-7}–10^{-9} s. In 1994, excimer and exciplex lamps were given the name excilamps [1].

The design of laboratory excilamps stimulated the research on their potential applicability in photochemistry. One of the first publications on the subject reports that "potential applications <of the excilamps> include the wavelength-selective investigation of photochemical or photobiological reactions, the stability testing of pharmaceuticals, cosmetics, dyes, detergents and food additives in solution, and for industrial photochemical processes" [2].

By now, there are numerous papers on designs and studies of excilamps [3-11] and less common publications demonstrate their applications in photochemistry [7,12-16].

In this paper, we describe several types of VUV/UV excilamps designed to date and reviews our seven-year laboratory studies of their applications in photochemistry.

2. Design And Characteristics Of Excilamps

Excilamps are a rather young family of light sources. First complete excilamps were developed by G.A. Volkova et al. at the Vavilov State Optical Institute in 1982-1984 [17]. These works pioneered the design of excilamps in Asea Brown Bovery (ABB, Switzerland). Since then, increasingly more and more excilamps are patented [18], and more and more firms and research groups involved directly with excilamps or employing them as part of equipment appear every year. Excilamps also attract the attention of known light source producers such as Heraeus Noblelight GmbH (Germany), Philips (Netherlands), Ushio Inc. (Japan), etc.

Excilamps can be driven in various ways. However, the longest lifetime of excilamps on inert gas – halogen mixtures is attained with electrodeless discharges: capacitive, barrier, and microwave discharges in which the working mixture is in a closed quartz bulb with a shape dependent on the discharge type. Other types of discharge (glow, spark and corona discharges, high-pressure pulsed space discharge with UV preionization) can also be used to excite excilamps, though the lifetime of the working matter of the thus driven excilamps is limited. In practical terms, the most attractive excilamps are those driven by barrier (BD) and capacitive (CD) discharges. The general properties and advantages of these excilamps are:

1. no self-absorption at operating wavelengths (except for excilamps on dihalides, e.g., I_2-excilamps);

2. mercury-free mixture (except for Hg-dihalides and Hg-inert gas excilamp mixtures), which better complies with ecological standards;
3. versatility of designs that for the first time allowed the optimization of lamp geometry for some photochemical reactors (discharge geometry determines the gas pressure: in glow and capacitive discharge excilamps the pressure is no greater than tens of Torr, but in the barrier discharge lamps it can exceed 760 mm Hg);
4. easy instant startup (BD and CD driven excilamps are powered by pulsed voltage of amplitude up to several kilovolts and frequency up to several hundreds of hertz, and hence these devices dispense with the need for special starters);
5. fairly long lifetime: t_r^{15-20} > 1000-2000 h for Cl-containing excilamps and t_r^{5-15} > 10000 h for inert gas excilamps (t_r^k is the time in which the excilamp radiant power decreases by k percents);
6. comparatively narrow UV or VUV band in which most of the radiant power is concentrated, as opposed to that in luminescent and thermal UV and VUV sources; for exciplexes, the band halfwidth is between 2 and 15 nm and for inert gas excimers, it is up to 30 nm (Fig. 1,2) [4,7];
7. moderate heating of bulbs for which excilamps sometimes are called cold emission sources;
8. high radiant power volume densities (W/cm^3), compared to that characteristic of low-pressure gas discharge lamps at resonance atomic transitions;
9. reasonable electrical safety (BD and CD excilamps use high-frequency power supplies, the discharge in a bulb is bounded by a dielectric barrier, and the discharge current is thus no greater than tens of milliamperes).

The fundamental principles of the UV and VUV generation by barrier and capacitive discharge plasmas are discussed elsewhere [9,11,16,19-22].

The characteristic optical spectra of some excilamps are shown in Figs. 1-3. It is seen in the figures that the optical spectra of Xe_2, Kr_2, and Ar_2 excilamps are continua with one peak. At very low pressures (fractions of a Torr), the Xe resonance line dominates in the spectrum and at pressures greater than 100 Torr the band with a peak at 172 nm dominates. Because of the higher intensity, the latter continuum is commonly used at high pressures. Typical spectra corresponding to this continuum are shown in Fig. 1. Among these excilamps, Xe_2-excilamps (172 nm) now find most extensive use.

Excilamps on inert gas halogenides are no less interesting for photochemical applications. The most intense band in the spectra of these excilamps is always the B–X transition of the working exciplex molecule. The spectra may also contain weak bands of other transitions (Table 1, Fig. 2) and low-intensity bands of halogen molecules. Most widespread in photochemistry applications are XeCl-, KrCl-, XeBr- and XeBr-excilamps. The emission spectra for these excilamps in barrier and capacitive discharges are shown in Figs. 2, 3. Because the pressures at which typical capacitive and barrier discharges operate are different (up to several Torr, and tens and hundreds of Torr, respectively), the spectra of these excilamps are different, too (Fig. 3). In the barrier discharge, the main continuum (the B–X transition) is much narrow than that in the capacitive discharge and the contribution of other transitions is less significant.

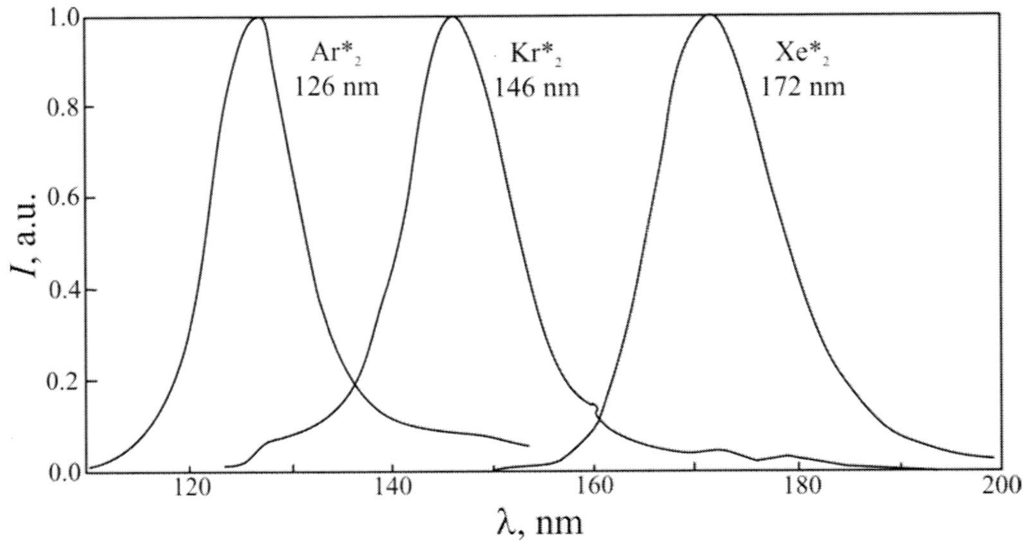

Figure 1. Emission spectra of some inert gas excimers in a barrier discharge.

Thus, the emission spectrum halfwidth can be specified by selecting one or another discharge type with due regard to particular requirements.

Table 1. Peak wavelength of RX* exciplex at different transitions

R	X	Peak wavelength of RX* exciplex at different transitions, nm			
		D1/2 → X1/2	B1/2 → X1/2	C3/2 → A3/2	D1/2 → A1/2
Ar	F	185	193	203	204
Ar	Cl		175		195
Ar	Br		165	172	183
Kr	F	220	248	275	272
Kr	Cl	200	222	240	235
Kr	Br		207	222	228
Kr	I		190	195	225
Xe	F	264	351	460	410
Xe	Cl	236	308	345	340
Xe	Br	221	282	300	325
Xe	I	203	253	265	320

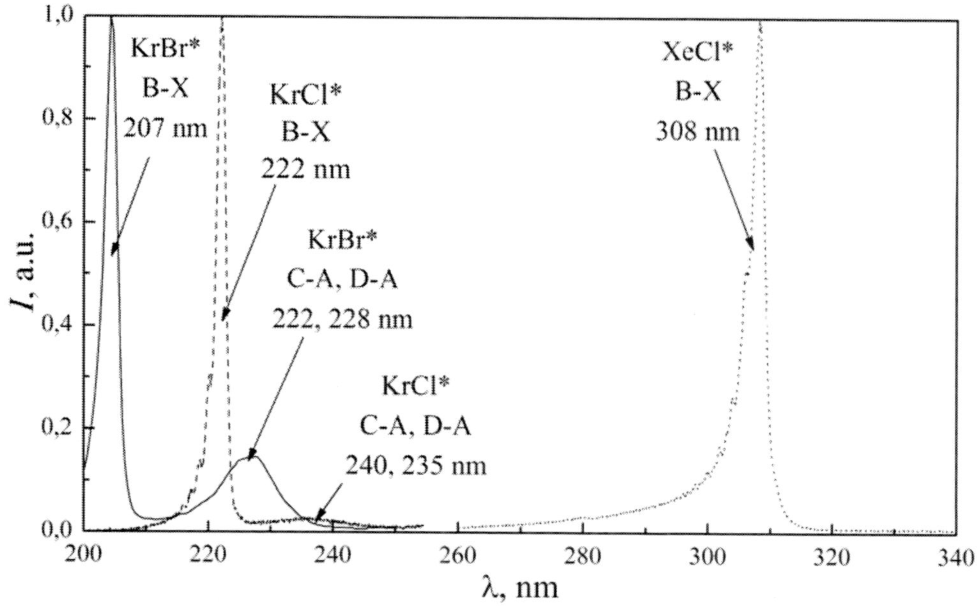

Figure 2. Emission spectra of some exciplex molecules in the barrier discharge.

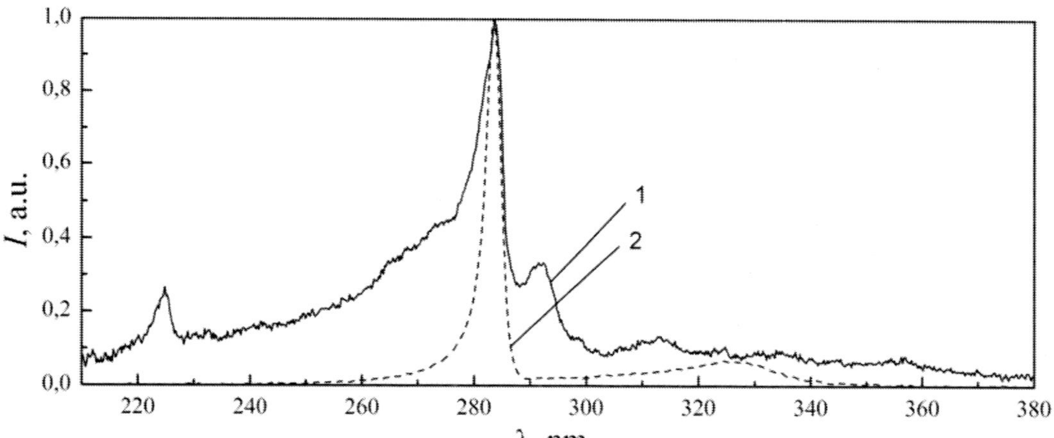

Figure 3. Emission spectra of XeBr* exciplexes in the capacitive (1) and barrier (2) discharges.

BD and CD excilamps differ in both the discharge form and the excilamp geometry (Fig. 4). The most-used excilamps are planar (Fig. 4(a, b)) and coaxial (Fig. 4(c)) BD excilamps and cylindrical (Fig. 4(d)) CD excilamps. In BD excilamps, the discharge, as a rule, is the multitude of chaotic microdischarges (Fig. 4 on the left), and in CD excilamps the discharge resembles a classical glow discharge (Fig. 4 on the right).

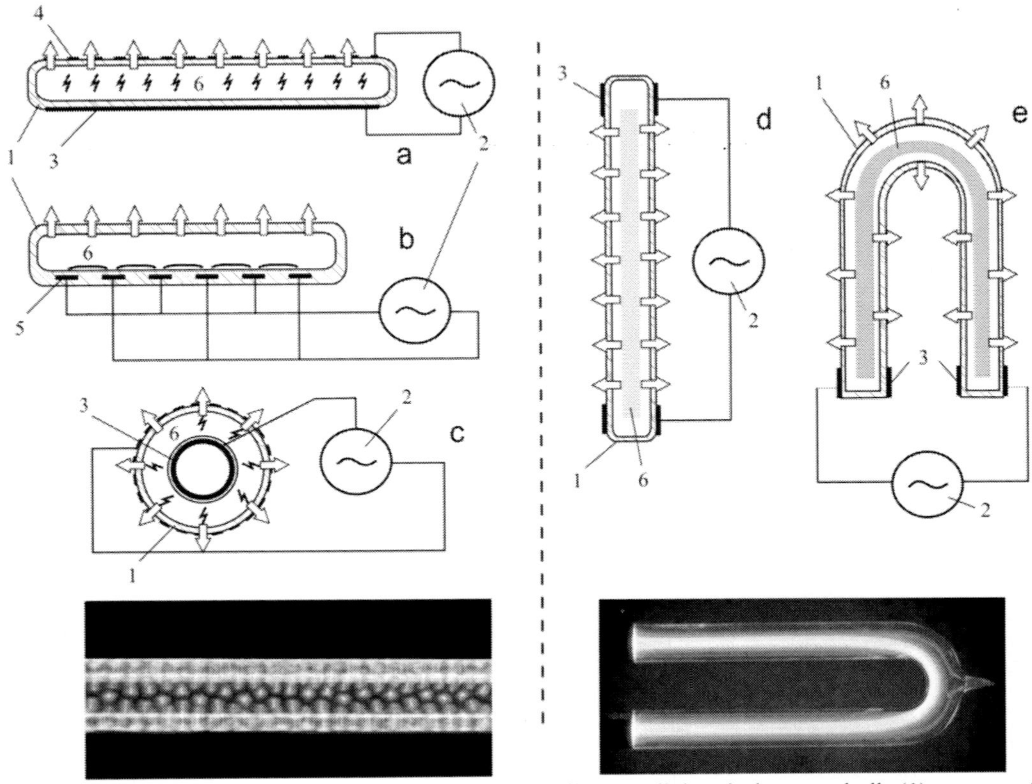

Figure 4. BD (on the left) and CD (on the right) excilamps: dielectric layer or bulb (1); ac or pulsed voltage source (2); solid and perforated electrodes (3, 4); electrode housed in the bulb wall (5); discharge gap (6). The radiation is shown by white arrows.

The variety of excilamp designs makes possible various photochemical reactors.

In 1998–2006, the team of the Laboratory of Optical Radiation (Institute of High Current Electronics, SB RAS) developed five series of BD and three series of CD radiation sources, and more then ten individual models, including photochemical reactors.

Below are examples of these radiation sources:

Sources of the series BD_P (barrier discharge, portable) are a compact portable device (dimensions 230×80×80 mm and weight 700 g) consisting of a housed coaxial excilamp with air cooling, power supply, and reflector (Fig. 5). The sources provide an exitance power up to 5-15 mW/cm^2 for the working molecules XeCl*, XeBr*, KrCl*, KrBr* and up to 35 mW/cm^2 for Xe_2*. The exitance power for the molecules XeI* and I_2*, Cl_2* is smaller than the above value and ranges to 1.5 mW/cm^2. The sources have found many research applications.

Sources of the series BD_L (barrier discharge, large) are a large device (dimensions 410×160×160 mm and weight 1.6 kg) consisting of a housed coaxial excilamp with air cooling, power supply, and reflector (Fig. 6). The radiant power is between 3 and 9 W and the working molecules are XeCl*, KrCl* and XeBr*.

One of the parameters that affect the excilamp efficiency and lifetime is temperature. Experiments show that increasing the bulb temperature decreases the excilamp efficiency and lifetime that necessitates forced cooling of the bulb. In practice, bulbs are normally water-cooled, which provides in an increase in the radiant power volume densities of KrCl- and XeCl- BD excilamps to 0.1 W/cm^3.

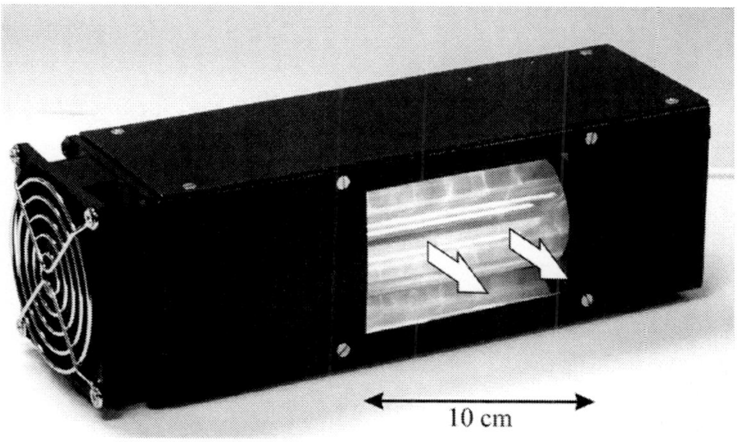

Figure 5. BD_P source. The output radiation is shown by arrows.

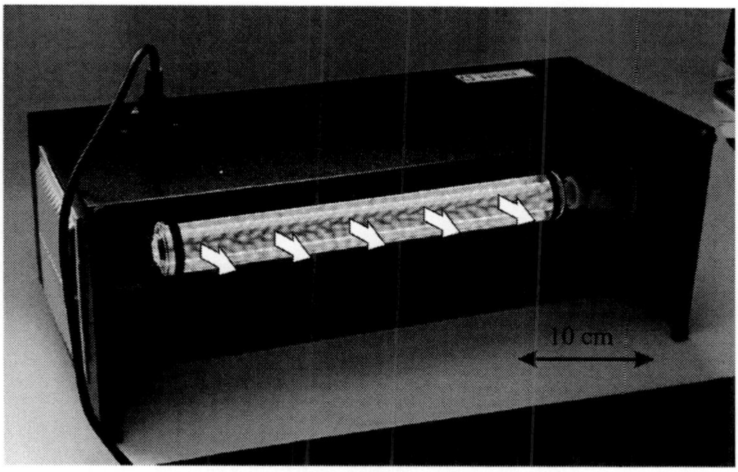

Figure 6. BD_L source. The output radiation is shown by arrows.

In the BD_L and BD_P sources, we employed simple air cooling of the coaxial excilamps. Remind that in both versions, the excilamp, the cooling system and the power supply are a single unit equipped with a reflector to increase the irradiance. A similar design was used in sources of the series CD_L (capacitive discharge, large), except that the reflector was located some distance away from the outer tube, whereas in BD_L and BD_E sources the reflector (foil) was attached directly to the outer tube surface.

For increasing the radiant power, sources BD_Cs (barrier discharge, cascade) were designed. The sources are a large device consisting of coaxial excilamps assembled in a cascade with air cooling and a common reflector. Figure 7 shows an example of the Xe_2_BD_Cs sources. The source provides a maximum emission at 172 nm, a radiant power of 50 W, an exitance power up to 120 mW/cm^2. The efficiency of the device is up to 5%.

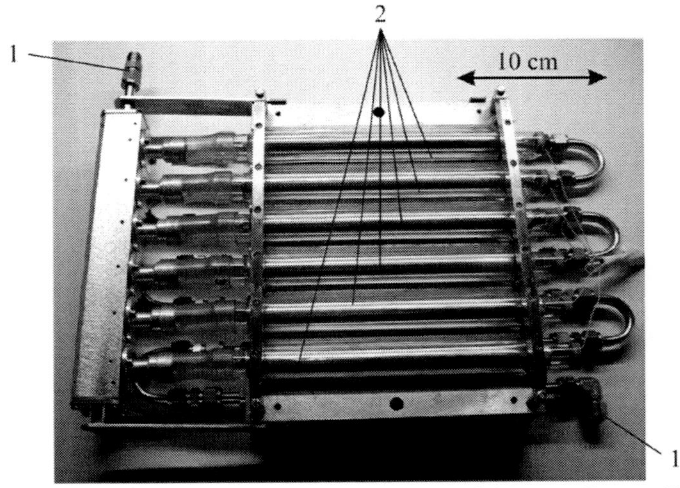

Figure 7. Xe2_BD_Cs source: water cooling unions (1); six coaxial BD excilamps (2).

In some cases, the excilamp bulb should be separated from the power supply. For this purpose, sources of the series CD_E (capacitive discharge, external) and BD_E (barrier discharge, external) were designed. An example of the sources is shown in Fig. 8.

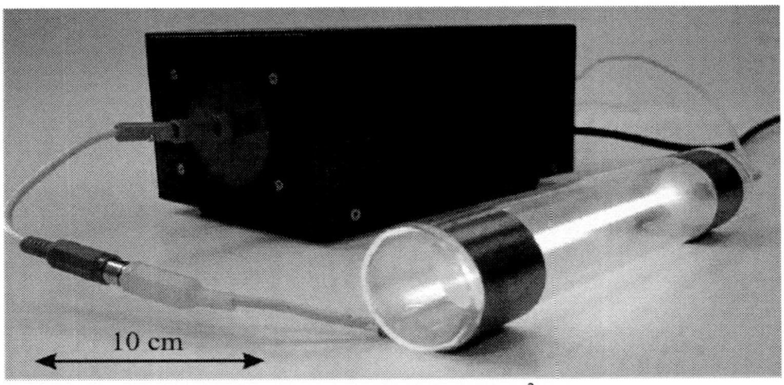

Figure 8. KrCl_CD_E source. The exitance power is 6.5 mW/cm^2.

Around sources of the BD_E series, it is rather easy to design systems for irradiating solutions and gases (Fig. 9). In 1996, the BD excilamp systems were given the name excilamp flow-through photoreactors [23].

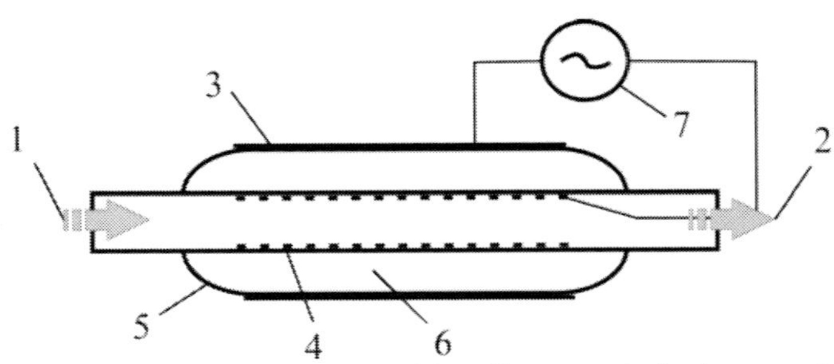

Figure 9. Flow-through photoreactor based on coaxial excilamps: solution/gas supply (1); irradiated solution/gas (2); outer electrode-reflector (3); inner grid/spiral electrode (4); bulb (5); discharge gap (6); power supply (7).

Note high-power pulsed excilamps [24-28] operating at low pulse repetition rates (little used in photochemistry) and windowless excilamps on xenon, krypton, and argon molecules [29-30], and multiwave lamps emitting on several working molecules [31-34] are omitted in this review.

3. Photolysis of Organic Matters

Ultraviolet radiation is fundamental for advanced oxidation technologies or advanced oxidation processes (AOPs) [13] used in purification of aqueous and gaseous media.

Traditionally, the range of wavelengths required for direct photolysis of dissolved organic matter (DOM) is 200–280 nm [35], since it is in this range that most of the components (dissolved organic and inorganic) contained in water absorb radiation.

Direct photolysis leads to excitation of singlet states of organic matters, which then relax to triplet states. These excited states may go through homolysis, heterolysis, and photoionization (Fig. 10).

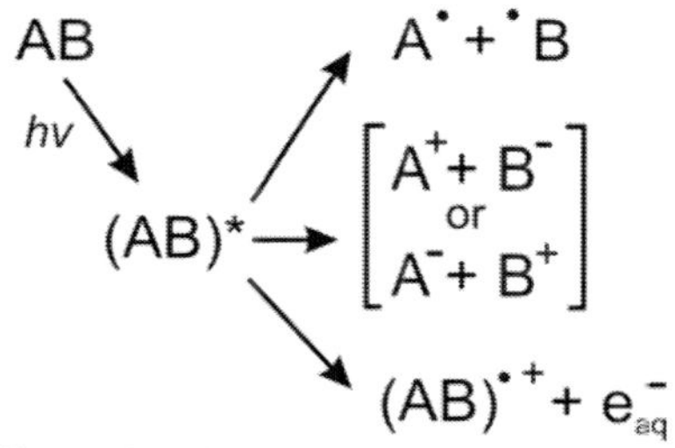

Figure 10. Possible pathways in direct photolysis.

The range of 100–200 nm wavelengths is very attractive for photolysis. Because the typical concentration of matters dissolved in water is moderate, VUV radiation is absorbed mainly by water whose concentration is 55.5 M. That is in typical conditions the amount of water is million times greater than that of the matter dissolved in it, and hence most of the VUV source energy is expended in homolysis of water ($H_2O - hv \rightarrow H_2O^* \rightarrow H^\bullet + {}^\bullet OH$) [12-14]. Further, the pollutant reacts with the resulting radicals and goes through mineralization:

$$C_nH_mX_z\,(hv, O_2) \rightarrow nCO_2 + (m\text{-}z)/2\; H_2O + zHX, \qquad (1)$$

where X is heteroatomic organic molecules which transform into the corresponding mineral acids HX (HNO_2, H_2SO_4, HCl, HNO_3 etc.). A similar reaction occurs on the interaction of the oxidant and UV radiation with metalloorganic complexes.

In all these AOPs, VUV and UV excilamps can successfully be used. Some results of their applications in AOP treatment of aqueous and gaseous media are indicated in Table 2.

Analysis of available papers shows that the most efficient AOP is realized with a Xe_2-excilamp (λ~172 nm) [13,36]. The trouble with VUV excilamps is a small penetration depth (microns) of VUV radiation in water. Therefore, this method for the moment finds commercial applications only for removal of organic carbon from ultrapure water or for treatment in combination with aeration.

The problem here is that water, which is often a solvent, prevents the VUV-DOM interaction and decreases the effective interaction volume.

This problem can be solved, e.g., by changing the state of aggregate state of water, as was done in [37]. In this work, water with a pollutant was first converted to vapor and then the vapor was irradiated by a Xe_2-excilamp. This transformation greatly increased the reactive volume and accelerated the photolysis of phenol. The process was called photoreactive distillation.

Table 2. Examples of VUV photolysis of organic matters in the liquid and gas phases

Substrate	Excilamp	Consumption power, W	Irradiated volume, V/L	Initial concentration, mg/L	Analytical method	Ref.
1,1,1-Trichloroethane	Xe2*	150	2.59	24.9	TOC	[14]
1-Butanol	Xe2*	155	gas phase	5	GC	[38]
2,4-Dichlorophenol	Xe2*	960	20	–	TOC	[39]
2-Chlorodibenzo-p-dioxin	KrCl* Xe2*	–	3-4 mL	–	GC, MS	[40]
3,4-Dimethoxynitrobenzen	XeCl*	–	3-4 mL	25	UVS	[41]
4-Chlorophenol	Xe2*	155	1	–	TOC	[42]
Alcohols (C1-C8)	Xe2*	155	1	–	TOC	[43]
Phenol	Xe2*	155	1	–	TOC	[43]
Acetone	Xe2*	155	1	–	TOC	[43]
Antrazine	KrCl*	80	5	10	GC,MS,nitrate-ions	[44]
Formaldehyde	Xe2*KrCl*	150	0.8	30	TOC	[45]
Humic acid	Xe2*	150	0.350	500	TOC	[46]
Tetrachloromethane	Xe2*	53	gas phase	326	GC, MS	[47]
Uridine	KrCl*	–	3-4 mL	0.03	UVS	[48]

Analytical method: total organic carbon (TOC); gas chromatography (GC); mass-spectroscopy (MS); UV spectroscopy (UVS).

Another way of solving the problem is to use systems in which the water content is small, but its positive or negative effect on the required process is quite considerable. Among these systems is low-water natural gas from gaseous condensate field. The results of studies of this system are described in Section 5.

Finally, we may resort to engineering to simply intensify the solution mixing in the reactive zone, e.g., by bubbling [49].

Moreover, VUV excilamps are applicable where the potentialities of direct photolysis are exhausted. The spectral absorbance of a molecule depends on its structure and interaction with a solvent. Most UV-absorbing molecules have either monovalent or conjugate double bonds, contain carbon, nitrogen, or oxygen atoms, and are characterized by delocalized π-electrons. These systems are called chromophores (Table 3). However, the majority of chromophore-containing environmental pollutants – alkenes (unsaturated hydrocarbons), aldehydes, aromatic and heterocyclic compounds, dyes, ketones, carboxylic acids, etc., absorb little or no radiation at $\lambda > 205–210$ nm. Thus, direct photolysis, in this case, fails, whereas VUV radiation does not.

The advantages of VUV photolysis are:

- extremely high concentration of hydroxyl radicals (locally);
- no need for additional oxidants (in situ H_2O_2 and O_3 generation);
- efficient photomineralization of organic matters in solution;
- lesser influence of bicarbonates and nitrate-ions on the photolysis rate.

By the time we began our research, most publications on photochemistry discussed direct photolysis, photocatalysis, and combined methods of treatment, and the subject matter of these works was to determine the resulting photoproducts and their toxicity. However, the processes occurring in the first phase of direct photolysis are photophysical processes of population and relaxation of energy levels of an organic molecule, rather than chemical reactions as such. These processes are indubitably affected by the optical parameters of exciting radiation. Therefore, our task was to find the most optimal (in photophysical terms) radiation parameters.

Table 3. Characteristics of typical organic chromophore functional groups [50]

Chromophore	λmax, nm	εmax, M-1·cm-1	Transition
C–H or C–C	< 180	1000	$\sigma \to \sigma^*$
C=C	180	10000	$\pi \to \pi^*$
C=C–C=C	220	20000	$\pi \to \pi^*$
Benzene	260	200	$\pi \to \pi^*$
Naphthalene	310	200	$\pi \to \pi^*$
Anthracene	350	10000	$\pi \to \pi^*$
Phenol	275	1500	$\pi \to \pi^*$
Aniline	290	1500	$\pi \to \pi^*$
C=O	280	20	$n \to \pi^*$
C=C–C=O	320	50	$n \to \pi^*$
C=N	< 220	20	$n \to \pi^*$
N=N	350	50	$n \to \pi^*$
N=O	300	100	$n \to \pi^*$
Ar–NO2	280	7000	$\pi \to \pi^*$

3.1. Effect of the UV source parameters on photolysis of phenol and its derivatives

The mechanisms of phenol photodegradation are critical for water purification technologies.

Phenols are aromatic compounds that contain one or several hydroxyl groups bound with a benzene ring (Fig. 11). Phenols, including nitro- and chlorophenols, are referred to most plentiful technosphere-produced environmental pollutants in the world. For example, water chlorination produces chlorophenols, and wood processing uses pentachlorophenol due to its insecticidal and bactericidal properties. Moreover, phenols and their derivatives are the resulting products of the bacterial degradation of aromatic acids in anaerobic conditions [22].

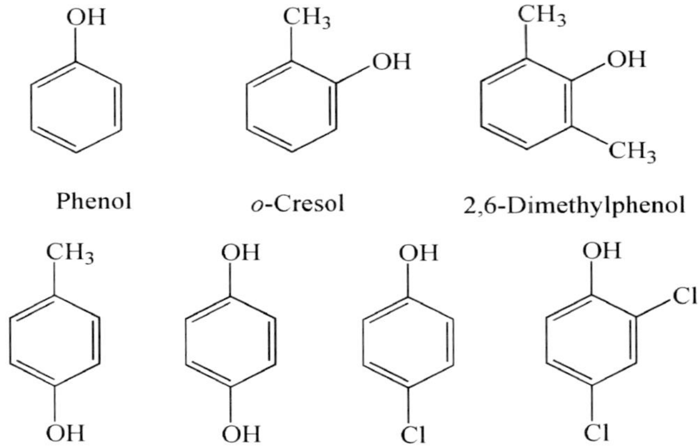

n-Cresol Hydroquinone 4-Chlorophenol 2,4-Dichlorophenol

Figure 11. Structural formula and names of some phenols and their derivatives,

Most of research on photolysis of phenols concerns direct photolysis [51] or advanced oxidation technologies (AOTs), including combined treatment by UV radiation and various oxidants (O_3, O_3/H_2O_2, Fe^{2+}) [52], ultrasound [53], and semiconductor catalysts, e.g., TiO_2 [54], $TiO_2 + H_2O_2$ [55].

Traditionally, photochemical research puts emphasis on what photoproducts and/or changes of toxicity are produced by one or another UV source, e.g., by a wide-band xenon flash lamp [56], nanosecond laser with λ=266 nm [57], or low-pressure mercury lamp [58]. However, in each particular case, the pulsewidth, the peak power, and the spectrum are different resulting in different scenarios of population of triplet and singlet states and, finally, in different kinetic pathways of photodegradation of organic matter in solution.

Therefore, our task was to find the most optimal parameters of the exciting UV radiation with respect to photophysical processes, using phenols as an example. First, we examined the effect of optical spectral distributions on the efficiency of phenol phototransformation (Sections 3.1.1, 3.1.2) and then compared the photodegradation of phenols irradiated by a nanosecond laser and by an excilamp with a relatively long pulsewidth (Section 3.1.3).

3.1.1. Photolysis of cresols with capacitive discharge excilamps

Para- and ortho-cresols with initial concentrations of 10^{-3} mol/liter were irradiated by KrCl- and XeBr- capacitive discharge excilamps (CD_E model, IHCE, SB RAS, Russia). The excilamps had U- and I-like shapes and emitted narrow bands (B→X transition) at 222 and 283 nm ($\Delta\lambda$ = 5–10 nm). The exitance power of the excilamps was 18 and 28 mW/cm^2, respectively.

The irradiated solutions were in 10 ml quarts bulbs located 0.5-1.5 cm from the excilamp surfaces such that the radiation took the path through one of the bulb walls. The rest planes of the bulb were non-transparent. During the test, the bulb and the excilamp were cooled with a fan to prevent heating of the solutions (and DOM volatilization). With the cresol concentration used, the contribution of oxygen dissolved in water to chemical processes could be disregarded [59].

It was not our aim to examine the products of photolysis, and the influence of the radiation on the DOM was judged from spectral and luminescence analyses (see, e.g., [60]), namely the cresol phototransformations and the photolysis efficiency were judged from the fluorescence spectra and the changes in the solution absorption spectra.

After irradiation of ortho-cresol, a decrease in the intensity of the main fluorescence band at 300 nm was observed (Table 4). At the same time, with two lamps operating at a time, the fluorescence intensity at $\lambda = 350$ nm increased. However, with the KrCl-excilamp, a 5 J/cm^2 radiation doze suffices for phototransformation of the intermediate photoproducts, whereas with the XeBr-excilamp, the doze must be at least 1.5 times greater to observe the effect.

Table 4. Maximum absorption wavelength and fluorescence of cresols

Matter	Maximum absorption wavelength, nm		Fluorescence, nm
Ortho-cresol	220	270	300
Para-cresol	220	280	305

Table 5. Quantum yield φ of cresol photolysis

Excilamp	D, J/cm2	φ	
		para-cresol	ortho-cresol
KrCl (λ ~ 222 nm)	2	0.5	0.57
XeBr (λ ~ 283 nm)	2	0.2	0.2

After irradiation of para-cresol, the intensity of the main fluorescence band at 305 nm decreased. The formation of the primary photoproduct with maximum fluorescence at ~410 nm was observed within the first minutes of irradiation by the excilamps. The decay of the product occurred at radiation dozes of 2 and 5 J/cm^2 for the KrCl- and XeBr-excilamps, respectively. The quantum yield of ortho- and para-cresol photolysis was calculated from the decrease in the intensity of the main band (Table 5).

Thus, the KrCl-excilamp provides higher efficiency of phototransformations, compared to the XeBr-excilamp.

This does not necessarily mean that the KrCl-excilamp is just the device to be used in the sewage treatment technology. For example, in deep water purification, the products of photolysis also display toxicity, and hence additional water treatment is required. The results of investigation of combined two-stage removal of chlorophenols from water are described in Section 3.2. At the first stage, the solutions were irradiated by the XeBr-excilamp and, at the second stage, the resulting decay products were "digested" by special microorganisms-destructors. The KrCl-excilamp treatment gives rise to more toxic products of photolysis, which oppress the vital activity of the microorganisms-destructors. Thus, the applicability of one or another excilamp for water purification should be judged not only from direct photolysis.

3.1.2. Photolysis of phenol and its bromine and chlorine derivatives with capacitive discharge excilamps

The phototransformation of phenol, 4-chlorophenol (n-chlorophenol), and 4-bromophenol (n-bromophenol) irradiated by KrCl- and XeBr- CD excilamps were studied under conditions similar to those described in Section 3.1.1.

First, for interpretation of the spectral and luminescent properties and elucidation of the mechanisms of photodissociation of the bonds O–H, C–Cl, C–Br of separate molecules of phenol, 4-chlorophenol and 4-bromophenol, we performed quantum-chemical calculations. The energy-level diagrams of electron-excited states of the molecules were calculated by the semiempirical method PNDO (partial neglect differential overlapping). The rate constants of photophysical intramolecular processes (internal and singlet-triplet conversions) were estimated and the photoreactions of the O–H and C–Cl bond dissociation were examined.

The obtained energy-level diagram of the electron-excited states of the examined molecules is shown in Fig. 12.

The substitution of Br and Cl atoms changes the pathway of phenol molecule photoprocesses, much as the heavy atom substitution does. The chlorine/bromine substitution causes a low-energy shift of the transition $S_0 \rightarrow S_1$ without changing its nature ($\pi\pi^*$-type transition). In the energy-level diagram of 4-bromophenol and 4-chlorophenol molecules, new S_2 and T_2 states of the $\pi\sigma^*$ type appear with the σ^*-orbital involving the C–Cl/C–Br bond. The energy of the S_2 and T_2 states of the halogen-substituted corresponds to the energy of dissociation of the C–Cl/C-Br bond. In a range up to 50 000 cm^{-1} of phenol, 4-bromophenol, and 4-chlorophenol molecules spectral absorption, states inherent to O–H bond are not found. This suggests a high probability of the C–Cl/C–Br bond photodissociation, which is what we observed in further experiment.

Comparison of the spectral and luminescent characteristics of phenols irradiated by the XeBr- and KrCl-excilamps reveals the following effects.

The harder quantum increases the phenol photodecay efficiency approximately two times. When excited by radiation at $\lambda \sim 222$ nm, the phenol molecule transforms into the state S_2. According to the calculation, the rate constant of the transition from this state into the photodissociative state is an order of magnitude higher than that of the transition from the state S_1. Thus, the more efficient photolysis of phenol on excitation at $\lambda{\sim}222$ nm, compared to that at 283 nm, is due to the increase in the probability of dissociation of the O–H bond on harder excitation.

There is no dependence like this for 4-chlorophenol and 4-bromophenol. Although the probability of photodissociation of the O–H bond for them also increases, the photostability of these molecules is governed by the disruption of the C–Cl/C–Br bond, for which the quantum energy at $\lambda \sim 283$ nm is sufficient. On harder excitation, long-wave photoproducts are more abundant (Figs. 13, 14).

Thus, the change of the excilamp wavelength from 283 to 222 nm significantly affects the photolysis of phenol, whereas changes but little the photodecay efficiency of 4-chlorophenol and 4-bromophenol, while being responsible for the formation of a large amount of fluorescent photoproducts in the long-wave spectral region. The obtained experimental data correlate with the results of quantum-chemical calculations.

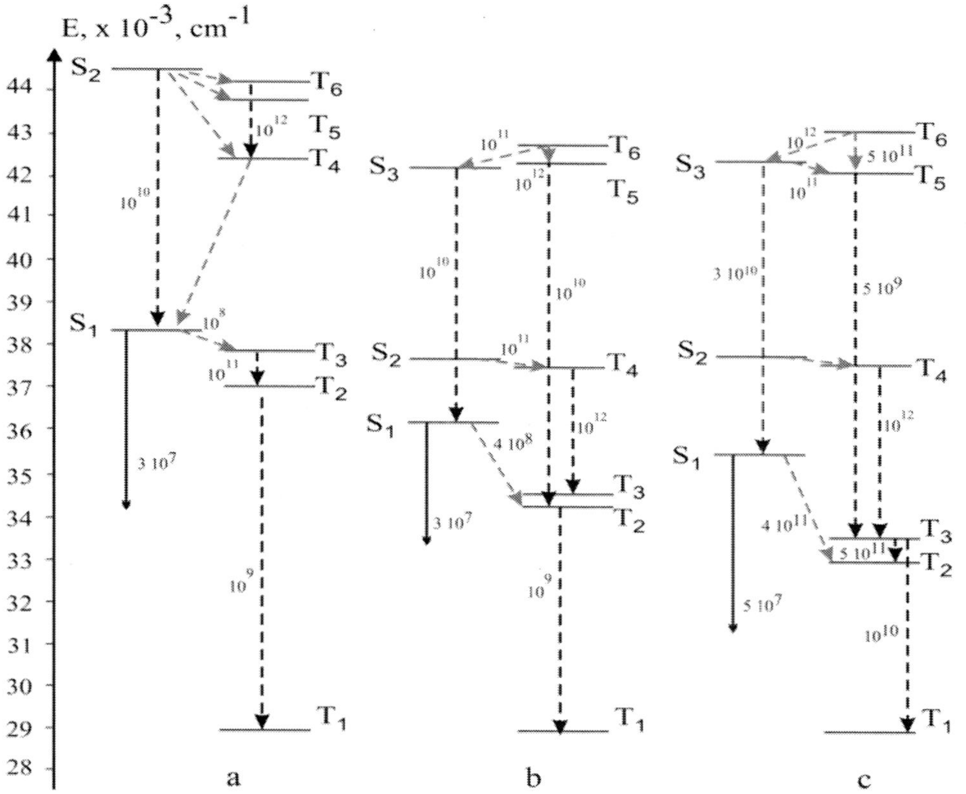

Figure 12. Energy-level diagram of the electron-excited states of phenol (a), 4-chlorophenol (b), and 4-bromophenol (c). Arrows indicate the energy relaxation pathways and figures near the arrows indicate the rate constants of these processes, s^{-1}.

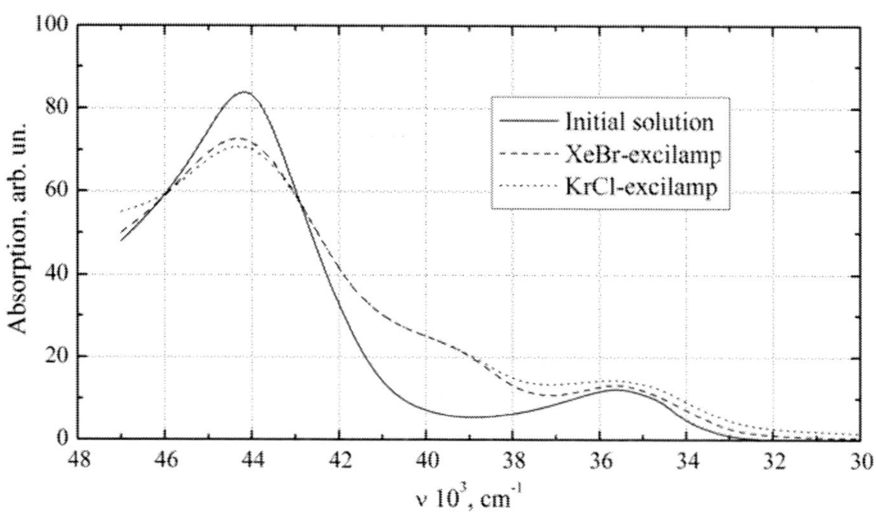

Figure 13. Absorption spectra of 4-bromophenol before and after irradiation by the excilamps. The radiation dose is 0.01 J/cm^2.

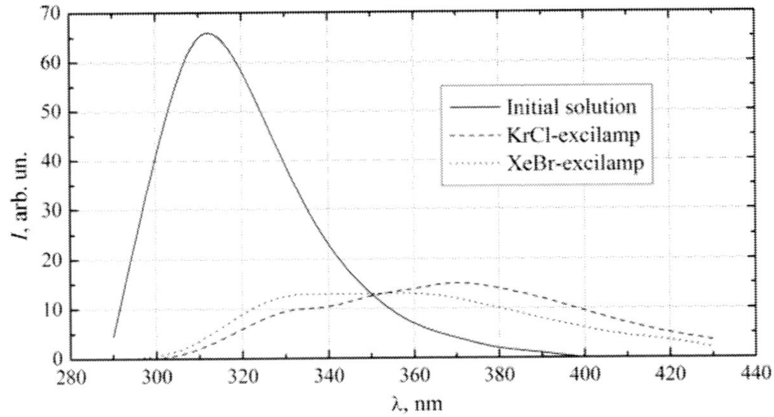

Figure 14. Fluorescence spectra of 4-chlorophenol before and after irradiation by the excilamps. The radiation dose is 0.1 J/cm^2.

3.1.3. Comparative analysis of KrCl-laser and KrCl-excilamp photolysis of phenol

For increasing the efficiency of photolysis and investigating the influence of the radiation parameters on the mechanisms of photolysis, it is appropriate to use radiation sources that allow selective excitation of electron states of compounds and variations in radiation pulse intensity and duration. Both narrow-band excilamps and exciplex lasers comply with the above requirements. A nanosecond exciplex laser makes it possible to compare the effect of the radiation pulsewidth on the photolysis of phenol [62].

In experiments, a Foton KrCl-laser with a wavelength of 222 nm ($\Delta\lambda < 1$ nm), a peak exitance power of 2 mW/cm^2, a frequency of 1 Hz, and a pulsewidth of 10 ns was used. The excilamp radiation pulsewidth was 1 μs. Thus, the laser had a peak power 10^8 times greater and a radiation pulsewidth two orders of magnitude smaller, compared to the excilamp.

It should be noted that the laser could operate at different wavelength (222, 266 and 308 nm) and at varying radiation pulsewidth (7-200 ns) and power density (0.4 kW/cm^2 – 35 MW/cm^2). However, earlier studies of photolysis of phenol and n-chlorophenol show that the most efficient photodecay of phenols occurs on excitation with 1 MW/cm^2 at $\lambda = 222$ nm [63]. It was in these conditions that we compared the efficiencies of photolysis of phenol and 4-chlorophenol with the laser and with the properly optimized KrCl-excilamp.

The solutions to be irradiated were neutral aqueous phenol and 4-chlorophenol solutions of concentration 10^{-3} mol/liter. The irradiation conditions were similar to those described in Section 3.1.1. The excilamp irradiated the solutions with a dose was up to 0.1 J/cm^2 were within the time 1 - 13 min. The irradiation by the laser with the same dose took 15–20 pulses.

Analysis of fluorescent spectra of the laser- and excilamp-irradiated solutions shows that the intensity of the main fluorescent band with at 296 nm decreases and that of the fluorescent band in the range of 350 and 390 nm increases with increasing the radiation dose. It is readily seen from Fig. 15 that with equal radiation doses, the excilamp provides more efficient phototransformations than the laser.

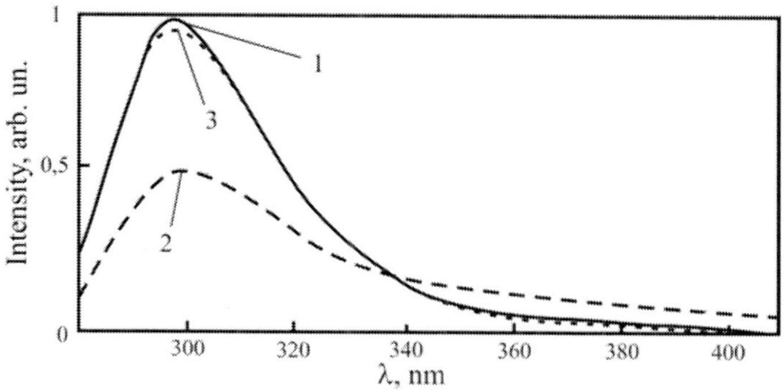

Figure 15. Fluorescent spectra of phenol before irradiation (1) and after excilamp (2) and laser (3) irradiation. The excitation energy (the radiation dose) is 0.1 J/cm^2 (2,3). The excitation wavelength is 270 nm.

The peculiarities found in the fluorescence spectra of the irradiated 4-chlorophenol solutions are the following:

– for excilamp irradiation with a dose up to 0.01 J/cm^2, a band with a maximum at λ = 330 nm, rather than the main band of 4-chlorophenol with a maximum at λ = 310 nm, is observed in the spectrum. To this band there corresponds the hydroquinone fluorescent. As the dose is increased, the intensity of this band decreases and that at λ = 380 nm increases;

– the lamp provides more efficient phototransformations than the laser. With a 0.1 J/cm^2 dose, the hydroquinone fluorescence is detected. As the dose is increased, the fluorescent intensity at 370 and 410 nm increases. With the excilamp, the band with a maximum at λ = 410 nm was not registered.

For interpretation of the results, we resort to the pathway of energy relaxation in the phenol molecule (Fig. 12(a)). When irradiated at λ = 222 nm (45000 cm^{-1}), the molecule transforms into the singlet state S_2 and then relaxes to lower triplet and singlet states of long lifetime within 0.1 ns. Hence, the irradiation of phenols by long pulses increases the probability of reabsorption in the channel of the excited states and population of higher photodissociative states, and also that of the initiation of radical mechanisms of photolysis [64].

A similar situation occurs in the 4-chlorophenol molecule. On excitation, the 4-chlorophenol molecule transforms into the state S_3 from which it rapidly relaxes to the T_2 (the photodissociative state) and T_1 levels. Increasing the pulsewidth increases the probability of both the population of the photodissociative triplet state, which corresponds to the dissociation of the C–Cl bond, and the reabsorption of lower photodissociative states that correspond to the dissociation of the O–H bond.

Thus, our experimental study demosntartes for the first time that the KrCl-excilamp irradiation with a pulsewidth of 1 μs ensures more efficient phototransformations than the KrCl-laser radiation with a pulsewidth of 10 ns.

3.2. Combined UV and biodegradation decomposition of chlorophenols

Like phenol, chlorophenols are toxic artificial or natural pollutants characterized by the ability for bioaccumulation and by the relative stability to decomposition in the environment.

In this section, we review our study undertaken after the detection of chlorophenols in the surface water of the river Selenga (the largest inflow of the lake Baikal) and in the lake Baikal in the region of waste water disposal of the Baikal pulp and paper plant (BPPP). The biodegradation of highly concentrated chlorophenols in pods-aerators of pulp and paper plants, as a rule, involves inhibition of growth of organisms and relatively low decomposition rates. Therefore, to increase the efficiency of chlorophenol biodegradation, many researchers propos to use combined methods of treatement [65-68].

For the biodegradation level to increase, we suggested preliminary UV irradiation of chlorophenol solutions by a XeBr-excilamp. The experimental setup is shown in Fig. 16. The setup includes a BD_P excilamp (Fig. 5) and a circulation system.

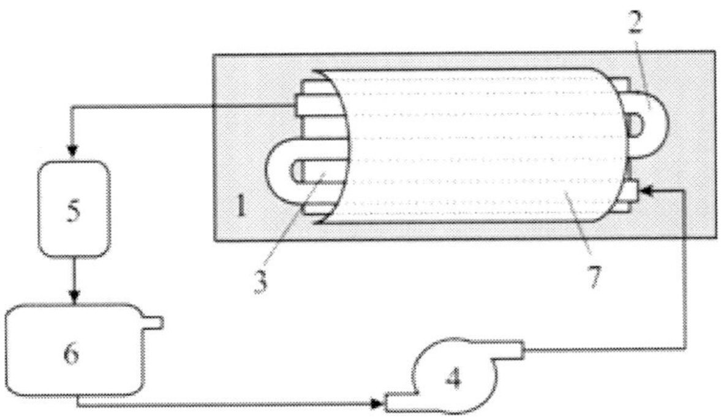

Figure 16. Experimental setup: excilamp (1); quarts tube (2); output window of the excilamp (3); pump (4); cooling chamber (5); sampling chamber (6); reflector (7).

The irradiated chlorophenol solution circulated with a rate of 66 liter/h through three quarts tubes located at the output excilamp window. On cooling, the solution was supplied to a reservoir for water sampling. The effective irradiance produced by the lamp at the output plane of the window was 18 mW/cm^2 that corresponded to 6.63×10^{-6} Einstein $l^{-1}s^{-1}$.

First, we irradiated 2-chlorophenol, 4-chlorophenol, and 2, 4-dichlorophenol aqueous solutions of initial concentrations 10, 20, 50 and 100 mg/liter and pH 5.5-6.2.

Then, we investigated the chlorophenol degradation with Bacillus cereus before and after the UV treatment. The culture was extracted from the silt in the BPPP pod-aerator and identified by sequencing 16S pDNA [69]. The aqueous chlorophenol solutions were preliminary UV-treated for the times $t_{1/2}$ and $t_{1/10}$ to attain residual chlorophenol concentrations of 50% and 10%. Subsequent cultivation of B. cereus in the irradiated solutions was realized after increasing pH of the medium to 7.0 and introducing the mineral salts described in [69]. The degradation ability of B. cereus before the UV treatment was also tested in a mineral medium containing 2-chlorophenol, 4-chlorophenol or 2,4-dichlorophenol as the only carbon and energy source.

The chlorophenol concentration during decomposition was controlled by the HPLC method. The intermediates and photoproducts of photolysis were identified by the GC-MS method. The cellular growth was controlled by measuring the optical density at 600 nm.

3.2.1. UV photolysis

The photolysis of chlorophenol is found to correspond to the equation of the first-order reaction. The maximum decomposition rate is attained for 4-chlorophenol that agrees with the data reported in [69-71]. In our experiments, the maximum initial concentration of 4-chlorophenol and 2,4-dichlorophenol at which the rate constant and the decay halfperiod vary, but little, were 20 and 50 mg/liter, respectively (see Table 5). The concentration threshold for 2-chlorophenol was lower and was approximately 10 mg/liter. Further increasing the initial concentration involves a consistent decrease in rate constant and an increase in half-decay period. Thus, the photolysis rate increases as follows: 2-chlorophenol < 2,4-dichlorophenol < 4-chlorophenol. This result confirms the data on the decrease in chlorophenol decay rate with increasing the number of the molecules in the benzene ring [70] or on the o-position o Cl atoms [71].

Table 5. First-order rate constants and chlorophenol half-decay periods for direct photolysis

Substance	C0, mg/liter	pH	I, Einstein l-1s-1	λ, nm	k0, min−1	t1/2, min	Reference
	50	2.5	6.45 × 10-7	250	8.76 × 10-3	25	[71]
	100	9.0	5.97 × 10-5	320-400	9.00 × 10-4	8.5	[71]
	100	7.1	no data	250-370	2.85 × 10-2	24.3	[71]
2-chlorophenol	10	5.6	6.63 × 10-6	283	0.100	6.9	o/d*)
	20	5.9	6.63 × 10-6	283	6.70 × 10-2	10.3	o/d
	50	5.7	6.63 × 10-6	283	5.89 × 10-2	11.8	o/d
	100	5.9	6.63 × 10-6	283	4.51 × 10-2	15.4	o/d
	50	2.5	6.45 × 10-7	250	2.52 × 10-2	27.3	[71]
	50	9.5	6.45 × 10-7	250	5.08 × 10-2	13.4	[71]
	100	6.0	5.97 × 10-5	320-400	9.32 × 10-3	8.6	[71]
4-chlorophenol	100	9.0	5.97 × 10-5	320-400	5.10 × 10-4	7.4	[71]
	10	6.2	6.63 × 10-6	283	0.777	0.9	o/d
	20	6.0	6.63 × 10-6	283	0.809	0.9	o/d
	50	5.8	6.63 × 10-6	283	0.729	0.9	o/d
	100	5.9	6.63 × 10-6	283	0.507	1.4	o/d
	50	2.0	4.80 × 10-4	185-436	3.80 × 10-2	17.5	[71]
	100	4.5-	2.5 × 10-6	350	5.0 × 10-5	>90	[72]
2,4-dichlorophenol	10	6	6.63 × 10-6	283	8.63 × 10-2	8.1	o/d
	20	5.6	6.63 × 10-6	283	8.46 × 10-2	8.1	o/d
	50	5.6	6.63 × 10-6	283	6.29 × 10-2	11.0	o/d
	100	5.5	6.63 × 10-6	283	4.69 × 10-2	14.7	o/d

*) o/d – our data

The obtained rate constants of photolysis are up to three orders of magnitude greater than those found earlier on direct UV photolysis initiated by other radiation sources (Table 5). Apparently, short-band UV radiation at λ ~ 282 nm is more efficient than long-band radiation or radiation of longer-wave sources. The rate constant of photolysis of 2,4-dichlorophenol of initial concentration 50 mg/liter (6.29×10^{-2} min^{-1}) without oxidants is comparable with that found in the following AOP-systems (pH = 2.0, λ = 185-436 nm): UV/H_2O_2 (4.40×10^{-2} min^{-1}), UV/Fenton (8.80×10^{-2} min^{-1}), and UV/O_3 (6.5×10^{-2} min^{-1}) [70]. With an initial 2,4-chlorophenol concentration of 100 mg/liter, the rate constant of direct photolysis (4.69×10^{-2}

min^{-1}) is comparable with that in the UV/Fenton system (5.7×10^{-2} min^{-1}) is two orders of magnitude greater than the rate constant found earlier in the systems UV/H_2O_2 (2.6×10^{-4} min^{-1}) and UV/Fe(III) (6.2×10^{-4} min^{-1}) (pH = 4.5-6, λ = 350 nm, I = 2.5×10^{-6} Einstein $l^{-1}s^{-1}$) [72]. The rate constant of photocatalytic decay of 2-chlorophenol with TiO_2 at λ = 365 nm [72] is 2.46 times lower than the rate constant of direct photolysis with the same initial concentration and pH of the medium at λ = 282 nm.

After irradiation, pH of the medium reached 3.50 due to the formation of intermediate oxygen products [73]. Table 6 shows identified intermediates after UV photolysis within the time $t_{1/2}$ and $t_{1/10}$. It is seen that the key stages of chlorophenol decay are the reactions of dechlorination and hydrooxidation of the benzene ring. Photolysis within $t_{1/10}$ results in a number of aromatic and organic acids and alcohols.

Table 6. Intermediates of chlorophenol photolysis identified by the GC-MS method

Substance	Time of UVtreatment	Intermediates
2-chlorophenol	t1/2 t1/10	Catechol, 2-chlorohydroquinone Phenol, benzyl alcohol, catechol, 2-phenoxyathanol, 4-chlororesorcin, phthalic acid
4-chlorophenol	t1/2 t1/10	п-benzoquinone, hydroquinone п-benzoquinone, hydroquinone, benzyl alcohol, 5-chlororesorcin, 2,5-cyclohexadiene-1,4-dion, benzoic acid, phthalic acid
2,4-dichlorophenol	t1/2 t1/10	Hydroquinone, 2-chlorohydroquinone Hydroquinone, benzyl alcohol, 2-chlorohydroquinone, 4-chlorocatechol, phthalic acid

3.2.2. Biodegradation of chlorophenol before and after UV photolysis

Earlier tests on biodegradation of chlorophenol without UV treatment show that with initial chlorophenol concentration up to 20 mg/liter, the chlorophenol solutions can be utilized directly by biooxidation in aerobic conditions using the culture B. cereus [71, 74]. However, with initial chlorophenol concentrations greater than 20 mg/liter, the solutions are difficult to biooxidize. In this case, the most efficient way of treatment is combined degradation.

First, chlorophenols are oxidized by XeBr-excilamp UV radiation to transform into relatively easy-to-oxidize states, and then they are biodegraded. With initial concentrations to over 50 mg/liter, we recommend fine preliminary UV treatment to remove 80-90% of the initial substrate. Although 4-chlorophenol displays relative stability to B. cereus, it easily decays under UV treatment.

Further biodegradation for 10 days provides utilization of the main photoproducts of the chlorophenols under study. In our experiments, the time it took for preliminary UV treatment was much shorter than that found earlier in [75] where the half-decay periods of chlorophenols of the same initial concentration were 2.2–54 h.

Thus, the study allows us to conclude that combined treatment using the XeBr-excilamp UV radiation and B. cereus can be used for developing efficient technologies of decomposition of chlorophenol of increased concentrations.

3.3. Effect of humic acids on the cresol mixture fluorescence produced by various UV sources

Humic substances (HSs) are present in water most of the time. They result from organic residua decomposition that is responsible for synthesis of numerous products of various reactions. Because of the natural selection, the most stable products are very complex in structure. The HMs are an important class of natural high-molecular compounds with an irregular structure that has been unknown so far. They comprise three fractions: fulvic acids (FA) dissolvable in water at all values of pH, humic acids (HA) dissolvable at pH > 2, and other water-insolvable HSs. According to modern concepts, the most representative group of humic substances is humic acids [76] and a HAs mixture, which also includes reactive components, may display stability for a long time. Apparently, a significant contribution to the HA stabilization is made by intermolecular interactions due to polar groups and peculiarities of carbon-carbonic bonds in cycles and linear structures.

However, exposure to light can disturb the natural stabilization and cause different types of efficient HA interaction with the surrounding, including xenobiotics present in aqueous media. The photochemical properties of HA are still little understood and their direct photolysis is rarely addressed in the literature. It is known that HAs can absorb light and transmit light energy to other components of aqueous solutions and may at time strongly affect the photolysis of xenobiotics [77]. Irradiated humic acids can produce active forms of oxygen [78] and induce phototransformations of herbicides [79]. Some data on the interrelated affect of HA and organic xenobiotics during photolysis are conflicting.

We studied the influence of the optical spectral distribution and the method of preliminary treatment of peat on the HA photodegradation. HA samples were taken from the valley peat of the Klyukvennoe field (Tomsk region). The samples were treated under different conditions of mechanical activation: HA-1 – grinded peat, HA-2 – mechanically activated peat, HA-3 – mechanically activated peat with celloveridin ferment [80]. Preliminary mechanical treatment increases the effective surface of HM components. Mechanical treatment using alkali allows chemical reactions with the formation of water-soluble compounds involving in the acid fraction. The introduction of celloverdine ferment into the system causes destruction of the cellulous structure and an increase in the yield of the products of mechanochemical reaction [81]. The HA concentration was 10^{-3} g l^{-1}. A mixture of methyl-substituted phenol and ortho- and para-cresols ($c=10^{-4}$ mol/liter) was used as benzene derivatives present in aqueous solutions. The solutions were irradiated in a quarts bulb 1 cm long. The irradiation time was varied and was 1, 2, 4, 8, 16, and 32 min. The UV source was BD_P excilamps (HCEI, Tomsk Russia) on the working molecules KrCl* ($\lambda = 222$ nm), XeBr* ($\lambda = 283$ nm) and XeCl* ($\lambda = 308$ nm). We also employed a DRT-240 high-pressure Hg lamp (Lisma-C, Novosibirsk, Russia). The absorption and fluorescence spectra were measured with a Specord M40 spectrophotometer and a CM2203 spectrofluorimeter ("Solar", Belarus).

The most efficient photodegradation of the cresol mixture was attained using the KrCl-excilamp ($\lambda = 222$ nm) with an irradiation time of 32 min (Fig. 17).

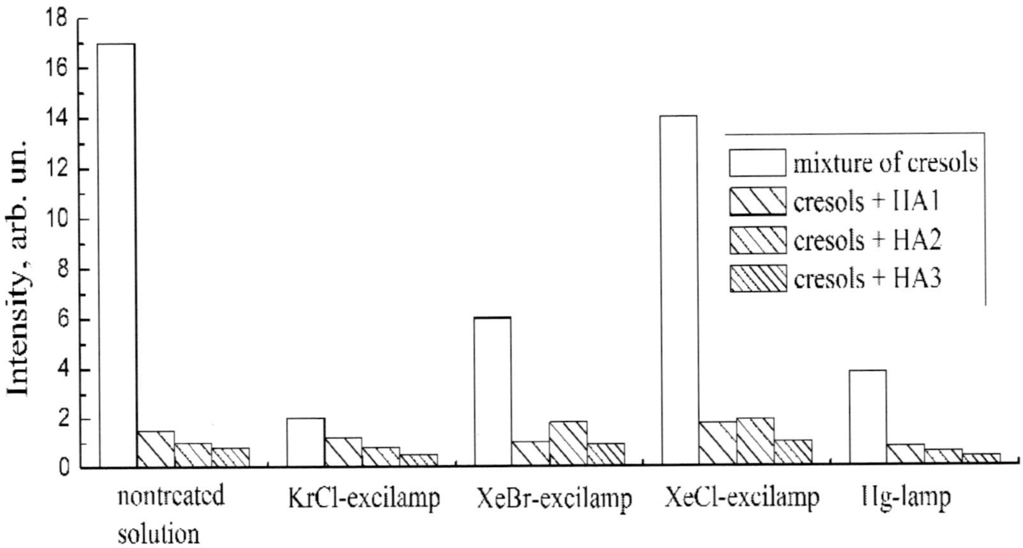

Figure 17. Fluorescence intensity of the cresol mixture photoproducts within 32 min after irradiation by different UV sources.

The fact is that among the sources employed, the KrCl-excilamp excites the matter to the highest states that often leads to disruption of the bonds and destruction of the cresol mixture. On addition of any of three HA samples, a decrease in the fluorescence intensity of the methyl-substituted phenols in the solution is observed. The most efficient attachment of cresols is ensured by the HA-3 both with and without photodegradation. In all likelihood, this is due to the celloverdin ferment used for preparing the HA samples from peat in the case of HA-3, which increases the content of actively interacted HA components (because of the destruction of hardly-hydrolyzed matters), whereas mechanical alkali treatment (HA-2) decreases it (because of the formation of water-soluble humates). The mineral composition in such treatments remains the same [6]. Mechanical treatment causes grinding and disordering of the structure of the material. It should also be kept in mind that after mechanical treatment of HAs, which are free radical sources, additional bonds are dissociated which increases their activity several times. It is because of this fact that fluorescent intensity of cresols with HAs is much lower than that without them no matter what the radiation source. After irradiation of the cresol mixture with humic matters, the most significant decrease in fluorescence was observed for the KrCl-excilamp and Hg-lamp. On addition of any of the HA samples, the fluorescence of the cresol mixture decreases due to complex formation. The most appreciable effect is found for the HA-3 sample both with and without irradiation. Our previous study of photoprocesses with the participation of HAs shows that their complicated influence on xenobiotics are associated, among other things, with the properties of HA triplet states of [82] and that the origin and degree of HA humification affect the photodecomposition of xenobiotics [83].

Thus, our work reveals the strong influence of humic acids on the fluorescence of the aqueous cresol mixture, which causes a considerable decrease in its intensity. It is found that the maximum degradation of the cresols mixture with humic acids occurs during the irradiation by the Hg-lamp or KrCl-excilamp. As for the effect of preliminary treatment (mechanical activation) of the HA samples on the photodegradation of cresols, an addition of any of three samples reveals a decrease in cresol fluorescence in the solution. In the presence of humic acids, the most efficient attachment of cresols is found for the HA-3 sample both with and without irradiation. For such complex and chemically irregular macromolecules like humic acids, it is difficult to determine more accurately what the exact way of the HA interaction with methyl-phenols. Apparently, this interaction is strong, and is responsible for significant changes in fluorescence intensity.

3.4. VUV and direct photolysis of carbamide

When used for direct photolysis, a light source must produce sufficiently intense radiation in the required wavelength range corresponding to the maximum DOM absorption and, on the hand, must provide negligible solvent absorption.

UV lamps employed in photochemistry produce nonpolychromatic light, except for lamps on metal vapors and low-pressure mercury lamp. To produce radiation in the required wavelength range, filter solutions, light filters, monochromatic filters, and monochromators are used. However, the use of these tools decreases the radiant power that makes the light source hardly applicable. Excilamps allows solving the problem by selectively affecting the DOM.

In most case, the DOM solvent is water, which displays strong absorbance in the short-wave UV spectrum (Table 7).

Table 7: Short-wave transmission cutoff for various water layer thicknesses

Thickness, mm	Optical transmission wavelength (nm)		
	50%	30%	10%
20	188	186	185
40	192	188	186
80	202	194	188

The most attractive source for direct photolysis is a KrBr-excilamp with a maximum at 206 nm. The radiation of the excilamp is transmitted through water, but falls within the absorption band of chromophore-containing environmental pollutants. In this connection, we can compare the effect of the KrBr-excilamp (direct photolysis) with that of the VUV excilamp (VUV-photolysis).

The setup used in the experiment is shown in Fig. 18.

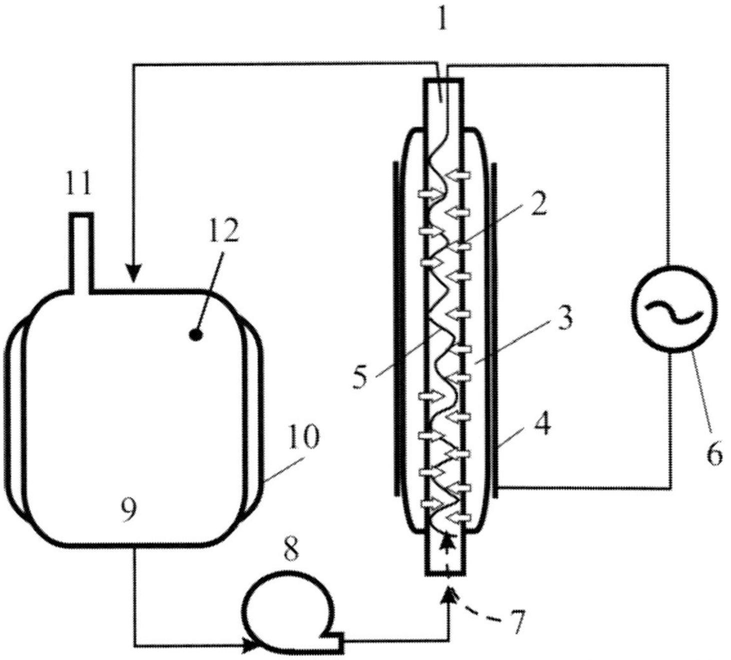

Figure 18. Experimental setup: excilamp (1); UV/VUV radiation (2); discharge gap (3); outer electrode-reflector (4); inner spiral electrode (5); power supply (6); air inlet (7); pump (8); reservoir (9); reservoir air cooling (10); sampling chamber (11); temperature and solution acidity control (12).

In the setup, narrow-band BC UV and VUV excilamps on the molecules Xe_2* (λ=172 nm) and KrCl* (λ=222 nm) (Heraeus Noblelight Co., Germany). The band halfwidth of the lamps were 14 and 4 nm, respectively. The the exitance power of the both excilamps was ~15 mW/cm^2. We also employed a BD KrBr-excilamp (model DB_E, IHCE, SB, RAS) with 5.6 mW/cm^2 at the internal excilamp surface. At the external excilamp surface, there was an aluminum reflector. Thus, the radiation was collected inside the bulb and, finally, interacted with the solution. The length and the diameter of the inner tube of the KrBr-, KrCl- and Xe_2-excilamps was 16, 90, 100 am and 2.1, 2, 2 cm, respectively. The spectra of these excilamps were similar to those shown in Figs. 1-2.

The excilamp cases were coaxial that allowed their use as flow-through photoreactors.

The solution circulated through the excilamp and reservoir with a velocity of 11 liter/min. The reservoir was used for cooling the solution and sampling. The reservoir volume was between 1 and 2 liter.

3.4.1. Photoreactor testing

The KrCl- and Xe_2-excilamps were already used earlier in AOP research (see, e.g., [14]), whereas the KrBr-excilamp required preliminary testing. For this purpose, we treated a 3 ml rhodamine 6G solution in 1 liter of water. The absorption spectra of the rhodamine solution have a maximum at 553 nm, which is fully reduced within 30 min of irradiation by the KrBr-excilamp (Fig. 19).

The KrCl-excilamp ensured the same result within 12 min. However, taking into account that the power of the KrCl-excilamp is an order of magnitude higher than that of the KrBr-excilamp, we can conclude that with comparable radiant powers, the KrBr-excilamp would provide more intense solution discoloration.

The Xe_2-excilamp ensures discoloration within 4 min. The peculiarity of the KrBr-photoreactor is that there is no need for bubbling and saturation of the solution with oxygen [49], since the water solution is transparent at $\lambda = 206$ nm and the dye absorption spectrum contains a strong short-wave absorption band at $\lambda < 220$ nm.

Thus, the KrBr-excilamp amplifies the direct photolysis, requiring no bubbling of the solution.

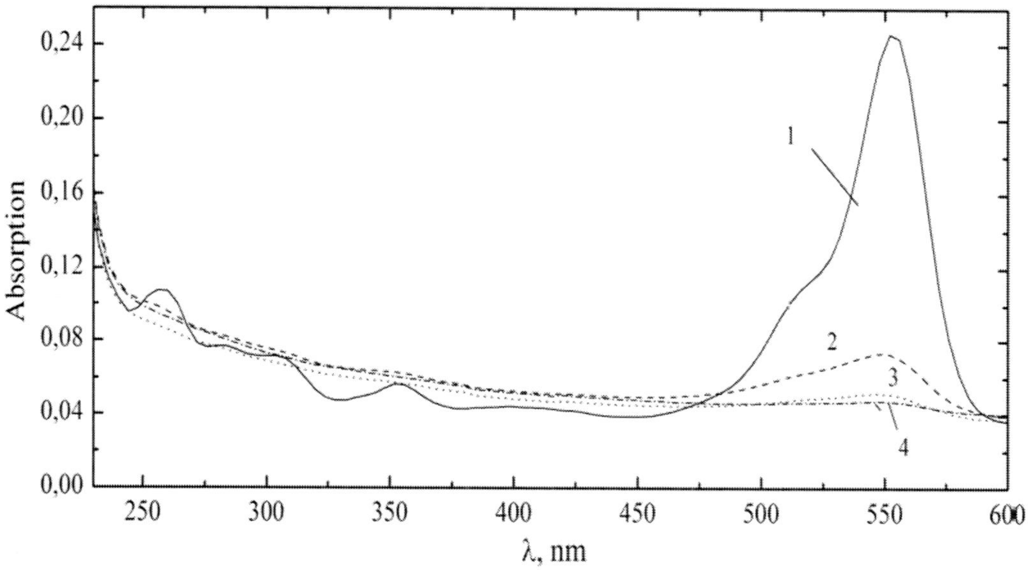

Figure 19. Absorption spectrum of the rhodamine 6G solution irradiated with the KrBr-photoreactor for 0 (1), 13 (2), 20 (3) and 30 (4) min.

3.4.2. Effect of the excilamp radiation on carbamide solutions

We referred to the problem of the UV stability of carbamide and attempted to determine the conditions under which its decay is most efficient [91]. Photostability of carbamide is a known empirical factor that could play an important role in bringing about the first organic molecules on the Earth.

Carbamide is contained in pesticides, making them stable to destruction, and hence the study of carbamide photostability appears to be urgent. Direct photolysis of carbamide by UV sources, as a rule, is inefficient, since carbamide solutions reveal no absorption at $\lambda > 240$ nm. Partial oxidation of carbamide was attained though photocatalysis (with Ag or TiO_2 at $\lambda > 350$ nm) [84, 85] and long pyrolysis [86]. However, in the latter case, ammeline and cyanuric acid resistant to UV radiation are mainly synthesized.

Finally, γ-irradiation of aqueous carbamide solutions also reveals «relative radiation stability» of carbamide [87].

The carbamide to be tested was CH_4N_2O (M = 60.06 g/mol, Fluka, frequency > 99.0%). The amount of carbamide dissolved in water using an ultrasound mixer was such that the concentration of the total organic carbon (TOC) was 40-50 mg/liter. For the initial experimental conditions to be alike, the reactor was filled two times with bidistilled. The solution temperature was varied depending on the power dissipated by the excilamp and was no greater than 55°C for the KrCl- and Xe_2-excilamps and 35°C for the KrBr-excilamp.

During the irradiation, samples (up to 5 ml) were taken from the reservoir to determine the total organic carbon content with a SHIMAZU TOC analyzer (5050 model). Each sample was analyzed three times. The standard measurement error of the TOC method was no greater than 1%. At the same time, the absorption spectra of the solutions in the range from 194 to 300 nm were fixed by a spectrophotometer. Changes in the spectra were indicative of the occurring photochemical processes.

Figure 20 shows the carbamide photodegradation under the action of different excilamps. With the KrBr- and KrCl-excilamps, a moderate effect comparable with an error of three repetitions of 15% was observed.

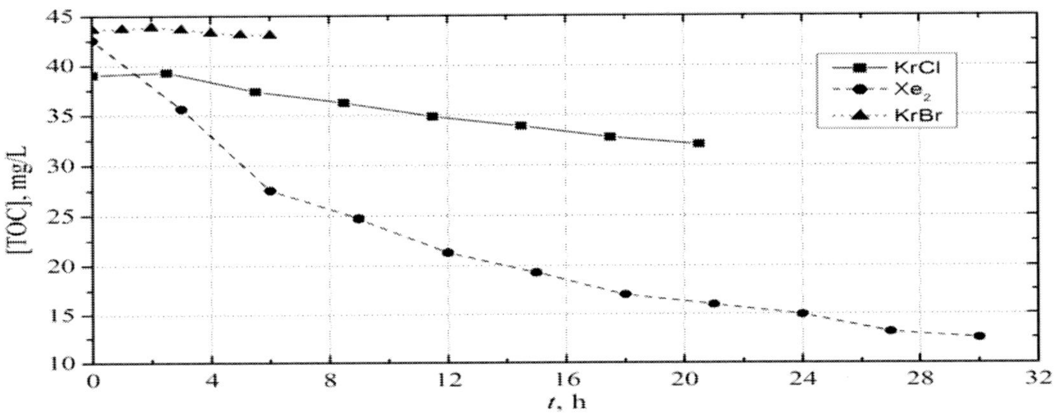

Figure 20. Photodegradation of carbamide.

The absorptions of the carbamide solution at λ=206 and λ=222 nm differ about three times (Fig. 21(1)). However, with the KrBr- and KrCl-excilamps no transformation in the absorption spectrum is found at both wavelengths. Efficient degradation with a 75% decrease in TOC is initiated by the Xe_2-excilamp (Fig. 20).

This suggests that direct photolysis is ineffective and the carbamide degrades exclusively through VUV photolysis. Hence, VUV photolysis appreciably changes the absorption spectrum of the solution (see Fig. 21(2,3)). It is seen that after long irradiation, NO_2^- and NO_3^- ions appear in the solution [88].

Taking into account the on •OH radical oxidation, a plausible mechanism for carbamide mineralization was proposed (Fig. 22).

For the process to be completed, three photon are required, hence the oxidation time is long.

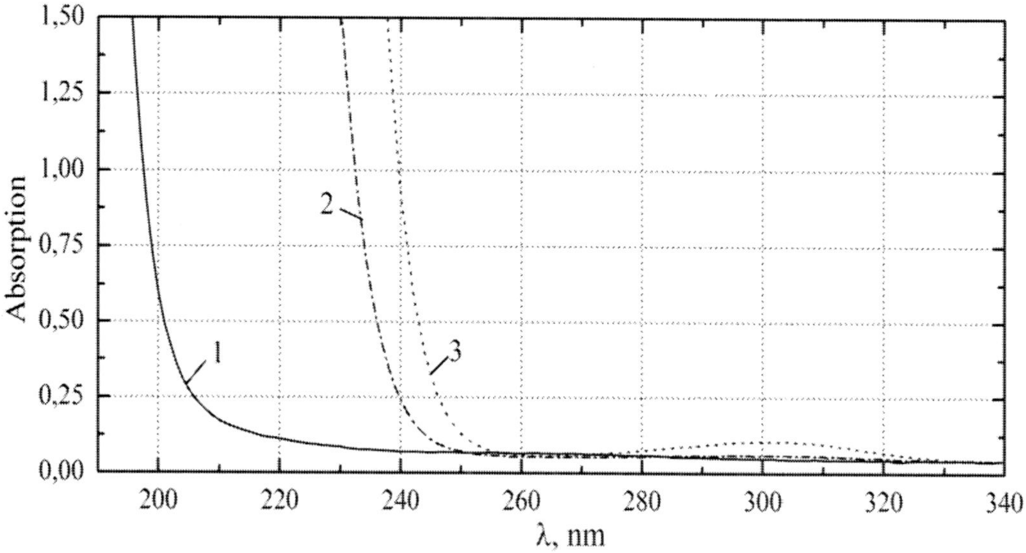

Figure 21. Absorption spectra of the initial carbamide solution (250 mg/liter) (1), after 16 (2) and 30 (3) h of irradiation by the Xe_2-excilamp (a weak maximum at λ = 300 nm is indicative of NO_3^- ions present in the solution).

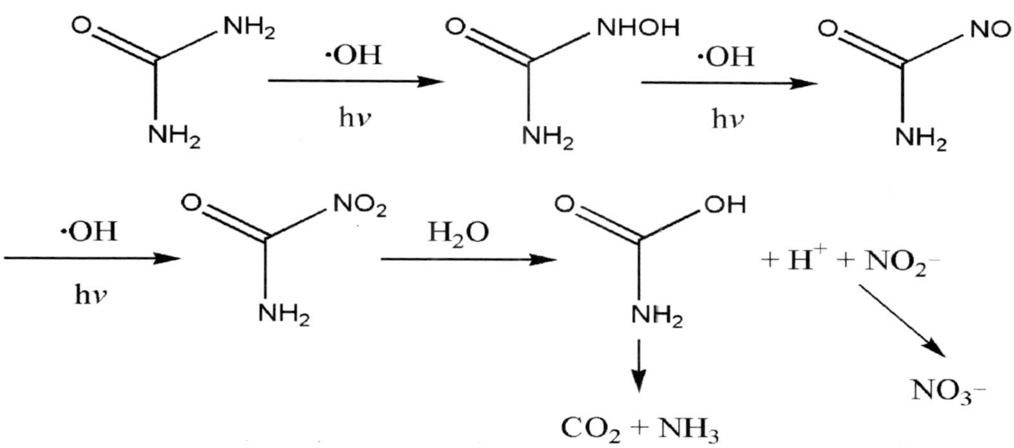

Figure 22. Carbamide oxidation.

The obtained data can be interpreted from photophysical standpoint. To answer the question why direct photolysis of carbamide at wavelengths 206 and 222 nm is ineffective, we performed quantum-chemical calculations by the PNDO method with spectroscopic parameterization [89, 90]. The calculations made it possible to determine the electron density distribution in the ground and excited states and the nature of electron-excited states of the molecule.

Figure 23 shows the calculated energy conversion rate constants.

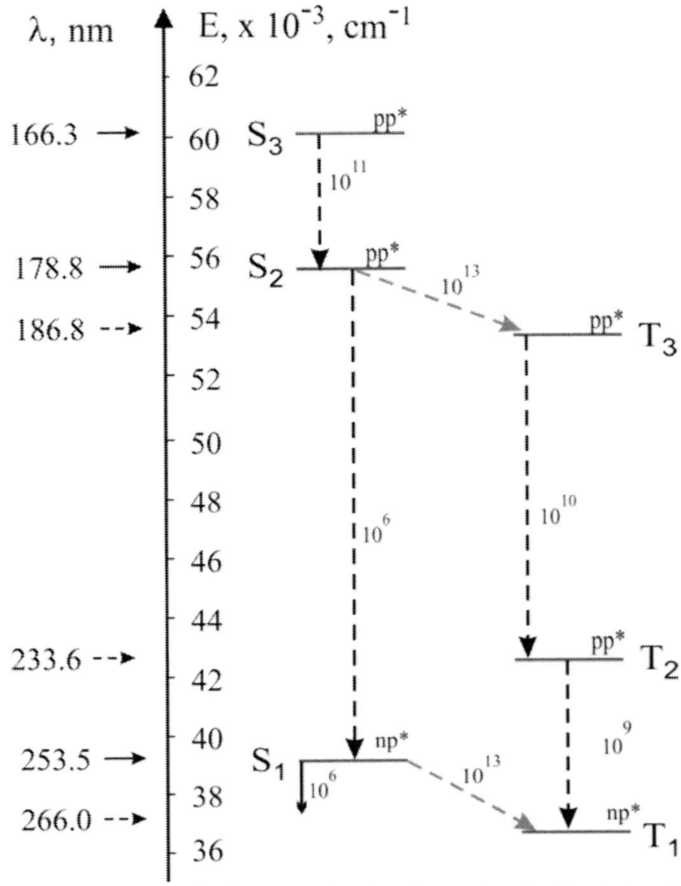

Figure 23. Paths of energy relaxation in the neutral carbamide molecule. The internal conversion and interconversions are indicated by vertical line and inclined arrows, respectively.

The lower excited states S_1* and T_1* develop with the participation of n-electrons of carbonyl oxygen. The presence of the nπ* states in the energy-level diagram makes the triplet channel of energy degradation dominating. That is the path of energy relaxation is such that the excitation at 222 and 206 nm leads to population of lower singlet oscillatory states. Therefore, within the measurement error, these two cases differ, but little, though the excilamp radiation intensities and the solution absorbance at these wavelengths differ significantly.

Thus, because of the photophysical peculiarities of the carbamide molecule, direct photolysis is ineffective and the carbamide degrades solely through oxidation by OH-radicals produced on Xe_2-excilamp VUV photolysis of water.

4. VUV Photolysis of Methanol in Actinometry

The data on excilamp-initiated DOM decay can be used to solve the inverse problem on determination of the excilamp radiation power, i.e., an actinometrical problem. This section briefly reviews one of our tests of electrochemical actinometers [92].

It was shown earlier that the Xe_2 VUV irradiation of methanol (CH_3OH) at λ=172 nm causes its photooxidation that may occur in parallel with other reactions [93, 94]. It was also shown that for concentrations $C(CH_3OH)$ = 0.1–0.2 mol/l, phototransformation of methanol into formaldehyde follows only one direction and we may speak about 100% of the methanol molecule transformation into formaldehyde.

In this case, controlling the formaldehyde yield by one or another method allows accurate estimation of the radiation dose absorbed by the solution from the quantum yield of the reaction. This procedure was used in designing and testing a methanol VUV actinometer [95].

The excilamp was a BC Xe_2-excilamp designed at the Laboratory of Optical Radiation (IHCE, SB RAS). The excilamp was built around a Radium Xeradex 20 bulb (OSRAM, Germany) [96]. The excilamp was a cylindrical quarts tube (outer diameter 31 mm, length 5 cm) with a transmission up to 70% at 180 nm. Inside the lamp, there was a spiral electrode and the external lamp surface was covered with a metal grid. During the operation of the lamp, a pulsed voltage was applied to the electrode and the grid was grounded. The xenon pressure in the bulb was 210 Torr. The maximum emission of the Xe_2^* molecule fell on 172 nm and the bandwidth at half height was ~20 nm.

For actinometrical study, we used the setup shown in Fig. 24. Lamp 1 was placed in reactor 2 filled with aqueous 0.1 M methanol solution 4. The solution was bubbled by nitrogen through tube 3 throughout the test and for 5 min before the experiment.

To determine the concentration of formaldehyde formed during the irradiation, a simple and accurate method of amperometric titration was used [97].

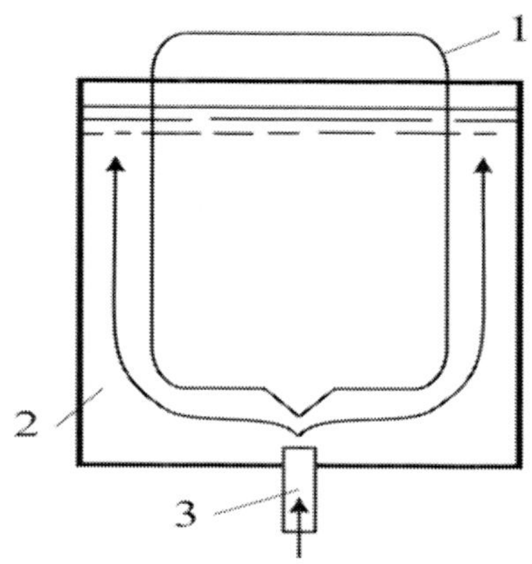

Figure 24. Experimental setup: excilamp (1); reactor with methanol solution (2); tube for nitrogen supply (3).

For this purpose, during irradiation we sampled 0.60 ml of the irradiated solution in a closed electrochemical cell at regular intervals (1 min), added 0.20 ml of I_2 (excess), diluted with water up to 10 ml, kept in the dark for 2 min, and after addition of water made titration of the iodine excess by the amperometric titration method. The measurements show that the dependence of the formaldehyde concentration on the irradiation time is a straight line. The tangent of the slope angle of this line gives the rate constants of the reaction k^o = 0.146 mmol/liter·min.

The excilamp power is equal to the absorbed photon flux. Calculations of the excilamp power were made according to the constants and formulae derived in [93,94] (Table 8):

$$P_{abs} = 1/0.946\ k^o\ V_R / (\Phi[H_2O]\cdot\xi[H_2O] + \Phi[MeOH]\cdot\xi[MeOH]) \qquad (2)$$

Table 8. Constants and experimental data for calculation of the excilamp power

Parameter	Quantity	value
[MeOH]0	Initial methanol concentration	0.10 mol/liter
kH2O = ε172H2O·[H2O]	Linear water absorptivity	550 cm–1
kMeOH = ε172MeOH·[MeOH]	Linear methanol absorptivity	4000 cm–1
VR	Volume of irradiated solution	0.125 liter
Φ[H2O](λ=172 нм)	Total quantum yield of water photolysis	0.47
Φ[MeOH](λ=172 нм)	Total quantum yield of methanol photolysis	0.88
ε 172(H2O)	Molar water absorptivity	10 liter/(mol·cm)
ε 172(MeOH)	Molar methanol absorptivity	162 liter/(mol·cm)
ξ[H2O] = kH2O / (kH2O + ε172MeOH·[MeOH]0)		0.286
ξ[MeOH] = ε172MeOH[MeOH]0/(ε172MeOH[MeOH]0 + kH2O)		0.9714

Substitution of the data from the table in formula (2) gives P_{abs} = 0.0194 mmol/min = 1.16 mmol/h. The power can also be expressed in kW/h. To do this, we use the formula:

$$P_{el} = P_{abs} \cdot E^{172} \qquad (3)$$

where $E^{172} = (N_A hc/\lambda) \cdot 2.778 \cdot 10^{-7} = 0.1932$ kW/mol, $N_A = 6.023 \cdot 10^{23}$ mol^{-1} and is Avogadro constant, c is the velocity of light and $h = 6.63 \cdot 10^{-34}$ J·s is Planck constant. Moreover, we take into account that 1 J = $2.778 \cdot 10^{-7}$ kW/h. Hence, P_{el} = 0.4685 W.

The thus obtained value corresponds to an excilamp power density of 5.1± 0.7 mW/cm^2.

For comparison, we measured the power of the same lamp with a HAMAMATSU physical photodetector (UV Power Meter C8026, Series H8025-185) immediately adjacent to the excilamp surface such that the spacing between them allowed argon blow. Measurements were taken at several point of the bulb. The average value was calculated to be 8.5±1 mW/cm^2.

The difference in physical and actinometrical measurements can be explained if we take into account that the Xe_2-excilamp radiation intensity depends on the cooling mode. In [98], a high-power Xe_2-excilamp was operated with minimum cooling and with gas cooling at a rate of 20 l/min. The decrease in excilamp wall temperature by 60 °C led to an increase in power by a factor of 1.6 and the discharge transformed from separate filaments to a near-space form.

In our tests, the lamp was in an aqueous medium that decreases the heat removal from the bulb and the discharge was far from the space form. Our experience of the Xe_2-excilamp irradiation of solutions shows that heating of the bulb can decrease about 1.5 times the power of lamps similar to that described in the paper, which agrees with data reported in [98]. This indicates that direct measurements with a physical photodetector cannot be compared with measurements of the lamp power in the reactor. In the last case, the power intensity of the lamp was lower.

Taking into account these remarks, the differences in the photodetector and actinometer measurements become clear. Moreover, when operating photodetectors with peculiar mode of aqueous solution circulation (and, thus, peculiar temperature mode), it is important to know the power absorbed by the solution. In these conditions, direct physical photodetector measurements give little information, whereas the proposed procedure allows fast determination of the VUV radiation dose.

5. VUV-Initiated Photodimerization and Removal of Water from Hydrocarbon Gas Mixtures

Oxidation and structural transformations of hydrocarbons are critical for modern plasmochemical technologies of treatment of hydrocarbon raw materials.

Water removal from natural gas and extraction of heavy hydrocarbon components from gas-condensate fields in Russia are realized using low-temperature processes in both field and plant conditions [99–101].

Early in the exploitation of gas-condensate fields, the gas flow in the field is cooled by the field gas pressure (excess pressure compared to the gas pressure in gas-main pipelines), i.e., by isoetalpic (throttling) or isentropic (cryogenic) expansion of the gas with subsequent recuperation of the gas expansion field in “gas–gas” and “gas–hydrocarbon condensate” heat exchangers. The fact is that early in the exploitation the gas pressure in the field is normally much greater than the operating pressure in the gas-main pipeline. As the pressure in the filed decreases to a certain level, it is required to complement the technological process with comprimation and to use external cooling of the gas flow. The temperature of the process on the extraction of C_{5+} hydrocarbons from the gas ranges from $-20\ ^{o}C$ to $-30\ ^{o}C$. At the same time, almost entire ethane and more than 50-60% of propane and butanes remain (i.e., are essentially lost) in the tank gas prepared for transportation.

In energy terms, the process under consideration is imperfect. For extraction of a small portion (up to a few percent) of heavy components and fractions of a percent of water vapors, the entire gas flow is subjected to the technological treatment. Moreover, low-temperature process involves cumbersome elaborate hardware and is thus too expensive. Therefore, the development of new technologies of the treatment of target components of natural gas from gas-condensate fields and that of physicochemical methods of intensification of low-temperature processes are urgent.

The ways of intensifying the low-temperature process are:

– acoustic treatment and corona discharge for intensifying the gas phase condensation;

– gas-dynamic separation that decreases the hardware bulkiness and increases the reliability of the gas preparation technology, particularly in marine conditions.

In this study, we consider fundamentally new trend toward photochemical processes in the gas flow with the use of excilamps. The study was conducted by request of Tomskgasprom in 2003.

5.1. Photolysis of water

In 2003, we proposed to realize water removal from natural gas through photolysis of water with subsequent bonding of $^{\bullet}OH$ radicals with the substrate (gas mixture components) [102]. The idea was to be tested experimentally.

Appreciable light absorption that leads to photolysis begins at wavelengths shorter than 200 nm (E_g > 6 eV). In water with impurities, e.g., with Fe ions, the photochemical decomposition proceeds with the participation of these impurities that can significantly decrease E_g [103].

In the gas phase, the absorption begins on the excitation of the lowest electron-excited state of water $^3B_1$ with an energy of ~7 eV which corresponds to ~ 180 nm [104]. This energy is sufficient for water photodissociation in the gas phase through heterolysis ($H_2O \rightarrow H_2$ + :O) and homolysis ($H_2O \rightarrow H^{\bullet} + {}^{\bullet}OH$). As shown in [105], the quantum yield of homolysis of the vapor water subjected to short-wave irradiation is unity.

For liquid water, the least energy-consumed process is the photoionization occurring at a quantum energy hν ~ 6.4 eV (λ = 195 ± 5 nm) [106]. Recent measurements of the quantum yield of liquid water homolysis give a refined value of 0.42 at λ=172 nm, which corresponds to the Xe_2 excilamp radiation maximum [94, 108]. This value is very high, and the contribution of other processes is thus negligible. Therefore, it was the Xe_2-excilamp that was selected for study.

5.2. Experimental verification and numerical simulation of the VUV effect on hydrocarbon mixtures

The design of the setup is shown in Fig. 25. In the study, we employed natural gas supplied by Tomskgasprom from the Myldzhinskoe gas-condensate field (Tomsk region, Russia). The gas from vessel 1 was reduced through a high-pressure redactor at a pressure of 5–7 MPa to a pressure slightly higher atmospheric pressure and then tuned accurately to atmospheric pressure, whereupon the gas was supplied in chamber 6 with an Xe_2-excilamp through quarts tube 4. The chamber 6 was paced in hermetically sealed cylindrical cell 13. The VUV-irradiated natural gas was analyzed using a Tsvet–500 gas chromatograph with a measurement error of the gas components of ± 2.5 %. In all experiments, the linear velocity of the natural gas flow through the photoreactor was 0.33 liter/min.

Table 9 presents the results of gas-chromatographic analysis of the natural gas and the results of experiments on their irradiation. It is seen that the initial gas contains 82.85 % of methane, 16.17 % of C_2 hydrocarbons and higher, and 0.25 % of water.

The table suggests that the most significant changes are those of the C_6–C_8 and H_2O compounds. The C_6–C_8 content increases about two times. The water vapor concentration decreases from 0.25 to 0.14 Mol. % (1.8 times). The absolute water vapor concentrations change significantly: $\Delta[H_2O]$ – 0.11% ($\Delta[C_6$–$C_8]$ – 0.05%).

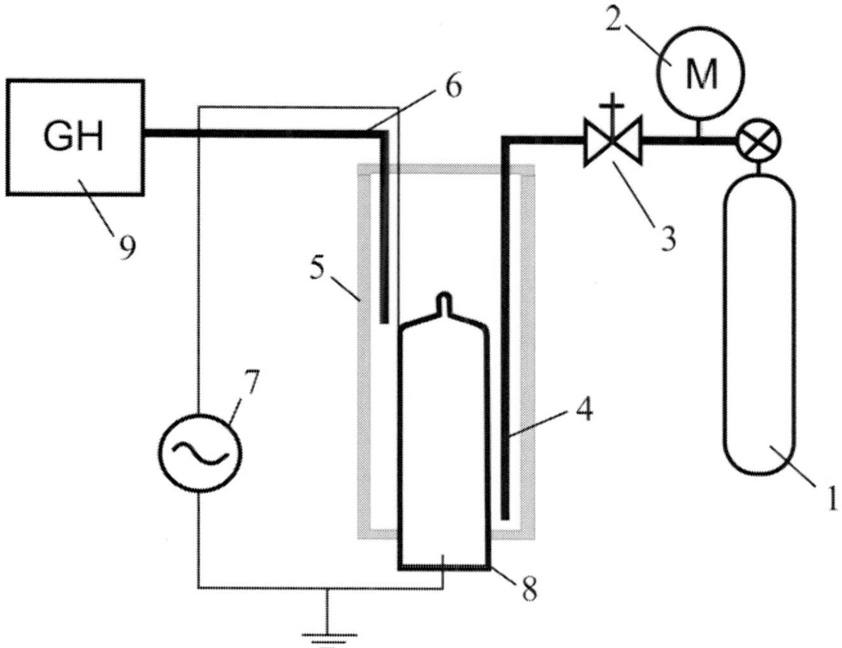

Figure 25. Xe_2-excilamp photoreactor: vessel with natural gas (1); manometer (2); valve (3); gas supply tube (4); chamber (5); gas sampling tube (6); power supply (7); excilamp (8); chromatograph (9).

Thus, the VUV irradiation causes not only the water vapor conversion, but also the hydrocarbon dimerization. The mechanism for hydrocarbon dimerization in BD plasmochemical reactors is well-understood [109]. The dimerization occurs through reactions with hydroxyl radicals:

$$C_nH_{2n+2} + {}^{\bullet}OH \rightarrow {}^{\bullet}C_nH_{2n+1} + H_2O \quad (4)$$

$${}^{\bullet}C_nH_{2n+1} + {}^{\bullet}C_mH_{2m+1} \rightarrow C_{(n+m)}H_{2(n+m+1)} \quad (5)$$

We simulated the mechanism of water removal from the VUV-irradiated natural gas. The initial concentrations of the above components were similar to those indicated in Table 9. The reactor model involved calculations of the $H^{\bullet}$, ${}^{\bullet}C_nH_{2n+1}$, and ${}^{\bullet}OH$ radical production under the action of UV radiation and constant rates of elementary processes with the participation of free radicals. The gas-phase kinetics of chemical transformations of the gas components was calculated using a kinetic model that included approximately 200 reactions [109-112].

In addition to the direct kinetic problem, the model allowed for solving the problem on the mechanism of a complex chemical reaction for which the most significant elementary stages were determined and a hierarchy of the processes from specified criteria was constructed.

Table 9. Natural gas composition before and after Xe2-excilamp irradiation (the gas flow rate 0.33 l/min, the VUV radiant power 2 W)

Component	Before irradiation, mol %	After irradiation, mol %
Methane	92.38	92.52
CO2	0.39	0.39
Ethane	3.48	3.48
H2O	0.25	0.14
Propane	2.10	2.04
i-butane	0.57	0.55
n-butane	0.52	0.52
i-Pentane	0.16	0.16
n-Pentane	0.11	0.11
C6–C8 isomers	0.04	0.09

According to the calculations, the pathway of the processes initiated the water removal from the mixture is the following.

When UV-irradiated, the natural gas, which contains water vapors, undergoes mainly water homolysis, resulting in highly reactive $^{\bullet}OH$ and $^{\bullet}H$ radicals:

$$H_2O + h\nu \rightarrow H_2O^* \rightarrow H^{\bullet} + {}^{\bullet}OH, \quad (6)$$

The radicals enter into chemical reactions with hydrocarbons. Reaction (4) occurs on the irradiation at $\lambda < 190$ nm, and hence the Xe_2-excilamp is an ideal source for its initiation.

The calculations show that the hydroxyl radicals produced in reaction (6) partially convert to hydrogen peroxide and H_2O:

$$^{\bullet}OH + {}^{\bullet}OH + M \rightarrow H_2O_2 + M, \quad (7a)$$
$$^{\bullet}OH + {}^{\bullet}OH \rightarrow H_2O + {}^{\bullet}O \quad (7b)$$

and a $^{\bullet}O$ radical appears, and the unspent $^{\bullet}OH$ radicals participate in the formation of new radicals from organic C_3–C_6 compounds in the reactions:

$$^{\bullet}O + C_nH_{2n+2} \rightarrow OH + и\text{-}{}^{\bullet}C_nH_{2n+1}, \quad (8)$$
$$^{\bullet}OH + C_nH_{2n+2} \rightarrow и\text{-}{}^{\bullet}C_nH_{2n+1} + H_2O. \quad (9)$$

The produced radicals then recombine with each other or with hydroxyl radicals:

$$^{\bullet}OH + {}^{\bullet}C_nH_{2n+1} + M \rightarrow C_nH_{2n+1}OH + M, \quad (10)$$
$$^{\bullet}C_nH_{2n+1} + {}^{\bullet}C_nH_{2n+1} + M \rightarrow и\text{-}C_{2n}H_{4n+2} + M, \quad (11)$$
$$C_nH_{2n+1}OH + h\nu \rightarrow C_nH_{2n+1}OH^* \rightarrow C_nH_{2n+1}{}^{\bullet} + {}^{\bullet}OH. \quad (12)$$

Thus, synthesis of dimers and alcohols takes place. The most efficient processes in the system are the reductions of water during the interaction of hydroxyl radicals and C_{3+} hydrocarbons. The rate constants of these processes are two-three orders of magnitude greater than the rate constants of similar $^{\bullet}OH$ interactions with methane and ethane molecules.

Therefore, despite the moderate concentration of the C_3–C_6 compounds in the natural gas composition, a hydrogen atom in the main breaks off these molecules, resulting in the $^{\bullet}C_nH_{2n+1}$ radical. Next, these radicals undergo dimerization and recombine with $^{\bullet}OH$. Thus, complex isomeric C_{6+} molecules and products of incomplete oxidation of alkanes (alcohols and aldehydes) are synthesized. Figure 26 shows experimental and numerical data on the concentrations of the components of the Xe_2-excilamp-irradiated natural gas.

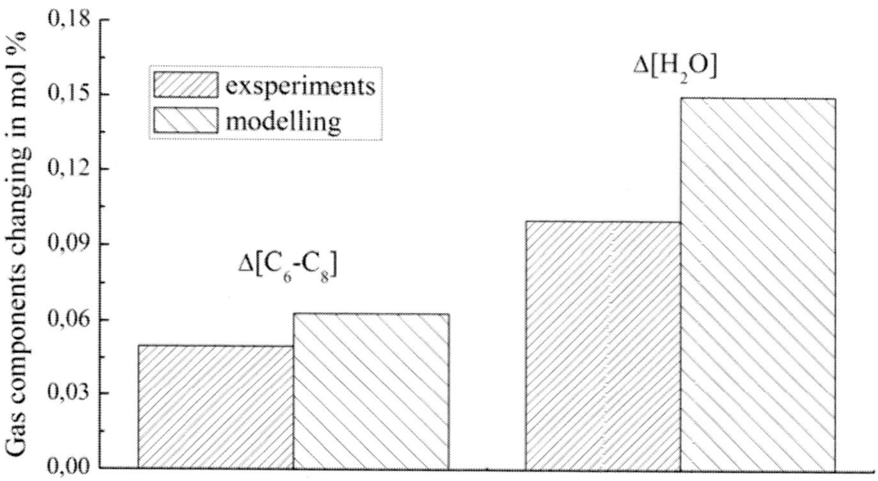

Figure 26. Experimental and numerical concentrations of the natural gas components after irradiation with the Xe_2-excilamp.

The increase in C_{6+} concentration (increase in gas condensate fraction) and the decrease in the water vapor concentration of the natural gas were found attractive for applications in gas industry. Therefore, high-power reactors around Xe_2-excilamps were designed [113]. For this purpose, at the Laboratory of Optical Radiation, special Xe_2-excilamps capable of holding an external pressure up to 60 atm were developed. The excilamp length was variable between 25 and 80 cm and the exitance power was variable between 19 and 25 mW/cm^2. In 2006-2007, the photoreactors were tested both in the pilot mode and directly on the site of Myldzhinskoe gas-condensate field [114–116].

A reactor of this type is shown in Fig. 27. The reactor has a volume of 760 cm^2 and allows irradiating aqueous and gaseous media in the flow-through mode.

The excilamps is sealed-off quarts tube 5 of outer diameter 22 mm. The length and the thickness of the tube wall are 1100 and 2.5 mm, respectively. The tube ends are near spherical. This design makes it possible to place the tube in reactor 1 at a pressure up to 40 atm. Cathode 2 can be either a spiral wound on the quarts rod or a zigzag wire with a step equal to the inner diameter of the bulb. The Xe_2-excilamp provides an exitance power between 12 and 17 mW/cm^2 [117].

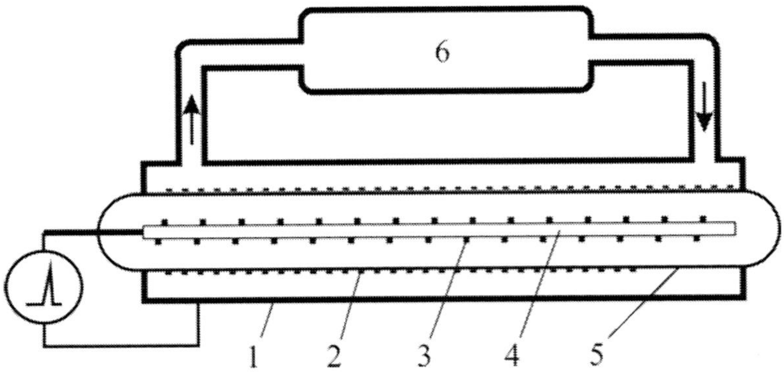

Figure 27. Schematic of the photoreactor: case (1); grid (2); metal spiral (3); quarts rod (4); excilamp case (5); pumping system (6).

Tests of the reactor at a pressure of 7.1 atm support the facts reveled earlier. In particular, it is shown that the lamp decreases the water content of the natural gas 2.5-times. The components most susceptible to photodimerization are propane and butane.

The reactor can be scaled, e.g., as shown in Fig. 28. This photoreactor was tested directly on the site of the Myldzhinskoe gas-condensate field. Analysis of chromatograms of the gas condensate obtained with and without treatment at a pressure gradient shows that an increase in condensate fraction is accompanied by an increase in C_4 and C_5 concentrations and by a decrease in C_6 concentration. Thus, retrograde condensation, i.e., precipitation of the liquid phase in the multicomponent gas system near its dew point, takes place with decreasing the isothermal pressure. Moreover, under hard exploitation conditions, the Xe_2-excilamp increases two-three times the gas condensate fraction in the gas.

The study gives grounds to suggest that the VUV irradiation of the natural gas is promising for gas industry. The process can be integrated with low-temperature gas separation to increase the yield of heavy hydrocarbons. Moreover, for "lean gases"[3] the irradiation can be considered as an original step of preliminary gas dewatering.

6. Conclusion

The applications of excilamps addressed in the paper are, to be sure, only part of the entire list of their use in photochemistry. Among the range of excilamp applications that are to be developed and/or being already developed are:

1) sampling preparation of various objects (ecological and bioobjects, foods, pharmaceutical substances, etc.) for inorganic analysis using both the UV irradiation alone or its combinations with oxidants (excited oxygen forms, OH-radicals, and hydrogen peroxide, persulphate, Ti dioxide, etc.) [15, 118–120];
2) determination of the total organic carbon, nitrogen and phosphorous in water;

[3] Lean gas: 1) gas containing methane, ethane, and propane after extraction of butane, propane and heavy homologues; 2) low-calorific gas.

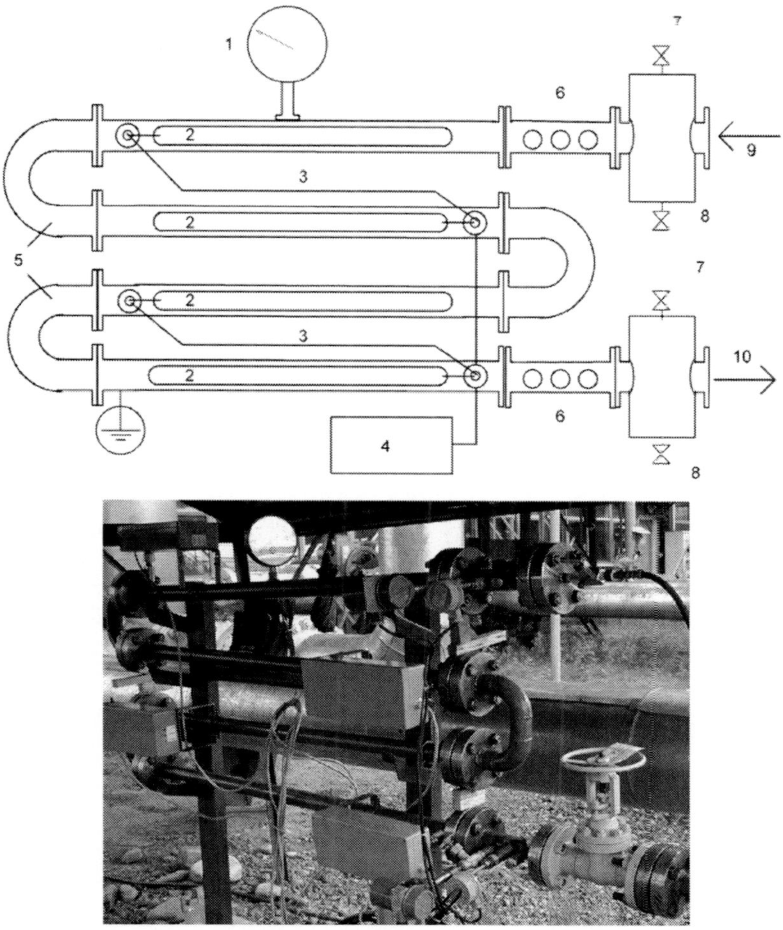

Figure 28. Pilot test setup: manometer (1), Xe_2-excilamp (2); high-voltage lead-ins (3); excilamp power supply (4); pipeline (5); moisture, temperature, and gas flow sensors (6); gas sampler chamber (7); gas condensate sampler (8); gas supply from the gas pipeline (9); exhaust (10).

3) oxidation/reduction of solutions (e.g., in photochemical titration) using the irradiation-produced oxidants and reducers;
4) photochemical reactors for detection of matters after chromatographic separation, e.g., photochemical production of matters that are electrically active in amperometric detectors;
5) inclusion of excilamp photoreactors in laboratory hardware, e.g., besides, mineralization, for inactivation of many toxic reagents – mycotoxins, cyanides in flow-through analysis (e.g., transformation of nitrates into nitrites);
6) development of novel actinometrical techniques for express-analysis of the UV and VUV intensities;

There is no doubt that the excilamps will take a prominent place in modern photochemistry.

References

1. Boichenko, A.M., Skakun, V.S., Tarasenko, V.F., Fomin, E.A., and Yakovlenko, S.I. (1994). Cylindrical excilamp pumped by a barrier discharge, *Laser Physics*, 4, 635-637
2. Oppenländer, T. (1994). Novel incoherent excimer UV irradiation units for the application in photochemistry, photobiology, photomedicine and for wastewater treatment, *Eur. Photochem. Assoc. Newsletter*, 50, 2-8.
3. Boyd, I.W., Zhang, J.Y., and Kogelschatz, U. (20030-). Development and Applications of UV Excimer Lamps, In *Photo-Excited processes, Diagnostics and Applications*; ed. by A. Peled, Kluwer Academic Publishers, the Netherlands, 161-199.
4. Gellert, B., and Kogelschatz, U. (1991). UV Generation of Excimer Emission in Dielectric Barrier Discharges, *Appl. Phys*, 52, 14-21.
5. Shevera, V.S., Shyaibov, A.K., Malinin, A.N., and Herz, S.Yu. (1980). The formation efficiency of inert gas monohalogenides in dielectric pulse discharge, *Opt. Specrtosc*, 49, 1205-1206 [in Russian].
6. Fortov, V.E., Yakovlenko, S.I. (2005). Eds. *Encyclopaedia of low-temperature plasma*, XI-4, Fizmatlit., Moscow, Chapter 3 [in Russian].
7. Lomaev, M.I., Sosnin, E.A., Tarasenko, V.F., Shits, D.V., Skakun, V.S., Erofeev, M.V., Lisenko, A.A. (2006). Capacitive and barrier discharge excilamps and their applications (Review), *Instrum. Exp. Tech*, 49, 595-616.
8. Sosnin, E.A. (2007). Excimer lamps and a new excilamp-based family of UV sources. *Light and Engineering*, 15, 49-56.
9. Lomaev, M.V., Skakun, V.S., Sosnin, E.A., Tarasenko, V.F., Schitz, D.V., Erofeev, M.V. (2003). Excilamps - efficient sources of spontaneous UV- and VUV-radiation, *Phys. Uspekh*, 173, 193-210.
10. Sosnin, E.A., and Tarasenko, V.F. VUV and UV excilamps and their applications, *High-Power Laser Ablation VI*, edited by C.R. Phipps, Proceedings of SPIE, Bellingham, WA, 2207, 626136.
11. Lomaev, M.I., Panchenko, A.N., Sosnin, E.A., Tarasenko, V.F. (1999). *Effective gas discharge UV sources: physics and engineering. Excilamps*; Tomsk State University Press, Tomsk, Russia [in Russian].
12. Gonzalez, M.G., Oliveros, E., Wörner, M., Braun, A.M. (2004). Vacuum-ultraviolet photolysis of aqueous reaction systems, *J. Photochem. Photobiol. C, Photochem. Reviews*, 5, 225-246.
13. Oppenlaender, T. (2003). *Photochemical Purification of Water and Air*, Weincheim, WILEY–VCH Verlag.
14. Baum, G., Oppenlaender, T. (1995). VUV-Oxidation of Chloroorganic Compounds in an Excimer Flow Through Photoreactor, *Chemosphere*, 30, 1781-1790.
15. Oppenlaender, T. (2003). Applications of excimer lamps in photochemistry and in photochemical technology, *Trends in Chemical Engineering*, 8, 123-136.
16. Sosnin, E.A., Oppenlaender, T., Tarasenko, V.F. (2006). Applications of Capacitive and Barrier Discharge Excilamps in Photoscience, *J. Photochem. Photobiol. C, Photochem. Reviews*, 7, 145-163.
17. Volkova, G.A., Kirillova, N.N., Pavlovskaya, E.N., Yakovleva, A.V. (1984). Barrier discharge VUV-lamps with inert gases, *J. Appl. Spectroscopy*, 41, 681-695 [in Russian].

18. Sosnin, E.A. (2004). *Principles of development of gas discharge spontaneous radiation sources: the developer manual.*, Tomsk State University Press, Tomsk, Russia, [in Russian].
19. Eliasson, B., Kogelschatz, U. (1988). UV Excimer Radiation from Dielectric-barrier Discharges, *Appl. Phys. B*, 46, 299-303.
20. Eliasson, B., Kogelschatz, U. (1991). Modeling and Application of Silent Discharge Plasmas, *IEEE Trans. Plasma Sci*, 19, 309-323.
21. Sosnin, E.A., Erofeev, M.V., Tarasenko, V.F. (2005). Capacitive discharge exciplex lamps, *J. Phys. D*, 38, 3194-3201.
22. Arnold, E., Driskemper, R., Reber, S. (1998). High-power excimer sources, *Proceedings of the 8th Intern. Symp. on the Science and Technology of Light Sources (LS-8)*, edited by G. Babucke, Greifswald, Germany, 90-98.
23. Oppenländer, T., Baum, G. (1996). Wasseraufbereitung mit Vakuum-UV/UV-Excimer-Durchflussphotoreaktoren (Water remediation with Vacuum-UV/UV-Excimer Flow-Through Photoreactors), *gwf-Wasser/Abwasser*, 137, 321-325 [in German].
24. Boichenko, A.M., Skakun, V.S., Sosnin, E.A., Tarasenko, V.F., Yakovlenko, S.I. (1996). Characteristics of an Exciplex KrCl Lamp Pumped by a Volume Discharge, *Quantum Electron*, 26, 336-340.
25. Erofeev, M.V., Tarasenko, V.F. (2006). Barrier-discharge XeCl-, KrCl-, XeBr-and KrBr-excilamps with a nanosecond pulse duration, *J. Phys. D*, 39, 3609-3614.
26. Erofeev, M.V., Lomaev, M.I., Sosnin, E.A., Tarasenko, V.F., Shitz, D.V. (2001). A 1-kW/cm^2 Flash KrCl Excimer Lamp, *Tech. Phys*, 46, 1341-1344.
27. Lomaev, M.I., Mesyats, G.A., Rybka, V.D., Tarasenko, V.F., Baksht, E.Kh. (2006). High-power short-pulse xenon dimer spontaneous radiation source, *Quantum Electron*, 37, 595-596.
28. Panchenko, A.N., Tarasenko, V.F., Belokurov, A.N., Mendoza, P. (2006). Planar KrCl* excilamp pumped by transverse self-sustained discharge with an optical system for radiation concentration, *Phys. Sc*, 74, 108-113.
29. Kogelschatz, U. (1992). Silent-discharge-driven excimer UV sources and their applications, *Appl. Surf. Sci*, 54, 410-423.
30. Lomaev, M.I., Tarasenko, V.F., Shitts, D.V., Lisenko, A.A. (2006). A windowless VUV excilamp, *Tech. Phys. Lett*, 32, 590-592.
31. Shuaibov, A.K., Dashchenko, A.I. (2000). Conditions for the simultaneous formation of Ar, Kr, and Xe chlorides in a multiwave radiation source pumped by a transverse discharge, *Quantum Electron*, 30, 279-281.
32. Shuaibov, A.K., Grabovaya, I.A. (2005). Electric-discharge source of steady-state UV-VUV radiation from iodine atoms and iodine-containing molecules, *J. Opt. Technol*, 72, 544-547.
33. Zhang, J.Y., Boyd, I.W. (2000). Multi-wavelength excimer ultraviolet sources from a mixture of krypton and iodine in a dielectric barrier discharge.. *Appl. Phys*, 71, 177-179.
34. Avdeev, S.M., Boichenko, A.M., Sosnin, E.A., Tarasenko, V.F., Yakovlenko, S.I. (2007). Barrier-discharge excilamp on a mixture of krypton and molecular bromine and chlorine, *Laser Phys*, 17, 1119-1123.
35. Phillips, R. (1983). *Sources and applications of ultraviolet radiation,* Academic Press, New York.

36. *Advanced oxidation processes for water and wastewater treatment*, edited by S. Parsons, IWA Publishing, Cornwall, UK, (2004).
37. Oppenländer, T., Fradl, M. (1999). TOC Destruction of a phenol/water Azeotrope by "photoreactive distillation" through an incoherent vacuum-UV excimer lamp, *Chem. Eng. Technol*, 22, 951-954.
38. Benoit-Marque, F., Wilkenhörner, U., Braun, A.M., Oliveros, E., Maurette, M.T. (1999). Lampes a excimere au Xe et XeCl adapties a la Photochimie,applications a la degradation de matiure organique en phase gaz, *J. Phys. IV France*, 9, 113-116.
39. Oppenländer, T. (1999). Photochemische Methoden in der Umwelttechnik: Vakuum-UV-Oxidation von Wasserinhaltsstoffen mit Xenon-Excimerlampen. *CLB Chem. Lab. Biotech*, 50, 175-179 [in German].
40. Nohr, N.S., MacDonald, J.G., Kogelschatz, U., Mark, G., Schuchmann, H.P., Von Sonntag, C. (1994). Application of excimer incoherent UV-sources as a new tool in photochemistry, Photodegradation of chlorinated dibenzodioxins in solution and absorbed in aqueous pulp sludge, *J. Photochem. Photobiol. A, Chem*, 79, 141-149.
41. Zhang, J.Y., Esrom, H., Boyd, I.W. (1999). UV intensity measurement of 308 nm excimer lamp using chemical actinometers, *Appl. Surf. Sci*, 138/139, 315-319.
42. Jacob, L., Hashem, T., Bürki, S., Guindy, N.M., Braun, A.M. (1993). Vacuum-ultraviolet (VUV) photolysis of water: oxidative degradation of 4-Chlorophenol, *J., Photochem. Photobiol. A, Chem*, 75, 97-103.
43. Oppenländer, T., Gliese, G. (2000). Mineralization of organic micropollutants (homologous alcohols and phenols) in water by vacuum-UV-oxidation (H_2O-VUV) with an incoherent xenon-excimer lamp at 172 nm, *Chemosphere*, 40, 15-21.
44. Rehberg, K., Mannschott, P., Erdiner, L., Sonntag, H.G., Schramm, B. (1994). Atrazinabbau durch UV-oxidation und identifizierung der photoabbauprodukte, *Vom Wasser*, 83, 139-156 [in German].
45. Oppenländer, T., Baum, G., Egle, W. (1995). Abwasserreinigung durch Vakuum-UV-Oxidation in Excimer-Durchflußreaktoren. *gwf-Wasser/Abwasse*, 136, 311-316 [in German].
46. Grichetschkina, M.V., Zaitsev, N.A., Braun, A.M. (1996). VUV-photolysis. Degradation of organic inhibiting the inverse-voltammetric of heavy metals 1. Humic substances, *Toxicol. Enviroment. Chem*, 53, 143-151.
47. Falkenstein, Z. (1997). Processing of C_3H_7OH, C_2HCl_3 and CCl_4 in flue gases using silent discharge plasmas (SDPs) enhanced by (V)UV at 172 nm and 253.7 nm, *J. Adv. Oxid. Technol*, 2, 223-238.
48. Zhang, J.Y., Boyd, I.W., Esrom, H. (1997). UV intensity measurement of novel 222 nm excimer lamp using chemical actinometer, *Appl. Surf. Sci*, 109/110, 482-486.
49. Oppenländer, T., Walldörfer, C., Burgbacher, J., Kiermeier, M., Lachner, K., Weinschrott, H. (2005). Improved vacuum-UV (VUV) photomineralization of organic compounds in a water photoreactor (Xe_2^* lamp, 172 nm) containing an axially centred ceramic oxygenator, *Chemosphere*, 60, 302-309.
50. Larson, R.A., Weber, E.J. (1994). *Reaction mechanisms in environmental organic chemistry*; CRC Press, Inc., Boca Raton, FL.
51. Chun, H., Yizhong, W., Hongxiao, T. (2000). Destruction of phenol aqueous solution by photocatalysis or direct photolysis, *Chemosphere*, 41, 1205-1209.

52. Esplugas, S., Gimenez, J., Contreras, S., Pascual, E., Rodrigues, M. (2002). Comparison of different AOPs for phenol degradation, *Water research*, 36, 1034-1042.
53 Wu, C., Liu, X., Wei, D., Fan, J., Wang, L. (2001). Photosonochemical degradation of phenol in water, *Water research*, 35, 3927-3933.
54. Akbal, F., Nur Onar, A. (2003). Photocatalytic degradation of phenol. *Environmental monitoring and assessment*, 83, 295-302.
55. Martyanov, I.M., Savinov, E.N., Parmon, V.N. (1997). A comparative study of the efficiency of photooxidation of organic contaminants in water solutions in various photochemical and photocatalytic systems. 1. Phenol photooxidation promoted by hydrogen peroxide in a flow reactor, *J. Photochem. Photobiol. A: Chem*, 107, 227-231.
56. THo, F.L., Bolton, J.R. (1998). Toxicity changes during the UV treatment of pentachlorophenol, *Water Research*, 32, 489-497.
57. Durand, A.P., Brown, R.G., Worall, D., Wilkinson, F. (1996). A nanosecond laser flash photolysis study of aqueous 4-chlorophenol, *J. Photochem. Photobiol. A, Chem*, 96, 35-43.
58. Audureau, J., Filiol, C., Boule, P., Lenaire, L. (1976). Photolyse et photo-oxydation du phenol en solution aqueeuse, *J. Chem. Phys*, 73, 513-620.
59. Arkhipova, M., Tereschenko, L., Arkhipov, Yu. (1995). Removal of phenols by photooxidative water cleaning, *J. Appl. Chem*, 8, 1563-1568.
60. Svetlitchnyi, V.A., Tchaikovskaya, O.N., Bazyl, O.K., Kuznetsova, R.T., Kopylova, T.N., Sokolova, I.V., Meshalkin, Yu.P. (2001). Phenol and p-chlorophenol photolysis under UV laser excitation, *High Energy Chemistry*, 35, 294-301.
61. Bazyl, O.K., Kopylova, T.N., Sokolova, I.V., Sosnin, E.A., Svetlitchnyi, V.A., Tarasenko, V.F., Tchaikovskaya, O.N. Improvement of photodecomposition methods of phenol-containing ecotoxicants in aqueous media, *Proceedings of Int. Conf. on Atomic and Molecular Pulsed Lasers IV*, edited by Tarasenko, V.F., Mayer, G.V., and Petras, G.G. (2002). Vol. 4747, Proceedings of SPIE; SPIE: Bellingham, WA, pp 240-244.
62. Sokolova, I.V., Chaikovskaya, O.N., Svetlichnyi, V.A., Kuznetsova, R.T., Kopylova, T.N., Maier, G.V., Sosnin, E.A., Lipatov, E.A., Tarasenko, V.F. (2002). Phenols phototransformation in water solutions at various excitation degree, *High Energy Chemistry*, 36, 272-275.
63. Svetlichnyi, V.A., Chaikovskaya, O.N., Bazyl', O.K., Kuznetsova, R.T., Sokolova, I.V., Kopylova, T.N., Meshalkin, Y.P. (2001). Photolysis of phenol and para-chlorophenol by UV laser excitation, *High Energy Chemistry*, 35, 258-264.
64. Oudjehani, K., Boule, P. (1992). Photoreactivity of 4-chlorophenol in aqueous solution, *J. Photochem. Photobiol, A: Chem*, 68, 363-373.
65. Pulgarin, C., Invernizzi, M., Parra, S., Sarria, V., Polania, R., Peringer, P. (1999). Strategy for the coupling of photochemical and biological flow reactors useful in mineralization of biorecalcitrant industrial pollutants, *Catalysis Today*, 54, 341-352.
66. Marco, A., Esplugas, S., Saum, G. (1997). How and why combine chemical and biological processes for wastewater treatment, *Wat. Sci. Technol*, 35, 321-327.
67 Amat, A.M., Arques, A., Beneyto, H., Garcia, A., Miranda, M.A., Segui, S. (2003). Ozonisation coupled with biological degradation for treatment of phenolic compounds: a mechanistically based study, *Chemosphere*, 53, 79-86.
68. Hörsch, P., Speck, A., Frimmel, F. (2003). Combined advanced oxidation and biodegradation of industrial effluents from the production of stilbene-based fluorescent

ahitining agents, *Water research*, 37, 2748-2756.

69. Matafonova, G.G., Shirapova, G.S., Giffhorn, F., Batoev, V.B., Kohring, G.-W. (2006). Degradation of 2,4-dichlorophenol by *Bacillus sp.* isolated from an aeration pond of Baikalsk pulp and paper mill (Russia), *J. International Biodeterioration and Biodegradation*, 58, 209-212.
70. Benitez, F.J., Beltran-Heredia, J., Acero, J.L., Rubio, F. J. (2000). Contribution of free radicals to chlorophenols decomposition by several advanced oxidation processes, *Chemosphere*, 41, 1271-1277.
71. Pera-Titus, M., García-Molina, V., Baños, M.A., Giménez, J., Esplugas, S. (2004). Degradation of chlorophenols by means of advanced oxidation processes: a general review. *Applied Catalysis B: Environmental*, 47, 219-256.
72. Al Momani, F., Sans, C., Esplugas, S. A. (2004). comparative study of the advanced oxidation of 2,4-dichlorophenol, *Journal of Hazardous Materials*, B107, 123-129.
72. Rao, N.N., Dubey, A.K., Mohanty, S., Khare, P., Jain, R., Kaul, S.N. (2003). Photocatalytic degradation of 2-chlorophenol: a study of kinetics, intermediates and biodegradability, *Journal of Hazardous Materials*, B101, 301-314.
73. Skurlatov, Yu.I., Ernestova, L.S., Vichutinskaya, E.V., Samsonov, D.P., Semenova, I.V., Rodko, I.Ya., Shvidky, V.O., Pervunina, R.I., Kemp, T.J. (1997). Photochemical transformation of polychlorinated phenols, *J. Photochem. Photobiol., A: Chemistry*, 107, 207-213.
74. Annachhatre, A.P., Gheewala, S.H. (1996). Biodegradation of chlorinated phenolic compounds, *Biotechnology Advances*, 14, 35-56.
75. Tamer, E., Hamid, Z., Magdy Aly, A., Tayeb Ossama, El., Bo, M., Benoit, G. (2006). Sequential UV-biological degradation of chlorophenols, *Chemosphere*, 63, 277-284.
76. Orlov, D.S. (1997). Humic substances in hydrosphere, *Soros Educ., J*, 2, 56-63 [in Russian].
77. Mekkaoui, M., Elizzouzi, M., Bouhaouss, A. et al. (2000). Photostability and photostabilizing effect of humic acids, *Intern. Journal of Photoenergy*, 2, 55-57.
78. Paul, A., Hackbarth, S., Vogt, R.D., Burnison, B.K., Steinberg. C.E. (2004). Photogeneration of singlet oxygen by humic substances: comparison of humic substances of aquatic and terrestrial origin, *Photochem. Photobiol. Sci*, 3, 273-280.
79. Stangroom, S.J., Macleod, C.L., Lester, J.N. (1998). Photosensitized transformation of the herbicide 4-chloro-2-methylphenoxy acetic acid (MCPA) in water, *Water Research*, 32, 623-626.
80. Sokolova, I.V., Tchaikovskaya, O.N., Tregubkina, M.V., Yudina, N.V., Maltseva, E.V. (2007). In *Hydrogeochemistry of Sedimentary Basins;* edited by Shvartsev, S.L. Ed. Proceedings of the Russian Scientific Conference, Scientific Technology Publishing House: Tomsk, pp 213-219 [in Russian].
81. Lomovskii, O.I., Ivanov, A.A., Rozhanskaya, O.A., Yudina, N.V., Korolev, K.G. (2004). Variation of the composition and properties of the water-soluble components of peat during mechanochemical treatment, *Chemistry for Sustainable Development*, 12, 355-361 [in Russian].
82. Chaikovskaya, O.N., Levin, P.P., Sultimova, N.B., Sokolova, I.V., Kuzmin, A.V. (2004). Triplet states of humic acids studied by laser flash photolysis using different excitation wavelengths, *Russian Chemical Bulletin*, 53, 313-317.

83. Chaykovskaya, O.N., Sokolova, I.V., Yudina, N.V. (2005). Fluorescence analysis of photoinduced degradation of ecotoxicants in the presence of humic acids, *Luminescence*, 20, 187-191.
84. Pelizzetti, E., Calza, P., Mariella, V., Maurino, V., Minero, C., Hidaka, H. (2004). Different photocatalytic fate of amino nitrogen in formamide and urea, *J. Chem. Soc. Chem. Comm*, 13, 1504-1505.
85. Hallman, M.M. (1996). *Photodegradation of water pollutants*; CRC Press, Boca Ration, N-Y, London, Tokio, 182-183.
86. Shaber, P.M., Colson, J., Higgins, S., Dietz, E., Thieilen, D., Brauer, J. (1999). Study of the urea thermal decomposition (pyrolysis) reaction and importance to cyanuric acid production, *American Laboratory*, August, 13-21.
87. Navarro-Gonzalez, R., Negron-Mendoza, A., Chacon, E. (1989). The γ-irradiation of aqueous solutions of urea, *Origins of life and evolution of biosphere*, 19, 109-118.
88. Mark, J., Bolton, J.R. (1999). Photochemistry of nitride and nitrate in aqueous solution: a review, *J. Photochem. Photobiol. A, Chem*, 128, 1-13.
89. Vasilyeva, N. Yu., Sokolova, I.V., Samsonova, L.G., Kopylova, T.N., Mayer, G.V. (1999). Nature of the electron-excited states and energy transfer in bichromophore-coumarin molecules, *J. Fluorescence*, 9, 17-25.
90. Artyukhov, V.Ya., Galeeva, A.I., Maier, G.V. (1997). Quantum-chemical study of triplet electronic energy transfer in biochromophore molecular systems, *Opt. Specrtosc*, 83, 685-690.
91. Sosnin, E.A., Groß, A., Bartnik, N., Oppenländer, T., Vasil'eva, N.Yu. (2006). Study of carbamide photodegradation in flow-through UV and VUV excilamp photoreactors, *Fundamental problems of modern technologies*, edited by Shipilov, S.E. Atmospheric Optics Institute Publishing House, Tomsk, pp 169-172 [in Russian].
92. Batalova, V.N., Sosnin, E.A., Zakharova, E.A., Tarasenko, V.F. (2003). An electrochemical ferrioxalate actinometer and its use in measuring the radiation intensity of excilamps, *Instrum. Exp. Tech*, 46, 73-76.
93. Oppenlaender, T., Schwarzwalder, R. (2002). Vacuum-UV oxidation (H2O–VUV) with a xenon excimer flow-through lamp at 172 nm, Use of methanol as actinometer for VUV intensity measurement and as a reference compound for OH-radical competition kinetics in aqueous systems, *J. Adv. Oxid. Technol*, 5, 155-163.
94. Heit, G., Neuner, A., Saugy, P.Y., Braun, A.M. (1998). Vacuum UV (172 nm) actinometry: The quantum yield of the photolysis of water, *J. Phys. Chem., A*, 102, 5551-5561.
95. Sosnin, E.A., Zakharova, E.A., Moskaleva, M.L., Batalova, V.N. (2006). An electrochemical methanol actinometer and its application to measurements of the vacuum UV Xe intensity, *Instrum. Exp. Tech*, 49, 101-105.
96. Radium XeradexTM 20 Excimer System: New surface treatment solution, fact sheet, Radium VF 505/4.99.
97. Songina, O.A., Zakharov, V.A. (1979). *Amperometric titration*, Chemistry, (Khimia), Moscow, [in Russian].
98. Matsuzawa, S., bSumimoto, T., Yoshioka, M., Matsuno, H., Hiramoto, T. (2004). *Proceedings of the 10th International Symposium on the Science and Technology of Light*

Sources (LS-10), edited by Zissis, G. IOP Conference Series 182, IOP Publishing: Bristol, Philadelphia, pp 175-176.
99. Grotsenko, A.I., Istomin, V.A., Kulkov, A.N., Suleimanov, R.S. (1999). Gas collection and preparation in Russian northern condensate fields, "Nedra" publishing, Moscow, [in Russian].
100. Bekirov, T.M., Lanchakov, G.A. (1999). Gas and gas-condensate treatment technology, Nedra- Business Center, Moscow, [in Russian].
101. Oskoblo, A.I., Molokanov, Yu. K., Vladimorov, A.I., Tschelkunov, V.A. (2000). Oil-gas and petrochemistry facilities, Nedra-Business Center, Moscow, [in Russian].
102. Sosnin, E.A., Erofeev, M.V. (2004). Natural gas drying and methanol photolysis in a flow-through photoreactor based on Xe_2-and KrCl-excilamps, *Proceedings of the 3rd workshop of young Russian scientists*, edited by Tulochonov, A.K. BNTz, SB RAS Publishing: Ulan-Ude, pp 247-248.
103. Solar energy photocatalytic transformations. Part 2, Molecular systems for water decomposition, Nauka, Novosibirsk, (1985) [in Russian].
104. Wang, H., Felps, W.S., McGlynn, S.P. (1977). Molecular Rydberg states. 7. Water., *J. Chem. Phys*, 67, 2614-2628.
105. Bolton, J.R., Cater, S.R. (1994). Homogeneous photodegradation of pollutants in contaminated water: an introduction. in *Aquatic and Surface Photochemistry*; edited by HelzEd, G.R. Lewis Publishers: Boca Ration, pp 467-490.
106. Nason, D., Fletcher, N.H. (1975). Photoemission from ice and water surfaces, *J. Chem. Phys*, 62, 4444-4449.
107. Heit Braun, A.M.G. (1997). Vacuum UV (172 nm) actinometry: The quantum yield of the photolysis of water, *Water Sci. Technol*, 35, 25-30.
108. Hu, W.P., Rossi, I., Corchado, J.C., Truhlar, D.G. (1997). Molecular modeling of combustion kinetics. The abstraction of primary and secondary hydrogen's by hydroxyl radicals, *J. Phys. Chem. A*, 101, 6911-6921.
109. Okazaki, K., Nozaki, T., Uemitsu, Y., Hijikata, K. (1995). Direct conversion from methane to methanol by silent pulsed discharge plasma, *Proceedings of 12th Int. Symp. on Plasma Chemistry*; Minnesota, USA, pp 581-586.
110. Tsang, W. (1988). Chemical kinetic data base for combustion chemistry. Part 3, *J. Phys. Chem. Ref. Data*, 17, 887.
111. Cohen, N. (1991). Are the reaction rate coefficients additive? Revised transition state theory calculations for OH + alkane reactions, *Int. J. Chem. Kinet*, 23, 397-417.
112. Akhmedov, A.Ya., Erofeev, V.I., Erofeev, M.V., Istomin, V.A., Korovin, S.D., Medvedev, Yu.V., Poligalov, Yu.I., Orlovskii, V.M., Sergeev, O.A., Sosnin, E.A., Stepanov, V.P., Tarasenko, V.F. (9 March 2005). Method of natural gas drying. Flow-through reactor for natural gas drying. Patent RU No. 2284850.
113. Bayerlen, A.N.,. Medvedev, Yr.V, Polygalov, Yr.I., Sergeev, O.V., Stepanov, V.P., Yakovlev, V.G., Alekseev, S.V., Orlovskii, V.M., Panarin, V.A., Suslov, A.I., Tarasenko, V.F. (2006). Effect of VUV Radiation on phase transition in natural gas. *Russian Physics Journal*, No. 10, application, 445-448.
114. Medvedev, Yr.V., Ivanov, V.G., Sereda, N.I., Polygalov, Yr.I., Erofeev, V.I., Korovin, S.D., Erofeev, M.V., Sosnin, E.A., Suslov, A.I., Tarasenko, V.F., Istomin, V.A. (2004). Interaction of high-power UV radiation on the natural gas in a flow-through reactor, *Science and techniques of gas industry*, No. 3-4, 83-87.

115. Kuvshinov, V.A., Orlovskii, V.M., Rozdestvenkii, E.A., Suslov, A.I., Tarasenko, V.F. (2006). Changes in the composition of the natural gas irradiated by an Xe_2-excilamp, *Atmos. Oceanic Opt*, 42, 815-818.
116. Alekseev, B.A., Kuvshinov, V.A., Lisenko, A.A., Lomaev, M.I., Orlovsky, V.M., Panarin, V.A., Rozhdestvensky, E.A., Skakun, V.S., Tarasenko, V.F. (2006). A photoreactor on the basis of a Xe_2-excilamp. *Instrum. Exp. Tech*, 49, 132-134.
117. Sosnin, E.A., Zakharova, E.A., Batalova. V.N. (2005). *Applications of excilamps in analytical chemistry*. Industrial Laboratory, 71, 18-24 [in Russian].
118. Sosnin, E.A., Batalova, V.N., Slepchenko, G.B., Tarasenko, V.F. (2002). Excilamp applications in the chemical sample pretreatment process, Proceedings of the *Int. Conf. on Atomic and Molecular Pulsed Lasers IV*, edited by Tarasenko, V.F., Mayer, G.V., Petrash, G.G. Vol. 4747, Proceedings of SPIE; SPIE: Bellingham, WA, pp 352-357.
119. Batalova, V.N., Bylatskaya, O.A., Sosnin, E.A. (2004). Biological objects pretreatment optimization using an XeBr-excilamp for mercury concentration control by the ASVA method, Proceedings of the *Conf. on Atomic and Molecular Pulsed Lasers*, edited by Tarasenko, V.F., Mayer, G.V., Petrash, G.G. Vol. 5483. Proceedings of SPIE; SPIE: Bellingham, WA, pp 323-327.

In: Photochemistry Research Progress
Editors: A. Sánchez, S. J. Gutierrez

ISBN: 978-1-60456-568-3

Chapter 8

PHOTOCHEMICAL PROCESSES AT THE AIR-WATER INTERFACE OF ATMOSPHERIC DROPLETS

Davide Vione, Claudio Minero and Valter Maurino
Dipartimento di Chimica Analitica, Università di Torino, Via Pietro Giuria 5, 10125 Torino, Italy

ABSTRACT

Photochemical processes play an important role in the atmospheric aqueous phase. They are involved in the transformation of organic and inorganic compounds, some of which have an impact on global climate. Atmospheric waters (cloud, fog, mist, rain), due to their elevated surface-to-volume ratio, are also a favourable environment for various kinds of interface phenomena. Photoinduced reactions at the air-water interface have been the object of much attention, because different kinds of surface effects could considerably enhance the reaction rates.

Early studies have been focused on the distribution of the light-field intensity inside disperse water droplets, starting from the assumption that the radiation intensity in the droplet could be considerably enhanced by cavity resonance effects compared to the surrounding gas phase. However, such phenomena would only involve peculiar combinations of radiation wavelength and droplet diameter and would play a minor role for polychromatic irradiation, as is the case of sunlight, and a continuous distribution of droplet diameters as is typical of the atmosphere. Multiple radiation reflections at the air-water interface would nevertheless increase the effective radiation intensity in the droplet by a factor of 1.3-1.8 compared to the gas phase. A number of additional effects are then able to substantially enhance the interface photochemical activity.

Many organic compounds are known to undergo accumulation at the air/water interface, and some of them can also interact with inorganic ions. Such an interaction, which can be approximately assessed by means of the Wagner-Onsager-Samaras approach, allows the concentration of photoactive anions near the surface to increase in the presence of organic compounds. For instance, the surface co-adsorption of nitrate and benzene considerably enhances the rate of the photochemical conversion of benzene into phenol upon nitrate photolysis. Just because of co-adsorption, in a spherical water drop of

1 μm radius around 15% of the cited reaction would take place in a surface layer of sub-nanometric thickness, which accounts for just 0.3% of the total volume.

Very important is also the enhancement of the photolysis quantum yields at the interface, where photoactive molecules are surrounded by an incomplete solvent cage. The presence of a surrounding solvent cage is able to considerably decrease the photolysis quantum yields by facilitating the recombination of photogenerated radical fragments. Molecules at interfaces would undergo the solvent-cage effect on photolysis to a very limited extent, which could considerably enhance their surface reactivity. The quantum yield of nitrate photolysis might be increased up to 6 times at the interface, and the combination of surface co-adsorption and increased quantum yield would cause over 50% of the photochemical conversion of benzene into phenol to occur in just 0.3% of the drop volume. This means that the reaction rate would be increased by a factor of two if a solution bulk is dispersed into 1 μm drops. Other photochemical reactions (e.g., the photolysis of $FeCl_2^+$) could be even more markedly enhanced. These findings could substantially modify the current knowledge concerning multiphase atmospheric chemistry.

INTRODUCTION

Photochemical processes play an important role in the transformation of organic compounds in the atmosphere. These reactions control the atmospheric lifetime of many species, including harmful pollutants and climate forcing gases [1]. The relevant reactions can take place in the gas phase, inside or at the surface of water droplets, and on airborne particulate matter. They can consist of direct or indirect photolysis processes [2]. In direct photolysis a sunlight-absorbing compound undergoes bond breaking (reaction 1) or photoionisation (reaction 2) upon photon absorption. Homolytic bond breaking with formation of radical species is favoured in apolar media or in the gas phase, but would also occur very often in the aqueous solution. In contrast, photoionisation reactions that yield charged species could involve organic compounds that are adsorbed on the surface of inorganic oxides, such as silica, which has environmental significance for molecules on mineral surfaces exposed to the atmosphere, including the crustal component of airborne particulate matter [3].

$$A{-}B + h\nu \rightarrow A^{\bullet} + B^{\bullet} \quad (1)$$

$$A{-}B + h\nu \rightarrow A{-}B^{+\bullet} + e^{-} \quad (2)$$

Concerning direct photolysis on particulate matter, an interesting finding has been that black carbonaceous particles absorb sunlight and can protect adsorbed compounds from photodegradation. Accordingly, the relevant phototransformation kinetics is significantly faster on white substrates such as silica and alumina, compared to darker compounds such as Fe oxides and carbonaceous material [4].

Indirect photolysis takes place in the presence of photosensitisers that absorb sunlight and produce reactive species that can induce the transformation of other molecules. The latter do not necessarily need to absorb solar radiation. Photogenerated reactive species can be the excited triplet states of aromatic ketones, quinones and of polyaromatic compounds, singlet oxygen ($^1O_2$) [5,6], and reactive radicals formed upon photolysis of the sensitisers [7]. In the case of the excited triplet states and of singlet oxygen, reactions (3-7) would involve the sensitiser S, oxygen, and an organic substrate M, where the superscripts ^{S1} and ^{T1} indicate the first excited singlet and triplet states, respectively [8,9].

$$S + h\nu \rightarrow S^{S1} \quad (3)$$
$$S^{S1} \text{ —(inter-system crossing)}\rightarrow S^{T1} \quad (4)$$
$$S^{T1} + O_2 \rightarrow S + {}^1O_2 \quad (5)$$
$$M + {}^1O_2 \rightarrow \text{Products} \quad (6)$$
$$S^{T1} + M \rightarrow \text{Products} \quad (7)$$

Transformation reactions induced by the photosensitisers would mainly take place on the surface of organic particles and in water droplets (mist, fog, cloud). The case of particles has been studied in more detail because many important organic components of airborne particulate matter (quinones, aromatic carbonyls, furans) are known photosensitisers [5,6]. These compounds are also present in disperse atmospheric droplets. Moreover, humic and fulvic substances play a major role in inducing photosensitised reactions in river and lake water [10], and somewhat similar compounds (Humic-Like Substances, HULIS) are also present in atmospheric aerosol, fog and cloud [11,12]. The possible photochemical properties of HULIS are completely unknown at the moment and would deserve dedicated studies. A simple transfer to HULIS of the known photochemical properties of humic substances is not allowed. Indeed, a major difference between surface-water humic acids and HULIS is that the latter have smaller average molecular weight and a lower degree of aromaticity [13]. These different features might have considerable impact on the photochemistry of HULIS vs. humic acids.

The environmental importance of ${}^1O_2$ as an oxidant has been a subject of controversy, but the determination of its photoproduction rate in fog water from California's Central Valley has for instance indicated that ${}^1O_2$ can be a significant sink of electron-rich organic compounds and of some amino acids [14,15].

Reaction (7) as directly involving the excited triplet states of the photosensitisers could involve electron or H abstraction, therefore transforming both sensitiser and substrate [16,17]. The oxidised substrate ($M^{+\bullet}$, $(M\text{-}H)^\bullet$) would undergo further transformation reactions, while the reduced sensitiser ($S^{-\bullet}$, $(S\text{+}H)^\bullet$) would in many cases be recycled back to S by molecular oxygen, yielding superoxide or hydroperoxide (reaction 8a,b) [8]. In such a case the whole process that involves S could be photocatalytic, and induce the degradation of the substrate without net transformation of the sensitiser.

$$S^{T1} + M \rightarrow S^{-\bullet} + M^{+\bullet} \quad (7a)$$
$$S^{T1} + M \rightarrow (S\text{+}H)^\bullet + (M\text{-}H)^\bullet \quad (7b)$$
$$S^{-\bullet} + O_2 \rightarrow S + O_2^{-\bullet} \quad (8a)$$
$$(S\text{+}H)^\bullet + O_2 \rightarrow S + HO_2^\bullet \quad (8b)$$

An important exception to reactions (8a,b) has been found in the case of the irradiation of PAHs and nitro-PAHs as sensitisers and of methoxyphenols as substrates. The addition of methoxyphenols has been found to enhance the photodegradation of the sensitisers, possibly because the reduced sensitisers undergo further transformation and net consumption instead of back oxidation by oxygen [18].

As far as the role of reactive radical species is concerned, in the gas phase the most important ones are $^{\bullet}OH$ and $^{\bullet}NO_3$. They are mainly generated by ozone, the former upon photolysis [19-21] and the latter after reaction with $^{\bullet}NO_2$ [22,23]. Further sources of $^{\bullet}OH$ are the photolysis reactions that involve H_2O_2 and HNO_2 [24]. The hydroxyl radical, due to photochemical formation and very fast reactivity is present in the atmosphere only during daytime. In contrast, the radical $^{\bullet}NO_3$ undergoes efficient photolysis under sunlight, and can be present at significant concentration in the atmosphere only during the night [25]. Note that ozone, besides being a source of reactive species, is also able to directly transform gas-phase alkenes [26].

$$O_3 + h\nu \rightarrow O_2 + O^* \quad (9)$$
$$O^* + H_2O \rightarrow 2\ ^{\bullet}OH \quad (10)$$
$$H_2O_2 + h\nu \rightarrow 2\ ^{\bullet}OH \quad (11)$$
$$HNO_2 + h\nu \rightarrow {}^{\bullet}OH + {}^{\bullet}NO \quad (12)$$
$${}^{\bullet}NO_2 + O_3 \rightarrow {}^{\bullet}NO_3 + O_2 \quad (13)$$
$${}^{\bullet}NO_3 + h\nu \rightarrow {}^{\bullet}NO_2 + O \quad (14)$$

Chlorine atoms, formed under irradiation upon heterogeneous reactions of sea-salt particles with ozone or nitrogen oxides [27], would also contribute to the gas-phase transformation of organic compounds in marine areas [28].

The radical $^{\bullet}OH$ also occurs in the atmospheric aqueous phase, because of transfer from the gas phase, reaction between O_3 and $O_2^{-\bullet}$, the Fenton reaction (Fe^{2+} + H_2O_2), and the photolysis of hydrogen peroxide, nitrate, nitrite, nitrous acid and Fe(III) hydroxocomplexes [7]. Table 1 reports the relative contribution of the different photochemical sources to $^{\bullet}OH$ photogeneration under conditions typical of continental clouds, in the presence of 0.5 μM Fe(III).

Table 1. Relative contribution of the photolysis of H_2O_2, NO_3^-, NO_2^-, HNO_2 and Fe(III) to the photochemical generation of $^{\bullet}OH$ in typical continental clouds [7]. The reference is the contribution of Fe(III), taken as 1.0.

Species	Concentration, M	Contribution to $^{\bullet}OH$, relative to Fe(III)
H_2O_2	5.8×10^{-5}	9.8
NO_3-	3.1×10^{-4}	4.2
NO_2-	1.9×10^{-7}	0.05
HNO_2	1.0×10^{-8}	0.05
Fe(III)	5.0×10^{-7}	1.0

The scenario can be significantly different under more polluted conditions. In the case of fog water from California's Central Valley the concentration of NO_2^-/HNO_2 reached up to 6×10^{-5} M, while nitrate had a maximum value of 3×10^{-3} M. Under these circumstances the photolysis of NO_2^-/HNO_2 accounted for 50-100% of $^{\bullet}OH$ photoproduction [14].

The occurrence of the radical $^{\bullet}NO_3$ in atmospheric waters would mainly be accounted for by transfer from the gas phase [29], but other reactive species can be formed upon photolysis in solution. Examples are $^{\bullet}NO_2$, formed upon nitrate photolysis or nitrite photooxidation [30,31], $Cl_2^{-\bullet}$, produced by chloride oxidation or by the UV photolysis of the complex $FeCl^{2+}$ [32,33], and $Br_2^{-\bullet}$, deriving from bromide oxidation or $FeBr^{2+}$ photolysis [34].

The relative importance of gas vs. aqueous-phase degradation processes depends on many factors, among which the liquid water content of the atmosphere, the partitioning of a given compound between the gas phase and the aqueous solution, and the reactivity of the same compound in each phase. As far as the latter issue is concerned, consider for instance the reaction of 2-propanol with $^{\bullet}OH$ in the gas phase (10^6 (molecules $^{\bullet}OH$) cm^{-3} on average) and in solution (average $[^{\bullet}OH] = 2 \times 10^{-14}$ M) [35,36]. From the reported reaction rate constants [25,37], the atmospheric lifetime of 2-propanol would be 2 days in the gas phase and only 7 hours in water droplets. For phenol the corresponding lifetimes would be 10 hours and 1 hour, respectively. Reactivity in water droplets could therefore be very important for the compounds that undergo efficient partitioning in the aqueous phase and can efficiently be transformed in that compartment.

As far as photochemistry is concerned, the atmospheric importance of reactions in water droplets would be influenced, and possibly enhanced, by some additional factors. They are the light-field distribution inside the droplets, the accumulation at the air-water interface of surface-active compounds, the surface co-adsorption of organic and inorganic species, and the enhanced photolysis quantum yields because of the incomplete solvent cage at the air/water interface. These issues could influence both the reaction rates in the whole droplet, and the rate distribution in the droplet volume. For instance, early studies of the light-field intensity in water droplets indicated that small spherical cavities could be subjected to considerable irradiation intensity, much higher compared to the surrounding gas phase, therefore enhancing photochemical reactions in solution [38,39]. Later studies have mitigated the early statements [40], leading to a better understanding of the light-field distribution inside the droplet volume [41].

The droplet surface could be a much more effective photochemical reactor than the bulk, because it could accumulate hydrophobic and surface-active organic compounds [42,43]. Additionally, the interaction between organic and inorganic species could lead to co-adsorption phenomena that would increase the surface concentration of photochemically active ions compared to the case where the organic compounds are absent [44]. Such a surface concentration effect could considerably enhance the rates of photochemical reactions at the air/water interface compared to the bulk, suggesting that a significant fraction of the total droplet reaction could take place in a very small volume near the surface [45]. The reaction in the whole droplet could be enhanced at a variable extent by the presence of a small, very reactive surface layer.

Finally, the air-water interface is different from the solution bulk because the surface molecules are only partially surrounded by the solvent. In other words, the solvent cage is incomplete at the droplet surface. This fact could considerably enhance the rates of photolysis because a major factor that limits the photolysis quantum yields is the recombination of photogenerated fragments inside the cage [46-49].

$$A{-}B + h\nu \rightleftarrows [A^{\bullet} + B^{\bullet}]_{cage} \rightarrow A^{\bullet} + B^{\bullet} \qquad (15)$$

The incomplete solvent cage at the air/water interface could lead to increased photolysis quantum yields and to higher rates of the photoinduced processes as a consequence [41]. Such an effect would be added to surface accumulation and co-adsorption phenomena, thereby enhancing the effectiveness of the droplet surface layer as a photochemical reactor compared to the bulk. The different effects that could increase the photoinduced reaction rates inside droplets and at the droplet surface are discussed in the paragraphs that follow.

ACTINIC FLUX IN WATER DROPLETS

Atmospheric waters modify significantly the flux of sunlight because of different phenomena. Clouds as well as aerosols are able to scatter and absorb radiation. As a consequence, the sunlight intensity is considerably increased on top of clouds and is usually reduced inside and below them [50,51]. This fact leads to a non-homogeneous distribution of the photochemical reaction rates inside the cloud. Understandably, photochemical processes would be favoured in the top layer and inhibited in the bottom one [52], but no average enhancement of cloud photochemistry can be expected because of this phenomenon.

However, spherical (or almost spherical) droplets are also able to refract and diffract the incident light. Incident radiation in a spherical droplet can be reflected and refracted at the air-water interface, and the refraction index n of water is in the 1.33-1.37 range between 800 and 300 nm. Due to the spherical symmetry, a series of reflection/refraction events would take place (see Figure 1), which increase the effective path length of the radiation in the droplet and therefore the probability of photon absorption by the dissolved compounds [40,53,54].

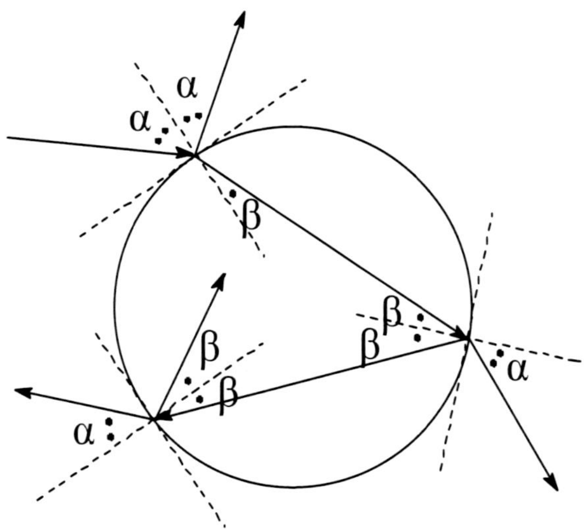

Figure 1. Multiple reflection/refraction phenomena at the air-water interface of spherical droplet. It is $\sin \alpha = n \sin \beta$, with n = 1.33-1.37 depending on the wavelength of the incident beam. Three events only are shown, but further ones would take place until the intensity of the light beam in the droplet vanishes because of absorption and refraction.

At each interface event a fraction R of the incident radiation is reflected, depending on the angles of incidence and of refraction. If the size of the droplet is much higher than the radiation wavelength, the laws of optical geometry apply and the value of R is given by equation (16), where α and β have the same meaning as in Figure 1 [55].

$$R = \frac{1}{2} \cdot \left[\left(\frac{\sin(\alpha - \beta)}{\sin(\alpha + \beta)} \right)^2 + \left(\frac{\tan(\alpha - \beta)}{\tan(\alpha + \beta)} \right)^2 \right] \tag{16}$$

A fraction 1-R of the incident radiation from the gas phase will therefore enter the droplet, and at each successive contact with the interface a fraction R will be reflected back into the droplet. Within successive reflections the radiation can be absorbed by the molecules that are dissolved in water. Figure 2 reports the trend of R as a function of the angle β, for n = 1.33-1.37. Note that the highest value of β that is possible inside a spherical water droplet is 0.26-0.27 π (47-49°), corresponding to α = π/2 (90°).

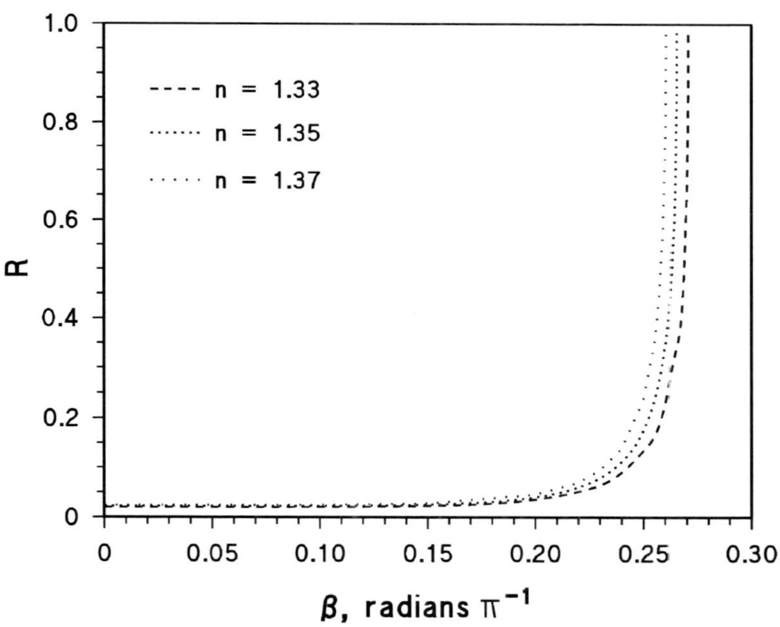

Figure 2. Fraction R of reflected radiation in a water droplet, as a function of the angle β and the refraction index n. Note that 0.25 on the X-axis scale means ¼ π (radians units).

A fraction of the incident radiation from the gas phase will remain inside the droplet till it is absorbed, and a part will be lost by refraction back into the gas phase. On average, the multiple reflections will increase the absorbed photon flux by a dissolved molecule M compared to the case where M is in the gas phase. The phenomenon is equivalent to an enhancement of the intensity of the radiation field inside the droplets compared to the surrounding gas phase [40].

Considering all the possible angles and positions of incidence of the radiation into the spherical droplet one obtains the actinic flux enhancement factor η, which represents the increase of the effective radiation intensity inside the water droplets compared to the surrounding gas phase. The value of η varies in the range 1.26-1.28 for a variation of the refraction index n between 1.33 at 800 nm and 1.37 at 300 nm [40,56].

The applicability of the laws of optical geometry depends on the ratio between the droplet size and the radiation wavelength, usually measured by the size parameter $x = 2 \pi R_D \lambda^{-1}$, where R_D is the droplet radius, and λ the radiation wavelength [55]. For cloud droplets with $R_D = 10$ μm, and λ in the 300-600 nm range, a size parameter x = 100-200 suggests that the optical geometry would just be a rough approximation. The exact treatment should consider the wave description of light and apply the Mie theory of refraction. The calculation of the actinic flux enhancement η can be carried out numerically by means of the appropriate codes, and the resulting values depend on both the radius of the droplet and the radiation wavelength [57,58]. Understandably, the η values derived from the Mie theory ($\eta(R_D,\lambda)$) tend to the optical geometry one ($\eta(\lambda)$) at large R_D, when the droplet size is much higher than the radiation wavelength. For atmospherically significant values of the size parameter ($x \approx 10$ or higher) it is $\eta(R_D,\lambda) > \eta(\lambda)$, but $\eta(R_D,\lambda)$ can be very variable. In most cases it is below 1.8, down to $\lim_{RD\to\infty}[\eta(R_D,\lambda)] = \eta(\lambda) = 1.26\text{-}1.28$ [40], but resonance effects are observed for certain values of (R_D,λ) or of x. In these cases the combination of droplet size and radiation wavelength increases the actinic flux up to values that can be even four orders of magnitude higher compared to the surrounding gas phase [39]. A substantial effect on η is thus observed for discrete values of the size parameter, but it is important to assess the impact of resonance for a continuous distribution of radiation wavelength and droplet size. In such a distribution [59] the effect of resonance appears as a series of spikes superposed to the continuum, and the numerical assessment of the spike importance is highly dependent on the resolution of x adopted in the numerical calculations. The Δx width of a single spike could be lower than 10^{-5}, e.g. totally included within x = 80.99427 and x = 80.99428, and the spike contribution could well be lost in low-resolution calculations that might completely miss a peak. In contrast, if a low-resolution calculation (assume 10^{-4} resolution) incidentally hits a peak, the peak value could be smeared out over a 10^{-4} Δx interval, thereby overestimating its importance by orders of magnitude. The application of high-resolution numerical calculations indicated that the resonance spikes would play on average a negligible role for distributions of R_D and λ of atmospheric significance [40]. Independently of the spikes, the actinic flux enhancement $\eta(R_D,\lambda)$ by the droplets due to multiple reflections would vary from 1.8 at λ = 300 nm and $R_D = 1$ μm, to 1.4 for λ = 300 nm and $R_D = 25$ μm. At 800 nm the corresponding $\eta(R_D,\lambda)$ would vary from 1.7 to 1.4 between 1 and 25 μm [40]. The rate of photochemical reactions in water droplets could thus be enhanced significantly compared to the surrounding gas phase, although not as much as the phenomenon of resonance might suggest.

Another interesting effect connected with the Mie theory is represented by the light-field distribution inside the droplet [60]. The relative intensity at the surface can be considerably enhanced compared to the bulk in the equatorial plane of the droplet (see Figure 3A), but in other spherical sections the surface intensity is lower than in the bulk (Figure 3B,C). The data of Figure 3 are referred to an environmentally significant irradiation spectrum of 290-600 nm, and $R_D = 1$ μm [41]. The absolute values of the average relative intensity are dependent on the irradiation wavelength, but the general framework would not change dramatically if a different spectrum is considered.

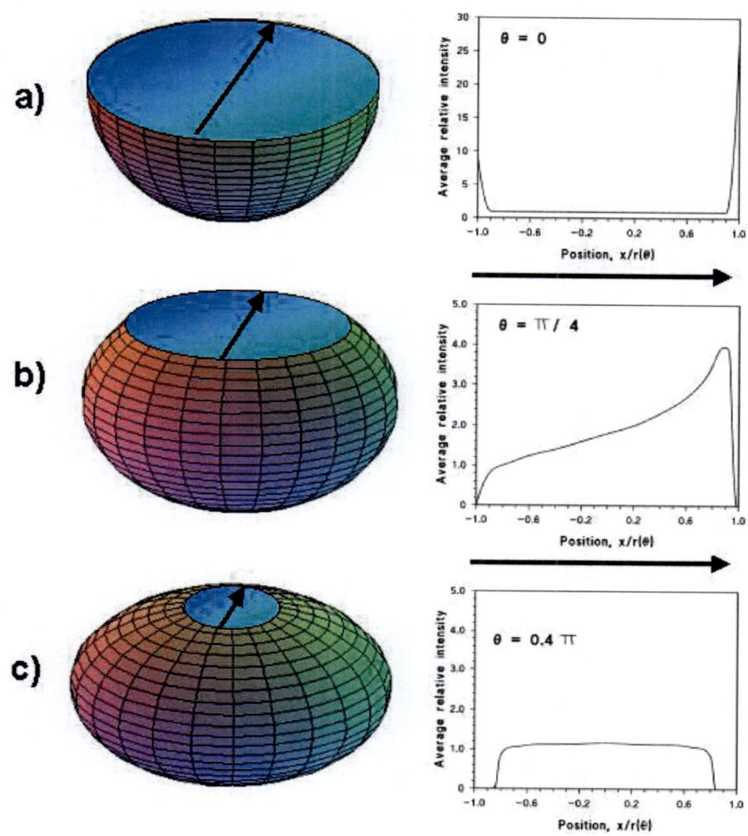

Figure 3. Light-field intensity (averaged over the 290-600 nm wavelength interval) in different sections of a spherical water drop (R_D = 1 μm). Each section is graphically depicted. The arrows represent both the direction of illumination and the path along which the average irradiation intensity is calculated.

In summary, the multiple reflections inside water droplets can lead to a 1.4-1.8 time enhancement of the irradiation intensity compared to the surrounding gas phase. Enhancement factors up to 10^4 would be observed, because of resonance effects, for certain combinations of radiation wavelength and droplet radius. However, resonance would only involve some discrete values of the size, and its role in a continuous size distribution of water droplets would be very minor [40]. The light field intensity within a droplet would not be uniform, and intensity at the surface can be considerably higher than in the bulk when considering the equatorial plane in the direction of illumination. The same would not be true of other planes, with the result that the average irradiation intensity at the surface would be a bit lower compared to the bulk [41]. Accordingly, the latest studies indicate that the most important factor to influence the light intensity in water droplets is the 1.4-1.8 times enhancement factor due to multiple reflections at the air-water interface.

SURFACE ACCUMULATION AND CO-ADSORPTION PHENOMENA

Many solutes are able to undergo accumulation at the air-water interface of atmospheric droplets, because of different processes. Amphiphilic compounds, such as fatty acids, dicarboxylic acids and some fractions of HULIS are well known to show significant surface activity. They can reach much higher concentration at the air-water interface compared to the solution bulk [42,61].

Surface accumulation is also operational for hydrophobic species such as aliphatic and aromatic hydrocarbons. They undergo poor solvation by the water molecules and tend to be "pushed" at the interface, where the incomplete solvent cage would make interaction with water energetically less expensive. In the case of aromatic hydrocarbons, the surface accumulation would increase with molecular size in the order benzene < naphthalene < anthracene [62]. For these compounds, surface accumulation would be an important phenomenon in the process that finally leads to volatilisation [44]. An increase of the interface concentration is also predicted for dissolved gases and small, neutral molecules and radicals such as N_2, O_2, O_3, $^{\bullet}OH$, $HO_2^{\bullet}$, and H_2O_2. All these species show in fact a free-energy minimum at the surface [63].

Recent studies indicate that the phenomenon of surface accumulation can also involve ionic species, depending on their interaction with water molecules and their polarisability. Such an effect would involve the anions rather than the cations, and within homologous series it is generally stronger with increasing atomic or molecular size. Larger size would generally mean that the ion fits less well inside the water cage, and tends to be shifted to a less hindered environment such as the air-water interface. For similar reasons, monoatomic cations are usually hydrated better than monoatomic anions due to their smaller diameter [27]. Among the anions that can undergo surface accumulation there are the halides, with the exception of fluoride, in the order $Cl^- < Br^- < I^-$, and this feature could have important consequence in the chemical and photochemical processes involving sea-salt droplets and aerosol [64,65]. The surface accumulation of chloride plays a significant role in the processes of uptake of reactive gas species and radicals (e.g. O_3, H_2O_2, $^{\bullet}OH$), with subsequent possible oxidation of chloride to photochemically active compounds such as Cl_2 and HClO. The associated photolysis processes are an important source of atmospheric Cl atoms [27].

Other anions that can undergo surface accumulation because of solvation issues are acetate (but not the smaller formate), benzoate, and phenolate [42]. A smaller but non-negligible effect would also be operational in the case of nitrate [62,66]. The surface accumulation of the cited species could have a significant influence on photochemical processes, in particular as far as photoactive compounds such as hydrogen peroxide and nitrate are concerned.

The surface accumulation of hydrophobic species and of photoactive compounds could co-operate in the enhancement of the photochemical reactions at the interface. For instance, hydrogen peroxide has a slight preference for the droplet surface layer, where it can be present at double concentration compared to the bulk [63]. Benzene would undergo much higher surface accumulation. Experimental data indicate that, in the presence of 1 μM bulk benzene, the concentration in the surface layer would be increased by a factor of 400 compared to the bulk. The thickness of such a layer would be just 0.5 nm, beyond which the concentration of benzene would rapidly fall down to the bulk value. Accordingly, for calculation purposes the distribution of benzene inside the droplet can be assumed as a slab profile with a concentration of 0.4 mM in the first 0.5 nm from the surface, and 1 μM at higher distance [44]. Note that a concentration of 1 μM is representative of the levels of many toxic aromatic compounds in polluted areas [36].

The interface accumulation of hydrogen peroxide is likely to involve a thin layer as well, and for sake of simplicity it can be hypothesised that the two layers have equal thickness. A concentration $[H_2O_2] = 5$ μM would be representative of atmospheric water in polluted areas [67].

The photolysis of hydrogen peroxide yields $^{\bullet}OH$ radicals (see reaction 11, which takes place both in the gas phase and in solution) [68], which can react with benzene to give phenol with 95% yield [69]. At 5 μM H_2O_2 and the small optical path lengths that are typical of atmospheric droplets, the photon flux absorbed by hydrogen peroxide (moles of photons L^{-1} s^{-1} units) would be directly proportional to $[H_2O_2]$. The rate of photolysis, which is the product of the absorbed photon flux times the photolysis quantum yield, would be proportional to $[H_2O_2]$ as well [67]. Additionally, benzene or otherwise another aromatic compound at typical atmospheric concentration values would not be the main $^{\bullet}OH$ scavenger in water droplets; compounds such as formate and formaldehyde would generally play the main role [70,71]. As a consequence, the formation rate of phenol (r_{Phenol}) would also be directly proportional to [Benzene]. One gets $r_{Phenol} \propto [Benzene]\,[H_2O_2]$, which varies with the distance from the droplet surface following the distribution of the two species. Consider a spherical layer located between x_i and $x_i + \Delta x$ from the surface. The number of moles of phenol produced in the layer would be given by $r_{Phenol}\, V_i$, where V_i is the layer volume. If Δx is very small, the layer volume would be $V_i = 4\,\pi\,(R_D - x_i)^2\,\Delta x$, where R_D is the radius of the droplet. Passing from finite to infinitesimal quantities [45], the fraction $F(x_i)$ of moles of phenol produced in a spherical layer of thickness x_i from the surface, relative to the whole droplet, would be given by the integral over x of the layer volume times the product [Benzene] $[H_2O_2]$, divided by the corresponding integral over the whole droplet (equation 17).

$$F(x_i) = \frac{\int_0^{x_i} (R_D - x)^2 \cdot [\text{Benzene}(x)] \cdot [H_2O_2(x)]\, dx}{\int_0^{R_D} (R_D - x)^2 \cdot [\text{Benzene}(x)] \cdot [H_2O_2(x)]\, dx} \qquad (17)$$

Note that x=0 corresponds to the air-water interface and x measures the distance from the interface. Obviously $F(R_D) = 1$, just meaning that 100% of the reaction takes place inside the whole droplet. The trend of $F(x_i)$ in the droplet is shown in Figure 4. Note that half of the reaction would take place in a surface layer of just 1 nm thickness, to be compared with R_D = 1 μm = 1000 nm. This practically means that the reaction in the surface layer has the same weight as that in the bulk, and that the surface accumulation of benzene and H_2O_2 could double the overall reaction rate in the droplet, compared to a scenario in which the concentration profiles are totally isotropic.

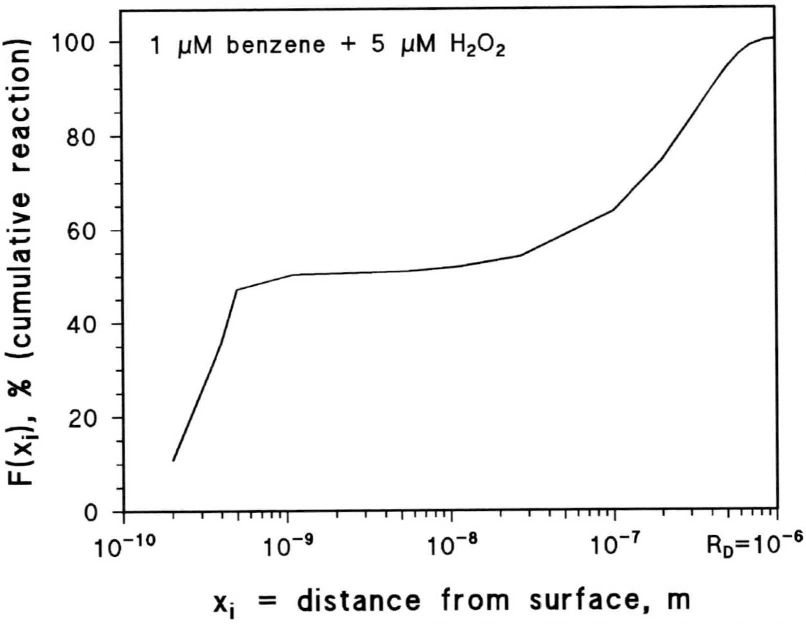

Figure 4. Trend of $F(x_i)$, expressed as percentage, as a function of the distance from the droplet surface.

A further effect that can increase the surface concentration of ionic species, both cations and anions in this case, is the interaction (co-adsorption) with some hydrophobic compounds that undergo surface accumulation [43,44]. This phenomenon can modify some features of the air-water interface, including most notably the dielectric constant, and the configuration water-organic layer-air can be more favourable to the occurrence of ionic species near the droplet surface compared to a water-air system [72]. Additionally, cations can interact with the delocalised π electrons of the aromatic rings and, if the aromatic compounds are accumulated at the surface, the interface concentration of the cations would also be increased. The surface concentration of the anionic counter-ions would be increased as a consequence upon interaction with the cations. To date, experimental data of co-adsorption between aromatic molecules and inorganic species are only available for the case of benzene and nitrate salts [44].

The phenomenon of surface co-adsorption can be treated approximately by means of the Wagner-Onsager-Samaras (WOS) formalism. This is an adaptation to the treatment of interfaces of the Debye-Hückel theory of electrolyte [73]. In this simplified theory the ions are described as impenetrable charged spheres in a solvent that is treated as a continuous and structureless dielectric. The same features are retained in the WOS formalism, the main difference being that the Debye-Hückel theory adopts a spherical symmetry because the solution bulk is isotropic [72]. This is reflected in the spherical symmetry form chosen for the Laplacian operator $\nabla^2 = (1/R)[\partial^2(R.)/\partial R^2]$:

$$\left(1/R\right)\left[\partial^2(R\,\psi)/\partial R^2\right] = \kappa^2\,\psi \qquad (18)$$

In contrast the WOS treatment describes an anisotropic surface layer, and the anisotropy is usually considered as localised in a direction perpendicular to the surface plane. Consequences of the anisotropy are the abrupt change of the dielectric constant at the interface and the cylindrical (instead of spherical) symmetry of the problem [73,74].

The Debye-Hückel theory considers the coulombic potential w_{ij} of the mean force acting between the ions i and j. To take into account the anisotropy, be i an ion of charge q at distance x from the interface. The WOS treatment considers the potential w(x) of the force exerted on the ion i, with charge q, by the image charge q' located at a distance 2x from q (see Figure 5).

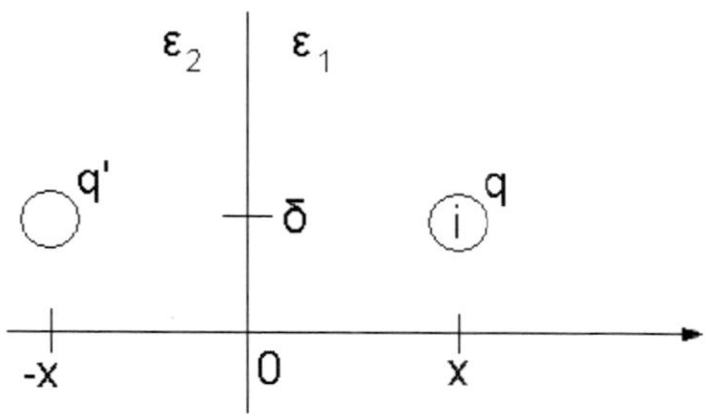

Figure 5. Schematic of the ion i (charge q) and of its image charge q', symmetric relative to the air-water interface. The position of the interface corresponds to the origin of the x axis. δ is the distance between the ion and the x axis. ε_1 and ε_2 are the dielectric constants of the two media (1 = water, 2 = air).

Be ε_1 and ε_2 the dielectric constants of the media containing q and q', respectively. The following relationships hold for q' and w(x):

$$q' = \left[(\varepsilon_1 - \varepsilon_2)\cdot(\varepsilon_1 + \varepsilon_2)^{-1}\right]\cdot q \tag{19}$$

$$w(x) = \left[(\varepsilon_1 - \varepsilon_2)\cdot(\varepsilon_1 + \varepsilon_2)^{-1}\right]\cdot q^2\cdot \exp(-2\kappa x)\cdot(16\pi\varepsilon_0\,\varepsilon_1\,x)^{-1} \tag{20}$$

In cylindrical symmetry one obtains equation (21), which leads to equation (22) for $w(x)=\frac{1}{2}\,q\,\psi$, and to equation (23) for the concentration $c_i(x)$ of the ion i as a function of the distance x from the interface [75]:

$$(\partial^2\psi/\partial x^2) + \delta^{-1}(\partial/\partial\delta)(\delta\,\partial\psi/\partial\delta) = \kappa^2\,\psi \tag{21}$$

$$w(x) = \left[(\varepsilon_1 - \varepsilon_2)\cdot(\varepsilon_1 + \varepsilon_2)^{-1}\right]\cdot\left[(1+\kappa a)^{-1}\cdot \exp(\kappa a)\right]\cdot q^2\cdot \exp(-2\kappa x)\cdot(16\pi\varepsilon_0\,\varepsilon_w\,x)^{-1} \tag{22}$$

$$c_i(x) = c_i(\infty)\cdot \exp(-w(x))\cdot(k_B\,T)^{-1} \tag{23}$$

δ is the distance between the charge q and the x axis (cylindrical symmetry, Figure 5), a is the radius of the ion, the charge of which is q, ε_0 is the dielectric constant of the vacuum, $c_i(\infty)$ the bulk concentration of the ionic species, k_B the Boltzmann constant, and T the absolute temperature. If 1 = water and 2 = air, equation (21) holds for a water-dissolved charged species near the interface with air, and foresees a lower interface concentration compared to the bulk.

In the presence of an organic compound that undergoes surface accumulation, it can be hypothesised that it forms a layer of thickness e and dielectric constant ε_{org} between water and air [45]. This is not unreasonable considering that the surface layer of atmospheric water droplets is more similar to an oily environment than to a simple air-water interface [76]. In such a case, for an ion of charge q at distance h from the water-organic interface one should consider the image charge q' inside the organic layer, and also the image q" of q' in the air phase (Figure 6).

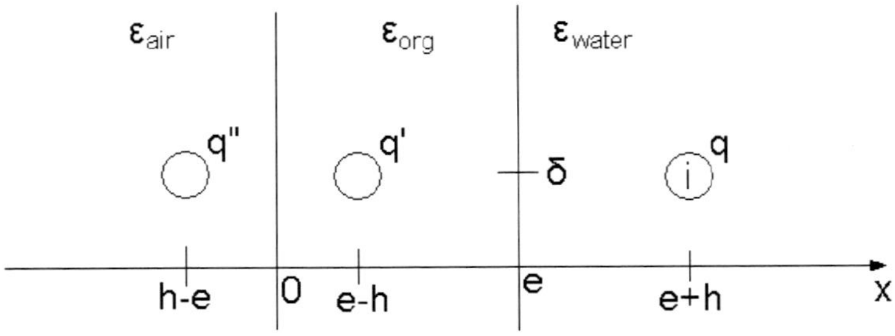

Figure 6. Schematic of an ion i (charge q) at distance h from the air-organic interface and of the image charges q' and q". δ is the distance of i from the x axis.

The following equations hold for q' and q" [45]:

$$q' = (\varepsilon_{water} - \varepsilon_{org}) \cdot (\varepsilon_{water} + \varepsilon_{org})^{-1} \cdot q \tag{24}$$

$$q'' = (\varepsilon_{water} - \varepsilon_{org}) \cdot (\varepsilon_{org} - \varepsilon_{air}) \cdot \left[(\varepsilon_{water} + \varepsilon_{org}) \cdot (\varepsilon_{org} + \varepsilon_{air})\right]^{-1} \cdot q \tag{25}$$

If org = benzene, from $\varepsilon_{water} = 78$, $\varepsilon_{benzene} = 2.2$ and $\varepsilon_{air} = 1$ one obtains q ≈ q' and q" ≈ 1/3 q. On these premises the potential w(x) is given by equation (26) which, substituted into equation (23), yields the concentration profile $c_i(x)$ of the ion i.

$$w(x) = \left(\frac{q^2}{16\pi\varepsilon_0\,\varepsilon_{water}}\right) \cdot \left(\frac{1}{x} + \frac{1}{3e}\right) \tag{26}$$

Figure 7 shows the concentration profiles of benzene (bulk concentration 1 μM) and of nitrate (0.1 mM) as a function of the distance from the interface, in a droplet of 1 μm radius. For benzene it is hypothesised an accumulation of 400 times in a 0.5 nm layer near the interface. The nitrate profile is reported both in the absence of benzene (equations 22,23) and in the presence of coadsorption of benzene and nitrate according to the WOS formalism (equations 23,26) [45].

The reported concentration profiles can have important consequence on photochemistry, considering that nitrate photolysis upon sunlight UV absorption yields $^{\bullet}OH$ radicals (reaction 27) [77], which can react with benzene to give phenol with 95% yield [69].

$$NO_3^- + h\nu + H^+ \rightarrow {}^{\bullet}OH + {}^{\bullet}NO_2 \tag{27}$$

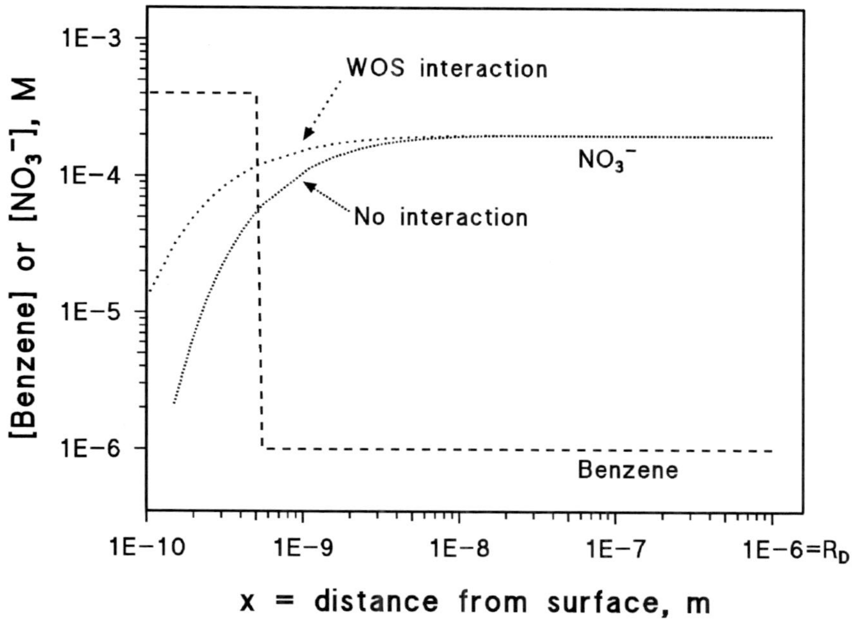

Figure 7. Concentration profiles of benzene and nitrate in a droplet of 1 μm radius (RD). The profile of [NO_3^-] is reported both when it is alone and upon interaction with benzene. It is evident that the co-adsorption enhances the concentration of nitrate near the interface.

As already seen for H_2O_2, with 0.1 mM nitrate in a small droplet the rate of NO_3^- photolysis would be directly proportional to [NO_3^-]. Benzene would not be the main $^{\bullet}OH$ scavenger in solution [70,71], thus it would be $r_{Phenol} \propto$ [Benzene] [NO_3^-]. In the absence of co-adsorption the reaction would be accelerated by the accumulation of benzene near the interface and, in a droplet of 1 μm radius, 5% of the total reaction would take place in a surface layer of 1 nm thickness, accounting for just 0.3% of the total volume. If the co-adsorption of benzene and nitrate is considered, in the same surface layer it would take place around 15% of the total reaction of the droplet [45]. These data confirm the potential importance of interfaces as efficient photoreactors in the atmosphere.

Solvent-Cage Effect on the Photolysis Quantum Yields

Many compounds can dissociate into fragments upon photon absorption. If the photolysis process occurs in the gas phase, the fragments would quickly diffuse away from each other and the probability of recombination would be very low. The situation is very different for a photolysis reaction involving a molecule in the solution bulk. In this case the molecule is surrounded by the solvent and, following the primary bond breaking, photogenerated fragments would be formed in the same "solvent cage" that was surrounding the molecule before photolysis. In most cases the photogenerated "fragments" are radical species. Under these circumstances two competitive processes can take place. The first is the diffusion of the fragments out of the cage, in which case the radicals can react with dissolved molecules in the solution bulk. However, radical diffusion into the solution can be a relatively slow process due to the need of passing the barrier represented by the surrounding solvent molecules. The process of diffusion out of the cage is therefore in competition with the recombination of the fragments inside the cage to give the initial molecule. Recombination would decrease the photolysis quantum yield, when the latter is measured from the reactivity of the radical species that exit the cage [77]. Note that the radical-radical recombination out of the cage is usually very unlikely because of the low steady-state concentration values of the radical species in the solution bulk. This consideration is particularly relevant if one of the photogenerated radicals is $^{\bullet}OH$, which is efficiently scavenged out of the cage by dissolved organic matter and by many inorganic compounds [71].

The solvent-cage effect plays an important role in many photochemical reactions. Because the rate of diffusion out of the cage is faster in low-viscosity solvents, the photolysis quantum yields of the reactions that are influenced by solvent-cage recombination are usually decreasing with increasing solvent viscosity [78,79]. In some cases the solvent-cage effect can result into a significant wavelength dependence of the quantum yield. For instance, the photolysis of aqueous OClO to yield $ClO^{\bullet}$ + O is favoured upon irradiation at shorter wavelength because of the increased kinetic energy of the photogenerated O atom, which makes its escape from the solvent cage easier [80].

In the case of the photolysis of H_2O_2 it has been found that the quantum yield in aqueous solution is around 0.5, much lower compared to the gas phase where it is near 1 [67,68]. It has also been found that about 64-68% of photogenerated hydroxyl radicals can escape the solvent cage [81]. It would imply a photolysis quantum yield in the primary process (before recombination) of about 0.75, much nearer to that in the gas phase.

The solvent-cage effect on photolysis can be very relevant to the photochemistry of atmospheric droplets, and in particular to the photochemistry at the interface, because at the droplet surface the solvent cage is incomplete and the photogenerated fragments can avoid recombination. Accordingly, photolysis processes can be enhanced at the air-water interface of atmospheric droplets compared to the solution bulk [46,47].

It has been found that the photodegradation of $Mo(CO)_6$ in 1-decene is increased by about 3 orders of magnitude in small liquid droplets (2 μm average diameter) compared to the bulk solution. The increase of the irradiation intensity in 1-decene droplets due to multiple reflections at the interface could at most be equal to a factor of 2, which leaves unexplained the largest part of the effect. The fast photodegradation in droplets has been explained with an increased photolysis quantum yield of $Mo(CO)_6$ at the air-water interface (incomplete solvent cage, see Figure 8) [41].

Various theoretical studies suggest that solvent-cage recombination of photofragments would be negligible or very low at the air-water interface. Obviously, surface effects of this kind would be much more important in small droplets compared to a macroscopic solution, due to the much larger surface-to-volume ratio [49].

Solvent-cage effects have also been reported on the photolysis of nitrate (reaction 27), based on the observation that the photolysis reaction is enhanced upon addition of $^{\bullet}OH$ scavengers such as 2-propanol and bromide [82,83]. The scavengers at elevated concentration would be able to penetrate the solvent cage and react with $^{\bullet}OH$ radicals in the cage, thereby preventing the recombination of $^{\bullet}OH$ and $^{\bullet}NO_2$. Under these circumstances one may expect an increased quantum yield of $^{\bullet}NO_2$ photogeneration.

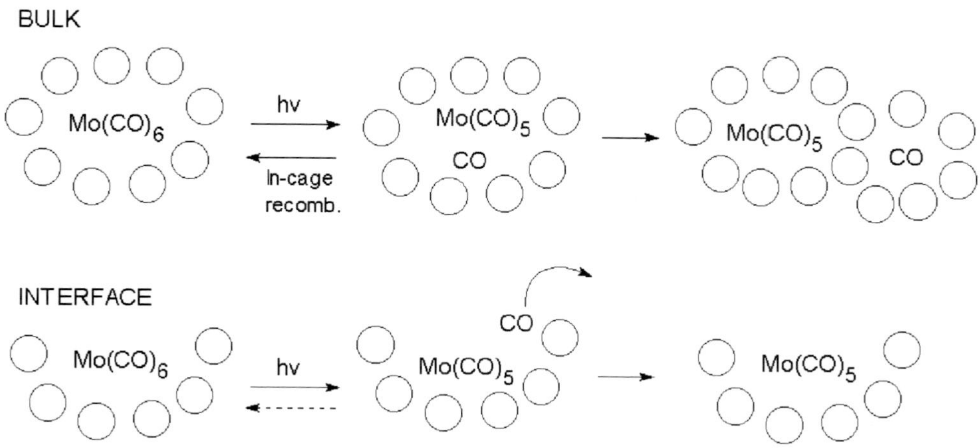

Figure 8. Schematic to explain the increased photodegradation kinetics of $Mo(CO)_6$ in small droplets, compared to the homogeneous solution.

The radical $^{\bullet}NO_2$ undergoes dimerisation and hydrolysis in water, yielding nitrate and nitrite [84]:

$$2\ ^{\bullet}NO_2 \rightleftarrows N_2O_4 \quad (28)$$

$$N_2O_4 + H_2O \rightarrow NO_2^- + NO_3^- + 2\ H^+ \quad (29)$$

The generation rate of nitrite upon nitrate irradiation would be proportional to that of $^{\bullet}NO_2$ and, under constant conditions (irradiation intensity, nitrate concentration), to the quantum yield of nitrate photolysis. It has recently been reported that the generation rate of nitrite upon nitrate irradiation increases in the presence of $^{\bullet}OH$ scavengers such as formate and formaldehyde [71]. The effect cannot be due to the inhibition of recombination between $^{\bullet}OH$ and $^{\bullet}NO_2$ in the solution bulk, which is negligible [71,85]. Interestingly the formation rate of nitrite reached a plateau with increasing scavenger concentration [71], which is reasonable under the hypothesis that the scavengers react with $^{\bullet}OH$ in the solvent cage. Indeed, sufficiently elevated scavenger concentration would be required for an effective penetration inside the solvent cage, but in the presence of an excess of scavengers the consumption of $^{\bullet}OH$ radicals in the cage would be practically quantitative. Limited effect would therefore be expected by further increasing the scavenger concentration.

The generation rate of nitrite upon nitrate irradiation, at the plateau caused by the presence of the scavengers in excess, was around 6 times higher than in the absence of the scavengers [71]. It is thus possible to assume that, in the absence of the solvent-cage effect, the quantum yield of nitrate photolysis would be six times higher than in the solution bulk. The solvent cage is incomplete at the air-water interface of atmospheric droplets, where the described enhancement of the photolysis quantum yield might be observed. This effect would be added to the possible co-adsorption phenomena at the interface, such as the already described interaction between benzene and nitrate [45]. The benzene-nitrate system would therefore be characterised by both surface co-adsorption and the enhancement of the quantum yield of nitrate photolysis. Within this scenario the fraction $F(x_i)$ of moles of phenol, produced upon reaction between benzene and photogenerated $^{\bullet}OH$ in a spherical layer of thickness x_i, would be proportional to the layer volume, the photolysis quantum yield, and the product of the concentrations of both benzene and nitrate:

$$F(x_i) = \frac{\int_0^{x_i} (R_D - x)^2 \cdot \Phi_{NO_3^-}(x) \cdot [\text{Benzene}(x)] \cdot [NO_3^-(x)]\, dx}{\int_0^{R_D} (R_D - x)^2 \cdot \Phi_{NO_3^-}(x) \cdot [\text{Benzene}(x)] \cdot [NO_3^-(x)]\, dx} \tag{30}$$

Under the arbitrary (but not too far from reality) hypothesis that both benzene accumulation and quantum yield enhancement would occur in a surface layer of 0.5 nm thickness, it is possible to calculate the corresponding values of the product $(R_D\text{-}x_i)^2\ \Phi_{NO3-}(x_i)$ [Benzene(x_i)] [$NO_3^-(x_i)$], and of $F(x_i)$, as a function of the distance x_i from the surface. They are reported in Figure 9 under three different scenarios: (i) no surface co-adsorption between benzene and nitrate, no quantum yield enhancement (just surface accumulation of benzene); (ii) surface co-adsorption but no quantum yield enhancement; (iii) surface co-adsorption and 6-fold surface enhancement of Φ_{NO3-}.

Figure 9 considers a water droplet of 2 μm diameter (R_D = 1 μm). It is apparent from the Figure that, when all the surface effects are considered together, around 55% of the total droplet reaction would take place in a surface layer of around 1 nm, which accounts for only 0.3% of the total droplet volume. This indicates the potential importance of the surface layer as an effective photochemical reactor. Additionally, the surface accumulation and co-adsorption phenomena would not modify at a significant extent the reaction rates in the solution bulk. The bulk processes would therefore take place at approximately the same rate that would be observed in the absence of surface phenomena. The latter, accounting for over 50% of the total reaction rate, would thus lead to a more than twofold increase of the rate in the droplet compared to a scenario in which only bulk processes were operational.

Current models of atmospheric chemistry and photochemistry only take into account the bulk reactions in droplets [7,86], and could therefore underestimate considerably the reactivity of atmospheric waters. It is thus very important to gain a more precise knowledge of the possible interface processes, their significance for a variety of chemical and photochemical reactions, and finally their impact on the reactivity of atmospheric droplets. The inclusion of such processes into the models of atmospheric reactivity would considerably increase the precision with which the models describe the real environment. The predictive capability of the models themselves would be increased as a consequence.

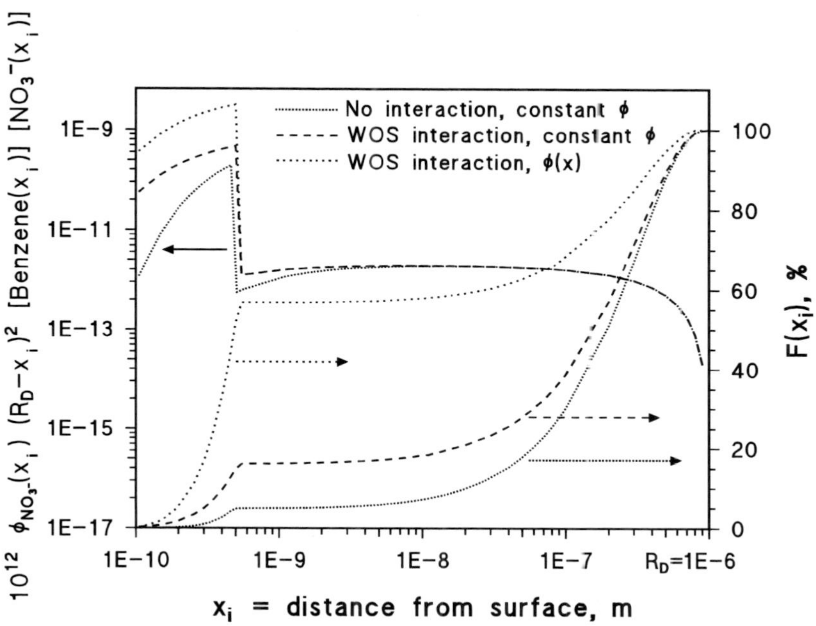

Figure 9. Effects of the surface co-adsorption of benzene and nitrate and of the surface enhancement of the quantum yield of nitrate photolysis on the production of phenol from benzene. It was assumed a 6-fold enhancement of Φ_{NO3-} in a 0.5 nm surface layer.

Conclusions

Photochemical processes at the surface of atmospheric droplets could be very important because different phenomena are likely to increase the effectiveness of the interface as a photochemical reactor, compared to the solution bulk. First of all, irrespective of interface effects, the droplet itself is a favourable environment for photochemistry due to increased irradiation intensity compared to the surrounding gas phase. The multiple reflections at the air-water interface of the droplets would in fact increase the sunlight intensity in the drop by a factor η = 1.3-1.8, depending on irradiation wavelength and droplet diameter. Figure 10 reports the value of η as a function of the droplet diameter for different values of the irradiation wavelength λ. The radiation of lower wavelength, which is more relevant to photochemical processes, shows more significant changes of η as a function of R_D [40].

From the Mie theory it is derived that multiple reflections would not increase the irradiation intensity by the same extent in the whole droplet. Accordingly, the intensity at the surface would be highly enhanced in the direction of illumination at the equatorial plane of the droplet, but considerably decreased at different angles. On average, the irradiation intensity would be slightly lower (around 25%) at the droplet surface compared to the bulk [41].

The irradiation intensity inside droplets could be enhanced at a far higher extent by morphology-dependent resonance phenomena, which are operational for certain combinations of droplet size and irradiation wavelength [39]. However, the resonance would give discrete and narrow spikes in the broader distribution of radiation wavelength and droplet size, and their average importance would be minimal [40,41].

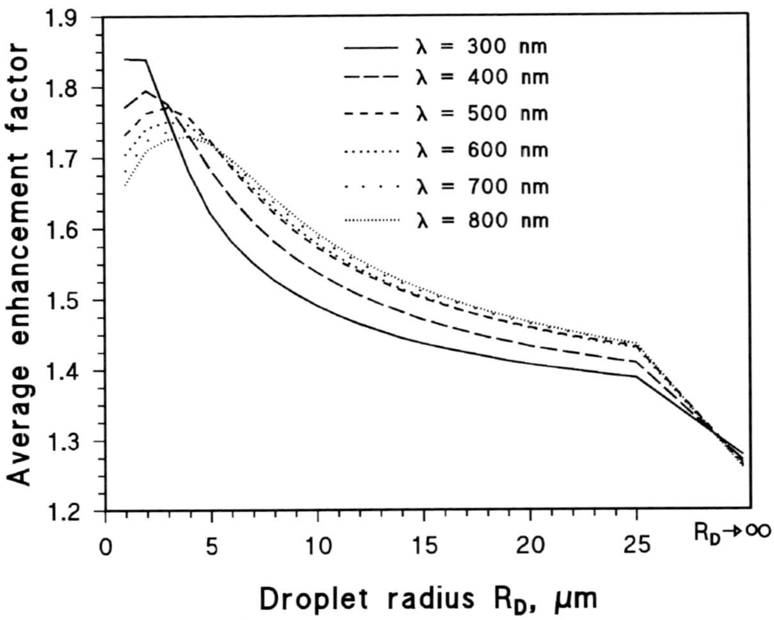

Figure 10. Average enhancement factor η of the irradiation intensity in water droplets, compared to the surrounding gas phase, as a function of the droplet radius R_D and the radiation wavelength λ.

Different phenomena could be able to enhance the rates of photochemical reactions at the interface, compared to the bulk of the droplet. First of all, a number of photochemically active species in water droplets have a greater affinity for the interface [87], as is the case of O_3 and H_2O_2. The latter would undergo an accumulation factor of two at the droplet surface [63], where also some organic substrates are more concentrated than in the solution bulk [44]. The combination of increased concentration of both photoinducer (H_2O_2) and substrate (benzene) would make the reaction in a thin surface layer as important as that in the bulk, the latter accounting for over 99% of the droplet volume.

Inorganic ions that do not undergo surface accumulation themselves can be present at higher concentration at or near the interface due to the interaction with hydrophobic molecules that are accumulated at the surface, as is the case of nitrate and benzene [44]. The so-called co-adsorption phenomenon could give a significant contribution to the interface photochemistry [45].

Finally, at the air-water interface the solvent cage surrounding the photoactive molecules is incomplete. It could lead to a considerable enhancement of photolysis processes at the interface because the recombination of photofragments would be less likely [77]. Nitrate photolysis could undergo a 6-fold enhancement at the interface due to the lack of solvent-cage inhibition [71]. The combined effects of co-adsorption and enhanced quantum yield of $^{\bullet}OH$ photogeneration at the interface could result into higher reactivity at the surface than in the bulk, where the vast majority of the droplet mass is concentrated.

The enhancement of the photolysis quantum yield due to the presence of an incomplete solvent cage could even be higher for other molecules. Enhancement by orders of magnitude has been observed in the case of $Mo(CO)_6$ in 1-decene [41], but a comparable situation could also be present in the environmentally more significant aqueous solution. Studies of laser flash photolysis have shown that the quantum yield of $FeCl^{2-}$ photocleavage is as high as 0.5±0.1 [88]. However, when measuring the reactivity of photogenerated $Cl^{\bullet}/Cl_2^{-\bullet}$ toward the degradation of phenol (C_6H_5-OH), one only finds an effective quantum yield of 4×10^{-3} for the photolysis of $FeCl^{2+}$ [89].

$$FeCl^{2+} + h\nu \rightarrow Fe^{2+} + Cl^{\bullet} \quad (31)$$

$$Cl^{\bullet} + Cl^{-} \rightarrow Cl_2^{-\bullet} \quad (32)$$

$$C_6H_5\text{-}OH + Cl_2^{-\bullet} \rightarrow C_6H_5\text{-}O^{\bullet} + 2\ Cl^{-} \quad (33)$$

The most likely explanation for the observed difference is the efficient recombination of Fe^{2+} and $Cl^{\bullet}$ inside the solvent cage, so that only a small minority of the photogenerated fragments would be able to diffuse into the solution and react with phenol.

Considering that the current models of atmospheric photochemistry only take the bulk reactions into account, they are likely to underestimate the importance of photochemical processes in atmospheric waters. Indeed, recent research advances in different fields suggest that surface accumulation phenomena, co-adsorption and enhanced photolysis quantum yields could make the reactions in a thin surface layer, accounting for less than 1% of the total volume, at least as important as those in the bulk.

Acknowledgments

Financial support by PNRA – Progetto Antartide, Inter-University Consortium of Chemistry and the Environment (INCA, Working group GLOB-CHEM), and CIPE Regione Piemonte (Project A142) is gratefully acknowledged.

References

[1] Atkinson, R. (1998). *Pure Appl. Chem*, 70, 1327-1334.

[2] Vione, D., Maurino, V., Minero, C., Pelizzetti, E., Harrison, M. A. J., Olariu, R. I., Arsene, C. (2006). *Chem. Soc. Rev*, 35, 441-453.

[3] Reyes, C. A., Medina, M., Crespo-Hernandez, C., Cedeno, M. Z., Arce, R., Rosario, O., Steffenson, D. M., Ivanov, I. N., Sigman, M. E., Dabestani, R. (2000). *Environ. Sci. Technol*, 34, 415-421.

[4] Dunstan, T. D. J., Mauldin, R. F., Jinxian, Z., Hipps, A. D., Wehry, E. L., Mamantov, G. (1989). *Environ. Sci. Technol*, 23, 303-308.

[5] Jang, M., McDow, S. R. (1995). *Environ. Sci. Technol*, 29, 2654-2660.

[6] Jang, M., McDow, S. R. (1997). *Environ. Sci. Technol*, 31, 1046-1053.

[7] Warneck, P. (1999). *Phys. Chem. Chem. Phys*, 1, 5471-5483.

[8] Anastasio, C., Faust, B. C., Rao, C. J. (1997). *Environ. Sci. Technol*, 31, 218-232.

[9] Feilberg, A., Nielsen, T. (2001). *Environ. Sci. Technol*, 35, 108-113.

[10] Southworth, B. A., Voelker, B. M. (2003). *Environ. Sci. Technol*, 37, 1130-1136.

[11] Chan, M. N., Chan, C. K. (2003). *Environ. Sci. Technol*, 37, 5109-5115.

[12] Feng, J. S., Moller, D. J. (2004). *Atmos. Chem*, 48, 217-233.

[13] Graber, E. R., Rudich, Y. (2006). *Atmos. Chem. Phys*, 6, 729-753.

[14] Anastasio, C., McGregor, K. G. (2001). *Atmos. Environ*, 35, 1079-1089.

[15] McGregor, K. G., Anastasio, C. (2001). *Atmos. Environ*, 35, 1091-1104.

[16] Alegrìa, A. E., Ferrer, A., Santiago, G., Sepúlveda, E., Flores, W. J. (1999). *Photochem. Photobiol. A: Chem*, 127, 57-65.

[17] Maurino, V., Borghesi, D., Vione, D., Minero, C. (2008). *Photochem. Photobiol. Sci*, 7, 321-327.

[18] Feilberg, A., Nielsen, T. (2000). *Environ. Sci. Technol*, 34, 789-797.

[19] Finlayson-Pitts, B. J., Pitts, Jr. J. N. (1997). *Science*, 276, 1045-1052.

[20] Atkinson, R. (1985). *Chem. Rev*, 85, 69-201.

[21] Atkinson, R., Arey, J. (1994). *Environ. Health Persp*, 102, 117-126.

[22] Atkinson, R., Aschmann, S. M., Arey, J. (1992). *Environ. Sci. Technol*, 26, 1397-1403.

[23] Barletta, B., Bolzacchini, E., Meinardi, S., Orlandi, M., Rindone, B. (2000). *Environ. Sci. Technol*, 34, 2224-2230.

[24] Lammel, G., Cape, J. N. (1996). *Chem. Soc. Rev*, 25, 361-369.

[25] Atkinson, R. J. (1994). *Phys. Chem. Ref. Data*, monograph n° 2.

[26] Atkinson, R., Arey, J. (2003). *Atmos. Environ*, 37, S197-S219 (Suppl. 2).

[27] Knipping, E. M., Lakin, M. J., Foster, K. L., Jungwirth, P., Tobias, D. J., Gerber, R. B., Dabdub, D., Finlayson-Pitts, B. J. (2000). *Science*, 288, 301-306.

[28] Wang, L., Arey, J., Atkinson, R. (2005). *Environ. Sci. Technol*, 39, 5302-5310.

[29] Bolzacchini, E., Bruschi, M., Hjorth, J., Meinardi, S., Orlandi, M., Rindone, B., Rosenbohm, E. (2001). *Environ. Sci. Technol*, 35, 1791-1797.

[30] Vione, D., Maurino, V., Minero, C., Pelizzetti, E. (2005). *Environ. Sci. Technol*, 39, 7921-7931.

[31] Minero, C., Maurino, V., Pelizzetti, E., Vione, D. (2007). *Environ. Sci. Pollut. Res*, 14, 241-243.

[32] Vione, D., Maurino, V., Minero, C., Calza, P., Pelizzetti, E. (2005). *Environ. Sci. Technol*, 39, 5066-5075.

[33] Chiron, S., Minero, C., Vione, D. (2006). *Environ. Sci. Technol*, 40, 5977-5983.

[34] Vione, D., Maurino, V., Cucu Man, S., Khanra, S., Arsene, C., Olariu, R. I., Minero, C. (2008). *ChemSusChem*, 1, 197-204.

[35] Umschlag, T., Zellner, R., Herrmann, H. (2002). *Phys. Chem. Chem. Phys*, 4, 2975-2982.

[36] Harrison, M. A. J., Barra, S., Borghesi, D., Vione, D., Arsene, C., Olariu, R. I. (2005). *Atmos. Environ*, 39, 231-248.

[37] Buxton, G. V., Greenstock, C. L., Helman, W. P., Ross, A. B. J. (1988). *Phys. Chem. Ref. Data*, 17, 513-886.

[38] Chan, C. K., Flagan, R. C., Seinfeld, J. H. (1991). *Appl. Opt*, 30, 459-467.

[39] Ray, A., Bhanti, D. (1997). *Appl. Opt*, 36, 2663-2674.

[40] Mayer, B., Madronich, S. (2004). *Atmos. Chem. Phys*, 4, 2241-2250.

[41] Nissenson, P., Knox, C. J. H., Finlayson-Pitts, B. J., Phillips, L. F., Dabdub, D. (2006). *Phys. Chem. Chem. Phys*, 8, 4700-4710.

[42] Minofar, B., Jungwirth, P., Das, M. R., Kunz, W., Mahiuddin, S. J. (2007). *Phys. Chem. C*, 111, 8242-8247.

[43] Acharid, A., Sadiki, M., Elmanfe, G., Derkao ui, N., Olier, R., Privat, M. (2006). *Langmuir*, 22, 8790-8799.

[44] Sadiki, M., Quentel, F., Elléouet, C., Huruguen, J. P., Jestin, J., Andrieux, D., Olier, R., Privat, M. (2003). *Atmos. Environ*, 37, 3551-3559.
[45] Vione, D., Minero, C., Hamraoui, A., Privat, M. (2007). Atmos. Environ, 41, 3303-3314.
[46] Graham, J. D., Roberts, J. T., Anderson, L. D., Grassian, V. H. J. (1996). *Phys. Chem*, 100, 19551-19558.
[47] Furlan, A. J. (1999). *Phys. Chem*. B, 103, 1550-1557.
[48] Benjamin, I. J. (1991). *Chem. Phys*, 95, 3698-3709.
[49] Winter, N., Benjamin, I. J. (2004). *Chem. Phys*, 121, 2253-2263.
[50] Madronich, S. J. (1987). *Geophys. Res*, 92, 9740-9752.
[51] Mayer, B., Fischer, C., Madronich, S. (1998). *Geophys. Res. Lett*, 25, 4321-4324.
[52] Ruggaber, A., Dlugi, R., Nakajima, T. J. (1994). *Atmos. Chem*, 18, 171-210.
[53] Ruggaber, A., Dlugi, R., Bott, A., Forkel, R., Herrmann, H., Jacobi, H. W. (1997). *Atmos. Environ*, 31, 3137-3150.
[54] Bott, A., Zdunkowski, W. J. (1987). *Opt. Soc. Am. A*, 4, 1361-1365.
[55] Kerker, M. (1969). *The scattering of light and other electromagnetic radiation*. Academic Press: NY, USA.
[56] Wiscombe, W. (1994). Refwat, a program to calculate the complex refraction index of pure liquid water for wavelengths between 0.01 microns and 10.0 m. ftp://climate.gsfc.nasa.gov/wiscombe.
[57] Bohren, C., Huffman, D. (1983). *Absorption and scattering of light by small particles*. Wiley: NY,USA.
[58] Wiscombe, W. (1980). *Appl. Opt*, 19, 1505-1509.
[59] Mayer, B., Schröder, M., Preusker, R., Schüller, L. (2004). *Atmos. Chem. Phys*, 4, 1255-1263.
[60] Barber, P. W., Hill, S. C. (1990). *Light Scattering by Particles: Computational Methods*.World Scientific: Singapore.
[61] Minofar, B., Vacha, R., Wahab, A., Mahiuddin, S., Kunz, W., Jungwirth, P. J. (2006). *Phys. Chem. B*, 110, 15939-15944.
[62] Vacha, R., Jungwirth, P., Chenb, J., Valsaraj, K. (2006). *Phys. Chem. Chem. Phys*, 8, 4461-4467.
[63] Vacha, R., Slavicek, P., Mucha, M., Finlayson-Pitts, B. J., Jungwirth, P. J. (2004). Phys. *Chem. A,* 108, 11573-11579.
[64] Jungwirth, P., Tobias, D. J. J. (2002). *Phys. Chem. B,* 106, 6361-6373.
[65] Hofft, O., Borodin, A., Kahnert, U., Kempter, V., Dang, L. X., Jungwirth, P. J. (2006). *Phys. Chem. B*, 110, 11971-11976.
[66] Salvador, P., Curtis, J. E., Tobias, D. J., Jungwirth, P. (2003). *Phys. Chem. Chem. Phys*, 5, 3752-3757.
[67] Vione, D., Maurino, V., Minero, C., Pelizzetti, E. (2003). *Ann. Chim.* (Rome), 93, 477-488.
[68] Finlayson-Pitts, B. J., Pitts, Jr. J. N. (1986). *Atmospheric Chemistry*. Wiley: NY.
[69] Deister, U., Warneck, P., Wurzinger, C. (1990). *Ber. Bunsenges. Phys. Chem*, 94, 594-599.
[70] Marinoni, A., Laj, P., Sellegri, K., Mailhot, G. (2004). *Atmos. Chem. Phys*, 4, 715-728.
[71] Minero, C., Maurino, V., Bono, F., Pelizzetti, E., Marinoni, A., Mailhot, G., Carlctti, M. E., Vione, D. (2007). *Chemosphere,* 68, 2111-2117.
[72] Berry, R. S., Rice, S. A., Ross, J. (1980). *Physical Chemistry*, Wiley. NY.

[73] Onsager, S., Samaras, N. N. T. J. (1934). *Chem. Phys*, 2, 528-536.
[74] Wagner, V. C. (1924). *Phys. Zeit*, 25, 474.
[75] Buff, F. P., Stillinger, Jr. F. H. J. (1956). *Chem. Phys*, 25, 312-318.
[76] McNeill, V. F., Wolfe, G. M., Thornton, J. A. J. (2007). *Phys. Chem*. A, 111, 1073-1083.
[77] Mack, J., Bolton, J. R. J. (1999). *Photochem. Photobiol. A: Chem*, 128, 1-13.
[78] Lima, J. F., Motonaga, S. N., Iha, N. Y. M. J. (1994). *Photochem. Photobiol. A, Chem*, 84, 177-182.
[79] Traylor, T. G., Magde, D., Marsters, J., Jongeward, K., Wu, G. Z., Walda, K. J. (1993). *Am. Chem. Soc*, 115, 4808-4813.
[80] Thogersen, J., Thomsen, C. L., Poulsen, J. A., Keiding, S. R. J. (1998). *Phys. Chem. A*, 102, 4186-4191.
[81] Crowell, R. A., Lian, R., Sauer, M. C., Oulianov, D. A., Shkrob, I. A. (2004). *Chem. Phys. Lett,* 383, 481-485.
[82] Warneck, P., Wurzinger, C. J. (1988). *Phys. Chem*, 92, 6278-6283.
[83] Bouillon, R. C., Miller, W. L. (2005). *Environ. Sci. Technol*, 39, 9471-9477.
[84] Grätzel, M., Henglein, A., Lilie, J., Beck, G. (1969). *Ber. Bunsenges. Phys. Chem*, 73, 646-653.
[85] Mark, G., Korth, H. G., Schuchmann, H. P., Von Sonntag, C. J. (1996). *Photochem. Photobiol. A: Chem*, 101, 89-103.
[86] Harrison, M. A. J., Heal, M. R., Cape, J. N. (2005). *Atmos. Chem. Phys*, 5, 1679-1695.
[87] Jungwirth, P., Tobias, D. J. (2006). *Chem. Rev*, 106, 1259-1281.
[88] Nadtochenko, V. A., Kiwi, J. (1998). *Inorg. Chem*, 37, 5233-5238.
[89] Khanra, S., Minero, C., Maurino, V., Pelizzetti, E., Dutta, B. K., Vione, D. (2008). *Environ. Chem. Lett*, 6, 29-34.

In: Photochemistry Research Progress
Editors: A. Sánchez, S. J. Gutierrez

ISBN: 978-1-60456-568-3

Chapter 9

PHOTOCHEMISTRY OF TETRAZOLE DERIVATIVES IN CRYOGENIC RARE GAS MATRICES

A. Gómez-Zavaglia[1,2], I. D. Reva [1], L. M. T. Frija [1,3], M. L. S. Cristiano [3] and R. Fausto[1]

[1]Department of Chemistry, University of Coimbra, P-3004-535, Portugal
[2]Faculty of Pharmacy and Biochemistry, University of Buenos Aires, C.P. 1113, Argentina
[3]Department of Chemistry and Biochemistry, F.C.T., and CCMAR, University of Algarve, Campus de Gambelas, 8005-039 Faro, Portugal

ABSTRACT

In this chapter, besides a brief description of the matrix isolation technique and discussion of its main advantages and drawbacks when applied to photochemical studies on organic molecules, a general overview of the photochemistry of tetrazoles and the general pattern of their photoreactions is presented.

The application of tetrazoles in agriculture and medicine is widespread and well known. In many drugs and drug-candidates, replacement of carboxylic acid groups, −COOH, by tetrazolic acid fragments, $-CN_4H$, is frequent, resulting in an increase in bioavailability and potency of the pharmacophore. Tetrazoles also have many other applications, such as in photography and photoimaging, and in the automobile industry as gas generating agents for airbags. These compounds also exhibit a very rich photochemistry, which is strongly influenced by the substituents present in the tetrazolic ring.

Several representative tetrazoles were trapped in a rigid environment of solidified noble gases (argon, xenon) at cryogenic temperatures (typically 10 K) and subjected to in situ photolysis. The range of possible rearrangements of the isolated species was controlled by choice of the excitation wavelengths. FT-IR spectroscopy provided experimental frequencies and intensities of characteristic absorptions of the isolated chemical species, both for reagents and photoproducts. The analysis of experimental data was assisted by their direct comparison with the vibrational spectra theoretically calculated for the single molecule in vacuum. This comparison was facilitated because the inert matrix only slightly affects the structure of the isolated molecules, and also owing to the high resolution of infrared matrix spectra (a few tenths of cm^{-1}). UV-excitation resulted in photofragmentation of the monomeric tetrazoles with a wide range of exit channels. Since the obtained fragments in general stay in the matrix cage where they are formed, no subsequent cross-reactions involving species resulting from photolysis of different reactant molecules can occur, strongly reducing the number of possible photoproducts in comparison with gas phase or solution studies. This fact introduced a useful simplification for the interpretation of the reaction mechanisms.

A number of relatively unusual or highly reactive molecules, such as antiaromatic azirines, azides, isocyanates or isothiocyanates, may be formed from photolysis of tetrazoles in a matrix. In some examples described in this chapter, the spectroscopic characterization of these molecular species was presented for the first time.

1. INTRODUCTION

Matrix isolation is a technique whereby gaseous atoms or molecules are trapped in an environment of solidified inert gases at temperatures close to absolute zero [1-4]. Under these conditions, intramolecular rearrangements are stopped for any process with an activation barrier larger than a few kJ mol^{-1}. On the other hand, because the matrix rigidity prevents diffusion of the sample, in general bimolecular reactions are also suppressed. Preservation for long time of unstable or usually short-lived species can then be achieved, which allows for their detailed study using conventional spectroscopic techniques [1].

One additional advantage of this methodology is that the technique of detection can easily be tailored to the problem under investigation, due to the spectroscopic transparency of the inert gas host. Vibrational spectroscopy, in particular, becomes a very powerful tool for studies of the structure and reactivity of moderate-size organic molecules when combined with low temperature matrix isolation. Infrared spectroscopy provides direct data on frequencies and intensities of characteristic absorptions of the isolated species, and matrix isolation facilitates obtaining this information, due to the high resolution of infrared matrix spectra and high sensitivity of the method. In the spectra of matrix-isolated compounds, the effect of inhomogeneous band broadening nearly vanishes in comparison with the condensed phase, and unlike the gaseous state the fine rotational structure is, in general, completely absent. Furthermore, unlike for solutions, it is possible to neglect the effect of the environment (inert matrix) on the structure of the molecules under investigation. These factors result in a unique narrowing of bands in vibrational spectra (down to a few tenths of cm^{-1}), which makes possible to detect the band shifts of a few wavenumbers that are characteristic for the finest changes of sample's structure. Moreover, because the inert matrix only slightly affects the structure of the isolated molecules it is possible to compare directly matrix experimental results and theoretically calculated spectra, which most of times are obtained for the single molecule in vacuum.

One of the most important advantages of matrix isolation spectroscopy is the possibility to follow photochemical reactions unambiguously. For a matrix isolated compound, in situ photolysis enables selective introduction of energy in the molecule under study, controlling the range of possible rearrangements of the isolated species. Thus, a wide range of intramolecular changes, up to molecule decomposition and formation of new entities, can be studied. In addition, contrarily to what occurs in the gaseous phase or in solution, where a multitude of possible photochemical reaction pathways leading to different products can be observed, in a matrix the processes are cage-confined (molecular diffusion is inhibited) and therefore, a useful simplification to the photochemical reactivity is obtained. For example, if photofragmentation of a matrix-isolated species occurs, most of the time the obtained fragments stay in the matrix cage where they are formed. Then, no subsequent cross-reactions involving species resulting from photolysis of different reactant molecules can occur, strongly reducing the number of possible photoproducts in comparison with gas phase or solution studies. This fact may considerably facilitate the interpretation of the reaction mechanisms.

We have been using the matrix-isolation method for many years, together with infrared spectroscopy, to study the photochemistry of several families of molecules with biological or industrial relevance. Among these, tetrazoles have been considered in great detail in the last few years.

The relevance of tetrazolic compounds is widespread because of their applications in fields such as agriculture and medicine [5-9]. This relevance is mainly due to the fact that the tetrazolic acid fragment, $-CN_4H$, has similar acidity and volume to the carboxylic acid group, $-COOH$, but it is metabolically more stable [5]. For this reason, the investigation of tetrazolic compounds has become a research area of major interest [6].

From a more fundamental point of view, tetrazoles are also particularly interesting molecules because they exhibit a very rich photochemistry [10-18]. Two main factors contribute to enrich photochemistry of tetrazoles: the possibility of tautomerism (which is associated with the presence in the molecule of labile hydrogen atoms) and conformational flexibility of the substituents.

Tautomerism has been found to be important in the tetrazole family in general. For example, hydrogen atoms directly bound to the tetrazole ring are labile and give rise to different tautomers, whose relative stability may strongly depend on the environment. Tetrazole itself occurs exclusively as the 1H-tautomer in the crystalline phase and mostly as the 2H-tautomer in the gaseous phase (~90% of the total population at 90°C); in solution, both tautomers coexist, with the population of the most polar 1H-form increasing with the polarity of the solvent [19-23]. The infrared spectrum of unsubstituted tetrazole (CN_4H_2) isolated in argon matrix (T= 10 K) was investigated in our research group, showing essentially the expected signature of the 2H-tetrazole tautomer [10]. The amount of the 1H-tautomer in the matrix was found to be ca. 10% of that of the most abundant form. This study has solved a long-term controversy regarding the existence of the 1H-tautomer of tetrazole in the gaseous phase [10, 24-26].

In the case of substituted tetrazoles, tautomerism may also involve substituent groups, as it was found, for example, in the case of 1-methyl-tetrazol-5-thione and 2-methyl-tetrazol-5-amine [13, 14]. In general, the presence of labile hydrogen atoms (either directly linked to the tetrazole ring or belonging to the tetrazole substituents) is a source of complexity in photochemical reactions, that opens additional reaction channels or allows for secondary photochemical reactions to take place concomitantly with the main primary photoprocesses [9, 13, 14, 27-29].

When substituents are linked to the tetrazole ring, the photochemistry of the molecule can also be influenced by the conformational flexibility of the substituents, which may favor or exclude certain reaction channels [13, 16]. For this reason, it has been shown that the properties of the substituents are very relevant in determining the precise nature and relative amount of the final photoproducts, then making tetrazoles a permanent challenge for investigation [9, 13, 14, 27-29].

The photochemistry of matrix-isolated unsubstituted tetrazole was studied for the first time by Maier and coworkers [9]. Upon photolysis with the 193 nm emission line of an ArF laser, rapid photocleavage of tetrazole was observed, leading to extrusion of N_2 and formation of several different photoproducts, including nitrilimine, carbodiimide, diazomethane, an HCN···NH complex and cyanamide [9]. By use of different excitation wavelengths, cyanamide and carbodiimide could be accumulated as final products. Meier's group study also revealed, for the first time, the vibrational signature of matrix-isolated nitrilimine [9]. In our more recent studies on tetrazoles, a general pattern of reaction of this family upon UV excitation could be defined [13-18]. Besides, the observation of a number of relatively unusual or highly reactive molecules, formed from photolysis of the studied tetrazoles, such as antiaromatic azirines, azides, isocyanates or isothiocyanates, led, in some cases, also to the first spectroscopic characterization of these species.

In the following sections, a general overview of the photochemistry of matrix-isolated tetrazole derivatives, based mainly on our previous studies on this family of molecules, is presented.

2. Photochemistry of Matrix-Isolated 5-Methoxy-1-Phenyl-1H-Tetrazole (5MPT)

Among the compounds investigated in our laboratories, 5-methoxy-1-phenyl-1H-tetrazole (5MPT) was found to exhibit a less complex photochemistry. This simplicity resulted essentially from two facts: 5MPT adopts only one tautomeric form and exists as a single conformer [16].

Upon UV irradiation (□>235 nm), 5MPT was found to react only according to two pathways (Figure 1): a) N2 elimination, with production of the antiaromatic 3-methoxy-1-phenyl-1H-diazirine (MPD), which was observed experimentally for the first time in our investigation [16]) and b) tetrazole ring-opening, leading to phenylazide and methylcyanate. The latter compound was found to quickly convert to its more stable isomer methylisocyanate, as it was observed in previous studies [30-33]. Phenylazide further eliminated molecular nitrogen to give phenylnitrene that underwent subsequent ring expansion to form the seven-membered ring 1-aza-1,2,4,6-cycloheptatetraene (ACHT, which can be easily identified by the observation of a very characteristic IR band at ca. 1913 cm−1). Identification of phenylazide was straightforward, since the vibrational spectrum of matrix-isolated phenylazide is well-known [27, 34, 35].

Due to this relative simple photochemistry, the photolysis of matrix-isolated 5MPT could be easily followed and investigated in detail. The obtained results [22] have been precious for the interpretation of the UV-induced reactivity of other tetrazole molecules, which exhibit much more complex photochemistries [14, 15, 17, 18].

In 5MPT, both observed photoprocesses involve the cleavage of the (N-3)–(N-4) bond. In channel a (Figure 1), the cleavage of the (N-1)–(N-2) bond occurs in addition. Channel b, which was found to correspond to the preferred reaction channel [16], also implies disruption of the (C-5)–(N-1) bond (Figure 1). According to DFT(B3LYP)/ 6-311++G(d,p) calculations performed on 5MPT, these are the three longest bonds in the tetrazole ring (calculated lengths are longer than 135.5 pm) [16]. In turn, the (N-2)=(N-3)

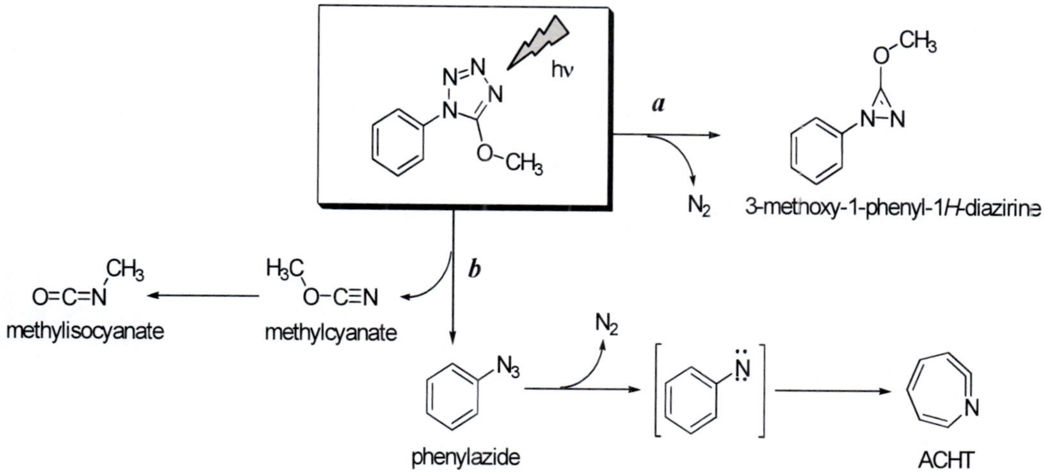

Figure 1. Photochemical reactions observed for 5-methoxy-1-phenyl-1H-tetrazole (5MPT) isolated in cryogenic inert matrices upon irradiation with UV (> 235 nm) light.

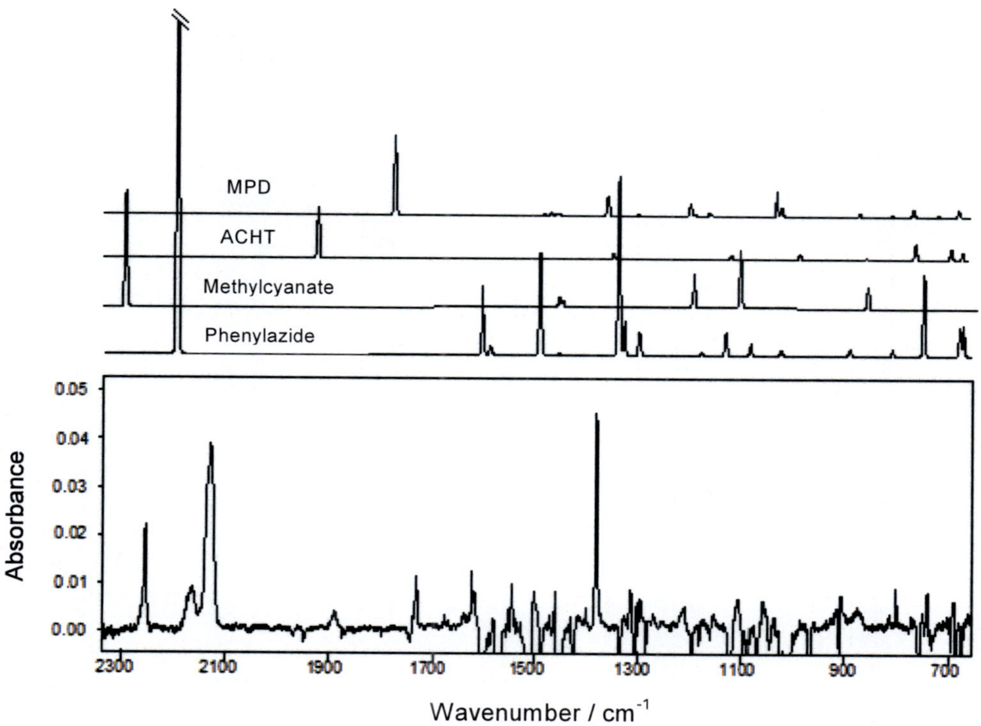

Figure 2. Bottom: Changes in infrared spectrum of 5MPT trapped in an argon matrix induced by UV (λ>235 nm) irradiation (spectrum of UV-irradiated (80 min) sample minus spectrum of freshly deposited matrix). Upper frames: B3LYP/6-311++G(d,p) calculated spectra of the observed photoproducts. In the theoretical spectra, intensities were scaled by different factors, in order to better simulate the experimental spectrum presented in the figure. Reproduced from [16] with permission.

and (C-5)=(N-4) bonds were calculated to be considerably shorter (128.0 and 131.3 pm, respectively). Hence, the bonds that are cleft in the photochemical reactions correspond to the weaker bonds, which are formally single bonds. Moreover, both the HOMO and the LUMO of 5MPT have -character (Figure 3). According to the DFT(B3LYP)/ 6-311++G(d,p) calculations, the vertical $\pi^* \leftarrow \pi$ transition corresponding to a single excitation from the HOMO to the LUMO requires a wavelength of ca. 220 nm (estimated from the LUMO−HOMO energy difference), which doubtlessly indicates that this corresponds to the reactive excitation. Very interestingly, contrarily to the HOMO orbital, which has a bonding character on the (C-5)–(N-1) bond, the LUMO is anti-bonding on this bond, thus explaining why it is so easy to break this bond upon photochemical excitation at $\lambda > 235$ nm. Note that the LUMO orbital does also have an anti-bonding character on the (N-1)−(N-2) bond, which is cleaved in the alternative reaction path instead of (C-5)−(N-1) (see Figure 3). The preference for the reaction path leading to phenylazide and methylcyanate, when compared with that leading to molecular nitrogen plus MPD, is also in agreement with the reaction energies (94.0 and 162.6 kJ mol^{-1}, respectively [16]), which clearly favor the first process.

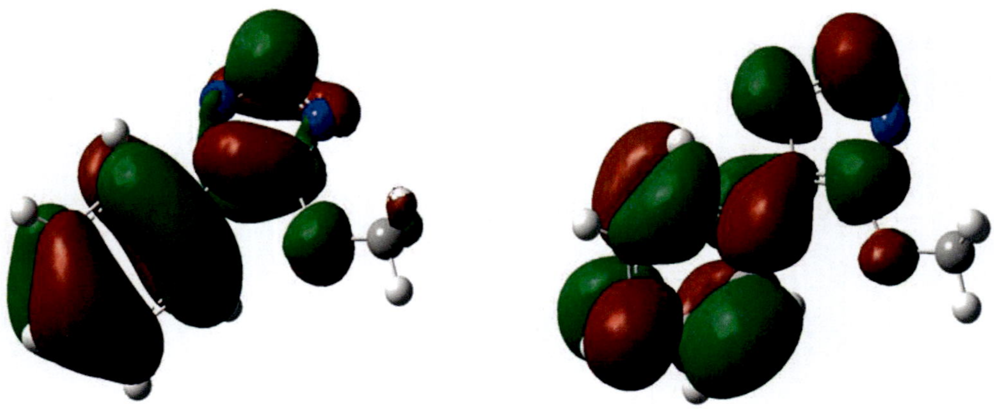

HOMO **LUMO**

Figure 3. HOMO and LUMO of 5MPT, calculated at the DFT(B3LYP)/6-311++G(d,p) level of theory. The isosurface corresponds to the electronic density of 0.02 e.

4. Photochemistry of Matrix-Isolated 5-Ethoxy-1-Phenyl-1H-Tetrazole (5EPT)

5-Ethoxy-1-phenyl-1H-tetrazole (5EPT) differs from 5MPT in the replacement of the methoxy by the ethoxy substituent. As it could be expected, the photochemistries of these two molecules are similar. On the other hand, the conformational flexibility of the ethoxy group of 5EPT could at first be thought to be a source of complexity to the analysis of the experimental data. Indeed, 5EPT possesses 3 conformational isomers (T, G and G'), which were predicted by DFT(B3LYP)/6-311++G(d,p) calculations [18] to have relative energies within the 0-3 kJ mol^{-1} range and, then, are all significantly populated in the gas phase prior to deposition. Fortunately, these conformers are separated by low energy barriers (less than 4 kJ mol^{-1}) on the ground-state potential energy surface and, during deposition of the matrix, the higher energy forms relax to the most stable conformer (this phenomenon is known as conformational cooling and has been studied in detail in our laboratory [36-38]). Hence, only the most stable conformer of 5EPT remains trapped in the as-deposited matrix. This conformer (T form) has the ethyl group placed nearly in the plane of the tetrazole ring and as far away as possible from the phenyl group, the dihedral angle between the two rings (phenyl and tetrazole) was calculated to be ca. 30° [18].

Similarly to 5MPT, UV irradiation (λ > 235 nm) of matrix-isolated 5EPT gives rise to different primary photochemical reactions, which are equivalent to those observed for the first molecule: a) N_2 elimination, leading to production of the anti-aromatic 3-ethoxy-1-phenyl-1H-diazirine (EPD), a photoproduct observed and spectroscopically characterized for the first time in our study [18] and b) tetrazole ring-opening, leading to ethylcyanate and phenylazide (as found for the experiments on 5MPT, phenylazide was also shown to subsequently release N_2, forming phenylnitrene that later rearranges to ACHT). As for 5MPT, the pathway leading to ethylcyanate and phenylazide prevails, and, as for 5MPT, only formally single bonds are cleft in the photoinduced reactions of 5EPT.

It is worth noting that both the diazirine and ethylcyanate photoproducts were observed to exist in the photolyzed matrix in a single conformation (the trans isomer, where the C−O−C−C dihedral is ca. 180°; see Figure 4) despite they possess more than one low energy conformational state. Indeed, ethylcyanate has one intramolecular degree of freedom, corresponding to internal rotation around the central (NC)O−C_2H_5 bond, which may result in the existence of different conformers (trans and gauche). However, DFT(B3LYP)/6-311++G(d,p) calculations indicate that the energy barriers separating the less stable gauche conformers from the conformational ground state are very low (less than 4 kJ mol^{-1} [18]). Then, similarly to the matrix-isolated reactant (5EPT), the photochemically produced ethylcyanate can also undergo conformational cooling, justifying the sole observation in the irradiated matrices of its most stable trans conformer.

Similarly to the general picture above described for ethylcyanate, three conformers separated by low intramolecular energy barriers (less than 3 kJ mol^{-1}) were also predicted by the DFT(B3LYP)/6-311++G(d,p) calculations for EPD. Once again, occurrence of conformational cooling in this species can easily explain the exclusive observation of the most stable conformer in the 5EPT photolyzed matrix.

It is also interesting to mention that ethylcyanate is known to isomerize readily at room or higher temperatures to ethylisocyanate [39] and the process is believed to be autocatalyzed [39, 40]. Because of its instability, ethylcyanate can only be stored at low temperature and its application in chemical reactions requires its generation in situ prior to use [39-42]. For the same reason, this compound had not been the subject of previous detailed experimental structural or vibrational studies and was only characterized in some detail in our investigation [18]. Very interestingly, no evidence of isomerization of ethylcyanate to ethylisocyanate was found in our study, indicating that under matrix-isolation conditions ethylcyanate is stable regarding the photochemical (λ>235 nm) isomerization to the isocyanate.

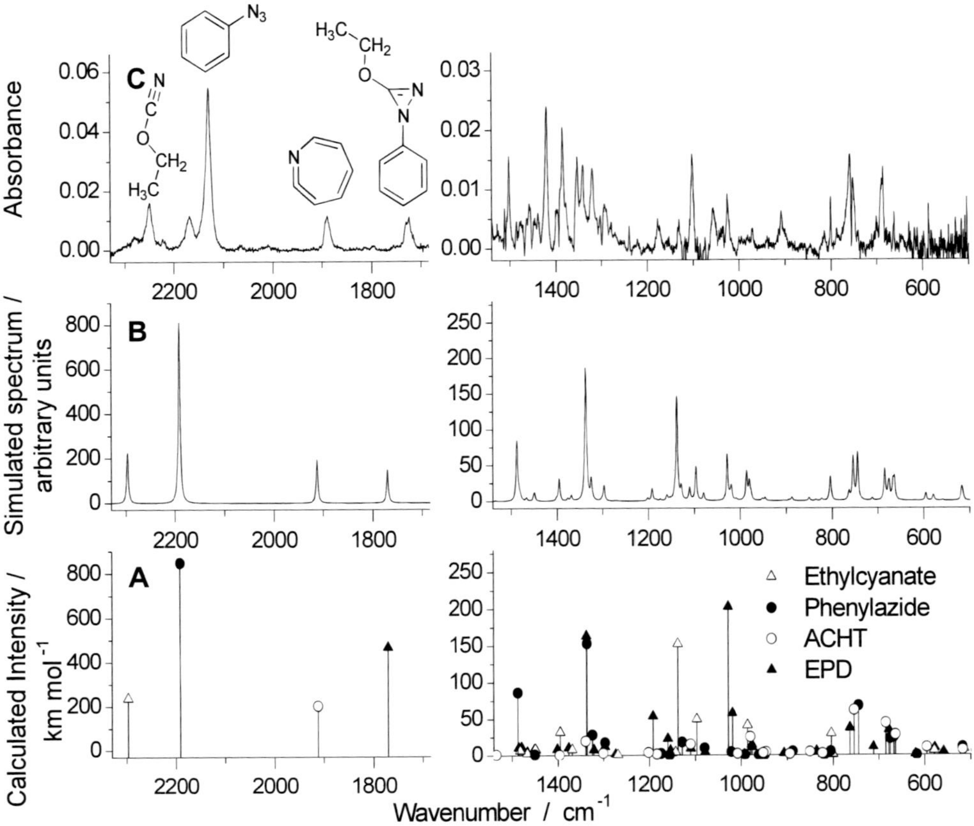

Figure 4. A: DFT(B3LYP)/6-311++G(d,p) calculated spectra of the observed photoproducts of 5EPT (scaled with a uniform factor of 0.978). B: Simulated spectrum for the mixture of photoproducts. The calculated intensities in the sum spectrum relate in the proportion 1:1:1:0.33, where 0.33 refers to EPD and unities refer to ethylcyanate, phenylazide and ACHT. C: Extracted spectrum of the photoproducts formed after 60 minutes of UV-irradiation (λ> 235 nm) of 5EPT trapped in an Ar matrix. The extracted spectrum was obtained by subtraction of the scaled spectrum of non-irradiated matrix from the spectrum of irradiated sample. The scaling factor was chosen so that the absorptions due to the originally deposited compound (5EPT) were nullified. Adapted from [18] with permission.

4. Photochemistry of Matrix-Isolated 1-Phenyl-Tetrazolone (PT)

1-Phenyl-tetrazolone (PT) can exist in five tautomeric forms: two keto tautomers, one mesoionic olate form and two hydroxy conformers (Figure 5). 1-phenyl-1,4-dihydro-5H-tetrazol-5-one (keto tautomer "A" in Figure 5) was predicted to be considerably more stable than all the remaining species, due to the existence in this tautomer of more favorable interactions between the phenyl hydrogen atoms ortho to the tetrazole ring and the N-2 and carbonyl oxygen atoms [15]. Because of the much greater stability of the "A" tautomer of PT, this is the only species populated in the gas phase at room temperature. Then, tautomer "A" is the unique species initially present in the matrix-isolated PT. This fact strongly simplifies the interpretation of the results of the subsequent photochemical experiments.

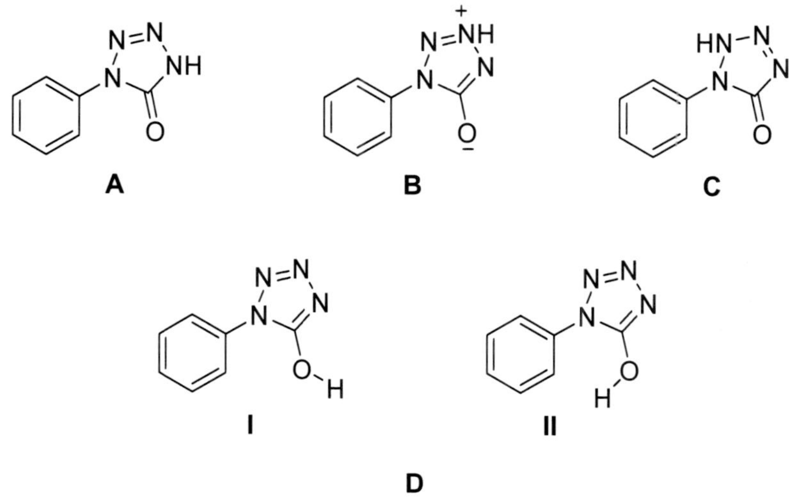

Figure 5. Tautomers of 1-phenyltetrazolone (PT). A: 1-phenyl-1,4-dihydro-5H-tetrazole-5-one; B: 1-phenyl-1H-tetrazole-3-ium-5-olate; C: 1-phenyl-1,2-dihydro-5H-tetrazole-5-one; D: 1-phenyl-1H-tetrazole-5-ol (two conformers, I and II).

Compared to 5MPT and 5EPT, the photochemistry of 1-phenyl-tetrazolone (PT) is rather complex. The main observed photoreactions induced by UV light (□>235 nm) in matrix-isolated PT are summarized in Figure 6. They are: a) extrusion of molecular nitrogen, giving 1-phenyl-diaziridin-3-one; b) elimination of HNCO with production of phenylazide; c) tetrazole ring-opening, leading to phenylisocyanate and azide; and d) loss of CO, leading to the final production of phenyldiazene and N_2, with all probability occurring via the phenyltetrazete intermediate.

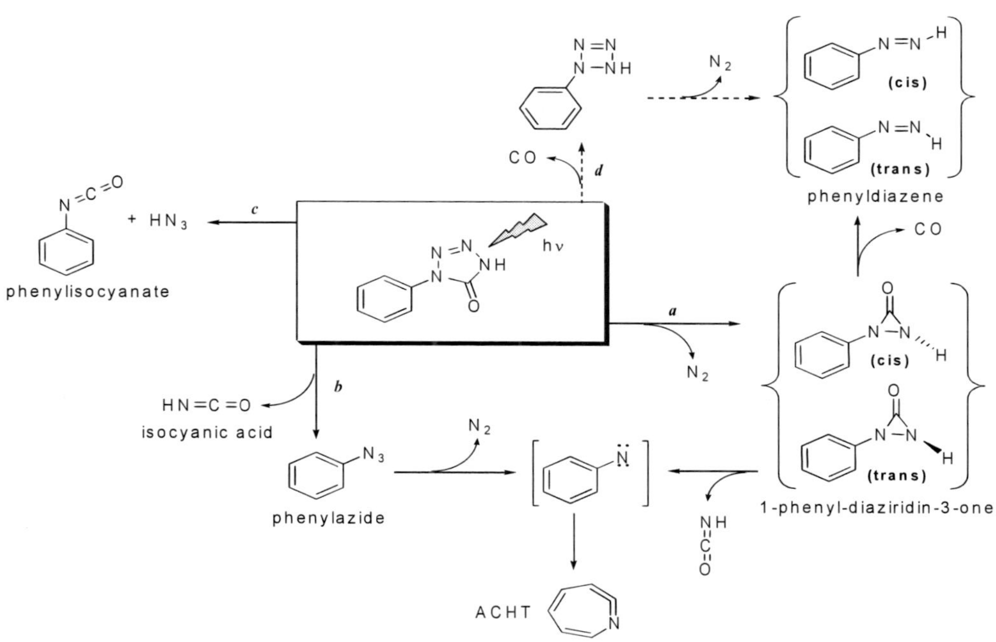

Figure 6. Photochemical reactions observed for 1-phenyl-tetrazolone isolated in cryogenic inert matrices upon irradiation with UV (λ> 235 nm) light.

Pathway a, leading to the production of 1-phenyl-diaziridin-3-one, is similar to pathway a in 5MPT and 5EPT. The photoproduced diaziridinone adopts two different conformers, which were characterized for the first time in our study [15]. Theoretical calculations indicated that the conformer where the phenyl group and the diaziridinone-ring hydrogen atom are trans to each other is more stable by 14.5 kJ mol^{-1} than the one where these groups are cis to each other. The most intense bands of the diaziridinone isomers correspond to the νC=O and νC−N antisymmetric stretching vibrations and one of the phenyl ring out-of-plane bending modes (γC−H ring 1). All the bands corresponding to these vibrations could be easily identified in the IR spectra of the photolyzed matrix [conformer trans: 1931.9/1926.0/1877.8/1874.3 cm^{-1} (site-split Fermi resonance doublets resulting from interaction of νC=O with the first overtone of νC−N, see Figure 7), 964.5/962.8 (νC−N) and 771.0 cm^{-1} (γC−H ring 1); calculated values: 1931.2, 945.5 and 756.2 cm^{-1}; conformer cis: 1913.1/1899.4/1865.8/1862.7 (Figure 7), 924.0/922.8 and 695.3 cm^{-1}, respectively; calculated values: 1921.2, 910.0 and 694.2 cm^{-1}].

The bands due to the diaziridinone are relatively easy to identify because their intensity increased only during the first stages of irradiation (i.e., until ca. 30 minutes of irradiation). Subsequent irradiation led to a decrease of intensity of these bands, which was accompanied by the appearance in the spectra of new bands (and subsequent progressive growth) that could be assigned to secondary photoproducts resulting from photolysis of the diaziridinone. Two different pathways could be postulated for photodegradation of the diaziridinone, the first one yielding phenyldiazene (in two different conformers) plus CO, and the second one leading to production of phenylnitrene and isocyanic acid. The phenylnitrene underwent the usual subsequent ring expansion to ACHT [27, 34, 35, 43-45]; which gives rise to the characteristic bands observed at 1894.2 and 1889.8 cm^{-1} [43-45] that grew mostly at later stages of irradiation (See Figure 7).

Pathway b leads to production of phenylazide and isocyanic acid (HNCO) and is similar to pathway b in 5MPT and 5EPT. The experimental identification of isocyanic acid was complicated because the most intense bands of this compound are nearly coincident with bands due to other photoproducts, in particular phenylisocyanate, which is formed in pathway c and is one of the major observed photoproducts. Nevertheless, bands assigned to HNCO were observed at ca. 2263 (buried under the intense group of bands due to νNCO as of phenylisocyanate), 824 and 630 cm^{-1}, in fairly good agreement with the DFT(B3LYP)/6-31++G(d,p) calculated frequencies for this molecule (2287, 850 and 627 cm^{-1}, respectively) [15]. Since the bands assigned to phenylazide still continued to grow considerably after 60 min of irradiation, it could be concluded that in this case phenylazide can undergo its usual subsequent reactions (as mentioned before, phenylazide tends to release N_2 and form phenylnitrene and then ACHT) only with relatively low efficiency.

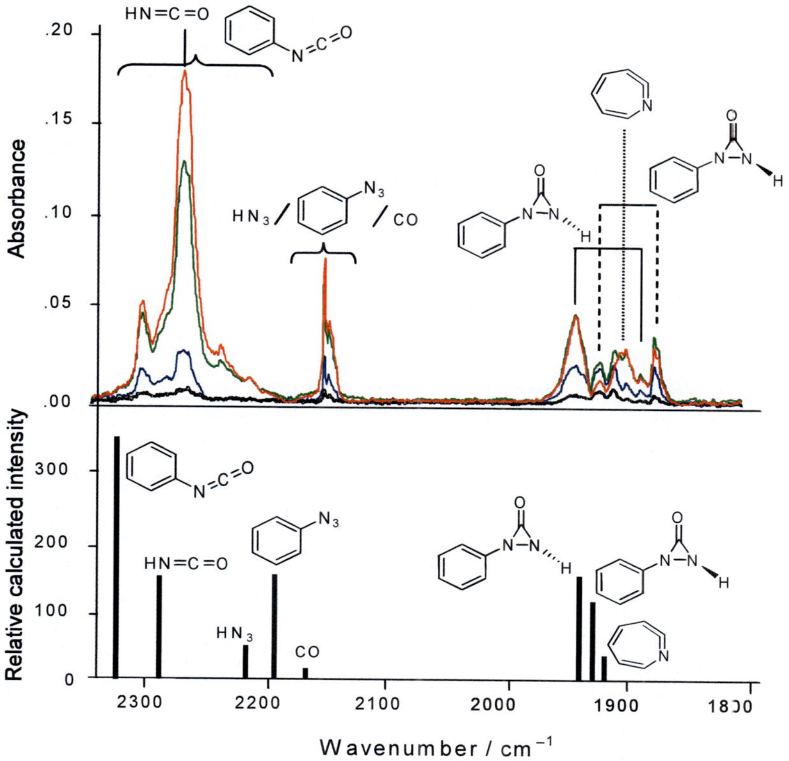

Figure 7. 2350-1800 cm–1 spectral region of the irradiated (λ> 235 nm) matrix of 1-phenyl-tetrazolone at different times of irradiation [2 (black line), 10 (blue), 30 (green) and 60 (red) minutes] and B3LYP/6-31++G(d,p) calculated spectra, in this spectral range, for the observed photoproducts.

Pathway c corresponds to tetrazole ring opening leading to phenylisocyanate and azide (see Figure 6). No indications were found that these compounds underwent additional reactions under the experimental conditions used [15]. The experimental spectrum of phenylisocyanate was easily identifiable because it shows a very intense multiplet in the 2290-2260 cm^{-1} region (Figure 7), due to the νNCO antisymmetric stretching vibration, that clearly dominates the spectrum of the photolyzed matrix. According to the theoretical predictions, the intensity of such unusually intense band represents about 90% of the total intensity of the spectrum of phenylisocyanate. The bands corresponding to azide could also be easily identified in the spectra of the irradiated matrix by comparison with the spectrum of the matrix-isolated azide previously reported by Himmel et al. [46].

A fourth possible reaction pathway (*pathway d*; see Figure 6) could not be eliminated (though it could not be doubtlessly proved in our study [15]). This reaction channel can lead to production of phenyldiazene through initial elimination of CO (to yield phenyltetrazete) followed by N_2 extrusion.

It is interesting to notice here that for 5MPT, 5EPT and PT all the primary photochemical reactions imply the simultaneous cleavage of two single bonds in the tetrazole ring of the reactant molecule. Compared to PT, the number of single bonds in both 5MPT and 5EPT tetrazole rings is reduced from 4 to 3, and the number of observed primary photochemical processes decreases from 3 (or 4, depending if pathway d in PT is effectively active or not) to

2. Indeed, as it will be shown in the next sections, this trend appears to be a general rule for tetrazole derivatives.

5. Photochemistry of Matrix-Isolated 2-Methyl-2H-Tetrazol-5-Amine (2MTA)

When the substituent at the tetrazole ring carbon atom is an amino or imino group, photolysis can result in the production of small C, N and H containing molecules of important astrophysical and/or industrial interest (e.g., CN_2H_2, CN_3H or CNH_3 isomers). Since in our studies we wanted to avoid the complexity introduced by the presence of a labile hydrogen atom at the tetrazole ring, 2-methyl-2H-tetrazol-5-amine (2MTA) was selected as a simple prototype molecule having the adequate characteristics for studying the photochemistry of amino/imino containing tetrazoles and their photoproducts as well as their structural and vibrational properties.

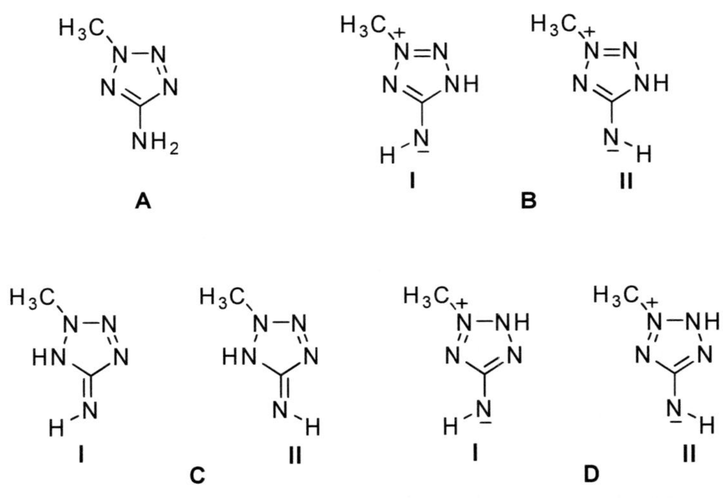

Figure 8. Structures of 2-methyl-2H-tetrazol-5-amine (A) and of its imine (aminide) tautomers: 3-methyl-1H-tetrazol-3-ium-5-aminide (B), 2-methyl-1,2-dihydro-5H-tetrazol-5-imine (C) and 2-methyl-3H-tetrazol-2-ium-5-aminide (D). In the case of the aminide tautomers, the placements of the charges on atoms correspond to one of the possible canonic forms for the mesoionic structures, which were also used to name the compound.

2MTA may exist in different tautomeric forms (Figure 8). The hydrogen shift from the amino group to three different positions of the tetrazole ring leads to three possible pairs of tautomers bearing an imine (or aminide) NH substituent at the position 5 [13]. Taking as reference the atom numbering of the tetrazole ring in 2MTA, the hydrogen involved in the tautomeric rearrangement can occupy positions 1, 3 and 4 of the ring. In the first case tautomerism leads to the imine species, 2-methyl-1,2-dihydro-5H-tetrazol-5-imine (structures C in Figure 8), which can exist in two different conformations, depending on the orientation of the hydrogen atom in the imino group at position 5 [13]. The remaining two cases lead to mesoionic-type structures, 2-methyl-3H-tetrazol-2-ium-5-aminide (Figure 8, structure D) and

3-methyl-1H-tetrazol-3-ium-5-aminide (Figure 8, structure B), respectively, that may also exist in two different stable conformations [13].

The most stable tautomer of 2MTA was found to have a planar tetrazole ring, with both the methyl carbon and amine nitrogen atoms lying in the plane of the heteroaromatic ring and the amine group pyramidalized [the calculated H−N(−C)−H dihedral angle was predicted to be ca. 136°]. One of the methyl hydrogens is also in the ring-plane, with the C−H bond eclipsing the (N-2)–(N-3) bond, whereas the two remaining methyl hydrogen atoms occupy nearly symmetric positions above and below the plane of tetrazole ring.

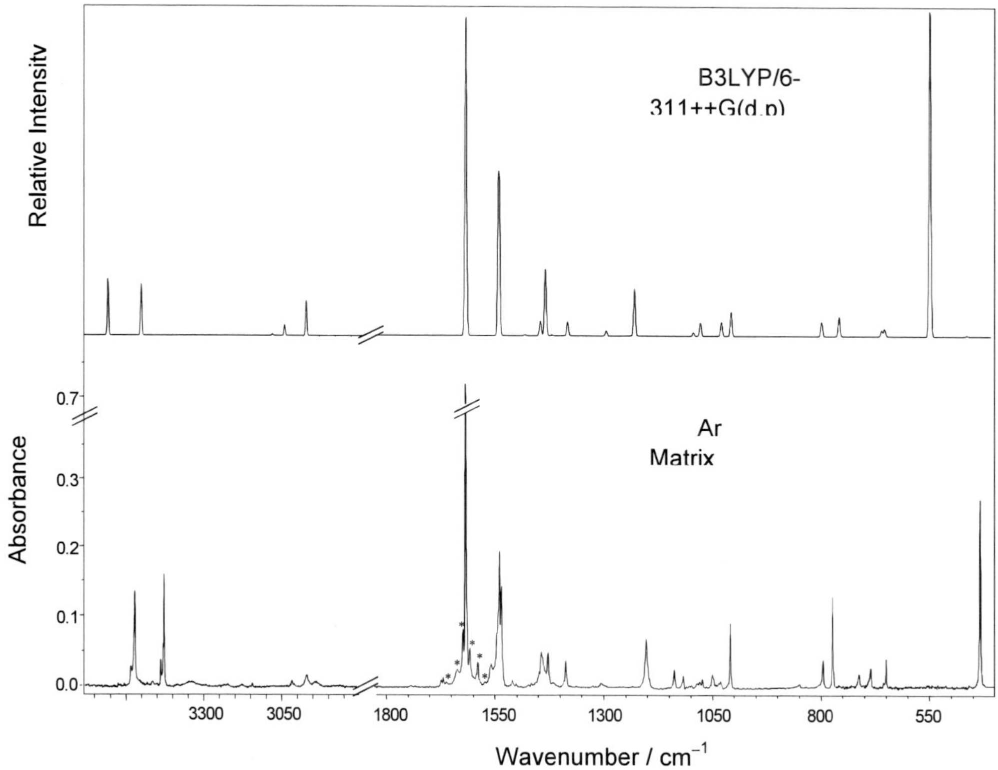

Figure 9. Infrared spectrum of 2MTA in argon matrix and DFT(B3LYP)/6-311+G(d,p) calculated spectrum. Calculated frequencies were scaled by 0.978. In the experimental spectrum, the asterisks indicate bands due to traces of isolated monomeric water. Reproduced from [13] with permission.

The experimental spectrum of 2MTA isolated in an argon matrix fits well the spectrum of tautomer A calculated at the DFT(B3LYP)/6-311++G(d,p) level of theory and no evidence of presence of any of the imine (aminide) tautomers was found (Figure 9). This observation is in agreement with the theoretically predicted relative energies of tautomers B, C, D: all of them are less stable than tautomer A by more than 100 kJ mol^{-1} and are not populated in the gaseous phase. [13]

In situ UV irradiation ($\lambda > 235$ nm) of matrix-isolated 2MTA induces three main primary photochemical processes. Two of them correspond to the prototype reactions found for other tetrazoles, i.e., a) N_2 elimination, with production of 1-methyl-1H-diazirine-3-amine, and b) ring cleavage, leading to production of methyl azide and cyanamide. The third observed reaction [pathway c); Figure 10] was tautomerization of 2MTA to mesoionic 3-methyl-1H-tetrazol-3-ium-5-aminide (tautomer B, Figure 8) via [1,3]-hydrogen shift.

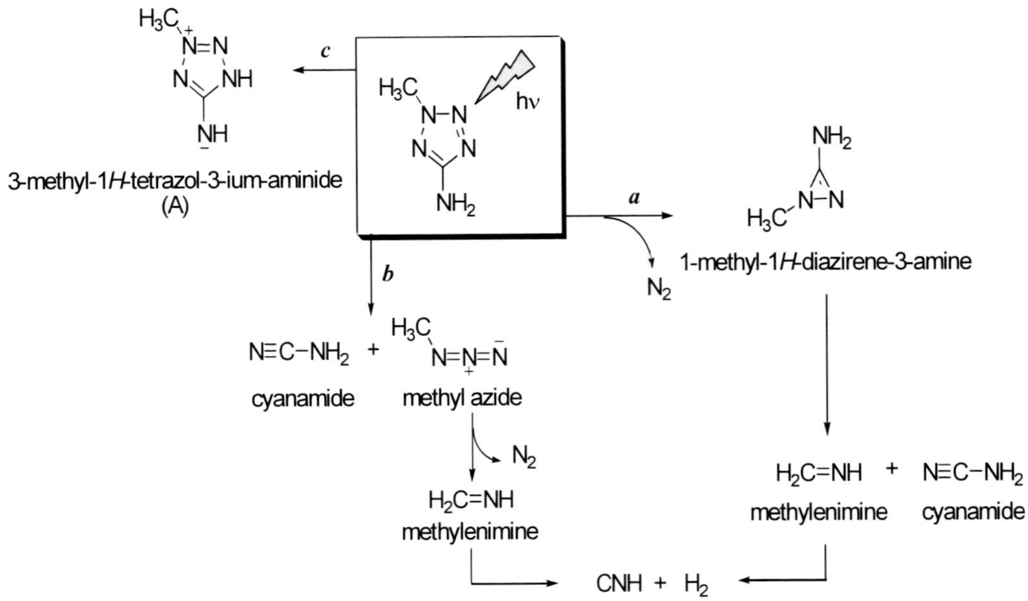

Figure 10. Photochemical reactions observed for 2-methyl-2H-tetrazol-5-amine isolated in cryogenic inert matrices upon irradiation with UV (□> 235 nm) light.

The photoproduced aminide tautomer BI is the lowest energy imine/aminide tautomer of 2MTA and, among the six possible imine/aminide tautomers, it is one of the two forms that can be directly produced from 2MTA as a result of an energetically accessible hydrogen-shift (see Figure 8). The other is conformer CII of 2-methyl-1,2-dihydro-5H-tetrazol-5-imine, which was not identified as contributing to the spectra of the irradiated matrix (this form is less stable than BI by ca. 20 kJ mol^{-1} and has a non-planar structure, then requiring a more important rearrangement of the matrix to be produced from 2MAT [13]). Once formed, the photoproduced aminide form stays stable in the matrix, since it can be comparatively less photoreactive than 2MTA because the two pathways similar to a) and b) would require an additional hydrogen atom migration.

Following the primary photoproducts, secondary reactions were observed, leading to spectroscopic observation of methylenimine and hydrogen isocyanide (CNH) [13]. As it can be easily recognized in Figure 10, these secondary reactions were also facilitated by the presence of labile hydrogen atoms in the primary photoproducts.

6. Photochemistry of Matrix-Isolated 5-Mercapto-1-Methyltetrazole (2MTT)

Like in the 2MTA molecule, tautomerism could be expected to be important in 5-mercapto-1-methyltetrazole (1-methyl-1H-tetrazole-5-thiol; MTT). In this case, tautomerism may be seen as mainly determined by the possibility of the sulphur atom bound to C-5 to adopt the thiocarbonyl or the mercapto form (Figure 11).

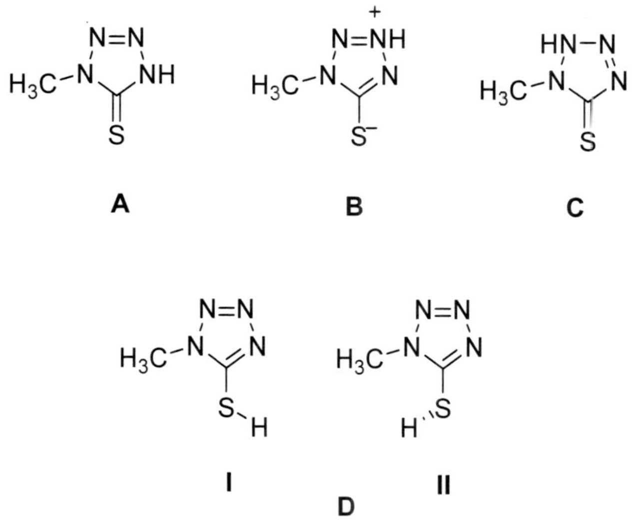

Figure 11. Tautomers of 5-mercapto-1-methyltetrazole: (A) 1-methyl-1,4-dihydro-5H-tetrazol-5-thione, (B) 1-methyl-1H-tetrazol-3-ium-5-thiolate; (C) 1-methyl-1,2-dihydro-5H-tetrazol-5-thione, (D) 1-methyl-1H-tetrazol-5-thiol (two conformers).

The 5-mercapto tautomer D (1-methyl-1H-tetrazol-5-thiol) may adopt two different conformers, I and II, with form I, with the methyl group and sulphydryl hydrogen atom oriented to opposite sides being a more stable thiol form [14]. Because of the destabilizing $CH_3 \cdots H(S)$ interaction in conformer II, this form has the thiol hydrogen atom considerably out of the tetrazole-ring plane (the calculated N=C-S-H dihedral angle is ca. 137°), whereas in conformer I this atom is in the ring plane. There are also two 5-thione tautomers, which differ in the position of the tetrazole-ring hydrogen atom (see Figure 11). The most stable thione tautomer A (1-methyl-1,4-dihydro-5H-tetrazol-5-thione) corresponds to the global minimum of the molecule in the gas phase and is considerably more stable than all other species {thiol tautomer DI is the second most stable form and has a relative energy of 32.5 kJ mol^{-1}, calculated at the DFT(B3LYP)/6-311++G(d,p) [14]}.

As expected taking into account the predicted relative stability of all the possible tautomers of MTT, only the most stable form (1-methyl-1,4-dihydro-5H-tetrazol-5-thione) was found in the as-deposited argon matrices. This is clearly demonstrated in Figure 12, by the pretty good agreement between the observed infrared spectrum of the as-deposited argon matrix of MTT and that calculated at the DFT(B3LYP)/6-311++G(d,p) level of theory for thione tautomer A (1-methyl-1,4-dihydro-5H-tetrazol-5-thione).

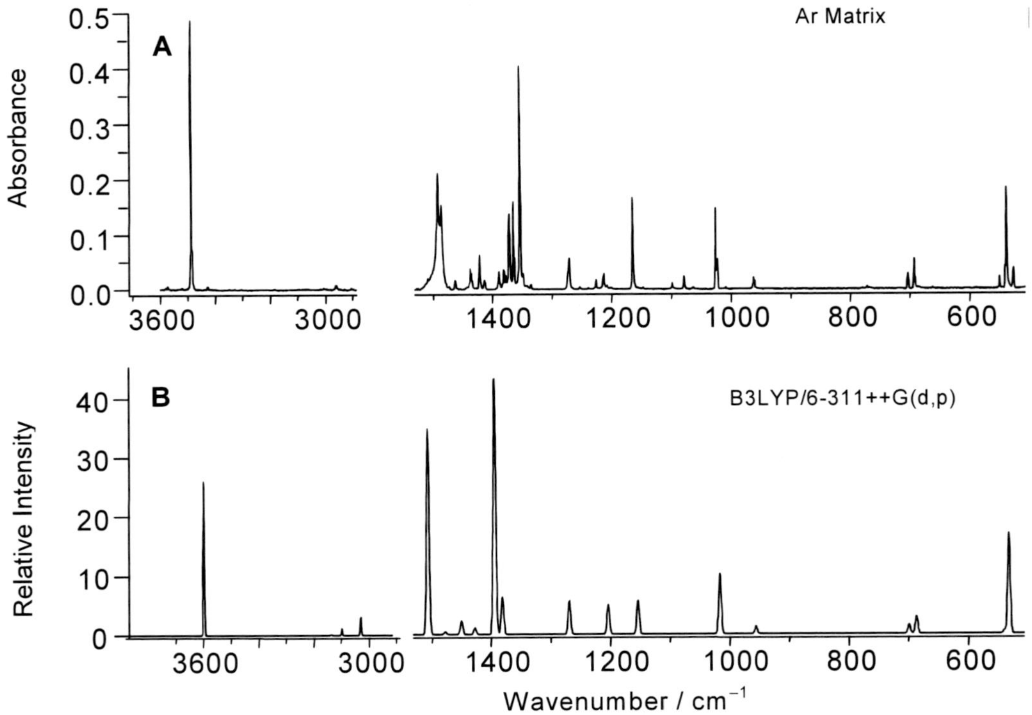

Figure 12. Infrared spectrum of MTT in argon matrix (A) and DFT(B3LYP)/ 6-311++G(d,p) calculated spectrum for its 1-methyl-1,4-dihydro-5H-tetrazol-5-thione tautomer (B). Adapted from [14] with permission.

Figure 13 depicts the proposed reaction pathways for the observed photochemistry of MTT. In general terms, the observed photochemistry of MTT follows closely those observed for other tetrazoles, in particular 2MTA and 1-phenyl-tetrazolone. Nevertheless, the presence of the sulphur atom in the molecule allows for additional reaction pathways. Indeed, in MTT molecular nitrogen can be eliminated by two different ways in a primary photolytic step: a) alone, with simultaneous production of 1-methyl-1H-diazirine-3-thiol (MDT, which is produced in two different conformers), and b) together with atomic sulphur, leading to the production of N-methyl-carbodiimide. Pathway a is similar to the reactions of 2MTA and PT leading to the formation of 1-methyl-1H-diazirine-3-amine (see Figure 10) and 1-phenyl-diazirin-3-one (see Figure 6), respectively, whereas pathway b has no counterpart in the photochemistries of those compounds. In addition to the above reactions, MTT can also undergo c) ring cleavage leading to production of methyl isothiocyanate and azide and d) elimination of carbon monosulphide leading to production of 1-methyl-1,2-tetrazete (which then eliminates N_2 to form methyl diazene) [14]. These two photolytic pathways are identical to pathways c and d shown in Figure 6 for PT.

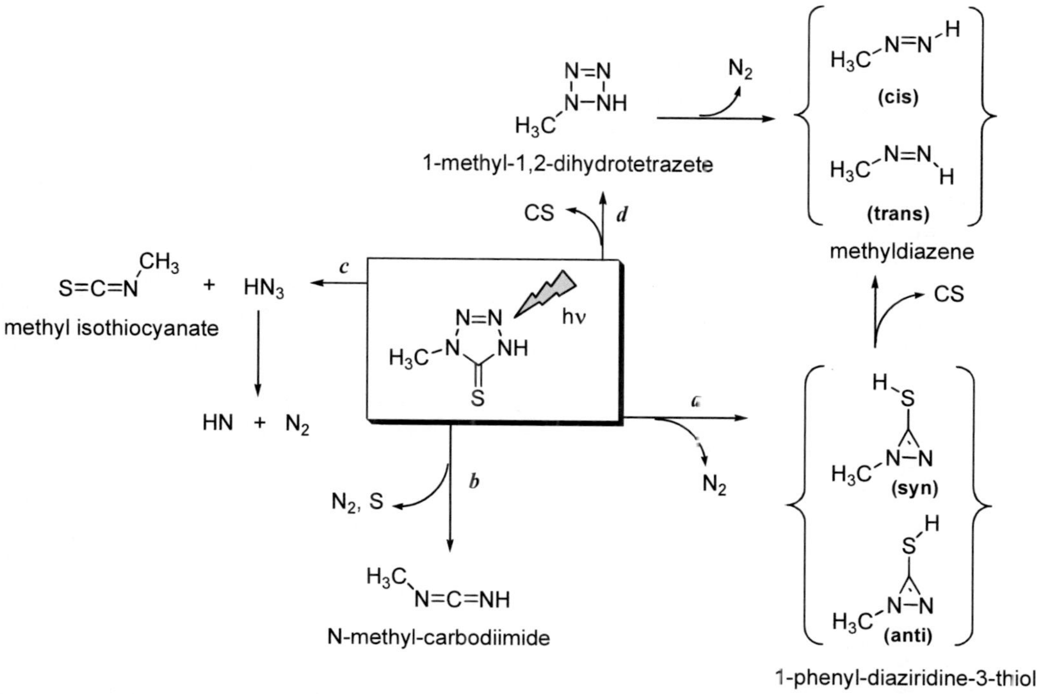

Figure 13. Photochemical reactions observed for 5-mercapto-1-methyltetrazole isolated in cryogenic inert matrices upon irradiation with UV (λ>235 nm) light.

Another pathway can also be postulated, involving cleavage of the (C-5)-(N-1) and (N-3)-(N-4) bonds and leading to methyl azide and HNCS, in a similar way to what occurs for 2MTA. Despite this pathway could not be discarded in a definitive way, no clear experimental evidence of its occurrence in the present case could be found.

Figure 14 shows the spectra of matrix-isolated MTT before and after irradiation, in the 2300-1500 cm^{-1} region, where the characteristic bands of the photoproducts were observed [14] as well as the calculated spectra of the photoproducts in the same spectral region. The dependence of the experimental spectra on the time of irradiation clearly reveals two types of variation patterns of band intensities: some bands increase in intensity mostly during the first 20 minutes of irradiation and decrease upon prolonged irradiation (group A bands), while intensity of other bands (group B) increases during the whole time of irradiation. The observation of bands of type A clearly indicates that there are primary photoproducts, which undergo secondary reactions. These bands could be identified as belonging to 1-methyl-1H-diazirine conformers, which undergo secondary photolysis to methyldiazene (with elimination of CS; see Figure 13). Bands of type B are given rise by all the other photoproducts and could be assigned based on results of theoretical calculations of the infrared spectra of the postulated reaction products [14].

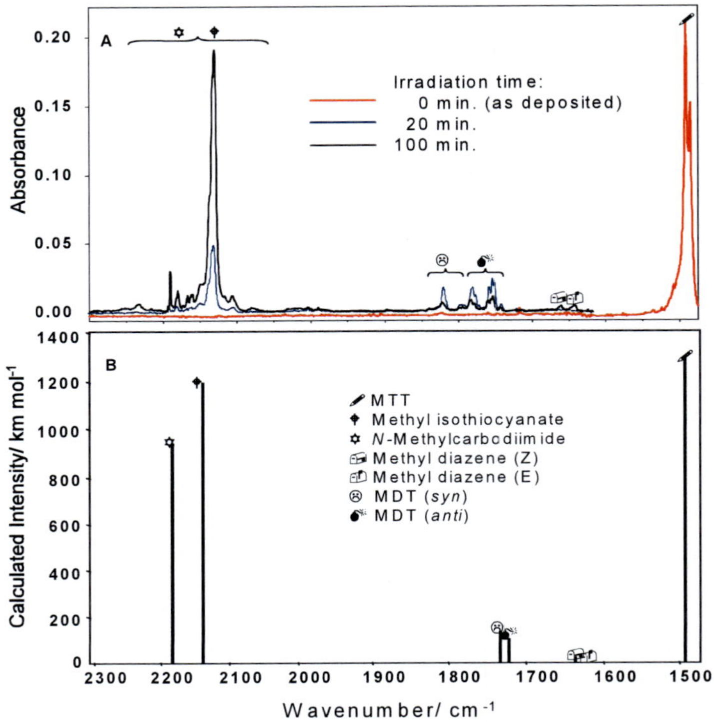

Figure 14. (A) Infrared spectra (2300-1500 cm−1 region) of MTT along the irradiation (λ>235 nm) process. (B) DFT(B3LYP)/6-311++G(d,p) calculated spectra for MTT and most of the observed photoproducts (among the identified photoproducts, only CS and HN do not absorb in the spectral region shown in the Figure). The intensity of the calculated band of MTT was multiplied by 7, in order to approximately preserve the relative band intensities in the experimental and calculated spectra shown in the figure. Adapted from [14] with permission.

It is interesting to note that MTT is a particularly reactive tetrazole when submitted to UV irradiation (λ>235 nm), under the experimental conditions used. In fact, the bands due to MTT reduce to ca. 60 % of their initial intensity after the first 20 min of irradiation and to less than 1 % after 100 min of irradiation, i.e., MTT almost totally converts into other species after 100 min of irradiation. It is also worth mentioning that at the end of the irradiation process, the total intensity of the spectrum was found to be considerably weaker (less than 30% of the initial total absorbance) than that of the as-deposited matrix. This result clearly demonstrates that some of the final products absorb only weakly in the infrared or do not absorb at all (among them, N_2 and S are certainly the dominant species; see Figure 13). The high observed photoreactivity is in keeping with the widespread usage of tetrazole thiones in imaging systems, namely as stabilizers in photography [47], as deep-UV resists in photolithographic applications [48] and as photoremovable capping agents in the semiconductor photocatalytic industry [49].

7. Photochemistry of Matrix-Isolated 1-Phenyl-4-Allyl-Tetrazolone (APT)

In studying the photochemistry of 1-phenyl-4-allyl-tetrazolone (1-phenyl-4-allyl-1,4-dihydro-5H-tetrazol-5-one; APT) we intended to look at a molecule without labile hydrogen atoms (thus simplifying the system in terms tautomerism) but with a conformationally flexible substituent (allyl group; $-CH_2-CH=CH_2$). Our interest in the investigation of 4-allyl-tetrazolones was also motivated by the fact that these compounds are important synthetically, as precursors of other biologically active heterocycles [50], and may be obtained from the corresponding 5-allyloxytetrazoles in quantitative yields through a thermally induced Claisen-type isomerisation [51]. By their side, 5-allyloxytetrazoles are important intermediates for hydrogenolysis of allylic alcohols through heterogeneous catalytic transfer reactions [52-56], presenting a practical and selective synthetic alternative to other methods [57-59]. Furthermore, allyltetrazolones have been found to be vital structural units of biologically active systems, acting as key intermediates in the biosynthesis of several important biomolecules [60-62].

As can be observed from Figure 15, the structure of APT just differs from that of 1-phenyl-tetrazolone (PT) in the substitution of the hydrogen linked to the tetrazole ring by an allyl group. Thus, one can expect that the observed differences in their photochemistry result essentially from this relevant structural modification.

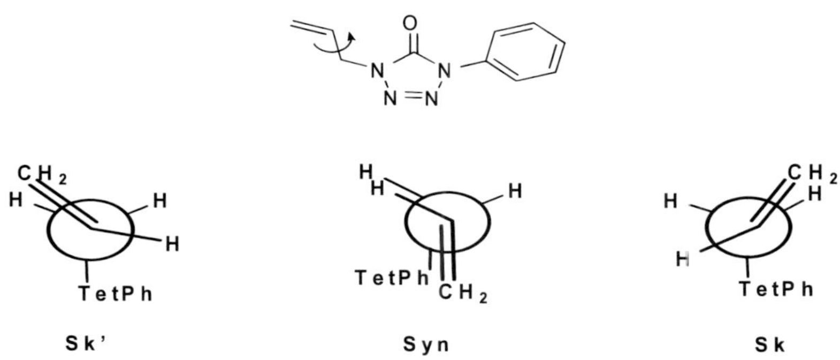

Figure 15. Minimum energy conformations of 1-phenyl-4-allyl-1,4-dihydro-5H-tetrazol-5-one, as calculated at the DFT(B3LYP)/6-31(d,p) level of theory [17]: Sk', Syn and Sk (represented as Newman projections). The lowest-energy structure was found to be conformer Sk'.

All minimum energy conformations of APT depend on the orientation of the $-CH=CH_2$ group on the allylic chain relatively to the tetrazole ring. Three conformers were found on the potential energy surface of the molecule (Figure 15). In these conformers, the $N-CH_2-CH=CH_2$ dihedral angle is ca. $-123°$ (Sk'), $121°$ (Sk) and $-2°$ (Syn) [17]. Their relative populations in the gas phase prior to deposition {as predicted by DFT(B3LYP)/6-311++G(d,p) calculations} are 41.8%, 38.8% and 19.4%, respectively [17].

Somewhat surprisingly, we found that the photochemistry exhibited by matrix-isolated APT is not significantly influenced by the conformation of the allyl group. Indeed, matching closely the general scheme for the photochemistry of phenyltetrazolic compounds, the following photolysis pathways could be identified for matrix-isolated APT (Figure 16): a) N_2 elimination with formation of 1-allyl-2-phenyldiaziridin-3-one (APD) which partially reacts further to form 1-allyl-1H-benzoimidazol-2(3H)-one (ABZ); b) ring cleavage to form allyl isocyanate and phenyl azide (and then ACHT) and c) ring cleavage to form phenylisocyanate and allyl azide. Interestingly, the allylic chain is maintained intact in all observed photochemical processes, indicating the photochemical stability of this residue under the experimental conditions used and justifying the lack of relevance of the presence in the matrix of several conformers to the observed photochemistry.

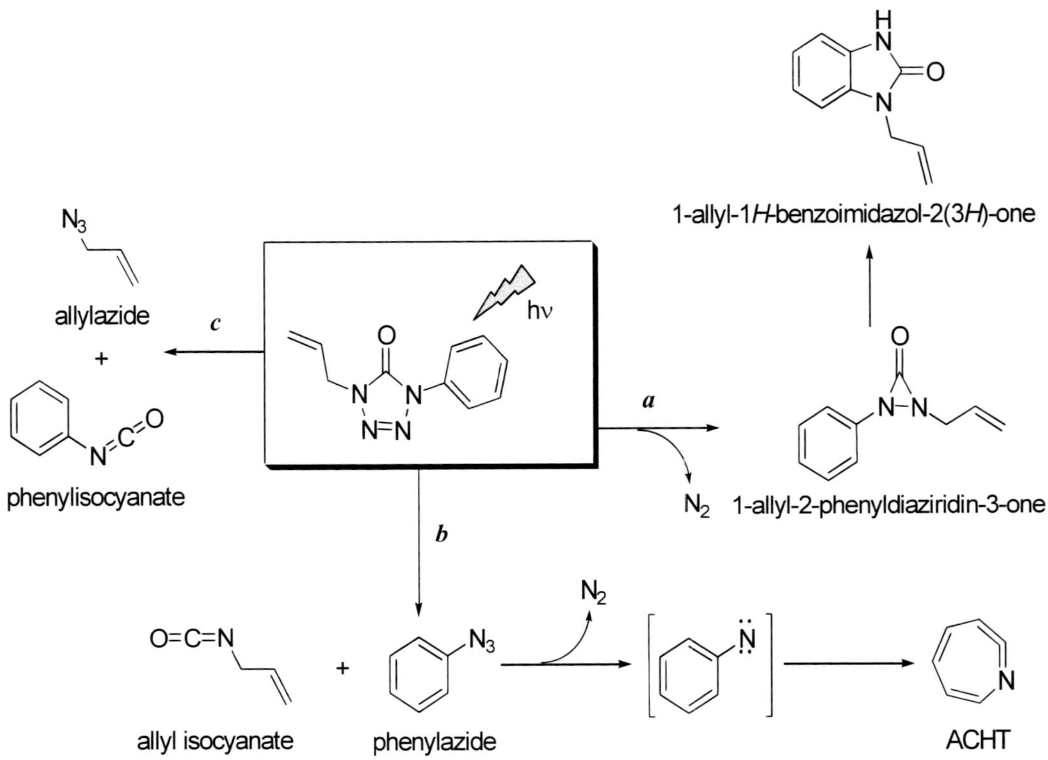

Figure 16. Photochemical reactions observed for 1-phenyl-4-allyl-1,4-dihydro-5H-tetrazol-5-one isolated in an argon matrix upon irradiation with UV (λ> 235 nm) light.

It is worth mentioning that for the conformationally flexible photoproducts all conformational possibilities were explored [17]. In the simulated spectra shown in Figure 17, only the most stable conformers for each photoproduct were considered. Further conformers were neglected in this analysis on the basis that their calculated spectra are very similar within the same photoproduct. In spite of the complexity due to the occurrence of different conformers among the photoproducts, the validity of such approach is demonstrated by a reasonable correspondence between the simulated and the observed spectra of photoproducts.

From the relative amounts of the photoproduced species extracted from the experimental spectra (calculated spectra were used as reference data for normalization of the intensities), it was also possible to determine the relative efficiencies of the different reaction channels as being b: c: a = 8: 16: 24 %. After 40 minutes of irradiation 48% of the reactant initially present in the matrix had been consumed (see Figure 17).

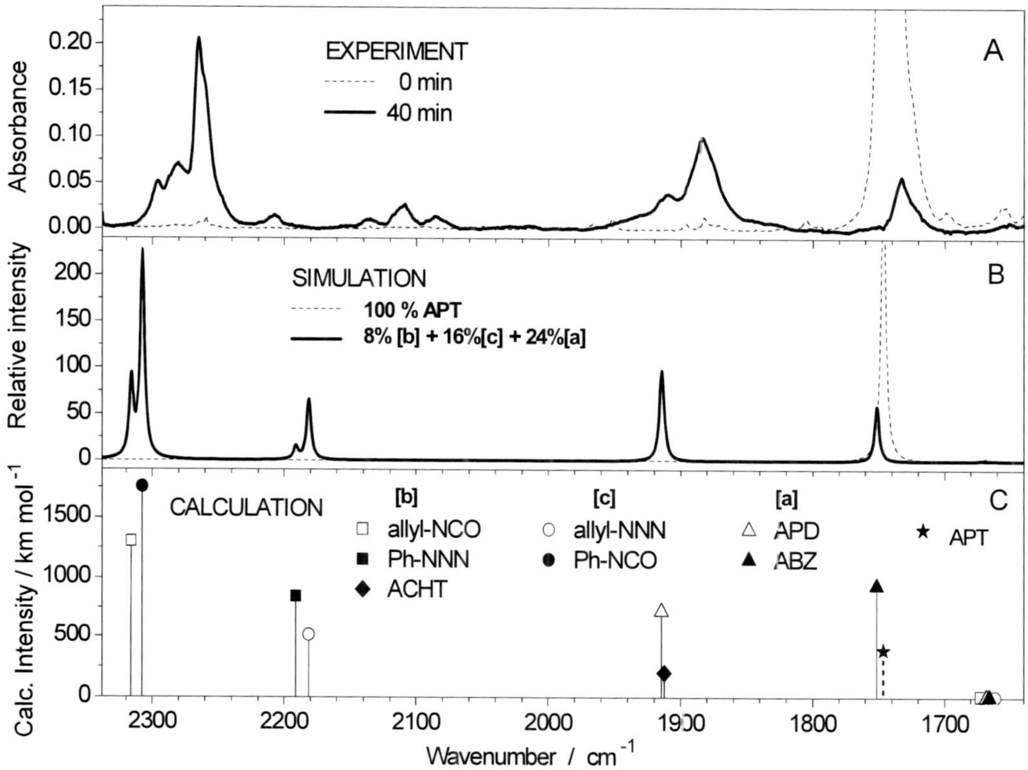

Figure 17. Infrared spectra of APT and the proposed photoproducts, in the carbonyl stretching region. (A): dashed line - FTIR spectrum of the as-deposited matrix of APT (argon, 10 K); solid line - extracted spectrum of the photoproducts formed after 40 min. of UV-irradiation (λ >235 nm). (B): simulated infrared spectra of APT (dashed line) and a mixture of photoproducts (solid line). The calculated intensities in the spectrum of APT were taken as 100%. The calculated intensities of photoproducts were scaled down to reproduce the relative intensities in the experimental spectrum of photoproducts and to obey the normalization condition: total amount of photoproducts is equal to the amount of consumed reagent (48%). (C): DFT(B3LYP)/6-311++G(d,p) calculated spectra of APT (dashed sticks) and possible photoproducts. The calculated wavenumbers were scaled with a uniform factor of 0.978. The calculated intensities were not scaled. Groups [a], [b] and [c] designate different photochannels.

Very interestingly, the observed photochemistry of the matrix-isolated APT is distinct from the preferred photochemical fragmentation in solution, where 3,4-dihydro-3-phenylpyrimidin-2(1H)-one, PDT, is produced as the major photoproduct [50, 63]. The non-formation of PDP for the matrix-isolated APT is doubtlessly related with the matrix medium where the reaction is suppressed. We believe that a strong restriction of conformational mobility in the argon matrix affects not only APT itself, but also its photoproducts. The design of PDP from APT (after extrusion of N_2) involves an approach between the terminal carbon of the allylic chain and the nitrogen atom N-1, in order to form a new C–N single bond. Such approach is strongly hindered in the rigid matrix, which consequently precludes formation of PDP. On the other hand, formation of ABZ appears as the main photoreaction channel for APT, in keeping with the results reported by Quast and Nahr [64].

Compared to 1-phenyl-tetrazolone, the photochemistry of matrix-isolated APT differs mainly in the secondary reactions, essentially because the labile (and much smaller, when compared with the allyl fragment) hydrogen atom in 1-phenyl-tetrazolone confers to the primary photoproducts produced from this compound more flexibility regarding possible subsequent photochemical reactions, in particular under volume-restricted conditions as it is the case of a solid matrix.

8. Photochemistry f Matrix-Isolated 5-Allyloxy-1-Phenyl-1H-Tetrazole (5APT)

The alkyl tetrazolyl ethers are synthetically important compounds. They are readily prepared from 5-chloro-1-aryltetrazol and easily converted in synthetically useful 4-alkyltetrazolones through a thermally induced Chapman-type isomerisation [50, 51, 63]. 1-phenyl-4-allyl-tetrazolone (APT), described in the previous section, was synthesized from 5-allyloxy-1-phenyl-1H-tetrazole (5APT). Figure 18 demonstrates that a moderate heating results in the quantitative transformation of 5APT into APT. This transformation is in a good agreement with the relative calculated energies of the two compounds: the most stable conformer of APT is above 80 kJ mol^{-1} more stable than that of 5APT (calculation at the DFT(B3LYP)/6-311++G(d,p) level).

In 5APT, there are four internal rotational axes that can give rise to nine different conformers, all of them with relative energies within a range of 5 kJ mol^{-1} [65]. However, the conformational effects of the allyl chain are of minor importance regarding the differences between the spectra of 5APT and APT. These spectra are dominated by the strong absorptions due to the stretching vibrations of (C-5)–O and C=O bonds, that appear in the spectra of 5APT and APT at ca. 1557 and 1727 cm^{-1}, respectively (see Figure 18). The frequencies of these characteristic vibrations appear very close in all conformers of the respective compounds and remain practically unaltered for the neat compounds in the bulk (Figure 18) and the monomers isolated in argon matrices (compare with Figure 17 for APT and Figure 19 for 5APT).

In the recent experiments, monomers of 5APT were isolated in xenon matrices and their UV-induced photochemistry was studied. The preliminary conclusions from these experiments will be analyzed in this section. There are experimental indications that 5APT does not undergo conformational cooling during deposition of the compound in xenon matrix at ca. 20 K. There is a good agreement between the experimental and calculated spectrum in the (C-5)–O stretching region where all conformers exhibit a strong absorption band at similar frequencies (see Figure 19). However, in the region where the vibrations of the CH and CH_2 groups appear (1450-1250 cm^{-1}), the experimental bands are broadened compared to the calculated spectrum of a single conformer. This is an indication of conformational multiplicity around the allyl fragment. Also, annealing of the matrix to the highest attainable temperatures (ca. 65 K) does not result in any changes ascribable to conformational isomerization. Therefore, there is more than one reactive conformational species of 5APT in the matrices, according to the conformational complexity existing in the gas-phase equilibrium.

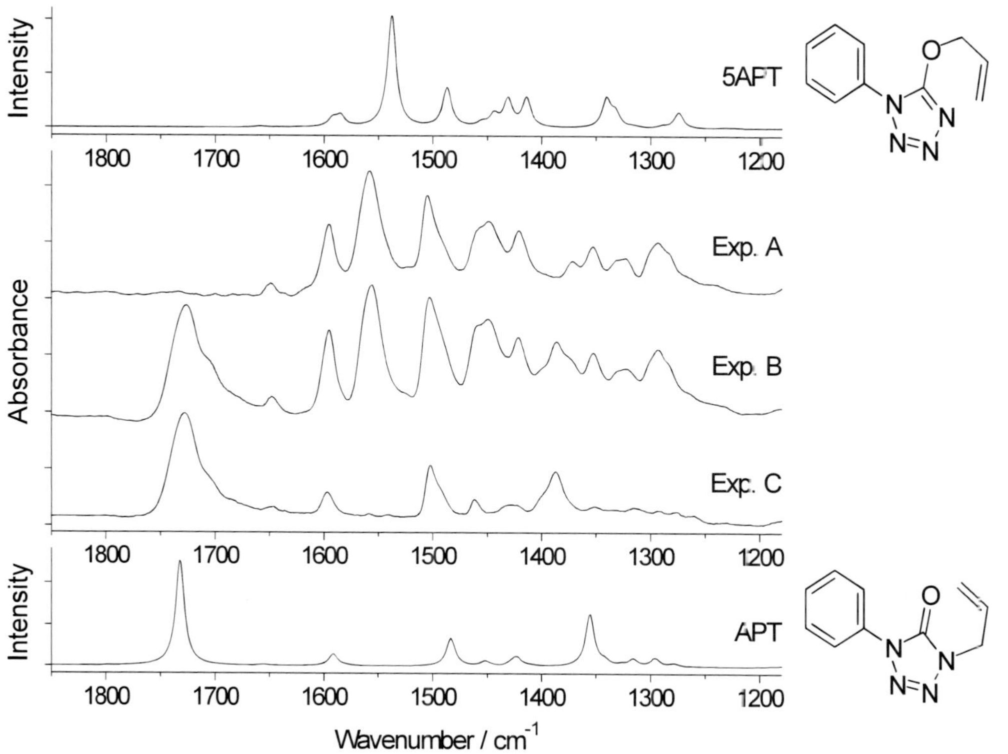

Figure 18. Experimental infrared spectra demonstrating thermal rearrangement of 5APT in APT. Upper and lower frame: theoretical spectra calculated at the DFT(B3LYP)/6-311+–G(d,p) level for the most stable conformer of 5APT and APT, respectively. The calculated wavenumbers were scaled with a uniform factor of 0.970, the absorption bands were simulated using Lorentzian functions centered at the calculated frequency with bandwidth at half-height equal to 10 cm-1. Middle frame, top trace (Exp. A): Experimental spectrum of 5APT layer over a KBr pellet at 24°C. Middle frame, bottom trace (Exp. C): Experimental spectrum of the same 5APT sample after heating at 150°C for 3 hours, showing complete transformation into APT. Middle frame, middle trace (Exp. B): Experimental spectrum at an intermediate stage of heating.

The simplicity of the analysis of the photochemistry of the simpler 5-alkoxy (methoxy and ethoxy) phenyltetrazoles, described in sections 2 and 3, is undoubtedly an adequate starting point for the investigation and further interpretation of the photochemistry of the more complex 5-allyloxy phenyltetrazoles isolated in cryogenic matrices. As it was shown for 5MPT and 5EPT, the absorptions of main the photoproducts appear as very characteristic bands in the 1800-2300 cm^{-1} region and can be easily identified also in the case of 5APT. Upon UV-irradiation, matrix-isolated 5APT exhibits a similar photochemistry to those observed for 5MPT and 5EPT [16, 18]: a) direct N_2 elimination, to give 3-(allyloxy)-1-phenyl-1H-diazirine (3APD; which could be identified for the first time in our investigation) and b) ring cleavage to phenylazide and allylcyanate. As usually, phenylazide further evolves to phenyl nitrene and, subsequently to ACHT, while allylcyanate rearranges to allylisocyanate in a similar way as methylcyanate was found to isomerize to methylisocyanate in the photolysis studies on 5MPT (see Figure 1) [16].

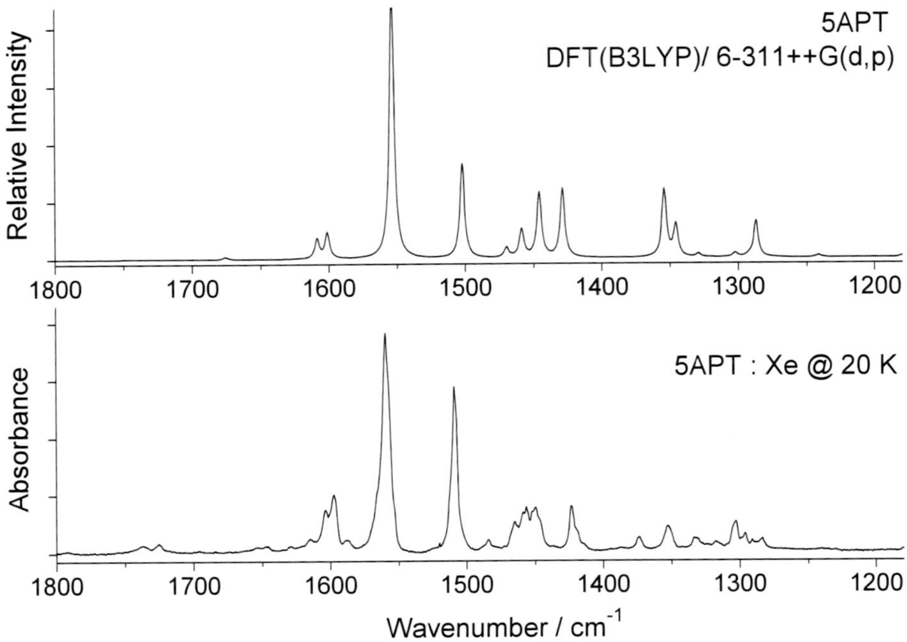

Figure 19. Fragments of the infrared spectra of 5APT. Lower frame - FTIR spectrum of 5APT monomers isolated in a xenon matrix at 20 K. Upper frame - simulated infrared spectrum of the most stable conformer. The calculated DFT(B3LYP)/6-311++G(d,p) wavenumbers were scaled with a uniform factor of 0.978, the absorption bands were simulated using Lorentzian functions centered at the calculated (scaled) frequency with bandwidth at half-height equal to 4 cm-1.

The conformational flexibility of the allyl group has some relevant consequences relative to the photoproduced diazirine (3APD). For simpler tetrazoles, like 5MPT and 5EPT [16, 18] (but also MTT [14]) only one or two conformers of the photoproduced diazirine can be observed. However, in 3APD, the presence of three internal rotational axes (CO–CC, OC–CH_2 and NC–CCH_2) leads to the existence of 6 conformers, from which 5 have relative energies within 1.4 kJ mol^{-1}. Besides, the barriers associated with the interconversion of these conformers are high enough (9-10 kJ mol^{-1}) to allow for observation of these forms in the irradiated matrix. In agreement with these expectations, the observed spectra of the 5APT irradiated matrix was found to exhibit clear signs of presence of several 3APD conformers in the matrix, including extensive broadening and splitting of the bands which correspond to vibrations of 3APD that are predicted by the calculations to occur at nearly the same frequency for the different conformers.

In the present study, the analysis of the products formed in the second observed photochannel received strong support from theoretical calculations, since the conformational flexibility of the allyl fragment also introduces some complexity in what concerns the structures of both the cyanate or isocyanate derivatives. According to DFT(B3LYP)/6-311++G(d,p) calculations, the two rotational axes of allylcyanate can give rise to 5 different conformers with relative energies within 0-5.5 kJ mol^{-1} and barriers for interconversion within the 8-9 kJ mol^{-1} range. However, all conformers of allylcyanate have a much higher energy than allylisocyanate (> 115 kJ mol^{-1}) and, if formed, they quickly isomerize to this latter species. This fact is evidenced by observation in the spectra of the photolyzed matrix of the characteristic intense band at 2164 cm^{-1} due to the asymmetric NCO stretching vibration of the isocyanate group from the very beginning of the irradiation process [64]. Very interestingly, allylisocyanate is produced in a single conformer, as all components of the structured intense □NCO as band, assigned to this compound, grow uniformly along the irradiation.

Allylisocyanate has been studied previously by infrared, Raman and microwave spectroscopy by Torgrimsen et al. [66]. These authors suggested that in the neat crystalline state this compound exists in a single conformer, which exhibits the cis arrangement of the allyl group, whereas in liquid phase the additional gauche conformers exist with significant population. In addition, they found that the bands corresponding to less stable conformers are either absent or very weak in the vapor spectra. The prevalence in the gas phase of the cis conformer was also proposed by Maiti et al. [67] from the analysis of the microwave spectrum of the compound. According to our DFT(B3LYP)/6-311++G(d,p) calculations, three different minima were found on the potential energy surface of allylisocyanate: a C_s symmetry form (cis: with C=CCN and CCN=C dihedral angles equal to 0 and 180°, respectively), and two gauche structures, which were found to be only slightly less stable than the cis conformer ($\Delta E_{gauche\text{-}cis}$ = 0.99 and 1.15 kJ mol^{-1}). In spite of having relatively low energies, the gauche forms were found to be separated from the most stable cis form only by very low energy barriers (< 0.5 kJ mol^{-1}), which explains their non-observation either in the spectroscopic studies carried out in the gaseous phase [66, 67] or in our investigation of the matrix-isolated compound.

The studies of photochemical and thermal reactivity of 5APT in solutions [65] revealed its propensity to a very effective isomerisation into APT (see Figure 20). It was very interesting to trace the possibility of such rearrangement for matrix-isolated 5APT monomers. Indeed, upon prolonged UV-irradiation (λ >235 nm) of the matrices, a new photoproduct band appeared at 1743 cm^{-1}. This absorption corresponds to the matrix isolated APT and no other putative photoproducts may absorb at this frequency. Moreover, the monomers of matrix isolated APT were found to absorb at exactly the same frequency in separate experiments [17]. It must be stressed that the conformation of the allyl group properly aligned for the rearrangement of 5APT in APT does not correspond to a stable conformer (it is not an energy minimum), and then such structure is unlikely to be present in the initially deposited matrix. The appearance of APT in the photolyzed matrix implies the occurrence of an intramolecular rotation of the allyl fragment induced by UV-irradiation. It is difficult to estimate the efficiency of this photochemical channel, since the photoproduced APT also undergoes UV-induced photodecomposition [17]. Currently, the possibility to occur in a matrix of other pathways characteristic of the photochemistry in solutions (like formation of oxazines, see Figure 20), is under analysis in our laboratories.

The possibility of thermal and photochemical rearrangement from 5APT to APT establishes a direct link between all the photochemistries of tetrazoles and tetrazolones reported in the sections 2-7. Such rearrangement also reveals the intrinsic complexity related with multitude of the exit channels in the photochemistry of alkyltetrazoles (in particular, 5APT).

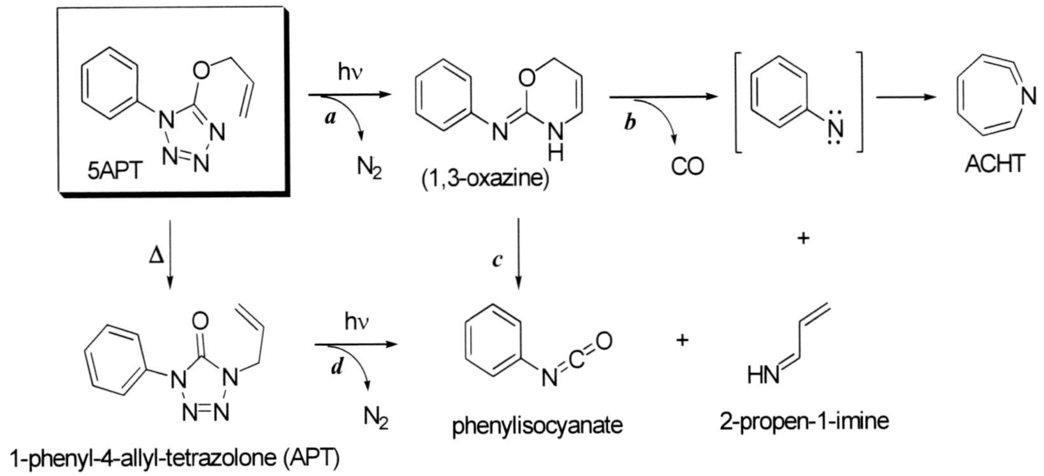

Figure 20. Thermal (□) and selected photochemical (h□, □> 235 nm) reactions observed for 5-allyloxy-1-phenyl-1H-tetrazole (5APT) in solutions [65].

9. Conclusion

The matrix isolation technique represents a powerful approach to investigate the photochemistry of tetrazole derivatives. The use of matrix isolation technique coupled to a suitable probing method, such as FTIR spectroscopy, represents an accurate strategy to understand the mechanisms involved in photochemical reactions, its main advantages being the simplification of the photochemical processes due to cage confinement of the reactions and the possibility to achieve a high spectroscopic resolution, increasing the probability to detect chemical species produced in low amounts during photolysis or those with intrinsically low intensity IR absorptions.

According to our results, the fundamental photochemistry of matrix-isolated tetrazoles involves fragmentation of the tetrazole ring. The specific substituents present in the ring determine in large amount the preferences for the different available primary photochemical reaction channels and strongly determine secondary processes in which the initial photoproducts participate. As a general rule, the tetrazole ring can undergo photolytic cleavage in three different ways: a) through the (N-1)–(N-2) and (N-3)–(N-4) bonds, releasing molecular nitrogen and forming a diazirine derivative; b) through the (N-1)–(C-5) and (N-3)–(N-4) bonds and c) through (N-1)–(N-2) and (N-4)–(C-5) bonds. In the two latter cases, the precise nature of the photoproducts varies depending on the substituents present in the tetrazole ring, but one of the products is always an azide, which then can undergo subsequent reactions, most of times eliminating N_2 to form the nitrene that then can further react to form the final product (e.g., ACHT, produced by ring expansion of phenyl nitrene).

The number of available reaction channels correlates with the number of formally single bonds in the tetrazole ring. When four single bonds are present (like in tetrazolones or tetrazole-thiones), the 3 photochemical fundamental types of reactions can take place, a fourth reaction path being sometimes also activated, corresponding to cleavage through the (N-1)–(C-5) and (N-4)–(C-5) bonds. For all molecules we investigated exhibiting only three formally single bonds in the tetrazole ring, only two reaction channels were found to be active, with that corresponding to direct N_2 elimination always playing the most important role.

The presence in the ring or in the substituents of labile hydrogen atoms always increases the complexity of the photochemistry, mainly due to the occurrence of different tautomeric forms, whereas the conformational flexibility of the substituent may have effects difficult to predict a priori. These are strongly determined by the number and relative energies of the possible conformers and conformational interconversion barriers.

The presence of a phenyl substituent at N-1 (or N-4) in alkyloxy tetrazoles results in strong activation of the channel leading to production of phenylazide and the corresponding alkyl-cyanate (that undergoes major isomerization to the isocyanate). In tetrazolones and tetrazole-thiones, a phenyl or alkyl substituent at N-1 (or N-4) favors the pathway leading to direct production of the corresponding isocyanate or thioisocyanate, which is in general inhibited in the alkyloxy and allyloxy tetrazoles.

The mechanism for direct elimination of molecular nitrogen from the tetrazole ring was not yet fully explained, but it seems to occur mainly through the biradical intermediate [50, 63, 65] that subsequently undergoes cyclization and forms diazirine. The biradical has been recently observed in an investigation carried out in our laboratory on substituted tetrazoles in solution, by laser flash-photolysis [65]. The mechanisms for the remaining two prototype primary ring-cleavage photochemical reactions of the tetrazole ring are still to be investigated.

It is fortunate that most of the prevalent photoproducts of the photochemistry of tetrazole derivatives can be easily identified by IR spectroscopy. In fact, three groups of bands, occurring in the usually clean spectral regions allow easy identification of the main products resulting from the three fundamental paths in tetrazoles' photochemistry: diazirines and diaziridinones give rise to characteristic bands in the 1800-1900 cm^{-1} spectral region, azides have their most intense band around 2100 cm^{-1} ($\nu N{=}N^{+}{=}N^{-}$ antisymmetric stretching) and isocyanates strongly absorb in the 2290–2300 cm^{-1} range ($\nu N{=}C{=}O$ antisymmetric stretching).

A final note shall be made regarding the possibility of using the in situ photolysis of matrix-isolated tetrazoles to produce novel molecular species, like for example antiaromatic diazirines or reactive isocyanates and azides (e.g., 3-methoxy-1-phenyl-1H-diazirine, 3-ethoxy-1-phenyl-1H-diazirine, 3-allyloxy-1-phenyl-1H-diazirine, 1-phenyl-diaziridin-3-one, 1-allyl-2-phenyldiaziridin-3-one, allylisocyanate, allylazide).

ACKNOWLEDGMENTS

The authors are grateful to Fundação para a Ciência e Tecnologia (FCT, Portugal) and FEDER for financial support (Projects POCI/QUI/59019/2004 and POCI/QUI/58937/2004) and grants SFRH/BD/17945/2004 (L.M.T.F.), SFRH/BPD/11499/2002 (A.G-Z.) and SFRH/BPD/1661/2000 (I.D.R.). AGZ thanks Consejo Nacional de Investigaciones Científicas y Técnicas (CONICET) and the Argentinian Agencia Nacional de Promoción Científica y Tecnológica (PICT 2006/68).

REFERENCES

[1] Bally. T. (2004). Chapter 17. Matrix Isolation. In *Reactive Intermediate Chemistry*, R. A. Moss; M. S. Platz and J. Maitland Jones, (Eds.) Wiley, New York, USA, pp 797-846.
[2] Whittle, E., Dows, D. A., and Pimentel, G. C. (1954). *Journal of Chemical Physics*, 22, 1943-1943.
[3] Becker, E. D., and Pimentel, G. C. (1956). *Journal of Chemical Physics*, 25 224-228.
[4] Norman, I., and Porter, G. (1954). *Nature*, 174 508-509.
[5] Singh, H., Charla, A. S., Kapoor, V. K., Paul, D., and Malhotra, R. K. (1980). In *Progress in Medicinal Chemistry*, p 151.
[6] Noda, K., Saad, Y., Kinoshita, A., Boyle, T. P., Graham, R. M., Husain, A., and Karnik, S. S. (1995). *Journal of Biological Chemistry*, 270, 2284-2289.
[7] Mavromoustakos, T., Kolocouris, A., Zervou, M., Roumelioti, P., Matsoukas, J., and Weisemann, R. (1999). *Journal of Medicinal Chemistry*, 42, 1714-1722.
[8] Desarro, A., Ammendola, D., Zappala, M., Grasso, S., and Desarro, G. B. (1995). *Antimicrobial Agents and Chemotherapy*, 39, 232-237.
[9] Maier, G., Eckwert, J., Bothur, A., Reisenauer, H. P., and Schmidt, C. (1996). *Liebigs Annalen*, 1041-1053.
[10]Bugalho, S. C. S., EMacoas, . M. S., Cristiano, M. L. S., and Fausto, R. (2001). *Physical Chemistry Chemical Physics*, 3, 3541-3547.
[11]Bugalho, S. C. S., Lapinski, L., Cristiano, M. L. S., Frija, L. M. T., and Fausto, R. (2002). *Vibrational Spectroscopy*, 30, 213-225.
[12]Bugalho, S. C. S., Serra, A. C., Lapinski, L., Cristiano, M. L. S., and Fausto, R. (2002). *Physical Chemistry Chemical Physics*, 4, 1725-1731.
[13]Gómez-Zavaglia, A., Reva, I. D., Frija, L., Cristiano, M. L., and Fausto, R. (2005). *Journal of Physical Chemistry A*, 109, 7967-7976.
[14]Gómez-Zavaglia, A., Reva, I. D., Frija, L., Cristiano, M. L., and Fausto, R. (2006). *Journal of Molecular Structure*, 786, 182-192.
[15]Gómez-Zavaglia, A., Reva, I. D., Frija, L., Cristiano, M. L., and Fausto, R. (2006), *Journal of Photochemistry and Photobiology A – Chemistry*, 179, 243-255.
[16]Gómez-Zavaglia, A., Reva, I. D., Frija, L., Cristiano, M. L. S., and Fausto, R. (2006). *Journal of Photochemistry and Photobiology A – Chemistry*, 180, 175-183.
[17]Frija, L. M. T., Reva, I. D., Gómez-Zavaglia, A., Cristiano, M. L. S., and Fausto, R. (2007). *Photochemical and Photobiological Sciences*, 6, 1170-1176.
[18]Frija, L. M. T., Reva, I. D., Gómez-Zavaglia, A., Cristiano, M. L. S., and Fausto, R. 2007). *Journal of Physical Chemistry A*, 111, (2879-2888.
[19]McCrone, W. C., Grabar, D., and Lieber, E. (1951). *Analytical Chemistry*, 23, 543-543.
[20]Putten van der, N., Heijdenrijk, D., and Schenk, H. (1974). *Acta Crystallographica Section C-Crystal Structure Communications*, 3, 321-322.
[21]Goddard, R., Heinemann, O., and Krüger, C. (1997). *Acta Crystallographica Section C-Crystal Structure Communications*, 53, 590-592.
[22]Butler, R. N., Garvin, V. C., Lumbroso, H., and Liègeois, C. (1984). *Journal of the Chemical Society-Perkin Transactions*, 2, 721-725.
[23]Zhaoxu, C., and Heming, X. (1998). *Theochem-Journal of Molecular Structure*, 453, 65-70.

[24]Krugh, W. D. and Gold, L. P. (1974). *Journal of Molecular Spectroscopy*, 49, 423-431.
[25] Razynska, A., Tempczyk, A., Malinski, E., Szafranek, J., Grzonka, Z., and Hermann, P. (1983). *Journal of the Chemical Society-Perkin Transactions*, 2, 379-383.
[26]Palmer, M. H., Simpson, I., and Wheeler, J. R. (1981). Zeitschrift Fur Naturforschung Section A - A *Journal of Physical Sciences*, 36, (1981). 1246-1252.
[27]Dunkin, I. R. (1986). *Spectrochimica Acta Part A-Molecular and Biomolecular Spectroscopy*, 42, 649-655.
[28]Dunkin, I. R., Shields, C. J., and Quast, H. (1989). *Tetrahedron*, 45, 259-268.
[29]Awadallah, A., Kowski, K., and Rademacher, P. (1997). *Journal of Heterocyclic Chemistry*, 34, 113-122.
[30]Jensen, K. A., Due, M., and Holm, A. (1965). *Acta Chemica Scandinavica*, 19, 438.
[31]Martin, D., and Mucke, W. (1965). *Chemische Berichte – Recueil*, 98, 2059-2062.
[32]Pasinszki, T., and Westwood, N. P. C. (1995). *Journal of Physical Chemistry*, 99, 1649-1654.
[33]Faustov, V. I., Baskir, E. G., and Biryukov, A. A. (2003). *Russian Chemical Bulletin*, 52, 2328-2333.
[34]Chapman, O. L., and Le Roux, J.P. (1978). *Journal of the American Chemical Society*, 100, 282-285.
[35]Hayes, J. C., and Sheridan, R. S. (1990). *Journal of the American Chemical Society*, 112, 5879-5881.
[36]Reva, I. D., Stepanian, S. G., Adamowicz, L., and Fausto, R. (2003). *Chemical Physics Letters*, 374, 631-638.
[37]Borba, A., Gómes-Zavaglia, A., Simões, P. N. N. L., and Fausto, R. (2005). *Journal of Physical Chemistry A*, 109, 3578-3586.
[38]Reva, I. D., Lopes Jesus, A. J., Rosado, M. T. S., Fausto, R., Eusébio, M. E., and Redinha, J. S. (2006). *Physical Chemistry Chemical Physics*, 8, 5339-5349.
[39]Martin, D. (1964). *Tetrahedron Letters*, 5, 2829-2832.
[40]Martin, D., and Weise, A. (1966). *Chemische Berichte – Recueil*, 99, 976-983.
[41]Holm, A., and Hugejens, E. (1974). *Acta Chemica Scandinavica Series B-Organic Chemistry and Biochemistry B*, 28, 705-710.
[42]Fischer, H., Zeuner, S., Ackermann, K., and Schubert, U. (1984). *Journal of Organometallic Chemistry*, 263, 201-212.
[43]Otsuka, S., Nakamura, A., and Yoshida, T. (1967). *Journal of Organometallic Chemistry*, 7, 339-344.
[44]Huisgen, R., Vossius, D., and Appl, M. (1958). *Chemische Berichte – Recueil*, 91, 1-12.
[45]Huisgen, R., and Appl, M. (1958). *Chemische Berichte – Recueil*, 91, 12-21.
[46]Himmel, H.J., Junker, M., and Schnöckel, H. (2002). *Journal of Chemical Physics*, 117, 3321-3326.
[47]Eachus, R. S., Muenter; A. A., Pawlik, T. D., and Lenhard, J. R. (2005). *Journal of Physical Chemistry B*, 109, 10126-10136.
[48]Taylor; J. W., Jiang, Y., and Bassett, D. R. (1991). The Use of Carbodiimide Chemistry in Photolithography. In *Polymeric Materials Science and Engineering*, p 50.
[49]Uchihara, T., Kamiya, T., Maedomari, S., Ganeko, M., Kinjo, S., Tamaki, Y., and Kinjo, A. (2000). *Journal of Photochemistry and Photobiology A – Chemistry*, 130, 63-69.
[50]Frija, L. M. T., Khmelinskii, I. V., and Cristiano, M. L. S. (2005). *Tetrahedron Letters*, 46, 6757-6760.

[51]Cristiano, M. L. S., and Johnstone, R. A. W. (1997). *Journal of the Chemical Society-Perkin Transactions*, 2, 489-494.

[52]Johnstone, R. A. W., Wilby, A. H., and Entwistle, I. D. (1985). *Chemical Reviews*, 85, 129-170.

[53]Araújo, N. C. P., Brigas, A. F., Cristiano, M. L. S., Frija, L. M. T., Guimarães, E. M. O., and Loureiro, R. M. S. (2004). *Journal of Molecular Catalysis A-Chemical*, 215, 113-120.

[54]Cristiano, M. L. S., Johnstone, R. A. W., and Price, P. J. (1996). *Journal of the Chemical Society-Perkin Transactions*, 1, 1453-1459.

[55]Hussey, B. J., Johnstone, R. A. W., and Entwistle, I. D. (1982). *Tetrahedron*, 38, 3775-3781.

[56]Boutros, A., Legros, J.Y., and Fiaud, J.C. (2000). *Tetrahedron*, 56, 2239-2246.

[57]Larock, R. C. (1999). *Comprehensive Organic Transformations: A Guide to Functional Group Preparations* (2nd ed). John WIley and Sons, Inc., New York, USA.

[58]Hayashi, M., Yamada, K., and Nakayama, S. (2000). *Journal of the Chemical Society-Perkin Transactions*, 1, 10, 1501-1503.

[59]Green, T. W., and Wutts, P. G. M. (1999).*Protecting Groups in Organic Synthesis*. New York, USA.

[60]Kumar, A., and Dittmer, D. C. (1994). *Journal of Organic Chemistry*, 59, 4760-4764.

[61]Adam, W., Peters, K., and Renz, M. (1994). Angewandte Chemie - International Edition in English, 33, 1107-1108.

[62]Bergmeier, S. C., and Stanchina, D. M. (1995). Tetrahedron Letters, 36, 4533-4536.

[63]Frija, L. M. T., Khmelinskii, I. V., and Cristiano, M. L. S. (2006). *Journal of Organic Chemistry*, 71, 3583-3591.

[64]Quast, H., and Nahr, U. (1985). *Chemische Berichte-Recueil*, 118, 526-540.

[65]Frija, L. M. T., Khmelinskii, I. V., Serpa, C., Reva, I. D., Fausto, R., and Cristiano, M. L. S. (2008). *Organic and Biomolecular Chemistry*, 6, Paper Ref : B718104C.

[66]Torgrimsen, T., Klaboe, P., and Nicolaisen, F. (1974). *Journal of Molecular Structure*, 20, 213-227.

[67]Maiti, S., Jaman, A. I., and Nandi, R. N. (1993). *Journal of Molecular Spectroscopy*, 158, 8-13.

In: Photochemistry Research Progress
Editors: A. Sánchez, S. J. Gutierrez

ISBN: 978-1-60456-568-3

Chapter 10

BENZOINDOLES AND PIRIDOINDOLES: REVIEWING THEIR PHOTOCHEMISTRY AND APPLICATIONS IN A COMPARATIVE WAY

Sergio M. Bonesi and Rosa Erra-Balsells
Departamento de Química Orgánica, Ciudad Universitaria, Buenos Aires, Argentina

ABSTRACT

The carbazole family is subjected to an overview in terms of the photophysics and photochemistry aspects. In this chapter, no attempts are made to include polymeric systems or macromolecules in which the carbazole nuclei are pendant groups. The photophysical behavior of carbazoles deals with the characterization of the electronic excited states (singlet and triplet states) as well as the derivation of the dynamic properties of the excited states in terms of radiative and radiationless rate constants by means of steady-state and time-resolved spectroscopy (absorption, emission fluorescence and phosphorescence spectroscopy [optical methods]; laser induced optoacoustic spectroscopy [LIOAS; optoaocousitc method]). This discussion is restricted to experiments done in solution at 298 K and in solid matrix at 77 K. The photochemistry of carbazoles can be grouped according to different classes of photoreactions in which these compounds are involved: photoinduced electron transfer process (PET), photosensitized process and photoinduced rearrangement process. In this regard, the photoreactions of carbazoles are discussed from preparative and mechanistic viewpoints.

INTRODUCTION

From the photochemical point of view, the aromatic compounds containing nitrogen atoms in their structure are quite interesting because after photon absorption π,π^* and/ or n,π^* singlet and/or triplet states can be populated. These electronic excited states show in general quite different behaviour. Thus, it is possible to find quite interesting applications for photochemical reactions.

In order to study the peculiar effect of the presence of nitrogen atoms on the photochemistry of aromatic compounds, two families of heterocyclic aromatic compounds have been studied in parallel way in our laboratory: compounds containing two nitrogen atoms in its aromatic structure: piridoindoles (carbolines) and compounds containing only a nitrogen in its aromatic structure: benzoindoles (carbazoles). Within the former family, β-carbolines (pirido[3,4-b]indoles) have been chosen because several members of this family, biomolecules present in plant tissues, are commercially available.

The photophysics studies deal with the characterization of the electronic excited states (singlet and triplet states) as well as the derivation of the dynamic properties of the excited states in term of radiative and radiationless rate constants by means of steady-state and time-resolved optic and optoacoustic spectroscopy.

The preparative photochemistry of carbazoles can be grouped according to different classes of photoreactions in which these compounds are involved: photoinduced electron transfer process (PET; electronic excited carbazoles working as electron donor and as electron acceptors), photosensitized process (carbazoles as photosensitizers), and photoinduced rearrangement process (photorearrangements of carbazoles). In this regard, the photoreactions of carbazoles will be discussed from preparative and mechanistic viewpoints. Prepartive photochemistry can be used as a tool in organic synthesis to prepare new compounds containing the carbazole moiety in its structure.

On the contrary, the photochemistry of β-carbolines and its applications is related with the absence of photochemical reactions and the radiative (fluorescence) and radiationless (heat release to the medium) processes as the predominant deactivation of the electronic excited states. Owing to this special photochemical stability and the strong basic character of the electronic excited state, some β-carbolines have shown to be outstanding photosensitizers (matrix) in UV-matrix assisted laser desorption/ionization (UV-MALDI) mass spectrometry. This soft ionization method described by K. Tanaka (Nobel Prize in Chemistry 2002) is one of the more challenge applications of β-carbolines photochemistry.

Both families will be subjected to overviews in terms of the photophysisc and photochemistry aspects due to the structural similarity that the carbazole and β-carboline nucleus have.

To begin with, the Part I of this series of planned reviews, titled photochemistry of benzoindoles (carbazoles) is here presented.

Results and Conclusions

The photochemistry of carbazoles has been studied from different points of view: (a) To begin with, the photophysics has been described carefully and carbazoles have been addressed as a good fluorescent chromophores [1,2], and depending on the nature of the substituted carbazole, derivatives show in some cases efficient phosphorescent emission [3] and radiationless deactivation characterized by Laser Induced Optoacoustic Spectroscopy (LIOAS) [4,5]; (b) carbazoles are also well known as electron donors and excellent reducing agents from the electronic singlet excited state. There are some nice reported examples in which carbazole and *N*-substituted carbazoles act as photosensitizers in photoinduced electron transfer (PET) processes; (c) Preparative photochemical reactions involving carbazoles have also been studied as a key step to introduce new functional groups in the carbazole moiety of use to synthesize more complex macromolecules.

(a) *Photophysics of carbazoles*

The electronic spectra and photophysics of carbazole and some carbazole derivatives have been mainly studied by using steady-state spectroscopic methods [6-21]. These heterocyclic compounds show special photophysical properties and are used as fluorescent chromophore in many biochemical and polymeric applications. For example, polymers containing carbazole as pendant groups show outstanding electrical and photoelectrical properties [1]. Also, carbazole-labeled phospholipids have been used to monitor the kinetics of thermotropic and inotropic lateral phase separations in vescicles in the presence of Ca^{2+} [22]. It is interesting to mention that by using carbazoles linked to a phospholipid as a fluorescent chromophore marker, the bioaccumulation of synthetic chlorinated pesticides into membranes have been estimated [23-26].

The electronic spectroscopy of a series of *N*- and C-substituted carbazoles have been studied with steady-state and time-resolved spectroscopic methods at room temperature and in solid matrix (77 K) [1]. Electronic spectra (absorption, fluorescence and phosphorescence emission spectra) of carbazole and carbazole derivatives have been recorded at 298 and 77 K and the solvent and substituent effects on the spectroscopic data and on the photophysical rate constants (k^0_f, k_{isc}, k^0_p and k^0_{nr}) have also been analyzed. The structure of the carbazoles studied are shown in Table 1.

The absorption spectra of carbazole and *N*- and C-substituted carbazoles were recorded in non-polar and in polar solvents at 298 K and they showed three general bands at 258, 292 and 335 nm. Previous studies led to unambiguous assignment of the lowest-lying excited electronic states of carbazoles as ^1L_a and ^1L_b in the C_{2v} symmetry point group with short z-axis and long y-axis, respectively, in the plane of the moleculel [2-15, 27-29]. Thus, comparing the position and oscillator strength of the bands located at 292 and 335 nm regions for carbazole derivatives with those of carbazole, these bands could be assigned to $^1L_b(S_1\text{-}S_0)$ and $^1L_a(S_2\text{-}S_0)$ electronic transitions, respectively.

Once the assignment of the lowest-lying bands of carbazole derivatives was accomplished, the substituent and solvent effects were studied. In general, the absorption spectra of the C- and *N*-substituted carbazoles recorded in cyclohexane at 298 K exhibit small bathochromic shift of the ^1L_b bands (2-16 nm) in comparison to carbazole. On the contrary, an hypsochromic shift of *N*-acetyl, *N*-benzoyl, 2-hydroxy and 2-acetyloxycarbazole was observed. Particularly, 3-nitrocarbazole show a large red shift for the ^1L_b band (33 nm) compared to carbazole. This noticiable red shift reflects the fact that in-plane conjugation of the nitro group has a significant effect on the absorption spectrum. The shifts observed for the

Table 1. Structure of the carbazoles studied

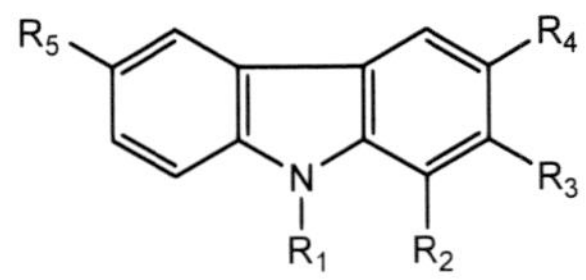

R_1	R_2	R_3	R_4	R_5
H	H	H	H	H
CH_3	H	H	H	H
Ph	H	H	H	H
C_2H_3	H	H	H	H
CH_3CO	H	H	H	H
PhCO	H	H	H	H
CH_3	H	CH_3O	H	H
H	H	HO	H	H
H	H	CH_3CO	H	H
H	H	H	Cl	H
H	H	H	Br	H
H	H	H	NO_2	H
H	H	H	Br	Br
H	H	H	PhCO	H
H	H	H	PhCO	PhCO
PhCO	H	H	Cl	Cl
H	NO_2	H	H	H
H	H	H	$(CH_3)_2N$	H
H	H	H	NH_2	H
H	H	H	CH_3	H
H	H	H	Cl	Cl

^1L_b transitions can be interpreted by mesomeric and/or inductive effects induced by the substituent attached to the carbazole moiety. It is interesting to point out that the absorption spectrum of substituted carbazoles shows no noticeable differences of ^1L_a electronic transition bands with respect to carbazole.

The spectroscopical behavior of the ^1L_a band with the substitution of the carbazole moiety has been also analyzed in terms of the Hammett relationship developed first by Kosower [30] and successfully used by Reichardt *et al.* [31,32]. In this regard, the substituent effect on the 0,0 electronic transition energy ($E_{0,0}$) of carbazole derivatives have been correlated with the Hammett parameters (σ_I, σ_p and σ_m) [33]. The results obtained from the linear correlations with σ_I lead to the following conclusions: i) weak electron-withdrawing substituents attached to the N atom of carbazole produces an important hypsochromic shift; ii) electron donor substituent in the N position show red shifts, iii) these red and blue shifts account for the fact that electron-donor substituents increase the electron charge density on the N atom while electron-acceptor substituents in the same position operates oppositely. Besides, the $E_{0,0}$ values of substituents attached to the C-3 position of carbazole correlate very good with σ_I and σ_p. The statistical parameters obtained, $\rho_I = 14.9$ for N-substituted and $\rho_I = -$

4.46 and ρ_p = - 6.12 for C-substituted carbazoles accounts for the red and blue shifts observed on the absorption spectrum.

The solvent effect on the absorption spectrum of carbazoles was analyzed. A small bathochromic shift was observed on going from non-polar (cyclohexane) to polar (ethanol) solvent. This behavior suggestes that the nature of the lowest singlet excited state is most likely to be π,π^*. This small solvating interaction agrees with the lower Stokes shift ($\Delta\nu s$) values calculated for carbazoles. Application of the Lippert-Mataga [34] linear correlation between the Stokes shift values of carabzole and its derivatives and the polarity-polarizability parameter ($\Delta f = f(\varepsilon) - f(\eta)$) gives the $\Delta\mu$ values which is the difference between the dipole moment of the excited state and the dipole moment of the ground state of carbazoles. These values ranged among 3.81 D and 12.06 D and indicate that the singlet excited state of carbazoles has a dipole moment higher than that of the ground state, which accounts for a more polarized electronic state, with a charge transfer character.

The fluorescence and absorption spectra of carbazoles measured at 298 K are found to display mirror symmetry with exact overlap of 0,0 band, as shown in Figure 1.

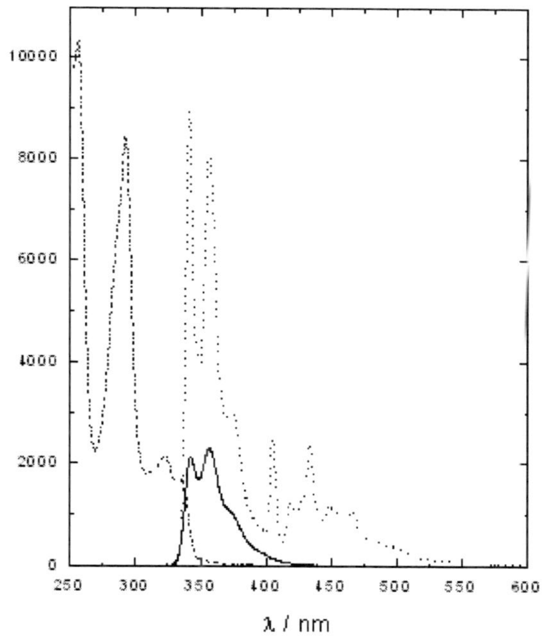

Figure 1. Electronic absorption spectra and fluorescence and phosphorescence emission spectra of carbazole. Dash line: absorption spectra of carbazole in acetonitrile at 298 K; solid line: fluorescence emission spectra of carbazole in acetonitrile at 298 K under nitrogen atmosphere; dot line: fluorescence and phosphorescence emission spectra (total luminiscence) of carbazole in solid matrix at 77 K.

This indicates that the geometries of the molecules are quite similar in their ground and first excited states. In a fixed solvent, *i.e.*, cyclohexane, λ_{max}(fluo) of carbazoles shows a bathochromic shift when the *N*-substituent is an electron acceptor group as well as the C-substituent is an electron-acceptor group. For compounds such as *N*-acetyl-, *N*-benzoyl-, 2-hydroxy- and 2-acetyloxycarbazoles, an hypsochromic shift is observed.

On changing the solvent polarity no significant change in the shape as well as in the λ_{max}(fluo) is observed, except a mere red shift of only 3 – 10 nm due to solvent polarity effect. It clearly diminishes the importance of specific interactions like solute-solvent hydrogen bonding in the excited state.

The dynamic properties of the lowest excited singlet and triplet states of carbazoles have been derived. The fluorescence lifetime τ_f, phosphorescence lifetime τ_p, fluorescence quantum yield ϕ_f and phosphorescence quantum yield ϕ_p of carbazole and carbazole derivatives have been measured in different solvents (cyclohexane, methanol and acetonitrile) at 298 K and in solid matrix (isopropanol – ethyl ether 1:1 (v:v)) at 77 K.

From these spectroscopic data different photophysical rate constants such as k_f, k^o_f, k_{isc}, ϕ_{isc}, k^o_p and k^o_{nr} were easily calculated. These photophysical rate constants were calculated under the assumption that non-radiative internal conversion (k_{ic}) is neglected owing to the planarity and the rigidity of the carbazole structures. Sarkar and Chakravorti have analyzed the photophysical properties of some heteroaromtic compounds, such as 9-phenylfluorene, triphenylamine, 9-phenylcarbazole and carbazole [35]. They have observed that the radiationless intersystem crossing process is favoured as the planarity of the molecule is increased. Also, the radiative singlet decay shows a similar trend on going from triphenylamine to 9-phenylcarbazole. Both processes increase at the expense of suppressed radiationless internal conversion. The susbstituent and solvent effects on the singlet excited state deactivation constants derived at 298 K have been analyzed. It was observed that *N*-substituted carbazoles and C-substituted carbazoles with electron-donor substituents showed a noticeable decrease of k^o_f and ϕ_f values while the τ_f values remain constant when were compared to carbazole. Besides, a significant increase in the k_{isc} and ϕ_{isc} was observed for the same substituted carbazoles.

In the case of *N*-acetyl-, *N*-benzoyl, 3-bromo- and 3-nitrocarbazole a very weak fluorescence emission was detected at room temperature and high values of k_{isc} and ϕ_{isc} were estimated which indicates that an intersystem crossing radiationless process deactivates efficiently the singlet excited state. This behavior can be ascribed to the spin-orbital coupling effect induced by the substituents.

The substituent effect was analyzed in a more quantitative way by plotting $\log[k^o_f(\text{X-CA}) / k^o_f(\text{CA})]$ versus the Hammett parameters (σ_I, σ_p and σ_m), where X denotes the substituent and CA the carbazole moiety. The results obtained show that the *N*-substitution has a longer effect on k^o_f, and thus on the singlet excited state, than the C-substitution. Also, a withdrawing substituent on the carbazole moiety causes a significant enhancement of the radiationless process which accounts for a destabilization of the intramolecular charge transfer singlet excited state which is of π,π^* type. The more relevant fact is the absolute value of the Hammett's slopes which account for a lower inductive effect (σ_I) on k^o_f than on k_{isc}. Comparing the slope values of k^o_f (ρ_I: - 3.22 for *N*-substitution; ρ_I: - 0.57 for C-substitution) and k_{isc} (ρ_I: 5.30 and ρ_I: 0.24, respectively), it was concluded that the larger the inductive effect on k^o_f, the higher is the probability of the deactivation of the singlet excited state via the intersystem crossing pathway.

The solvent effect on k^o_f and k_{isc} values was also analyzed selecting the $E_T(30)$ and the Z values as the solvent polarity parameters. Linear correlations of k^o_f were obtained for carbazole and *N*-methylcarbazole with both solvent parameters. The k_{isc} values show a parabolic trend as the solvent polarity increases. This behavior is due to a mild intermolecular heavy atom effect (HAE) because ethyl acetate and dichloromethane operate simultaneously as solvent and as a quencher of the singlet excited state. In this regard, the deactivation of the excited state through the intersystem crossing channel operates. Also, a similar trend was observed when the solvent effect was analyzed on the fluorescence quantum yield.

Carbazole and its derivatives exhibit structured fluorescence and phosphorescence emission spectra when measurements were performed in solid matrix at 77 K with excitation into the S_1 level. The shape and λ_{max} of the fluorescence emission spectra of carbazoles do not change noticeably on going from 298 K (ethanol) to 77 K which accounts for the fact that the geometries of the first-excited state of carbazoles are quite similar at both temperatures.

The experimentally obtained lowest radiative triplet transition of carbazole is found to range from 406 to 417 nm. As an exception, 3-nitro- and 1-nitrocarbazole show their lowest radiative triplet transitions at 472 nm which is due to mesomeric effect of this strong electron acceptor group on the lowest triplet excited state. Another interesting feature is the long lifetime of phosphorescence (τ_p) of carbazole and carbazole derivatives which are higher than 1 s except for 1-nitro-, 3-nitro-, 3-benzoyl- and 3,6-dibenzoylcarbazole whose lifetime values are shorter than 1 s. These results suggest that the nature of the lowest triplet excited state in each case is most likely to be π,π*, but for nitrocarbazoles as well as 3-benzoyl- and 3,6-dibenzoylcarbazole the lowest triplet excited state would be an n,π* electronic state.

The k_{isc} and k^o_p values of carbazoles decrease significantly relative to diphenylamine and its *N*-methylderivative owing to the increased planarity of the carbazole molecules. *N*-phenylcarbazole shows a k_{isc} and k^o_p slightly enhanced with respect to carbazole due to a forced out-of-plane of the phenyl group from the carbazole moiety. *N*-Acetyl-, *N*-benzoyl-, 3-chloro-, 3-bromo- and 3,6-dibromocarbazole show a noticeable enhancement of k_{isc} and k^o_p rate constants due to a spin-orbital coupling between the S_1 and T_1 promoting the phosphorescence emission process. This behavior is due to two known effects: the heavy atom effect which is enhanced by the presence of halogen groups attached to the carbazole moiety and the n,π* character of the electronic excited state involved when acyl groups are attached to the carbazole moiety. In general, the ϕ_p and ϕ_{isc} measured at 77 K are practically unaltered when methyl, phenyl, methoxy and acetyloxy groups are introduced at the carbazole moiety. In a clear way when the acyl groups, the nitro groups and the halogen groups are atteched to the carbazole moiety the ϕ_p and ϕ_{isc} noticeable increase. Also, the oscillator strength $f(T_1\text{-}S_0)$ for these series of carbazoles were calculated from the radiative phosphorescence rate constants (k^o_p) and the values range between 10^{-7} and 10^{-9} which are characteristic of spin forbidden transitions. Therefore, taking into account the $f(T_1\text{-}S_0)$, the k_{isc} and k^o_p values and the shape of the phosphorescence emission spectra, two dintinctly type of lowest triplet excited state can be addresed for the series of carbazoles studied. Those carbazoles with substituents such as *N*-methyl, *N*-phenyl, *N*-vinyl, 3-bromo, 3,6-dibromo, 3-chloro, 2-methoxy-*N*-methyl and 2-acetyloxy show ^{3}π,π* lowest triplet excited state whereas those carbazoles with C-substituents such as nitro, acetyl and benzoyl show ^{3}*n*,π* lowest triplet excited states.

The HOMO and LUMO energies and the LUMO-HOMO energy difference (ΔE) of the carbazoles studied have been calculated by using the semi empirical PM3 method. These values have been also related to the ^1L_b electronic transition values measured in acetonitrile at room temperature. The results showed that for almost all the carbazoles studied acceptable linear regressions are obtained. This fact means that the contribution of the substituents to the

HOMO orbital is properly described by the semi empirical method used which has the advantage of having the corresponding parameter for halogen substituents included. Additionally, the HOMO energy, the ΔE energy and the I_a values, which were calculated by semi empirical PM3 method, were correlated to the one-electron potential ($E_{p/2}$) of carbazoles. The plots show a linear correlation with positive slopes. The trend observed may be interpreted as an enhancement of the one-electron reductive capability of carbazoles which depends on the substituents. Thus, electron donor and electron acceptor substituents bonded to the *N*- and C-atoms of carbazole moiety produce an increase or a decrease of the $E_{p/2}$ value, respectively. This observation agrees with the fact that the significant change on the $E_{p/2}$ values of carbazole derivatives compared to that of the $E_{p/2}$ value of carbazole is consistent with the inductive and mesomeric effects of the substituents bonded to the carbazole moiety. It is interesting to point out that there are some carbazoles whose $E_{p/2}$ are not known or difficult to be measured by cyclic voltammetry owing to their high irreversibility [36,37]. Therefore, the above linear correlations may be used for estimating the $E_{p/2}$ values of carbazoles from thermodynamic parameters, such as E_{HOMO}, ΔE or I_a, which are easily calculated by semiempirical method.

Bonesi and Erra-Balsells have also studied the intramolecular heavy atom effect on the spectroscopic data and photophysical rate constants of a series of mono-, di-, tri- and tetra-chloro *N*- and C-substituted carbazoles in acetonitrile at 298 K and in solid matrix at 77 K [1]. These data, which were measured with steady-state and time-resolved spectroscopic techniques, have been compared to the corresponding *N*- and C-substituted carbazoles and the effect of incorporation of the chlorine atom to the carbazole moiety proved to quench efficiently the fluorescence emission by spin-orbital coupling. The structures of the series of chlorocarbazoles and the non-chlorinated carbazole derivatives, studied in comparative way, are shown in Table 2.

Table 2. Structure of the chlorocarbazoles and the non-chlorinated carbazole derivatives studied

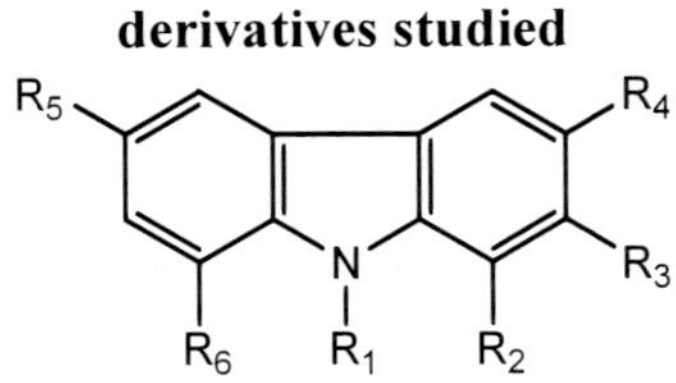

R_1	R_2	R_3	R_4	R_5	R_6
H	H	H	H	H	H
H	H	H	Cl	H	H
H	Cl	H	H	Cl	H
H	H	H	Cl	Cl	H
H	Cl	H	Cl	Cl	H
H	Cl	H	Cl	Cl	Cl
H	H	HO	H	H	H
H	H	HO	Cl	H	H
H	Cl	HO	Cl	H	H
H	H	CH_3CO	H	H	H
H	H	CH_3CO	Cl	Cl	H
CH_3	H	CH_3O	H	H	H

CH_3	H	CH_3O	Cl	H	H
CH_3	H	CH_3O	Cl	Cl	H
CH_3	Cl	CH_3O	Cl	Cl	H
H	H	H	Br	H	H
H	H	H	Br	Br	H
H	H	H	NO_2	H	H
H	H	H	NO_2	Cl	H
Ph	H	H	H	H	H
Ph	H	H	Cl	H	H
CH_3	H	H	H	H	H
CH_3	Cl	H	H	Cl	H
CH_3	H	H	Cl	Cl	H
CH_3	Cl	H	Cl	Cl	Cl
CH_3CO	H	H	H	H	H
CH_3CO	H	H	Cl	H	H
CH_3CO	H	H	Cl	Cl	H

The absorption spectra of carbazoles and chlorocarbazoles were recorded in acetonitrile at 298 K and generally they showed three bands at 260, 298 and 350 nm. The bands located in the 298 and 350 nm region were assigned as ^1L_b ($S_1 - S_0$) and ^1L_a ($S_2 - S_0$) electronic transitions, respectively. It was observed that the introduction of one to four chlorine atoms as substituents in the carbazole moiety produced no change on the nature of the excited singlet state which is a π,π^* excited state. The absorption spectra of the chlorocarbazoles recorded in acetonitrile at room temperature exhibit a noticeable bathochromic shift of the ^1L_b ranging between 10 and 22 nm compared to the non-chlorinated carbazoles. This large red shift accounts for the in-plane conjugation of the chlorine substituents. Also, it can be interpreted by the mesomeric effect induced by the chlorine atoms bonded to the carbazole moiety. In general, the absorption spectrum of chlorocarbazoles show a small bathochromic shift of the ^1L_a electronic transition band from 2 to 10 nm as the number of chlorine atoms is increased.

The oscillator strength $f(S_1 - S_0)$ associated to ^1L_b electronic transition of carbazoles were calculated from the radiative fluorescence rate constant ($k^o{}_f$) and the wavenumber of the absorption – fluorescence crosspoint. These parameters depends on the number of chlorine substituents attached to the carbazole moiety and the values are similar or lower than that of the corresponding non-chlorinated carbazoles. This dependence was attributed to a change in the charge transfer character of the ^1L_b electronic transition, and therefore, a noticeable change of the transition dipole moment ($\Delta\mu$) associated to these electronic transition could occur.

Systematically, the fluorescence and absorption spectra of the carbazoles under study recorded in acetonitrile at 298 K display excellent symmetry with exact overlap of 0,0 band. This behavior implies that the geometries of the molecules are quite similar in ground and excited state. Also, the λ_{max}(fluo) of the chlorocarbazoles show a significant bathochromic shift (10 to 24 nm) as the number of chlorine atoms attached to the carbazole moiety increases. This trend is due to an in-plane conjugation of the chlorine atom with the carbazole moiety. An interesting feature to be pointed out is the intramolecular quenching of the fluorescence emission of chlorocarbazoles as the number of the chlorine atoms attached to the carbazole moiety is increased. Thus, the fluorescence quantum yield reflects quantitatively the above mentioned trend and the values decrease 10 to 50 times as the number of chlorine atoms is increased. The fluorescence lifetimes of chlorocarbazole derivatives was also measured in acetonitrile at room temperature and the values decrease significantly as the number of chlorine atoms are increased, *e.g.*, 2-hydroxycarbazole, 12.3 ns; 3-chloro-2-hydroxycarabzole, 2.47 ns and 1,3-dichloro-2-hydroxycarbazole, 0.20 ns.

The individual rate constants of the radiative and radiationless processes from S_1 (k_f, $k^o{}_f$ and k_{isc}) were derived from the measured ϕ_f and τ_f. These data clearly show that the $k^o{}_f$ values of chlorocarbazoles decrease sligthly with respect to those of the non-chlorinated carbazoles. As expected, the efficient deactivation process of the singlet excited state of chlorocarbazole derivatives through the intersystem crossing channel is highly favored owing to both rigidity and planarity of the carbazole structures and the spin-orbit coupling process between the singlet and triplet excited states [38,39].

For the case of 3-chloro- and 3,6-dichloro-*N*-acetylcarbazole it was observed that $k^o{}_f$ and ϕ_f increase sligthly in comparison to *N*-acetylcarbazole while k_{isc} decrease. The particular trend is due to a change in the geometry of S_0 and S_1 as well as for an n,π^* and π,π^* mixture character of the excited states. Also, this "inverse" heavy atom effect is explained in terms of the influence of chlorine substitution on the position of T_2. Thus, the T_2 electronic state is lowered in energy so that it falls below S_1 and a S_1 – T_2 radiationless mechanism for intersystem crossing becomes available.

The total luminscence spectra (fluorescence and phosphorescence emission spectra) of carbazoles and chlorocarbazoles were measured in solid matrix at 77 K. These compounds exhibit structured fluorescence and phosphorescence emission spectrum with excitation to the S_1 level. The shape and λ_{max}(fluo) of the fluorescence emission spectra do not change on going from 298 to 77 K. The phosphorescence emission spectra of chlorocarbazoles shows λ_{max}(phosph) in the region of 410 to 434 nm while 6-chloro-3-nitrocarbazole shows λ_{max}(phosph) at 470 nm. This noticeable red shift is due to a significant mesomeric effect of the nitro group on the lowest triplet excited state.

The phosphorescence lifetimes (τ_p) of chlorocarbazoles range between 2.75 s to 30 ms and are lower than those of the non-chlorinated carbazoles. For the case of 6-chloro-3-nitrocarbazole it was observed that the introduction of a chlorine atom to 3-nitrocarbazole has no effect on the τ_p value. The ϕ_p values show a noticeable enhancement as the number of chlorine atom are attached to the carbazole moiety which reflect a greater efficiency of population of T_1 (k_{isc} is enhanced) and a greater efficiency of emission from T_1. A comparative analysis of the photophysical rate constant $k^o{}_p$, k_{isc} and k_{nr}, the shape of the phosphorescence emission spectra and the values of $f(T_1\text{-}S_0)$ let to conclude that the lowest triplet excited state for the chlorocarbazoles is most likely to be a $^3\pi,\pi^*$ electronic state while for the 3-chloro-*N*-acetyl- and 3,6-dichloro-*N*-acetylcarbazole the T_1 electronic state is to be likely an $^3n,\pi^*$.

The values of the HOMO and LUMO energy and the LUMO-HOMO energy difference (ΔE) of the chlorocarbazoles calculated by using the semiempirical PM3 method were related to the ^1L_b electronic transition values measured in acetonitrile. These correlations show good linear regressions that suggest a significant contribution of the chloro substituents in the HOMO orbital and, therefore, the red shift observed on the absorption spectrum accounts for this contribution. Also, the adiabatic ionization potential (I_a) calculated by semi empirical PM3 methods are related to the one-electron potential ($E_{p/2}$) values and good linear correlation is obtained. From the slope of this correlation the free solvation energy ($\Delta G(s)$) values of carbazole radical-cations can be estimated which are intermediates that are formed during electrochemical and photochemical processes.

The preparation, isolation and characterization (elemental analysis, M.p., ^{1}H-NMR, ^{13}C-NMR and MS) of the bromo derivatives of *N*-substituted carbazoles (*N*-methylcarbazole (**1**), *N*-phenylcarbazole (**2**), *N*-benzylcarbazole (**3**), 2-methoxy-*N*-methylcarbazole (**4**)) and *C*-substituted carbazoles (2-acetyl- oxycarbazole (**5**) and 3-nitrocarbazole (**6**)) was recently reported by Erra-Balsells et al. [3].

Table 3. Structure of the bromocarbazoles studied

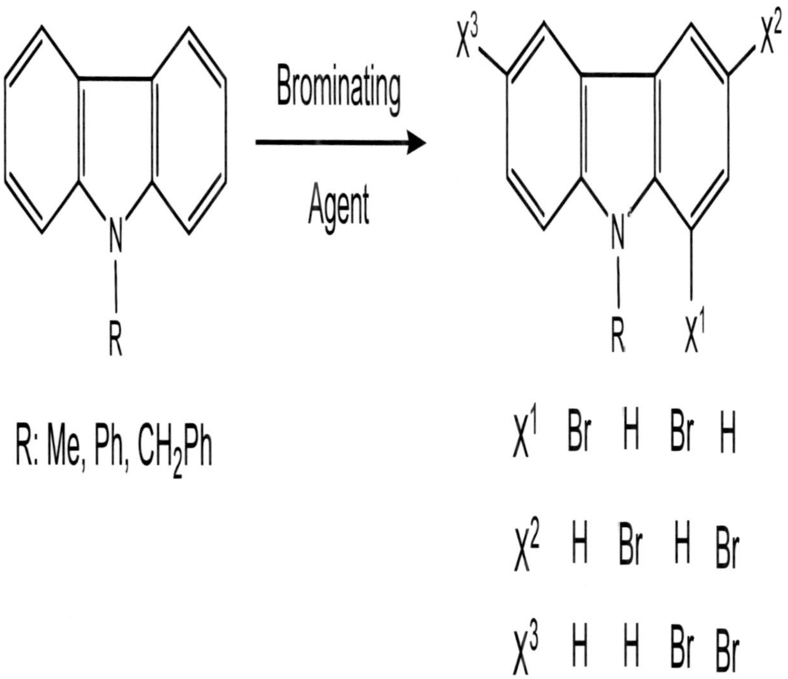

The brominating reaction of carbazoles **3**, **4**, **5** and **6** and the characterization of the corresponding bromo derivatives were described for the first time. Semi empirical PM3 calculations were performed in order to predict reactivity of *N*-substituted and *C*-substituted carbazoles and their bromo derivatives. Electronic spectra (UV-absorption, fluorescence and phosphorescence emission spectra) of bromocarbazole derivatives in acetonitrile solution at 298 K and in solid matrix at 77 K were recorded and compared with those of the corresponding carbazoles **1** – **6**. The dynamic properties of the lowest excited singlet and triplet states (τ_f, τ_p, ϕ_f and ϕ_p) have been measured in the same experimental conditions. The bromine and nitro intramolecular spin – orbital coupling effect on the spectroscopic data, photophysical parameters and on the photoreactivity were also analyzed briefly.

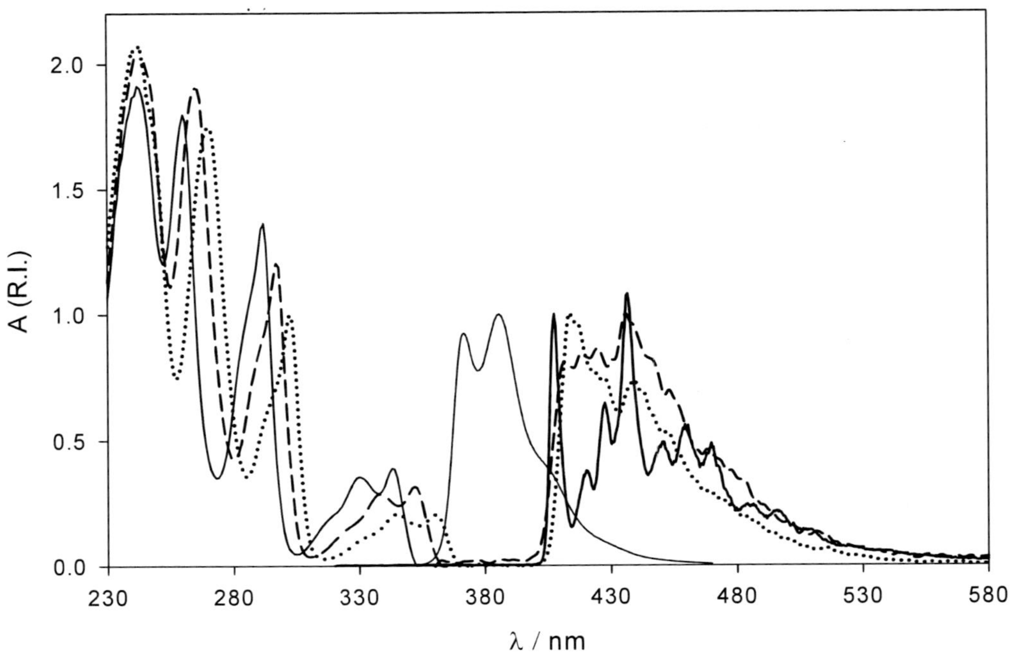

Figure 2. Electronic Absorption and fluorescence emission spectra (acetonitrile; 298 K) and phosphorescence emission spectra (solid matrix; 77 K) of *N*-methylcarbazole (1) (solid line), 3-bromo-*N*-methylcarbazole (1b) (dashed line) and 3,6-dibromo-*N*-methylcarbazole (1c) (dotted line). Concentration: 1.0 x 10^{-5} mol.dm^{-3}; λ_{exc}: 320 nm.

The spectroscopic studies performed on carbazoles **1** – **4** and their bromo derivatives (**1b**, **1d**, **2a**, **2b**, **3a**, **3b**, **4b** and **4c**) showed a noticeable bathochromic shift on the absorption and phosphorescence emission spectra at 298 K and 77 K as the number of bromine atoms attached to the carbazole moiety increased (Figure 2). This red shift is due to mesomeric and / or inductive effects owing to the higher electronic density of the bromine compared with hydrogen as substituent at C(3) and / or C(6). Also, these bromo carbazole derivates are shown to be good phosphorescent chromophores in solid matrix at 77 K due to a high spin – orbital coupling mechanism between the bromine atoms and the carbazole moiety with ϕ_P values in the range of 0.23 to 0.99 (see Figure 2 and data in Table 4).

The shape of the phosphorescence emission spectra of these compounds are "naphthalene – like" and taking into account the phosphorescence lifetime values the lowest triplet excited state is most likely to be π,π^* electronic transition. In the case of compounds **6** and **6a** the "benzophenone – like" shape of the phosphorescent emission spectra indicates that the lowest triplet excited state is a n,π^* electronic transition. The effect of bromine on the n,π^* transitions is very minor because compounds **6** and **6a** showed similar absorption and phosphorescence spectra (Table 4).

Table 4. Spectroscopic data (λ_{max}), fluorescence quantum yield (ϕ_f) and phosphorescence quantum yield (ϕ_p) of carbazoles and bromo carbazoles in MeCN at 298 K and in solid matrix at 77 K

Compound	λ_{max}(abs) /nm [a])	λ_{max}(fluo) /nm [a])	λ_{max}(phosph) /nm [b])	ϕ_f [a])	τ_f / ns [c])	ϕ_p [b])
1	345	356	408	0.43	14.7	0.24
1b	355	NF [d])	414	NF	-	0.56
1d	362	NF	417	NF	-	0.74
2	342	358	407	0.34	11.4	0.35
2a	352	NF	415	NF	-	0.69
2b	361	NF	424	NF	-	0.88
3	342	355	410	0.40	12.3	0.47
3a	351	NF	415	NF	-	0.70
3b	357	NF	425	NF	-	0.99
4	330	350	412	0.45	14.3	0.23
4b	342	NF	426	NF	-	0.77
4c	349	NF	428	NF	-	0.94
6	365	NF	480	NF	-	0.04
6a	365	NF	481	NF	-	>0.001

[a]) Measured in MeCN at 298 K. [b]) Measured in solid matrix (isopropanol – diethylether (1:1)) at 77 K. [c]) Values taken from [26a]. [d]) NF: no fluorescent detected.

In conclusion, the regioselectivity of the brominating reaction provides an alternative method for the synthesis of selected bromo heterocyclic compounds which are not commercially available. These compounds are interesting examples to be tested as substrates for the study of the photoinduced heterolytic cleavage reaction of the C – Br bond. In this regard, a previous study in terms of spectroscopic and photophysical analyses of bromo carbazole derivatives was needed in order to asses which lowest excited state (singlet or triplet state) is efficiently populated. Therefore, from this study is possible to address that compounds **1b**, **1d**, **2a**, **2b**, **3a**, **3b**, **4b**, **4c** and **6a** populate their lowest triplet excited states efficiently and, from these results, it is expected that the photoinduced heterolytic cleavage reaction would take place as well.

Mesaros *et al.* have studied the photophysical behavior of 1-nitrocarbazole and 3-nitrocarbazole in different organic solvents using spectroscopic, luminiscence and photoacoustic techniques [4]. Absorption spectra and triplet state transient spectra were recorded and formation quantum yields and lifetimes determined. Also, phosphorescence emission at 77 K and laser induced luminiscence in acetonitrile solution at 298 K for the nitrocarbazoles were studied. It was observed that the nitro group attached to the carbazole moiety induces significant changes in the photophysical behavior of those compounds.

Since carabzole is a good fluorescent chromophore in acetonitrile at room temperature (ϕ_f: 0.62 and τ_f: 15.1 ns), 1-nitro- and 3-nitrocarbazole do not show any fluorescent emission under similar experimental conditions. Thus, from the electronic excited singlet state one or two radiationless deactivation processes could be operating: (i) an efficient intersystem crossing process and/or (ii) an efficient release of heat to the medium. In this regard, the ϕ_{isc}

(or ϕ_T) of the nitrocarbazoles were measured by laser flash photolysis at λ = 355 nm in acetonitrile, under Ar atmosphere at 298 K. In these experiments, benzophenone was selected as a reference and as a triplet energy donor whose ϕ_T value is known (ϕ_T: 1.00) while 1-cyanonaphthalene was selected as a triplet energy acceptor. The ϕ_T values thus obtained are as follow: carbazole, 0.38; 1-nitrocarbazole, 0.40; 3-nitrocarbazole, 0.50. The high ϕ_T values obtained indicates that *ca.* 50% of the initial electronic excited state S_1 population is involved in the intersystem crossing process $S_1 \rightarrow T_1$. Taking into account that the nitrocarbazoles are not fluorescent chromophores, it was clearly concluded that the other 50% of the absorbed photons should be released to the medium as heat through a radiationless internal conversion pathway from the S_1 state. The triplet excited state of nitrocarbazoles were characterized by laser flash photolysis and the electronic absorption of the triplet transients were recorded. Also, the triplet lifetime (τ_T) of 3-nitrocarbazole was measured at λ: 500 and 650 nm. At both wavelengths, the triplet transient decay was monoexponential giving a value of 500 ns. For the case of 1-nitrocarbazole, no transient triplet absorption spectra was obaitned and a τ_T lesser than 10 ns was estimated.

Photoacoustic measurements were performed at 298 K in order to determine the prompt heat released to the medium by the nitrocarbazoles after excitation. In these experiments, 2-hydroxybenzophenone was used as a calorimetric reference, the solutions were electronically excited with a laser beam (λ: 355 nm) and the photoacoustic signals were detected with a piezoelectric transducer (PZT crystal, resolution time: 1.5 μs) and a PVF_2 film (resolution time: 150 ns). Correlation of the photoacoustic signal amplitude (H) of the calorimetric reference and the samples (nitrocarbazoles) with the excitation fluence (F) of the laser beam gives nice linear regressions with slopes equal to 1 when a PZT transducer was used as a detector. The slope of this linear relationships is known as the ratio of the fraction of the energy released in the medium as prompt heat by the calorimetric reference (α_R) and the sample (α) within the time resolution of the experiments. Since the calorimetric reference has an α_R = 1, from the slope the sample α value is easily obtained. Thus, for the 1-nitrocarbazole α = 1 was obtained when the experimental time resolution ranged from 1.5 μs to 150 ns. The same value was obtained when the experiments were carried out in different organic solvents of different polarity. As a result, it can be concluded that 1-nitrocarbazole releases to the medium all the absorbed energy as heat in a time shorter than 150 ns. This result agree with those obtained by LFP. These experiments were also performed for 3-nitrocarbazole and the results obtained were quite different. For both resolution time the H_S/H_R ratio were 0.88 (1.5 μs) and 0.77 (150 ns) and this implies that a transient species with a lifetime between 150 ns and 1.5 μs is formed and it stores part of the absorbed energy. Then, a deconvolution method was used in order to estimate the prompt heat released to the medium and the lifetime of the transient. The results thus obtained give an α value in the range 0.65 – 0.74 and a lifetime of the transient between 350 – 650 ns. The last estimation was in complete agreement with the triplet state value τ_T = 500 ns determined by laser flash photolysis experiments.

The measured lifetime of the lowest triplet state of 3-nitrocarbazole is *ca.* 50 times higher than that of 1-nitrocarbazole. This noticeable behavior can be rationalized due to the different location of the nitro group as substituent in the carbazole moiety, that allows the formation of an intramolecular hydrogen bond only in 1-nitrocarbazole through a keto-enol-like equilibrium (Figure 3). Then, a preferential deactivation pathway for the T_1 state of 1-nitrocarbazole is through radiationless mechanism. On the contrary, for the T_1 state of 3-

nitrocarbazole, radiative deactivation through phosphorescence emission is a quite efficient process.

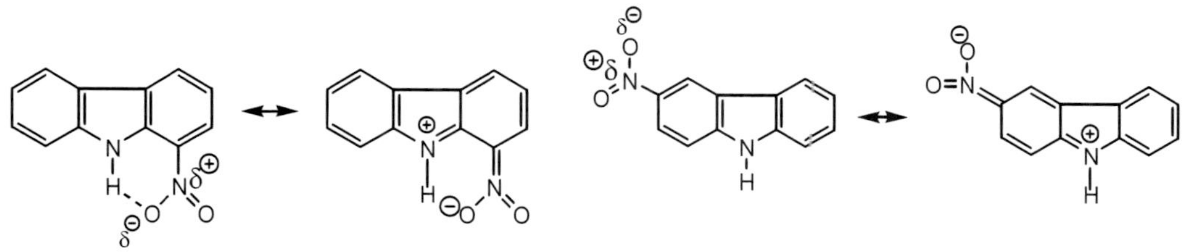

1-Nitrocarbazole 3-Nitrocarbazole

Figure 3. Keto-enol-like equilibrium for 1- and 3-nitrocarbazole.

Recently, Bonesi et al. [5] have carried out a comparative study of a series of nitrocarbazoles such as 1-nitro-, 3-nitro, 1,6-dinitro- and 3,6-dinitrocarabzole in terms of the absorption and emission properties in steady-state conditions and the calorimetric behavior and the triplet state properties using photoacoustic measurements and laser flash photolysis spectroscopy.

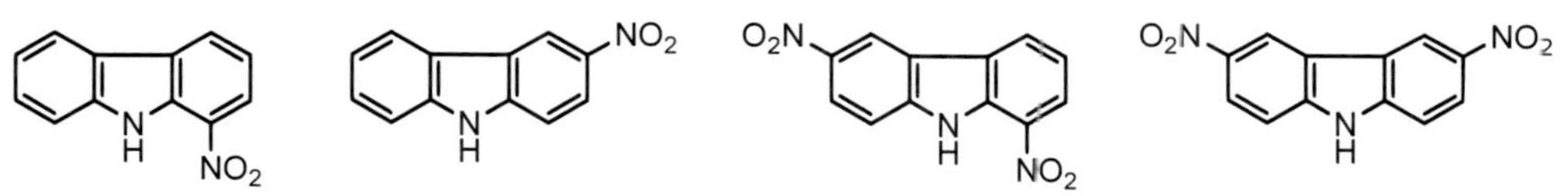

1-Nitrocarbazole ***3-Nitrocarbazole*** ***1,6-Dinitrocarbazole*** ***3,6-Dinitrocarbazole***

Figure 4. Nitrocarbazoles studied.

The absorption spectra of carbazole and nitrocarbazoles were recorded in acetonitrile at 298 K and the bands located in the 300 and 350 nm regions as ^1L_a (S_2-S_0) and ^1L_b (S_1-S_0) electronic transitions, respectively, were unambiguously assigned. The nitrocarbazoles showed a noticeable bathochromic shift of the ^1L_b band (26 to 70 nm) compared to that of carbazole, whereas, the ^1L_a band showed a small bathochromic shift ranging between 8 and 28 nm.

The oscillator strength $f(S_1\text{-}S_0)$ and the dipole moment associated to the ^1L_b electronic transition of the nitrocarbazoles are higher than that of carbazole, suggesting that a significant change in the charge transfer character of the ^1L_b transition accounts for this behavior. Instead, the S_2-S_0 electronic transitions of nitrocarbazoles show $f(S_2\text{-}S_0)$ and $\Delta\mu$ values lower than that of carbazole.

The nature of the lowest singlet excited state of nitrocarbazoles are most likely to be π,π^* as in the case of carbazole due to a small bathochromic shift of the ^1L_b electronic transition on going from non-polar to polar solvents.

Since it was pointed out that carbazole is a good fluorescent chromophore in acetonitrile at 298 K, the mono- and dinitrocarbazoles did not show any fluorescent emission in similar experimental conditions. Thus, the singlet excited state of these compounds deactivates through an efficient intersystem crossing process and/or an efficient internal conversion. The ϕ_T values for the nitrocarbazoles were measured by laser flash photolysis at λ_{exc}: 355 nm in acetonitrile under Argon atmosphere at 298 K. The triplet lifetimes of these compounds were also measured under similar experimental conditions and show monoexponential decays. The ϕ_T and τ_T values are shown in Table 5.

Table 5. ϕ_T, τ_T and ϕ_{IC} values calculated for carbazole and nitrocarbazoles studied

Compound	ϕ_T	τ_T / ns	ϕ_{IC}
Carbazole	0.38	----	----
1-Nitrocarbazole	0.40	> 10	0.60
3-Nitrocarbazole	0.50	500	0.50
1,6-Dinitrocarbazole	0.67	130	0.33
3,6-Dinitrocarbazole	0.70	390	0.30

The high ϕ_T values indicate that the intersystem crossing process ($S_1 \rightarrow T$) is involved in the radiationless deactivation of the S_1 state. Also, this process increases as the number of nitro groups attached to the carbazole moiety is increased since the well-known intramolecular spin-orbital coupling effect of the nitro group operates efficiently. Taking into account that the nitrocarbazoles are not fluorescent chromophores, the radiationless internal quantum yield (ϕ_{IC}) can be easily derived ($\phi_{IC} = 1 - \phi_{ISC}$). These calculated values are among 0.30 and 0.60 and suggest that the internal conversion deactivates the S_1 state of nitrocarbazoles and competes efficiently with the intersystem crossing process.

The triplet transient lifetimes (τ_T) of mono- and dinitrocarbazoles show noticeable differences and the trend observed is due to the different location of the nitro group as substituent in the carbazole moiety. A rationalization of this trend was interpreted in terms of the facile intramolecular hydrogen bond formation between the NH group and the nitro group attached to the C-1 of the carbazole moiety, which are spatially vecinal each other, for 1-nitrocarbazole and 1,6-dinitrocarbazole (Figure 5). Of course, it is not possible for 3-nitro- and 3,6-dinitrocarbazole since the nitro group is attached to different positions in the carbazole moiety. The optimized geometry obtained at HF/3-21G level clearly shows that this hydrogen bond is possible because of the distance between the involved groups is *ca* 2.4 Ü. The out-of-plane fixed position of the nitro group with respect to the plane of the carbazole moiety could cause the triplet excited state to favor its deactivation and hence, shorter triplet transient lifetimes (τ_T) are observed.

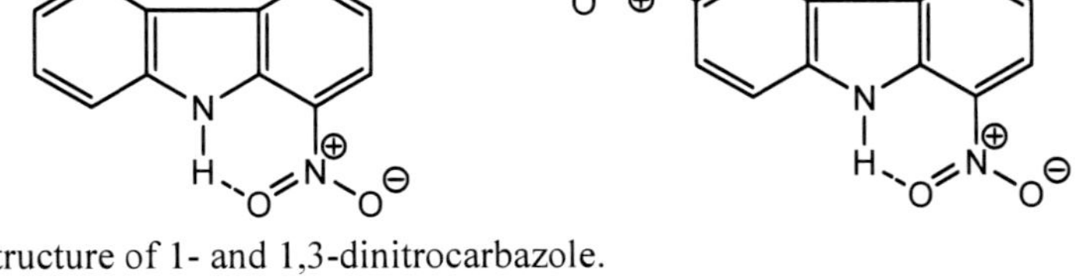

Figure 5. Keto-structure of 1- and 1,3-dinitrocarbazole.

The phosphorescence emission spectra of nitrocarbazoles were measured in solid matrix at 77 K. The shape of these spectra are very similar, showing three broad bands which resembles to that of the "benzophenone-like" molecules, but are quite different than those observed in the structured phosphorescence spectrum of carbazole. Also, the λ_{max}(phosph) of the nitrocarbazoles show a noticeable red shift compared to that of carbazole and mesomeric effect of the nitro group accounts for this behavior.

The phosphorescence lifetime (τ_p) and the phosphorescence quantum yield (ϕ_p) were measured in similar experimental conditions and the dynamic properties of the lowest triplet excited state (k^o_p and k_{nr}) were easily calculated. These photophysical data show that carbazole and 3-nitrocarbazole deactivate the lowest triplet excited state (T_1) through an efficient phosphorescence emission process, whereas, 1-nitro-, 1,6-dinitro- and 3,6-dinitrocarbazole deactivate the T_1 state through a radiationless pathway. Therefore, a comparative analysis of k^o_p, τ_p, $f(T_1\text{-}S_0)$ and the shape of the phosphorescnce spectra suggest that the lowest triplet excited state of nitrocarbazoles are to be an n,π^* electronic state. On the contrary, the same analysis for carbazole suggests that the lowest T_1 state is a π,π^* electronic state.

In order to study the thermal deactivation process of the nitrocarbazoles photoacoustic experiments were conducted in acetonitrile at room temperature under N_2 atmosphere. The solutions were irradiated with a pulse of λ_{exc}: 355 nm and the acoustic signal were detected with a PZT transducer set-up (τ_R: 1.5 μs) and with a PVF_2 transducer set-up (τ_R: 150 ns). 2-Hydroxybenzophenone was used as a calorimetric reference. The results obtained from these experiments are collected in Table 6.

Table 6. α and τ_T values calculated for nitrocarbazoles studied

Compound	α (τ_R: 150 ns)	α (τ_R: 1.5 μs)	τ_T / ns
1-Nitrocarbazole	1	1	> 30
3-Nitrocarbazole	0.77	0.88	$350 < \tau_T < 650$[a]
1,6-Dinitrocarbazole	0.76	1	$30 < \tau_T < 300$[b]
3,6-Dinitrocarbazole	0.76	1	$30 < \tau_T < 300$[b]

[a] Estimated from a deconvolution procedure. [b] Estimated from the resolution of the experiments

These results show that the dinitrocarbazoles have a value of α < 1 measured with a faster PVF_2 set-up which means that in this conditions there is an energy storing species present, *i.e.*, the triplet state. Therefore, 1,6-dinitrocarbazole and 3,6-dinitrocarbazole release a fraction of the absorbed energy to the medium as prompt heat. This calorimetric behavior is similar to that observed for 3-nitrocarbazole while 1-nitrocarbazole release all the absorbed energy as heat, also with the faster PVF_2 set-up transducer.

(b) *The use of carbazoles as photosensitizers*

The direct photoreduction of aromatic sulfoxides (λ_{exc}: 254 nm) in the presence of sodium methoxide in methanol has been studied in detail by Kropp et al. from mechanistic and preparative viewpoints [40]. Latter, Cubbage et al. and Jenks et al. studied the photosensitized reduction of the same family of compounds using carbazole, *N*-methylcarbazole and *N*-ethylcarbazole as photosensitizers [41,42]. In these experiments carbazoles can be used as photosensitizers because of they absorb considerably to the red (λ_{exc}: 350 nm) than the aromatic sulfoxides. Different sulfoxides have been subjected to react in the presence of carbazoles: diphenylsulfoxide, di(*p*-fluorophenyl)sulfoxide, dibenzothiophene sulfoxide and dibenzyl sulfoxide. The photoreactions were carried out in methanol, as the best solvent of choice among those investigated (methanol, acetonitrile and THF) under argon atmosphere. Also, the photoreaction is not interfered with by many reasonable starting materials or products and quantitative sulfide yields can be obtained (yields higher than 90%). Carbazole derivatives do not degrade significantly during the course of the reaction whereas anilines, such as *N*-methylaniline and *N,N*-dimethylaniline, used as sensitizers instead of carbazoles, tend to degrade somewhat during the course of the reaction and higher energy irradiation must be used. Comparing different photoinduced reduction reactions of arylsulfoxides: direct excitation of arylsulfoxides in the presence of methoxide, photosensitization with anilines and photosensitization with carbazoles, seem to make the last photosensitized reaction the method of choice.

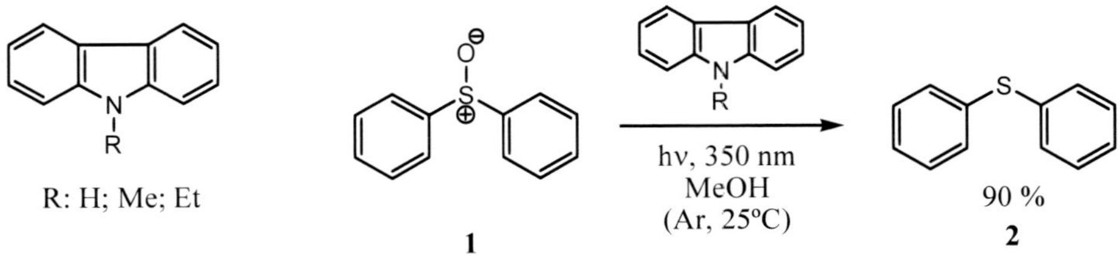

Scheme 1. Photosensitized reduction of arylsulfoxides.

Thus, the quantum yield of appearence of arylsulfide in the *N*-methylcarbazole photosensitized reduction of arylsulfoxide increases as the concentration of the sulfoxides changes in a range of 10 – 200 mM. Also, singlet quenching rate constants (k_q) were determined by monitoring the fluorescence lifetime of the carbazoles as the function of sulfoxide concentration, using the time-correlated single photon counting method. The k_q values are diffussion controlled ($\sim 10^9$ $M^{-1}s^{-1}$) for all the arylsulfoxide studied in the presence of carbazoles as the fluorescent chromophores. Triplet sensitization by energy transfer by carbazoles is not plausible, given the triplet energy values of carbazole ($E_T \sim 70$ kcal/mol) [33] and simple arylsulfoxides ($E_T \sim 80$ kcal/mol) [42]. Addition of isoprene ($E_T \sim 60$ kcal/mol) [33] at concentrations up to 10 mM did not affect the quantum yield of the photosentitized reduction, thus eliminating triplet-based chemistry.

Evidence of the photoinduced reduction of arylsulfoxides points toward a hydroxysulfuranyl radical (Scheme 2, **4**) as the key intermediate, and solvent effects point to heterolysis (***path b***), rather than homolysis (***path a***), as the step that breaks the S–O bond (Scheme 2).

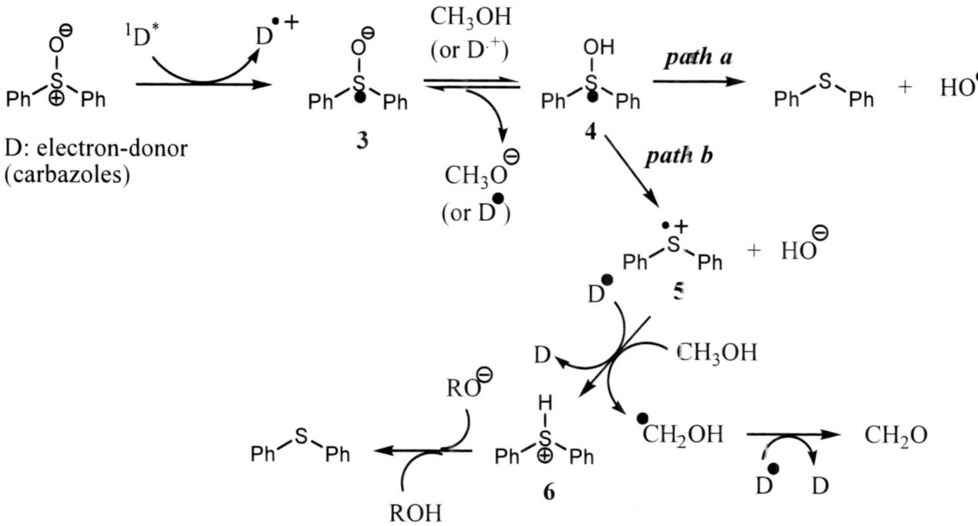

Scheme 2. Reaction steps for the photosensitized reduction of arylsulfoxides.

Dialkylhydroxysulfuranyl radicals (*e.g.*, **4**) have been the subjects of study in other contexts. In the gas phase, reaction of dimethyl sulfide with $HO^{\cdot}$ produces $(CH_3)_2SOH^{\cdot}$. While there has been controversy over the topic, both experimental and computational approaches have recently converged on an S–O homolytic bond dissociation energy of about 11 kcal/mol [43-49]. Although, the universally reported fate that $(CH_3)_2SOH^{\cdot}$ in aqueous solution suffers an heterolytic dissociation to $(CH_3)_2S^{\cdot+}$ and HO^{-}. Thermodynamic estimations provide a ΔG^{o} value for this process of about 5 kcal/mol. Therefore, the difference between the gas and solution phase decomposition routes is directly related to the solvation of ions. It is reasonable to postulate that in alcoholic solvents, the difference between homolysis and heterolysis may be smaller than in water, and homolysis may be favored in some of the solvents tested, such as THF.

It was shown that the fluorescence quenching of carbazoles by arylsulfoxides is diffusion controlled so an electron transfer process was considered thermodynamically feasible even if the back-electron transfer process could significantly compete. As is shown in Scheme 2, after electron transfer, proton transfer from the N–H or C–H groups of the carbazole radical cations ($D^{\cdot+}$) to **3** is expected, given the reduced pK_as of radical cations [50,51]. Thus, if heterolytic dissociation (***path b***) of **4** is rapid, then reaction between methanol and the radicals derived from *N*-methyl- or *N*-ethylcarbazole (*e.g.*, $D^{\cdot}$) is likely to produce $^{\cdot}CH_2OH$, while regenerating the sensitizer D. The N–H bond strengths of carbazole are in the same range, though the exact literature values differ slightly [50-52]. This transient involvement of sensitizer-derived radicals is also consistent with the gradual decomposition of the sensitizers.

Another hint to be pointed out regarding the low consumption of the sensitizer (carbazoles) is the ease of its regeneration. In the following scheme, Scheme 3, is outlined the regeneration process of the sensitizer (D).

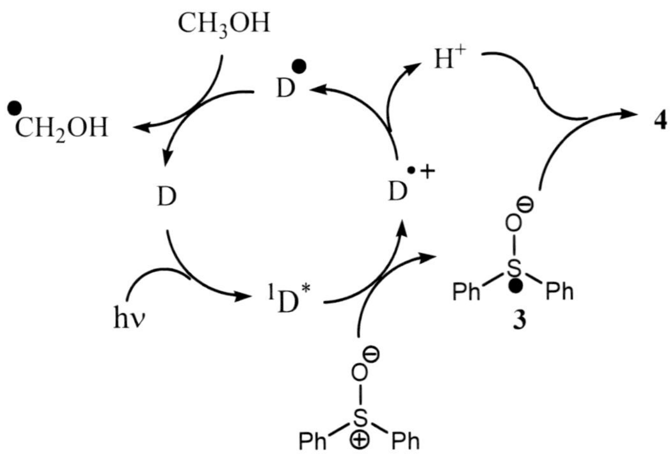

Scheme 3. Photosensitized reduction of arylsulfoxides, Regeneration of the sensitizer (D).

A general and practical usefull method for selective deoxygenation of secondary alcohols via benzoates by using of PET reactions has been studied by Saito *et al.* [53]. *N*-Methylcarbazole have been used as photosensitizer and the photoirradiations were carried out in THF – H_2O (10:1) solution under nitrogen through a Pyrex filter (λ_{exc} > 300 nm) (Scheme 4).

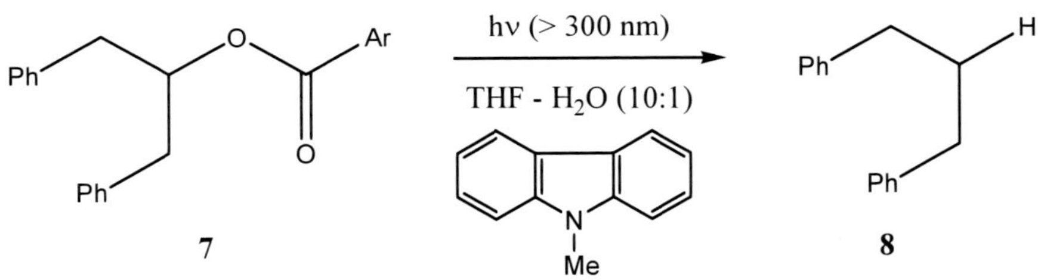

Ar. (a) *m*-$CF_3C_6H_4$;
(b) *p*-ClC_6H_4; (c) C_6H_5

Scheme 4. Deoxygenation of secondary alcohols photosensitized by N-methylcarbazole.

Compound **8** is obtained in 91%, 68% and 83% yield, respectively, depending on the nature of the benzoates **7a** – **c** used as starting material. The photosensitized reaction works well when water is used as co-solvent and no reaction occurs in the dark or in the absence of the photosensitizer (*N*-methylcarbazole). The free energy (ΔG) values associated with the electron-transfer process, calculated by using the Rehm-Weller equation, are large negative for all the benzoates **7a** – **c** studied. The fluorescence emission of *N*-methylcarbazole is quenched by the aryl esters **7a** – **c** giving near diffusion-controlled quenching rate constants (k_q ~ $4.5x10^9$ $M^{-1}.s^{-1}$). These results suggest an electron-transfer from the singlet *N*-methylcarbazole to the esters **7a** – **c**. It is interesting to point out that the photosensitized deoxygenation reaction of these benzoates is also chemoselective. As is shown in Scheme 5, Irradiation of compound **9** in a (10:1) mixture of isopropanol – water in the presence of *N*-methylcarbazole as photosensitizer led to the selective reduction of the ester of the tertiary alcohol to give a (1:1) mixture of compounds **10** and **11** in 85% yield. This result indicates the inertness of the ester of a primary alcohol toward the photosensitized reduction.

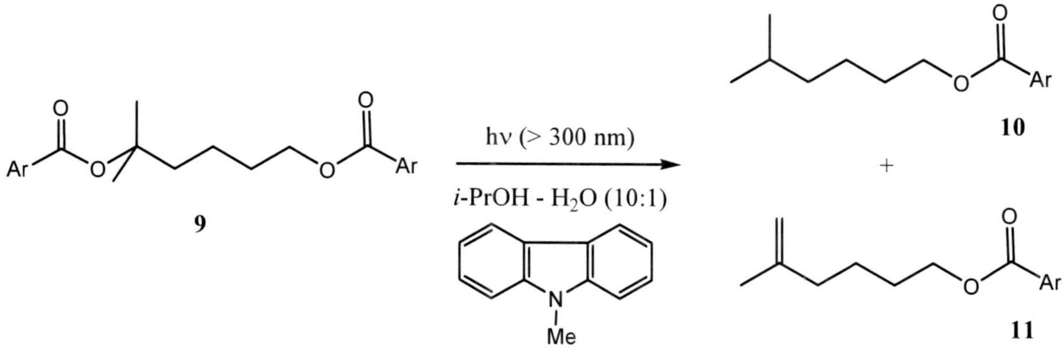

Scheme 5. Chemoselective deoxygenation of tertiary alcohols photosensitized by N-methylcarbazole.

The success of the selective photosensitized deoxygenation of secondary alcohols via benzoates (or aryl esters) in the presence of *N*-methylcarbazole led to the useful synthesis of 2',3'-dideoxyadenosine **13** (Scheme 6). Irradiation of 2',3',5'-tri-*O*-[*m*-(trifluoromethyl)benzoyl]-adenosine (**12**) in a (1:1) mixture of isopropanol – water in the presence of *N*-methylcarbazole gives compound **13** in 73% yield.

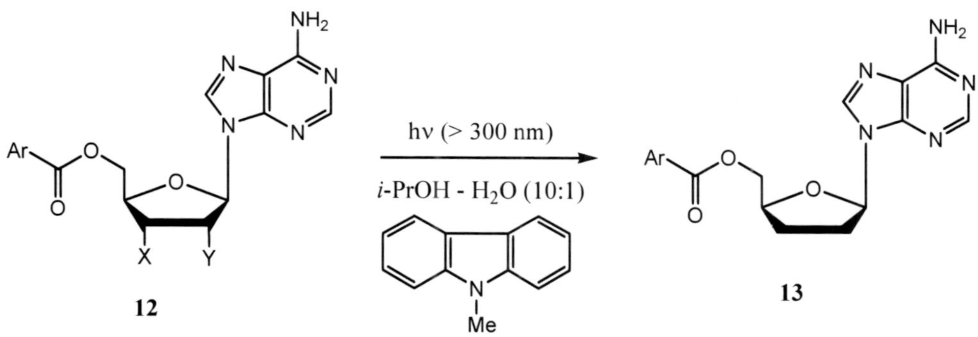

X = Y = OCOAr
(Ar: *m*-$CF_3C_6H_4$)

Scheme 6. Chemoselective deoxygenation of nucleosides photosensitized by N-methylcarbazole.

The result also indicates a higher reactivity of the C-2' and C-3' esters compared to the C-5' ester. Again, a noticeable chemoselectivity is observed.

It is interesting to point out that the sensitizer, *N*-methylcarbazole, was mostly recovered unchanged indicating a recycling of the sensitizer probably via the electron-transfer from the solvent radical ($S^{\cdot}$) to the sensitizer radical-cation ($D^{\cdot +}$) as is depicted in Scheme 7.

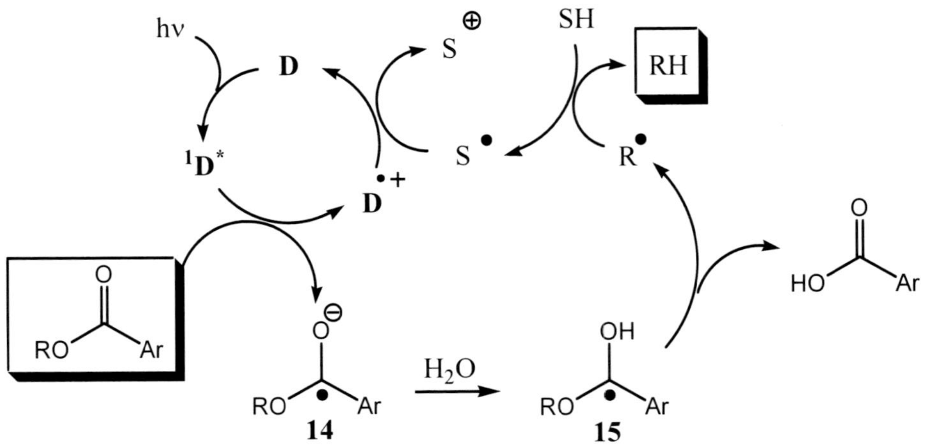

Ar: *m*-$CF_3C_6H_4$; *p*-ClC_6H_4; C_6H_5
SH: solvent

Scheme 7. Mechanism for deoxygenation of alcohols photosensitized by N-methylcarbazole (D).

Ten years latter, Park and Rizzo developed an stereocontrolled *de novo* synthesis of β-2'-deoxyribonucleosides [54]. Nucleosides and nucleoside analogs have long been an important class of medicinal agents possessing anticancer and antiviral activity. Modified nucleosides have also played a central role in the development of genetic therapy such as triplex (antigene) and antisense strategy [55-63]. The synthetic strategy for the *de novo* synthesis of 2'-deoxynucleosides is outlined in the Scheme 8, providing a valuable method for the synthesis of natural and unnatural 2'-deoxynucleosides of biomedical interest.

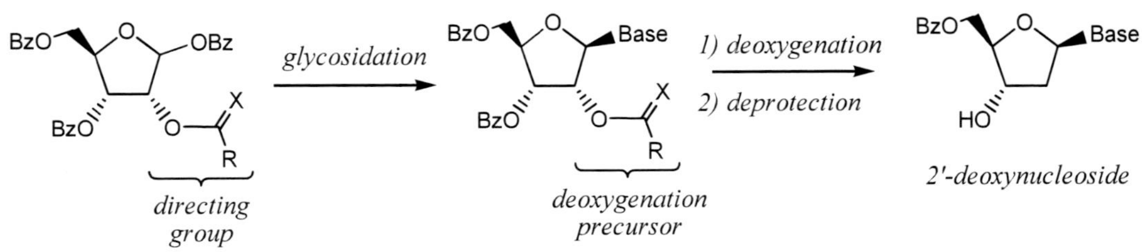

Scheme 8. Synthesis of 2'-deoxynucleosides.

This methodology is more efficient and involves minimun use of protecting groups, making the procedure simpler and more economical. It is interesting to point out that the deoxygenation step, the key step toward the target 2'-deoxynucleoside, is the photosensitized reduction of esters developed previously by Saito where *N*-methylcarbazole is used as a photosensitizer. Rizzo introduced a modification to Saito's methodology which is to carry out the photosensitized reaction in the presence of magnesium perchlorate (1.4 mM) (Scheme 9). The presence of the salt promoted the transformation of the contact ion pair (*CIP*) into free ion pair. Therefore, protonation of the free anion-radical **14** (Scheme 7) gives radical **15** (Scheme 7) which undergoes β-scission to give the deoxygenated radical (*e.g.*, R˙) (Scheme 7). Another hint to be pointed out in Rizzo's procedure is the direct competition, in the same molecule, between two secondary benzoate groups: the *m*-trifluoromethylbenzoates are deoxygenated faster than the benzoates.

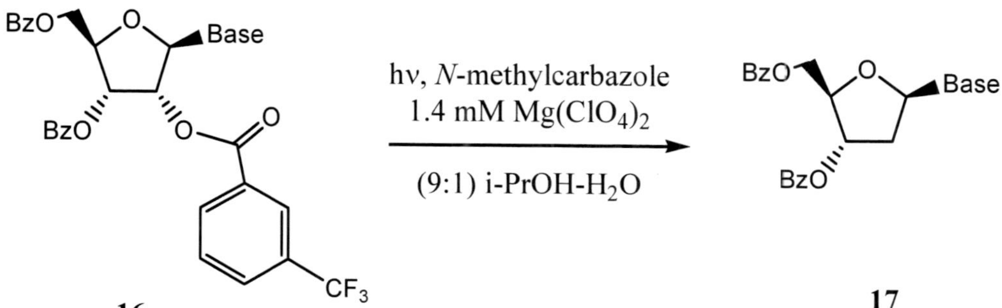

16 17

Scheme 9. Photosensitized synthesis of 2'-deoxynucleosides in the presence of $Mg(CLO_4)_2$.

Prudhomme and Rizzo have further studied the photosensitized reduction of benzoates using 3,6-dimethyl-*N*-ethylcarbazole as a sensitizer [64] (Scheme 10). The turnover properties of this carbazole photosensitizer in the deoxygenation of benzoyl derivatives have been improved because no significant change in the redox potential niether in the excited-state electron-transfer properties is observed. Also, the increased lifetime of the radical cation of 3,6-dimethyl-*N*-ethylcarbazole, by blocking the 3- and 6-positions of the carbazole moiety, allows for the photosensitizer to be regenerated and thus it can be used in substoichiometric amounts. This feature is losed when *N*-methylcarbazole was used as a sensitizer and explain why a *ca* 10% degradation of the photosensitizer was observed during the photoirradiations of the benzoates. Also, stoichiometric amounts of *N*-methylcarbazole are used in the photoirradiations and, in the case of long photolysis times, fresh sensitizer needed to be added after about 3 – 4 h of irradiation.

An advantage of the photodeoxygenantion over the Barton and related thermal deoxygenation reactions is that toxic tin species are avoided [65-67]. Also, the photoinduced electron-transfer reaction, where carbazoles (*N*-methylcarbazole and 3,6-dimethyl-*N*-ethylcarbazole) are used as sensitizers, is an attractive synthetic step for the reduction of secondary and tertiary arylesters and can be used widely as a general reduction reaction. In the following scheme are outlined some interesting examples of the photosensitized reduction reactions of arylesters, where the aryl moiety is m-$CF_3C_6H_4$.

Finally, a unified synthetic strategy for the synthesis of β-2'-deoxyribonucleosides, α-2'-deoxyribonucleosides, β-3'-deoxyribonucleosides and β-2',3'-dideoxynucleosides is shown in the Scheme 11 and involved the preparation of strategically protected furanose tetra-ester glycosidation precursors. The steroechemistry of the glycosidation is directed by a C2-ester of the glicosylation precursor. The key step in the sequences is the selective deoxygenation of a benzoyl or 3-(trifluoromethyl)benzoyl group of the ribo-, arabino- or xylo-nucleoside via a PET mechanism using carbazoles as the photosensitizer to give the desired deoxynucleoside [68].

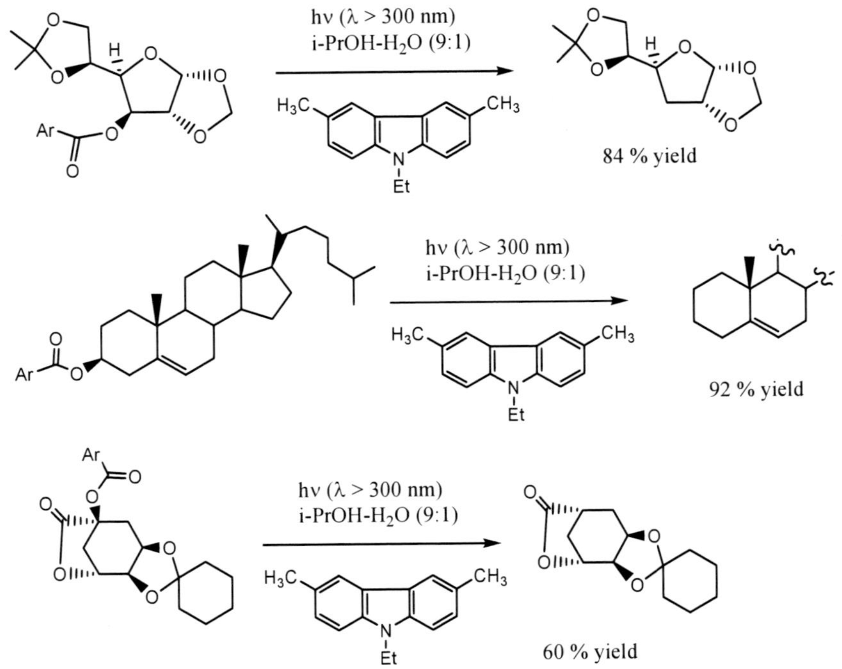

Scheme 10. Photosensitized reduction of benzoates using 3,6-dimethyl-N-ethylcarbazole.

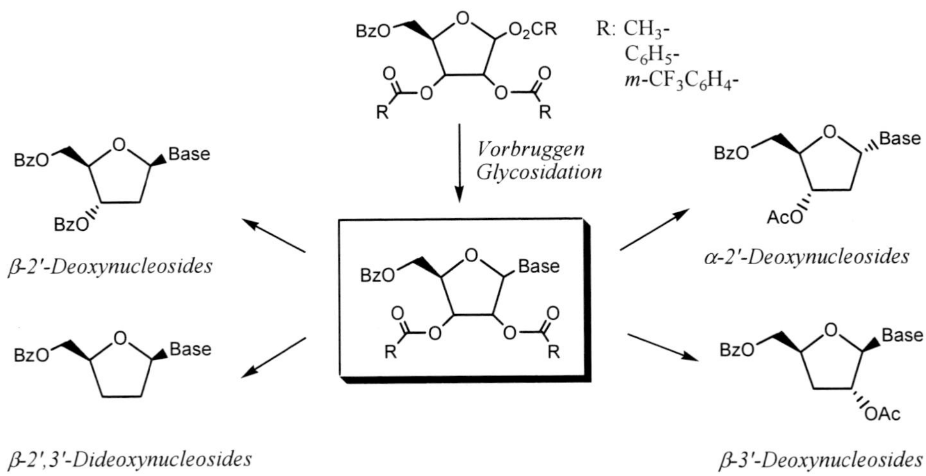

Scheme 11. Photosensitized synthesis of 2'- and 2',3'-deoxynucleosides.

A nice application of photosensitized reactions by carbazoles in biochemistry was reported by Yoshimura and Fujimoto with regards to the catalytic repair of thymine dimer in DNA via carbazole nucleoside [69] (Scheme 12). The catalytic repair of a thymine dimer incorporated in a DNA duplex via oligodeoxynucleotide (ODN) containing carbazole nucleoside (**K**) (Schemes 12 and 13) takes place through a photoinduced electron transfer process between **K** and thymine dimer under 366 nm irradiation for 30 minutes. The incorporation of carbazole nucleoside to DNA was achieved according to standard phosphoramidite chemistry and modified ODN containing **K** are as follows: ODN(K) 5'-d(ACTGTCACGC**K**TCACAT)-3'; ODN (KK) 5'-d(ACTGTCACGC**KK**TCACAT)-3'; ODN(KA) 5'-d(ACTGTCACGC**K**ATCACAT)-3' (Scheme 12). The phosphoramidite of **K** is outlined in the scheme, starting from carbazole and Hoffer's α-chlorosugar to give carbazole nucleoside **K** and further protection and deprotection reactions rend the target molecule.

Synthesis of phosphoramidite K

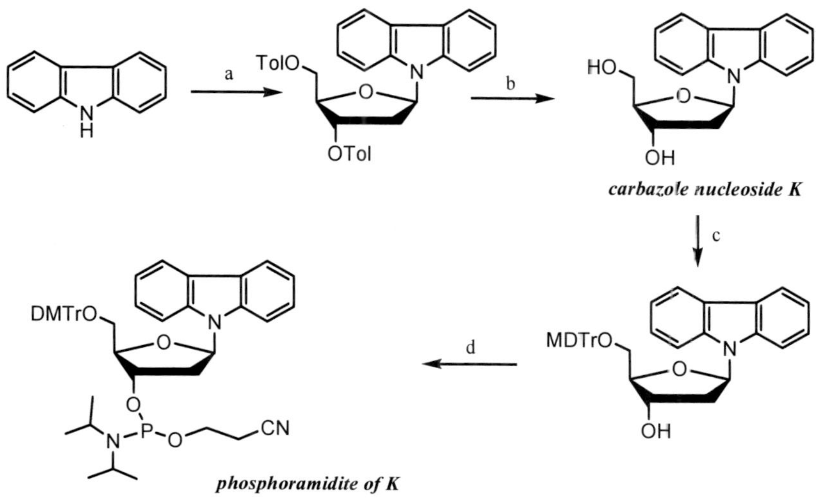

a: KOH, TDA-1, Chlorosugar, MeCN, rt, 20 min, 59%; **b**: NaOMe, MeOH, rt, 2 h 86%; **c**: 4,4'-dimethoxytrityl chloride, DMAP, Pyridine, rt, 15 h, 36%; **d**: (i-$Pr_2N)_2PO(CH_2)_2CN$, 1*H*-tetrazole, MeCN, rt, 2 h, quant.

Photochemistry of ODNs.

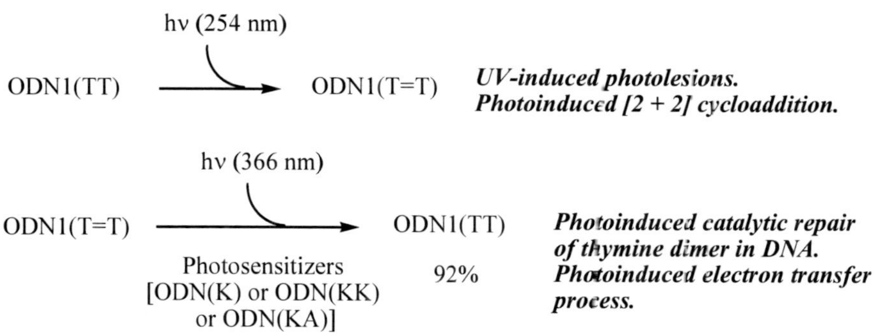

Schemes 12 and 13. Synthesis of phosphoramidite K and repairing thymine dimmer in DNA photosensitized by carbazole nucleoside.

After incorporation of phosphoramidite K to DNA, the thymine dimer formation in synthetic ODNs was performed photochemically under UV irradiation which induces a [2 + 2] cycloaddition of pyrimidines located above each other in the DNA double strand [70]. Then, the feasibility of the photochemical repair of a thymine dimer in DNA via ODN containing K was confirmed. When ODN(K) was irradiated at 366 nm for 30 minutes in the presence of ODN1(T=T) 5'-d(ATGTGA**T=T**GCGTGACAGT)-3', a thymine dimer, the appearence of a new signal determined by HPLC was asigned to the ODN1(TT) 5'-d(ATGTGA**TT**GCGTGACAGT)-3' photochemically repaired and it was obtained in 92% yield (Scheme 13). MALDI-TOF-MS and enzymatic digestion with AP and P1 nuclease at 37°C for 4 h of isolated ODN1(TT) showed that it was a repaired product formed from

ODN1(T=T). The quantum yield for the photochemical repair by using ODN(K) was estimated (□ = 0.014) at 365 nm, based on the desappearence of ODN1(T=T) by employing valerophenone as an actinometer [71,72]. Also, similar preparative results were obtained when ODN(KK) and ODN(KA) were used as repair of thymine dimer and ODN1(T=T) was obtained in 94 and 93 %, respectively. In order to infer that a photoinduced electron transfer takes place during the catalytic repair of thymine dimer in DNA, fluorescence emission quenching analyses of carbazole moiety in ODN(K) by ODN1(T=T) have been carried out. The results show an efficient quenching of the carbazole fluorescence emission in ODN(K) at □$_{exc}$: 330 nm. Additionally, the quenching of the fluorescence emission of ODN(K) was also studied with ODN composed of mismatch bases, ODN2(T=T) 5'-d(GCACAGT=TATACAGAGAG)-3', which was used as a quencher. In these experiments, scarcely quenching of the fluorescence emission of ODN(K) was observed. Furthermore, when ODN2(T=T) was used in repair, the repaired product of ODN(T=T) was scarcely observed which reinforces the fluorescence quenching results above mentioned. This nice application demostrates that the catalytic repair of a thymine dimer in DNA via ODN containing K is a viable repairing process of damage DNA through UV-induced photolesions, which can be both mutagenic and cancirogenic. Also, the catalytic repair process illustrates the use of carbazole incorporated in ODN as an excellent photosensitizer for promoting the reductive photoinduced electron transfer of thymine dimer.

Furthermore, carbazole-photosensitized dechlorination of pentachlorophenol could be used as a photochemical method for the destruction of impurities present in waterwaste [73]. Polychlorophenols represent an important class of compounds because of their widespread practical applications. They are very usefull as pesticides, germicides and fungicidal agents. Unfortunately, they are very toxic and their accumulation in the environment is a growing problem [74]. It is well known that the toxicity of polychlorinated aromatics, e.g. polychlorophenols, decreases with a decrease in the amount of chlorine substituents present in the molecule. Thus, photosensitized dechlorination of polychlorophenols by carbazole could be an alternative approach to the problem. In this regard, a preparative and mechanistic analyses have been carried out. Electronic absorption spectra of pentachlorophenol and carbazole in acetonitrile and 3-methylpentane show that the incident radiation (□$_{exc}$: 320-360 nm) is almost exclusively absorbed by carbazole. The photosensitization reaction of pentachlorophenol with □$_{exc}$: 350 nm in MeCN under Ar atmosphere in the presence of carbazole rendered 2,3,4,5-tetrachlorophenol as the main photoproduct (Scheme 14). Prolonged irradiation time results in the formation of trichlorophenol isomers. Unexpectedly, no less chlorinated phenols nor unsubstituted phenol were detected under this experimental conditions. Two side products are also formed in the photosensitized reaction. 2-Methyl-4,5,6,7-tetrachlorobenzoxazole is one of these side products and it is probably formed by reaction of tetrachlorophenol radical with the reaction solvent (MeCN). Coupling of chlorophenol radical with the sensitizer (carbazole) also takes place as a side reaction rendering a mixture of coupled products which were not carefully identified.

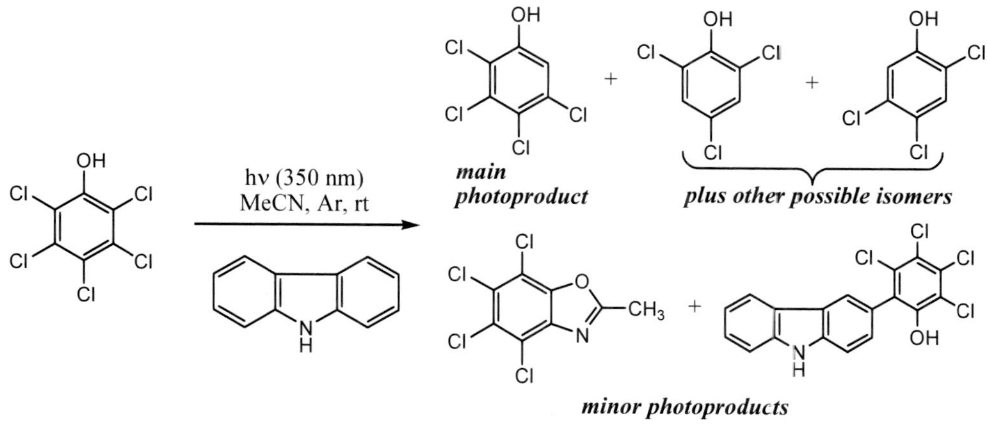

Scheme 14. Photosensitized dechlorination of pentachlorophenol.

This photosensitized reaction occurs through an electron transfer process from the singlet excited state of carbazole to the ground state of pentachlorophenol, which is the electron acceptor, to rend efficiently the carbazole radical-cation and the pentachlorophenol radical-anion in the solvent-cage. The radical-anion looses chloride ion to give the tetrachlorophenol radical which can abstract a H-atom from the solvent (main secondary pathway) or interact with the solvent or the photosensitizer (carbazole) (Scheme 15).

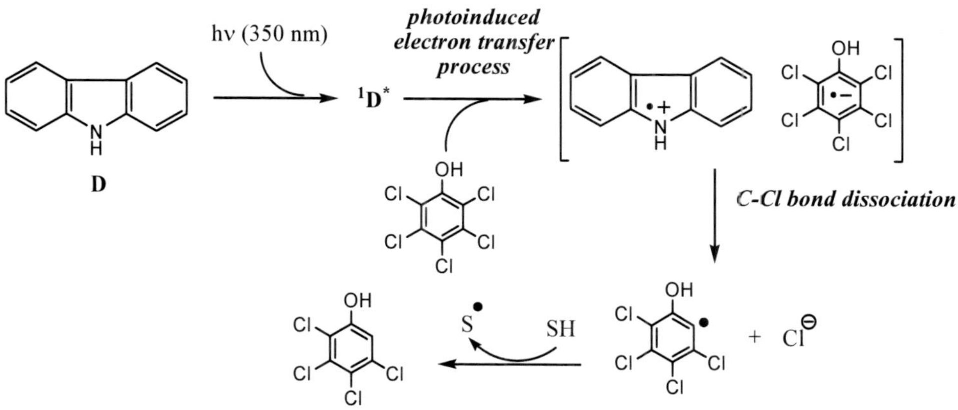

Scheme 15. Mechanism for photosensitized dechlorination of pentachlorophenol.

Quenching of the fluorescence emission of carbazole by pentachlorophenol have been performed by steady-state and time-resolved spectrsocopy. Stern-Volmer analyses provided rate constants of quenching of carbazole in a diffusion-controlled limit (1.6×10^{10} $M^{-1}.s^{-1}$ in heptane and 8.2×10^{9} $M^{-1}.s^{-1}$ in MeCN) leading to the conclusion that the proposed electron-transfer mechanism is feasible. Comparing the energies of the first excited singlet state of pentachlorophenol (93.3 kcal/mol) and carbazole (84.8 kcal/mol), the energy transfer process is negligible for this photosensitized reaction. Since, in addition to tetrachlorophenols, trichlorophenols are also produced at longer irradiation times, this mechanism may also be valid for less chlorinated phenols.

(c) *Preparative photochemical reactions of carbazoles.*

Another interesting PET reactions that involves electron excited carbazoles as electron donors are the photochemical nitration and halogenation reactions of carbazoles. The photochemical nitration takes place when carbazoles are subjected to react with tetranitromethane (TNM, electron acceptor species) while the photochemical halogenation takes place when tetrachloro- or tetrabromomethane (electron acceptor species) react with carbazoles. These photochemical reactions occur through a dissociative electron-transfer process, according to Saveant's model [75-77]. In this regard, the anion radical of the electron acceptor, viz., $C(NO_2)_4^{\cdot -}$ and $CX_4^{\cdot -}$ (X: Cl or Br), which is formed in the solvent cage with a fleeting lifetime, dissociates to lead $^-C(NO_2)_3$ and $NO_2^{\cdot}$ from TNM and X^- and $CX_3^{\cdot}$ from CX_4 [78-80].

The photochemical nitration of *N*-ethylcarbazole and *N*-isopropylcarbazole with TNM gives the corresponding 3-nitro-*N*-ethyl- and 3-nitro-*N*-isopropylcarbazoles in > 90 % yield when the mixture is irradiated with λ_{exc} > 300 nm in benzene for 90 minutes [78]. Slightly lower yields were obtained in MeOH and cyclohexane, under identical conditions. In these photoreactions nitroform is the other main product and is readly removed by extraction with water. Thermal nitration reaction is very slow both in acetic acid and dichloromethane requiring 4 weeks at room temperature to reach a 90% yield. In MeOH no thermal nitration reaction takes place.

Carbazoles and TNM form an electron-donor-acceptor complex in the ground state. Absorption spectroscopy of the mixture reaction shows a new band shifted to the red which is asigned to a charge-transfer electronic transition. The complex formed between *N*-ethylcarbazole and TNM shows a λ_{max} = 410 nm in cyclohexane with a $K_c = 7\ M^{-1}$. Selective irradiation of the charge-transfer band gives rise to the photochemical reaction, so a triad is inmediately formed after efficient dissociative electron transfer process takes place in the solvent cage. The photochemical nitration reaction mechanism proposed is outlined in the Scheme 16.

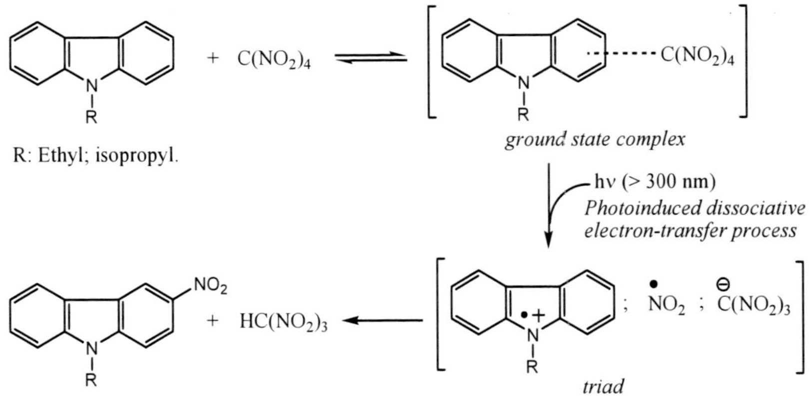

Scheme 16. Mechanism for photonitration of carbazoles.

The photohalogenation of carbazoles can be achieved by irradaition of carbazoles in the presence of halomethanes in protic and aprotic media under Ar atmosphere [80]. The photoreaction is initiated by a PET process from the singlet excited state of carabzole to the halomethane (RX). Depending on the nature of the halomethane, chlorocarbazoles and bromocarbazoles can be easily prepared (Scheme 17). The chloro- and bromocarbazole derivatives (**19a-f** and **20a-f**) are obtained in good chemical yield. Halogenation at C-3 position gives the 3-halocarbazole derivatives as the main photoproduct while halogenation at C-1 position takes place in lower yield. The quantum efficiency of conversion (ϕ_{reac}) of carbazoles **18a-f** in ethanol in the presence of CCl_4 ranged from 0.055 to 0.110, while in the presence of CBr_4 the ϕ_{reac} values ranged between 0.36 and 0.54. This behavior indicates that the photoreaction is more efficient as the electron affinity of the halomethane increases.

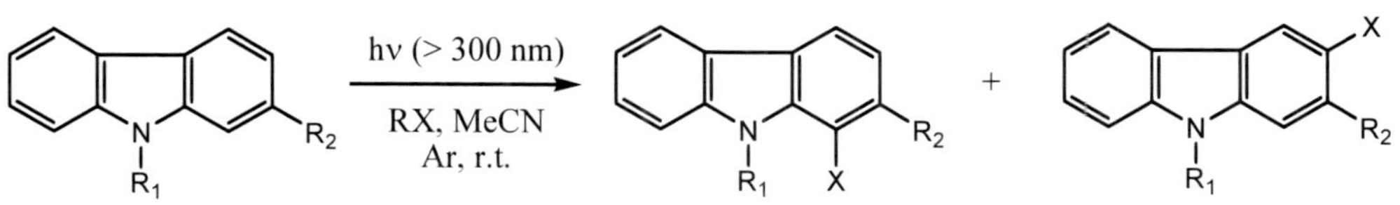

18 **19** **20**

18	R1	R2		RX		X
a	H	H		$CHCl_3$; CCl_4		Cl
b	Me	H		CH_2Br_2; $CHBr_3$; CBr_4		Br
c	COMe	H				
d	COPh	H				
e	H	OH				
f	Me	OMe				

Scheme 17. Photohalogenation of carbazoles.

Solvent effects on the photohalogenation reaction of carbazoles have been analyzed by means of the Reichardt parameter ($E_T(30)$) and was concluded that the photoreaction is efficient when polar protic and aprotic solvents are used [79].

It is interesting to point out that the polyhalomethanes quench the fluorescence emission of carbazoles **18a-f** and, by a Stern-Vomer analyses and carbazoles lifetime measurements by TC-SPC spectroscopy, is shown that the fluorescence quenching process is diffusional-controlled (kq ~ 0.048 – $1.5x10^9$ $M^{-1}.s^{-1}$). Laser flash photolysis experiments confirmed the formation of the *N*-acetyl- (**18c**) and *N*-benzoylcarbazole (**18d**) radical-cations when solutions of these carbazoles in CCl_4 under argon atmosphere were electronically excited with a beam of λ_{exc} 266 nm [79]. From a mechanistic point of view, the photohalogenation reaction of carbazoles **18a-f** is initiated by a dissociative photoinduced electron-transfer process. This process involves the reduction of the halomethane to the halomethane radical-anion, which has a fleeting lifetime, and the heterolytic dissociation of the halomethane radical-anion in a concerted step according to Saveant's model occurs.

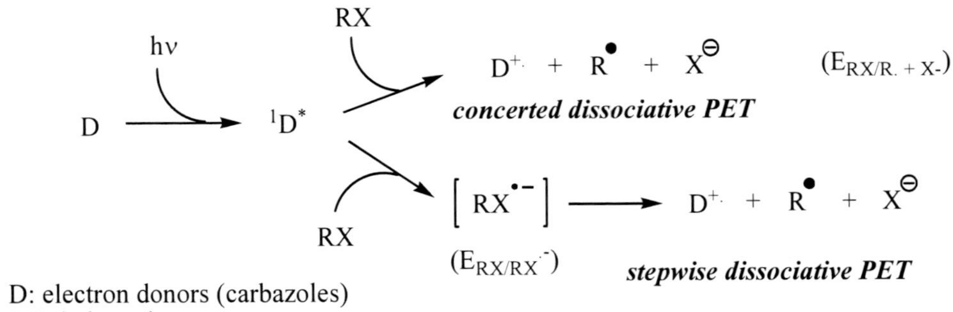

Scheme 18. Mechanism for photohalogenation of carbazoles.

The Rehm-Weller Gibbs energy relationship was applied to determine the fundamental parameters for the one-electron reduction, i.e., the one-electron reduction portentials ($E_{RX/RX.-}$) of the halomethanes and the intrincic barrier for the electron transfer reduction ($\Delta G^{\ddagger}(0)$. The $E_{RX/RX-.}$ values obtained were related to the concerted electron transfer-bond breaking reduction potentials ($E_{RX/R.+X-}$) and a linear relation was observed [79].

It is interesting to point out that Zelent and Durocher described some aspects of the photochemistry of carbazole (**18a**) in the presence of CCl_4 in three solvents: ethanol, 3-methylpentane and pure CCl_4 [81-83]. As is shown in Scheme 19, irradiation of carbazole with CCl_4 in ethanol gives **21** and **22**.

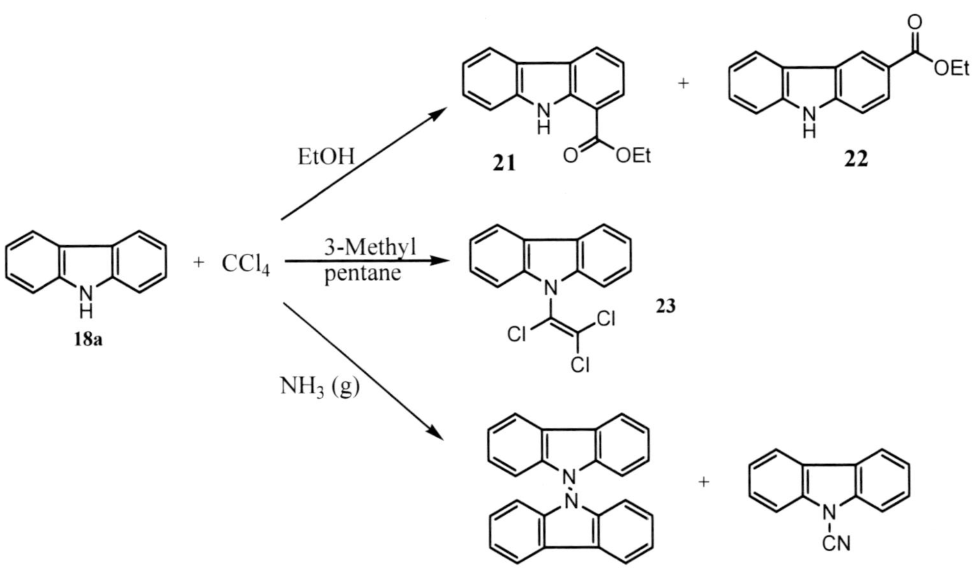

Scheme 19. Photohalogenation of carbazole with CCl_4 in 3-methylpentane.

In 3-methylpentane the photochemical reaction of carbazole gives *N*-(trichloroethylene)carbazole (**23**) as the main product. In pure CCl_4 saturated with ammonia gas, *N,N*-dicarbazyl (**24**) and *N*-cyanocarbazole (**25**) have been obtained. In each of the reaction media studied, hexachloroethane was also formed.

It is noteworthy to mention that these results do not match with those reported by Bonesi and Erra-Balsells [79] As is shown in Scheme 20, in this paper was reported that irradiaton of carbazole in a CCl_4 – EtOH (1:1) mixture under N_2 atmosphere, gives **19a** and **20a** as the main products with traces of compounds **21** and **22** (yield > 0.1%). Under O_2 atmosphere, the yields of compounds **21** and **22** increases (yield ~ 13%) but compound **20a** is the main photoproduct. When hexane was used as solvent, which is quite similar to 3-methylpentane, no reaction takes place meaning that a PET process between carbazole and CCl_4 is not feasible. When the photoreaction was carried out in methanol or in isopropanol in the presence of CCl_4, the carbomethoxy- (**26** and **27**) and carboisopropoxycarbazoles (**28** and **29**) were obtained. The formylcarbazoles (**30** and **31**) were easily obtained when the photoreaction of carbazole was carried out in the presence of $CHCl_3$ instead of CCl_4 under O_2 atmosphere and ethanol as the reaction solvent. It is noteworthy to mention that these kind of photoproducts are formed in 5 – 10% yield.

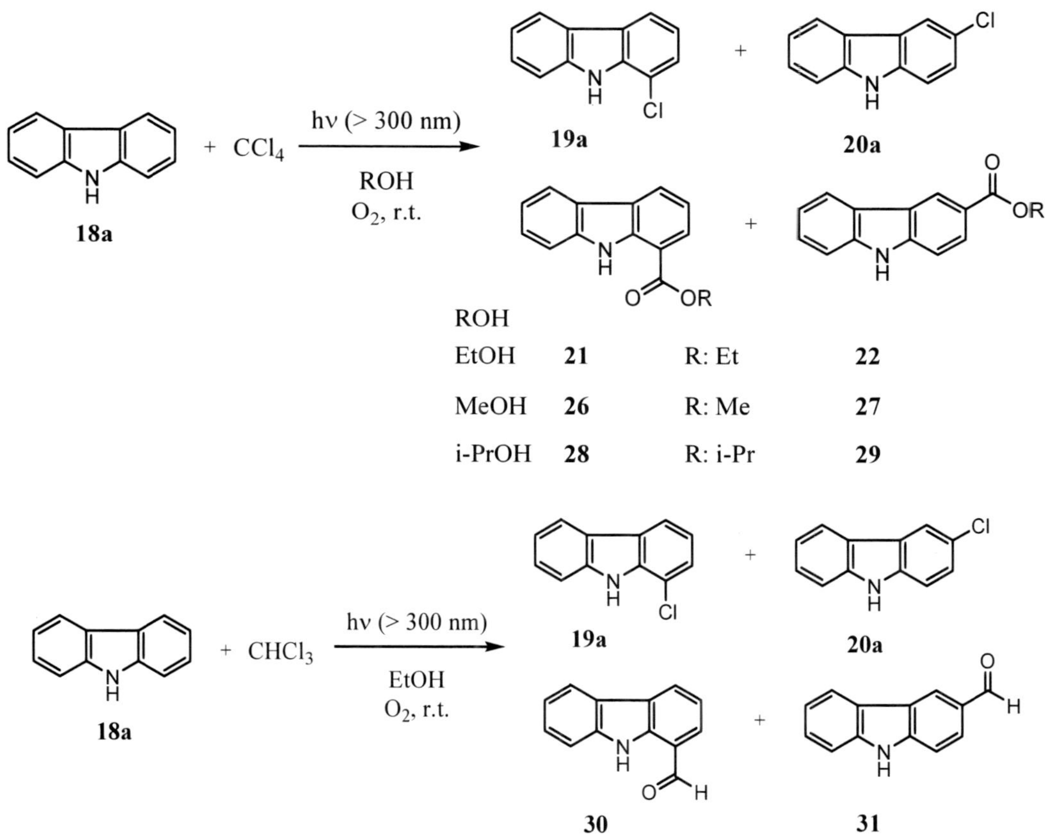

Scheme 20. Photohalogenation of carbazole with CCl_4 and $CHCl_3$ in ROH.

Further, Durocher *et al.* studied the one-electron photooxidation of carbazole in the presence of CCl_4 as a reaction media, using water, ethanol or ammonia after irradiation in order to isolate and identify the photoproducts that arise from secondary photochemical reaction [82,83]. The primary photochemical reaction in carbazole – CCl_4 system gives, in the solvent cage, the intermediates shown in Figure 6:

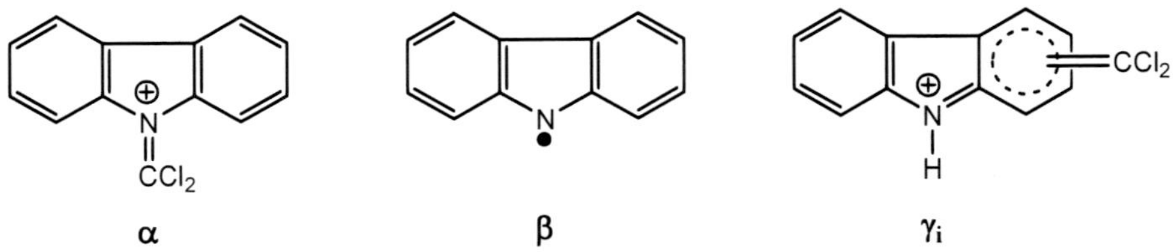

Figure 6. Intermediates formed by photochemical reaction of carbazole-CCl_4 system.

The probability of formation and further transformations of these transient products **α**, **β** and **γ**$_i$, depends strongly on the nature of the reaction media. Treatment of the photolyzed solution with water where transients show significant stability in pure CCl_4, undergo easy hydrolysis resulting in the corresponding (carbo-*N*-carbazyl)carbazole and carbazole carboxilic acids (Figure 7). These kind of photoproducts arise from the hydrolysis of intermediates **α** and **γ**$_i$.

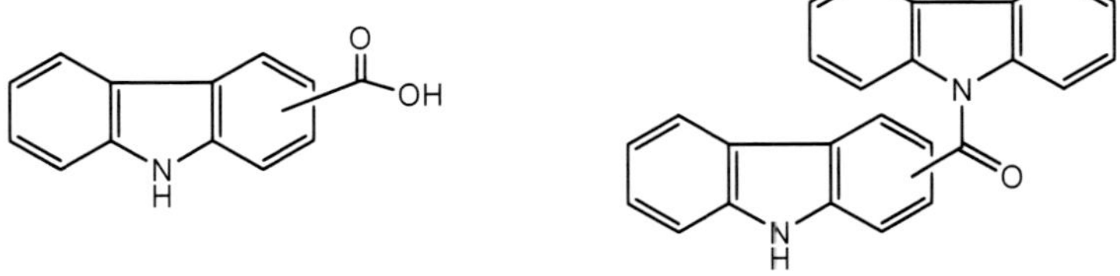

carbazole carboxilic acids ***(carbo-N-carbazyl)carbazoles***

Figure 7. Products formed by adding H_2O to the photochemical reaction of carbazole-CCl_4 system.

Furthermore, disubstituted carbazole carboxilic acids and di(carbo-*N*-carbazyl)carbazole can be observed by TLC of the photolyzed solution after water addition.

If ethanol is added to the photolyzed solution instead of water, the transient products formed in pure CCl_4 undergo reaction with ethanol and mono- and di-substituted (carboethoxycarbazoles) are formed. Since ethanol contains traces of water, hydrolysis of transients **α** and **γ**$_i$ take palce rendering the photoproducts already described in significant low yield.

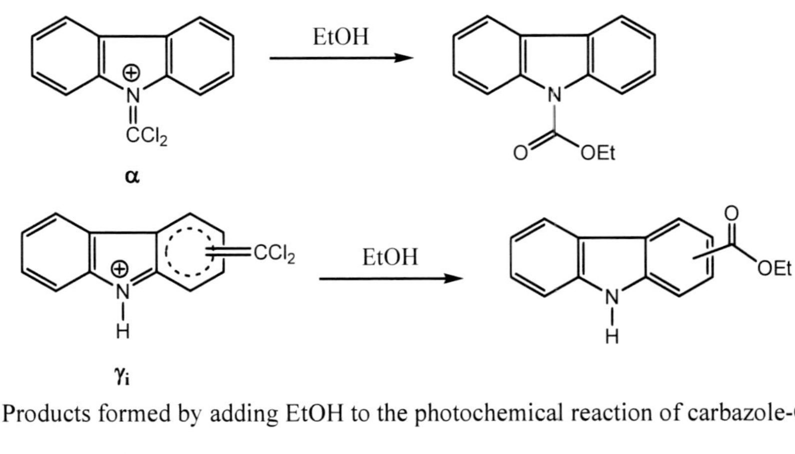

Scheme 21. Products formed by adding EtOH to the photochemical reaction of carbazole-CCl_4 system.

When ethanol is used as a solvent of carbazole in the presence of CCl_4, the secondary photochemical transformation is significantly changed in comparison with the reactions in pure CCl_4 above described. The polar medium favours the formation of ionic intermediates in the solvent cage. Thus, intermediates α and γ_i which are formed during irradiation can inmediately undergo further reactions which result in the formation of all the mono-substituted (carboethoxy)carbazoles and no di-substituted derivatives are detected.

Saturation of the photolysis solution by ammonia gas gives the corresponding cyano derivatives of carbazoles in addition to ammonium chloride. The main sources of these cyano derivatives are the intermediates α and γ_i transients, which undergo reaction with ammonia resulting in the corresponding (carboimino-*N*-carbazyl)carbazoles (Scheme 22).

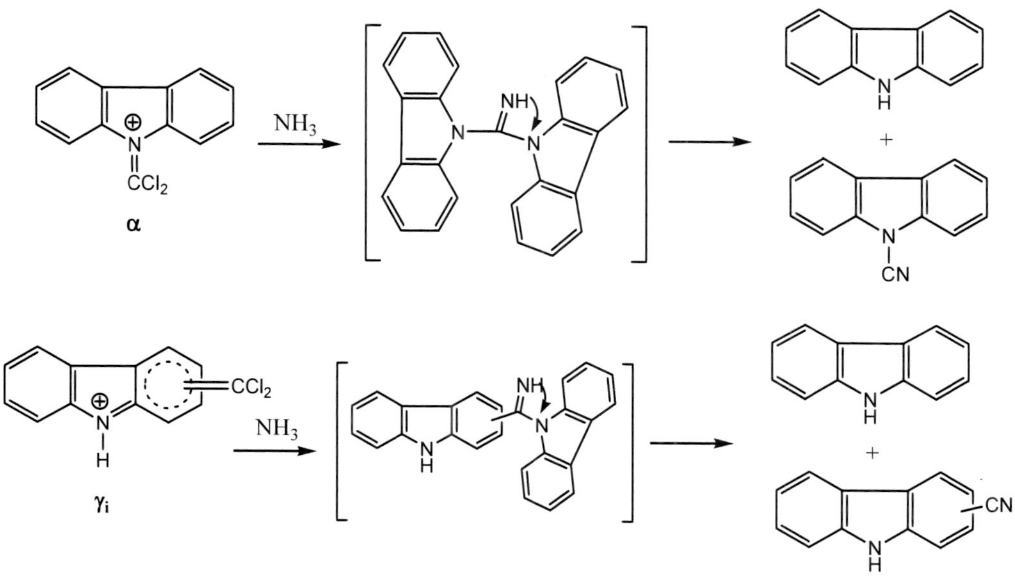

Scheme 22. Products formed by adding NH3 to the photochemical reaction of carbazole-CCl4 system.

The course of photochemical transformations in carbazole-CCl_4 system, is radically changed in the presence of ammonia and the photoproducts formed come from thermal secondary reaction of carbazole radical, the β intermediate. The main photoproducts obtained in these experiments are *N,N'*-dicarbazyl, 1-(*N*-carbazyl)carbazole and 3-(*N*-carbazyl)carbazole (Scheme 23).

The mechanism of the photodecomposition of *N*-ethylcarbazole in the presence of CCl_4 has been discussed on the basis of the photoproducts identified [84]. Photodisocciation of the *N*-ethyl bond and the one-electron transfer in the transiently formed excited charge-transfer complex between *N*-ethylcarbazole and CCl_4, have been proposed as the primary photochemical process involved in the singlet excited *N*-ethylcarbazole molecule. The latter process, treated as the main process, leads to the radical cation of *N*-ethylcarbazole, chloride ion and trichloromethyl radical in the solvent-cage. The other reactions in the system studied are analysed following the decomposition of the cation-radical in the presence of Cl^- and $CCl_3^{\cdot}$ which can occur by the *N*-ethyl group and (or) by aromatic ring. The formation of intermediate products such as those shown in Figure 8 gives rise to secondary photochemical reactions in the system studied. The reaction media strongly influence the nature of the secondary photochemical transformations and this photochemical reaction was studied in pure CCl_4, in ethanol as a solvent in the presence of CCl_4 and also, in pure CCl_4 in the presence of ammonia gas. In general, the comparison of the product distribution of the photochemical reaction of *N*-ethylcarbazole with those obtained for carbazole in the same experimental conditions, are quite similar. Thus, mono- and di-(carboethoxy)-*N*-ethylcarbazoles are formed when the photoreaction was carried out in CCl_4 – ethanol as solvent mixture; *N*-cyanocarbazole, 1-cyanocarbazole, 3-cyanocarbazole and 3-cyano-*N*-ethylcarbazole are formed in pure CCl_4 in the presence of ammonia gas and aslo, after treatment with ammonia gas. *N*-(trichlorovinyl)carbazole, *N*-β-dichlorovinyl)carbazole and 1,1-di(*N*-carbazyl)ethane were also formed as side products in CCl_4 in the prescence of NH_3.

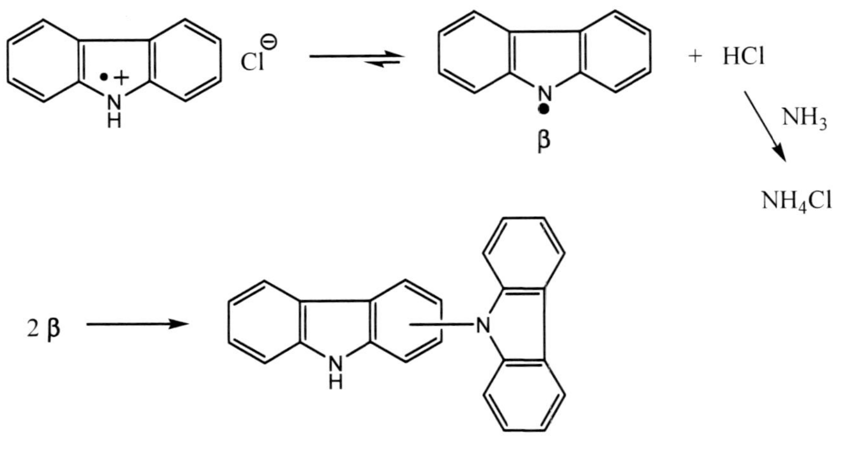

Scheme 23. Mechanism for the photochemical reaction of carbazole-CCl_4 system adding NH_3.

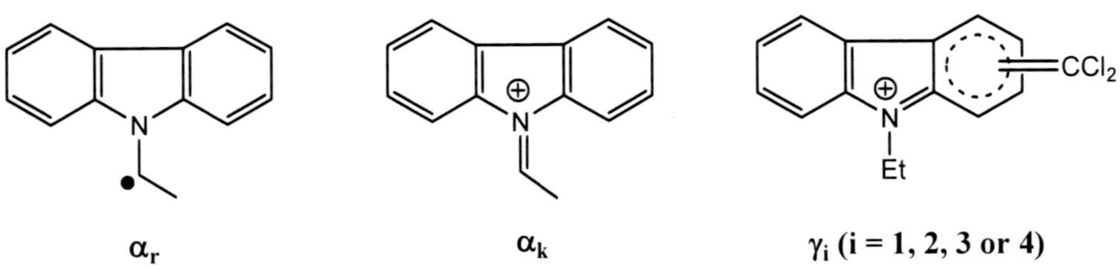

Figure 8. Primary products formed in the photochemical reaction of N-ethylcarbazole-CCl_4 system.

Erra-Balsells and Frasca studied the photochemical reaction of carbazole and some carbazole derivatives in dichloromethane [85,86]. Irradiation of carbazole in dichloromethane with a high-pressure Hg lamp gave a mixture of several products: bis(9-carbazyl)methane, *N*-methylcarbazole, *N*-ethylcarbazole and 3-chlorocarbazole. The conversion of the starting material was 50% after 18 h of irradiation (Scheme 24).

This photochemical reaction takes place through a dissociative PET reaction where the radical-cation of carbazole, the chloride ion and the chloromethyl radical ($ClCH_2^{\cdot}$) are formed in the solvent cage. The carbazole radical-cation reacts with chloride anion and the chloromethyl radical to give the 3-chlorocarbazole and the cation **A**. This intermediate is the source of the formation of bis(9-carbazyl)methane, *N*-methylcarbazole and *N*-ethylcarbazole (Figure 9).

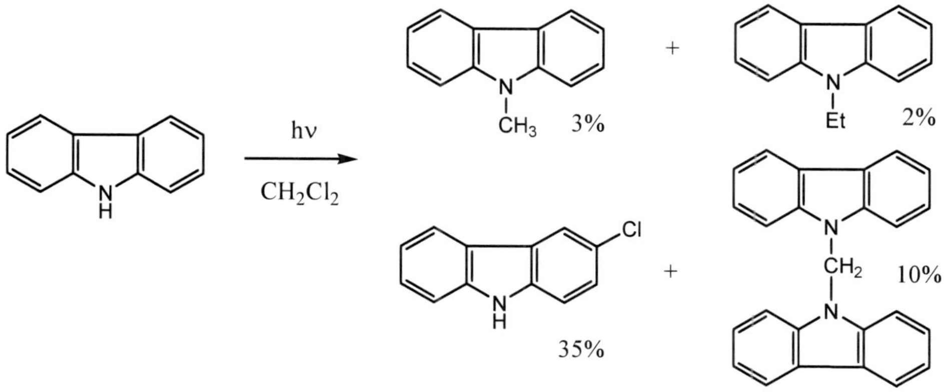

Scheme 24. Photochemical reaction of carbazole-CH_2Cl_2 system.

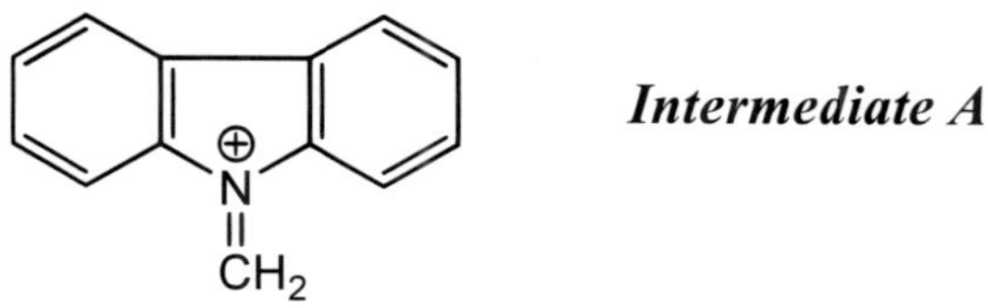

Figure 9. Intermediate formed in the photochemical reaction of carbazole-CH_2Cl_2 system.

The irradiation of 3,6-dichlorocarbazole and 3,6-dibromocarbazole in dichloromethane in the same experimental condition above mentioned gave the 9,9'-bis(3,6-dichlorocarbazyl) and 9,9'-bis(3,6-dibromocarbazyl) as the only products, respectively. It is noteworthy that the photodimerization occurs through a different mechanism from that already discussed. The photodimerization of these 3,6-chloro- and 3,6-dibromocarbazole takes place through the fragmentation of the N-H bond in the electronic excited state rendering the 9-aminyl radical as intermediate which dimerizes to give the corresponding biscarbazyl derivatives. Also, this photochemical behavior is similar to that described for 9H-pyrido(3,4-b)indoles [87]. As it can be seen, the photochemical reaction pathway depends on the nature of the substituent present in the carbazole ring. Thus, carbazole reacts with the solvent (dichloromethane) through a dissociative PET reaction while the 3,6-dichloro- and 3,6-dibromocarbazoles, which are not good electron donor as carbazole is, favor the radical dimerization pathway. It is interesting to mention that both reactions, dissociative PET reaction and radical dimerization reaction, are not concurrent, at least, in any of the examples already studied. Irradiation of carbazole in hexane with a high-pressure Hg lamp gave only the dimer 9,9'-biscarbazyl (Scheme 25).

These result confirms that 9-aminyl radicals are the key intermediates that are formed through the proposed photodimerization pathway when solvents with high electron affinity, such as dichloromethane or tetrachloromethane, are not used.

3-Nitrocarbazole was also irradiated in dichloromethane with a high-pressure Hg lamp and from the photolyzed solution an asimetric dimer, the 1,9'-bis(3-nitrocarbazyl), was detected as the only product (Scheme 26). This interesting result can be rationalized taking into account that the radical intermediate, the 9'-carbazyl, reacts with another molecule of 3-nitrocarbazole at the C-1 position, which is an activated position of the substituted carbazole moiety. This behavior is due to the presence of a withdrawing substituent attached to the carbazole moiety, such as the nitro group, that confers a σ character to the 9'-aminyl radical and no delocalization of the unpair electron takes place. Therefore, no C-radicals are formed during the photodimerization reaction of substituted carbazoles [88].

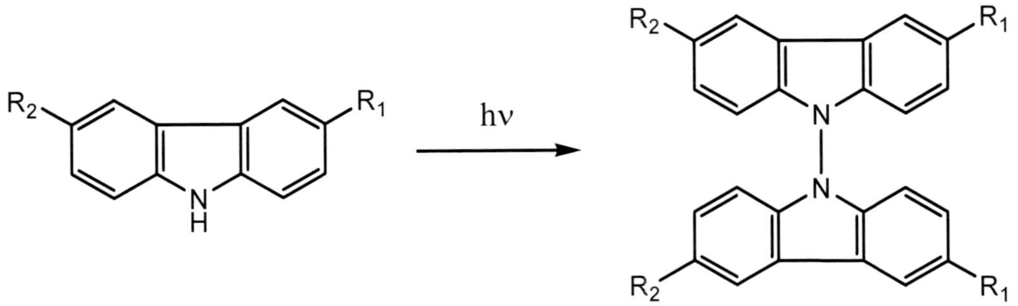

	Solvent	
$R_1 = R_2 = H$	Hexane	$R_1 = R_2 = H$
$R_1 = R_2 = Cl$	Dichloromethane	$R_1 = R_2 = Cl$
$R_1 = R_2 = Br$	Dichloromethane	$R_1 = R_2 = Br$

Scheme 25. Photochemical reaction of 3,6-dihalocarbazoles.

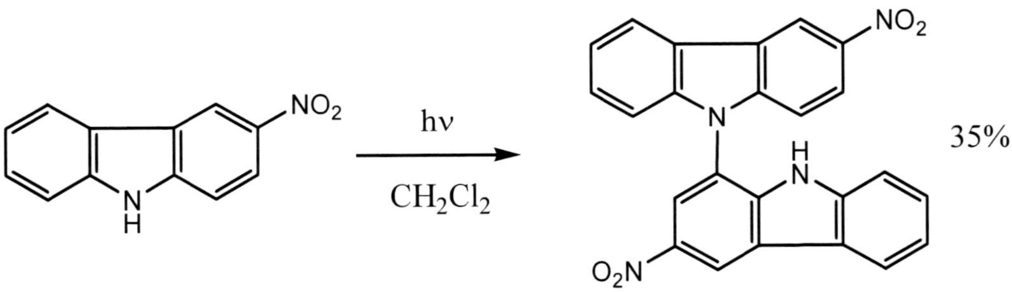

Scheme 26. Photochemical reaction of 3-nitrocarbazole.

N-Acyl- and 2-acylcarbazoles undergo photo-Fries rearrangement and the reaction products are the expectable *ortho*- and *para*-rearranged isomers and the corresponding amine or phenol, which are formed through the deacylation of the original carbazoles. The rearrangement photoproducts are obtained in good chemical yields and the photoreaction takes place with moderate to good quantum efficiency. The photo-Fries rearrangement has received considerable attention since its discovery in 1960 [89], and the studies have been devoted mainly to elucidating the mechanism of the reaction rather than its use in synthesis [90-98]. However, this photochemical reaction has been successfully used in the synthesis of natural products such as griseofulvin, daunomycine and flavonoids [99-102]. Recently, Miranda has extensively reviewed the photo-Fries rearrangement reaction [103].

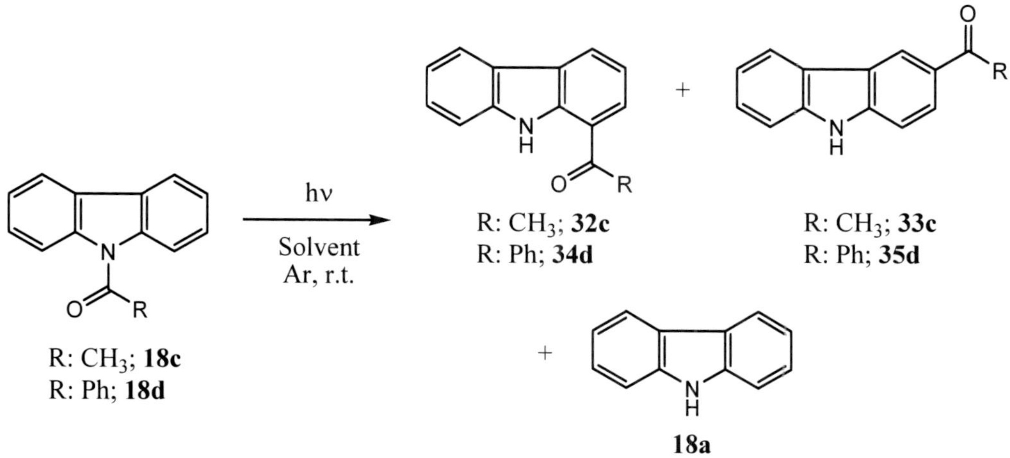

Scheme 27. Photochemical reaction of N-acylcarbazoles.

Zander described briefly the photorearrangement of *N*-acetylcarbazole in cyclohexane claiming that 1-acetylcarbazole was the only product [104]. Shizuka *et al.* have studied the photo-Fries rearrangement of *N*-acetyldiphenylamine and *N*-acetylcarbazole at 254 nm in cyclohexane and in ethanol [105]. The photochemical reaction of *N*-acetylcarbazole have been discussed on the bases of relationship between the isolated photoproducts and the odd π electron densities obtained from the Hückel molecular orbital (MO) calculations. The quantum yield of the photoproduct formation (ϕ) in the photo-Fries rearrangement are related to the odd π electron densities and was also observed that the ϕ does not show any dependence on irradiation time and on the initial *N*-acetylcarbazole concentration. The authors also stated that 1-acetylcarbazole (**32c**) (Scheme 27) and 3-acetylcarbazole (**33c**) (Scheme 27) were obtained with higher quantum yield of the former than the latter and the quantum yields for product formation were higher in cyclohexane than in EtOH. Bonesi and Erra-Balsells, have studied the photolysis of *N*-acetylcarbazole in ethanol and dichloromethane [106]. They showed that this photochemical reaction depends on the initial concentration of the *N*-acylcarbazole derivative. Thus, as the concentration of *N*-acetylcarbazole is increased in the range of $5x10^{-4}$ to $5x10^{-2}$ M, the quantum yield of conversion of *N*-acetylcarbazole of the photo-Fries reaction diminishes significantly. Also, two new products are observed when the photochemical reaction is carried out at higher initial concentration: 4-acetylcarbazole and 3,*N*-diacetylcarbazole. The former substitution pattern was also observed when *N*-ethylcarbazole was irradiated in a CCl_4 – EtOH mixture, and hence, 4-(carboethoxy)-*N*-ethylcarbazole was formed.

A comparative and systematic study of the photo-Fries rearrangement of *N*-acetyl- (**18c**) and *N*-benzoylcarbazole (**18d**) has also been described [107] (Scheme 27). In this paper was reported the photochemistry of **18c** and **18d** in a wide range of organic solvents and in different experimental conditions. The photochemical reaction of both *N*-acylcarbazoles under Ar, air and O_2 atmospheres gives the same product distribution with a quite similar chemical yields of the expectable photoproducts. Irradiation of compounds **18c** and **18d** at two different wavelengths (254 and 313 nm) gives the same product distribution with quite similar chemical and quantum yields. These results lead to state that O_2 does not quench efficiently the electronic excited state and the S_1 state is the electronic excited photoreactive state which is to be likely a π,π^* electronic state.

Photosensitization of **18c** and **18d** with xanthone (triplet energy donor) does not give any product while normal photo-Fries rearrangement was observed in the presence of tetramethyldiazetine dioxide (triplet energy acceptor).

The solvent effect on the photoreaction was also studied and no noticeable dependence on the photoproduct chemical yields was observed as the solvent was changed from cyclohexane to DMF. Also, the photo-Fries rearrangement of compounds **18c-d** was not modified when radical scavengers, such as cyclohexene or 1-dodecylmercaptane, are used. These results suggested that the formation of the radical species (acyl radical and carbazolyl radical) and hence, the photo-Fries rearrangement, take place in the solvent cage. Besides, carbazolyl radicals do not diffuse freely in the medium, probably reacting with hydrogen donor species which are close to the solvent cage.

It is interesting to point out that the photo-Fries rearrangement of *N*-acetyl- and *N*-benzoylcarbazole is quenched when halomethanes, such as $CHCl_3$ and CCl_4, are used as solvents (Scheme 28). In this regard, chloro-*N*-acylcarbazoles are formed in low yields while in CCl_4 – EtOH (1:1) mixture is formed in 10 – 15 % yield (conversion of the starting material ~ 15%). The fluorescence quenching experiments of carbazoles **18c-d** with halomethanes lead to conclude that the electronic excited state responsible for the photo-Fries rearrangement, the S_1 state, is efficiently quenched by halomethanes. Also, laser flash photolysis experiments in CCl_4 at 266 nm show the formation of the *N*-acylcarbazole radical cation with transient absorption spectra at 500 nm for **18c** and λ_{max}: 395 nm for **18d** and lifetime values of 15.2 and 16.8 ns, respectively [107].

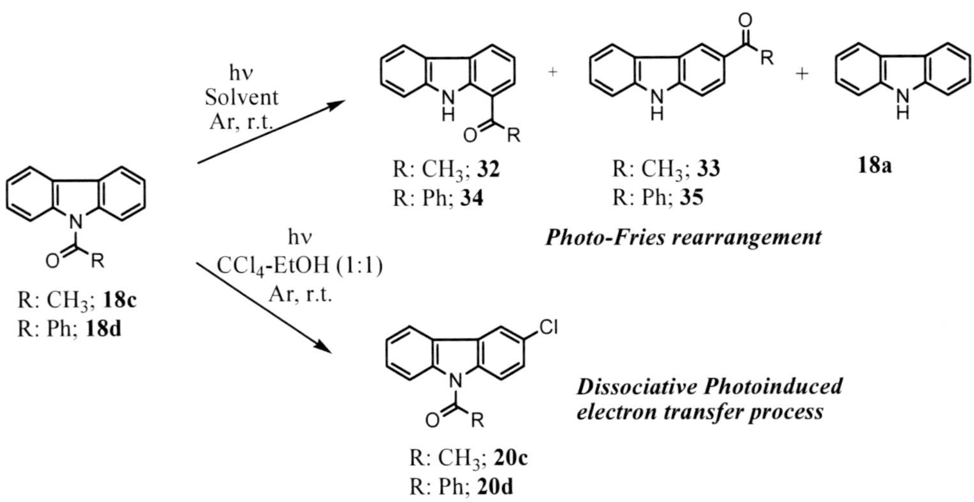

Scheme 28. Photochemical reaction of N-acylcarbazoles in the presence of CCl_4.

Ghosh *et al.* have studied the photochemical rearrangements of *N*-aroylcarbazoles using I_2 as an oxidant agent [108]. This study showed that the photo-Fries rearrangement is wavelength dependent. Irradiations carried out in methanol at 254 nm afforded **18a** and 3-aroylcarbazoles (**37a-c**) with variable chemical yields (Scheme 29). When the photochemical reaction was performed in methanol but with λ_{exc}: 366 nm, compounds **18c**, **37a-d** and **38a-d** are formed.

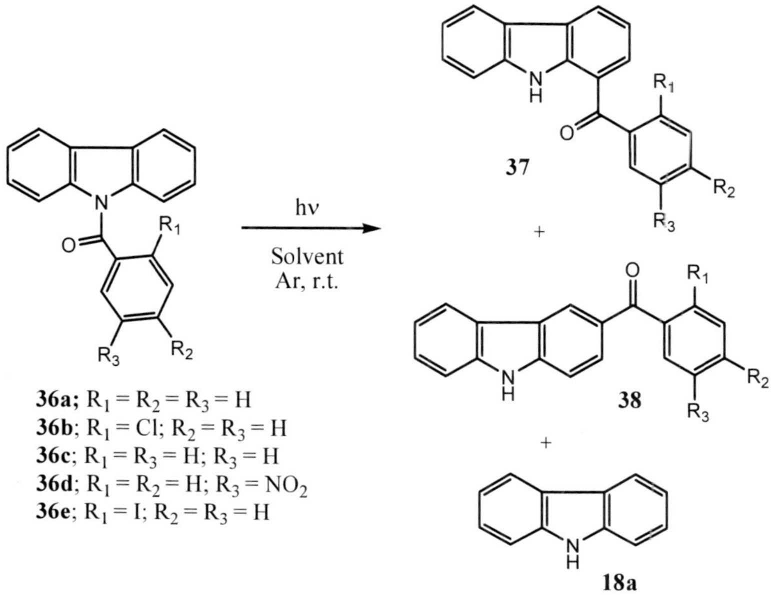

36a; $R_1 = R_2 = R_3 = H$
36b; $R_1 = Cl$; $R_2 = R_3 = H$
36c; $R_1 = R_3 = H$; $R_3 = H$
36d; $R_1 = R_2 = H$; $R_3 = NO_2$
36e; $R_1 = I$; $R_2 = R_3 = H$

Scheme 29. Photochemical reaction of N-aroylcarbazoles.

It is interesting to note that *N*-(3-nitrobenzoyl)carbazole do not rearrange at 254 nm while at 366 nm 3-(3-nitrobenzoyl)carbazole was obtained in very low yield (~ 6%). This behavior could be rationalized taking into account that the nitro group favors the intersystem crossing process ($S_1 \rightarrow T_1$) through spin-orbital coupling so the triplet excited state (T_1) deactivates very efficiently through a radiationless pathway that competes with the photochemical reaction.

Irradiation of *N*-(2-chlorobenzoyl)carbazole (**36b**) in benzene at 254 nm or *N*-(2-iodobenzoyl)carbazole (**36e**) in MeOH / I_2 at 254 nm afforded phenanthridone **39** in 23 % and 67 % yield, respectively (Scheme 30). This distinctly behavior reflects a greater tendency of the iodine-derivative (**36e**) for radical formation after excitation in comparison to the chlorine-derivative (**36b**). The ease of formation of intermediate (**I**), which depends on the C-X (X=Cl, I) bond energy, favors the photocyclisation to give phenanthridone **39** [109].

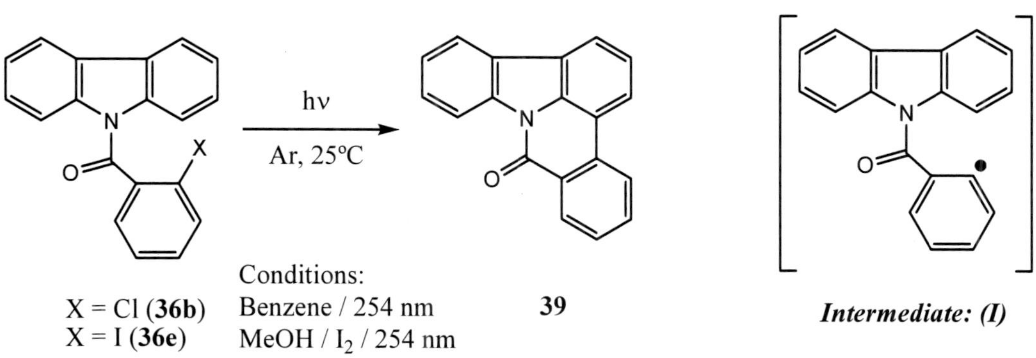

Scheme 30. Photochemical reaction of N-(2-chlorobenzoyl)carbazole.

N-Sulphonylcarbazoles also undergo photo-Fries rearrangement [110]. Irradiation of *N*-benzenesulphonyl-, *N*-(*p*-methylbenzene)sulphonyl- and *N*-mesylcarbazole in benzene at 254 nm afforded the [1,3]- and [1,5]-migrated photoproducts **41a-c** and **42a-c** and carbazole (Scheme 31). In all cases, starting material was recovered. Systematically, the photochemical reaction of *N*-sulphonylcarbazoles have been carried out also in methanol and two different wavelengths: 254 and 366 nm for each solvent were used. The results obtained showed that the photoreaction was efficient and gave the same product distribution.

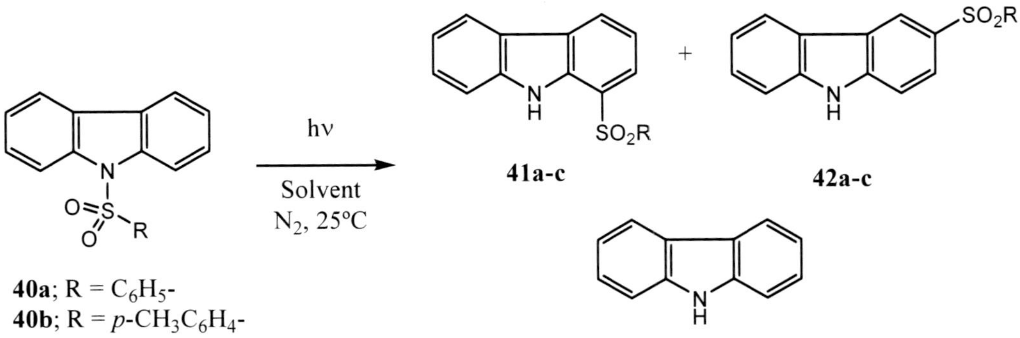

40a; R = C_6H_5-
40b; R = *p*-$CH_3C_6H_4$-
40c; R = CH_3

Scheme 31. Photochemical reaction of N-sulphonylcarbazoles.

Thus, compounds **41a-b** and **42a-b** are formed in a (3:4) molar ratio in both solvents and excitation wavelengths used. In the case of *N*-mesylcarbazole (**40c**) the photolyzed mixture rendered compounds **41c** and **42c** in a (5:3) molar ratio when the solutions in benzene or methanol were irradiated at 254 and 366 nm. In general, in these irradiations the conversion of the starting material reached *ca.* 85 – 90% yield and was observed that the photoreactions are more rapid at λ_{exc}: 366 nm than at λ_{exc}: 254 nm. Therefore, the same photoreactive excited state is populated efficiently and undergoes a fast homolytic cleavage of the S-N bond in *N*-sulphonylcarbazoles thereby generating the solvent-cage intermediates: carbazolyl radical and sulphonyl radical. These species may undergo a) cage recombination to starting material; b) intramolecular [1,3]- and [1,5]-migration of sulphonyl radical or c) escape of carbazolyl radical from the solvent-cage and hydrogen abstraction from neighbour molecules thereby producing carbazole.

Irradiation of 2-acyloxycarbazoles and 2-alkylsulphonyloxy- and 2-arylsulphonyloxycarbazoles also undergo photo-Fries rearrangement [111,112]. The photochemistry of 2-acetyloxy- and 2-benzoyloxycarbazole in different solvents have been studied. Irradiation of the 2-acyloxycarbazoles in organic media at 254 and 313 nm yield the [1,3]-migrated photoproducts: 1.acyl-2-hydroxycarbazole (**44a-b**), 3-acyl-2-hydroxycarbazole (**45a-b**) and 2-hydroxycarbazole (**46**) (Scheme 32). The effects of the solvent, the atmosphere and the intensity of the ligth source on the photochemistry of 2-acyloxycarbazoles have been studied.

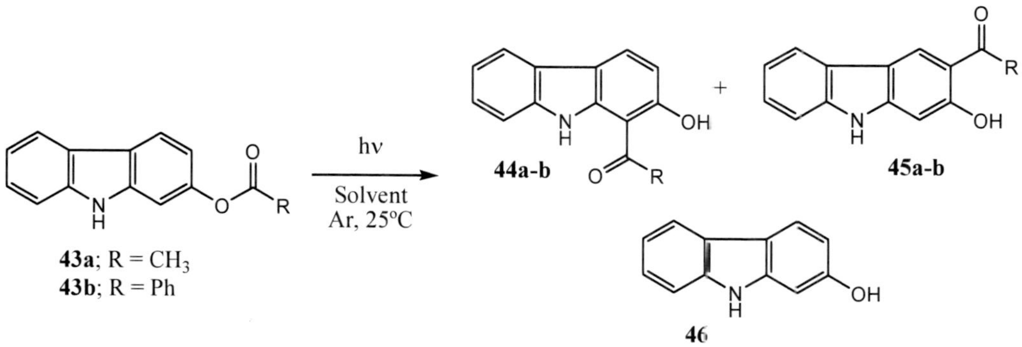

Scheme 32. Photochemical reaction of 2-acyloxycarbazoles.

The product distribution and the chemical yield of formation of the photoproducts **44**, **45** and **46** remain practically constant when the photoreaction is carried out in different solvents (cyclohexane, THF, acetonitrile and alcohols) and at different excitation wavelengths. Thus, the ratio of photoproducts **44** to **45** is *ca.* 2. Although, it was observed that in alcoholic media the conversion of the starting material is 100 % whereas in apolar and polar aprotic media the conversion varied from 33 % to 57 % for a fixed irradiation time (150 minutes).

The photoreactive excited state of 2-acyloxycarbazoles (**43a-b**) have been characterized by means of steady-state and time-resolved spectroscopy. The results suggested that the photoreactive excited state is the lowest singlet excited state (S_1) and it is mostly likely to be π,π*. The fluorescence emission of **43a** and **43b** was efficiently quenched by *p*-dicyanobenzene, which is a good and known electron-acceptor quencher of aromatic and polynuclear aromatic compounds [113]. Also, the photochemical reactions of 2-acyloxycarbazoles were quenched by *p*-dicyanobenzene showing that the S_1 state is the photoreactive excited state. These results were confirmed when photosensitization with xanthone of the reaction of 2-acyloxycarbazole were carried out and no photoproducts were detected. It is noteworthy that the multiplicity and the photoreactive excite state of 2-acyloxycarbazoles **43a-b** are similar to those obtained for the excited state of *N*-acetyl- and *N*-benzoylcarbazole. This means that the population of the singlet excited state, which is the photoreactive excited state, is independent of the position where the acyloxy or sulphonyloxy groups are attached to the carbazole moiety.

The photochemical reactions of **43a** and **43b** show quantum yields of conversion (Φ_r) between 0.020 and 0.181 and are independent of the atmosphere (Ar or air), whereas it is observed that an increase in the solvent polarity the Φ_r value increases due to the estabilization of the S_1 excited state. Also, the effective rate constants of the photochemical reaction (k_r) were calculated for compounds **43a** and **43b** in different solvents. The k_r values ranged between $2.6x10^6$ s^{-1} to $2.9x10^7$ s^{-1} and clearly show that the photo-Fries rearrangement is an important deactivation pathway of the lowest singlet excited state under continuous excitation of solutions of 2-acyloxycarbazoles. Also, the photochemical reaction competes significantly with the fluorescence emission process in non-polar solvents, while in polar protic solvents, competition occurs to a great extent.

Systematic studies on the photo-Fries rearrangement of different 2-sulfonyloxycarbazoles have shown that this type of conversion can be readily used for the preparative-scale introduction of alkyl- or arylsulfonyloxy groups into polycyclic aromatic compounds under very mild conditions. Also, this photochemical method showed that it is advantageous in comparison to other thermal methods where Lewis acids and hazardous reagents are used (S_2Cl_2; S / $AlCl_3$; ArSCl / Fe°; S_2Cl_2 / NaOH).

Irradiation of 2-sulfonyloxycarbazoles **47a-c** in benzene, cyclohexane, MeCN or MeOH at 313 nm under Ar atmosphere at room temperature afforded the *ortho*-migrated products **48a-c** and **49a-c** together with **46** and the corresponding sulfonic acids (Scheme 33) [112]. The pair of *ortho*-rearranged products, *i.e.*, **48a/49a**, **48b/49b** and **48c/49c** were all obtained in a molar ratio of *ca.* 2:1. The C-1 rearrangement products **48a-c** were the main products formed in yields of *ca.* 60% while compounds **49a-c** were side products (*ca.* 25%).

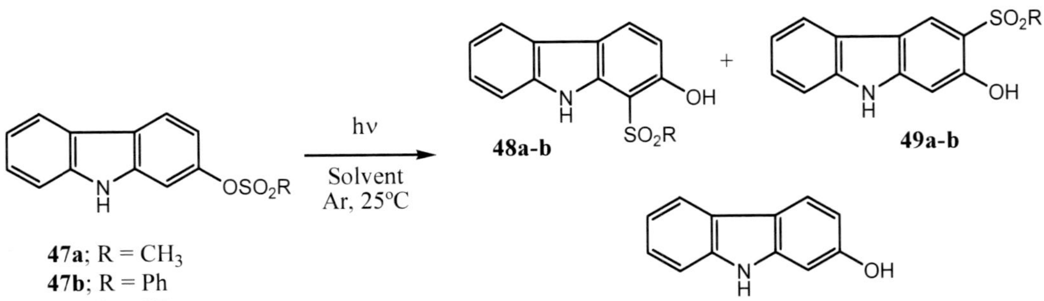

47a; R = CH_3
47b; R = Ph
47c; R = CF_3

Scheme 33. Photochemical reaction of2-sulfonyloxycarbazoles.

Rationalization of the observed regioselectivity in the above transformations, semi-empirical and *ab-initio* optimizations of 2-hydroxycarbazole radical were performed. The results showed that the charge density at C-1 was found to be sligthly higher than at C-3. Also, the carbazole NH might assist the acyl rearrangement though H-bonding, a process that, for geometric reasons, is possible only for the rearrangement to C-1, not to C-3. Therefore, both factors, accounts for the observed regioselectivity in favor of the 1-substituted product.

Effects of irradiation wavelength, solvent polarity, presence or absence of O_2 and photosensitizers on the photochemical reaction were studied in detail. Also, the photophysical properties of compounds **47a-c** in terms of absorption and fluorecence and phosphorescence emission spectroscopy under steady-state and time-resolved techniques have been studied.

The analyses of the photophysical properties, the photochemistry in solution and photosensitization experiments, clearly show i) that compounds **47** populated efficiently the S_1 state by direct absorption of ligth; ii) that the photoreactive excited affording the *ortho*-rearrangement products is the S_1 state (π,π^*); iii) that the photosensitized formation of the triplet-excited state does not yield any stable photoproduct; indeed, this triplet-excited state should be activated efficiently at room temperature through radiationless pathways. The quantum yields (ϕ) for the photo-Fries rearrangement of **47a-c** did not depend significantly on the presence or absence of O_2 while a moderate dependence on solvent polarity was observed.

With regards to the mechanism of the photo-Fries rearrangement of carboxilic esters, sulfonic esters, amides and sulfonamides of carbazoles, a [1,*n*]-sigmatropic rearrangement mechanisms takes place from the electronic excited state to afford the carbazole radical and the acyl (or sulfonyl) radical in the solvent-cage. These species live long enough in order i) to recombine forming the starting material; ii) to rearrange and afford the *ortho*- and *para*-migrated photoproducts, depending on the nature and substituents attached to the carbazole moiety and iii) to escape from the solvent-cage, trap an hydrogen atom to give carbazole and the acid (RCOOH or RSO_3H). The mechanistic consideration are outlined in a generalized form for *N*-acylcarbazoles in the Scheme 34.

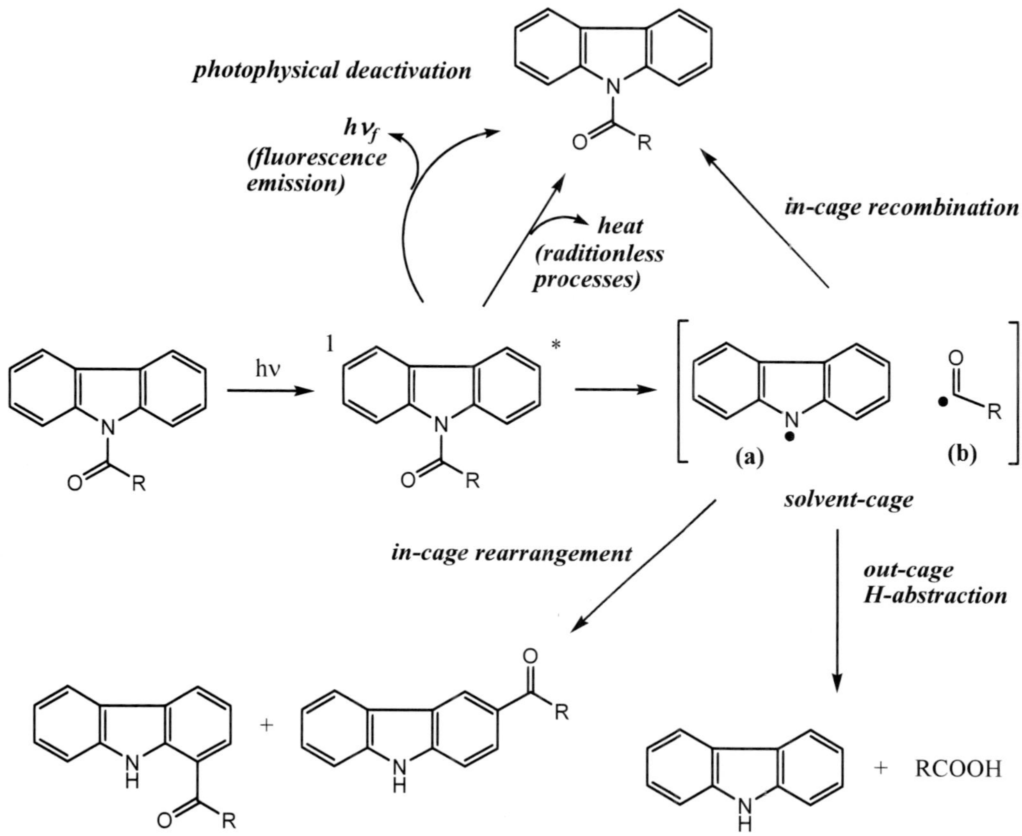

Scheme 34. Photorearrangement mechanism for N-acylcarbazoles.

Since Ghosh *et al.* [108,109] have proposed that the regioselectivity of the photo-Fries reaction of some *N*-aroylcarbazoles in MeOH / I_2 could proceed through a concerted reaction pathway, now-a-days it is accepted that this type of photo-rearrangements take place through a radical pathway involving radical species **(a)** and **(b)** [109,114]. The whole results reported clearly show that these radical species are formed in the solvent-cage and can react through a rearragement pathway to afford the migrated-photoproducts or escape from the solvent-cage. It is noteworthy to mention that radical scavangers do not affect the chemical yield of the migrated photoproducts (formed in-solvent-cage) and, at the same time, the formation of carbazole (formed out-solvent-cage), as a side product, is independent of the nature of the solvent, *i.e.*, poor or good H-donors like benzene or alcohols, respectively.

Finally it is interesting to mention that, in connection with carbazoles chemistry and photochemistry, among different studies related with synthesis of benzoyl- [115,116], and iodocarbazoles [118, 119] and spectroscopical carcaterization (Mas Spectrometry [120,121], ^{1}H and ^{13}C NMR [122-124]), Erra-Balsells *et al.* [125] has recently reported and interesting photohemical reduction of nitroaromatic compounds in the presence of acetic acid.

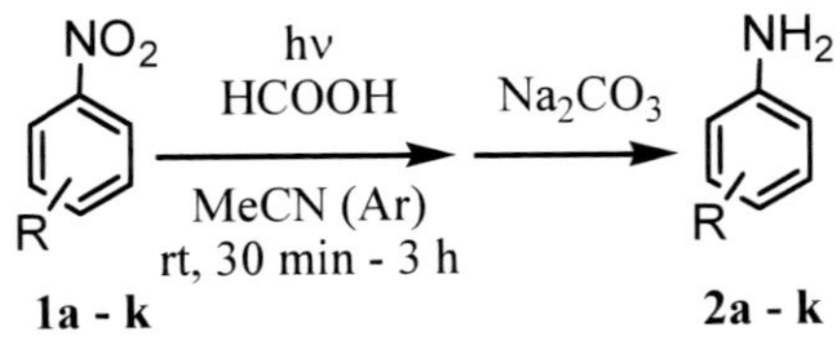

Scheme 35. Formic-mediated photochemical reduction of nitroaromatic compounds.

After the photon absorption, the electronic excited aromatic species can take and electron from the acetic acid starting the reduction process, as is shown in Scheme 36.

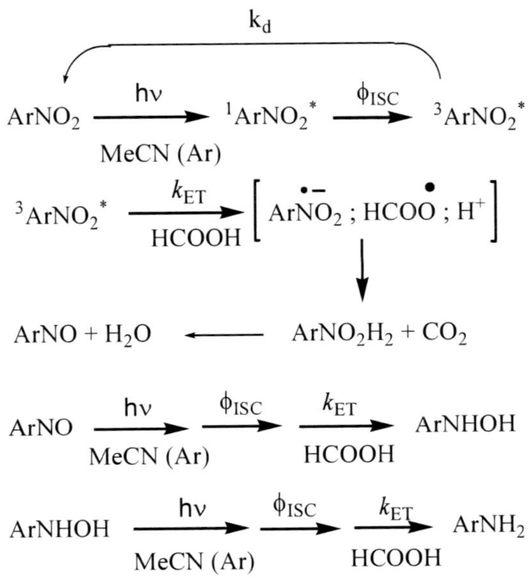

Scheme 36. Mechanism proposed for photochemical reduction of nitroaromatic compounds.

Among the different nitroarenes used, nitrocarbazoles were also included in this study. Chemical conversions were quite high, e.g. 1-nitrocarbazoles was phototransformed into the corresponding 1-aminecarbazole with 96% yield (irradiation time 45 min). In this case, a PET process is occuring but now the carbazoles moiety is working as the electron acceptor species.

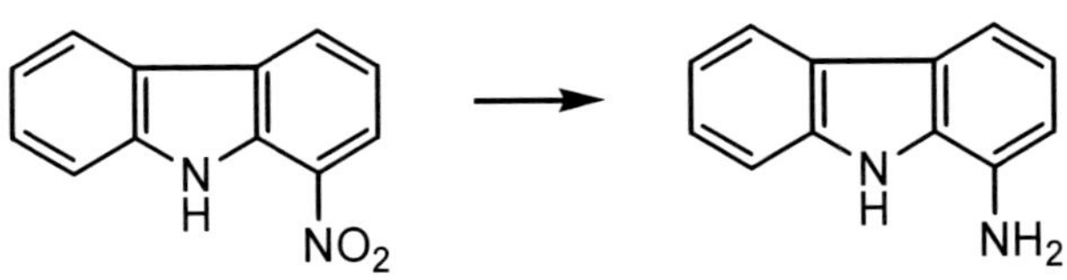

Figure 10. Formic acid-mediated photochemical reduction of nitrocarbazoles.

This mild photoreduction can be accomplished in high yield, with wide functional group tolerance and short reaction time (30 min to 1 hour), and allow a very clean method for the conversion of nitro arenes, nitro polycyclic and heterocylic arenas to the correspnding amine atructure.

Acknowledgments

The authors thank Universidad de Buenos Aires (X022), CONICET (PIP05/5443), ANPCyT (PICT 6-12312/02 and 0615/06) and Fundación Antorchas for finantial support. R. Erra-Balsells and S. M. Bonesi are research members of CONICET.

References

[1] Bonesi, S.M., Erra-Balsells, R. (2001). *J. Lumin.*, 93, 51.
[2] Bonesi, S.M., Erra-Balsells, R. (2002). *J. Lumin.*, 97, 83.
[3] Ponce, M. B., Cabrerizo, F. M., Bonesi, S. M., Erra-Balsells, R. (2006). *Helv. Chim. Acta,* 89, 1123.
[4] Mesaros, M., Bonesi, S. M., Ponce, M. A., Erra Balsells, R., Bilmes, G. M. (2003). *Photochem. Photobiol. Sci.*, 2, 808.
[5] Bonesi, S. M., Mesaros, M., Cabrerizo, F. M., Ponce, M. A., Bilmes, G. M., Erra-Balsells, R. (2007). *Chem. Phys. Lett.*, 446, 49.
[6] Ellinger, L. P. in: Pasika, W. M. (Ed.) (1968). *Advances in Macromolecular Chemistry*, Academic Press, London, pp. 169.
[7] Sharp, J. H. (1966) *J. Phys. Chem.*, 70, 584.
[8] Siminosescu, Cr. J., Orecec, V. (1979). *Rev. Roum. Chim.*, 24, 171.
[9] Landan, V., Ledwith, A., Marsh, D. G., Williams, D. J. (1976). *Macromolecules,* 9, 833.
[10] Solaro, R., Chiellini, E., Ledwith, A. (1980). *J. Am. Chem. Soc*, 58, 583.
[11] Pinkhan, C. A., Wait Jr., S. C. (1968). *J. Mol. Spectrosc.*, 27, 326.
[12] Bree, A., Zwarich, R. (1968). *J. Chem. Phys.*, 49, 3344.
[13] Bree, A., Zwarich, R. (1968). *J. Chem. Phys.*, 49, 3355.
[14] Witanoswski, M., Stefaniak, L., Januszewski, H., Grabowski, Z. R., Webb, G. A. (1972).*Tetrahedron*, 28, 637.
[15] Murk, D., Nitzche, L. E., Christoffersen, R. E. (1978). *J. Am. Chem. Soc*, 100, 1371.
[16] Bigelow, R. W., Ceasar, G. P. (1979). *J. Phys. Chem.*, 83, 1750.
[17] Haevey, P. D., Zelent, B., Durocher, G. (1983). *Spectroscopy*: Int. J., 2, 128.
[18] Johnson, G. E. (1974). *J. Phys. Chem.*, 78, 1512.
[19] Zelent, B., Harvey, P. D., Durocher, G. (1983).*Can. J. Spectrosc.*, 28, 188.
[20] Harvey, P. D., Zelent, B., Durocher. (1983).*Can. J. Spectrosc.*,29, 23.
[21] Zander, M. (1968). *Ber. Bunsenges. Phys. Chem.*, 72, 1161.
[22] Silvius, J. R. (1990).*Biochemistry*, 20, 2930.
[23] Omann, G. M., Glazer, M. (1984). *Biochemistry*, 23, 4962.
[24] Omann, G. M., Lakowicz, J. R. (1977). *Science*, 23, 465.
[25] Omann, G. M., Lakowicz, J. R. (1982). *Biochem. Biophys. Acta*, 83, 684.
[26] Abrams, F. S., Chattopadhyay, A. E. (1992). London, *Biochemistry*, 31, 5322.
[27] Mataga, N., Toriashi, Y., Ezumi, K. (1964). *Theor. Chim. Acta*, 2, 158.
[28] Huber, J. R., Adams, J. E. (1974). *Ber. Bunsenges. Phys. Chem.*, 78, 217.
[29] Chakravorty, S. C., Ganguly, S. C. (1970).*J. Chem. hys.*, 52, 2760.
[30] Kosower, E. M.,. Hofman, D., Wallenfels, H. (1962). *J. Am. Chem. Soc.,* 84, 2755.

[31] Reichardt, C., Grahn, W. (1970).*Chem. Ber.*, 103, 1072.
[32] Reichardt, C., Muller, R. (1976). *Liebigs. Ann. Chem.*, 1937.
[33] Murov, L. S., Carlmichael, I., Hug, G. L. (1993). *Handbook of Photochemistry*, Marcel Dekker, New York.
[34] Reichardt, C. (1979). *Solvent effect in Organic Chemistry*, Verlag Chemie, New York.
[35] Sarkar, A., Chakravorti, S. C. (1998).*J. Lumin.*, 78, 205.
[36] Ambrose, J. F., Nelson, R. F. (1975). *J. Electrochem. Soc.*, 122, 876.
[37] Ambrose, J. F., Nelson, R. F. (1968). *J. Electrochem. Soc.*, 115, 1159.
[38] Turro, N. J. Modern Molecular Photochemistry, Benjamin, Menlo Park, CA., 1978.
[39] Adams, J. F., Mantulin, W.W., Huber, J. R. (1973). J. *Am. Chem. Soc.*, 94, 5477.
[40] Kropp, P. J., Fryxell, G. E., Tubergen, M. W., Hager, M. W., Harris Jr, G. D.,McDermott Jr., T. P., Tornero-Velez, R. J. (1991). *J.Am. Chem. Soc.*, 113, 7300.
[41] J. W. Cubbage, T. A. Tetzlaff, H. Groundwater, R. D. McCulla, M. Nag, W. S. Jenks, J. (2001). *Org. Chem.*, 66, 8621.
[42] Jenks, W.S., Lee,W. Shutters, D. (1994). *J. Phys. Chem.*, 98, 2282.
[43] Martin, J. C., Perkins, C. W., Clarkson, R. B. (1986). *J. Am. Chem. Soc.*, 108, 3206.
[44] Margaretha, P. (1999). in *S-Centered Radicals*, Alfassi, Z. B., Ed. John Wiley and Sons: New York, pp 277-88.
[45] Hynes, A. J., Stoker, R. B., Pounds, A. J., McKay, T., Bradshaw, J. D., Nicovich, M., Winw, P. H. (1995). *J. Phys. Chem.*, 99, 16967.
[46] Barone, S. B., Turnipseed, A. A., Ravishankara, A. R. (1996). *Faraday Discuss.*, 100, 39.
[47] Barone, S. B., Turnipseed, A. A., Ravishankara, A. R. (1996). *J. Phys. Chem.*, 100, 14694.
[48] Turecek, F. (2000). *Collect. Czech., Chem. Commun.*, 65, 455.
[49] McKee, M. L. (1993). *J. Phys. Chem.*, 97, 10971.
[50] Bordwell, F. G., Zhang, X., Cheng, J. P. (1991). *J. Org. Chem.*, 56, 3216.
[51] Bordwell, F. G., Zhang, X., Cheng, J. P. (1993). *J. Org. Chem.*, 58, 6410.
[52] Jonsson, M., Lind, J., Eriksen, T. E., Merenyi, G. (1994). *J. Am. Chem. Soc.*, 116, 1423.
[53] Saito, I., Ikehira, H., Kasatami, R., Watanabe, M., Matsuura, T. (1986). *J. Am. Chem. Soc.*, 108, 3115.
[54] Park, M., Rizzo, C. J. (1996). *J. Org. Chem.*, 61, 6092.
[55] Chu, C. K., Baker, D. C. Eds. (1993). *Nucleosides & Nucleotides as Antitumor and Antiviral Agents*. Plenum Press: New York,
[56] Heildelberg, C. (1965).*Prog. Nucleic Acid Res. Mol. Biol.* 4, 1.
[57] Cooke, S. T., Lebleu, B. Eds. (1993). *Antisense Research and Applications*. CRC Press: Boca Raton, Fl.
[58] Milligan, J. F., Matterucci, M. D., Martin, J. C. (1993). *J. Med. Chem.*, 36, 1923.
[59] Uhlmann, E., Peyman, A. (1990).*Chem. Rev.*, 90, 543.
[60] Bishofberger, N., Shea, R. G. (1992). in *Nucleic Acid Targeted Drug Design: Propst.* C. L.; Perum, T. J. Eds. Marcel Dekker: New York, pp 579-612.
[61] Varma, R. S. (1993). *Synlett,* 713.
[62] Cohen, J. S.,Hogan, M. E. (1994). *Sci. Am.,* 76.

[63] DeMesmaeker, A., Haner, R., Martin, P., Moser, H. E. (1995). *Acc. Chem. Res*., 28, 366.

[64] Prudhomme, D. R., Wang, Z., Carmelo, C. J. (1997). *J. Org. Chem*., 62, 8257.

[65] Barton, D. H. R., McCombie, S. W. (1975). *J. Chem. Soc, Perkin Trans I*, 1574.

[66] Hartwig, W. (1983). *Tetrahedron*, 39, 2609.

[67] McCombie, S. W. in *Comprehensive Organic Synthesis*, Trost, B. M., Fleming, I., Eds. Pergamon Press: Oxford, 1991, pp 811 -833.

[68] Wang, Z., Pridhomme, D. R., Buck, J. R., Park, M., Rizzo, C. J. (2000). *J. Org. Chem*., 65, 5969.

[69] Yoshimura, Y., Fujimoto, K. (2006). *Chem. Lett.,* 35, 386.

[70] Dandlinker, J. P., Nuñez, M. E., Barton, J. K. (1998). *Biochemistry* 37, 6491.

[71] Wagner, P. J., Kochevar, I. E., Kemppainen, A. E. (1972). *J. Am. Chem. Soc*., 94, 7489.

[72] Wagner, P. J., Thomas, M. J.,. Harris, E. (1976). *J. Am. Chem. Soc*., 98, 7675.

[73] Nowakowska, M., Szczubialka, K. (1995). *J. Photochem. Photobiol A: Chem*., 91, 81.

[74] Crosby, D. G., Beynon, K. L., Greve, P. A., Korte, F., Still, G. G., Vonk, J. W. (1981). *Pure Appl. Chem*., 53, 1051.

[75] Saveant, J. M. (1993). *Acc. Chem. Soc*., 26, 455.

[76] Saveant, J. M. in Dissociative Electron Transfer, in Advances in Electron Transfer Chemistry, Mariano, P. S. Ed. JAI Press, New York, 1994, vol. 4, pp 53-116.

[77] Andrieux, C. P., Saveant, J. M., Tardy, C. *J. Am. Chem. Soc*., 119, 11546. (1997) and references cited therein.

[78] Iles, D. H., Ledwith, A. (1969). *Chem. Comm*, 364.

[79] Bonesi, S. M., Erra-Balsells, R. (1999). *An. Soc. Quim. Argent*., 87, 127.

[80] Bonesi, S. M., Erra-Balsells, R. (2000). *J. Chem. Soc., Perkin Trans*. 2, 1583.

[81] Zelent, B., Durocher, G. (1981). *J. Org. Chem*., 46, 1496.

[82] Zelent, B., Durocher, G. (1982).*Can. J. Chem.,* 60, 945.

[83] Zelent, B., Durocher, G. (1982). *Can. J. Chem*., 60, 2442.

[84] Zelent, B., Durocher, G. (1985).*Can. J. Chem*., 63, 1654.

[85] Erra-Balsells, R., Frasca, A. R. (1984). *Tetrahedron Lett*, 25, 5363 – 5366.

[86] Erra-Balsells, R., Frasca, A. R. (1985). *An. Asoc. Quim. Argent*., 73, 207.

[87] Erra-Balsells, R., Frasca, A. R. (1983). *Tetrahedron*, 39, 33.

[88] Danen, W.C., Neugebauer, F.A. (1975). *Angew. Chem., Int. Ed*., 14, 783.

[89] Anderson, J. C., Reese, C. B. (1960). *Proc. Chem. Soc*. 217.

[90] Bellus, D. (1971). *Adv. Photochem*. 8, 140.

[91] Stratenus, J. L., Havinga, E. (1966). *Rec. Trav. Chim*. 85, 434.

[92] Snell, B. K. (1968). *J. Chem. Soc. C*, 2367.

[93] Sandner, M. R., Hedaya, E., Tecker, D. J. (1968). *J. Am. Chem. Soc*., 90, 7249.

[94] Finnegan, R. A., Kunston, D. (1968).*Tetrahedron Lett*., 3429.

[95] Plank, D. A. (1968). *Tetrahedron Lett*., 5423.

[96] Meyer, J. W., Hammond, G. S. (1972). *J. Am. Chem. Soc*., 94, 2219.

[97] Kalmas, C. E., Hercules, D. M. (1974). *J. Am. Chem. Soc*., 96, 449.

[98] Adam, W. J. (1974). *Chem. Soc., Chem. Comm*., 289.

[99] Taub, D., Kuo, C. H., Slates, H. L., Wendler, N. L. (1963). *Tetrahedron*, 19,1.

[100] Kende, A. S., Belletrie, J., Bently, T. J., Hume, E., Airey, J. (1975). *J. Am. Chem. Soc.*, 97, 4425.

[101] Ramakrisham, V. T., Kagan, J. *J. Org. Chem.*, 35, 2901. (1970).

[102] Obara, H., Takahashi, H., Hirano, H. *Bull. Chem. Soc. Jpn.*, 42, 560. (1969).

[103] Miranda, M. A. in *Organic Photochemistry and Photobiology*. Horspool, W. M.; Song, P. S., Eds. CRC Press: Boca Raton; 1995; Chapter 47; pp 570 – 578.

[104] Zander, M. (1981). *Chem. Ber.*, 114, 176.

[105] Shizuka, H., Kato, M., Ochiai, T., Matsui, K., Morita, T. (1970).*Bull. Chem. Soc. Jp.*, 43, 67.

[106] Bonesi, S. M., Erra-Balsells, R. (1991). *J. Photochem. Photobiol., A: Chem.*, 56, 55.

[107] Bonesi, S. M., Erra-Balsells, R. (1997). *J. Photochem. Photobiol., A: Chem.*, 110, 271.

[108] Ghosh, S., Das, T. K., Datta, D. E., Mehta, S. (1987). *Tetrahedron Lett.*, 28, 4611.

[109] Ghosh, S., Datta, D. B., Data, I., Das, T. K. (1989). *Tetrahedron*, 45, 3775.

[110] Chakrabarti, A., Biwas, G. K., Chakraborty, D. P. (1989).*Tetrahedron*, 45, 5059.

[111] Bonesi, S. M., Crevatin, L. K., Erra-Balsells, R. (2004*). Photochem. Photobiol. Sci.*, 3, 381.

[112] Crevatin, L. K., Bonesi, S. M., Erra-Balsells, R. (2006). *Hel. Chim. Acta*, 89, 1147.

[113] Fox, M. A., Chanon, M. (1988). *Photoinduced Electron Transfer*. Elsevier Science, Amsterdam.

[114] Bouchet, P., Jonchery, G., Jacquier, R., Elguero, J. (1978). *J. Heterocycl. Chem.*, 15, 625.

[115] Bonesi, S. M., Erra-Balsells, R. (1991). *J. Heterocycl. Chem.*, 28, 1035.

[116] Bonesi, S. M., Erra-Balsells, R. (1991). *An. Asoc. Quim. Argentina*, 79, 113.

[117] Bonesi, S. M., Erra-Balsells, R. (1997). *J. Heterocycl. Chem.*, 34, 891.

[118] Bonesi, S. M., Erra-Balsells, R. (2001). *J. Heterocycl. Chem.*, 38, 77.

[119] Monge, M. E., Bonesi, S. M., Erra-Balsells, R. (2002). *J. Heterocycl. Chem.*, 39, 933.

[120] Erra-Balsells, R. (1988). *J. Heterocycl. Chem.*, 25, 221.

[121] Nonami, H., Fukui, S., Erra-Balsells, R. (1997). *J Mass Spectrom.*, 32, 287.

[122] Erra-Balsells, R. (1988). *Magnetic Res. Chem.*, 26, 1109.

[123] Bonesi, S. M., Ponce, M. A., Erra-Balsells, R. (2004). *J. Heterocyl. Chem.*, 41, 161.

[124] Bonesi, S. M., Ponce, M. A., Erra-Balsells, R. (2005). *J. Heterocyl. Chem.*, 42, 295.

[125] Cors, A., Bonesi, S. M., Erra-Balsells, R(2008). *Tetrahedron. Lett.*, in press.

In: Photochemistry Research Progress
Editors: A. Sánchez, S. J. Gutierrez

ISBN: 978-1-60456-568-3

Chapter 11

2+2 CYCLOADDITION ON FURAN DERIVATIVES

Maurizio D'Auria, Lucia Emanuele and Rocco Racioppi
Dipartimento di Chimica, Università della Basilicata, Italy

ABSTRACT

The Paternò-Büchi reaction on furan derivatives is reviewed. The photochemical reaction of furan derivatives in the reaction with carbonyl compounds gives the corresponding monoadduct with high regio- and stereoselectivity. When the reaction is carried out in benzene, a large amount of the diadduct can be obtained. The exo stereoselectivity at C-6 was explained assuming the presence of a secondary orbital effect on spin orbit coupling. Regio- and stereoselectivity of the reaction of carbonyl compounds with 2,3-dihydrofuran is described. Applications of the reaction to the synthesis of 2- and 3-furylmethanol derivatives is reported. The synthesis of asteltoxin, avenaciolide, and oxetanocin is also reported. Asymmetric reactions on chiral glyoxylate is reported. The low diastereoselectivity of the reaction with chiral ketones and aldehydes is discussed. Finally, the high diastereoselectivity observed in the reaction with 2- and 3-furylmethanol derivatives is reported.

INTRODUCTION

The Paternò-Büchi reaction is a milestone in organic photochemistry (Scheme 1a). Paternò and Chieffi reported obtaining oxetanes from the photocycloaddition of ketones to olefins in 1909, but this reaction was recognized as an important synthetic reaction only after the work of Büchi [1,2].

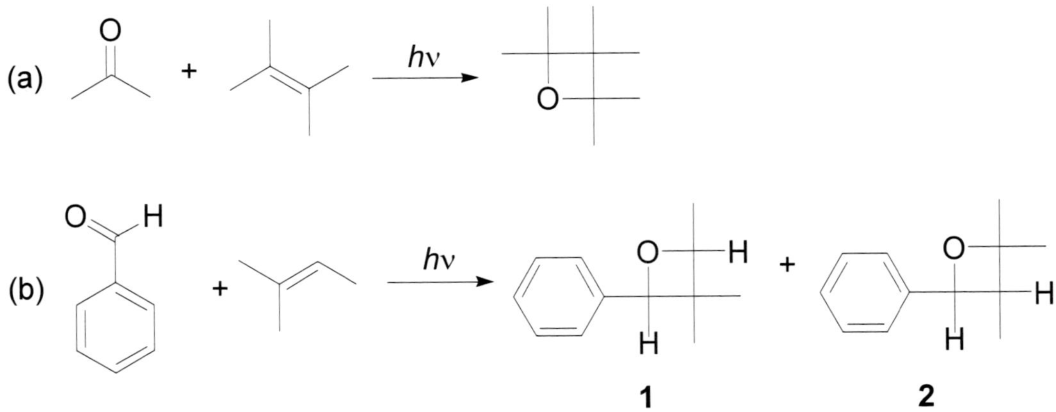

Scheme 1.

The regiochemistry of this reaction appeared to be a difficult problem to solve from the first. Paternò performed a reaction between 2-methyl-2-butene and benzaldehyde, and he could not assign the exact structure of the product. He was not able to distinguish between **1** and **2** [Scheme 1b] [1]. Büchi solved the problem showing that **1** was the actual product [2]. Furthermore, in this reaction the stereocontrol was not examined, but several stereoisomers could be obtained.

After these pioneering works, several reviews cover the enormous number of papers published in this field [3-7].

The synthesis of oxetanes can have a great importance in the development of compounds with relevant biological properties: oxetane ring is present in the skeleton of Taxol (**3**), an important drug used in the treatment of ovarian cancer [8], merrilactone A (**4**), a new sesquiterpene dilactone with neurotrophic activity [9], and several antiviral oxetanes, such as **5**, **6**, and **7**, have been described in literature (Scheme 2) [10-12].

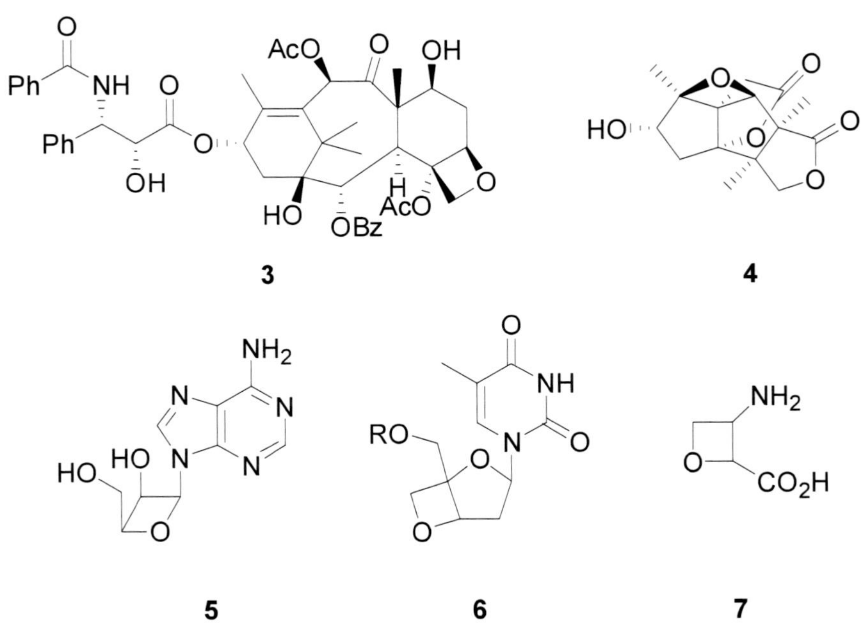

Scheme 2.

The Paternò-Büchi reaction is a photocycloaddition reaction of a n,π* carbonyl compound to an alkene in the ground state from either the S_1 or the T_1 state. The reaction can occur through the initial C-O bond formation or through a previous formation of the C-C bond.

Frontier orbitals approach can be used to explain the formation of oxetanes. We can observe HSOMO-LUMO interaction in which the half-occupied π* carbonyl orbital interacts with the unoccupied π* molecular orbital of an electron deficient alkene and a C,O-biradical is formed. The LSOMO-HOMO interaction in which the half-occupied n orbital of the carbonyl O atom interacts with the π orbital of an electron-rich alkene and a C,C-biradical is formed [13,14].

More recently a theoretical study of the Paternò-Büchi reaction has been written. This study showed that there are two conical intersection points located near the C-C and C-O bonded biradical regions of the ground state. These two conical intersections support a mechanism where the decay from the excited state is accompanied by a geometric rotation of the terminal group, in the case of C-O attack, and by an orbital rotation at the oxygen center, in the case of C-C attack. Furthermore, for C-O attack, the triplet surface must cross the singlet to reach a diradicaloid minimum. For C-C attack, the triplet biradical minimum is located at the same geometry as the conical intersection between the two singlet states, and the efficiency of the intersystem crossing will be determined by the nature of the spin-orbit coupling. Thus, for the triplet, the reaction path can be predicted by the most stable biradical rule [15]. A conformational analysis of the biradicals also appeared [16]. The biradical intermediate in the reaction between benzophenone and electron-rich alkene has been determined by using laser flash photolysis. An absorption with λ_{max} 535 nm has been observed (Scheme 3) [17,18].

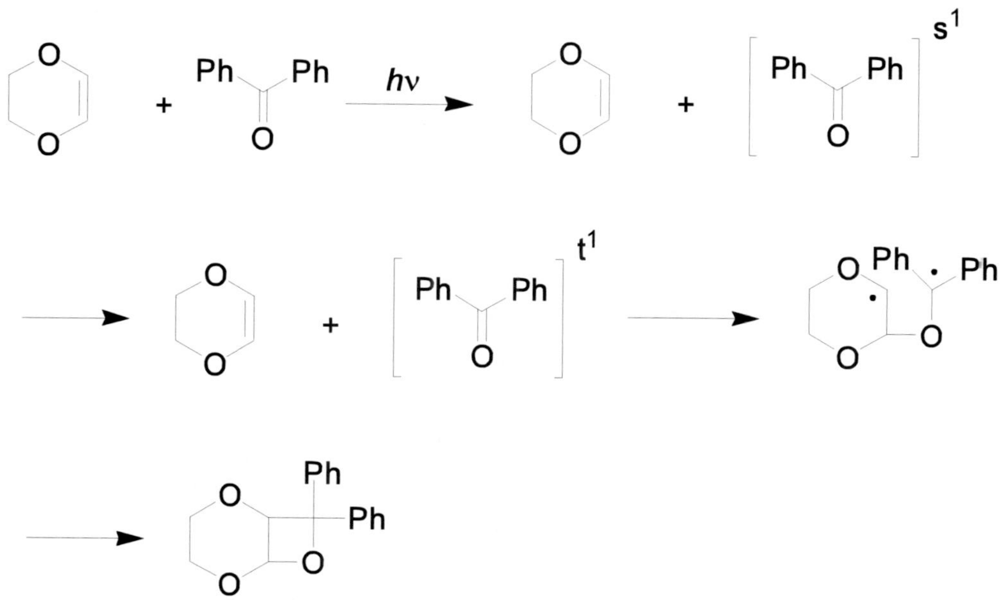

Scheme 3.

Furthermore, in the reaction of furan with benzophenone a transient species was observed [19]. Also in this case the transient species was identified as the triplet biradical and it showed an absorption at 531 nm.

We performed DFT calculations at B3LYP/6-31G+(d,p) level on the triplet biradical intermediates of these two reactions by using Gaussian 03 [20]. Figure 1 shows the optimized structure of the biradical intermediate in the reaction between 1,4-dioxene and benzaldehyde.

Table 1. Energy and shape of suitable molecular orbitals in the biradical intermediate obtained in the reaction between 1,4-dioxene and benzaldehyde

Molecular orbital	Energy (H)	Shape
54 (NLUMO)	-0.002	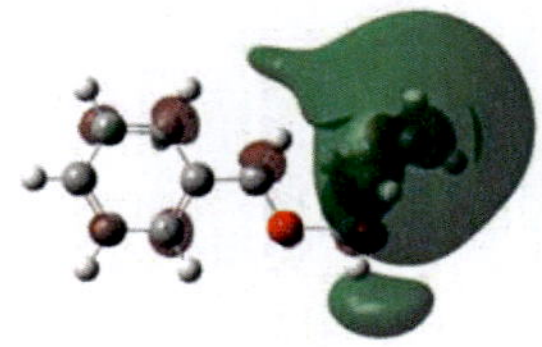
53 (LUMO)	-0.007	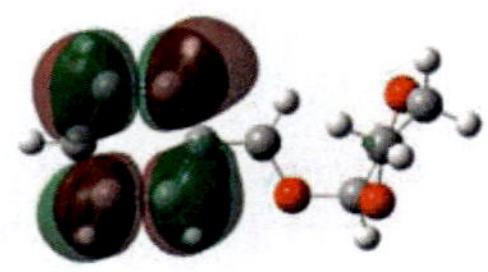
52 (HSOMO)	-0.076	
51 (LSOMO)	-0.090	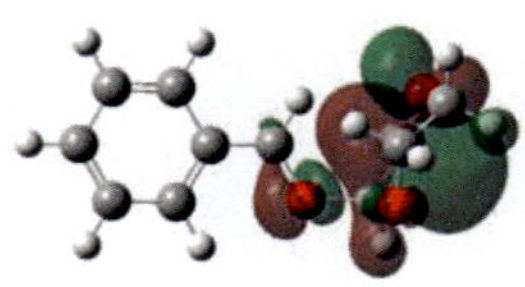
50	- 0.249	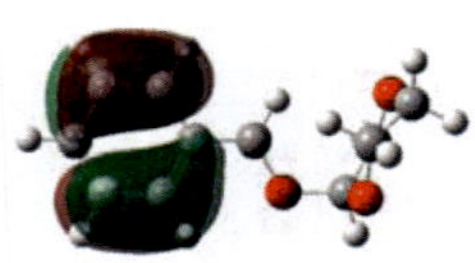

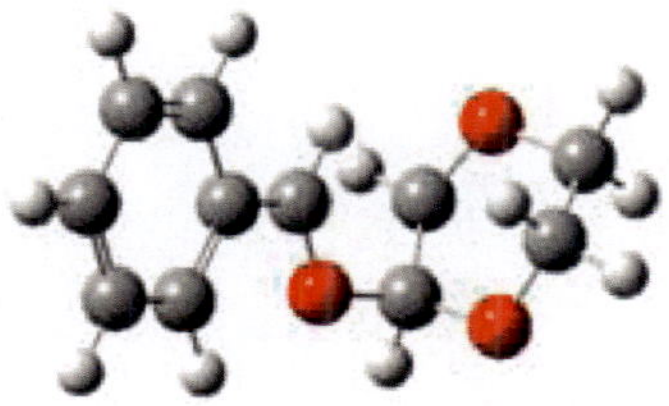

Figure 1. 1,4-Dioxene-benzaldehyde triplet adduct.

The structure is characterized by the orthogonal position between the *p* orbitals at the radical centers. This behavior is in agreement with geometry able to give an intersystem crossing process to the corresponding singlet biradical able to give the cyclized product. In fact, the perpendicular relation between the *p* orbitals increases the spin-orbit coupling. Furthermore, the radical centers are in an eclipsed conformation able to cyclized immediately after the intersystem crossing.

In Table 1 we have collected the energy and the shape of some molecular orbitals of the biradical.

The data collected in Table 1 gave us the indication that the only transition able to give the observed transient absorption is that from LSOMO to LUMO (549 nm) and that it is a $\pi \rightarrow \pi^*$ transition.

On the basis of these data we can conclude that the observed transient absorption can be ascribed to a π,π^* transition of the supposed biradical intermediate and that these data are in agreement with the presence of this biradical intermediate.

The same treatment can be performed on the supposed biradical intermediate whose transient absorption was observed in the reaction between furan and benzophenone. In Figure 2 we reported the optimized structure for this biradical intermediate in the reaction between furan and benzaldehyde.

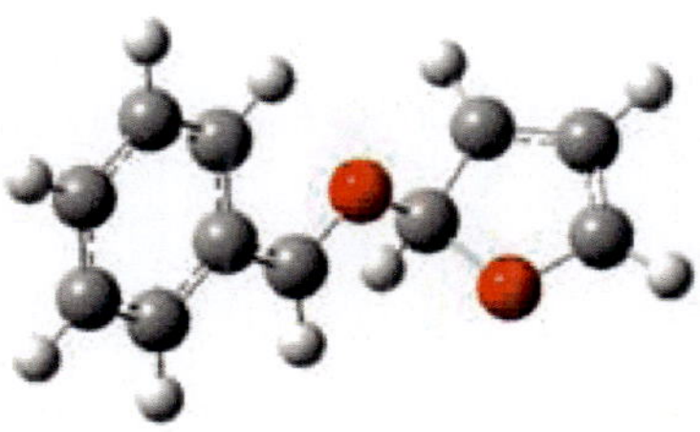

Figure 2. Furan-benzaldehyde triplet adduct.

Also in this case, the structure is characterized by the orthogonal position between the *p* orbitals at the radical centers. This behavior is in agreement with a geometry able to give an intersystem crossing process to the corresponding singlet biradical able to give the cyclized product. In this case the radical centers are not eclipsed. This behavior is in agreement with previous works in this field, showing the presence of two conformational minima for the biradical [19].

In Table 2 we have collected the energy and the shape of some molecular orbitals of the biradical.

Also in this case the only transition able to give the observed transient absorption is a LSOMO → LUMO transition. As in the case of 1,4-dioxene biradical, the transition is a $\pi \rightarrow \pi^*$ transition at 555 nm, in good agreement with the experimental results.

Table 2. Energy and shape of suitable molecular orbitals in the biradical intermediate obtained in the reaction between furan and benzaldehyde

Molecular orbital	Energy (H)	Shape
49 (NLUMO)	0.003	
48 (LUMO)	-0.008	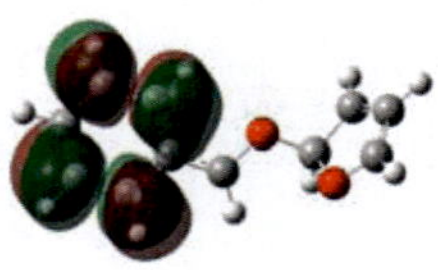
47 (HSOMO)	-0.078	
46 (LSOMO)	-0.090	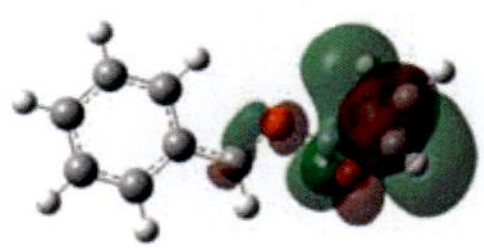
45	-0.251	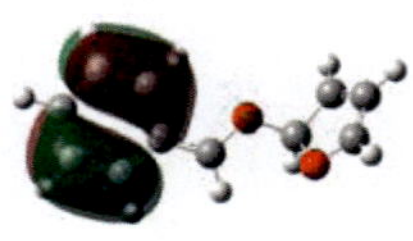

In conclusion all the above reported results are in agreement with the presence of a biradical intermediate in the Paternò-Büchi reaction between electron-rich alkenes and aromatic carbonyl compounds. We were able to find the transition responsible for the transient absorptions observed in these reactions. All the transitions are $\pi \rightarrow \pi^*$ ones, and corresponds to LSOMO → LUMO transitions.

The formation of exciplex is used to explain the reaction behaviour of simple alkene. However, evidence of monoelectron transfer processes are reported for electron-rich alkenes [21].

THE PATERNO-BÜCHI REACTION ON FURAN DERIVATIVES

The first report on the photochemical reactivity of furan towards carbonyl compounds appeared in 1966. Schenck reported that the irradiation of benzophenone in furan gave the corresponding adduct in 94% yield (Scheme 4) [22]. The structure assignment was confirmed by Gagnaire et al. on the basis of the NMR analysis of both the products of the same reaction and of the reaction between 2,5-dimethylfuran and benzophenone [23].

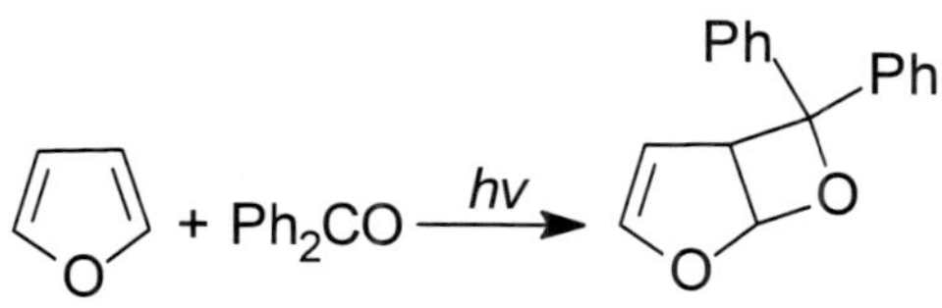

Scheme 4.

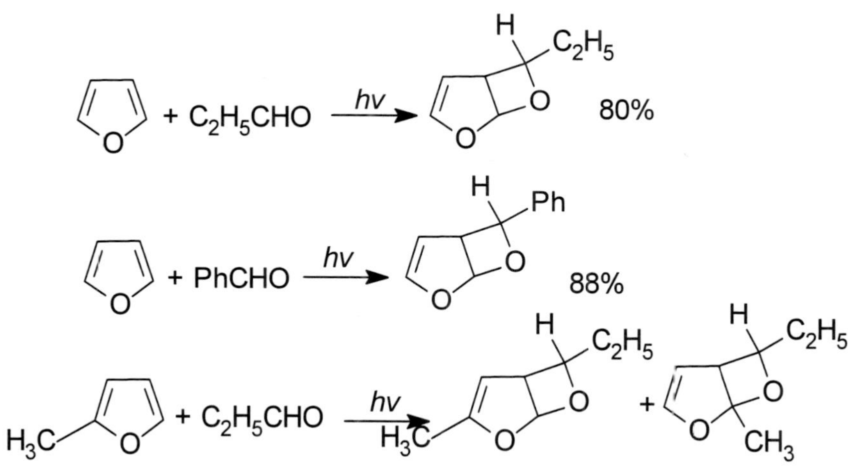

Scheme 5.

Two years later furan and 2-methylfuran were found to react with propanal and benzaldehyde. NMR spectral data supported the formation of the same type of products (Scheme 5) [24].

The structure of the products was confirmed also by NOE analysis [25]. The yields were very high. In this case, the authors did not specify the *exo* or *endo* stereochemistry at C-6 on the dioxabicyclo[3.2.0]heptene skeleton. This problem was solved some years later assigning the *exo* configuration to this carbon [26].

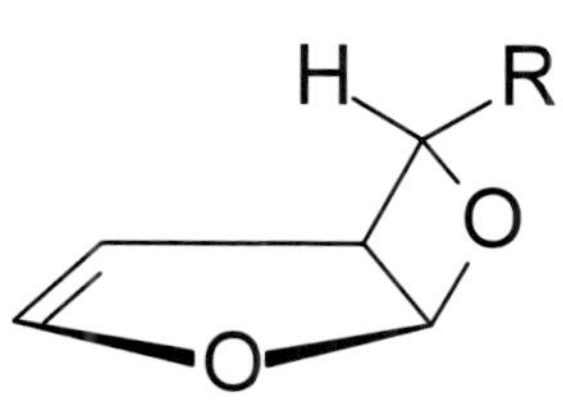

When 2-methylfuran was used, a mixture of regioisomeric products was obtained, but the authors did not report the regioisomeric ratio. In contrast with this result, other authors reported a complete regioselectivity in the reaction between substituted furans and benzophenone: in these reactions the coupling occurred on the most hindered side of the molecule (Scheme 6) [27,28].

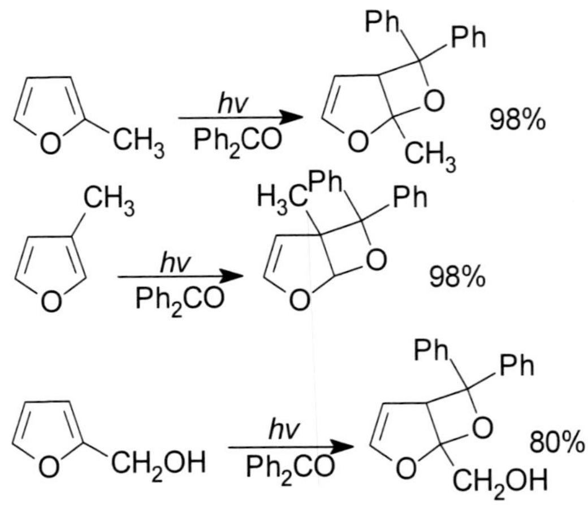

Scheme 6.

The Japanese group supposed that the reaction involved the formation of the biradical **8**, due to the attack of the n,π* triplet oxygen on the site of higher free valence or higher electron density of the furan ring.

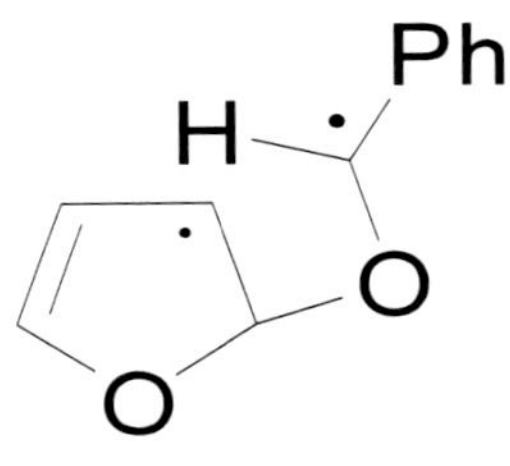

8

Afterwards, the same group tested the reaction on a large group of aliphatic and aromatic aldehydes and ketones [29]. They found that ketones reacted giving lower yields than the corresponding aldehydes: in particular acetone, butanone, and acetophenone gave yields in the range of 0.9-1.7%; furthermore, the oxetane yields increased with the number of carbon atoms: with acetaldehyde they obtained 15% yield, while cyclohexylcarbaldehyde gave 27% yield and benzaldehyde 35%.

The kinetics of the reaction is in agreement with a mechanism involving the formation of the biradical **8** following the scheme:

$$B \xrightarrow{h\nu} B^s$$

$$B^s \xrightarrow{k_1} B$$

$$B^s \xrightarrow{k_2} B^t$$

$$B^t \xrightarrow{k_3} B$$

$$B^t + F \xrightarrow{k_4} B + F$$

$$B^t + F \xrightarrow{k_5} (B\text{-}F)^*$$

$$(B\text{-}F)^* \xrightarrow{k_6} Ox$$

$$(B\text{-}F)^* \xrightarrow{k_7} B + F$$

Where B is benzophenone, F is furan, (B-F)* is the biradical **8** and Ox is the oxetane [30].

More complex carbonyl compounds were used to give the corresponding adducts. *n*-Butyl glyoxylate gave the adduct with furan in 77.3% yield, while diethyl ketomalonate gave the corresponding product in 30% yield [31].

Good regioselectivity was observed using silyl and stannyl furan derivatives: the reaction in this case occurred on the less hindered side of the molecule (Scheme 7) [32].

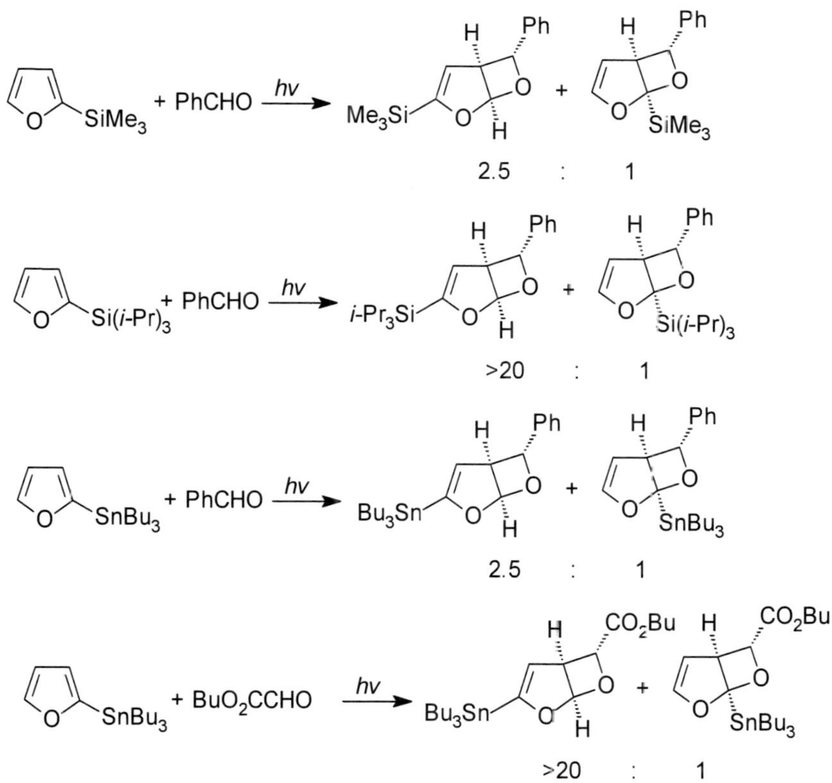

Scheme 7.

On the contrary, 2-furylmethanol and the corresponding silyl ether gave low regioselectivity [33,34].

The high *exo* stereoselectivity of the reaction has been extensively studied: the formation of the product occurs on a triplet 1,4-biradical which must be converted into the singlet biradical to give the product. Three mechanisms operate for the interaction between singlet and triplet states of 1,n-biradicals: electron-nuclear hyperfine coupling (HFC), spin-lattice relaxation (SLR), and spin-orbit coupling (SOC). HFC is an important control factor for biradicals with long carbon chains between the radical centers. SOC plays a dominant role in biradicals with shorter distances between the radical centers, whereas SLR seems to contribute only marginally. SOC is strongly dependent on the geometry of the triplet biradicals: in particular, SOC decreases with increasing distance between the two spin-bearing atoms. The best geometry to have SOC requires that the axes of the *p* orbitals at the radical centers are perpendicular to each other. In order to explain the pronounced *exo* stereoselectivity, a secondary orbital effect can be postulated: an interaction between the rather flexible α-oxy radical center and the allyloxy ring localized radical in **9**, likely plays a major role (Scheme 8) [35,36].

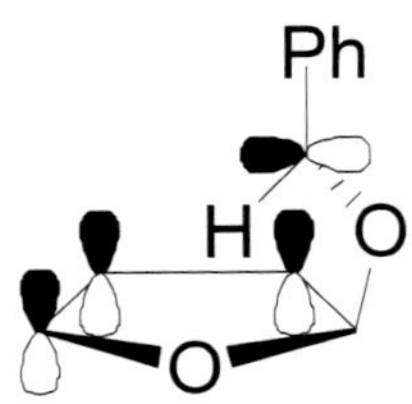

9

Scheme 8.

We examined the regioisomeric biradical intermediates **A** and **B** resulting from the head-to-head and head-to-tail addition, respectively (Figure 3) [37].

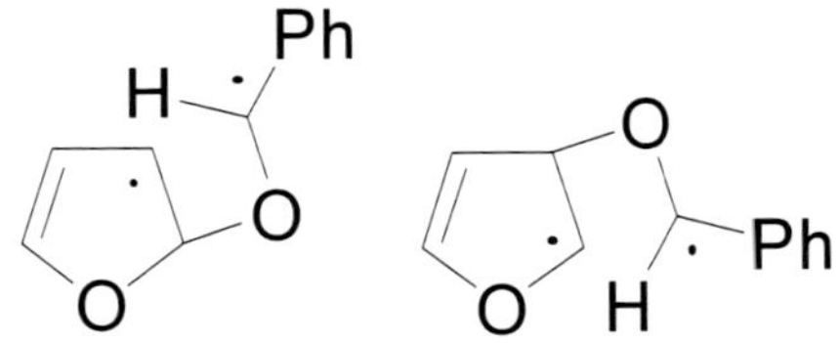

A **B**

Figure 3. Possible biradical intermediates in the reaction of furan with benzaldehyde.

The biradical **A** is more stable than **B** by 16.5 kcal mol^{-1}. The biradical **A** exists as two conformers and we considered only that conformer able to give the ring closure. The HSOMO and LSOMO of the biradical **A** are reported in Figure 4.

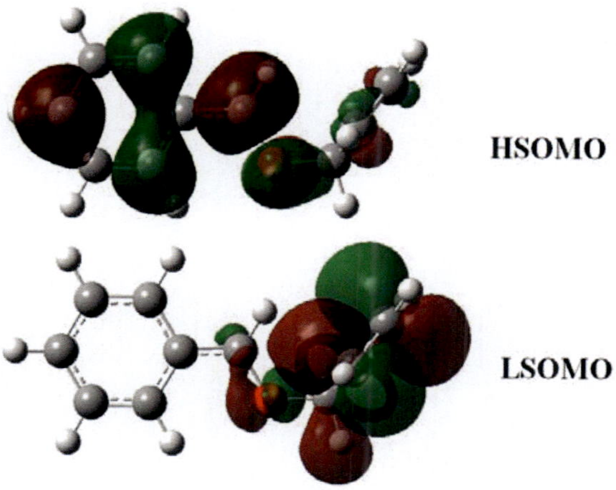

Figure 4. HSOMO and LSOMO of biradical intermediate A.

Also in this case the HSOMO is mainly localized on the benzaldehyde fragment of the biradical while the LSOMO is mainly localized on the furanoid part of the molecule. The coupling between the radical carbons in these two orbitals, considering that the atomic orbitals involved can superimpose themselves, can give only the *exo* isomer, in agreement with the experimental results (Figure 5).

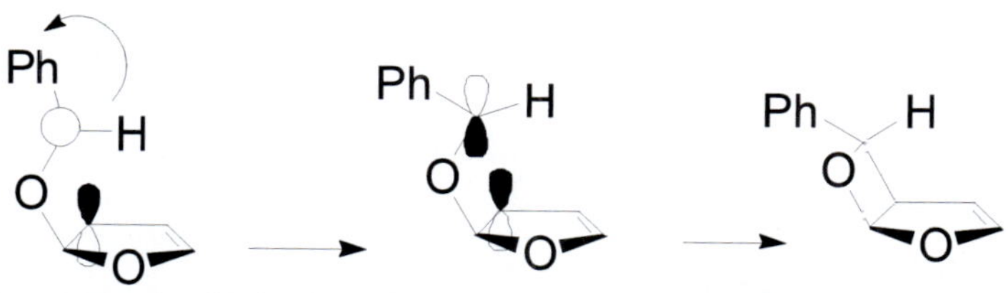

Figure 5. Ring closure reaction in the formation of the exo isomer of the adduct between furan and benzaldehyde.

When the reaction was carried out in benzene, new products were obtained. Ogata reported that two 2:1 adducts are the product of the reaction (Scheme 9) [38]. When 2-methylfuran was irradiated in the presence of benzophenone the 1:1 adduct was obtained with two 2:1 adducts (Scheme 9). On the contrary, 2,5-dimethylfuran gave only one 2:1 adduct (Scheme 9); benzaldehyde showed minor selectivity and this behaviour was attributed to the higher triplet energy of benzaldehyde (71 kcal mol^{-1}) in comparison with that of benzophenone (68.7 kcal mol^{-1}): the more energetic excited benzaldehyde would not entirely discriminate between the two double bonds.

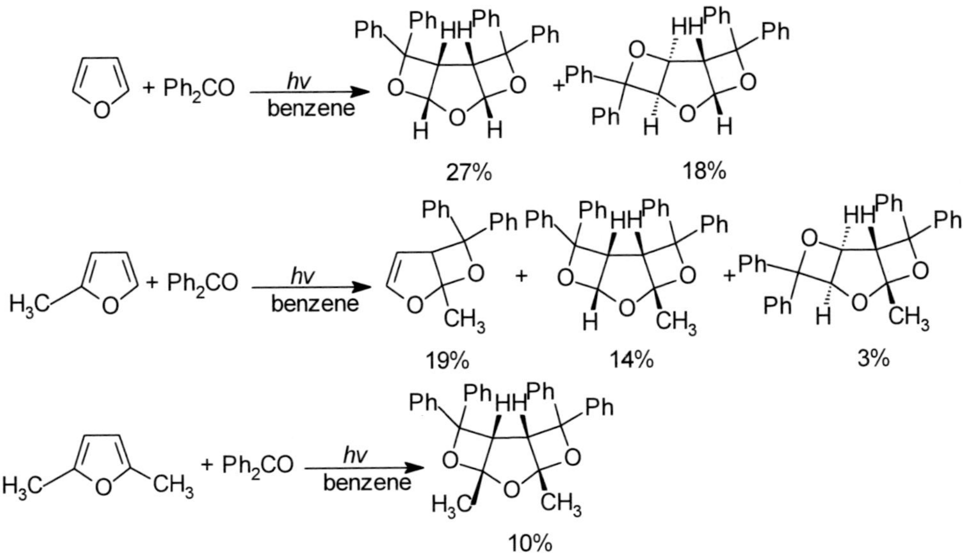

Scheme 9.

The formation of the major 2:1 adduct in the reaction of Ogata was immediately questioned by Leitich [39]. He considered that structure **10** was more acceptable.

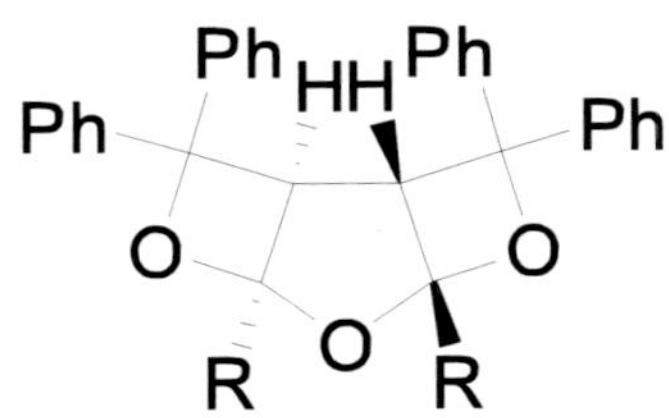

10

Evanega and Whipple made the reaction obtaining the 1:1 adduct, the adduct **10** in 29% yield, and the other 2:1 adduct reported in Scheme 9 in 18% yield [40]. Finally, the structure of the adduct **10** was confirmed through X-ray analysis [41].

Furan reacts with chloral to give unexpectedly the corresponding 2-furyl carbinols [42]. Furthermore, 2-cyanofuran did not react, while 2-furfural diacetate and furfural ethylene acetal showed low reactivity: in particular, furfural diacetate reacted with benzophenone in benzene, giving the monoadduct with high regioselectivity (only the adduct on the most hindered side of the molecule was obtained) but in low yield (21%), and in the presence of significant amounts of diadducts (two diadducts were obtained in 21 and 18% yields, respectively). Furthermore, furfural ethylene acetal reacted with benzophenone in benzene showing high regioselectivity because the ketone attacks the furan derivative only on the side bearing the substituent, but in low yield (22%) and in the presence of diadducts in 28 and 15% yields [42].

The cycloaddition reaction can be performed on esters. In this case, the adducts can be obtained in a few cases. In most of the examples, they underwent a cycloreversion reaction to give the ring opening products (Scheme 10) [43]. More recently, this result has been questioned and a 95:5 mixture of stereoisomeric adducts was identified as the product [36].

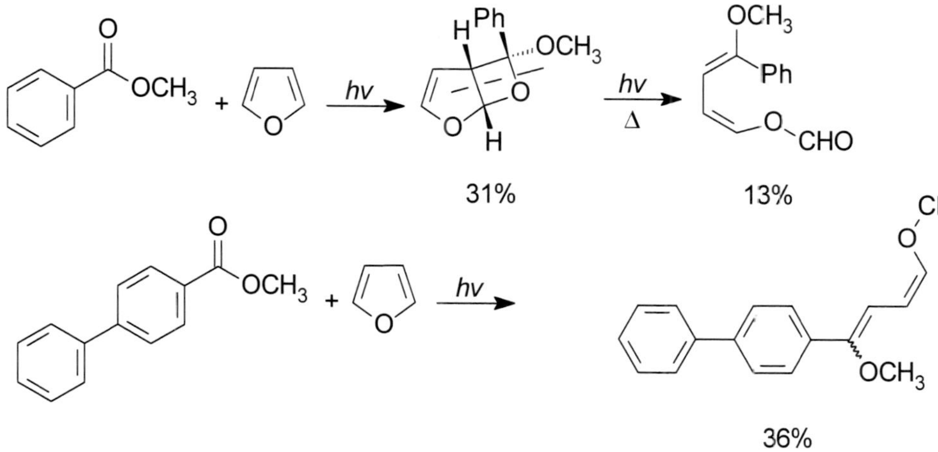

Scheme 10.

Coupling products can be obtained carrying out the reaction between an amide and furan: also in this case the cycloadduct can not be isolated, but the subsequent decomposition products can be isolated (Scheme 11) [44].

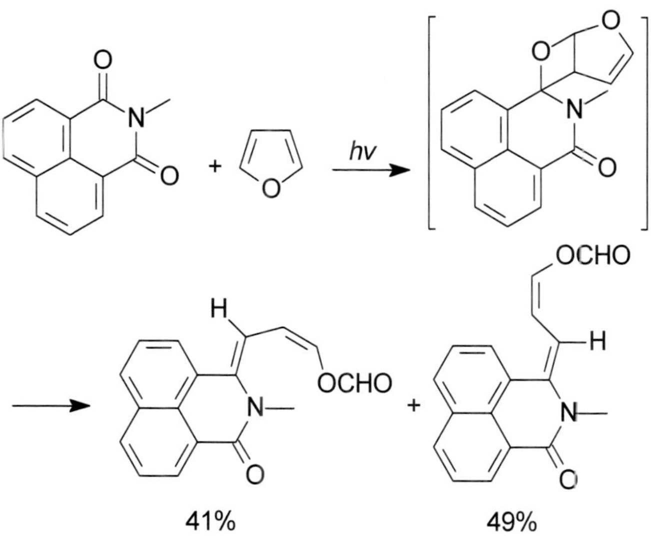

Scheme 11.

Furan quenches the fluorescence of the substrate, while a small new emission at 500 nm appears: this evidence is in agreement with a mechanism involving a reaction in the excited singlet state via the formation of an exciplex.

Schenck reported also the reaction of some aromatic carbonyl compounds (benzophenone, benzaldehyde) with benzofuran (Scheme 12); he showed that when high triplet energy compounds were used, dimers of benzofuran were obtained, while oxetanes were the products of the reaction when low triplet energy carbonyls were used [45]. As in the case of furan, he obtained only one product showing that the reaction has a high regioselectivity. Reinvestigation of the reaction of benzofuran with acetophenone or propiophenone showed that, also in this case, oxetanes were obtained [46].

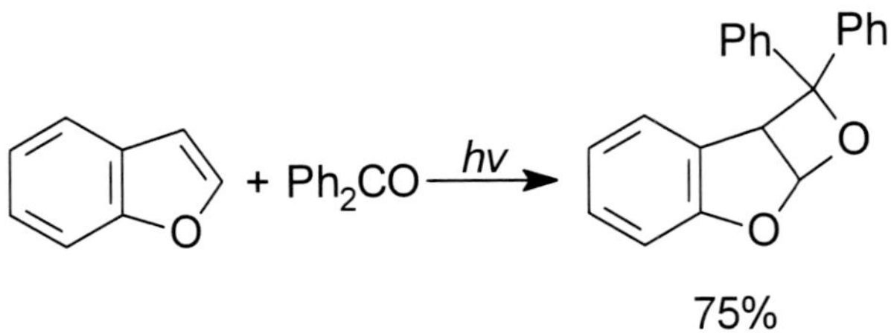

Scheme 12.

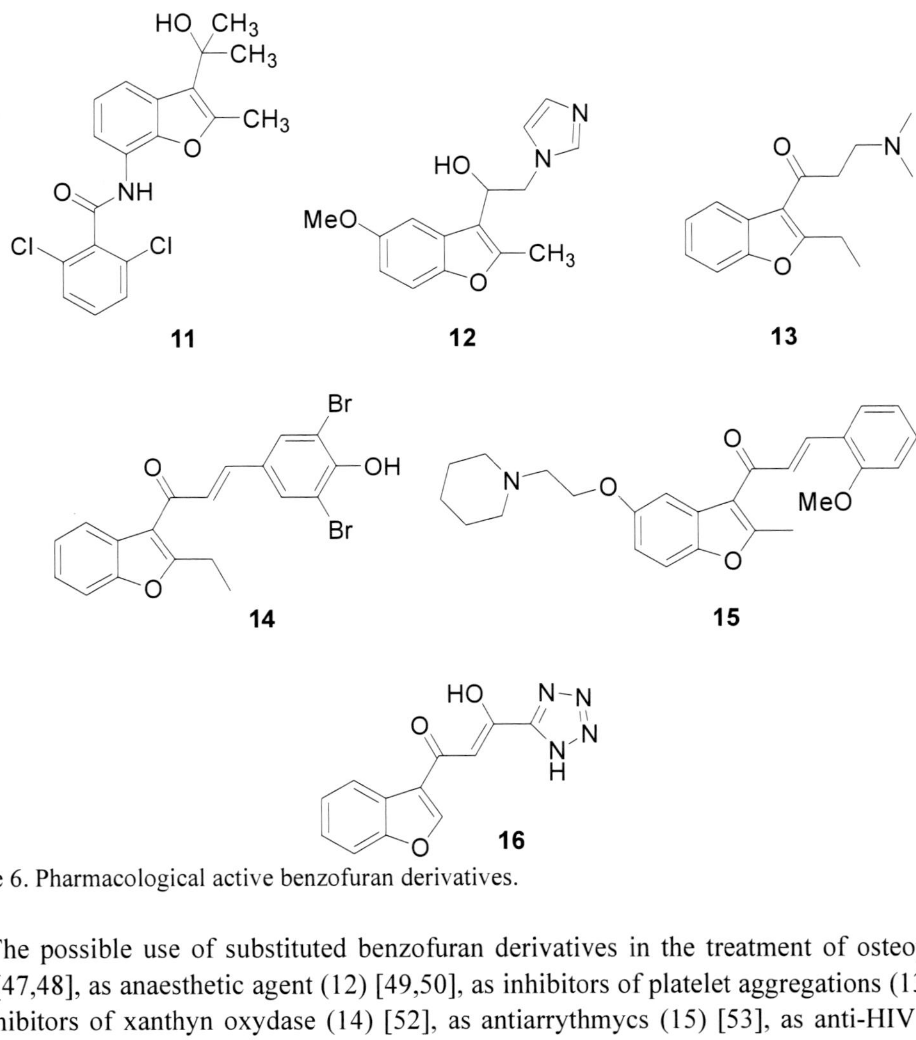

Figure 6. Pharmacological active benzofuran derivatives.

The possible use of substituted benzofuran derivatives in the treatment of osteoporosis (11) [47,48], as anaesthetic agent (12) [49,50], as inhibitors of platelet aggregations (13) [51], as inhibitors of xanthyn oxydase (14) [52], as antiarrythmycs (15) [53], as anti-HIV agents (16) [49], as agonist of the benzodiazepines [54], as inhibitors of ATPase (11) [55] has been described (Figure 5).

Most of these compounds are benzofuran derivatives substituted in position 3 with a side chain containing in position an alcoholic function or a carbonyl group.

In order to synthesize benzofuran derivatives with potential biological properties we used as starting material 4-amino-2,7-dimethylbenzo[b]furan (20) and 4-chlorobenzo[b]furan (23) (Scheme 13).

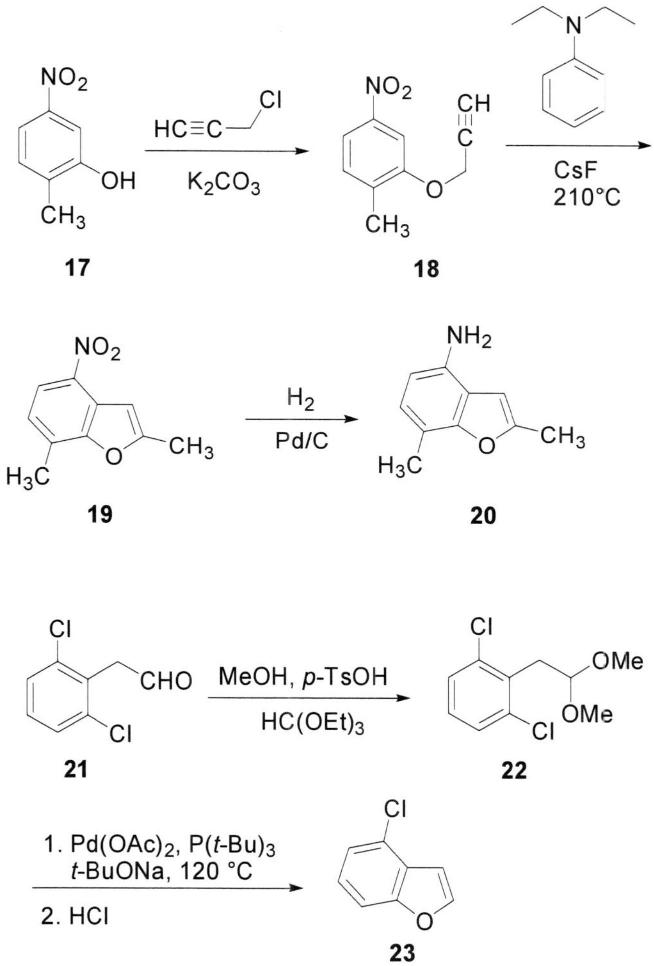

Scheme 13.

The selection of these substrates was due to our intention to give an answer to two questions in the field of the Paternò-Büchi reaction. First, we want to verify if the presence of both photolabile substituents, such as chlorine, and substituents able to induce electron transfer reactions, such as the amine, was able to modify the photochemical behaviour of the furan ring. Then, we wanted to verify if the Paternò-Büchi reaction on benzofuran could be used in the synthesis of potential biological properties. In fact, the amine group is present (as an amide) in compound 11 [47,48,55] while the chlorine substituent is present in some compounds claimed for their anaesthetic properties [49].

The synthesis of 20 started from 2-methyl-5-nitrophenol that was alkylated with propargyl chloride in basi medium: the resulting ether 18 was converted in the benzofuran derivative 19 through reaction with CsF in the presence of N,N-diethylaniline. The reduction of the nitro group afforded 20 (Scheme 13). The synthesis of 23 was performed starting from 2,6-dichlorophenylacetaldehyde (21) that was converted in the corresponding acetal through reaction with methanol in acidic condition and in the presence of triethylorthoformate; the acetal 22 gave 4-chlorobenzo[b]furan (23) via a condensation reaction in basic conditions in the presence of palladium (0) (Scheme 13).

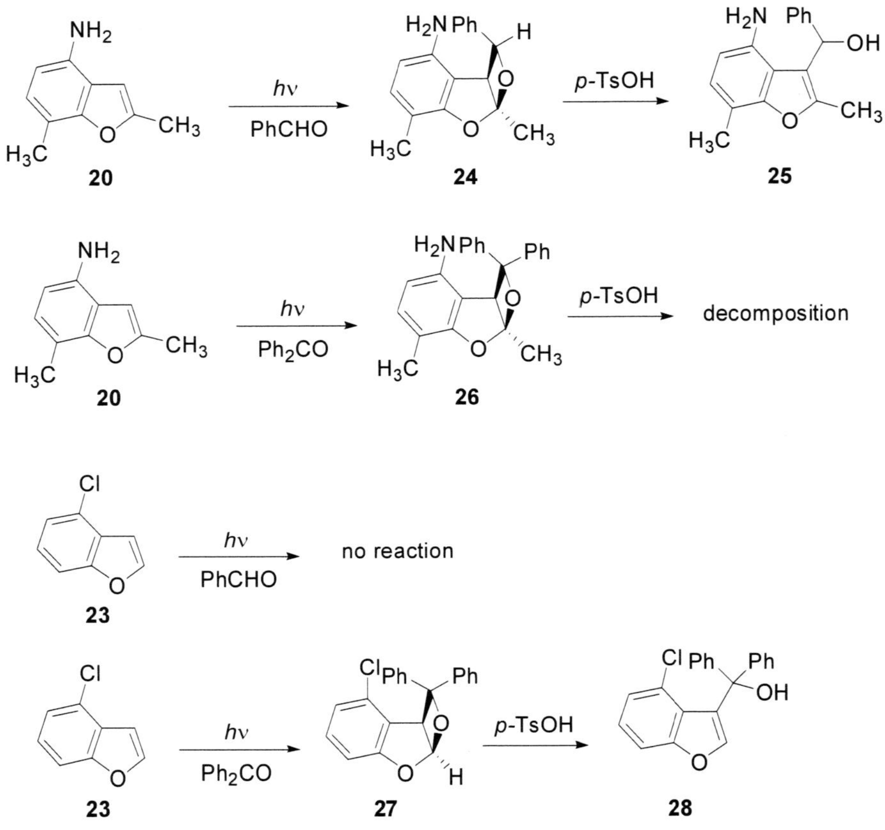

Scheme 14.

Table 3. Paternò-Büchi reaction on benzofuran derivatives and acid hydrolysis of the oxetanes

Benzofuran	*Carbonyl compound*	*Oxetane*	*3-benzofurylalcohol*	**Yield [%]**
20	benzaldehyde	**24**		66
		24	**25**	77
20	benzophenone	**26**		94
		26	-	-
23	benzaldehyde	-		-
23	benzophenone	**27**		40
		27	**28**	66

The irradiation of 20 in the presence of benzaldehyde in benzene gave the corresponding adduct 24 in good yields (Scheme 14, Table 3).

It is interesting to note that the reaction allowed the formation of the endo isomer. The reaction between benzo[b]furan and benzaldehyde gave an adduct whose stereochemistry has not been described [45]. However, in the NMR spectrum the chemical shift of the proton at C-6 (5.45 ppm) was in agreement with an exo stereochemistry. In our case, we observed that the proton at C-6 had a chemical shift at 5.96 while the proton at C-5 showed a chemical shift of 4.18. These data are in agreement with an endo stereochemistry: in fact, in the endo adduct between furan and benzaldehyde, the proton at C-6 had a chemical shift at 6.07 while the proton at C-5 showed an peak at 4.21 [36].

The hydrolysis of the adduct 24 was performed in ethyl ether in the presence of a catalytic amount of p-TsOH. This way, the 3-benzofurylmethanol derivatives 25 was obtained in good yields (Scheme 14, Table 3). The reaction of 20 with benzophenone gave the corresponding adduct 26 in very good yields (Scheme 14, Table 3). Unfortunately, this compound was labile and after a night at 5 °C it was completely decomposed. Compound 23 did not react with benzaldehyde (Scheme 14), while it reacted with benzophenone to give the corresponding adduct 27 (Scheme 14, Table 3). This compound, treated with p-TsOH, gave the 3-benzofurylmethanol derivative 28 (Scheme 14, Table 3).

The lack of reactivity of 23 towards benzaldehyde has to be explained. We assume that the Paternò-Büchi reaction occurs following the scheme depicted in the Scheme 15.

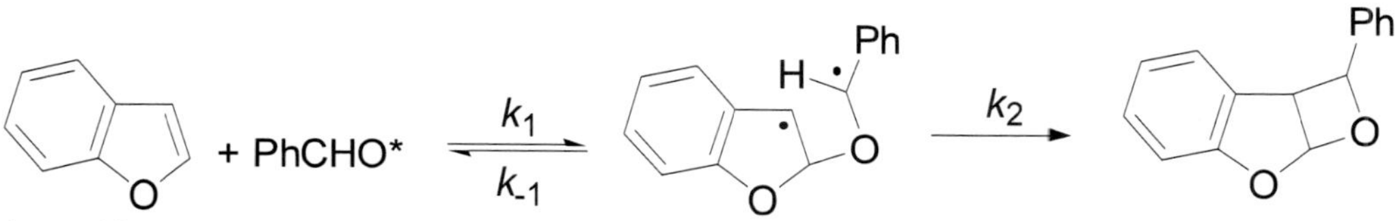

Scheme 15.

In this scheme we have a reversible addition of the carbonyl compound to the double bond to give the corresponding biradical, followed by an irreversible ring closure reaction. The lack of the reactivity observed in the case of the reaction between 23 and benzaldehyde can not be attributed to an inefficient attack of the excited triplet state of the benzaldehyde on the benzofuran double bond (k1). There is not significant difference between the benzophenone and benzaldehyde triplet energy (286 and 301 kJ mol-1, respectively) to explain the observed photochemical behaviour. We think that the observed lack of reactivity can be due to the relative rate between the retrocleavage (k-1) and ring closure (k2) reactions. We investigated on the presence of some features able to explain the different reactivity in the biradical intermediates 29 and 30 (Figure 7) obtained in the reaction of 23 with benzaldehyde and benzophenone, respectively.

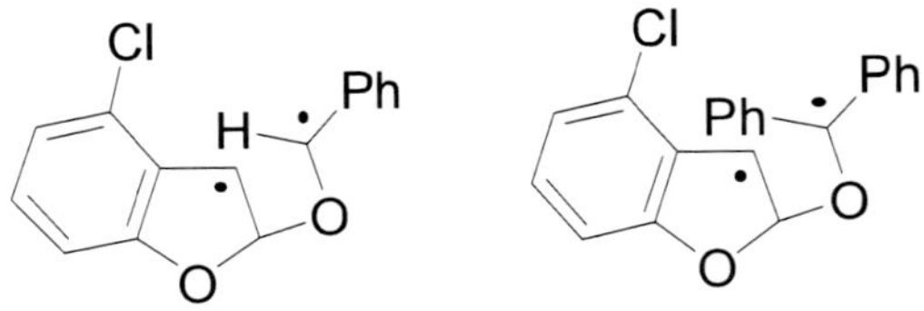

29 **30**

Figure 7. Biradical intermediates in the reaction of benzaldehyde and benzophenone with **23**.

We performed some DFT calculations with hybrid functional B3LYP and using 6-31G+(d,p) basis set. The biradicals are very similar: the only difference we found is in the charge on the radical carbon deriving from the carbonyl compound. The charge on the radical carbon on the benzofuran ring is -0.2. In the case of 29 the benzylic radical carbon has a charge of -0.1 while, in the case of 30, the charge on the diphenylmethyl radical carbon is 0.00. The repulsion between to similar charge can be responsible of the low k2 value, driving the reaction towards the retrocleavage.

2,3-Dihydrofuran

The irradiation of 2,3-dihydrofuran with benzophenone in benzene gave adduct **31** (Scheme 16). It should be noted that in this case a completely different regioselectivity was obtained in the coupling reaction [38].

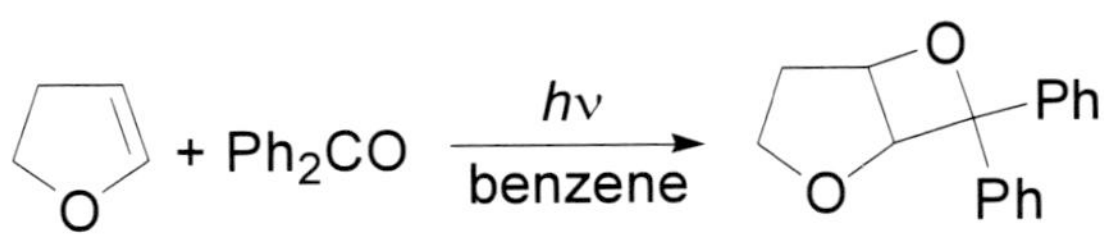

31

Scheme 16.

The same behavior was observed irradiating 2,3-dihydrofuran with aliphatic aldehydes and ketones. By using acetone the adducts were obtained in 52% yield in a 200:1 isomeric ratio (Scheme 17) [56]. The reaction of 5,5-dimethyl-2,3-dihydrofuran with both aromatic and aliphatic aldehydes gave the corresponding adducts with high *endo* selectivity [57]. When acetaldehyde was used the adduct was obtained in 63% yield as a mixture of stereoisomers. The selectivity depended on 2,3-dihydrofuran concentration: this behavior was in agreement with a switch from a triplet mechanism to a singlet mechanism at higher concentration [58-61].

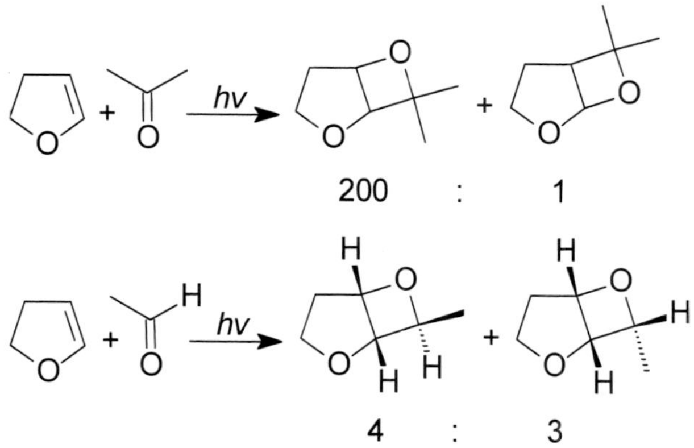

Scheme 17.

When benzaldehyde was used as carbonyl compounds the adducts were obtained with an overall yield of 98% as a >98:2 regioisomeric mixture. The major isomer is 88:12 *endo/exo* mixture. The reaction showed a good regio- and stereo-selectivity (Scheme 18) [62-64]. This stereoselectivity was partially lost when benzaldehyde reacts with 2-methyl-2,3-dihydrofuran [63].[46]

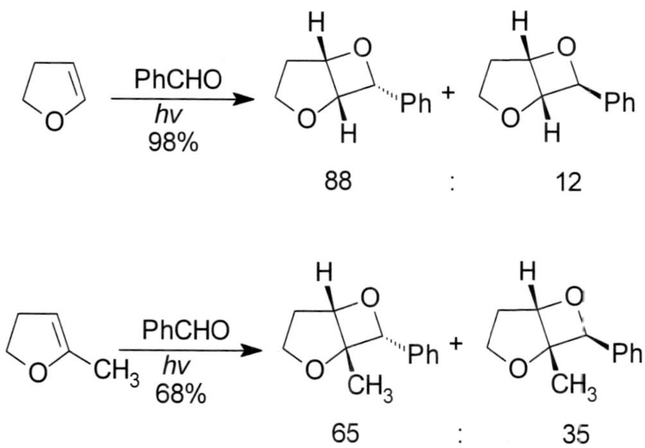

Scheme 18.

A change of regioselectivity was observed when the reaction with benzaldehyde was performed in high polar solvent in agreement with an electron transfer mechanism (Scheme 19) [58,59].

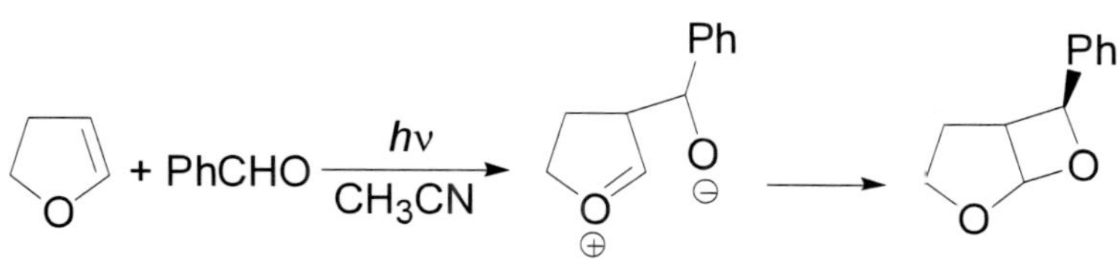

Scheme 19.

Benzaldehyde reacts in its triplet state. This way, a triplet biradical is formed as an intermediate. For the formation of the products, intersystem crossing into the singlet manifold is necessary. The most important factor influencing an intersystem crossing for flexibile triplet biradicals is spin-orbit coupling. The angle between p orbitals at the radical centers is approximately 90° for maximun spin orbit coupling.

For the pronounced endo selectivity in the reaction between aromatic aldehydes and 2,3-dihydrofuran, we can consider the two biradical conformers 32 and 33 to be responsible, with the alkyloxy substitutent localized in a pseudoequatorial position and 32 being more populated because of fewer steric interactions (Scheme 20). When a methyl group is present, the increasing gauche interactions with the -alkyloxy substituent lead to a certain concentration of 34 and 35, again with 34 being preferred because of fewer steric interactions.

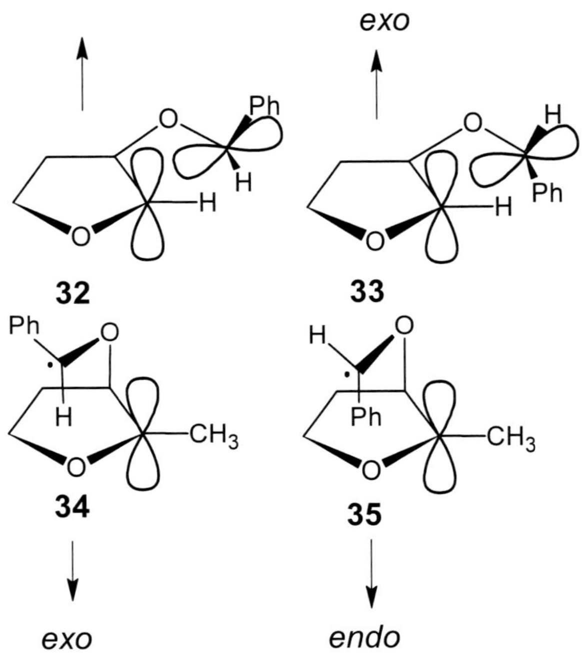

Scheme 20.

α– and β-naphthaldehydes, on the contrary, gave high *exo* selectivity. To understand this behaviour we have to consider that they gave cycloadducts also in the presence of triplet quenchers, while the authors observed fluorescence quenching in the presence of dihydrofuran: in this case, singlet excited states are responsible for the high *exo* selectivity [35,65].

We examined the regiochemistry of the reaction, and we tested the possibility that the regiochemical behavior of the reaction is determined by the formation of the more stable biradical intermediate. We performed DFT/B3LYP calculations using 6-31G+(d,p) basis set on Gaussian 03. We examined the structures **36** and **37** (Figure 8).

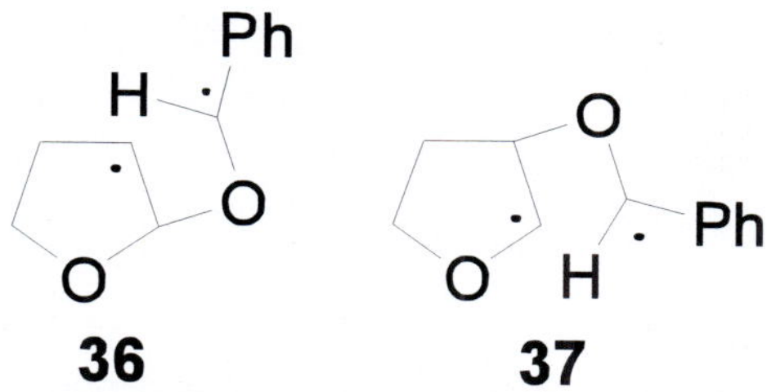

Figure 8. Possible biradical intermediates in the reaction of 2,3-dihydrofuran with benzaldehyde.

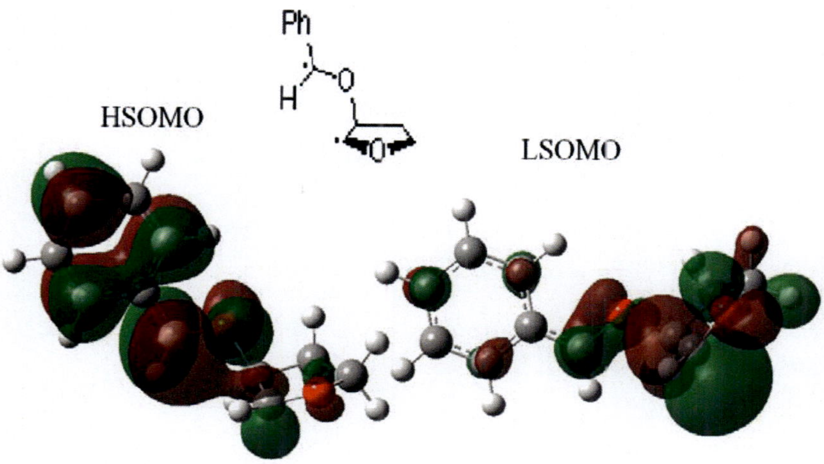

Figure 9. HSOMO and LSOMO of the biradical intermediate **37**.

The biradical **37** is more stable than **36** for 1.49 kcal mol^{-1}. This result is in agreement with the experimental result. On the basis of this result the stability of the biradical intermediates can account for the regiochemical behavior of the reaction. We can see also why the *endo* isomer is favored in the reaction. In Figure 9 we have collected the HSOMO and the LSOMO of the biradical **37.**

The HSOMO is mainly localized on the aromatic ring and it is extended until the radical carbon. The LSOMO is mainly localized on the dihydrofuran ring. The coupling between the radical carbons in these two orbitals is possible (the atomic orbitals involved can superimpose themselves) only if the *endo* isomer is formed (Figure 10).

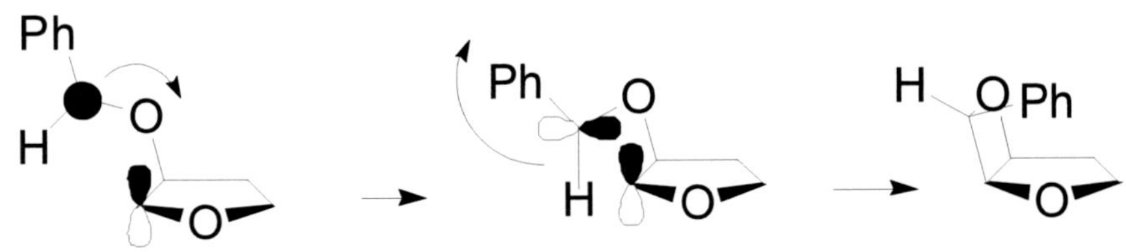

Figure 10. Ring closure reaction in the formation of the endo isomer of the adduct between 2,3-dihydrofuran and benzaldehyde.

When 2,3-dihydrofuran derivatives react with α,β-unsaturated carbonyl compounds, 2+2 cycloaddition between the olefins occurs (Scheme 21) [66].

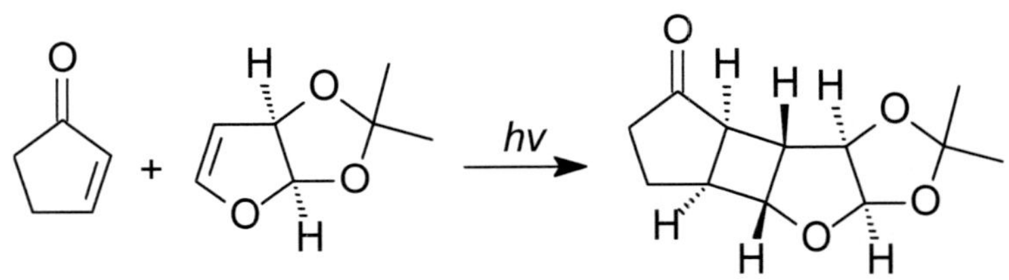

Scheme 21.

Synthetic Applications

Oxetanes can be converted into the corresponding 3-furylcarbinol through reaction with TsOH (Scheme 22) [31]. The formation of **38** is in agreement with the high negative entropy of the activation found (□$S^{\neq}$ = - 141.6 J K^{-1}) [67]. If the irradiation was performed in the presence of an acid, 3-furylcarbinol can be obtained in a one-step procedure in better yields [68].

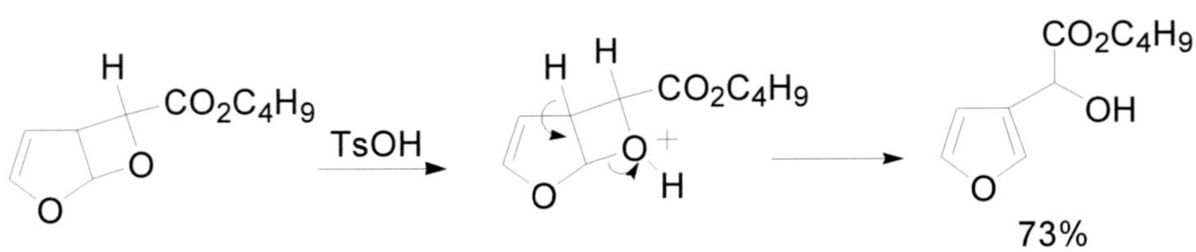

38

Scheme 22.

This procedure has been used in the synthesis of perillaketone, a naturally occurring 3-substituted furan [68]. Irradiation of furan with 4-methyl-pentanal in the presence of MsOH gave a mixture of **39** (18%) and **40** (5%). Oxidation of **39** with Collins reagent gave perillaketone (Scheme 23).

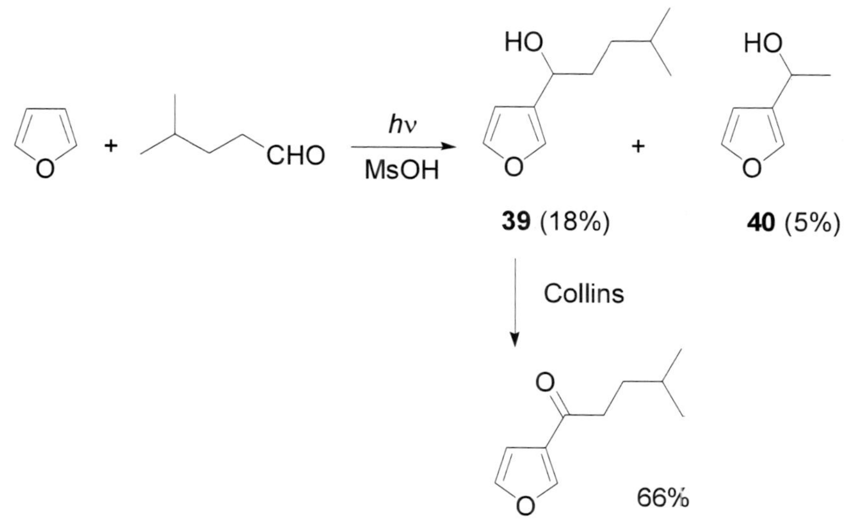

Scheme 23.

It is noteworthy that this is the first report on a probable Norrish type II on the carbonyl compound In fact, this type of reaction can be responsible in the formation of **40**: the excited aldehyde can undergo 1,5-hydrogen shift to give acetaldehyde, and this aldehyde, through the cycloaddition reaction, gives the 3-furylcarbinol **40**.

On the contrary, Lewis acids catalysed a different behavior: the treatment with $BF_3 \cdot Et_2O$ gave only 3-substituted furan in THF and 89% of 2-substituted furan in acetonitrile (Scheme 24) [67]. This behavior is in agreement with the formation of a carbocation **41** which can be stabilized by splitting off a proton to form 3-substituted furan or can undergo a 1,2-rearrangement to a more stable carbocation which splits off a proton to form 2-substitued furan (Scheme 24).

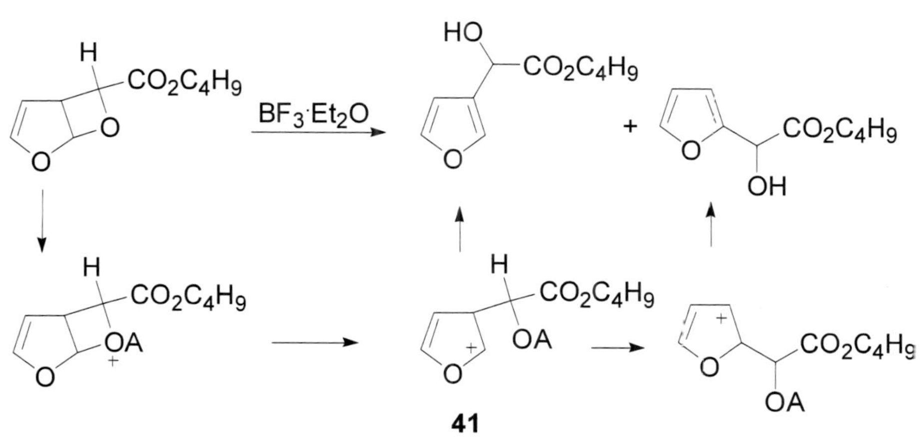

Scheme 24.

The oxetane derivatives can be treated also with $KMnO_4$ and the resulting *cis* diol was treated with acetone in the presence of an acid. This catalysis induces a ring opening reaction giving two products (Scheme 25) [69]. The major product was obtained through a spontaneous epimerisation of intermediate **42** in these reaction conditions. This product was 3-deoxy-1,2-O-isopropylidene-β-DL-streptose.

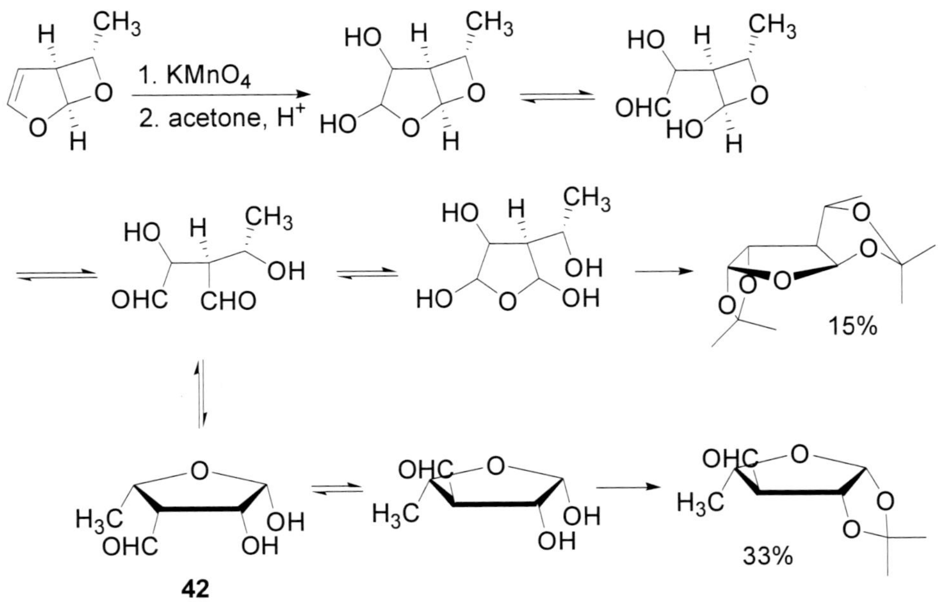

Scheme 25.

The same epimerisation reaction occurs on the *trans* diol obtained through treatment of the oxetane with MCPBA (Scheme 26) [69].

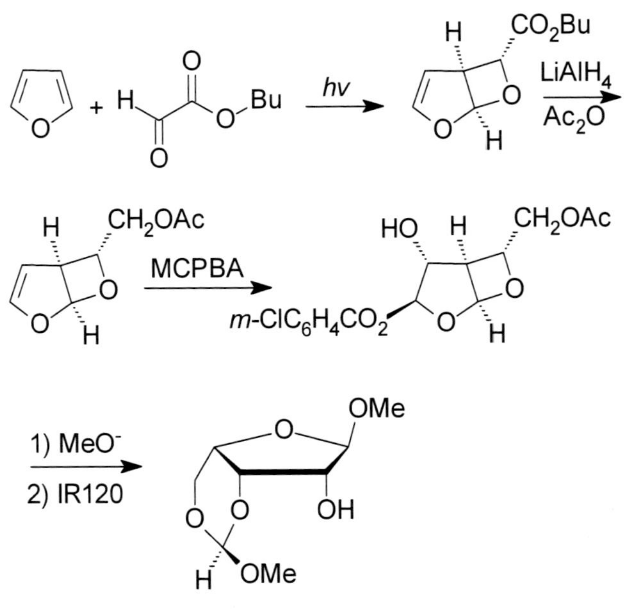

Scheme 26.

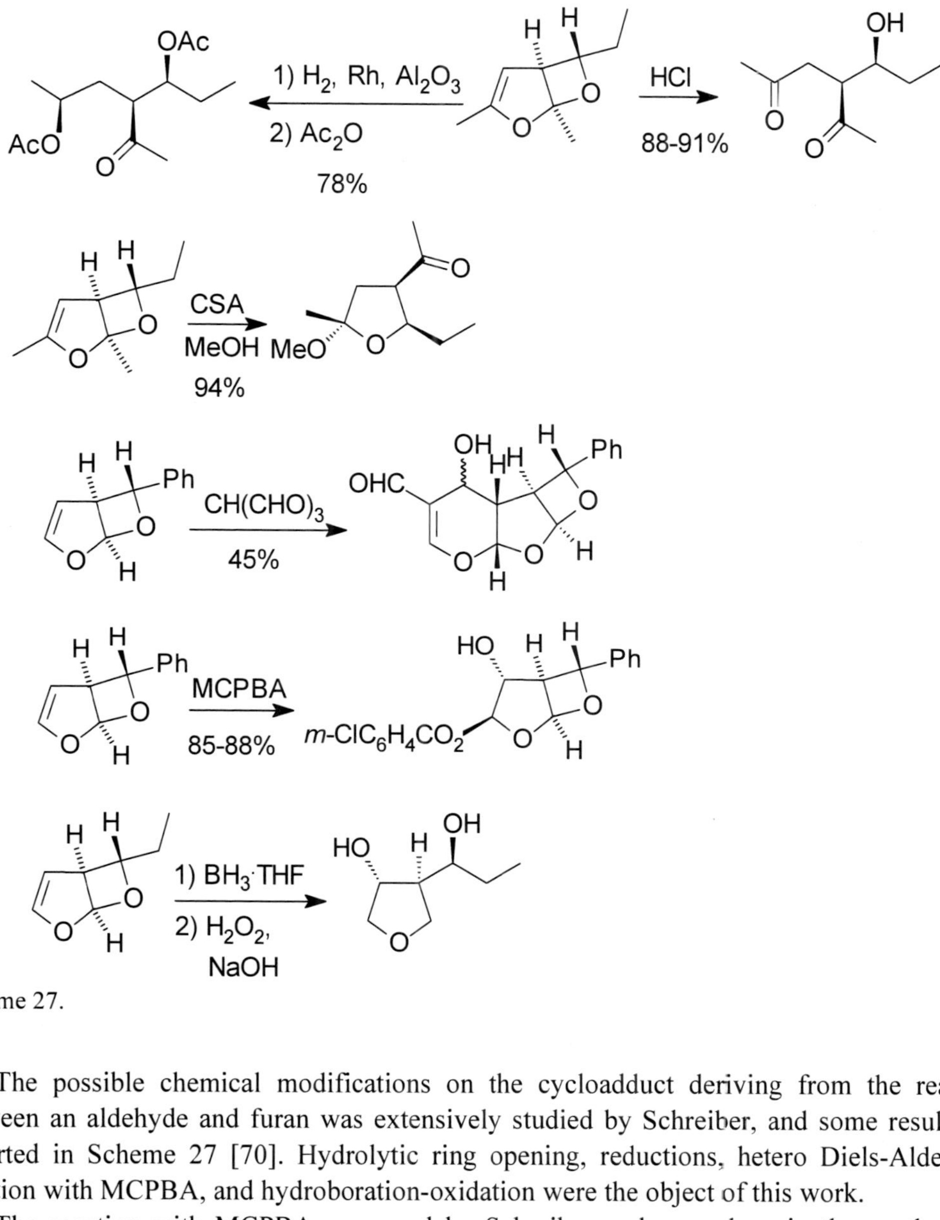

Scheme 27.

The possible chemical modifications on the cycloadduct deriving from the reaction between an aldehyde and furan was extensively studied by Schreiber, and some results are reported in Scheme 27 [70]. Hydrolytic ring opening, reductions, hetero Diels-Alder, the reaction with MCPBA, and hydroboration-oxidation were the object of this work.

The reaction with MCPBA was used by Schreiber and coworkers in the synthesis of Asteltoxin **43**, a potent inhibitor of *E. coli* BF_1-ATPase activity used as a fluorescent probe for mitocondrial F_1- and bacterial BF_1-ATPase [71].

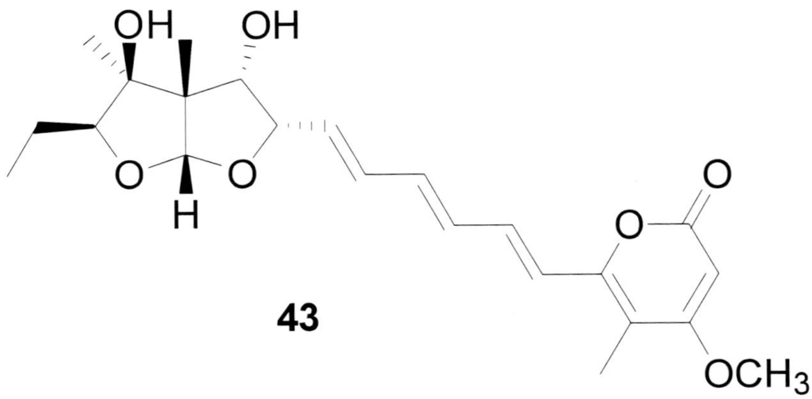

43

The synthetic sequence implies the photochemical coupling of 3,4-dimethylfuran with a functionalised aldehyde to give the corresponding adduct in 63% yield. The subsequent reaction with MCPBA gave the protected *trans* diol that was converted into **44** through acid hydrolysis. Then, the aldehyde was converted into the corresponding hydrazone and this substrate was treated with EtMgBr to give **45** with complete stereochemical control obtained through chelation. The subsequent conversion of the benzyl ether into a seleno derivative and the elimination of the selenoxide gave **46**. Compound **46** could be treated with ozone to give the corresponding aldehyde that was then converted into Asteltoxin (Scheme 28) [72].

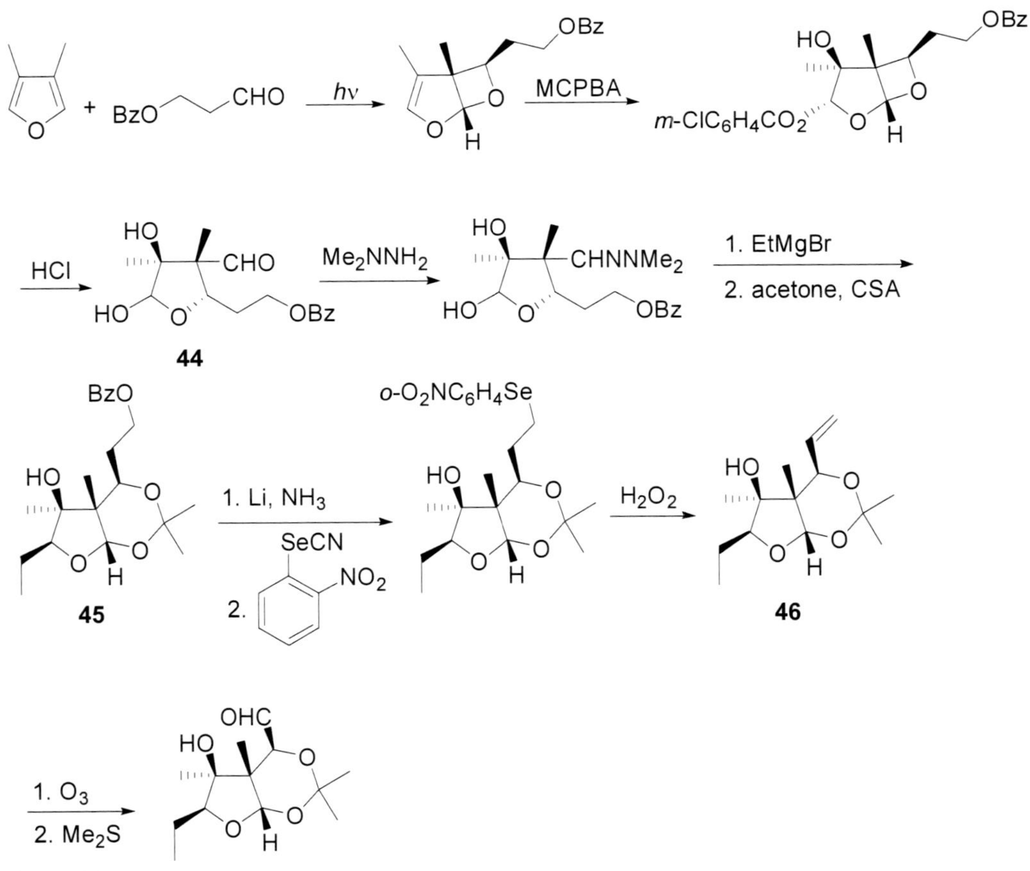

Scheme 28.

In fact, the aldehydic product could be alkylated with a vinyl anion and the subsequent double [2,3] sigmatropic rearrangement at room temperature furnished **47** as a 3:1 mixture of epimers. Camphorsulfonic acid induced the transposition of the substrate into a bis(tetrahydrofuran) derivative. Pummerer rearrangement and hydrolysis gave aldehyde **48**. Aldol condensation gave **49** and elimination gave Asteltoxin (Scheme 29) [73].

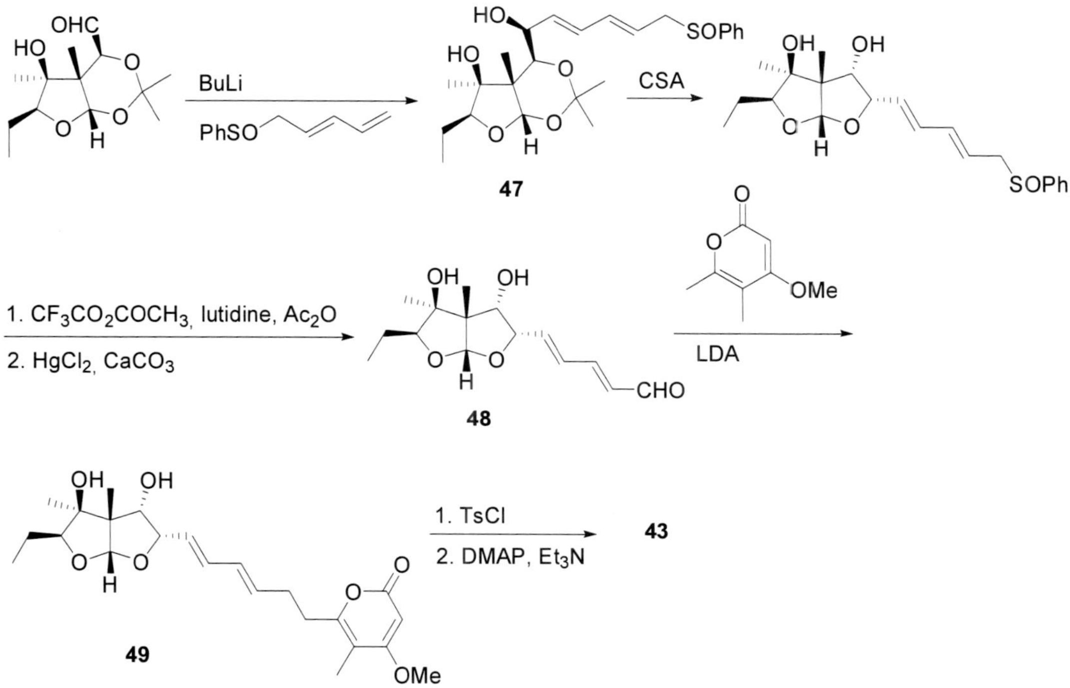

Scheme 29.

The same research group reported a formal synthesis of Avenaciolide, an antifungal metabolite (Scheme 30) [74]. In this case the oxetane (obtained in multigram quantities in high yields and with complete stereochemical control) was treated with hydrogen to give the saturated compound. The key step in this synthetic procedure is a reaction with ozone followed by a base catalysed epimerisation with potassium carbonate and cyclization in acidic medium.

The reaction of tributylstannylfuran with butyl glyoxylate [32] was used in the synthesis of a ginkgolide B-kadsurenone hybrid of two inhibitors of a platelet activating factor (Scheme 31). In the obtained oxetane the stannyl substituent was used to give a coupling Stille reaction; the subsequent hydrolysis gave the corresponding carbonyl derivative that was reduced with $NaBH_4$ in the presence of $CeCl_3$: this reaction allowed for the selective reduction of the aryl ketone in the presence of the aldehyde. The use of distannoxane transesterification catalyst gave a 1:1 mixture of the hybrid.

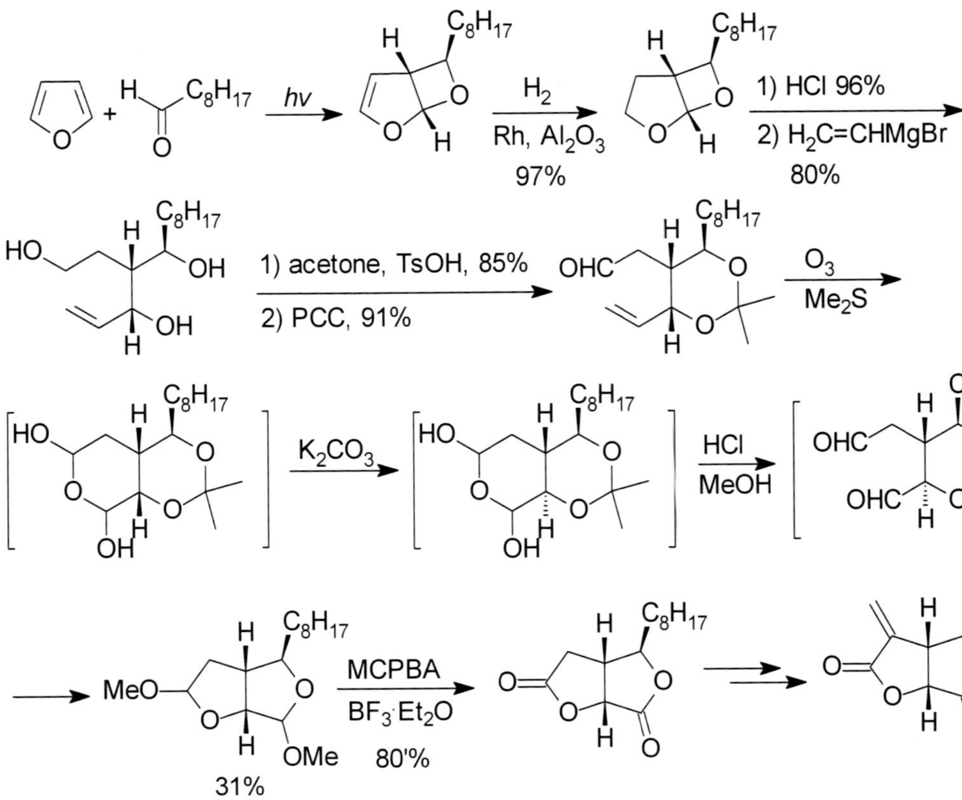

Scheme 30.

Further elaborations of the stannyl substituents in the adduct can be obtained *via* reaction with RCOCl in the presence of Pd(0) to give the corresponding ketone, or *via* reaction with BuLi and reaction of the corresponding anion with PhCHO to give the corresponding alcohol [75]. Furthermore, the cycloadduct obtained from the reaction between furan and an aldehyde can be treated with an excess of Schlosser's base (BuLi, *t*-BuOK) to give the corresponding anion which can react with carbonyl compounds or alkyl halides (Scheme 32) [75].

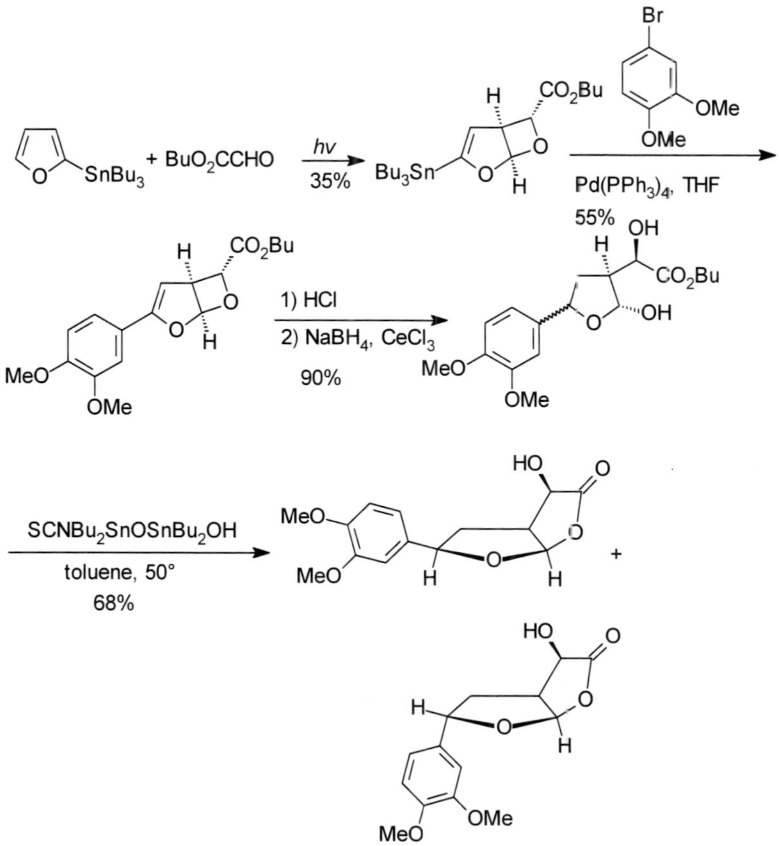

Scheme 31.

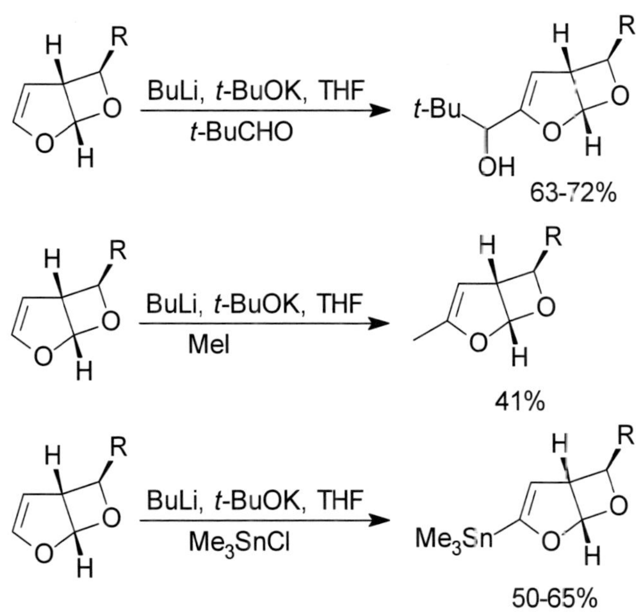

Scheme 32.

A possible further application of the Paternò-Büchi reaction between carbonyl compounds and furan could be related to the use of the oxetane ring. The most important target in this field was oxetanocin (**50**), a nucleoside isolated from *Bacillus megaterium* NK 84-0218 showing anti-HIV activity [76,77].

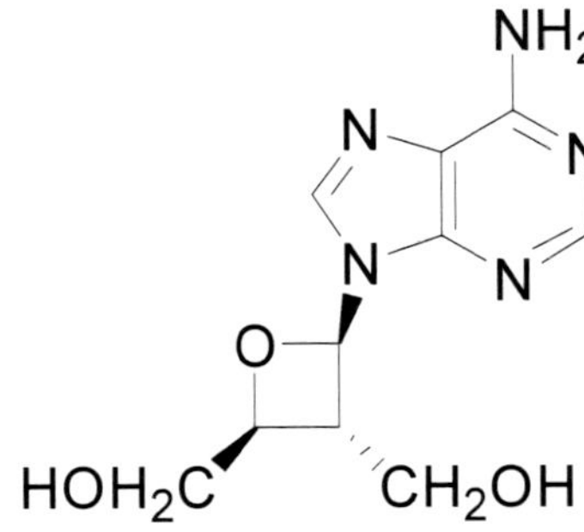

50

The treatment of oxetane **51** with *N*-iodosuccinimmide in the presence of methallyl alcohol gave the corresponding iodoacetals **52** (Scheme 33) [78]. The subsequent treatment of **52** with iodonium di-*sym*-collidine perchlorate (IDCP) [79] gave **53** in low yields as a relatively unstable mixture of diastereoisomers. Alternatively, treatment of **54** with dimethyl dioxirane gave the corresponding epoxide and the treatment with methallyl alcohol gave the hydroxy acetal **55**. The treatment of **55** with IDCP gave **56** in moderate yields.

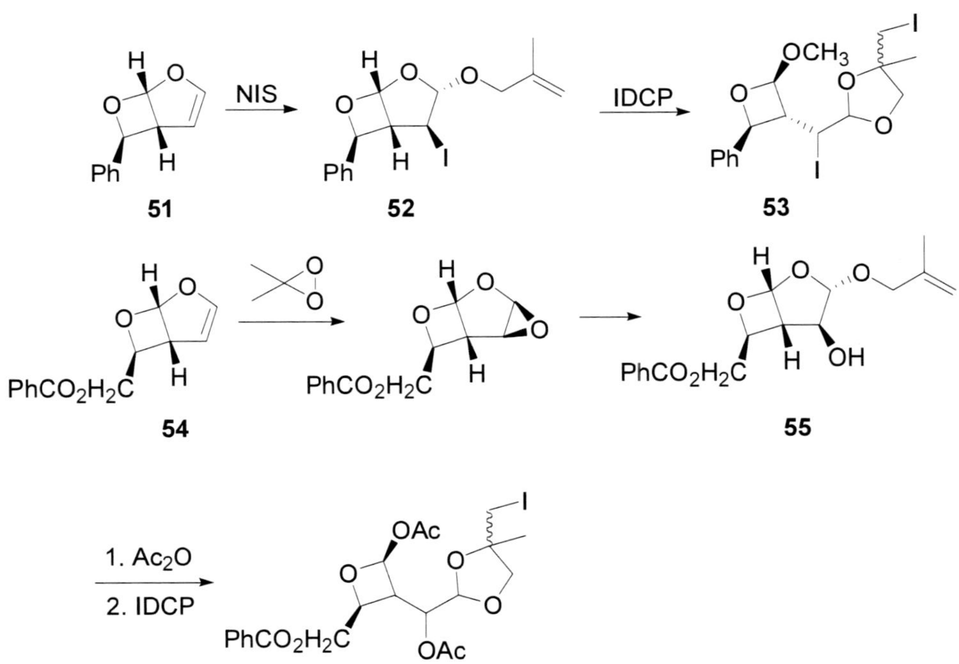

Scheme 33.

Oxetanocin was obtained carrying out the reaction between 2-methylfuran and benzoyloxyacetaldehyde. The corresponding adduct was treated with ozone and the product was reduced with $NaBH_4$. The obtained alcohols were protected. Product **57** was treated with N-benzoyl-disilyladenine and $SnCl_4$ to give **50** (Scheme 34) [80].

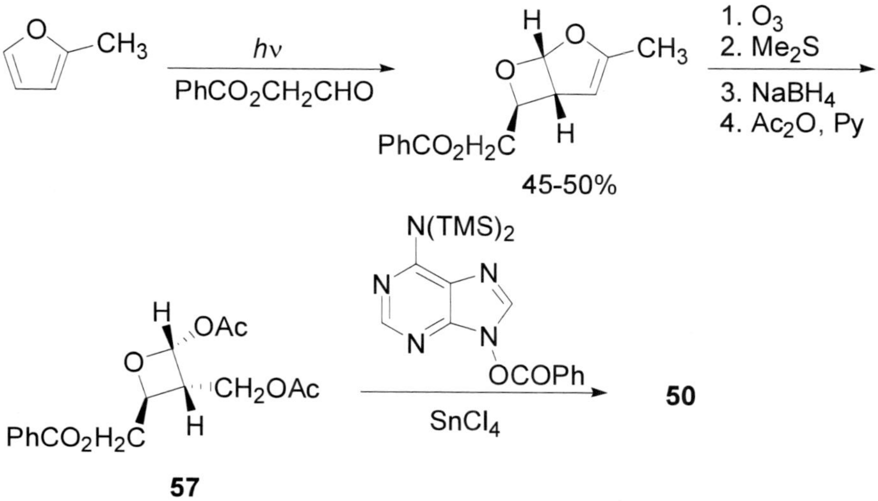

Scheme 34.

Irradiation of the oxetane **58** in the presence of 1-methoxynaphthalene and 2,7-dimethoxynaphthalene as sensitizers gave an efficient carbonyl-alkene metathesis of bicyclic oxetanes through an electron transfer process (Scheme 35) [81].

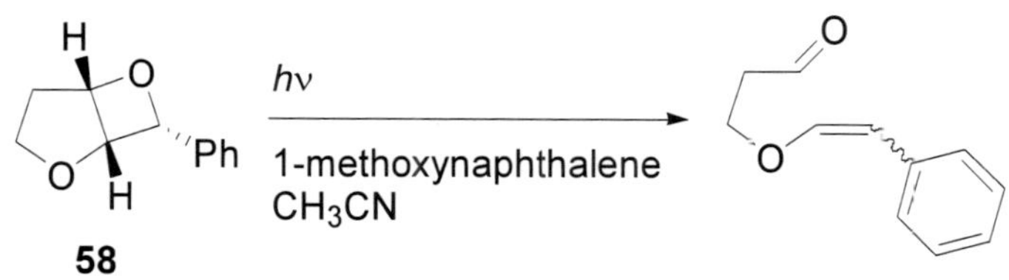

Scheme 35.

Asymmetric Reactions

The reaction of glyoxylates with furan can be performed also using chiral glyoxylate. In particular, the use of *R*-(-)-menthol, 2-octanol, and 2,2-dimethyl-3-butanol as chiral auxiliaries gave the corresponding oxetanes in high yields. These compounds can be converted into the corresponding 3-substituted furans. These furans showed low enantiomeric excess (Scheme 36, Table 4) [82].

On the basis of the experimental data the configuration of the oxetane was 1*R*, 5*S*, 6*S*.

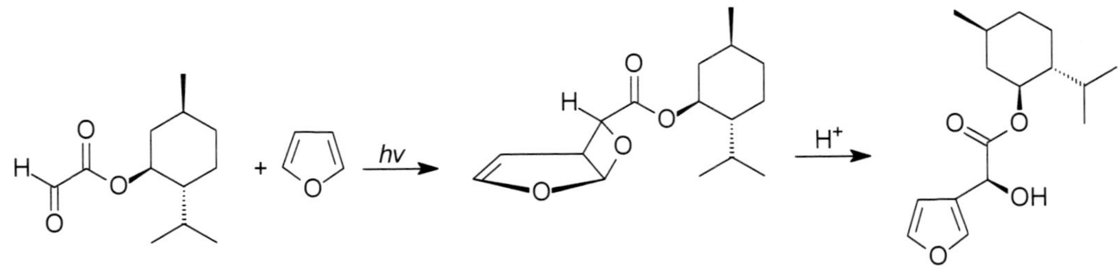

Scheme 36.

The low optical purity was explained considering that two conformers of the glyoxylate can approach furan.

Table 4. Reaction of chiral alkyl glyoxylate with furan

Glyoxylate	Oxetane [yield %]	3-furylcarbinol [yield%]	Optical purity [%]
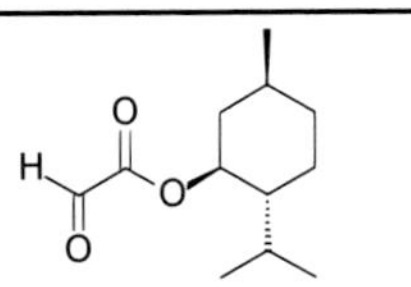	80	70	7.27
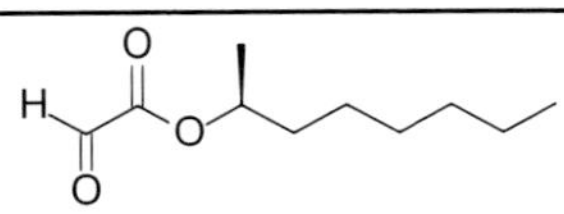	71	75	4.74
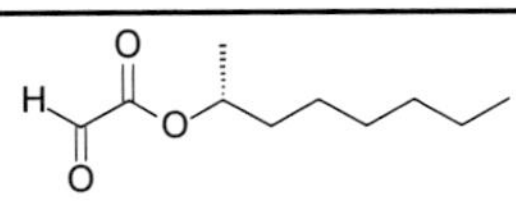	70	74	4.84
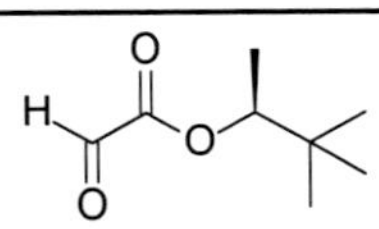	67	75	2.28
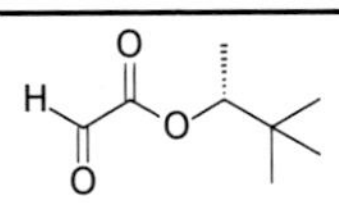	66	77	2.42

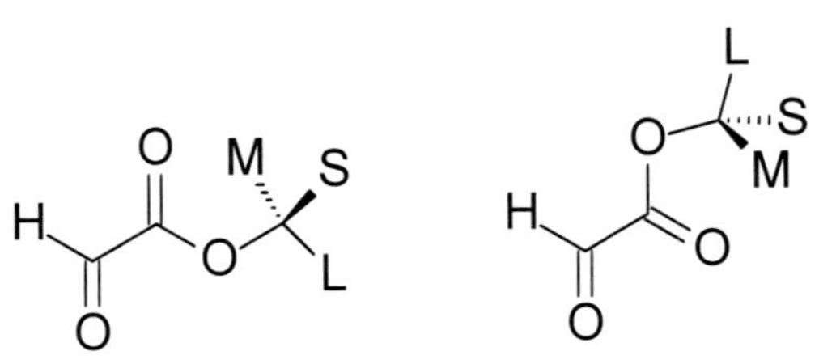

S-trans S-cis

The use of chiral phenylglyoxylate gave better results. The use of chiral alcohols gave diastereoisomeric excess in the range of 4-80% (Scheme 37, Table 5) [83-85].

Table 5. Reaction of chiral phenylglyoxylate with furan

OR*	Yield %	de %
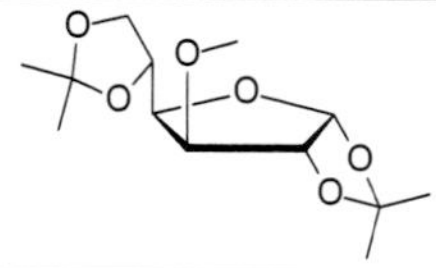	79	48
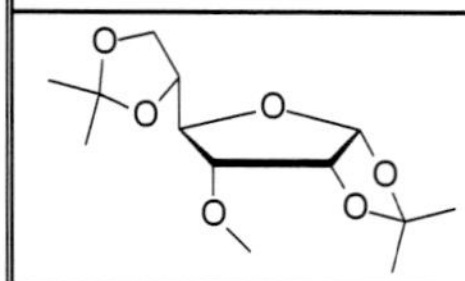	74	44
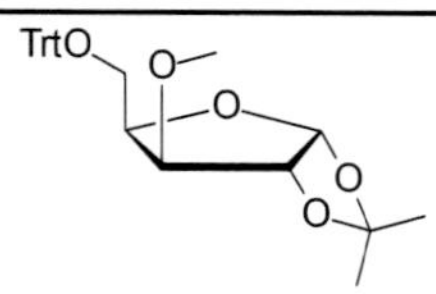	85	4
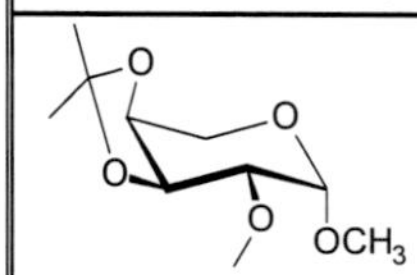	80	42
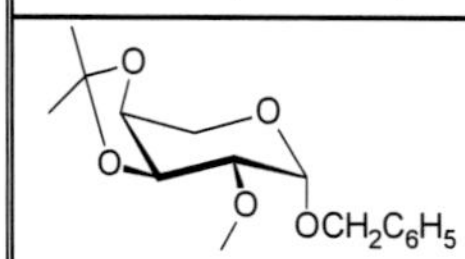	78	43
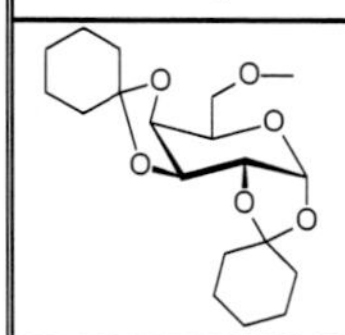	84	6
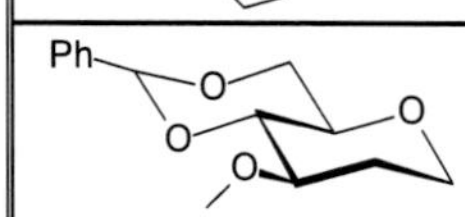	80	80

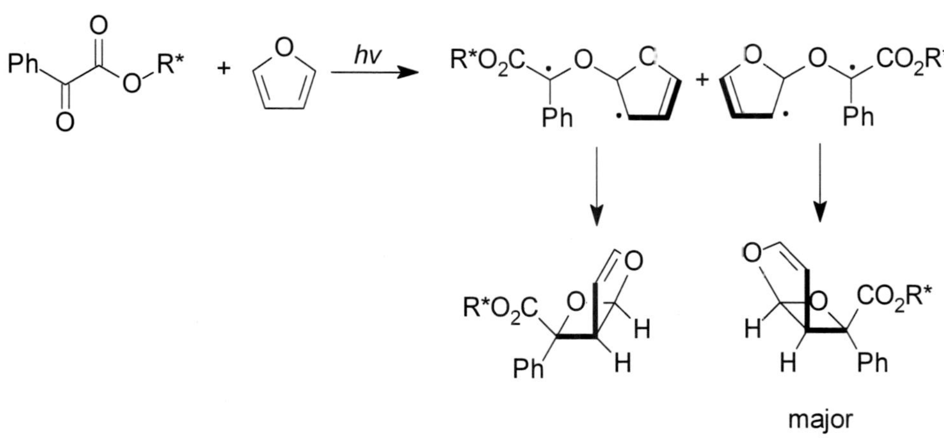

Scheme 37.

The authors observed also an important variability of the diastereoisomeric excess in function of the temperature, with the presence of an inversion temperature. These results were explained assuming that the diastereoselection is produced on two levels: 1) the preferred formation of the diastereoisomeric 1,4-biradical intermediate **59** and 2) the preferred retrocleavage of the energetically unfavored diastereoisomeric intermediate **60** to the starting materials [86].

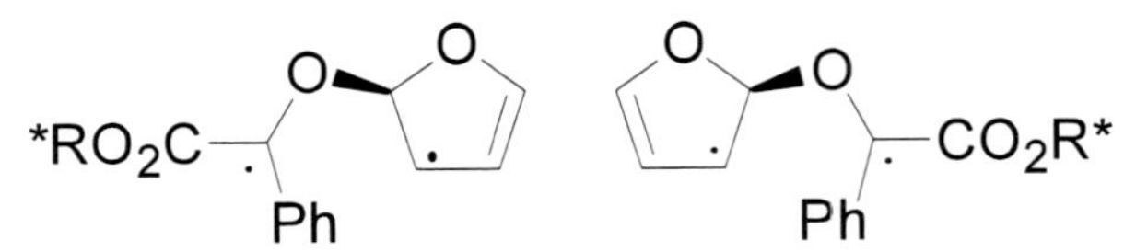

59 **60**

The diastereoisomeric excess in the high-temperature region ($T>T_{inv}$) is dominantly controlled by steric effects of the chiral auxiliaries, whereas in the low-temperature region ($T<T_{inv}$), the nature of the olefin has a dominating influence.

When the reaction is carried out on 2-methylfuran, a 2:1 regioisomeric mixture was obtained with a very high diastereoisomeric excess (Scheme 38) [87].

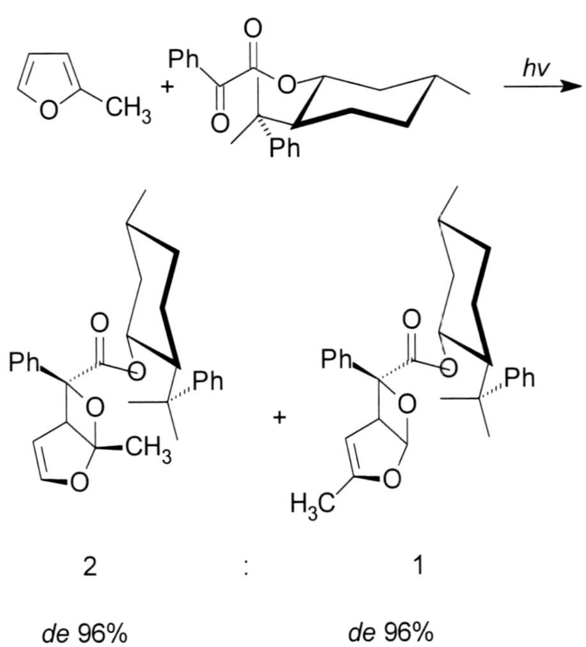

Scheme 38.

We studied the reactivity of some chiral phenylglyoxylates [88].We used three phenylglyoxylate esters: the compound **61**, obtained from (*S*)-1-methyl-1-propanol, the compound **62**, obtained from (*S*)-2-methyl-1-butanol, and the compound **63**, obtained from (-)-8-phenylmenthol (Figure 11).

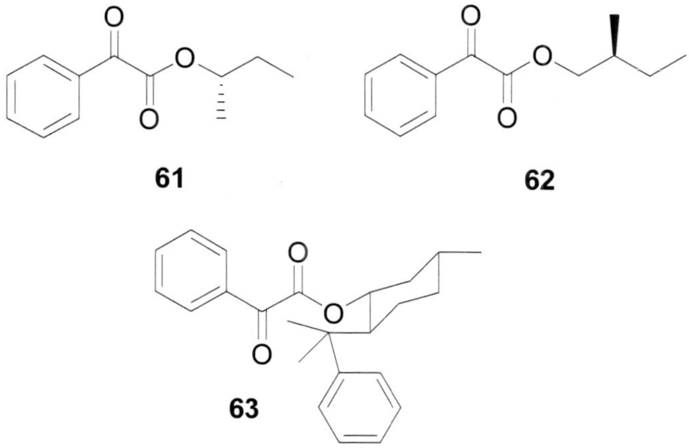

Figure 11. Phenylglyoxylates used in the study.

All the reactions gave in good yields the corresponding oxetanes. Considering the stereochemical behavior of the reaction, while the reaction with 8-phenylmenthol glyoxylate gave a high diastereoisomeric excess, the reaction with the glyoxylate ester **61** gave only 15% *de* and the reaction with the glyoxylate ester **62** resulted in no diastereoselectivity (Figure 12).

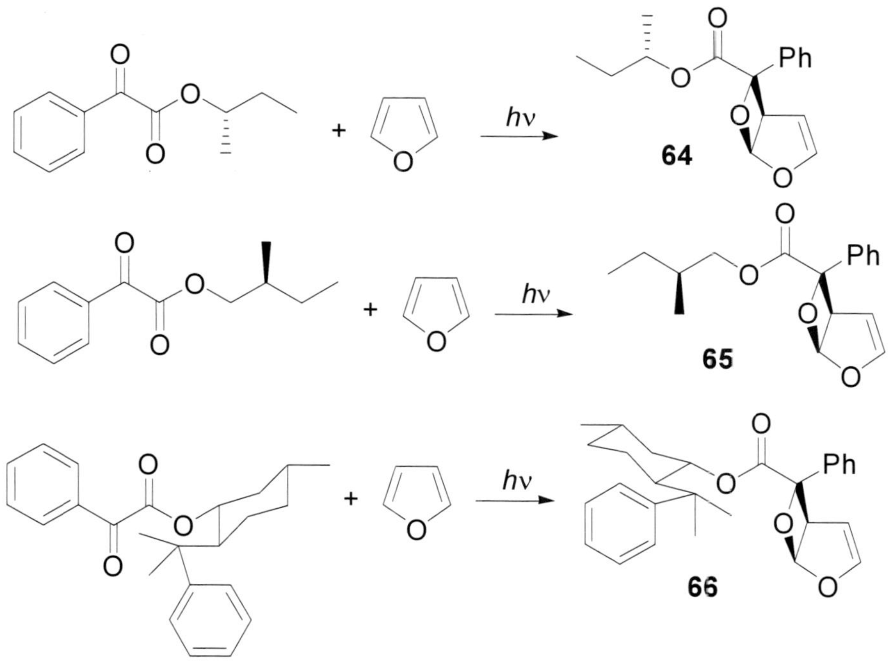

Figure 12. The reaction of chiral phenylglyoxylates with furan.

The best results in order to justify the observed stereoselectiivty can be obtained considering the energy of the triplet biradical intermediates in the reaction of 61-63 with furan (Figure 13).

Calculations on these biradical intermediates showed that 67a (the precursor of 64) was more stable than 67b by 0.73 kcal mol^{-1}. Furthermore, 68a and 68b differed by only 0.02 kcal mol^{-1}, in agreement with the observed no stereoselectivity of the reaction. Finally, the biradical intermediate 69a showed to be more stable than 69b by 21.9 kcal mol^{-1}: also this result is in agreement with the observed high diastereoisomeric excess.

In conclusion, the stereochemical behaviour of the Paternò-Büchi reaction between 61-63 and furan can be explained considering the relative stability of the possible biradical intermediates.

To improve diastereoselectivity, we tried to carry out a reaction in an organized medium. Selective absorption on the surface of the solid could improve diastereoselectivity. In the case of 66, obtained in solution with de = 95%, the use of the zeolite (NaY) increases the diastereoisomeric excess reaching the value of 98%.

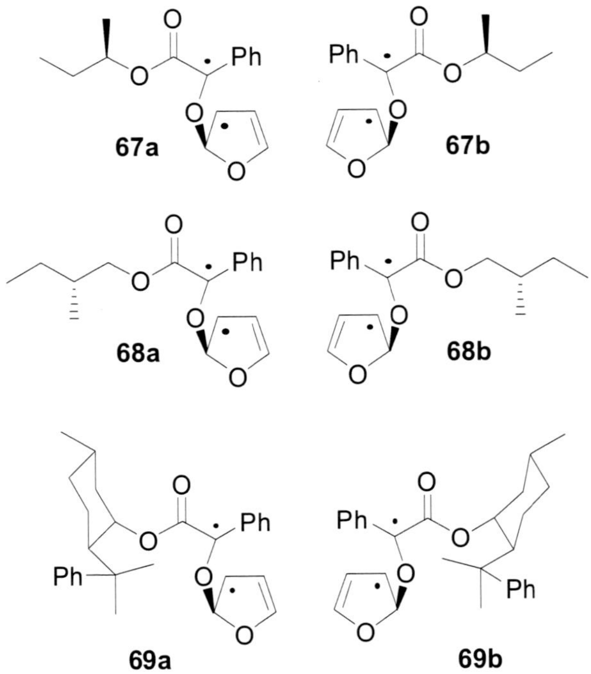

Figure 13. Radical intermediates in the reaction of 7-9 with furan.

In another study we used the chiral phenylglyoxylate esters **61** and **62**, obtained from (*S*)-1-methyl-1-propanol and (*S*)-2-methyl-1-butanol, respectively (Scheme 39). The esters were irradiated in furan within a zeolite. The results of these reactions are reported in Table 6.

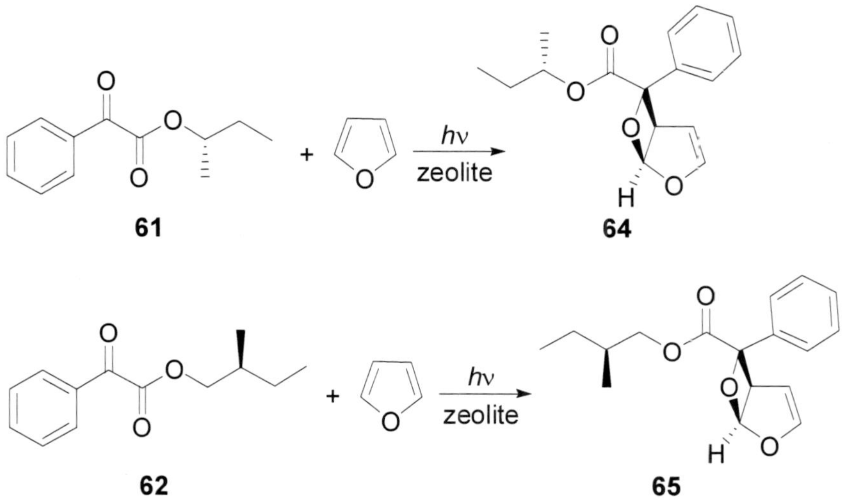

Scheme 39.

Table 6. Reaction of chiral phenylglyoxylate esters with furan within zeolites

Entry	Ester	Product	Zeolite	Yield (%)	de (%)
1	61	64	-	66	15
2	61	64	NaY	30	37
3	61	64	LiY	35	30
4	61	64	KY	39	6
5	62	65	-	66	0
6	62	65	NaY	58	18
7	62	65	LiY	63	17
8	62	65	KY	49	22
9	62	65	CsY	41	20
10	62	65	RbY	40	19

In Table 6 we reported also our previous reported results on the reaction in solution. The obtained results are in agreement with the observation that (with the exclusion of the reaction of entry 4) the performing of the reaction within a zeolite increases the stereoselectivity.

The origin of this behaviour can be explained considering the possible effect of the confinement on the reaction. In the Paternò-Büchi reaction on furan derivatives the biradical intermediate **70** can assume two possible comformations [19,37]. One of this has the radical carbons in almost *anti* conformation, while the other is almost a *syn* conformation (Scheme 40).

The *syn* conformation is in a steric situation where the following cyclization, after the intersystem crossing from the triplet to the singlet state, is a favoured process. In the *anti* conformation the most favoured process is the retrocleavage to the starting material.

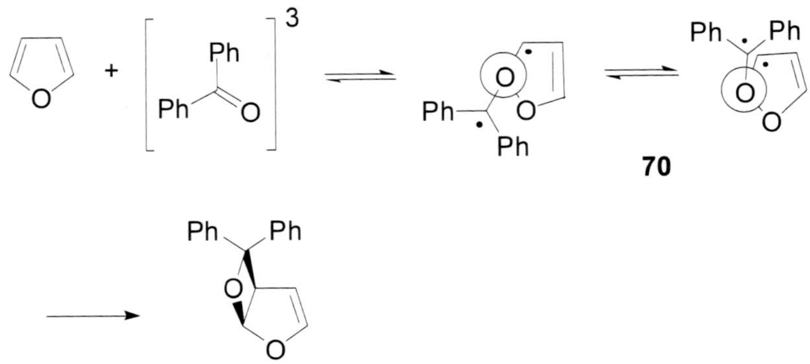

Scheme 40.

We performed DFT calculations on the possible biradical intermediates in the reaction of **61** and **62** with furan. We performed calculations at B3LYP level using 6-31G+(d,p) basis set on Gaussian 03. The results are collected in Tables 7 and 8.

The results showed in Tables 7 and 8 show that, with the exception of the *SS* biradical intermediate obtained from the phenylglyoxylate **1**, the *anti* conformations are more stable than the *syn* conformation. This is in agreement with previous reported results [19,37]. Furthermore, considering that the equilibrium between the two conformers is shifted towards the *anti* conformer and, then, towards the retrocleavage reaction, the data are in agreement with the hypothesis that the equilibrium reaction between triplet excited carbonyl compounds and furan is the rate determining step of the reaction. In the case of phenylglyoxylate ester **61**, the energy difference between the *syn* conformations of the possible biradical intermediates, is 2.57 kcal mol^{-1}.

In the reaction of the phenylglyoxylate **62** with furan, the energy difference between the *syn* conformations of the possible biradical intermediates is 2.92 kcal mol^{-1}: these results are not in agreement with the observed stereoselectivity in solution (Table 6). However, the *SS-syn* conformer derived from **61** was preferentially obtained while, in the case of *RS* conformers, the *anti* conformer was preferentially obtained. The preferential formation of the *syn* conformer in the case of the *SS* biradical intermediate favours the reaction because the ring closure can occur without the *anti-syn* equilibration required for the other stereoisomer. This effect can account for the observed stereoselectivity in solution. The *SS* conformer of the biradical intermediate is a precursor of the major stereoisomer obtained in the reaction.

Finally, Tables 2 and 3 reported also the dimension of the unitary cell for these intermediates. All these structures can enter into the supercage of the zeolites [89]. We can see that, with the exception of the *RS* biradical intermediate obtained from the phenylglyoxylate **62**, the *syn* conformations take up a smaller volume than the *anti* one.

The preferential formation of a stereoisomer in the reaction within a zeolite can be explained: in the case of the biradical intermediates obtained from the phenylglyoxylate **61**, the *SS-syn* conformer is obtained preferentially, while the *RS* conformer is obtained preferentially in *anti* conformation. The latter conformer has to rotate along the C-O bond to give one able to cyclize.

Table 7. Energy and volume of the possible biradical intermediates in the reaction of phenylglyoxylate 61 with furan

Conformer		Energy (H)	Volume (Å^3)
SS-syn	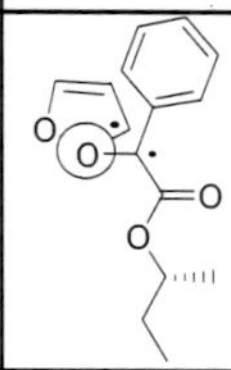	-916.43180528	496
SS-anti	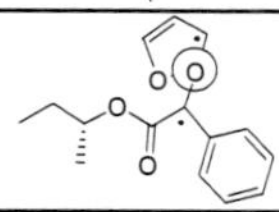	-916.42784617	556
RS-syn	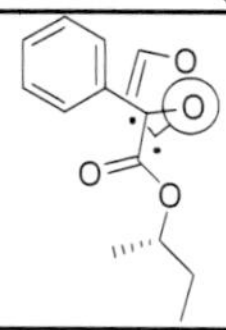	-916.42770328	449
RS-anti	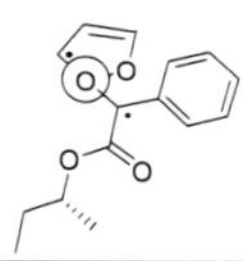	-916.43380265	644

Table 8. Energy and volume of the possible biradical intermediates in the reaction of phenylglyoxylate 62 with furan

Conformer		Energy (H)	Volume (Å)
SS-syn	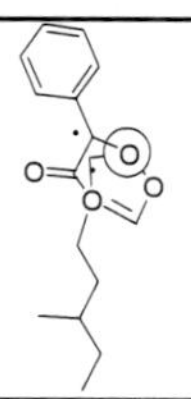	-955.52724261	550
SS-anti	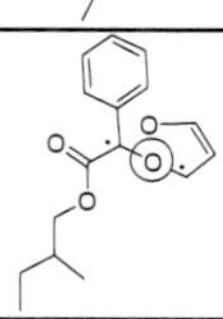	-955.53341862	728
RS-syn	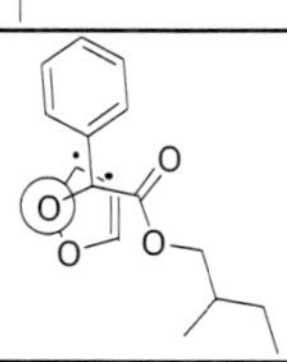	-955.52259314	713
RS-anti	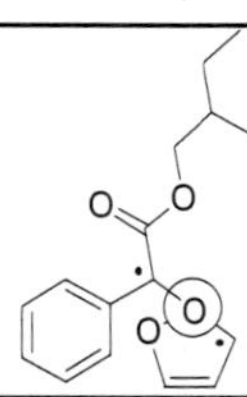	-955.53184611	667

Zeolites can act as agents able to reduce the conformational mobility of the intermediates, inhibiting the rotation of this conformer and, then, favouring the formation of the product deriving from the *SS* biradical intermediate. In the case of the biradical intermediates obtained from the phenylglyoxylate **62**, the *RS-syn* conformer occupies a larger volume than the *RS-anti* one. This fact prevents the *anti-syn* equilibration between these two conformation, favouring the formation of the *SS* conformer.

If the Paternò-Büchi reaction between chiral phenylglyoxylates and furan within zeolites depends on the capability of zeolites to inhibit the conformational equilibration of the biradical intermediates, we would expect that the presence of larger cations induces, reducing the available space in the supercage and, then, the mobility of the intermediates, an increase of stereoselectivity. The results, reported in Table 6, are not in agreement with this hypothesis. The reaction of **61** with furan showed a decrease in the stereoselectivity passing from Na to Li and K zeolites. Furthermore, the presence of different cation does not change the stereoselectivity in the reaction between **62** and furan.

In a recent work in this field the authors concluded the discussion of their results on the effect of alkali metal ions on the stereoselectivity of some photochemical reactions within zeolite claiming that "it is important to recognize that it is not only the cation size that controls the extent and direction of diastereoselectivity but also the binding ability and charge density of cations play an important role. Overall, too many factors control the chiral induction process within a zeolite making predictions a bit risky" [90]. In this work the authors tested the possibility that the coordination with an alkali metal ion induces "rigidification" of the substrates. To test this hypothesis we performed some calculations on the biradical intermediates involved in our reactions in the presence of an alkali metal ion (Li^+ and Na^+). We performed calculations on the lithium and sodium complexes of the biradicals obtained from the phenylglyoxylate **61**. The results are reported in Table 9. We considered only the biradical intermediates showing to be the most stable ones in Table 7.

All the structures showed a "rigidification" due to a complexation with the oxygen atom on the carboxylic group and with the aromatic ring. Furthermore, in the case of lithium complexes the distance between the alkali ion and the oxygen atom in the *SS-syn* stereoisomer is lower than in the *RS-anti* stereoisomer showing a higher "rigidification". However, this type of complexation is observed also by using sodium alkali ions, but in this case the lowest distance between Na^+ and the carboxylic oxygen atom was observed in *RS-anti* stereoisomer: this result is not in agreement with the experimental results. Furthermore, the difference between the distances of M^+ and O atoms in the different stereoisomers was higher by using lithium than in the case of sodium. These results should be in agreement with a higher "rigidification" effect in the case of lithium than when we used the sodium zeolite. However, these conclusions are not in agreement with the experimental results.

Finally, the energy difference between the *SS-syn* biradical intermediate obtained from **61** and the *RS-anti* isomer is higher when lithium is present: also this datum is in contrast with the experimental results.

Table 9. Lithium and sodium complexes of the biradical intermediates obtained from the phenylglyoxylate 61

Biradical intermediate	Alkaly ion	Structure	M^+ - O distance [Å]	Energy [H]
SS-syn	Li^+	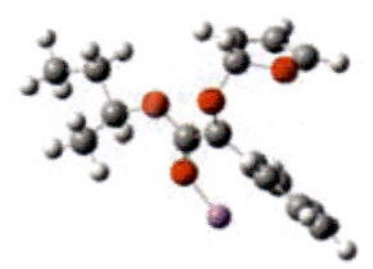	1.73919	-9.23.98645760
RS-anti	Li^+	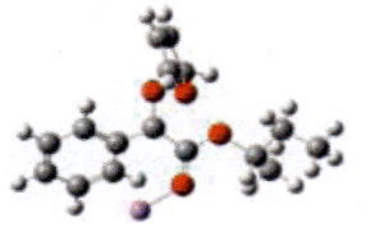	1.78591	-923.98077414
SS-syn	Na^+	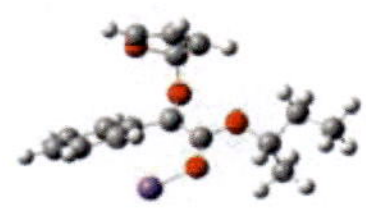	2.13952	-1077.85003987
RS-anti	Na^+	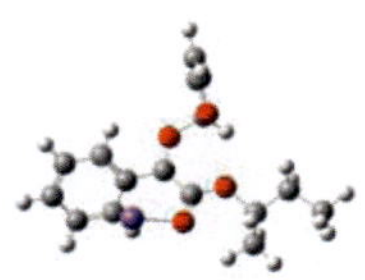	2.11897	-1077.64403658

On the basis of these controversial results we searched for a new explanation of the effect of the alkali ions in the stereoselectivity of our reactions. We tested the possibility that HOMO-LUMO interactions could play a role. The LUMO of Li^+ is at -0.254 H, that of Na^+ is at -0.256 H, while the LUMO of K^+ showed an energy of -0.220 H. In figure 14 we reported the HSOMOs of the biradical intermediates *SS-syn* and *RS-anti* obtained from the phenylglyoxylate **61**.

Both the orbitals are π orbitals with a different shape: that of the *SS-syn* stereoisomer is mainly localized on the furan ring, while that of the *RS-anti* stereoisomer is extended to both the carboxylic function and the phenyl ring.

The HSOMO of the *SS-syn* stereoisomer showed an energy of -0.196 H, while the other one has an energy of -0.192 H. Then, the interaction with the LUMO of the alkali ions is different for the two stereoisomers and has a maximum when Na^+ is used, considering that it has the lowest energy for the LUMO. Then, the intercation lowers when Li^+ and K^+ were used. This trend is in agreement with the experimental results.

Finally, we examined also the HSOMOs of the *SS-anti* and *RS-anti* stereoisomers of the biradical intermediate obtained from **62**. The orbitals are showed in figure 15.

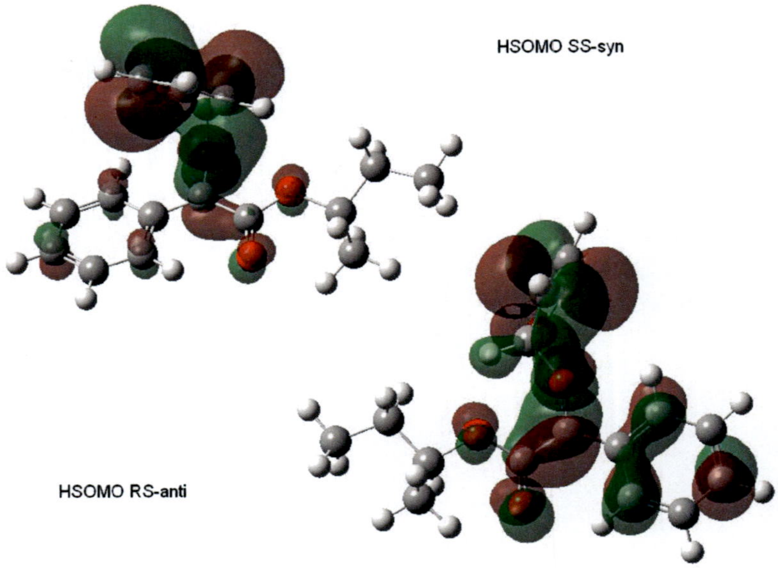

Figure 14. HSOMOs of the SS-syn and RS-anti biradical intermediates from the phenylglyoxylate 61.

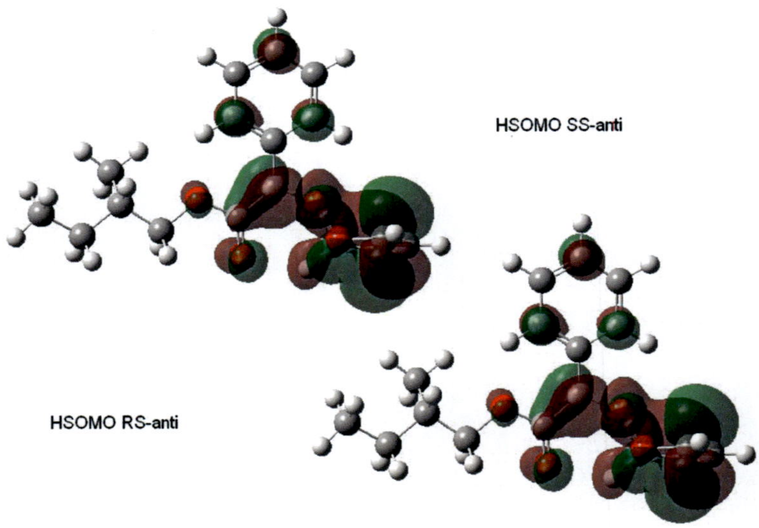

Figure 15. HSOMOs of the SS-anti and RS-syn biradical intermediates from the phenylglyoxylate 62.

Also in this case both the orbitals are π orbitals and they are very similar. Also the energy is the same, -0.190 H. Then the interaction with the LUMO of the alkaly ions is the same for both the stereoisomers and it is for this reason that we did not observe an alkaly ions effect in the Paternò-Büchi reaction of phenylglyoxylate ester **62** when the reaction is performed in different MY zeolites.

Finally, we tested the reactivity of phenylglyoxylate esters **61** and **62** when they were absorbed on β-cyclodextrin. The results are reported in Table 10.

Table 10. Reaction of chiral phenylglyoxylate esters with furan within β-cyclodextrin

Ester	Product	Yield [%]	de [%]
61	64	33	8
62	65	24	0

We can note that in this case the reaction occurs with lower yields than within zeolites and with a very low diasteroisomeric excess (zero in the case of the ester **62**). The effect of β-cyclodextrin on the stereoselectivity of the Paternò-Büchi reaction of phenylglyoxylate esters can be explained considering the presence on the β-cyclodextrin of twenty-one hydroxyl groups. The presence of a lot of hydroxy group in the microenviroment of cyclodextrin could induce the impossibility of the carbonyl compound to recognize a prochiral face of the alkene.

Isatine derivatives gave the corresponding cycloadducts with high stereoselectivity when irradiated in the presence of furan and benzofuran [91].

The reaction of furan with acyl cyanides yields the corresponding oxetanes but both diastereoisomeric *endo*- and *exo*-oxetanes are formed (Scheme 41). When chiral acyl cyanides are used low asymmetric induction is observed [92].

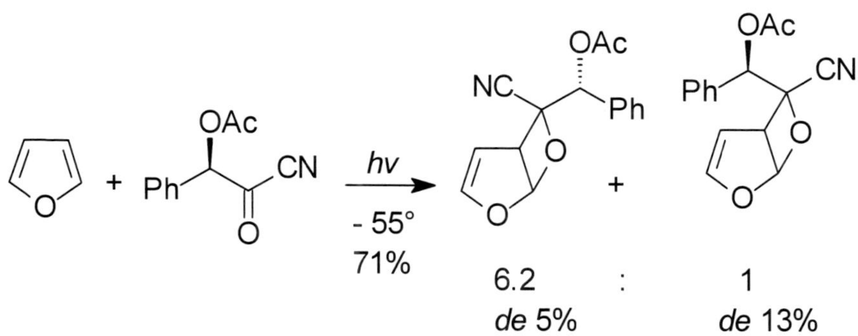

Scheme 41.

Furan reacts also with chiral ketones. In this case, an α-cleavage reaction before the 2+2 cycloaddition modifies the expected products (Scheme 42). When (-)-menthone was used as a substrate a chiral product was obtained as 2:1 diastereoisomeric mixture where the most abundant product has 1*R*, 3*R* configuration [93].

When the reaction was performed on carbohydrate **71**, a complex reaction mixture was obtained (Scheme 43) [93,94].

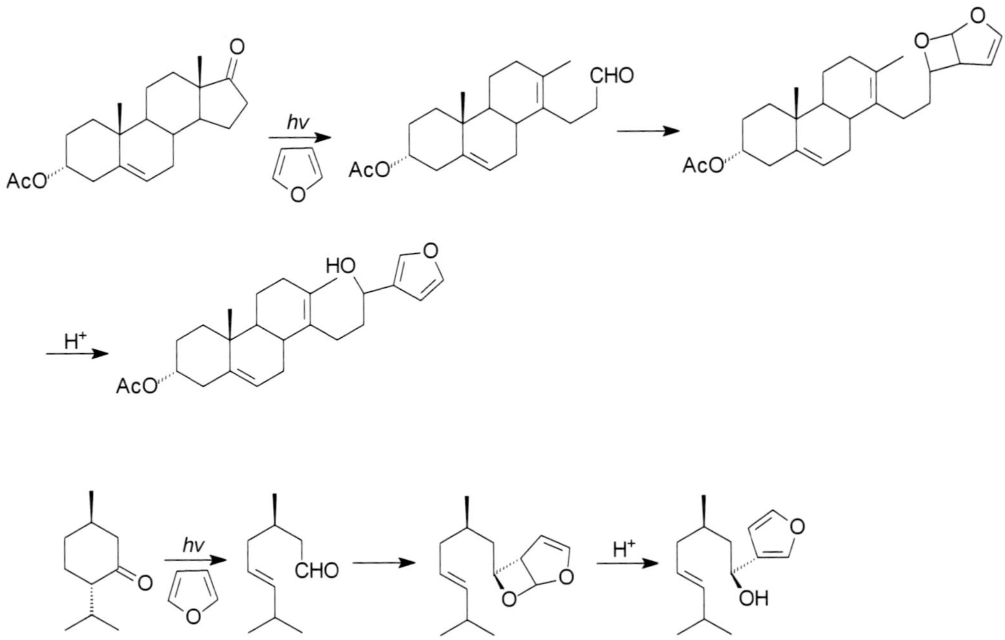

Scheme 42.

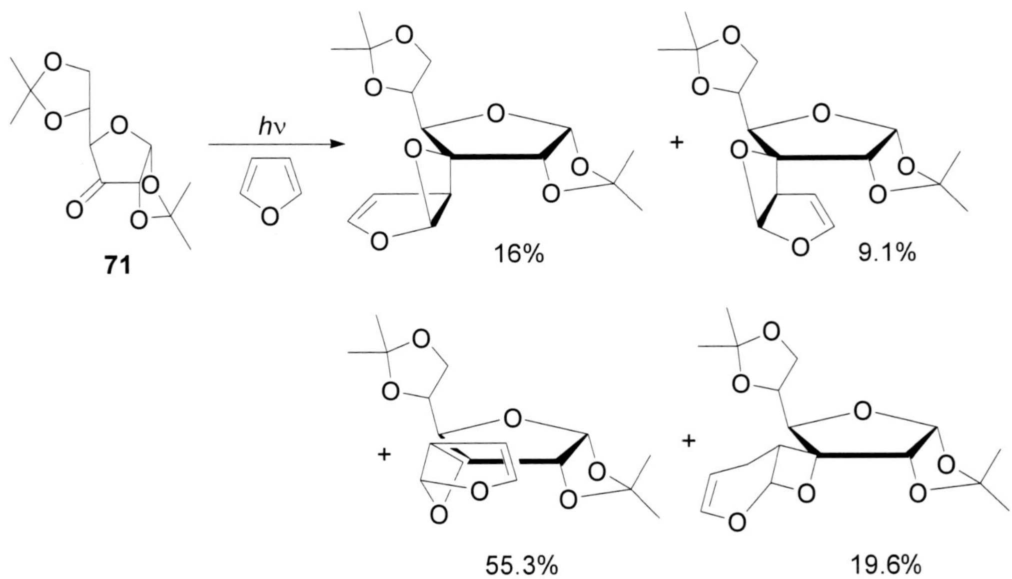

Scheme 43.

Attempts to obtain stereoselective Paternò-Büchi reactions were performed carrying out the reaction between 3,4-dimethylfuran and *R*-isopropylidene glyceraldehyde. The coupling products were obtained with an overall yield of 35% as a 1.2:1 mixture of diastereoisomers. Furthermore, the compound **72** was obtained with 54% ee (Scheme 44) [95,96].

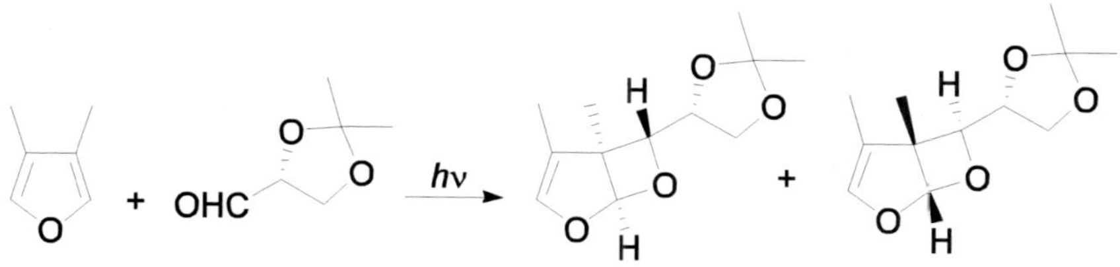

72

Scheme 44.

This behavior suggests the operation of a mechanism that is insensitive to the substitution pattern of chiral aldehydes. Reaction between an excited aldehyde (singlet or triplet state) and furan proceeds with initial carbon-oxygen bond formation to produce either of the two biradical species. The stereocenter adjacent to the carbonyl is now in a 1,4-relationship to the newly formed stereocenter at the acetal carbon and is expected to exert little influence as a stereocontrol device [95]. The extensive racemization observed probably reflects the photolability of the aldehydes towards racemization under the conditions of the reaction [96].

Nevertheless, compound 72 can be used in a chiral synthesis of the bicyclic part of asteltoxin confirming the assigned absolute configuration [96].

If the chiral center is near the reaction site the diastereoselectivity increases. Benzoin reacted with furan to give the corresponding adduct in acceptable yield (56%) and de = > 98% (Figure 16) [97].

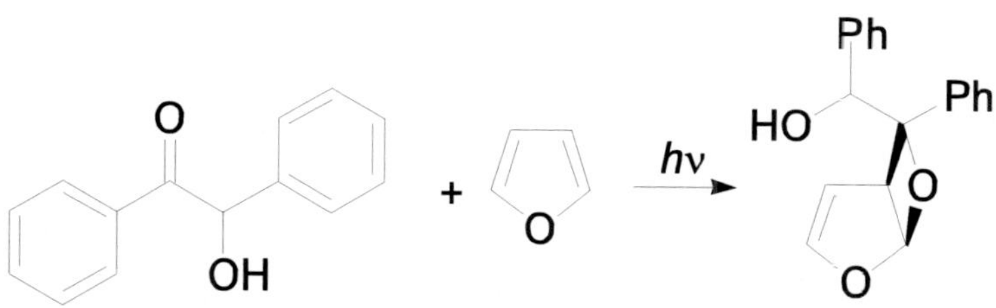

Figure 16. Photochemical reaction between benzoin and furan.

Also in this case we calculated the energy of the intermediates 73a and 73b for the reaction of benzoin with furan. The structures are depicted in Figure 17. The biradical 73a was more stable than 73b for 3.66 kcal mol-1, in agreement with the observed stereoselectivity.

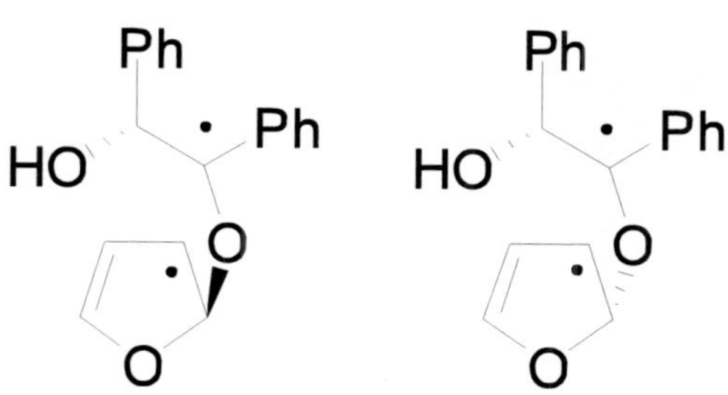

73a 73b

Figure 17. Biradical intermediates in the reaction of benzoin with furan.

When chiral ketones were used as substrates the Norrish Type II reaction is the In order to avoid Norrish Type II reaction **we used a substrate without** □hydrogen. We used the phenyl derivative 74 (Figure 18). In this case we observe the formation of the corresponding adduct [98].

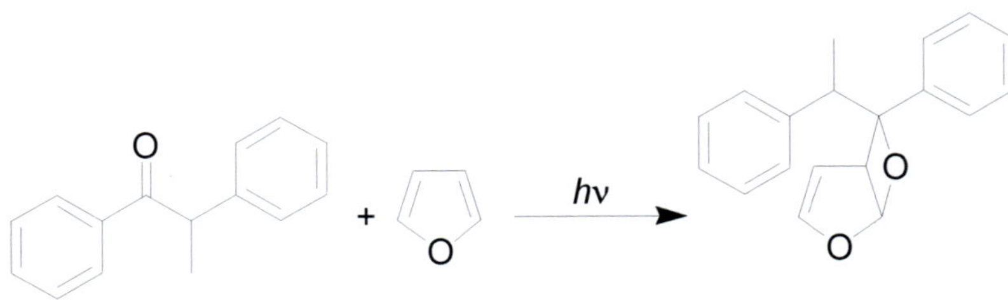

74 **75**

Figure 18. The reaction between chiral methyl ketone and furan.

The NMR data for the compound 75 are in agreement for an exo configuration for the phenyl group at C-6. Chiral HPLC (Chiralcel OJ, = 235 nm, 99:1 n-hexane/isopropanol) showed a de >98%.

We calculated the total energy of the possible intermediates in the reaction of 74 with furan. The structures of these compounds (76a, 76b) are reported in Figure 19.

The R,S conformer 76b is more stable than the S,S one by 1.1 kcal mol-1 and this difference can account for the observed diastereoselectivity.

Furthermore, the analysis of the HSOMO and LSOMO in this biradical intermediate (Figure 20) shows that the coupling can occur to give the product with the phenyl group in the exo configuration.

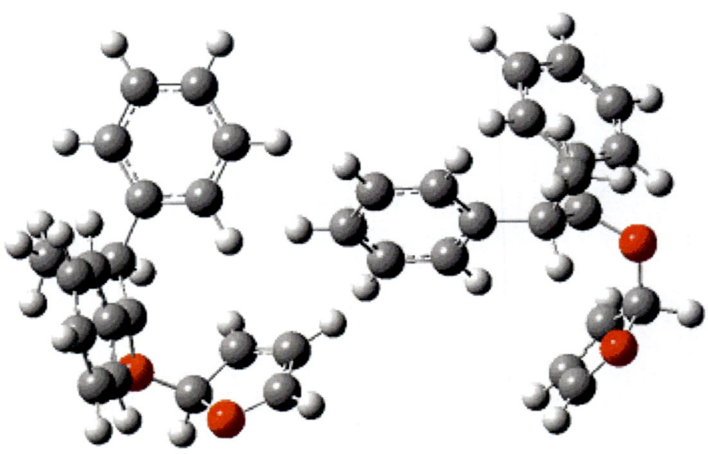

S,S ***R,S***

76a **76b**

Figure 19. Biradical intermediates in the reaction between 74 and furan.

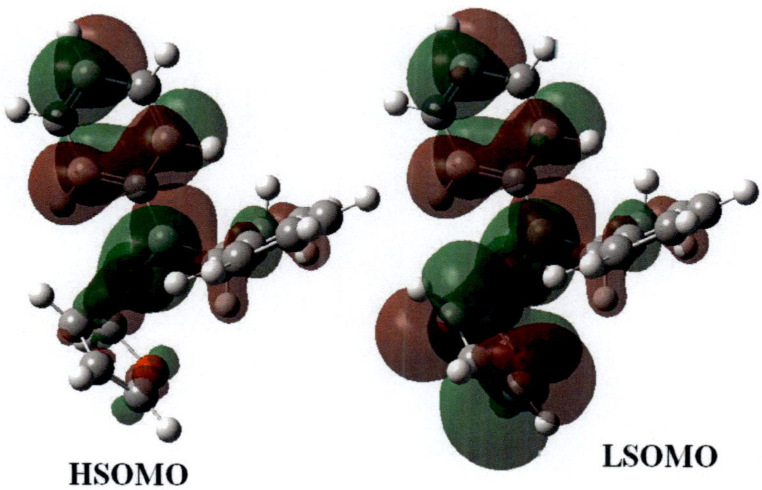

Figure 20. HSOMO and LSOMO of the biradical intermediate involved in the reaction between 74 and furan.

In 1990 Griesbeck found that the reaction of benzaldehyde with homoallylic alcohols did not show diastereoselectivity [62]. Ten years later Adam showed that allylic alcohols reacted with benzophenone to give the corresponding adducts with high regio- and diastereo-selectivity (Scheme 45) [99-101].

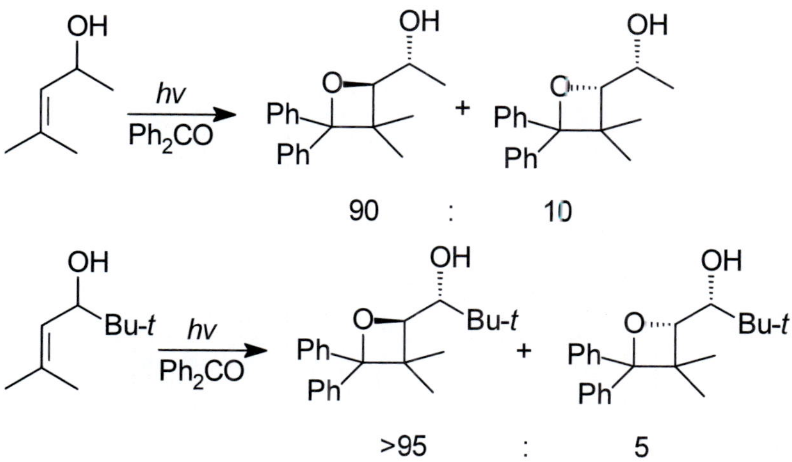

Scheme 45.

The diastereoselectivity dropped drastically in presence of protic methanol and totally disappeared for the corresponding silyl ethers. These data are in agreement with the presence of an hydroxy directing effect in the Paternò-Büchi reaction. Threo stereoisomer can be favoured through the formation of an hydrogen bond between triplet excited benzophenone and the substrate in the exciplex, while the formation of the erythro stereoisomer would be less favoured due to allylic strain (Scheme 46).

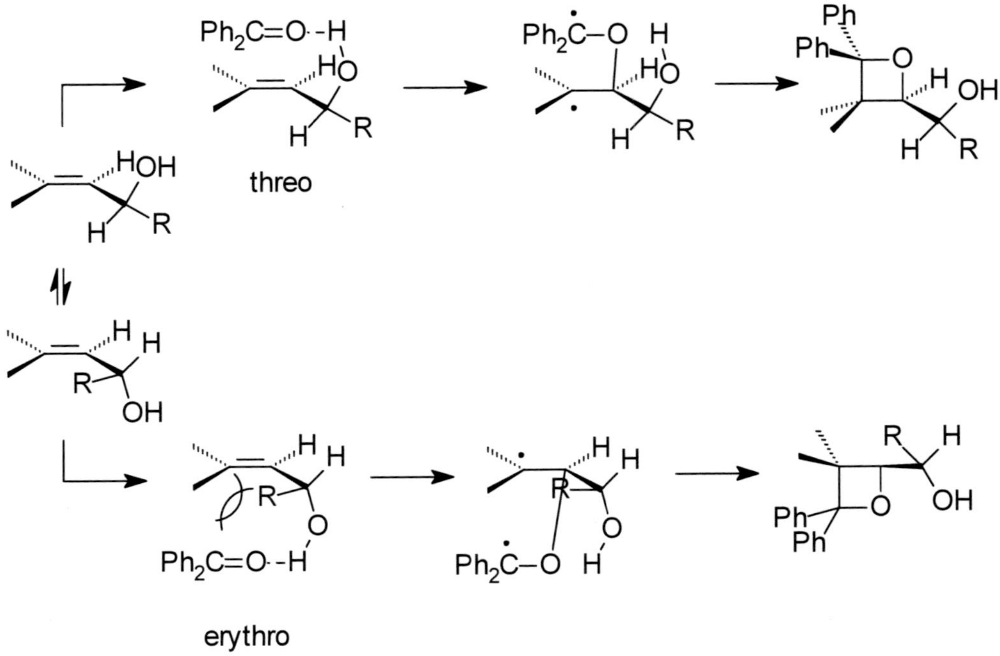

Scheme 46.

The formation of hydrogen bond to direct the Paternò-Büchi reaction has been considered by other researchers. Diastereoselective cycloaddition has been obtained using chiral enamide [102,103], or in the reaction of allylic alcohols with naphthalene rings [104].

When unsymmetrical carbonyl partners such as acetophenone or benzaldehyde were used high diastereoselectivity was observed to give the corresponding cis isomer. The regioselectivity was high with acetophenone but lower with benzaldehyde [100].

Cis diastereoselectivity can be explained by using the Griesbeck rule on the possible triplet biradicals formed in the reaction. Steric interactions are minimized when the biradical assumes the optimal conformation and this conformation accounts for the formation of the observed stereoisomer [105].

When chiral allylic alcohols were used as substrates in the reaction cis diastereoisomers were formed. Furthermore, also in this case, a pronounced threo diastereoselectivity was observed, in agreement with a less pronounced hydroxy directing effect when acetophenone and benzaldehyde were used [100,105]. Chiral allyl ether gave the corresponding adduct with high diastereoselectivity [106].

The reaction of allylic alcohols with carbonyl compounds was tested also on a particular type of allylic alcohol such as 2-furylmethanol derivatives. While the reaction of 2-furylmethanol with benzophenone showed low regioselectivity, the presence of larger substituents on the carbon bearing the alcoholic function allows a high regioselectivity (Scheme 47) [107]. Furthermore, when 2-furylethanol (77) was used as substrate a 1:1 mixture of stereoisomers was obtained, while, when 1-(2-furyl)-benzylic alcohol (78) was the substrate, only one diastereoisomer was obtained (Scheme 47) [107].

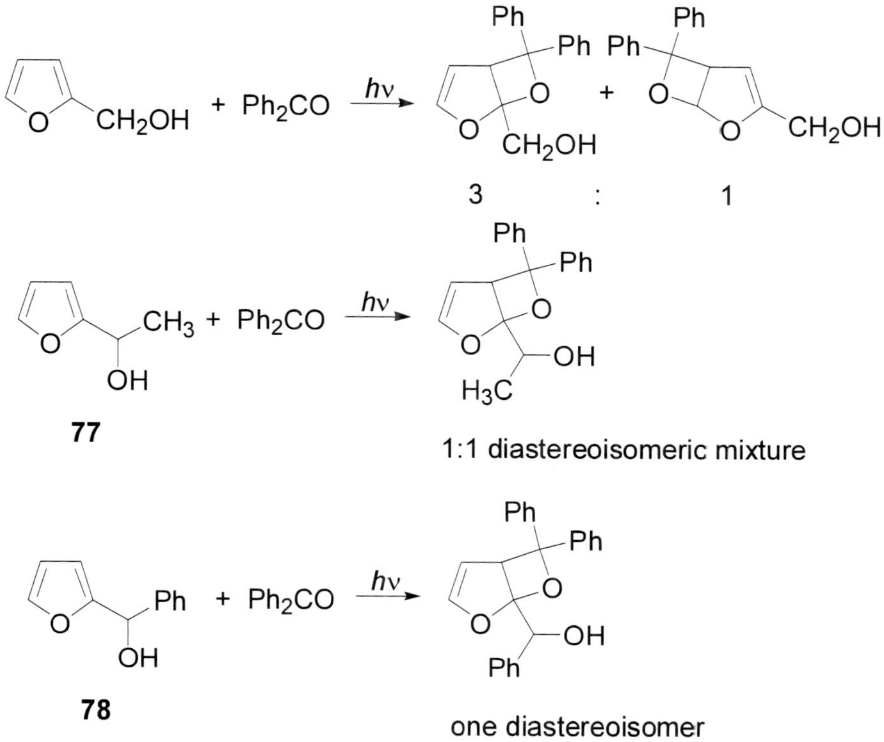

Scheme 47.

NOE experiments are in agreement with a 1RS, 1'RS, 5RS configuration. The high diastereoselectivity observed was confirmed using optically active 78.

When 5-methyl-2-furyl derivatives were used as substrates a different regioselectivity was observed. Compound 79 gave a 1:1 mixture of regioisomers, when irradiated in the presence of benzophenone, and a single regioisomer in the presence of benzaldehyde (Scheme 48) [108]. In agreement with the results obtained with 2-furyl derivatives, the products deriving from the attack on the side bearing the alcoholic function were obtained as a single diastereoisomer, while those deriving from the attack on the side bearing the methyl group were obtained as a mixture of diastereoisomers.

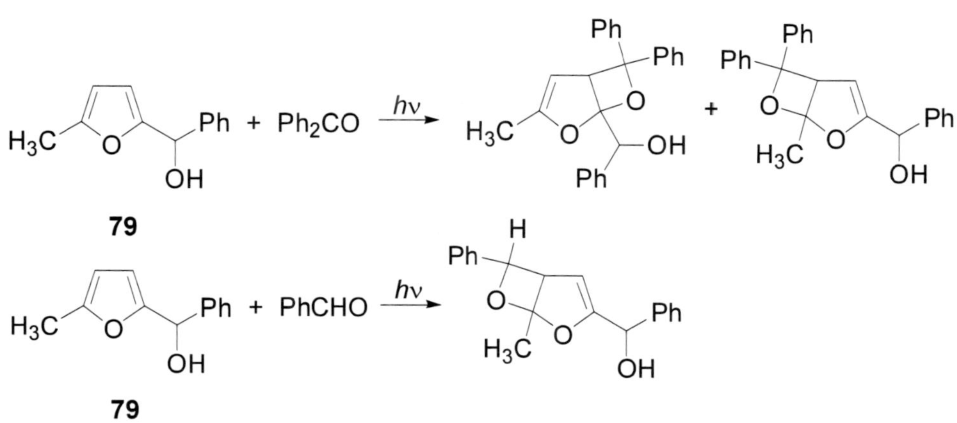

Scheme 48.

The reaction of 2-furylmethanol derivatives with aliphatic aldehydes and ketones gave the corresponding adducts with high regioselectivity but no diastereoselectivity (Scheme 49) [109].

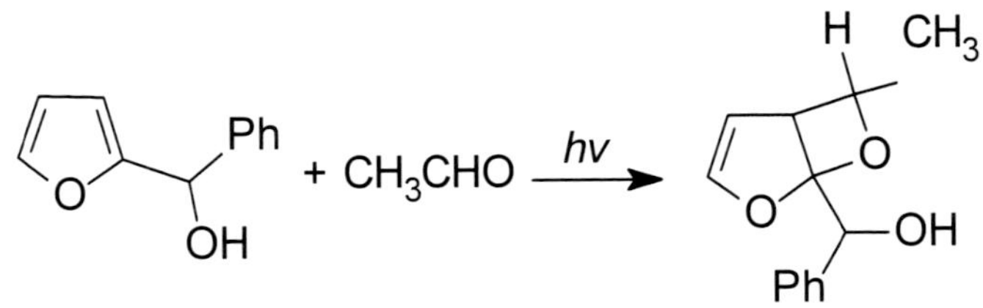

Scheme 49.

The regioselectivity of the reaction was explained on the basis of the relative stability of the biradical intermediates. Theoretical study (DFT) showed that the biradical obtained on the most hindered side of the molecule was more stable than the other [108]. The nature of the intermediate was in agreement with the observed value in a Hammett free energy correlation.

The observed diastereoselectivity in the reaction with aromatic carbonyl compounds clearly shows that it increases in relation to the nature of the substituents on the carbon bearing the alcoholic function as described by Adam. However, while Adam considers the allylic strain with a methyl group as the driving force for the diastereoselectivity, in this case, the methyl group is not present. Therefore, the allylic strain can not be used to explain diastereoselectivity.

In order to have more data to explain the observed stereoselectivity we studied the photochemical behavior or tertiary 2-furylcarbinols [110]. The photochemical reaction of 1-methyl-1-phenyl-1-(2-furyl)methanol with benzaldehyde gave a mixture of two regioisomeric product.

The regioisomer on the most hindered side of the molecule was obtained in low yield but it showed a complete diastereoisomeric control. On the contrary the main product was a mixture of four diastereoisomeric products (Scheme 50). The reaction of the same compound with benzophenone gave only the product deriving from the attack on the most hindered side of molecule. This compound was obtained with 48% diastereoisomeric excess.

The reaction of 1-methyl-1-t-butyl-1-(2-furyl)methanol with benzaldehyde and benzophenone gave complete regioselectivity. In all the experiments only the products deriving from the attack on the most hindered side of the substrate were observed (Scheme 51).

The regioselectivity was explained, as described above, considering the relative stability of the biradical intermediates. In this case, in the reaction of the 1-methyl-1-phenyl-1-(2-furyl)methanol with benzaldehyde the biradical obtained on the less hindered side of the substrate was more stable than the other one by 18.03 kJ mol-1.

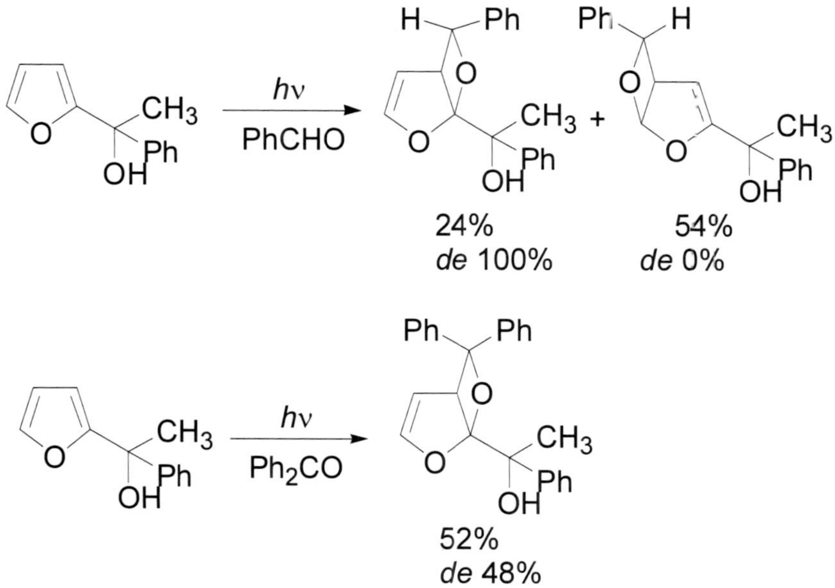

Scheme 50.

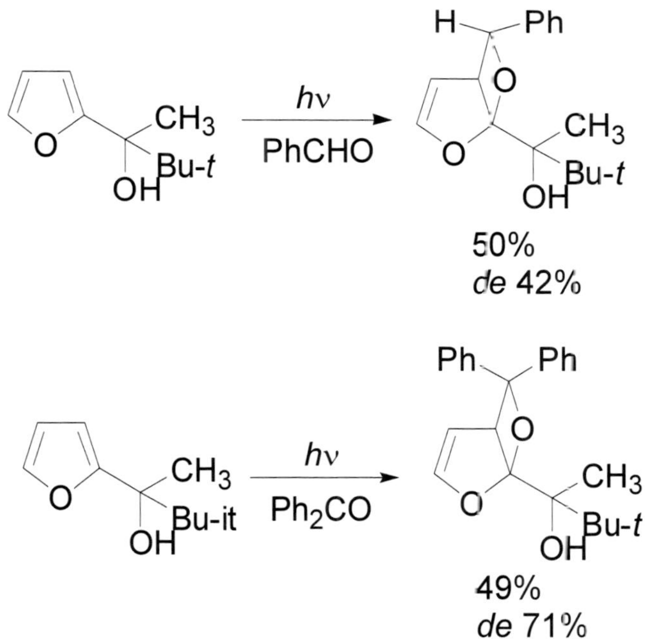

Scheme 51.

On the basis of these results we can attempt an explanation of the stereochemical behavior. 1-Methyl-1-phenyl-1-(2-furyl)methanol showed three conformations (Figure 21). All three conformers were in the range of 1.97 kJ mol-1 and they did not show a preference.

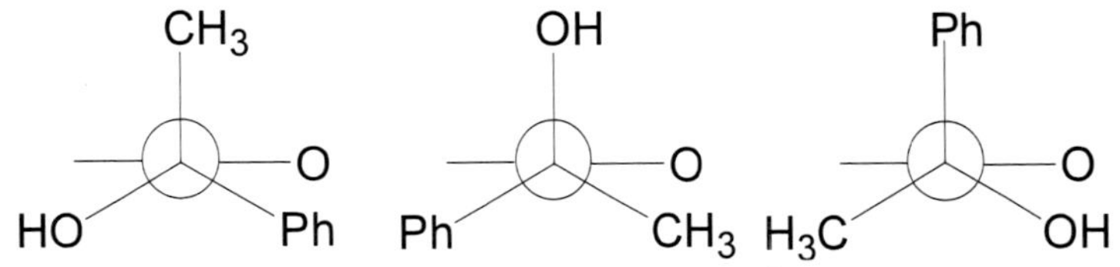

Figure 21. Conformers of 1-methyl-1-phenyl-1-(2-furyl)methanol.

To attempt to solve the problem we tested the following hypothesis: the directing effect exerted by the hydroxyl group is due to the formation of a hydrogen bond between the hydroxyl group and the oxygen of the excited carbonyl compound, or it is due to the formation of a complex. This type of interaction could favour the formation of a preferential conformation in the biradical intermediate where the hydroxyl group and the oxygen of the carbonyl compound are near. These conformations could have different energies for different diastereoisomeric biradicals, giving an explanation of the observed behaviour. The above described hypothesis requires that the biradical intermediates have a very short life enabling them to equilibrate to the most stable one. In the case of 1-methyl-1-phenyl-1-(2-furyl)methanol, if the hydroxyl group drives the attack of the oxygen of the carbonyl group, we obtained the conformations of the biradical intermediate represented in Figure 22.

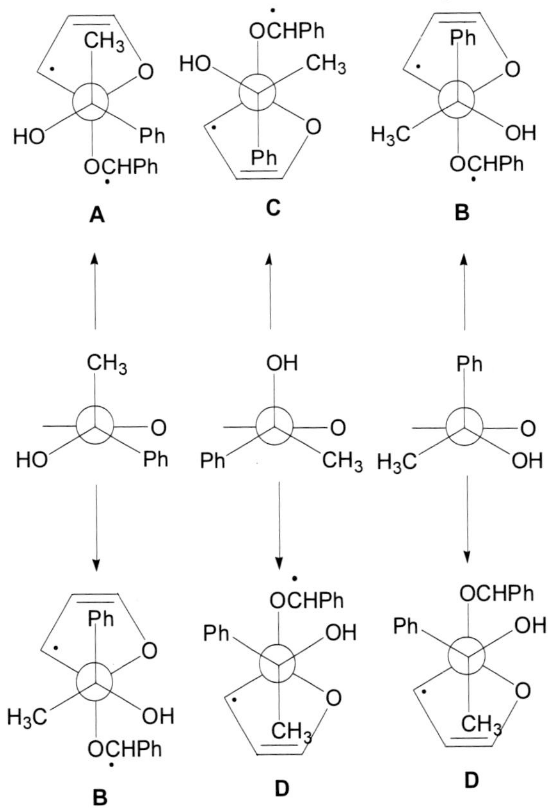

Figure 22. Possible conformations of the biradical intermediate from the reaction of 1-methyl-1-phenyl-1-(2-furyl)methanol with benzaldehyde.

We can observe that B and D are the preferential conformation: calculations on these conformations showed that there is a difference of 13.26 kJ mol-1 between the energies of these two conformations. This difference can account for the observed complete diastereoselectivity of the reaction. In the reaction of the same substrate with benzophenone the corresponding conformers B and D show a difference energy of 7.79 kJ mol-1: this difference is in agreement with the observed diastereoselectivity.

Figure 23 reports the possible conformations of 1-methyl-1-t-butyl-1-(2-furyl)methanol.

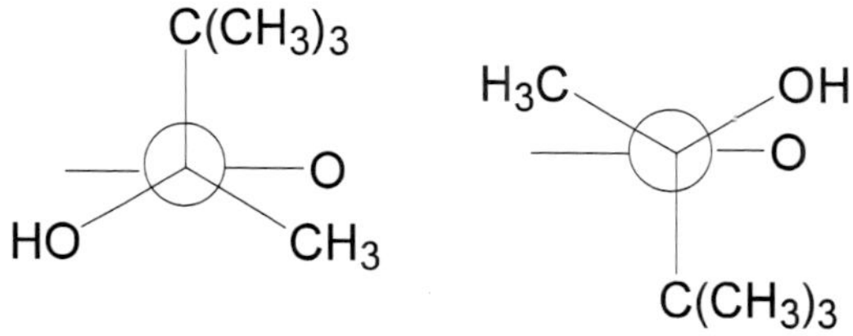

Figure 23. Conformers of 1-methyl-1-t-butyl-1-(2-furyl)methanol.

On the basis of these conformers, if the attachment of the excited carbonyl compound occurs on the same side of the hydroxyl group, the conformers described in Figure 24 can be obtained.

The conformers of Figure 24 differ by 6.90 kJ mol-1, in agreement with the observed diastereoisomeric excess.

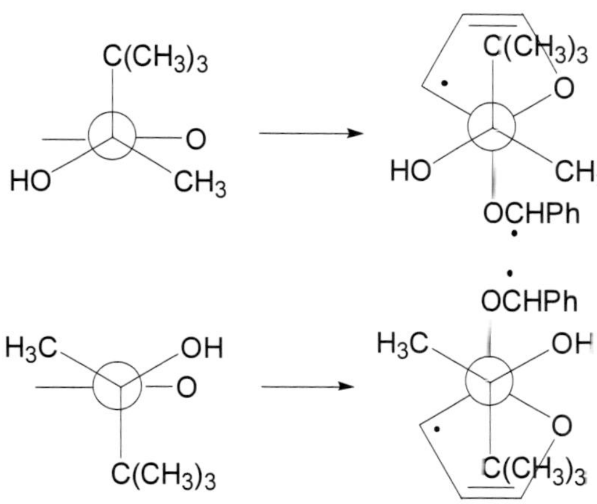

Figure 24. Possible conformations of the biradical intermediate from the reaction of 1-methyl-1-t-butyl-1-(2-furyl)methanol with benzaldehyde.

To test the above described hypothesis we examined also the results obtained on 2-furylmethanol derivatives. 2-Furylphenylmethanol gave complete diastereoisomeric control when it reacted with benzophenone [107]. The conformers of this substrate are reported in Figure 25.

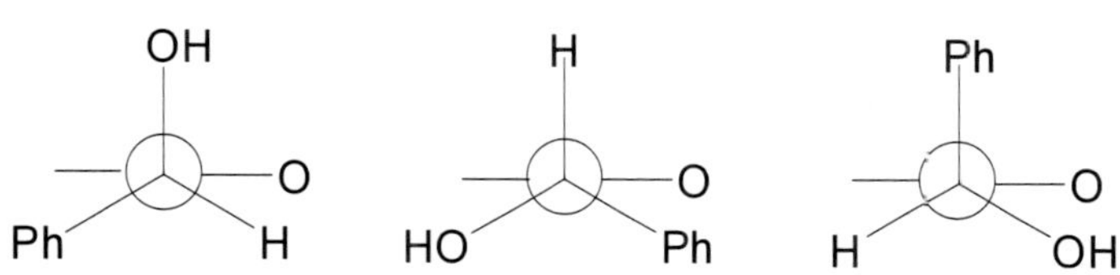

Figure 25. Conformations of 2-furylphenylmethanol.

The conformers deriving from the attack of the excited carbonyl compound on the same side of the hydroxyl group are described in Figure 26.

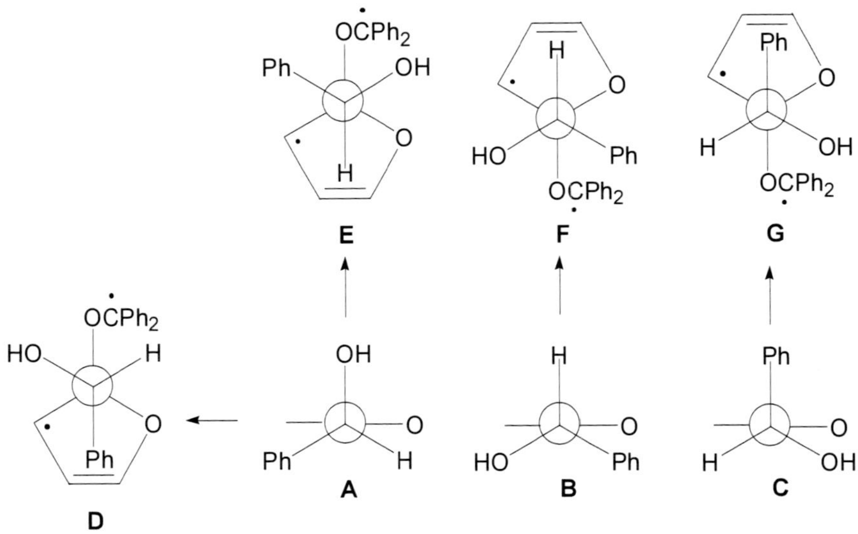

Figure 26. Possible conformations of the biradical intermediate from the reaction of 2-furylphenylmethanol with benzophenone.

Conformer **D** is the most stable one for the formation of (*RR*)* biradical, while conformer **F** is the most stable for the formation of (*RS*)* biradical. The conformer **F** is more stable than **D** by 16.57 kJ mol^{-1}, in agreement with the observed high diastereoselectivity. Furthermore, the same compound reacted with acetone to give the corresponding adduct without diastereoselectivity. Also in this case conformers **D** and **F** are those showing lowest energy but the difference between them is 1.09 kJ mol^{-1}, in agreement with the observed lack of diastereoselectivity.

The same approach can be used to justify the stereochemical behaviour of the reaction of allylic alcohols with benzophenone [111].

The irradiation of 3-furylmethanol **80** in the presence of benzophenone with a 125 W high pressure mercury arc gave the corresponding adduct **81** (Scheme 52, Table 11): the reaction occurs only at the C2/C3 double bond. This behaviour is in agreement with our previous reported results in this field. The regioselectivity of the reaction depended on the relative stability of the biradical intermediate. In this case, addition of an axcited carbonyl compound to C2 forms a tertiary radical site at C3, while such addition to C5 would generate only a secondary radical at C4. Unfortunately, the reaction of **80** in the presence of benzaldehyde did not occur (Scheme 52, Table 11). Probably, considering the mechanism of this reaction (see below), in this case, the retrocleavage reaction of the biradical intermediate does not allow the isolation of an oxetane.

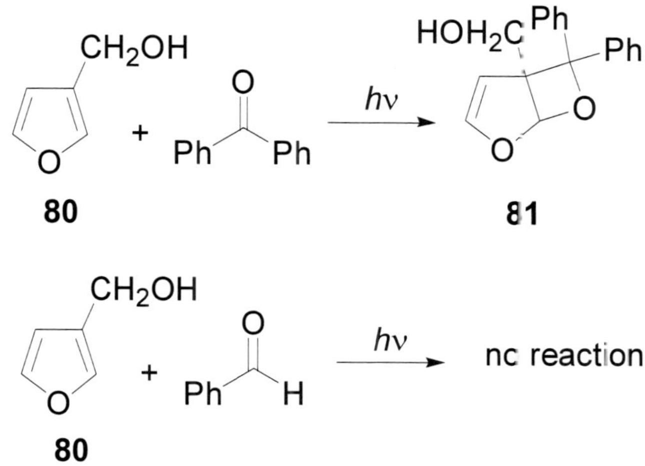

Scheme 52.

The irradiation of 1-(3-furyl)-*n*-heptanol **82** in the presence of benzophenone gave the adduct **83** (Scheme 53, Table 11). In this case the reaction could be performed on enantiomerically enriched **82**. The kinetic resolution of **82** by using Sharpless oxidation gave (*R*)-**82** in 30% yield with an *e.e.* = 60% ($[\alpha]_D$ = -20.4 (c 1, $CHCl_3$)).

The irradiation of (*R*)-**82** in the presence of benzophenone gave the (1*S*,5*S*,1'*R*)-**83** (Scheme 53, Table 11). For the stereochemical assignment see below. Chiral HPLC (Chiralcel OD) gave a *d.e.* = 23%. The irradiation of **82** in the presence of benzaldehyde gave a mixture of regioisomeric products **84** and **85** (Scheme 53, Table 11). The formation of regioisomers is due to the relative stability of the regioisomeric biradical intermediates. The same reaction was performed on (*R*)-**82** showing that (1*S*,5*S*,6*S*,1'*R*)-**84** was obtained with a *d.e.* = 43% while **85** was obtained as a mixture of diastereoisomers (Scheme 53. Table 11).

Table 11. Paternò-Büchi reaction on 3-furylmethanol derivatives

Furan	Carbonyl compound	Irradiation time [h]	Product	Yield [%][a]
80	Ph_2CO	44	**81**	42
80	PhCHO	27	-	-
82	Ph_2CO	18	**83**	73
(*R*)-**82**	Ph_2CO	18	(1*S*,5*S*,1'*R*)-**83**	56
82	PhCHO	24	**84**	42
			85	27
(*R*)-**82**	PhCHO	24	(1*S*,5*S*,6*S*,1'*R*)-**84**	54
			85	36
86	Ph_2CO	24	**87**	58
(*R*)-**86**	Ph_2CO	24	(1*S*,5*S*,1'*S*)-**87**	60
86	PhCHO	18	-	-

a) All the yields refer to isolated chromatographically pure products.

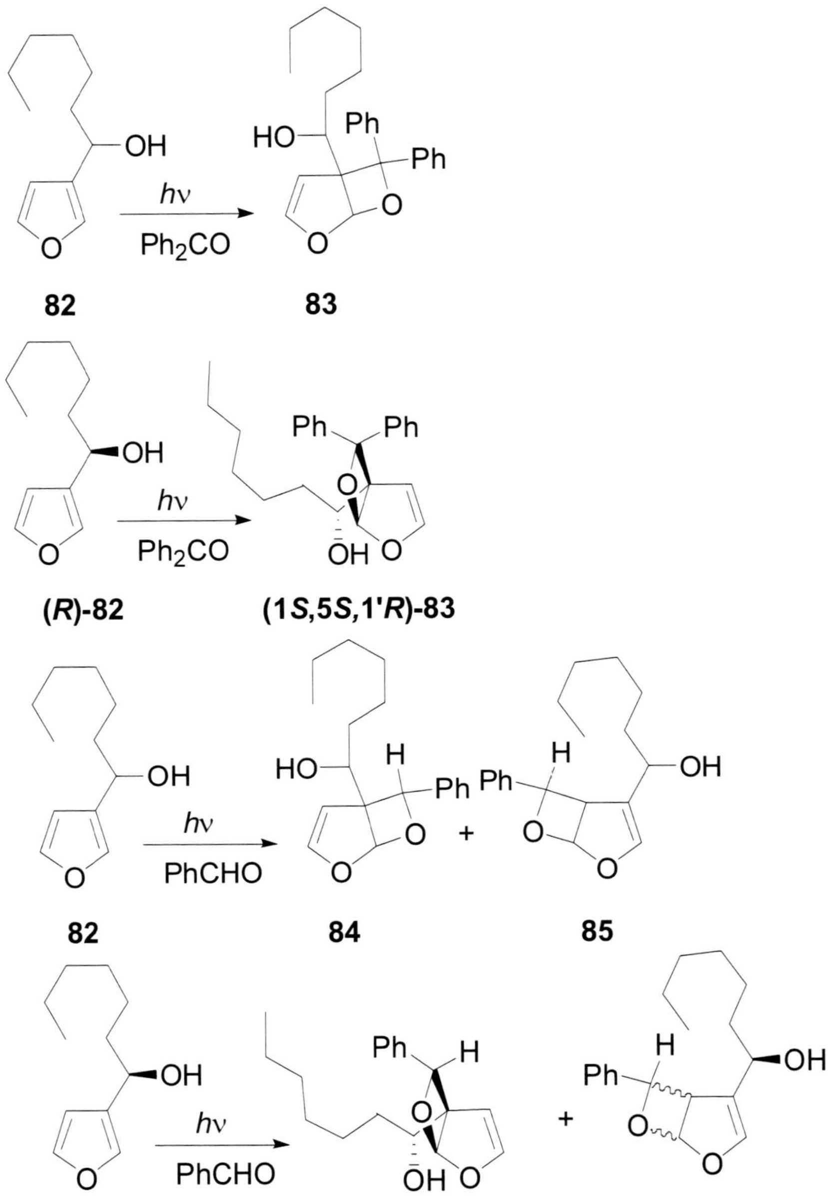

Scheme 53.

The stereochemical assignment of the products could be obtained on the basis of NOE experiments on (1*S*,5*S*,6*S*,1'*R*)-**84**. In fact, the irradiation on the singlet at δ 5.77 ppm (proton at C-6) gave a positive NOE effect (7%) on the *ortho* aromatic protons at δ 7.48 ppm. Furthermore, we observed a negative NOE on the C*H*OH proton at 3.60 ppm. PM3 optimization of the structures showed that the most stable conformation of **84** with 1*R*,5*R*,6*R*,1'*R* configuration should not present this negative NOE effect (the distance between the *ortho* aromatic protons ant C*H*OH was 4.76 Å) while this effect should be observable in the 1*S*,5*S*,6*S*,1'*R* diastereoisomer (the distance between the *ortho* aromatic protons ant C*H*OH was 2.48 Å).

The irradiation of the 3-furylmethanol derivative **86** in the presence of benzophenone gave the adduct **87** (Scheme 54, Table 11). Kinetic resolution of **86** gave (*R*)-**86** in 20% yield and *e.e* = 90% ($[\alpha]_D$ = -9.6 (c 1, $CHCl_3$)). The reaction of (*R*)-**86** with benzophenone gave the (1*S*,5*S*,1'*S*)-**87** diastereoisomer with a *d.e.*> 99% (Chiralcel OD) ($[\alpha]_D$ = 67.0 (c 1, $CHCl_3$)) (Scheme 54, Table 11). The relative configuration was assigned on the basis of the NOE experiments on **84**. The irradiation of **86** in the presence of benzaldehyde gave only decomposition products (Scheme 54, Table 11). The reaction occurs but we were not able to isolate the product after chromatography. Probably, it decomposed during the purification process.

To confirm the presence of a hydroxyl derecting effect in this reaction, we performed the reaction of the acetyl derivative of the compound **86** with benzophenone (Scheme 55). The reaction of **88** with benzophenone gave the corresponding adducts **89** and **90** with high regioselectivity. However, it was a 1:1 mixture of two diastereoisomeric compounds. This result is in agreement with the presence of a hydroxyl directing effect: if the hydroxy group is not present no diastereoselectivity was observed.

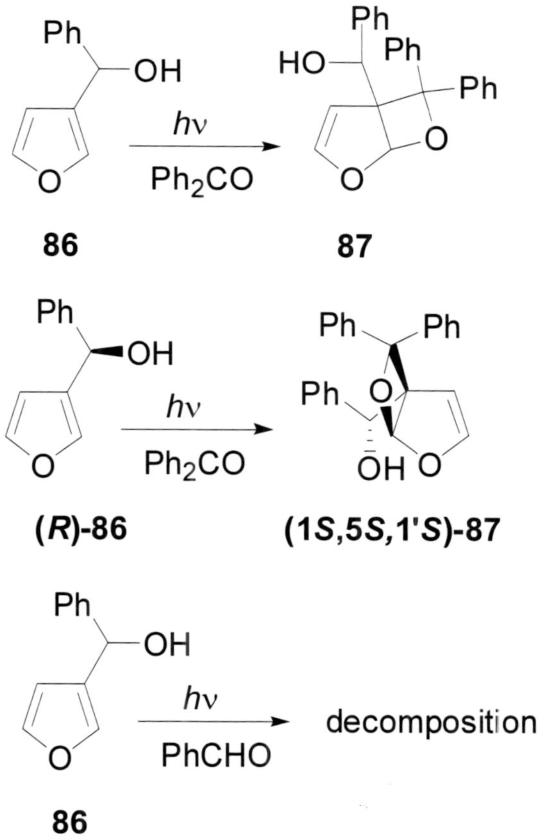

Scheme 54.

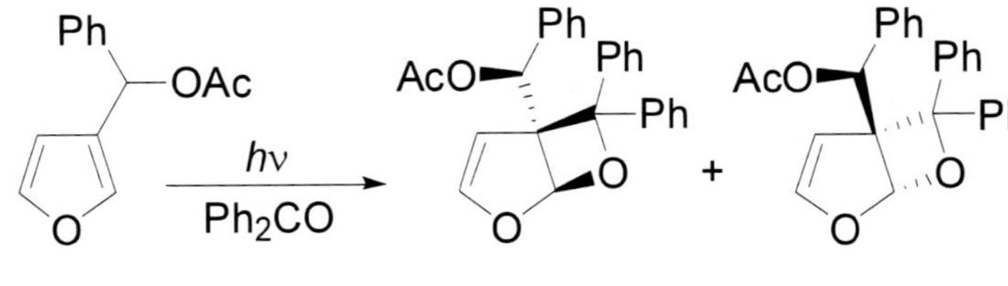

88 **(±)-89** **(±)-90**

Scheme 55.

As reported above the observed stereoselectivity in the reaction of 2-furylmethanol derivatives was explained on the basis of the possible conformers of the biradical intermediates obtainable admitting a directing effect of the hydroxyl group through coordination or hydrogen bond with the carbonyl compound. Recently, a transient absorption with λ_{max} at 531 nm has been observed in the reaction between furan and benzophenone in agreement with the presence of a biradical intermediate [19]. The presence of the biradical intermediate is in agreement with the mechanism depicted in Scheme 56. We followed the reaction kinetics in order to unravel whether the kinetics of the reactions of 3-furylmethanol derivatives should be compatible with this mechanism. The reaction of **82** with benzophenone follows a second order kinetics (k = $1.6 \cdot 10^{-3}$ $M^{-1}s^{-1}$, Figure 27) in agreement with a mechanism depicted in Scheme 56, where $k = k_1k_2/k_{-1}$.

In fact, for a sequence

$$A + B \underset{k_{-1}}{\overset{k_1}{\rightleftharpoons}} AB \xrightarrow{k_2} C$$

Assuming that the reaction of A with B follows a second order kinetics, we have

$$\frac{d[AB]}{dt} = k_1[A]\cdot[B] - k_{-1}[AB] - k_2[AB]$$

$$\frac{d[C]}{dt} = k_2[AB]$$

Applying the steady state approximation

$$k_1[A]\cdot[B] = (k_{-1} + k_2)\cdot[AB]$$

$$[AB] = \frac{k_1}{k_{-1} + k_2}[A]\cdot[B]$$

Then,

$$\frac{d[C]}{dt} = \frac{k_1 k_2}{k_{-1} + k_2}[A]\cdot[B]$$

Assuming $k_{-1} >> k_2$, we obtain

$$\frac{d[C]}{dt} = \frac{k_1 k_2}{k_{-1}}[A]\cdot[B]$$

These results are in agreement with the presence of a biradical intermediate.

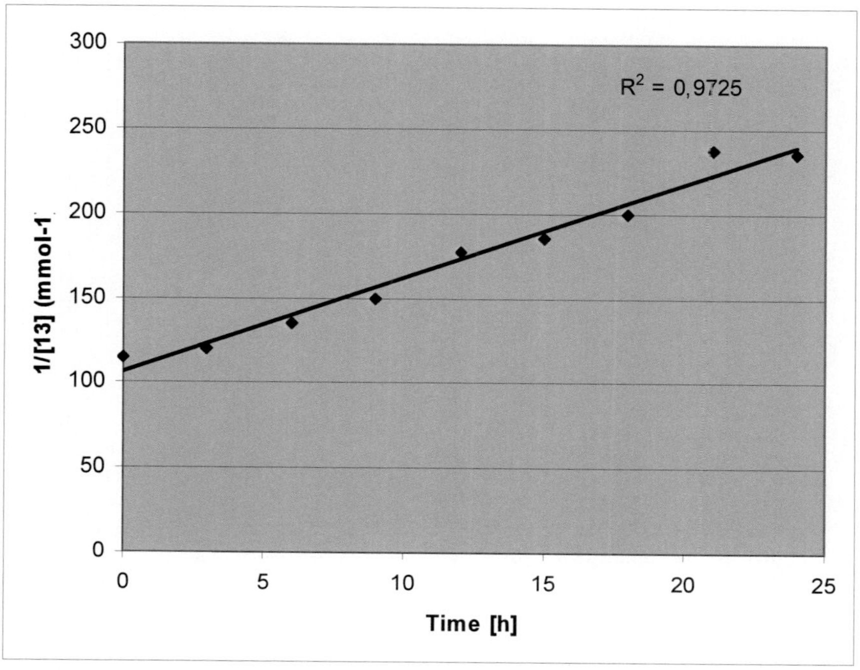

Figure 27. Kinetics of the reaction of **82** with benzophenone.

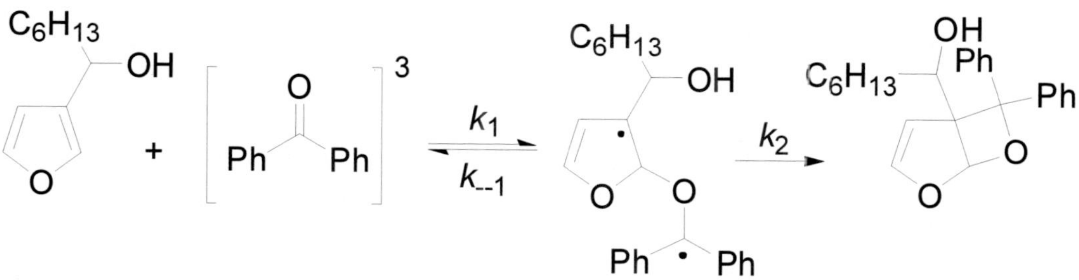

Scheme 56.

To explain the stereochemistry, we performed some calculations (DFT/B3LYP/6-31G+(d,p)). We have only two conformers, **86A** and **86B**. **86A** is more stable than the other one for 1.18 kcal mol^{-1} (Figure 28). Then, there is not a preferential conformation.

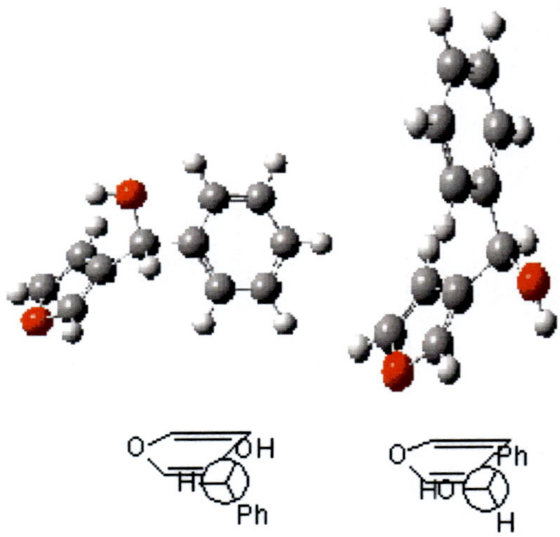

86A **86B**

Figure 28. Conformers of **86**.

The directing effect exerted by the hydroxyl group is due to the formation of a hydrogen bond between the hydroxyl group and the oxygen of the excited carbonyl compound, or it is due to the formation of a complex. This type of interaction could favour the formation of a preferential conformation in the biradical intermediate where the hydroxyl group and the oxygen of the carbonyl compound are near.

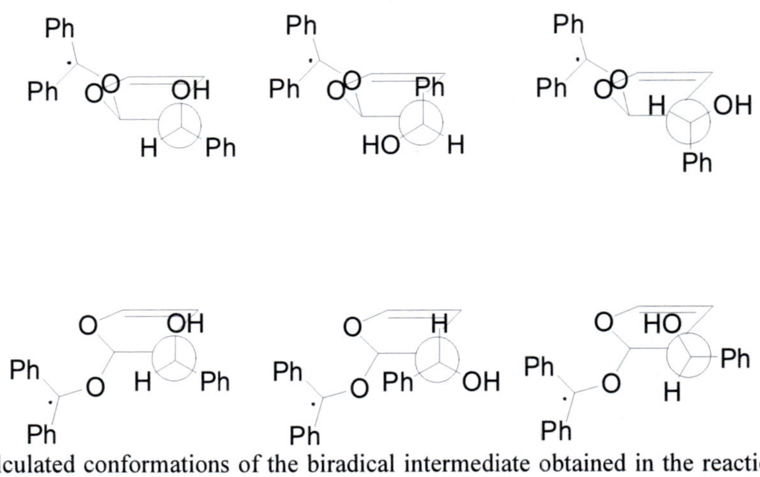

Figure 29. Calculated conformations of the biradical intermediate obtained in the reaction between **86** and benzophenone.

These conformations could have different energies for different diastereoisomeric biradicals, giving an explanation of the observed behaviour. The above described hypothesis requires that the biradical intermediates have a very short life enabling them to equilibrate to the most stable one. The attack of benzophenone on **86** allows the formation of a biradical intermediate. In Figure 29 we reported the calculated conformations of the biradical intermediate.

The above described conformations of **86**, when attacked by benzophenone and assuming the formation of a hydrogen bond between the hydroxy group and the oxygen atom of the excited carbonyl compound, can give the biradical intermediate conformation described in Figure 30. The conformation of the biradical intermediate is able to induce the observed stereoselectivity. In fact, it induces the formation of 1*S*,5*S* stereoisomer, if we start from (*R*)-**86**, and the 1*R*,5*R* stereoisomer starting from (*S*)-**86**.

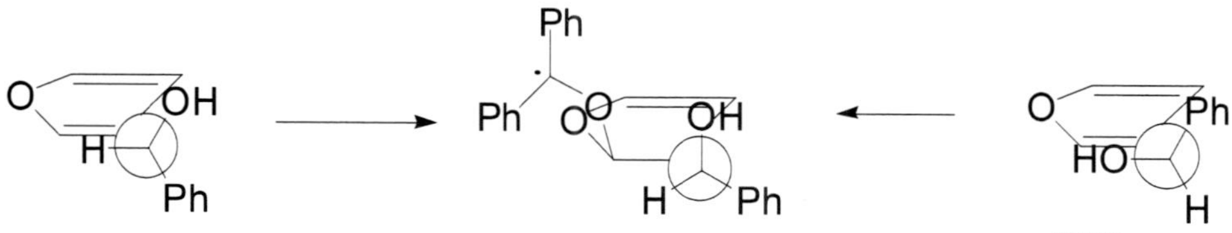

86A **86B**

Figure 30. Biradical intermediate obtained from the calculated conformations of **86**.

In the case of the compound (*R*)-**82** we observed the presence of two conformers **82A** and **82B** (Figure 31). Also in this case, there is not a prefential conformation (**82A** is more stable than **82B** for 0.7 kcal mol^{-1}).

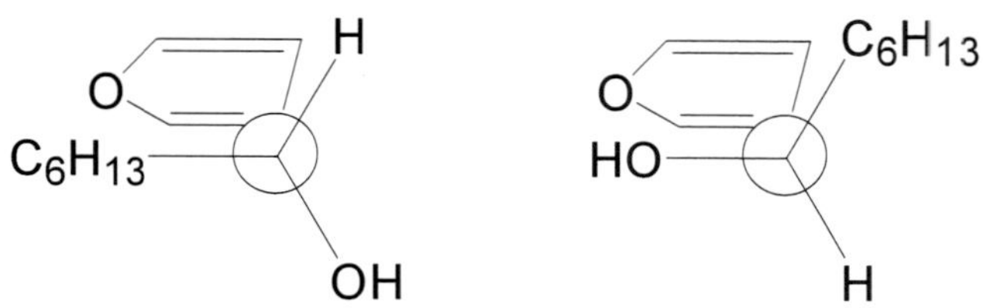

82A **82B**

Figure 31. Conformers of 82.

In the reaction of (*R*)-**82** with benzophenone six conformers of the biradical intermediates could be orbtained. These coformers are reported in Figure 32.

Assuming the presence of a hydroxy directing effect, benzophenone can give only four conformers of the biradical intermediates, two of them are *RS* conformers (**91**, **92**) while the other two are *RR* ones (**93**, **94**) (Figure 32).

The energy of the conformers **91-94** are reported in Table 12. We can see that *RS* conformers are more stable than the *RR* ones. Furthermore, the energy difference between the most stable *RS* conformer (**91**) and the most stable *RR* conformer (**93**) is 10.88 kcal mol^{-1}. This energy difference is in agreement with the observed diastereoisomeric excess.

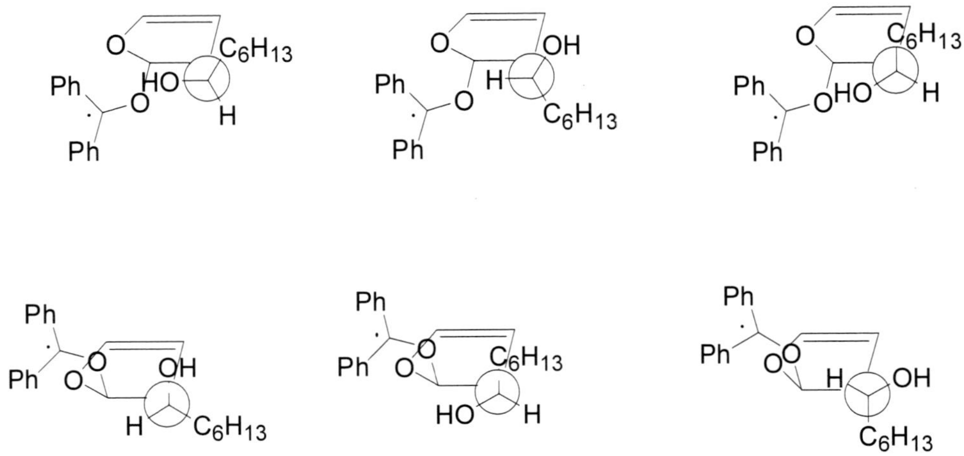

Figure 32. Calculated conformations of the biradical intermediate obtained in the reaction between **82** and benzophenone.

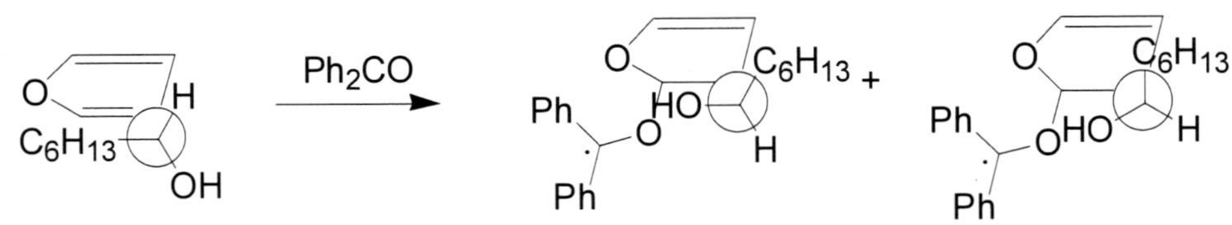

82A **91** **92**

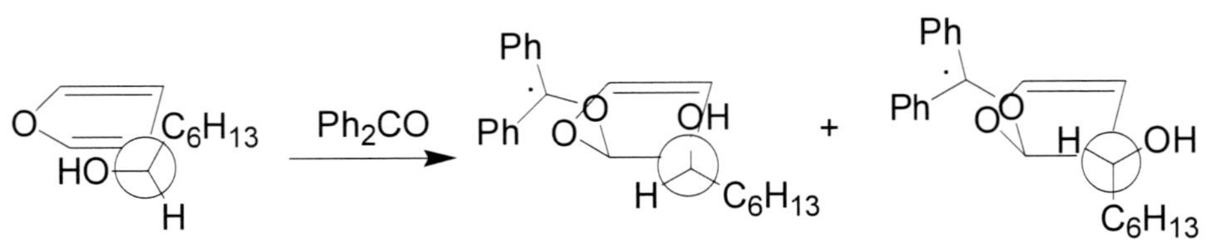

82B **93** **94**

Figure 33. Conformers of the possible biradical intermediate from 82 anf benzoprhenone.

We showed that the Paternò-Büchi reaction also on 3-furylmethanols can occur with high regioselectivity and with a good stereoselectivity, depending on the hydroxyl directing effect.

Table 12. Energy of the possible conformers obtained in the reaction between 82 and benzophenone

Conformer	Energy (H)
91	-1157.07698862
92	-1157.07585758
93	-1157.07284048
94	-1157.07134174

REFERENCES

[1] Paternò, E., Chieffi, G. (1909). *Gazz. Chim. Ital,* 39, 431

[2] Büchi, G., Inman, C. G., Lipinsky, E. S. (1954). *J. Am. Chem. Soc,* 76, 4327.

[3] Arnold, D. R. (1968). *Adv. Photochem,* 6, 301.

[4] Jones, G. (1981). *Org. Photochem,* 5, 1.

[5] Carless, H. A. J. (1984). In *Synthetic Organic Photochemistry*; Horspool, W. M. Ed. Plenum: New York, USA, p. 425.

[6] Porco, J. A., Schreiber, S. L. (1991). In *Comprehensive Organic Synthesis*; Trost, B. M., Fleming, I., Paquette, L. A. Ed. Plenum: New York, USA, Vol. 5, p. 151.

[7] Griesbeck, A. G. (1994). In *Handbook of Photochemistry and Photobiology*; Horspool, W. A., Song, P. S. Eds. CRC Press: Boca Raton.

[8] Kingston, D. G. I. (1991).*Pharmac. Ther,* 52, 1.

[9] Huang, J., Yokoyama, R., Yang, C., Fukuyama, Y. (2000).*Tetrahedron Lett,* 41, 6111.

[10] Wang, Y., Fleet, G. W. J., Storer, R., Myers, P. L., Wallis, C. J., Doherty, O., Watkin, D. J., Vogt, K., Witty, D. R., Wilson, F. X., Peach, J. M. (1990). *Tetrahedron asymmetry*, 1, 527.

[11] Yang, C. O., Kurz, W., Eugui, E. M., Mc Roberts, M. J., Verheyden, J. P. H., Kurz, L. J., Walker, K. A. M. (1992). *Tetrahedron Lett*, 33, 41.

[12] Kawahata, Y., Takatsuto, S., Ikekawa, N., Murata, M., Omura, S. (1986). *Chem.Pharm. Bull,*34, 3102.

[13] Kopecký, J. (1992). *Organic Photochemistry,* VCH, New York, USA, p. 126.

[14] Turro, N. J., Dalton, J. C., Dawes, K., Farrington, G., Hautala, R., Morton, D., Niemczyk, M., Schore, N. (1972). *Acc.Chem.Res,*5, 92.

[15] Palmer, I. J., Ragazos, I. N., Bernardi, F., Olivucci, M., Robb, M. A. (1994). *J.Am.Chem. Soc,*116, 2121.

[16] Kutateladze, A. G. (2001). *J. Am. Chem. Soc,*123, 9279.

[17] Freilich, S. C., Peters, K. S. (1981). *J. Am. Chem. Soc,*103, 6225.

[18] Freilich, S. C., Peters, K. S. (1985). *J. Am. Chem. Soc,* 107, 3819.

[19] Abe, M., Kawakami, T., Ohata, S., Nozaki, K., Nojima, M. (2004). *J. Am. Chem. Soc,*126, 2838.

[20] Gaussian 03, Revision A.1, M. J. Frisch, G. W. Trucks, H. B. Schlegel, G. E. Scuseria, M. A. Robb, J. R. Cheeseman, J. A. Montgomery, Jr., T. Vreven, K. N. Kudin, J. C. Burant, J. M. Millam, S. S. Iyengar, J. Tomasi, V. Barone, B. Mennucci, M. Cossi, G. Scalmani, N. Rega, G. A. Petersson, H. Nakatsuji, M. Hada, M. Ehara, K. Toyota, R. Fukuda, J. Hasegawa, M. Ishida, T. Nakajima, Y. Honda, O. Kitao, H. Nakai, M. Klene, X. Li, J. E. Knox, H. P. Hratchian, J. B. Cross, C. Adamo, J. Jaramillo, R. Gomperts, R. E. Stratmann, O. Yazyev, A. J., Austin, R., Cammi, C., Pomelli, J. W., Ochterski, P. Y., Ayala, K., Morokuma, G. A., Voth, P., Salvador, J. J., Dannenberg, V. G., Zakrzewski, S., Dapprich, A. D., Daniels, M. C., Strain, O., Farkas, D. K., Malick, A. D., Rabuck, K., Raghavachari, J. B., Foresman, J. V., Ortiz, Q., Cui, A. G., Baboul, S., Clifford, J., Cioslowski, B. B., Stefanov, G., Liu, A., Liashenko, P., Piskorz, I., Komaromi, R. L., Martin, D. J., Fox, T., Keith, M. A., Al-Laham, C. Y., Peng, A., Nanayakkara, M., Challacombe, P. M. W., Gill, B.,

Johnson, W., Chen, M. W., Wong, C., Gonzalez, and Pople, J. A. Gaussian, Inc., Pittsburgh PA, 2003.
[21] Mattay, J. (1987). *Angew. Chem., Int. Ed. Engl,* 26, 825.
[22] Schenck, G. O., Hartman, W., Steinmetz, R. (1963). *Chem. Ber,* 96, 498.
[23] Gagnaire, D., Payo-Subiza, E. (1963). *Bull. Soc. Chim. Fr,* 2623.
[24] Toki, S., Shima, K., Sakurai, H. (1965). *Bull Chem. Soc. Jpn,* 38, 760.
[25] Nakano, T., Rivas, C., Perez, C. (1973). *J.Chem. Soc., Perkin Trans,* 1, 2322.
[26] Whipple, E. B., Evanega, G. R. (1968). *Tetrahedron,* 24, 1299.
[27] Rivas, C., Payo, E. (1967). *J. Org. Chem,* 32, 2918.
[28] Rivas, C., Bolivar, R. A., Cucarella, M. (1982). *J. Heterocyclic Chem,* 19, 529.
[29] Shima, K., Sakurai, H. (1966). *Bull.Chem. Soc. Jpn,* 39, 1806.
[30] Toki, S., Sakurai, H. (1967). *Bull. Soc. Chim. Jpn,* 40, 2885.
[31] Zamojski, A., Koźluk, T. (1977). *J.Org. Chem,* 42, 1089.
[32] Schreiber, S. L., Desmaele, D., Porco, J. A. (1988). *Tetrahedron Lett,* 29, 6689.
[33] Carless, H. A. J., Halfhide, A. F. (1992). *J. Chem. Soc., Perkin Trans,* 1,1081.
[34] Abe, M., Torii, E., Nojima, M. (2000). *J.Org. Chem,* 65, 3426.
[35] Griesbeck, A. G., Mauder, H., Stadtmüller, S. (1994). *Acc.Chem. Res,* 27, 70.
[36] Griesbeck, A. G., Buhr, S., Fiege, M., Schmickler, H., Lex, J. (1998). *J.Org. Chem,* 63, 3847.
[37] D'Auria, M., Emanuele, L., Racioppi, R. (2006). *Lett Org. Chem,* 3, 244.
[38] Ogata, M., Watanabe, H., Kano, H. (1967). *Tetrahedron Lett,* 533.
[39] Leitich, J. (1967). *Tetrahedron Lett,* 1937.
[40] Evanega, G. R., Whipple, E. B. (1967). *Tetrahedron Lett,* 2163.
[41] Toki, S., Sakurai, H. (1967). *Tetrahedron Lett,* 4119.
[42] Sekretar, S., Rudā, J., Štibranyi, L. (1984). *Coll. Czech. Chem. Commun,* 49, 71.
[43] Cantrell, T. S., Allen, A., Ziffer, H. (1989). *J.Org. Chem,* 54, 140.
[44] Kubo, Y., Suto, M., Tojo, S., Araki, T. (1986). *J. Chem. Soc. Perkin Trans,* 1, 771.
[45] Krauch, C. H., Metzner, W., Schenck, G. O. (1966). *Chem. Ber,* 99, 1723.
[46] Farid, S., Hartman, S. E., DeBoer C. D. (1975). *J. Am. Chem. Soc,* 97, 808.
[47] Kaway, Y., Yamazaki, H., Kayakiri, N., Yoshihara, K., Yatabe, T., Oku, T. (1996). *PTC Pat. Appl.* 9529907, 1995 [*Chem. Abstr.*124, 202000].
[48] Bryant, H. U., Dodge J. A. (1996). *US Pat.* 5523309, 1996 [*Chem. Abstr.* 125, 142536].
[49] Fujishita, T., Yoshinaga, T., Sato, A. (2000). *PTC Pat. Appl.* 2000039086, 2000 [*Chem. Abstr.* 133, 89529].
[50] Zotova, S. A., Shvedov, V. I. (1994). *Khim. Farm. Zh,* 28, 22.
[51] Robba, M. F., Aurousseau, M. (1982). *Fr. Pat.* 2488893, 1982 [*Chem. Abstr. 97,* 38847].
[52] Deltour, G., Broekhuysen, J., Ghislain, M., Burgeois, F., Binon, F. (1967). *Arch. Int. Physiol,* 165, 25.
[53] Imbert, T., Lacour, A., Turin M. (1980). *Fr. Pat.* 2431491, [*Chem. Abstr,93,* 186160].
[54] Da Settimo, F., Lucacchini, A., Marini, A. M., Martini, C., Primofiore, G., Senatore, G., Taliani, S. (1996). *Eur. J. Med. Chem,*31, 951.
[55] Niikura, K., Takano, M., Sawada, M. (2004). *Br. J. Pharmacol,*142, 558.
[56] Carless, H. A., Haywood, D. J. (1980). *J.Chem. Soc. Chem. Commun,*1067.

[57] Bondock, S., Griesbeck, A. G. (2005). *Int. J. Photoenergy*, 7, 23.
[58] Griesbeck, A. G., Fiege, M., Bondock, S., Gudipati, M. S. (2000). *Org. Lett*,2, 3623.
[59] Griesbeck, A. G. (2003). *J. Photoscience*,10, 49.
[60] Griesbeck, A. G. (2003). *Synlett*, 451.
[61] Griesbeck, A. G., Abe, M., Bondock, S. (2004). *Acc. Chem. Res*,37, 919.
[62] Griesbeck, A. G., Stadtmüller, S. (1990). *J. Am. Chem. Soc*,112, 1281.
[63] Griesbeck, A. G., Stadtmüller, S. (1991). *J. Am. Chem. Soc*,113, 6923.
[64] Griesbeck, A. G., Stadtmüller, S. (1990). *Chem. Ber*, 123, 357.
[65] Griesbeck, A. G., Mauder, H., Peters, K., Peters, E.M., von Schnering, H. G. (1991). *Chem. Ber*, 124, 407.
[66] Smith, A. B., III, Sulikowski, G. A., Sulikowski, M. M., Fujimoto, K. (1992). *J.Am. Chem. Soc*, 114, 2567.
[67] Jarosz, S., Zamojski, A. (1979). *J.Org. Chem*, 44, 3720.
[68] Kitamura, T., Kawakami, Y., Imagawa, T., Kawanisi, M. (1977). *Synth. Commun*,7, 521.
[69] Kozluk, T., Zamojski, A.(1983). *Tetrahedron*, *39*, 805.
[70] Schreiber, S. L., Hoveyda, A. H., Wu, H. J. (1983). *J. Am. Chem. Soc*,105, 660.
[71] Satre, M. (1981). *Biochem. Biophys. Res. Commun*, 100, 267.
[72] Schreiber, S. L., Satake, K. (1983). *J. Am. Chem. Soc*, 105, 6723.
[73] Schreiber, S. L., Satake, K. (1984). *J. Am. Chem. Soc*, 106, 4186.
[74] Schreiber, S. L., Hoveyda, A. H. (1984). *J. Am. Chem. Soc*, 106, 7200.
[75] Schreiber, S. L., Porco, J. A. (1989). *J. Org. Chem*, 54, 4721.
[76] Shimada, N., Hasegawa, S., Harada, T., Tonisawa, T., Fujii, T. (1986). *J. Antibiot*, 39, 1623.
[77] Hoshino, H., Shimizu, N., Shimada, N., Takida, T., Takeuchi, T. (1987). *J. Antibiot*,60, 1077.
[78] Hambalek, R., Just, G. (1990). *Tetrahedron Lett*, 31, 4693.
[79] Mootoo, D. R., Date, V., Fraser-Reid, B. (1988). *J. Am. Chem. Soc*, 110, 2662.
[80] Hambalek, R., Just. G. (1990). *Tetrahedron Lett*, 31, 5445.
[81] Pérez-Ruiz, R., Miranda, M. A., Alle, R., Meerholz, K., Griesbeck, A. G. (2006). *Photochem. Photobiol. Sci*, 5, 51.
[82] Jarosz, S., Zamojski, A. (1982). *Tetrahedron*, 38, 1447.
[83] Pelzer, R., Jütten, P., Scharf, H. D. (1989). *Chem. Ber*, 122, 487.
[84] Pelzer, R., Scharf, H. D., Buschmann, H., Runsink, J. (1989). *Chem. Ber*, 122, 1187.
[85] Hu, S., Neckers, D. C. (1999). *J.Chem. Soc., Perkin Trans*, 2, 1771.
[86] Buschmann, H., Scharf, H.D., Hoffmann, N., Plath, M. W., Runsink, J. (1989). *J. Am. Chem. Soc*, 111, 5367.
[87] Buschmann, H., Scharf, H.D., Hoffmann, N., Esser, P. (1991). *Angew. Chem., Int. Ed. Engl*, 30, 477.
[88] D'Auria, M., Emanuele, L., Racioppi, R. (2003). *Photochem. Photobiol. Sci*, 2, 904.
[89] Ramamurthy,V.(1991). In *Photochemistry in organized and constrained media*. Ramamurthy, V., Ed. VCH Publishers, Inc., New York, USA, pp. 429-493.
[90] Natarajan, A., Ramamurthy, V. (2006). *Org. Biomol. Chem*, 4, 4533.
[91] Zhang, Y., Xue, J., Gao, Y., Fun, H.K, Xu, J.H. (2002). *J.Chem. Soc., Perkin Trans*, 1, 345.
[92] Žagar, C., Scharf, H. D. (1991). *Chem. Ber*, 124, 967.

[93] Jarosz, S., Zamojski, A. (1982). *Tetrahedron*, 38, 1453.
[94] Tronchet, J. M. J., Baehler, B.(1974). *J.Carbohydrates Nucleosides Nucleotides,* 1, 449.
[95] Schreiber, S. L. (1985). *Science,* 227, 857.
[96] Schreiber, S. L., Satake K. (1986). *Tetrahedron Lett,* 27, 2575.
[97] D'Auria, M., Emanuele, L., Racioppi, R. (2004). *Tetrahedron Lett,* 45, 3877.
[98] D'Auria, M., Emanuele, l., Pace, V., Racioppi, R. (2006). *Lett. Org. Chem,* 3, 350.
[99] Adam, W., Peters, K., Peters, E. M., Stegmann, V. R. (2000). *J. Am. Chem. Soc*, 122, 2958.
[100] Adam, W., Stegmann, V. R. (2001). *Synthesis*, 1203.
[101] Adam, W., Prein, M. (1996). *Angew. Chem., Int. Ed. Engl,*35, 477.
[102] Bach, T., Schröder, J., Harms, K. (1999). *Tetrahedron Lett,* 40, 9003.
[103] Bach, T., Bergmann, H., Harms, K. (1999). *J. Am Chem. Soc,*121, 10650
[104] Yokohama, A., Mizuno, K. (2000). *Org. Lett,* 2, 3457.
[105] Griesbeck, A. G., Bondock, S. (2001). *J. Am. Chem. Soc,* 123, 6191.
[106] Bach, T. (1997). *Liebigs Ann,* 1627.
[107] D'Auria, M., Racioppi, R., Romaniello, G. (2000). *Eur. J. Org. Chem,*3265.
[108] D'Auria, M., Racioppi, R. (2000). *Arkivoc*, 1, 145.
[109] D'Auria, M., Emanuele, L., Poggi, G., Racioppi, R., Romaniello, G. (2002). *Tetrahedron,*58, 5045.
[110] D'Auria, M., Emanuele, L., Racioppi, R. (2004). *Photochem. Photobiol. Sci,*3, 927.
[111] D'Auria, M., Emanuele, L., Racioppi, R. (2005). *Lett. Org. Chem,*2, 132.

In: Photochemistry Research Progress
Editors: A. Sánchez, S. J. Gutierrez

ISBN: 978-1-60456-568-3

Chapter 12

CHLOROPHYLL FLUORESCENCE: A USEFUL TOOL IN BARLEY PLANT BREEDING PROGRAMS

***Hazem M. Kalaji*[1] * *and Peiguo Guo*[2#] ***

[1] Department of Plant Physiology, Warsaw University of Life Sciences - SGGW, Warsaw, Poland

[2] College of Life Science, Guangzhou University, Guangzhou, China.

ABSTRACT

Plant growth and development are dynamic processes which are continuously changed by environmental conditions. Nowadays, with the availability of performance instruments, many physiological parameters, which correlate with plant growth, development and economic yield under biotic and abiotic stresses, are exploited as indicators in plant breeding programs. Chlorophyll *a* fluorescence is a fast, non-invasive, non-destructive and informative tool to study physiological state of photosynthetic apparatus of any photosynthesizing material. It gives the ability to analyze the individual life processes and to develop an understanding of the plant's response to fluctuations in environmental conditions. In this paper, we discuss responses of photosynthetic apparatus of barley plants growing under various abiotic stresses: high and low temperature, high and low light, salinity, heavy metals, nutrient deficiencies and drought. In addition, measurement of basic chlorophyll fluorescence, and quantitative analysis of the fluorescence transient (JIP-test) are discussed. This test helps to explain the stepwise flow of energy through PSII: absorption, trapping and electron transport. This sort of analysis enables an understanding of the relationships between the biophysical side of photosynthesis, signals of the fluorescence and its analytical expressions. The results show the differentiation in extracted and technical fluorescence parameters, quantum efficiencies, specific and phenomenological fluxes of barely plant growing under

---

* Both authors contributed equally to this work

Corresponding author: Tel +86-20-88531563, email guopg@gzhu.edu.cn

unfavorable growth conditions. This work proves that chlorophyll *a* fluorescence is one of the vital tools to monitor PSII reaction to different abiotic stresses. Moreover, it allows stress types that plants experience in certain environmental growth conditions to be distinguished. Different chlorophyll afluorescence parameters are discussed and proposed to be used as trustworthy indicators in selection for abiotic stress tolerance in barley germplasm.

1. Introduction

Plant physiology offers many physiological parameters to be exploited in different plant science fields like agriculture, agronomy, forestry and horticulture to study abiotic and biotic stress effects on plant growth and final yield. In the last few years, some physiological issues such as the photosynthetic efficiency of plants have started to be investigated within plant breeding programs to evaluate the performance of crops when growing under various unfavorable environmental conditions. One of these parameters is chlorophyll a fluorescence which indicates the capacity of plants to convert light energy to biochemical energy during the photosynthetic process. The advantages of this technique are that, it is non-invasive, non-destructive and rapidly measured using highly portable equipment. A typical measurement of one second duration provides a lot of information related to the transient fluorescence induction curve, which is observed when a plant is strongly illuminated after dark adaptation. This allows large scale screening of crops at throughput rates of many hundreds of samples per hour. Many breeders have already applied this technique in their research as a biomarker, bioindicator or selection criterion and proved that it provides reliable information about plant photosynthetic efficiency which is highly correlated to the final yield of crops. However, there are many different chlorophyll fluorescence parameters which may be used as reliable indicators and the objective of this work is to discuss some of those parameters and their use in selection for abiotic stresses tolerance in barley germplasm.

2. Chlorophyll Fluorescence Measurements

Chlorophyll fluorescence allows us to study the different functional levels of photosynthesis indirectly (processes at the pigment level, primary light reactions, thylakoid electron transport reactions, dark enzymatic stroma reactions and slow regulatory processes). It is useful to study the effects of environmental stresses on plants since photosynthesis is often reduced in plants experiencing adverse conditions, such as water deficit, temperature, nutrient deficiency, polluting agents, and attack by pathogens (Araus et al., 1998). Therefore, analysis of chlorophyll fluorescence parameters is considered as an important approach for evaluating the health or integrity of the internal apparatus during photosynthetic process within a leaf (Krause and Weiss, 1991; Clark et al., 2000), and provides a rapid and accurate technique for detecting and quantifying the tolerance of plants to stress (Li et al., 2006). Moreover, a new trend in plant science (neural networking) showed that, measurement of chlorophyll fluorescence induction curves is one of the most promising techniques for automatic identification of plant species and should also be useful with other pattern recognition problems where a priori unknown characteristics are extracted from a large feature space (Codrea et al., 2003). chlorophyll *a* fluorescence imaging can be used also on a microscopic level or by contrast in large scale remote sensing (Lawson et al., 2002; Moya and Cerovic, 2004).

Chlorophyll fluorescence is a re-emission of light energy absorbed by chlorophyll molecules or re-emitted red/far red light (mostly from PSII). The extent of chlorophyll fluorescence emission is determined by the two factors: the amount of light absorbed by the sample, competition for absorbed energy by other energy dissipation processes (mainly photochemistry and emission of heat). A reduction in energy use by one dissipation pathway enables greater energy dissipation by alternative pathways. Hence changes in fluorescence emission reflect changes in the efficiency of photochemistry (Hansatech Instruments, 2007). This can be expressed by the figure 1:

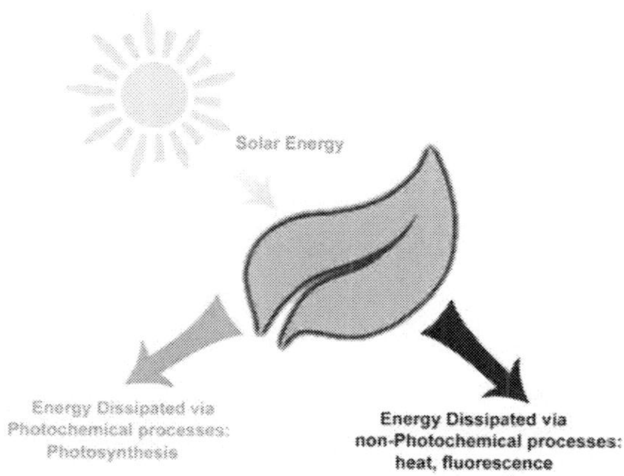

Figure 1. The fate of the absorbed light by photosynthesizing samples: heat, chlorophyll fluorescence and photochemistry output (Courtesy of Hansatech Instruments Ltd., Narborough Road, Pentney, King's Lynn, Norfolk PE32 1JL, England, www.hansatech-instruments.com).

In the last decades, the development of electronics has allowed the introduction of many techniques to measure chlorophyll fluorescence signals from any photosynthesizing material (mainly from chlorophyll a and Photosystem II). Most of the techniques allow investigation in vivo and are considered to be both non-destructive and non invasive. Typically, commercial instrumentation follows one of two approaches:

2. 1 Continuous excitation systems

A continuous excitation fluorimeter is designed to measure the Kautsky Induction or Fast Fluorescence Induction (Kautsky and Hirsch, 1931). In such systems the sample must be shielded from ambient light to prevent the red/far red component of daylight interfering with the measurement. In this case, measurements of chlorophyll fluorescence could be provided after sample dark adaptation in a dark room with very low light or a green safe light or by covering samples with leaf clips for 20-30 minutes. The measured parameters are as follows (Fig. 2):

2.1.1 Fo

The Fo parameter is thought to represent emission by excited chlorophyll a molecules in the antennae structure of Photosystem II. The true Fo level is only observed when the first stable electron acceptor of Photosystem II called Qa is fully oxidised. This requires thorough dark adaptation. Fo occurs at time base 0. It is the almost instantaneous (nanoseconds range) rise to an origin level of fluorescence upon illumination using the fluorimeter.

2.1.2 Fp or Fm

This is the maximum fluorescence value obtained for a continuous light intensity. This parameter may only be termed the maximum fluorescence (Fm) if the light intensity is fully saturating for the plant and the electron acceptor Qa is fully reduced. If the light intensity used for the recording is not sufficiently high, the plant may not be fully saturated in all circumstances and the peak value is obtained (Fp).

2.1.3 Fv

The Fv parameter indicates the variable component of the recording and relates to the maximum capacity for photochemical quenching. It is calculated by subtracting the Fo value from the Fm value.

2.1.4 Fv/Fm

Fv/Fm is a parameter widely used to indicate the maximum quantum efficiency of Photosystem II. This parameter is widely considered to be a sensitive indication of plant photosynthetic performance with healthy samples typically achieving a maximum Fv/Fm value of approx. 0.85. Values lower than this will be observed if a sample has been exposed to some type of biotic or abiotic stress factor which has reduced the capacity for photochemical quenching of energy within PSII. Fv/Fm is presented as a ratio of variable fluorescence (Fv) over the maximum fluorescence value (Fm).

2.1.5 Fv/Fo

It represents the maximum primary yield of photochemistry of photosystem II, and could provide an estimation of leaf photosynthetic capacity.

2.1.6 Tfm

Tfm is a parameter used to indicate the time at which the maximum fluorescence value (Fm) was reached. This parameter may be used to indicate sample stress which causes the Fm to be reached much earlier than expected.

2.1.7 Area

The area above the fluorescence curve between Fo and Fm is proportional to the pool size of the electron acceptors Qa on the reducing side of Photosystem II. If electron transfer from the reaction centres to the quinone pool is blocked such as is the mode of action of the photosynthetically active herbicide DCMU, this area will be dramatically reduced.

The Area measurement is a very useful parameter as it highlights any change in the shape of the induction kinetic between Fo and Fm which would not be evident from the other parameters e.g. Fo, Fm, Fv/Fm which only express changes of amplitude of the extreme Fo and Fm. An example of its use would be following the time dependence of herbicide penetration into the leaf by following changes in the induction kinetic with time (Fig. 2).

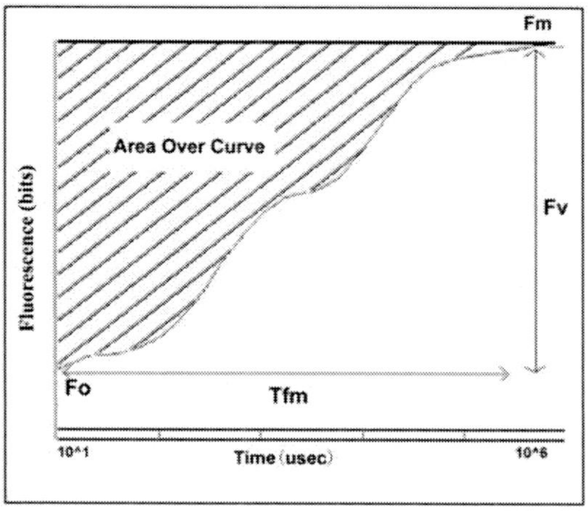

Figure 2. Main chlorophyll fluorescence parameters measured by the use of continuous excitation systems (Courtesy of Hansatech Instruments Ltd., Narborough Road, Pentney, King's Lynn, Norfolk PE32 1JL, England www.hansatech-instruments.com).

2.1.8 Additional Time Mark Parameters

Values of Fluorescence at specific times in the fluorescence induction rise curve have been used by Strasser et al. (2000) with great success for further analysis of the fluorescence data (Fig. 3). Typical times are:

T1= 50 microseconds.
T2= 100 microseconds.
T3= (K step) 300 microseconds.
T4 = (J step) 2 milliseconds.
T5 = (I step) 30 milliseconds.

Further biophysical parameters can also be calculated which quantify the photosystem II behavior for the specific energy fluxes per reaction centre (RC):

Absorption (Abs/RC)
Trapping (TR/RC)
Dissipation (DI/RC)
Electron transport (ET/RC)

or per sample cross section (CS)

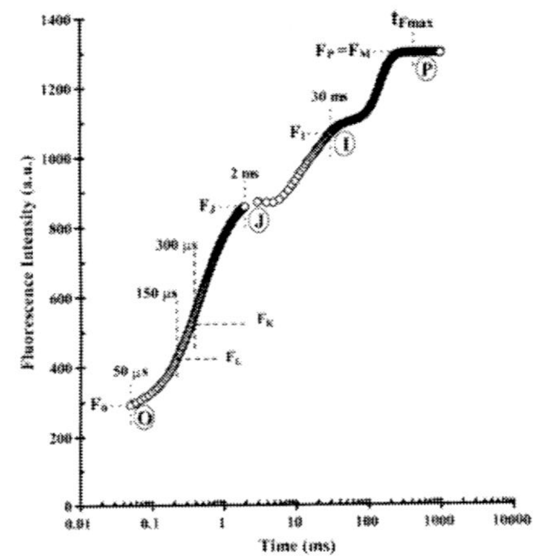

Figure 3. A typical chlorophyll a polyphasic fluorescence rise OJIP, exhibited by plants. The transient is plotted on logarithmic time scale from 50 µs to 1 s. The marks refer to the selected time points used by the JIP-test for the calculation of structural and functional parameters. The signals are: the fluorescence intensity F_O (at 50 µs); the fluorescence intensity F_L(at 150 µs); the fluorescence intensity F_K (at 300 µs); the fluorescence intensities F_J (at 2 ms) and F_I (at 30 ms); the maximal fluorescence intensity, $F_P = F_M$ (at time denoted as t_{FM}) (Strasser et al., 2000 and 2004).

Absorption (Abs/CS)
Trapping (TR/CS)
Dissipation (DI/CS)
Electron transport (ET/CS)

The above parameters and more than 40 others ones were also provided by Strasser et al. (1995, 1999, 2000, 2004) and described as OJIP analysis (JIP-test).

2.1.9 Performance Index Parameter

The Performance Index (PI) is essentially an indicator of sample vitality. It is an overall expression indicating a kind of internal force of the sample to resist constraints from outside. PI (on an absorption basis or on a chlorophyll basis) has three components. The first component shows the force due to the concentration of active reaction centers. The second component is the force of the light reactions, which is related to the quantum yield of primary photochemistry and the third component is the force related to the dark reactions (after Qa^-). These are normal redox reactions in the dark.

Usually, the instruments used to measure this chlorophyll fluorescence are called "plant stress meters" and are typically portable, light and are programmable to analyze measured signals such as Pocket PEA Chlorophyll Fluorimeter from Hansatech Instruments (Photo 1).

Photo 1. The Pocket PEA Chlorophyll Fluorimeter provides an affordable & convenient method of rapidly screening samples using 1 second measurements of the fast chlorophyll fluorescence kinetics (Courtesy of Hansatech Instruments Ltd., Narborough Road, Pentney, King's Lynn, Norfolk PE32 1JL, England www.hansatech-instruments.com).

2. 2. Pulse modulated systems

This technology provides the ability to electronically separate the fluorescence signal in full sunlight or a range of instrument controlled light intensities, thus providing a facility to monitor plant photochemistry under natural field conditions or an instrument controlled light regime to identify performance in a range of likely field conditions (Hansatech Instruments 2007).

In addition to the parameters measured by continuous excitation systems, pulse modulated fluorimeters can typically also measure the following parameters (Table 1):

Table 1. Some measured chlorophyll fluorescence parameters by the use of pulse modulated fluorimeters. (Courtesy of Hansatech Instruments Ltd., Narborough Road, Pentney, King's Lynn, Norfolk PE32 1JL, England, www.hansatech-instruments.com)

Parameter	Measurement	Units	Derivation
PAR	Incident photosynthetically active radiation	$\mu molm^{-2}s^{-1}$	
Temp	Temperature	°C	
Fs	Steady state fluorescence yield	Bits	
Fm'	light-adapted fluorescence maximum	Bits	
ϕ_{PSII}	quantum efficiency of PSII	No units	= (Fm'-Fs) / Fm' (Genty et al. 1989)
qP	photochemical quenching co-efficient	No units	= (Fm'-Fs) / (Fm'- Fo)
qNP	Non-photochemical quenching co-efficient	No units	= (Fm-Fm') / (Fm-Fo)
NPQ	Alternative definition of non-photochemical quenching	No units	= (Fm-Fm') / Fm'
ETR	Electron transport rate	No units	=PAR * 0.5 * ϕ_{PSII} * 0.84

Some of those parameters are measured and the others are calculated and described in the typical experimental trace (Fig 4).

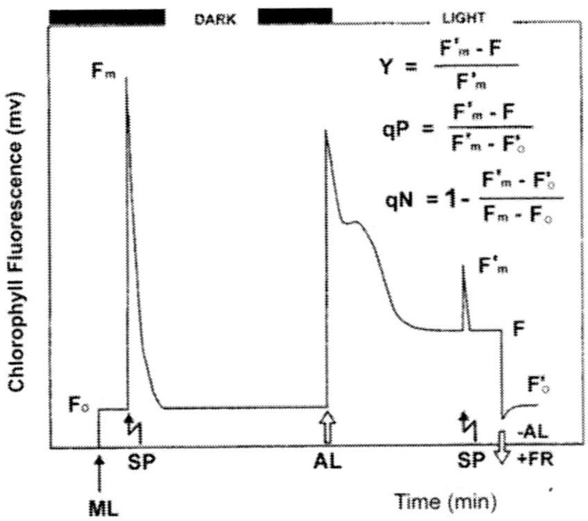

Figure 4. Chlorophyll fluorescence parameters measured by the use of pulse modulated systems. Fo- fluorescence yield of dark-adapted sample, Fm- maximal fluorescence yield of dark-adapted sample (reached during a saturation pulse), F- fluorescence yield observed at any time during illumination; Fm′- maximal fluorescence yield of illuminated sample (reached during a saturation pulse), Fo′- minimal fluorescence yield observed shortly after darkening of illuminated sample and reoxidation of the PS II acceptor side, qP- coefficient of photochemical quenching, qN- coefficient of nonphotochemical quenching, ML- Measuring light, SP- Saturation light pulse, AL- Actinic light, FR- Far red light (van Kooten and Snel, 1990).

The instruments for chlorophyll fluorescence measurements by these types of techniques are called pulse-modulated fluorimeters. A typical example is the chlorophyll fluorescence monitoring system FMS-2 from Hansatech Instruments Ltd. (Photo 2).

Photo 2. FMS-2 is a lightweight, field portable chlorophyll fluorimeter using the pulse-modulated principle. It is suitable for non-invasive sampling under natural light conditions & may be combined with other techniques such as infra-red gas analysis (IRGA) for simultaneous measurements. (Courtesy of Hansatech Instruments Ltd., Narborough Road, Pentney, King's Lynn, Norfolk PE32 1JL, England www.hansatech-instruments.com).

2. 3. The application of chlorophyll fluorescence

According to Strasser et al. (2000), there are four basic groups of questions in which the fluorescent transient can be used. Here the following are applicable:

2.3.1 Chemical stress

Well-known plants under well-defined conditions to analyze the response to different chemicals. Examples of these types of experiments would be effects of allelochemicals, herbicides, exogenous phytoregulators, fertilisers, CO_2, O_2, O_3, SO_2 or other gases concentrations, etc.

2.3.2 Physical stress

Well characterized plants under changing physical environmental parameters. Examples of this would be effects of quality or intensity of light or temperature.

2.3.3 Combined chemical and physical stress

Well-known plant materials can be used to analyze synergisms and antagonisms in the effects of chemical and physical co-stress. These are very relevant types of study in plant ecophysiology.

2.3.4 Biosensing

In these kinds of studies, the environmental conditions are maintained constant and fluorescence transient tests are used to:

2.3.4.1 Describe a plant (or even an ecosystem) in terms of vitality, productivity, and response to stress.
2.3.4.2 Study the relation structure-function in transgenic plants.
2.3.4.3 Study complex systems (from trees to ecosystems).

Strasser et al. (2000 and 2004) even noticed other possible or actual uses of fluorescence techniques ranging from testing of agricultural productivity as influenced by style of agriculture, herbicides, pesticides, hormones, cultivars, environmental factors that can cause stress to even behavior of plants at the ecosystem level to global climatic changes. Other cited uses are related to optimization of culture conditions, testing the degree of fruit ripening or measuring the effects of pollution.

In the last few years, plant breeders have started to be interested to apply chlorophyll fluorescence tools in their research and breeding programs. This will be described in detail in the next paragraphs.

3. Chlorophyll Fluorescence Parameters as Selection Criteria for Cereal Plants

Since breeders started to develop barley plants many physiological, morphological, phenological, or those directly related to final yield components traits were suggested as selection criterion (Slafer et al., 2005; Araus et al., 2002; Atanassov et al., 1999; González et al., 1999; Voltas et al., 1998; Isla et al., 1998; Ellis et al., 1997; Ceccarelli et al., 1991).

Different chlorophyll fluorescence parameters were proposed for selecting stress-tolerant progeny in breeding programs in cereal plants. Munns and James (2003) showed that, leaf injury of wheat plant can be measured chlorophyll fluorescence. They found that the dark-adapted fluorescence parameter Fv/Fm and the light-adapted photochemical quenching parameter NPQ was a sensitive index of salt injury. Research provided by Elhani et al. (2000) suggested that, recording of Fo-temperature curves could provide a rapid means of determining heat tolerance in durum wheat. In particular, the Fv/Fm ratio measured at 60°C could be a rapid and sensitive test to identify genotypes highly tolerant to terminal drought conditions typical of Mediterranean environments (Fig. 5).

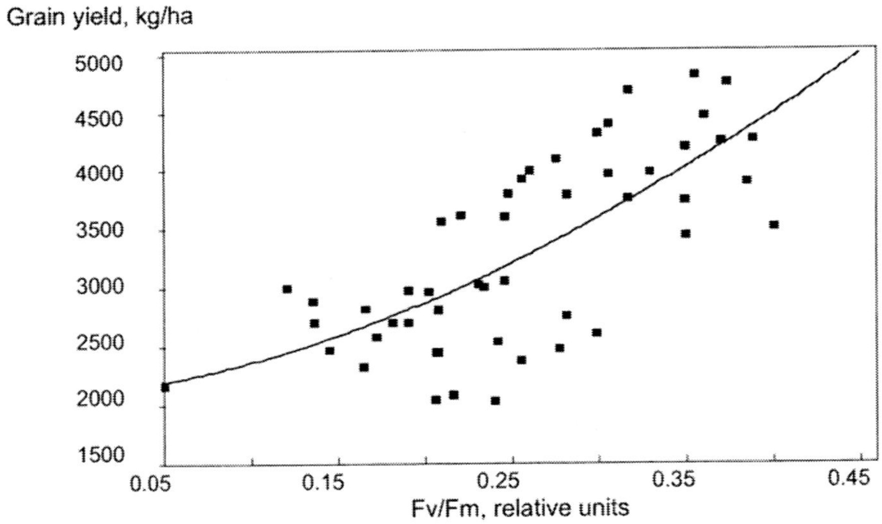

Figure 5. Regression of grain yield on Fv:Fm values measured on the flag leaves of 25 genotypes of durum wheat subjected to 60°C during 20 min in a growth chamber ($Y = 2070 + 10794\ X^{1.64}$, $r = 0.721^{***}$, $n = 50$) (Elhani et al., 2000).

Rapacz (2007) revealed that, chlorophyll *a* fluorescence measurement is a good evaluator of freezing tolerance in winter wheat plants. Analyzing the polyphasic transience (from 50 μs to 1 s) of chlorophyll *a* fluorescence he was able to get detailed studies of the progressive energy flows and efficiencies within the photosystem 2 (PS2) complex that ensue following initial exposure to freezing temperatures right through to the plant recovery stage. Freezing damages were reflected in decreased the Performance Index parameter based on equal absorption that characterizes all energy bifurcations in PS2. In contrast, the potential quantum yield of PS2 (Fv/Fm), which characterizes efficiency of energy trapping in the PS2 reaction centres (RCs), was only different in cultivars with contrasting freezing-tolerance during the recovery stage.

Some other chlorophyll fluorescence parameters along with chlorophyll content were also recommended as tools for improving the efficiency of durum wheat selection under Mediterranean conditions. Royo et al. (2000) indicated that, chlorophyll fluorescence analysis may provide a sensitive indicator of stress conditions in plants. It can also be used to estimate the activity of thermal energy dissipation in photosystem II, which protects photosynthesis from the adverse effects of light and heat stress. In their study, genotypic differences were found for all of the chlorophyll fluorescence parameters measured during the grain filling period, with the exception of the time of emission of Fm, which only varied significantly between environments and years. These two factors explained most of the variability found in the chlorophyll fluorescence parameters (Table 2). In this study, variations in the values of Fv (and therefore of the relationship Fv/Fm) were more associated with variations of Fm than of Fo, thus indicating the possible existence of a photoprotector mechanism in genotypes that are more productive under drought conditions. In addition, the significant relationship found between grain production and Fm, Fv and the relationship Fv/Fm suggest that these parameters could be used as easy and rapid tools for evaluating yield performance of durum wheat genotypes under Mediterranean conditions.

Table 2. Percentage of the sum of squares of the analyses of variance for different physiological characteristics measured at grain-milk stage, explained by the main factors in the analyses.

Factor	Canopy Temperature	Chlorophyll Content (SPAD)	Chlorophyll fluorescence*			
			Fo	Fm	Fv	Fv/Fm
Year	0-15	4-24	50	1	0	31
Environment	7-58	9	21	64	64	26
Genotype	1-2	20-35	2	2	2	4

*only determined at Granada site

Canopy temperature was measured with an infrared thermometer; chlorophyll content was determined with the SPAD, and chlorophyll fluorescence parameters were determined from data recorded with a fluorimeter. Measurements were made in 25 durum wheat genotypes at Lleida during three years and at Granada during two years (Royo et al. 2000)

Studying the effects of soil drought on crop yield of 4 strains and 7 cultivars of spring triticale Grzesiak et al., (2003) observed large differences in chlorophyll fluorescence parameters among studied forms. Screening of Fv/Fm values changes was one of the most suitable techniques for estimation of degree of drought tolerance for triticale. The statistically significant correlation between leaf water potential (ψ) was observed only with leaf fluorescence (Fv/Fm). Moreover, they proved that, changes of gaseous exchange parameters were not useful as screening test in this research.

The research of Moustakas et al. (1993) related to the study of the effects of aluminum on the photosynthetic apparatus of different cereals showed that, measurements of chlorophyll fluorescence *in vivo* could be used to monitor injury caused by "Al-stress", and thus they may serve as a rapid screening test for Al tolerance. Chlorophyll fluorescence kinetics confirmed that the soft wheat 'BHG' cultivar has the potential for growth on acid soils while triticale cultivars 'Niovi' and 'Dada' appeared to be relatively tolerant. The percentage decrease in Fv/Fm of the less tolerant cultivars after Al-treatment indicated a decrease in the efficiency of the primary photochemistry of PS II, while the decrease in the ratio Fv/Fo suggested that exposure of the cultivars 'Dio' and 'Appulo E' to aluminum caused injury to the thylakoid structure.

Zhao et al. (2007) indicated that there is great variability for salt tolerance among naked oat germplasms, and greater photosynthesis capacity based on chlorophyll fluorescence measurements appeared to be associated with salt tolerance in naked oats while Morant-Manceau et al. (2004) confirmed that, chlorophyll fluorescence along with gas exchange measurements are good indices for a hexaploid triticale and its parental species screening under salt stress conditions.

Rizza et al. (2001) used chlorophyll fluorescence to evaluate the cold acclimation and freezing tolerance of winter and spring oats. They measured the efficiency of the excitation capture by open Photosystem II (PSII) reaction centres by the Fv/Fm ratios. A significant reversible decrease in Fv/Fm was found in all genotypes during acclimation to low, non-freezing temperatures. This research underlined that, Fv/Fm analysis appears to be an attractive test for the evaluation of frost tolerance in oats, being rapid, non-invasive and capable of monitoring a trait related to a crucial stage in the acquisition of frost tolerance. Moreover, they proved that, chlorophyll fluorescence is a proper tool for stress detection in early stages of stress and it is more sensitive and precise than other standard methods and highly correlated with field-evaluated frost damage.

Gray et al. (1997) proved that, some measured chlorophyll fluorescence such as: Fm, Fm' (maximum fluorescence yield in the dark-adapted and light-adapted states, respectively); Fo, Fo' (minimal fluorescence yields in the dark-adapted and light-adapted states, respectively); Fs (steady-state fluorescence yield); Fv, Fv' (variable fluorescence yield in the dark-adapted and light-adapted states, respectively); Fv/ Fm, maximal photochemical yield of PSII in the dark-adapted state; qN (coefficient of nonphotochemical quenching); q_P (coefficient of photochemical quenching); 1-q_P (the relative reduction state of PSII) are useful for screening cold Acclimation and freezing tolerance in winter rye.

4. Chlorophyll Fluorescence Parameters as Selection Criteria for Barley

In the last few years, many works showed that, some physiological traits which are related to photosynthetic activity of the plant could be also used in barley selection programs (Kalaji and Loboda, 2007; Tóth et al., 2007; Ashraf, 2004; Kalaji and Pietkiewicz, 2004; Munns and James, 2003; Araus et al., 1998). In the frame of these researches it was proved that, some of chlorophyll fluorescence parameters are good indicators of adaptation of barley to different unfavorable growth conditions thus; could be treated as reliable selection element during barley plant breeding programs (Kalaji and Loboda, 2007; Oukarroum et al., 2007; Li et al., 2006; Jiang et al., 2006; Bort et al., 1998).

Planchon et al. (1989) recommended chlorophyll fluorescence transient to be used as a genetic marker of productivity in barley. They studied the photosynthetic characteristics of the penultimate leaf in barley genotypes (reciprocal F1 combinations and varieties) at the flowering stage in order to analyze the relationship between these traits and the biological yield and the economic yield. They found that, the biological yield was correlated with the leaf area whereas the economic yield was more dependent on the green area duration of the leaf or the combination of the two factors, leaf area and green area duration. Moreover, the net photosynthesis per unit leaf area, determined at anthesis, was not a factor of productivity. On the other hand, the chlorophyll fluorescence (P), which was negatively correlated with the leaf area (+0.5) and with the combination of leaf area and green area duration (+0.5), was found to be a good marker of the economic yield. In view of its high heritability, it may provide an effective selection criterion (data not shown).

Kalaji and Loboda (2007 and un-published data) conducted many experiments on barley seedlings (two Syrian landraces: Arabi Aswad and Arabi Abiad) to underline the proper chlorophyll fluorescence parameters adequate for plant selection towards tolerance to different abiotic stresses.

Beside measurements of transient chlorophyll signals and basic chlorophyll fluorescence parameters, to get more information about energy flux and its fate in photosystem II (PSII), JIP-test was applied (see above paragraph 2.1). Plants photosynthetic activity under different abiotic stresses (high and low temperature, light, salinity, heavy metals, and nutrient deficiency) was studied (Table 3). This was done to explain the stepwise flow of energy through PSII at the reaction centre (RC) level (ABS/RC, TR/RC, ET/RC and DI/RC) as well as at the level of a measured cross section (CS) (ABS/CS, TR/CS, ET/CS and DI/CS). Pipeline and leaf models and radar illustrations were applied to show barley plant reactions to different unfavorable growth conditions.

Table 3. Description of 14 different applied abiotic stresses and the way to achieve the requested level of each of them which caused about 50% of growth reduction after 14 days from germination

Stress group	Stress	Level of stress
Temperature	Low Temperature	10/6°C for day/night
	High Temperature	40/20°C for day/night
Light	Low PAR	200 μmol photon m^{-2} s^{-1}
	High PAR	1800 μmol photon m^{-2} s^{-1}
Salinity	Salt Stress	NaCl added to final concentration of 120 mM
Heavy Metals	Cadmium	$CdCl_2$ added to final concentration of 25 μM
	Lead	$Pb(NO_3)_2$ added to final concentration of 50 μM
Nutrients deficiency	Calcium	Modified Hoagland nutrients -Ca
	Sulphur	Modified Hoagland nutrients -S
	Magnesium	Modified Hoagland nutrients -Mg
	Potassium	Modified Hoagland nutrients -K
	Nitrogen	Modified Hoagland nutrients -N
	Phosphorus	Modified Hoagland nutrients -P
	Iron	Modified Hoagland nutrients -Fe

Table 4. Some chlorophyll fluorescence parameters of two barley cultivars (cv. Arabi Aswad and cv. Arabi Abiad) grown under different abiotic stresses as % of control after 24 hours of stress application (see paragraph 2.1 for description of abbreviations)

Stress	cv. Arabi Aswad					cv. Arabi Abiad				
	Tf(max)	Area	Fv/Fm	Fv/Fo	PI(abs)	Tf(max)	Area	Fv/Fm	Fv/Fo	PI(abs)
Low Temp.	100.00	96.09	97.61	90.01	65.41	64.29	60.77	93.64	75.85	36.72
High Temp.	100.00	98.34	99.54	97.95	94.59	3.57	0.63	21.71	5.58	0.68
Low PAR	142.86	69.95	94.16	78.04	63.81	85.71	71.97	99.21	96.40	84.48
High PAR	103.57	94.83	97.09	88.04	70.21	64.29	63.99	95.92	83.33	49.33
+NaCl	85.71	85.25	99.04	95.79	82.68	100.00	98.82	98.87	94.87	91.19
+Cd	0.57	0.01	0.63	0.14	0.00	0.82	0.01	0.71	0.15	0.00
+Pb	96.43	54.85	95.72	83.11	48.51	96.43	76.30	97.18	87.99	74.44
-Ca	89.29	65.94	95.74	83.22	56.55	178.57	84.85	97.19	88.06	76.01
-S	96.43	70.45	95.70	83.08	55.65	107.4	87.80	96.21	84.38	72.19
-Mg	103.00	61.87	94.20	78.16	52.20	285.71	84.29	96.36	84.93	66.90
-K	85.71	64.14	97.78	90.64	62.08	85.88	72.21	95.64	82.39	52.67
-N	92.86	89.20	97.97	91.42	86.27	100.00	87.91	97.98	91.14	76.10
-P	103.00	78.37	97.70	90.37	65.11	107.14	81.72	96.42	85.12	58.17
-Fe	178.57	53.41	95.24	81.49	50.27	250.00	87.77	96.07	83.90	64.28

Table 5. Some chlorophyll fluorescence parameters of two barley cultivars (cv. Arabi Aswad and cv. Arabi Abiad) grown under different abiotic stresses as % of control after 7 days of stress application (see paragraph 2.1 for description of abbreviations)

Stress	*cv.* Arabi Aswad					*cv.* Arabi Abiad				
	Tf(max)	Area	Fv/Fm	Fv/Fo	PI(abs)	Tf(max)	Area	Fv/Fm	Fv/Fo	PI(abs)
Low Temp.	103.57	85.12	95.26	81.47	73.18	178.57	69.36	94.34	78.78	58.08
High Temp.	178.57	30.04	31.48	9.13	1.57	285.00	80.98	53.64	20.50	6.12
Low PAR	107.14	17.34	38.91	12.23	2.12	142.86	61.79	91.20	69.78	37.39
High PAR	96.43	100.51	97.35	88.92	80.62	285.71	40.32	55.93	22.05	6.32
+NaCl	214.29	65.16	90.08	66.53	40.19	67.86	8.67	24.58	6.77	0.69
+Cd	-	-	-	-	-	1.04	0.01	0.91	0.21	0.00
+Pb	82.14	18.63	50.38	18.18	3.49	100.00	68.78	93.90	77.43	51.79
-Ca	78.57	13.67	74.56	39.08	7.02	103.57	30.74	67.39	31.53	4.31
-S	103.00	87.82	89.77	65.76	50.63	103.00	85.95	95.29	81.87	67.96
-Mg	100.00	13.23	53.58	20.16	3.20	92.86	43.91	84.22	54.33	23.65
-K	92.86	7.43	32.94	9.70	0.62	107.14	15.88	41.47	13.64	1.16
-N	142.86	88.19	89.61	65.36	51.69	178.57	51.20	76.36	41.86	12.34
-P	96.43	74.10	91.96	71.47	50.44	100.00	86.84	94.02	77.78	56.65
-Fe	85.71	4.12	25.24	6.88	0.50	100.00	43.28	80.27	47.57	16.32

The obtained results after 7 hours and 7 days of stress application showed different values of extracted and technical fluorescence parameters (Tables 4 and 5), quantum efficiencies, specific and phenomenological fluxes for each cultivar.

All applied stresses decreased PSII activity of barley seedlings. However, this depended on stress type and its duration. Phenomenological parameters (energy absorption, trapping and electron transport calculated on the basis of cross section) were shifted shortly after abiotic stresses application while after 7 days of stresses fluorescence ratios and yield values were mainly altered. The earliest effects of stresses on PSII activity were observed in the case of high temperature and cadmium treatment (Fig. 6). The most pronounced changes in most of studied chlorophyll fluorescence parameters were found for potassium deficiency. Both cultivars were relatively sensitive to potassium, calcium, magnesium and iron deficiencies. Cv. Arabi Aswad was more tolerant to high photosynthetic active radiation (PAR) and salinity while cv. Arabi Abiad was more tolerant to low PAR stress (data not shown).

The experiments allowed selection of those chlorophyll a fluorescence parameters which were changed most significantly under a specific stress and in consequence those that could be useful as selection criteria in barely plant breeding programs.

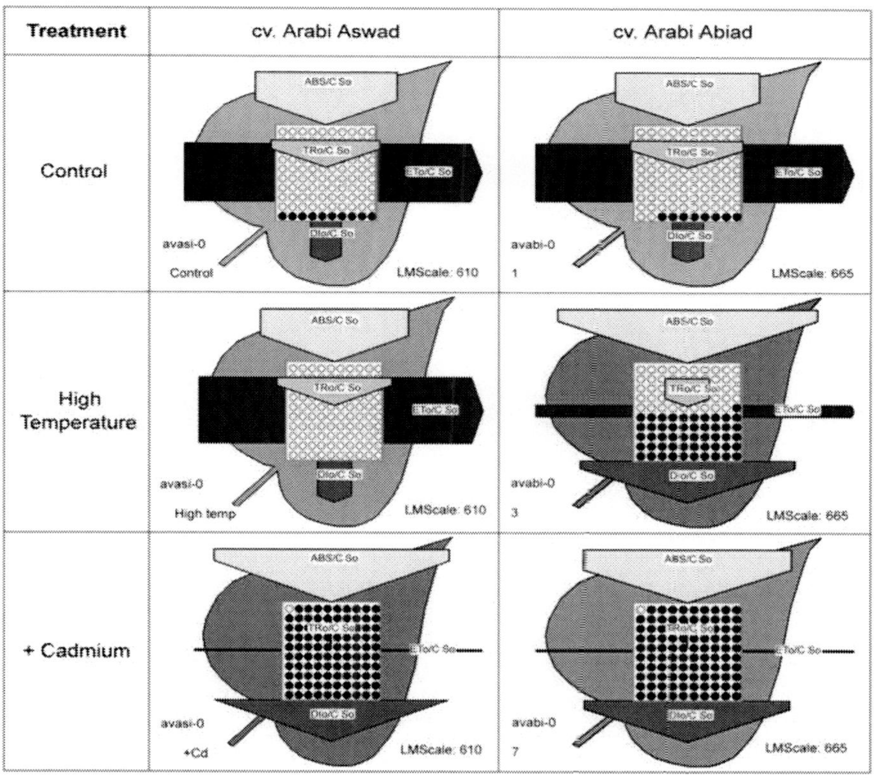

Figure 6. Leaf model showing phenomenological energy fluxes per excited cross section (CS) after 24 hours for barley cvs. Arabi Abiad and Arabi Aswad grown without stress (control) and under high temperature and cadmium stresses at t=0. ABS/CS -absorption flux per CS approximated by Fo, TR/CS - trapped energy flux per CS, ET/CS - electron transport flux per CS, DI/CS - dissipated energy flux per CS. Each relative value is represented by the size of proper parameters (arrow), empty circles represent reducing Q_A reaction centers (active), full black circles - non-reducing Q_A reaction centers (inactive or silent). Green color intensity of leaves is proportional to its chlorophyll content calculated by Biolyzer software (Bioenergetics laboratory, Geneva University, Switzerland).

Oukarroum et al. (2007) evaluated the effects of drought and re-watering on 10 varieties of barley (*Hordeum vulgare* L.) originating from Morocco. The polyphasic OJIP fluorescence transient was used to evaluate photosystem II (PSII) criteria at the end of the first week of drought stress (moderate drought), at the end of the second week (severe drought) and the end of the recovery phase. Drought and re-watering had little effect on the maximum quantum yield of primary photochemistry (Fv/Fm). The photosynthetic performance index (PI) is the product of an antenna, reaction center and electron transport dependent parameter. It revealed differences between varieties as a function of drought and re-watering. They denoted that, fluorescence rise during the first 300 μs (L-band) can give information on the energetic connectivity between PSII units whereas changes in the rise during the first 2 ms (K-band) offer information on developing limitations on the donor side of PSII. Changes in respectively the L and K-bands of the fluorescence transients OJIP were shown to have predictive value with respect to the vitality of leaves and the tolerance of the varieties to drought stress (Fig. 7).

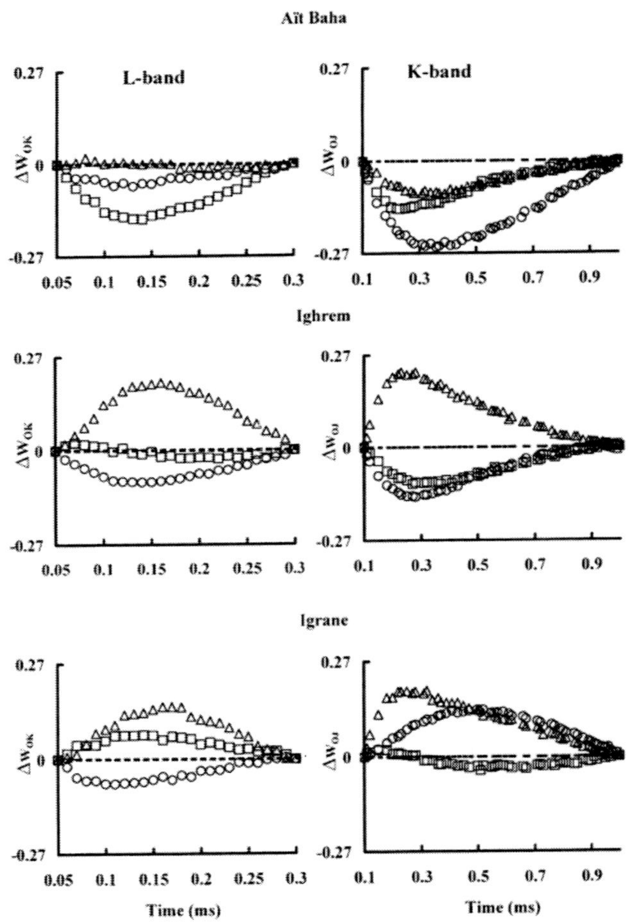

Figure 7. Change in the shape of the chlorophyll a fluorescence transient curves normalized between Fo and F_K expressed as W_{OK} (W_{OK} = (Ft −Fo)/(Fk −Fo)) and between Fo and F_J expressed as W_{OJ} (W_{OJ} = (Ft −Fo)/(F_J −Fo)). □W_{OX} =$W_{OX(treatment)}$ −$W_{OX(control)}$ where treatment is control (dash), moderate drought (square), severe drought (triangle), recovery (circle). (Oukarroum et al., 2007).

The results obtained by Pshybytko et al. (2003) led to conclusion that there are various mechanisms of heat damage to photosynthetic processes in barley seedlings of different ages. According to the data obtained the heating of young leaves inhibited the photosynthesis by limitation of the electron transport rate while in old leaves the enzymes of Calvin cycle were the most vulnerable (Table 6). They indicated that, chlorophyll fluorescence measurements help to identify the effect of heat on barley plants which cane be exploited in barely breeding programs.

Table 6. Effect of high temperature on the parameters of chlorophyll fluorescence (modified after Pshybytko et al., 2003)

Age (days)	Variant	Fv/Fm	Φ_{PSII}	qP	NPQ
4	Control	0.79	0.67	0.79	0.74
	Stressed	0.76	0.57	0.78	0.57
	% to control	97	86	98	77
7	Control	0.82	0.63	0.82	0.92
	Stressed	0.80	0.59	0.80	0.68
	% to control	98	93	98	74
11	Control	0.80	0.60	0.81	1.09
	Stressed	0.78	0.59	0.79	1.45
	% to control	98	97	98	133

The change of chlorophyll fluorescence parameters in froze leaves of 3 leaf-age seedlings were examined using two winter barley cultivars (Chumai 1 and Mo 103) differing in cold tolerance to investigate physiological response to low temperature as affected by cold (Dai et al., 2007). The results showed that non-lethal freezing shock (exposed to −8 °C for 18 h) did not obviously affect maximum quantum efficiency in photosystem II (PSII), but dramatically increased non-photochemical quenching and reduced effective quantum yield in PSII. Cold acclimation significantly improved stability of photosynthetic function of leaves after freezing stress through buffering excessive energy and alleviating photoinhibition during recovery, indicating it increased recovery ability of barley plants from freezing injury. High irradiance was quite harmful to the stability of PSII in barley plants during recovery from freezing injury. The electron transport rate of PSII varied with cold-acclimation, irradiance and genotype. Cold acclimation caused significant increase in electron transport rate of PSII for relatively tolerant cultivar Mo 103, but not for relatively sensitive cultivar Chumai 1 (Fig. 8 and 9). They concluded that some chlorophyll fluorescence parameters during recovery from freezing shock may be used as the indicators in identification and evaluation of cold tolerance in barley.

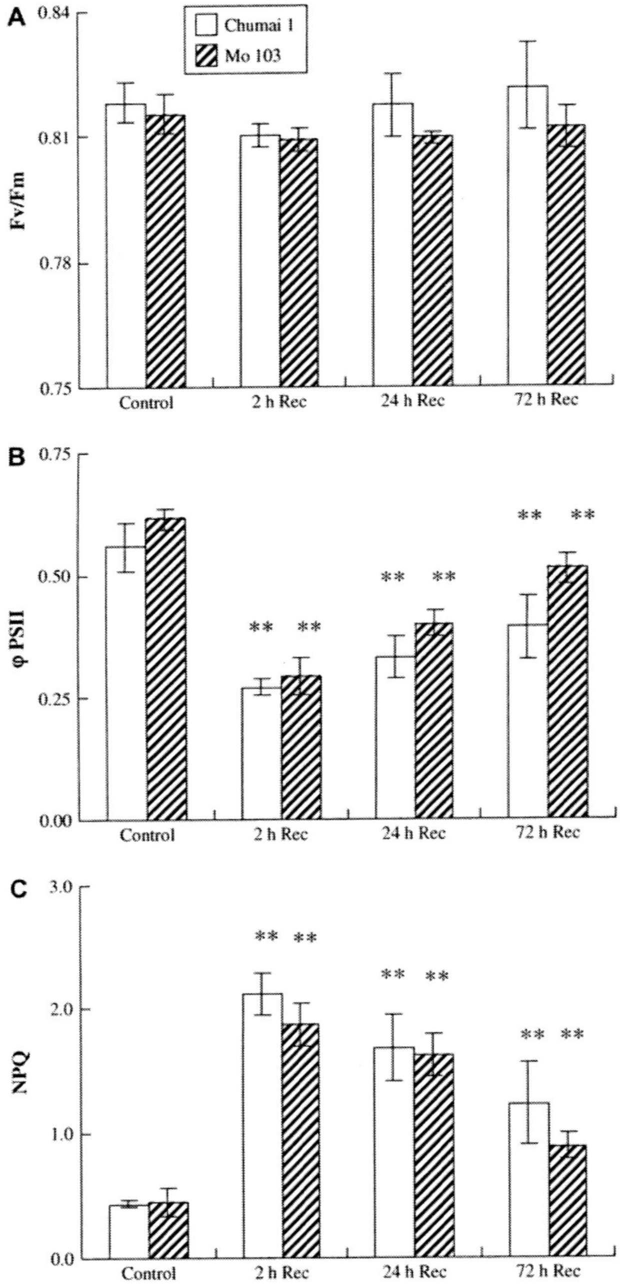

Figure 8. Effects of freezing treatment (exposed to -8 °C for 18 h) on chlorophyll *a* fluorescence parameters of barley leaves after freezing treatment and during recovery process. Control: no freezing treatment; Rec: recovery; ** significant at the 0.01 probability levels as compared to the control; Bars indicate S.D. (n=3). (Dai et al., 2007)

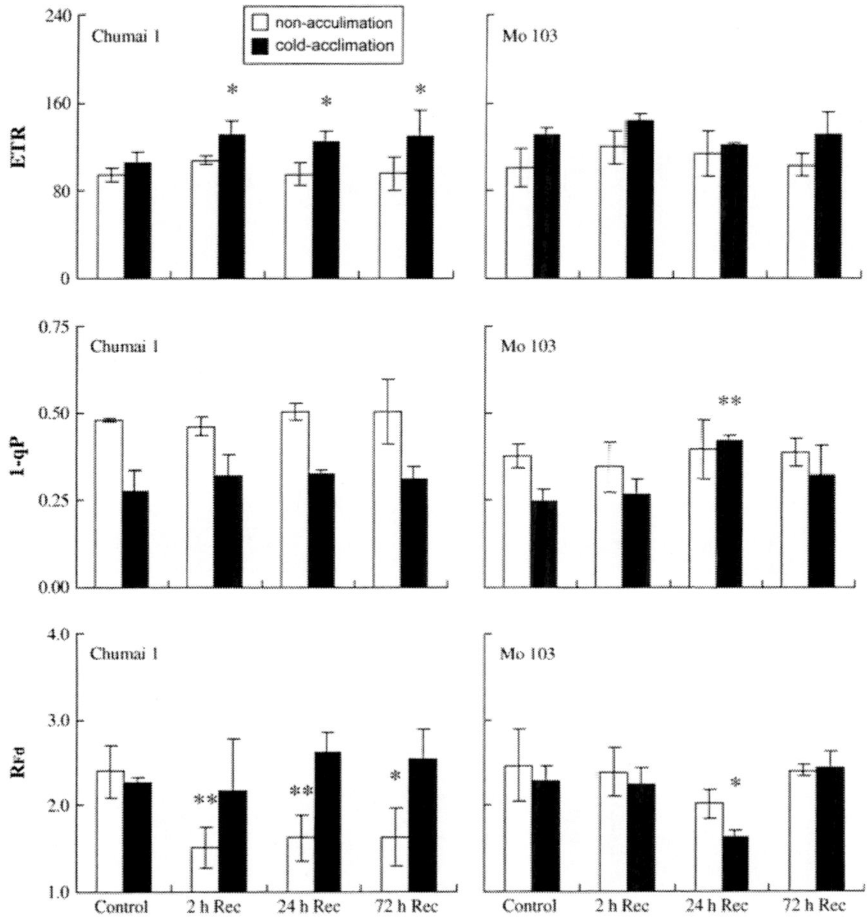

Figure 9. Effects of freezing treatment (exposed to -8 °C for 18 h) on chlorophyll *a* fluorescence parameters of cold- and non-acclimated barley leaves under low irradiance (PPFD: 225 ± 25 mmol m^{-2} s^{-1}). Control: no freezing treatment, Rec: recovery; *, ** significant at the 0.05 and 0.01 probability levels as compared to the control, respectively. Bars indicate S.D. (n =3). (Dai et al., 2007)

Li et al. (2006) identified some indicators related to drought tolerance through analysis of photosynthetic traits in barley. These traits included chlorophyll content, initial fluorescence (Fo), maximum primary yield of photochemistry of photosystem II (Fv/Fo) and maximum quantum yield of photosystem II (Fv/Fm). Four genotypes (Tadmor, Arta, Morocco9-75 and WI2291) variable in drought tolerance were used to investigate the correlation between these traits and drought tolerance. The obtained results showed that all of these traits were affected negatively in the four genotypes at different levels of post-anthesis drought stress, but the decrease in drought tolerant genotypes was much less than that of drought sensitive genotypes. Components of the photosynthetic apparatus could be damaged significantly in drought sensitive genotypes, while drought tolerant genotypes were relatively less affected. On the other hand, the values of chlorophyll content, Fo, Fv/Fo and Fv/Fm in drought tolerance genotypes were significantly higher than those in drought sensitive genotypes under drought stress (Table 7). They concluded that above mentioned parameters could be considered as reliable indicators in screening barley germplasm for drought tolerance.

Table 7. Effect of drought stress on chlorophyll content and chlorophyll fluorescence parameters in barley (Li et al. 2006)

Genotype	Chlorophyll content (mg/g fresh weight)		Fo		Fv/Fo	
	Control	Drought	Control	Drought	Control	Drought
Tadmor	2.42±0.11a	2.16±0.15b (*)	241.1±23a	237.5±29a (ns)	3.76±0.31a	3.23±0.28a (*)
Arta	2.51±0.09a	2.47±0.21a (ns)	233.4±19a	238.2±31a (ns)	3.91±0.41a	3.84±0.33a (ns)
Morocco9-75	2.33±0.09a	1.6±0.18c (**)	237.2±26a	192.8±21b (**)	3.96±0.22a	1.85±0.32c (**)
WI2291	2.46±0.13a	1.72±0.22c (**)	228.7±21a	186.5±27b (**)	3.83±0.27a	2.35±0.37b (**)

Letters a, b, c in same column indicate significances of Duncan test with P< 0.05 in genotypes under well-watered (control) and drought stress conditions, respectively. * and ** and ns with braces indicate significance for each trait within each genotype between control and drought stress in the level of 0.05, 0.01 and no significance, respectively. The data are the mean value and standard deviation of 6 duplicates.

Table 8. Main characteristics of QTLs with an LOD score > 2.5 for fluorescence parameters and chlorophyll content (Chl) in plants grown under well-watered conditions (Guo et al., 2008)

Trait [a]	QTL	Chromosome number	RP [#]	Peak *	Nearest Marker	LOD	% Expl. **	Additive effect [b]
Fo	$QcF_o1.1$	1H	RP1/RP2	76.1/75	e35m47-02	2.6/2.55	6.6/6.2	-6.89/-6.13
	$QcF_o2.1$	2H	RP1	38	BMag0378	2.5	6.4	6.69
	$QcF_o7.1$	7H	RP2	148.1	e35m47-01	2.5	6.2	6.21
Fm	$QcF_m2.1$	2H	RP1	1	e38m59-04	3.5	9.5	-57.33
	$QcF_m4.1$	4H	RP1	33.7	e32m48-08	2.8	7.1	-49.77
	$QcF_m5.1$	5H	RP1/RP2	13.7/15.7	p71m82-01	2.7/3.1	8/8.2	53.35/55.6
Fv	$QcF_v2.1$	2H	RP1	1	e38m59-04	3.5	9.4	-59.35
	$QcF_v5.1$	5H	RP1/RP2	15.7/13.7	p71m82-01	2.8/3.1	7.4/8.3	53.78/66.74
	$QcF_v6.1$	6H	RP1	98.3	e35m48-08	2.5	8.4	-56.04
Fv/Fm	$QcF_v/F_m2.1$	2H	RP1	1	e38m59-04	2.6	7	-0.012
	$QcF_v/F_m2.2$	2H	RP1	30.8	BMag0140	3	9.9	-0.014
	$QcF_v/F_m5.1$	5H	RP1/RP2	15.9/14.9	p71m82-01	2.5/2.8	6.3/7.6	0.011/0.015
Chl	QcChl2.1	2H	RP1	35.1	BMag0140	3.3	9.8	1.01
	QcChl2.2	2H	RP1/RP2	115/115	HVM54	4.4/4.8	12.8/13.6	-1.15/-1.34
	QcChl4.1	4H	RP1/RP2	82.9/81.9	p71m88-02	3.5/3.4	10.2/9.6	-1.03/-1.11

a: Fo = initial fluorescence (fluorescence level when plastoquinone electron acceptor pool (Qa) is fully oxidized), Fm = maximal fluorescence (fluorescence level when Qa is transiently fully reduced), Fv = variable fluorescence (Fm- Fo), Fv/Fm = maximum/potential quantum efficiency of PSII.

b: Negative values indicate effects from *H. spontaneum* 41-1; positive values indicate effects from Arta

#: RP presents replicate experiment.

*Peak = position of the peak of the QTL in centiMorgans

**%Expl. = % of phenotypic variance explained by the QTL

Table 9. Main characteristics of QTLs with an LOD score > 2.5 in plants grown under drought conditions (Guo et al., 2008)

Trait[a]	QTL	Chromosome number	RP[#]	Peak *	Nearest Marker	LOD	% Expl. **	Additive effect [b]
Fo	$QtF_o2.1$	2H	RP1/RP2	116/115.5	HVM54	3.9/3.7	13.5/12.6	-14.85/-13.9
	$QtF_o4.1$	4H	RP1	0	HVM40	3.0	7.9	11.23
	$QtF_o6.1$	6H	RP1/RP2	86.5/87.5	e38m59-06	2.8/2.6	7.0/7.4	-10.62/-11.73
	$QtF_o7.1$	7H	RP1	85.2	e38m59-17	3.3	9.0	-12.02
Fm	$QtF_m2.1$	2H	RP1/RP2	115/116	HVM54	5.4/5	15.1/15.7	85.24/105.75
	$QtF_m2.2$	2H	RP1/RP2	135.7/137.7	p71m24-01	2.6/2.51	9.2/8.8	66.35/65.73
Fv	$QtF_v2.1$	2H	RP1/RP2	115/116	HVM54	5.7/5	16.3/15.9	94.16/103
	$QtF_v2.2$	2H	RP1/RP2	135.7/137.7	p71m24-01	2.7/2.55	9.3/8.6	71.14/70.52
Fv/Fm	$QtF_v/F_m2.1$	2H	RP1/RP2	116/116	HVM54	5.2/3.6	15.5/15.2	0.069/0.067
	$QtF_v/F_m2.2$	2H	RP1/RP2	135.7/137.7	p71m24-01	2.8/2.61	9.3/8.9	0.053/0.051
Chl	QtChl2.1	2H	RP1/RP2	114/116	HVM54	3.6/2.7	9.9/8	1.93/1.67
	QtChl2.2	2H	RP2	137.7	p71m24-01	2.6	8.8	1.82

a: Fo = initial fluorescence (fluorescence level when plastoquinone electron acceptor pool (Qa) is fully oxidized), Fm = maximal fluorescence (fluorescence level when Qa is transiently fully reduced), Fv = variable fluorescence (Fm- Fo), Fv/Fm = maximum/potential quantum efficiency of PSII, chl = chlorophyll content (mg/g fresh weight)

b: Negative values indicate effects from *H. spontaneum* 41-1; positive values indicate effects from Arta

#: RP presents replicate experiment.

* Peak = position of the peak of the QTL in centiMorgans

** %Expl. = % of phenotypic variance explained by the QTL

Since the identification of molecular markers linked to genes controlling drought tolerance in barley is one way to improve breeding efficiency (Guo et al., 2008) analyzed the quantitative trait loci (QTL) controlling chlorophyll content and chlorophyll fluorescence in 194 recombinant inbred lines (RILs) developed from the cross between the cultivar 'Arta' and Hordeum spontaneum 41-1. Five traits, chlorophyll content, and four chlorophyll fluorescence parameters, namely initial fluorescence (Fo), maximum fluorescence (Fm), variable fluorescence (Fv), and maximum quantum efficiency of PSII (Fv/Fm) which are related to the activity of the photosynthetic apparatus, were measured under well-watered and drought stress conditions at post-flowering stage. QTL analysis identified a total of nine and five genomic regions, under well-watered and drought stress conditions, respectively, that were significantly associated with the expression of the five target traits at post-flowering stage. No common QTL was detected except one for chlorophyll content, which was identified in both growth conditions, demonstrating that the genetic control of the expression of the traits related to photosynthesis differed under different water conditions. A QTL for Fv/Fm, which is related to the drought tolerance of photosynthesis, was identified on chromosome 2H at 116 cM in the linkage map under drought stress. This QTL alone explained more than 15% of phenotypic variance of maximum quantum yield of PSII, and was also associated with the expression of four other traits. In addition, another QTL for Fv/Fm was also located on the same chromosome (2H) but at 135.7 cM explaining around 9% of the phenotypic variance under drought conditions. The obtained result suggested that two major loci, located on chromosome 2H, are involved in the development of functional chloroplast at post-flowering stage for drought tolerance of photosynthesis in barley under drought stress (Table 8 and 9). They concluded that, if validated in other populations, chlorophyll fluorescence parameters could be used as selection criteria for drought tolerance.

5. Conclusion

All the above researches confirm that chlorophyll fluorescence is a reliable method for determining selection criterion in barley breeding programs.

Acknowledgments

The authors gratefully acknowledge the assistance of Hansatech Instruments Limited, Narborough Road, Pentney, King's Lynn, Norfolk PE32 1JL, England (www.hansatech-instruments.com) for the supply of photographs and images.

References

Araus, J.L., Slafer, G. A., Reynolds, M. P., Royo, C. (2002).Plant breeding and drought in C_3 cereals: What should we breed for? *Ann. Bot*, 89, 925-940.

Araus, J.L.,Voltas, J.A., Nakkoul, H., Nachit, M. M. (1998). Chlorophyll fluorescence as a selection criterion for grain yield in durum wheat under Mediterranean conditions. *Field Crops Res*, 55, 209—223.

Ashraf, M. (2004). Some important physiological selection criteria for salt tolerance in plants. Flora - Morphology, Distribution, Functional Ecology of Plants, 199, 361-376.

Atanassov, P., Zaharieva, M., Merah, O., Monneveux P. (1999). Genetic and environmental variation of useful traits in a collection of naked barley I. Yield components and water use efficiency. *Cereal Res. Comm*, 27, 315-322.

Bort, J., Araus, J.L., Hazzam, H., Grando, S., Ceccarelli, S. (1998).Relationships between early vigour, grain yield, leaf structure and stable isotope composition in field grown barley. *Plant Physiol. Biochem*,36, 889-897.

Ceccarelli, S., Acevedo, E., Grando, S. (1991). Breeding for yield stability in unpredictable environments: single traits, interaction between traits, and architecture of genotypes. *Euphytica*, 56, 169-185.

Clark, A.J., Landolt, W., Bucher, J.B., Strasser R.J. (2000). Beech (Fagus sylvatica) response to ozone exposure assessed with a chlorophyll fluorescence performance index. *Environmental Pollution*, 109, 501-507.

Codrea, C.M., Aittokallio, T., Keränen, M., Tyystjärvi, E., Nevalainen, O.S. (2003). Feature learning with a genetic algorithm for fluorescence fingerprinting of plant species. Pattern Recogn. *Letters,* 24, 2663-2673.

Elhani, S, Rharrabti, Y., Roca, L.S., García del Moral, L.F. (2000). Evolution of chlorophyll fluorescence parameters in durum wheat as affected by air temperature.*Options Méditerranéennes* , Ser. A, 40,275-277.

Dai F., Zhou M., Zhang G. (2007). The change of chlorophyll fluorescence parameters in winter barley during recovery after freezing shock and as affected by cold acclimation and irradiance. *Plant Physiol.Biochem*, 45, 915-921

Elhani, S., Rharrabti, Y., García del Moral, L.F., Roca, L.F. (2000). *Evolution of chlorophyll fluorescence parameters in durum wheat as affected by air temperature*. In Royo C. (ed.), Nachit M.M. (ed.), Di Fonzo N. (ed.), Araus J.L. (ed.) . Durum wheat improvement in the Mediterranean region: New challenges. Zaragoza: CIHEAM-IAMZ. pp.275-277

Ellis, R.P., Forster, B.P., Waugh, R., Bonar, R., Handley, L.L., Robinson, D., Gordon, D.C., Powell, W. (1997). Mapping physiological traits in barley. *New Phytol*, 137, 149–157.

González, A., Martín, I., Eyerbe, L. (1999). The influence of precocity, osmotic adjustment and stomatal conductance. *Field Crops Res*, 62, 23-34.

Gray, G.R., Cauvin, L.P., Sarhan. F., Huner. N.P.A. (1997). Cold acclimation and freezing tolerance. A complex interaction of light and temperature. *Plant Physiology*, 114, 467–474.

Grzesiak, S., Grzesiak, M.T., Filek, W., Stabryła, J. (2003). Evaluation of physiological screening tests for breeding drought resistant triticale (x *Triticosecale* Wittmack). Acta Physiol. *Plantar*, 25, 29-37.

Guo, P., Baum, M., Rajeev, K., V., Graner, A., Grando, S., Ceccarelli, S. (2008). QTLs for chlorophyll and chlorophyll fluorescence parameters in barley under post-flowering drought. *Euphytica*, DOI: 10.1007/s10681-007-9629-6

Hansatech Instruments (2007). *Chlorophyll fluorescence guide*.(www.hansatech-instruments.com).

Isla, R., Aragüés, R., Royo, A. (1998). Validity of various physiological traits as screening criteria for salt tolerance in barley. Field Crops Res, 58, 97-107.

Jiang, Q., Roche, R., Monaco, T.A., Durham, S. (2006). Gas exchange, chlorophyll fluorescence parameters and carbon isotope discrimination of 14 barley genetic lines in response to salinity. *Field Crops Res*, 96, 269–278

Kalaji, M.H., Pietkiewicz, S. (2004). Some physiological indices to be exploited as a crucial tool in plant breeding. *Plant Breed and Seeds Sci*, 49, 19-39. Review.

Kalaji, H.M., Loboda, T. (2007). Photosystem II of barley seedlings under cadmium and lead stress. Plant Soil Environ, 53, 511-516.

Kautsky, H., Hirsch, A. (1931). Neue Versuche zur Kohlensäureassimilation. *Naturwissenschaften*, 19(48), Page 964.

Krause, G. H., Weiss. E. (1991). Chlorophyll fluorescence and photosynthesis: the basics. *Annual Review of Plant Physiology and Plant Molecular Biology*, 42, 3 13-349.

Lawson, T., Oxborough, K., Morison, J.I.L., Baker, N.R. (2002). Responses of photosynthetic electron transport in stomatal guard cells and mesophyll cells in intact leaves to light, CO_2, and humidity. *Plant Physiol.* 128, 52–62.

Li, R., Guo, P., Baum, M., Grande, S., Ceccarelli, S. (2006). Evaluation of chlorophyll content and fluorescence parameters as indicators of drought tolerance in barley. *Agricul. Sci. China*, 5, 751-757.

Morant-Manceau, A., Pradier, E., Tremblin, G. (2004).Osmotic adjustment, gas exchanges and chlorophyll fluorescence of a hexaploid triticale and its parental species under salt stress. *J. Plant Physiol*, 161, 25-33.

Moustakas, M., Ouzounidou, G., Lannoye, R. (1993). Rapid screening for aluminium tolerance in cereals by use of the chlorophyll fluorescence test. *Plant Breeding*, 111, 343–346.

Moya, I., Cerovic, Z.G. (2004). *Remote sensing of chlorophyll fluorescence: instrumentation and analysis*. In: Papageorgiou, G.C., Govindjee (eds). Chlorophyll *a* Fluorescence: A Signature of Photosynthesis, Advances in Photosynthesis and Respiration. Springer, Dordrecht,19, 429-445.

Munns, R., James, R.A. (2003). Screening methods for salinity tolerance: a case study with tetraploid wheat. *Plant Soil*, 253. 201–218.

Oukarroum, A., El Madidi, S., Schansker, G., Strasser, R.J. (2007). Probing the responses of barley cultivars (*Hordeum vulgare* L.) by chlorophyll *a* fluorescence OLKJIP under drought stress and re-watering . *Environ. Exp. Bot*, 60, 438-446.

Planchon, A., Sarrafi, A., Ecochard, R. (1989). Chlorophyll fluorescence transient as a genetic marker of productivity in barley. *Euphytica*, 42, 269-273.

Pshybytko, N.L., Kalituho, L.N., Kabashnikova, L.F. (2003).The various mechanisms of photosynthesis limitation in heat-treated barley seedlings of different ages. *Bulg. J. Plant Physiol*, Special Issue, 304–313.

Rapacz, M. (2007). Chlorophyll a fluorescence transient during freezing and recovery in winter wheat. *Photosynthetica*, 45 (3), 409-418.

Rizza, F., Pagani, D., Stanca, A.M., Cattivelli, L. (2001).Use of chlorophyll fluorescence to evaluate the cold acclimation and freezing tolerance of winter and spring oats. *Plant Breeding*, 120, 389–396.

Royo, C., García del Moral, L.F., Aparicio, N., Villegas, D., Casadesús, J.,Araus, J.L. (2000). *Tools for improving the efficiency of durum wheat selection under Mediterranean conditions*. In Royo C. et al. (eds). Durum Wheat Improvement in the Mediterranean Region: New Challenges. Zaragoza, Spain, pp. 63-70.

Slafer, G.A.,Araus, J.L., Royo, C., Garcia del Moral, L.F. (2005). Promising eco-physiological traits for genetic improvement of cereal yields in Mediterranean environments. *Ann. Appl.Biol*, 146, 61–70.

Strasser, B.J., Strasser, R.J. (1995). *Measuring fast fluorescence transients to address environmental questions: the JIP Test*. In: Mathis, P. (Ed.), Photosynthesis: From Light to Biosphere, vol. V. Kluwer Academic Publisher, The Netherlands, pp. 977–980.

Strasser, R.J., Srivastava, A., Tsimilli-Michael, M. (1999). *Screening the vitality and phytosynthetic activity of plants by fluorescence transient*. In: Behl, R.K., Punia, M.S., Lather, B.P.S. (Eds.), Crop Improvement for Food Security. SSARM, Hisar, India, pp. 72–115.

Strasser, R.J., Srivastava, A., Tsimilli-Michael, M. (2000). *The fluorescence transient as a tool to characterise and screen photosynthetic samples*. In: Yunus, M., Pathre, U., Mohanty, P. (Eds.), Probing Photosynthesis: Mechanisms, Regulation and Adaptation. Taylor & Francis, London, GB, pp. 445–483 (Chapter 25).

Strasser, R.J., Srivastava, A., Tsimilli-Michael, M. (2004). *Analysis of the chlorophyll a fluorescence transient*. In: Papageorgiou, G., Govinjee, Xu C.H. (Eds.), Advances in Photosynthesis and Respiration, Chlorophyll Fluorescence a Signature of Photosynthesis, vol. 19. Kluwer Academic Publishers, The Netherlands, pp. 321–362.

Tóth, S.Z., Schansker, G., Garab, G., Strasser, R.J. (2007). Photosynthetic electron transport activity in heat-treated barley leaves: The role of internal alternative electron donors to photosystem II. Biochim. *Biophys. Acta*, 1767 295-305.

Van Kooten, O., Snel, J.F.H. (1990). *The use of chlorophyll fluorescence nomenclature in plant stress physiology*. In: Snel, J., van Kooten, O. (Eds.), Photosynthetic Research, vol. 25, special issue, Kluwer Academic Publishers, Netherlands. pp. 147-150.

Voltas, J., Romagosa, I., Muñoz, P., Araus, J.L. (1998). Mineral accumulation, carbon isotope discrimination and indirect selection for grain yield in two-rowed barley grown under semiarid conditions. *Eur. J. Agronomy*, 9, 147-155.

Zhao, G.P., Ma, B.L., Ren. C.Z. (2007). Growth, gas exchange, chlorophyll fluorescence, and ion content of naked oat in response to salinity. *Crop Sci*, 47, 123-131.

In: Photochemistry Research Progress
Editors: A. Sánchez, S. J. Gutierrez

ISBN: 978-1-60456-568-3

Chapter 13

MARINE PHOTOCHEMISTRY IN EARTH SYSTEM MODELS

Scott Elliott [*]

Los Alamos National Laboratory, Los Alamos, New Mexico, U.S.A.

ABSTRACT

Environmental systems models now simulate major element geocycles spanning the ocean and periodic table, but development is lagging for aqueous photochemistry mechanisms which interrelate biogeochemical tracers in most surface waters. This chapter reviews knowledge of critical open marine photochemical channels for the major atom types. Potential feedbacks to climate change are described, occurring initially via the generation of small radicals but leading to low molecular weight volatiles such as carbon monoxide, carbonyl sulfide, dimethyl sulfide and hydrocarbons. Examples are provided of box and one-dimensional models developed over the last decade or so which entail relatively complete energy transfer and reaction schemes. Only a handful of large-scale, three-dimensional approaches have appeared. The available global photochemical codes are surveyed, with emphasis on studies conducted recently in the oceanic component of the Community Climate System Model. Numerical and computational issues arising are enumerated in all cases. Coupling with atmospheric chemistry and aerosol-consistent radiative transfer must be achieved within the model frameworks. Stiff reaction systems will ultimately result as the pelagic diel cycle is resolved. Multiple scattering algorithms must be incorporated to augment purely absorptive schemes within the water column. The dissolved organic matter must be segregated into tracer bins of graded lability in order to deal with bleaching and aging. Aspects of biogeochemical complexity are mentioned to conclude. It is noted that the composition of photochemical source material implies a rich reaction web involving polymers, colloids, surfactants and trace metals. Complete mechanism development will thus be challenging.

* Email: sme@lanl.gov

1. INTRODUCTION

The climate simulation community is currently moving rapidly toward planetary scale systems modeling in the areas of ecodynamics and biogeochemistry (IPCC, 1996; Sarmiento and Gruber, 2006). Multiple atom types, organisms and trophic levels are now carried and transformed in codes developed for general circulation model grids. A typical global biogeochemistry treatment might encompass nitrogen, phosphorus, iron, silicon and carbon flow through half a dozen phytoplankton classes. These may be processed by zooplankton in varying size groups and generate dissolved organic matter which is in turn remineralized by heterotrophic bacteria, represented at the very least as a temperature dependent time constant and sometimes as a set of real populations (Moore et al. 2002; Christian and Anderson, 2002).

Photochemistry and photobiological processes are subsumed by such models at several levels. Visible and ultraviolet photons are lumped together into a broad wavelength set described simply as the photosynthetically available radiation or PAR. Multiple spectral bands thus implicitly drive primary production and light-sensitive chemical processes in tandem. Depth of penetration is often determined solely by the chlorophyll distribution so that proportionality is presumed for other chromophores. In fact the dissolved organics generated by marine ecosystems contain functional groups which absorb exclusively in the ultraviolet A and B portions of the spectrum (Leifer, 1988; Blough, 2001; Benner, 2002). This particular energy input supports radicals, excited state electronic species and a complex interactive photochemistry, distributed across all sunlit surface waters. A significant fraction of carbon remineralization passes through such channels (Mopper and Kieber, 2002). Small greenhouse and climate-active gases are generated which may be oxygenated and sometimes sulfurous. They flow through/from the extensive interface comprising the pelagic microlayer. Its permeability is determined in turn by surfactants which have been aged not only microbially but photolytically (Liss and Duce, 1997). Highly reduced, multiple carbon species are also among the compounds under photochemical control in the marine mixed layer (Plass Dulmer et al. 1995; Toole et al. 2008), and moving into the atmosphere they may influence both gas phase oxidation potential and aerosol/cloud reflectivities (IPCC, 1996; Seinfeld and Pandis, 2006). Trace metals entering the sea within the dust aerosol undergo organic chelation and photoreduction processes which are critical to their bioavailability and residence time (Morel and Price, 2003).

There has been preliminary exploration of the channels listed above in low dimensionality models – see for example a seminal series from the Najjar group (Najjar et al. 1995 to Van Hobe et al. 2003). But aqueous photochemistry has only begun to appear in full, general circulation frameworks alongside the fundamental ecology and geocycling (Chu et al. 2008). In the present chapter, samples of box and column models are first described which deal in increasing detail with global change relevant, light-driven chemical reactions of the open ocean. Examples from among a relatively few global scale analogs are then listed, and the requirements for future additions and improvements are commented upon. It is concluded that a strong need exists for the extension of earth system models in the direction of coupled, synthetic marine photochemistry mechanisms.

2. Background

Autotrophic organisms of the ocean harvest incoming solar radiation via chlorophyll and related pigments in absorption bands order tens of nanometers in breadth, spread throughout the visible spectrum (Jumars, 1993). Much of the bioorganic material produced by photosynthesis is polysaccharide in nature, but sugar groups do not interact with light. A suite of free chromophores occurs as a byproduct, however, and includes not only generalized pigment fragments but also individual aromatics such as phenols/quinones and assorted, conjugate double bonded structures (Benner, 2002; Nelson and Siegel, 2002). Sometimes these entities couple with one another in surface water to form charge transfer complexes. Processing through the food web transforms operational biological compounds into detritus. Microbial action then leads to a conversion of particles, colloids and gels toward small molecular units. In the sense of chemical composition, there is a smearing of identities from familiar cell constituents to poorly identified polymers, peptides, nucleic acids and beyond. The kinetics are sometimes modeled in bulk with lability tracking the decrease in size (Christian and Anderson, 2002), but in fact details of the chemistry remain poorly known. During the breakdown, electronic absorption spectra become averaged and shift into a broad continuum rising steadily in the ultraviolet.

Bulk carbonaceous material in the open sea is often referred to as particulate or dissolved organic matter, depending on its characteristic sedimentation rate (P- or DOM: IPCC, 1996; Moore et al. 2002). The ultraviolet-absorbing portion then becomes CDOM, with the C signifying "chromophoric". Some phytoplankton synthesize and release specialized light interacting molecules directly. The mycosporine-like amino acids (MAA) constitute a notorious example and may serve as a microbial sunscreen, but they usually contribute only a minor fraction to overall attenuation. Riverine and coastal sources yield refractory substances such as lignins and their contribution to the remote marine situation continues to be debated. They may, for example, lead to differential photochemical intensities per unit organic mass in the Atlantic versus Pacific basins (Zafiriou, 2003). Emphasis here will be placed upon the representative, global central ocean.

The energy associated with CDOM absorption ultimately dissipates as heat, but at intermediate stages a variety of low molecular weight oxidants is supported at steady state (Plane et al. 1987; Blough, 2001; Mopper and Kieber, 2002). Radicals such as the superoxide and aqueous hydroxyl figure prominently among this collection of reactive species. Hydrogen peroxide, by contrast, is a highly oxidizing but stable compound formed through the overall photochemical activity. Multiple redox states of inorganic nitrogen contribute as well. A variety of the more complex oxidant structures/radicals are based on organic backbones. Within this chemically reactive environment high energy photon absorption leads ultimately to decomposition of the DOM, and several of the low molecular weight products have strong biogeochemical relevance. The present chapter will deal mainly with the one and two carbon gases CO, OCS, and DMS (carbon monoxide, carbonyl sulfide and dimethyl sulfide). They form strong climate teleconnections to atmospheric greenhouse and albedo systems (IPCC, 1996). Production and removal mechanisms remain poorly understood in most cases.

Climate and Earth System Models must ultimately parameterize all the above absorption, photochemical and intermediate steps in some form. Numerical tracers representing the dissolved organics must enfold chromophores with spectrally dependent cross sections. They must undergo aging and bleaching commensurate with their structure (Christian and Anderson, 2002). Radical systems must lead at some level of complexity to oxidation and then release of the climate relevant volatiles. Classes of surfactant must be generated simultaneously, leading to coatings of the sea surface with distributions dependent on wave action (Duce and Liss, 1997). It follows that the coatings will also modulate transfer of the gases, and this effect could well evolve during the era of global change. Photolysis of the teleconnection agents themselves must be simulated on top of all else (Toole et al. 2004 and 2008). Presently, very few of these processes are incorporated globally at the systems level (Elliott et al. 2008). The plan in this chapter is to review detailed treatments appearing in low dimensionality, and to propose their extrapolation to full three dimensional frames.

Optical depths tend to be less than unity in remote mixed layer waters at central wavelengths, but they then rise dramatically moving into the blue (Nelson and Siegel, 2002). As a reference point for the reader it may be noted that typical ecodynamic models set the chlorophyll absorption at about 0.03 ($1/m(mg/m^3)$, averaged over the PAR which is here taken to lie between roughly 400 to 700 nm (Sarmiento and Gruber, 2006). A median chlorophyll density for the remote mixed layer is about 0.1 mg/m^3 but background phenomena such as particulate interactions also remove light, with an e-fold of several tens of meters. UVA and UVB portions of the spectrum are generally defined by aquatic chemists as the regions 400 to 320 and 320 to 280 nm (Leifer, 1988). Tests conducted in remote areas such as the Sargasso Sea show that overall organic absorption coefficients rise from roughly 0.03 (1/m) to 0.3 (1/m) from the A to the B bands, almost without exhibiting spectral structure (Nelson and Siegel, 2002). The challenge for photochemistry modelers is to associate such values with local and dynamic CDOM distributions at the global scale. This must be accomplished through a combination of station measurements with simulations of the distribution of organic material. Some fraction of deposited incoming energy is siphoned into photochemical systems as a function of decreasing wavelength. The concept of quantum yield thus becomes the means for parameterizing much of the complex organo-radical chemistry lying between absorption and release steps.

3. Low Dimensionality Simulations

Equations fundamental to the modeling of aqueous photochemistry were laid out several decades ago in references ranging from short summaries (Zepp and Cline, 1977) to the textbook level (Leifer, 1988). Such treatments logically begin with discussions of the downwelling radiation input. Both visible and ultraviolet photons reach the sea surface after passing through absorbing and scattering layers of the atmosphere (which are ozone and aerosol laden). Energy losses are particularly strong in the UVB. Stratospheric ozone and tropospheric pollution dramatically decrease fluxes in this regime, with a sharp wavelength dependence. More generally, airborne particles near the Earth's surface are responsible for a partial rejection back toward space (Seinfeld and Pandis, 2006) and clouds have a small, wavelength independent effect (Zafiriou et al. 2003). Early research tended to account for both stratospheric and tropospheric attenuation through analytical models with idealized atmospheric chemical compositions inserted in appropriate altitude layers (Leifer, 1988).

More recently it has become possible to use satellite data to achieve the same purposes (Arrigo et al. 2003).

Reflection of light from the sea surface itself is non-negligible but is often treated as a rough 5 percent, "off the top" decrement in such calculations. Refraction is also significant and pulls any light stream entering the water column closer to the vertical. A common approach to simplifying the atmospheric situation, whether in the pre-satellite or contemporary era (Leifer, 1988; Van Hobe et al., 2003), has been to divide the ocean interface radiation source into two major fractions –the direct beam and then a total sky diffuse contribution. The latter is often modeled simply as a luminous hemisphere of uniform brightness. The hemispheric fraction of total input increases with sun angle whether the variation is daily or seasonal. Average entrance trajectories are usually quite steep, and with the added influence of refraction come close to being overhead on a net basis. Integrated angle adjustments may nonetheless be applied and where detailed data comparisons are available for the trace photochemicals, they may be quite important (Kettle, 2000).

Early models of seawater photochemistry were confronted with large uncertainties in the spatial distribution of both precursors and products. Basic issues such as the depth of radiation penetration were addressed initially, along with the capacity for CDOM to support the principal absorption, sensitization and reaction sequences. There was a tendency to focus on the small inorganic radicals, usually oxygenated (Plane et al. 1987), or else upon individual stable oxidants such as hydrogen peroxide (Sikorski and Zika, 1993). The latter compound is readily measurable by standard wet chemical techniques, and so provides a powerful entry point for validation. Locations for data comparison were often relatively accessible sites distributed through the subtropical Atlantic, Caribbean or Gulf of Mexico. In many cases the models so developed would not be applicable to the remote open ocean, and this is due in part to basin-peripheral geochemical effects such as sediment or river input. Historically coastal waters of the North Atlantic have received a great majority of attention from photo-geochemists, but they may not be globally representative (Zafiriou et al. 2003).

Global change studies require that linkages be made from the background activity of small energetic species to generation or consumption of key teleconnection agents. The first such substance to be simulated was carbon monoxide. This simple diatomic compound constitutes a high throughput intermediate and indicator for the surface ocean carbon cycle (Mopper et al. 1991). It then crosses into the troposphere to influence concentrations of the HOx-type radicals which determine gas phase oxidizing capacity (Logan et al. 1981; Seinfeld and Pandis, 2006). In connection with depth profile studies conducted in the equatorial Pacific, Johnson and Bates (1996) constructed analytical models for the diurnal cycling of dissolved CO concentrations. But these calculations were largely empirically based, with production rates drawn directly from seawater sample analysis. The BATS site (Bermuda Atlantic Time Series) has provided analogous data sets but with numerical simulations as follow up. Kettle (2000) computed downward energy fluxes to the surface of an idealized Sargasso Sea location, mixed layer penetration of the diffuse radiation with the slight angle factor included, then attenuation by multiple agents in the water column but with CDOM dominating. A portion of the energy was channeled into production of the selected trace gas, consistent with laboratory and field measurements of its quantum yield. This approach appears to be the most generally useful to earth system modelers. Differential energy deposition is converted into production of a direct/indirect greenhouse species via known or expected yield factors.

Carbonyl sulfide is a close molecular analog of the greenhouse gas carbon dioxide, but it is produced in the sea almost exclusively by photochemical processes (Ferek and Andreae, 1984). It may be thought of as deriving from light-driven decomposition of phytoplanktonic intracellular sulfur, occurring in shallow waters during remineralization. Potential precursor compounds include several amino acids and organo-sulfur frameworks lying along the DMS metabolism. OCS hydrolyzes rapidly at the seawater pH to give monothiocarbonates and then carbonic acid, but it nevertheless transfers into the atmosphere in part. There it is sufficiently long lived to supply much of the sulfur present in the stratospheric aerosol layer (Seinfeld and Pandis, 2006). Many groups have undertaken local scale simulation of the steady state supported by photochemistry in the marine mixed layer. Hydrolysis is a purely physicochemical removal process, with temperature/ionic strength dependent rate constants known to order ten percent (Elliott et al. 1989). Concentration measurements may thus be readily transposed into throughputs. Van Hobe et al. (2003) applied methods essentially identical to the above in order to study depth profiles compiled during ACE campaign ship cruises. Source/sink relationships were computed both in point or box models and in one dimensional numerical columns. Major conclusions were that the depth dependent data demand at least a single model dimension for study, that the selection of vertical physics applied within the mixed layer is crucial, and that the lack of CDOM distribution information can be a limiting factor. These authors state several times that photochemical mechanisms should be scaled upward into regional and global chemistry-transport models.

Dimethyl sulfide is perhaps the second most important agent of global sea-air exchange after carbon dioxide. It crosses the ocean interface to provide the source of tropospheric aerosol sulfate over most of the planet. Exceptions occur only downwind of polluted continental areas in the Northern Hemisphere. Atmospheric sulfate mass determines regional numbers of cloud condensation nuclei through Kohler curve effects. Strong influence follows on the brightness of low level clouds (Seinfeld and Pandis, 2006). DMS is produced as a stress-reducing metabolite for multiple reasons by multiple classes of phytoplankton. It appears to be utilized by autotrophic cells mainly to deal with internal radical oxidants and cryo-pressure, but is simultaneously released into the water column during standard trophic level interactions (grazing, senescence etc.; Simo, 2004). In most surface waters removal occurs mainly by bacterial consumption, but photolysis forms a ten percent level background channel. Near the atmospheric light source and in some special polar situations, photolysis may actually dominate. For example, prior to the strong spring bloom in Antarctic coastal waters, shallow residual CDOM or MAA drives oxidation before microbial populations can become established (Toole et al. 2004). There may also be connections with aqueous photochemistry of the nitrogen oxides, since waters of the Southern Ocean are iron limited and hence nitrate rich.

The initial product of DMS breakdown appears to be dimethyl sulfoxide, but little else is known regarding the mechanism. None of the critical sensitizers or intermediates have been identified (Mopper and Kieber, 2002). Nevertheless multiple efforts have been made to incorporate the photolysis reaction into low dimensionality sulfur cycle models. In the pioneering work of Gabric and company, a limited data set for light driven losses in coastal waters of the North Sea was extrapolated into box models, then applied to a variety of open ocean ecosystems (e.g. Gabric et al. 1996). The original studies were those of Brimblecombe and Shooter (1986), who were able to determine that the active wavelengths were coincident with PAR. Removal in the Gabric models was portrayed simply as a time constant, adjusted locally to match varying energy inputs. The approach thus lacked resolution for any of the major physical variables -depth, chromophore identity, wavelength and time were all subsumed.

More recently the group Toole et al. (2008) have undertaken a detailed simulation of sulfur photochemistry near the BATS location, in association with a series of Sargasso Sea marine optics expeditions. All the above processes were included in a one dimensional simulation driven by standard vertical physics. For the most part, however, pivotal quantities such as phytoplankton and bacterial densities were incorporated as measured data (assimilated). Atmospheric attenuation was set at a seasonal constant, a two percent sea surface reflection was assumed, and CDOM distributions were keyed to remotely sensed absorptions at 443 nm. The remainder of the spectrum extending into the ultraviolet was reconstructed using a standard slope. A relatively tight wavelength resolution of 5 nm was adopted. The authors find that seasonal changes in radiation penetration are critical to the simulation. They also note somewhat tangentially that the distinction between UVA and UVB will play into DMS sourcing via photobiology of the oxidant stress. Their simulations did not attempt to deal with dynamics of the DOM distribution.

4. Global Simulations

Several groups have taken intermediate steps toward global scale photochemical modeling by computing individual marine trace gas production fields. CO has again been a major player due to its centrality within the carbon cycle. Zafiriou et al. (2003) made measurements of CDOM absorption and apparent quantum yield from basin spanning north-south Pacific cruises and constructed latitudinally dependent inputs. Their chief assumption was that overall energy deposition in the upper ocean could be estimated as the difference between incident radiation and escape from the mixed layer into the thermocline. These authors were able to show that approximate balance is achieved with bacterial removal, and they also tabulated sea to air fluxes which were distributed regionally. Fichot and coworkers (Miller and Fichot, 2006) estimated ultraviolet attenuation from remote sensing data using multivariate statistical techniques. The algorithms involved are similar to those applied to deduce pigment distributions on broad scales. Results were validated in a relatively complete, columnar model of radiation transfer below the sea interface, then augmented with the quantum yields determined and reviewed in Zafiriou et al. (2003) to give three dimensional maps of global carbon monoxide production. The work is presented as a one-molecule example of the importance of ultraviolet penetration to overall planetary biogeochemistry.

Simulations based closely on ship or satellite energy deposition data lack predictive capability until they are applied to the validation of dynamic approaches. More complete models of the global photochemistry were pioneered by Najjar and coworkers. The paper Najjar et al. (1995) serves as an instructive preview. Both carbon monoxide and carbonyl sulfide are considered as sample compounds. The latter is taken as the primary case because certainty of the hydrolytic removal rates permits attention to be focused upon CDOM-ultraviolet interactions. The authors compute time dependent OCS distribution changes based on stratospheric ozone trends provided by TOMS (the Total Ozone Mapping Spectrometer). The methods are clearly extensible to future systems studies. It is also determined that dark reactions may be required in some areas to balance the overall sulfur budget. These would occur toward the bottom of the mixed layer, at depths to which the UV does not penetrate. Preiswerk and Najjar (2000) followed up on this work with a mobile one dimensional model of photochemical production, air-sea gas exchange, vertical turbulent mixing and hydrolysis. The code was transferred around the ocean such that zonally averaged production contours could be plotted. The authors confirmed the necessity of including a dark reaction and explored the seasonality of energy inputs and trace gas release. CDOM concentrations were constructed as correlations with satellite chlorophyll data.

At least one group has begun to couple photochemical production mechanisms for the climate active trace gases into true Earth System Models. This body of research takes carbon monoxide as the initial case but draws CDOM distributions from dynamic biogeochemistry routines (Chu et al. 2008). Coding has been developed in a framework provided by the Community Climate System Model, a joint venture of the National Science Foundation and Department of Energy (CCSM- Blackmon et al. 2001; Collins et al. 2006). CCSM has evolved over the last decade or so as a combination of the high performance Parallel Ocean Program (POP) with atmospheric general circulation models. One major direction of development has been the addition of increasingly complex and coupled chemistry/biology simulators within the sea. The intent is to asymptote toward a full marine systems approach to climate modeling. Accordingly ecodynamics are now simulated globally within POP-CCSM, driven by the chemistry and transport of a multi-elemental array of nutrients (Moore et al. 2002 and 2004). Both particulate and dissolved organic matter are treated as advected/diffused tracers, in order to recycle material requisite to the biota. In appending the small molecule CO, Chu et al. (2008) parameterize CDOM as a globally averaged, constant fraction of the total nonsinking organics. Incident ultraviolet is taken from offline data sets consistent with contemporary atmospheric conditions, but within the sea attenuation is accomplished dynamically through absorption by the implied local chromophore content. Resolution is essentially two-band, consisting of UVA and UVB. With the exception of certain high latitude complexities such as the trade off between ice coverage and atmospheric transfer, the authors were able to reproduce major ship track data (Bates et al. 1995; Zafiriou et al. 2003). Large scale distributions and budgets are all consistent with earlier estimates, some of which were based on the more rudimentary forms of global modeling.

A major conclusion from this exercise is that aging and bleaching of CDOM must be accounted within photochemical systems models in the near term future (Chu et al. 2008). The authors demonstrate that a time/space varying chromophore fraction could dramatically improve agreement with real carbon monoxide data, particularly at the periphery of the major ocean basins. Ultimately the dissolved organics must be decomposed numerically into bins of varying lability and absorptive properties. This will mean mapping in some way onto chromophoric functional groups or collections thereof, as well as the incorporation of bacterial processing dynamics (Christian and Anderson, 2002). It is not for the moment clear whether such operations will be best performed within the base ecodynamics or as a separate set of routines. Chu et al. (2008) also apply their model system to explore carbon monoxide removal processes, which are primarily biotic. Grazing dependent kinetics are sensitivity tested for the dominant consumer microbes. This sub-investigation shows that evolution of surface CO fields during the era of global change will depend critically on life histories of the relevant bacteria. Chemoautotrophic organisms which may specialize on CO will behave quite differently from heterotrophic generalists.

The same CCSM development group has expanded their photochemistry into a modular code suite tailored for the simulation of generic marine trace gases (Elliott et al. 2008). Carbonyl sulfide distributions have been reported at the level of a demonstration of capability and embody major features from the disperse, one dimensional calculations (Preiswerk and Najjar, 2000). Photolytic removal of dimethyl sulfide from the water column is included in all organo-sulfur simulations. Baseline mechanisms are still derived from the early work of Brimblecombe and Shooter (1986), though there has been a certain amount of testing on global photosensitizer distributions (Elliott et al. 2007). Plass Dulmer et al. (1995) reviewed concentration data and information relating to the photolytic production of nonmethane hydrocarbons in surface sea water (NMHC). Ethylene and isoprene appear to be exuded directly by the phytoplankton, but all other compounds arise from an interaction of ultraviolet radiation with CDOM. Idealized alkanes, alkenes and alkynes have been simulated in the POP model as well. Concentration patterns display major seasonal and geographic features apparent in the data sets. The NMHC feed back onto atmospheric global change by providing multi-carbon backbones and the associated ozone source/sink potential to the marine troposphere (Seinfeld and Pandis, 2006).

5. Summary and Discussion

Sunlit upper layers of the ocean cover the vast majority of our planet, but a rich photochemistry implied by this situation is only beginning to enter systems level models of global change. Multiple climate feedback loops can be identified, beginning with the absorption of radiation by aqueous organic chromophores and leading to greenhouse, albedo and atmospheric chemical effects (Najjar et al. 1995; IPCC, 1996; Mopper and Kieber, 2002; Seinfeld and Pandis, 2006). Additionally, photodegradation of detrital material constitutes a significant channel in the marine carbon cycle (Mopper et al. 1991). Low molecular weight gases often act as the agents of teleconnection from aqueous photochemistry up to large scale geocycles. Examples include carbon monoxide, carbonyl sulfide, dimethyl sulfide and a suite of pure hydrocarbons (Najjar et al. 1995; Plass Dulmer et al. 1995; Elliott et al. 2008). A brief history of the simulation of surface water photochemistry has been provided here, along with some notion of the community's present capability to explain distributions of precursor and climate active species. Successes include the representation of vertical distributions of certain inorganic oxidants and carbon monoxide in subtropical waters (e.g. Sikorski and Zika, 1993; Kettle, 2000) plus global sea-air fluxes of CO and carbonyl sulfide constrained by measured removal rates (e.g. Preiswerk and Najjar, 2000; Zafiriou et al. 2003). Planetary biogeochemistry codes are now available which can provide dynamic simulations of dissolved organic concentrations (Moore et al. 2002). The potential for coupling is already being exploited in the Community Climate System Model or CCSM (Chu et al. 2008; Elliott et al. 2008). But considerable effort must still be expended before aqueous photochemical aspects of global change will be faithfully simulated.

Critical remaining issues are now addressed to conclude. Coupled ocean-atmosphere climate models such as CCSM compute surface radiation downwelling consistent with atmospheric chemical and aerosol properties of the overhead column. Interactive, global radiation fields must be substituted for the offline inputs currently incorporated into most marine photochemistry studies (Kettle, 2000; Chu et al. 2008). The local, low dimensionality models often resolve diel cycling (Sikorski and Zika, 1993; Kettle, 2000), but initial attempts at systems simulation utilize monthly average radiation fields (Chu et al. 2008). Several groups have pointed out that the phasing of vertical turnover and radiation input to the mixed layer can influence production rates (Doney et al. 1995; Kettle 2000). If computational intensities are to be kept manageable, it seems likely that some means must be found for parameterizing the solar cycle retroactively into detailed photochemistry. Sun angle calculations could be performed intermittently with the consequent production rates ingested into the trace gas module. The ultraviolet spectrum has been resolved at the nanometer level in some of the regional and coarser studies (Kettle et al. 2000; Zafiriou et al. 2003), but in the new systems models distinction is limited to UVA versus B. Binning will likely have to be set finer than ten nanometers at some point in the near future. Light penetration is currently computed even in the most complete of the local codes solely in a single scattering mode –the first interaction of a photon with a reflecting particle or absorber removes it (Kettle et al. 2000). Multiple scattering algorithms should be adopted below the ocean interface, just as they are currently applied above.

It may prove necessary during the upcoming era of global change to push at least some of the mechanisms for trace gas generation/loss back toward the fundamental radical systems which dominate processing at the microscale. Realistic connections will thus be required with the functional group composition of dissolved organic matter (Blough, 2001; Benner, 2002; Mopper and Kieber, 2002; Zafiriou et al. 2003). At this point aqueous photochemists may find themselves confronting analogs of the stiffness problem so familiar to their atmospheric counterparts (Seinfeld and Pandis, 2006). As the morning sun rises over the open ocean and UVA/B along with their intermediate wavelengths begin to penetrate into the mixed layer, reactive species will enter into kinetic relationships characterized by extremely rapid time constants. It will not be possible to resolve all of the chemical interactions temporally. They will have to be handled implicitly or through molecular family groupings. While the computational techniques for dealing with these difficulties are well developed in other disciplines, they will be novel to the marine systems community.

Perhaps the greatest challenge awaiting marine photochemists in the next few decades will be comprehension of the sheer complexity of their particular Earth system component. Extremely shrewd modeling strategies must be exploited if the interaction of ultraviolet radiation with organics and the subsequent connections to climate are to be represented accurately. Close couplings must first be considered with global scale photobiology. Ecosystems which enfold multiple organism classes and trophic levels harvest light through processes not even listed here, then channel the energy into a panoply of organic structure. Phytoplankton, zooplankton, bacteria and viruses are all involved (Jumars, 1993; IPCC, 1996; Sarmiento and Gruber, 2006). The dynamic ecology is intricate in and of itself but serves merely as the backdrop. Reduced carbon compounds synthesized by the biota enter into a full marine detrital suite. Particles, colloids, gels and finally true dissolved forms all harbor chromophores and their chemical composition varies widely (Benner, 2002). Trace metals such as iron and manganese are coordinated by both free and bound carbonaceous ligands (Wells et al. 1995; Morel and Price, 2003). Oxidation states are determined in part by photoreduction, and they go on to play intimately into bioavailability and metabolic function. It is likely that many of these biogeochemical channels will have systems level import and demand detailed photochemical treatment in three dimensional ocean models. Current generation routines are difficult to alter or enhance such that organic mass and energy consistency are maintained (Elliott et al. 2008). An option which should be explored for dealing with this explosion of complexity is to automate photochemical and general biogeochemical code construction.

ACKNOWLEDGMENTS

The authors thank the U.S. Department of Energy Office of Science and Biology/Environmental Research SciDAC project for support -SciDAC stands for Scientific Discovery through Advanced Computing.

REFERENCES

Arrigo, K.R., Lubin, D., van Dijken, G.L., Holm-Hansen, O., and Morrow, E. (2003). Impact of a deep ozone hole on Southern Ocean primary production. *J. Geophys. Res*, 108, doi:10.1029/2001JC001226.

Bates, T.S., Kelly, K.C., Johnson, J.E., and Gammon, R.H. (1995). Regional and seasonal variations in the flux of oceanic carbon monoxide to the atmosphere. *J. Geophys. Res*, 100, 23092-23101.

Benner, R. (2002). *Chemical composition and reactivity*. In: Hansell, D.A., and C.A. Carlson (Eds.), Biogeochemistry of Marine Dissolved Organic Matter. Academic Press, New York, USA.

Blackmon, M., and 25 others. (2001). The Community Climate System Model. *Bull. Am. Met. Soc*, 82, 2357-2376.

Blough, N.V. (2001). *Photochemical processes*. In: Steele, J., S. Thorpe, and K. Turekian (Eds.), Encyclopedia of Ocean Sciences. Academic Press, New York, USA.

Brimblecombe, P., and Shooter, D. (1986). Photo-oxidation of dimethylsulfide in aqueous solution. Mar. Chem. 19, 343-353.

Christian, J.R., and Anderson, T.R. (2002). *Modeling DOM biogeochemistry*. In: Hansell, D.A., and C.A. Carlson (Eds.), Biogeochemistry of Marine Dissolved Organic Matter. Academic Press, New York.

Chu, S., Elliott, S., and Erickson, D. (2008). Basin scale carbon monoxide distributions in the Parallel Ocean Program. *Earth Interactions*, 11, 1-30.

Collins, W.D., and 14 others. (2006). The Community Climate System Model (CCSM3). *J. Climate*, 19, 2122-2143.

Doney, S.C., Najjar, R.G., and Stewart, S. (1995). Photochemistry, mixing and diurnal cycles in the upper ocean. *J. Mar. Res*, 53, 341-369.

Elliott, S., Lu, E., and Rowland, F.S.(1989). Rates and mechanisms for hydrolysis of carbonyl sulfide in several types of natural water. *Environ. Sci. Tech*, 23, 458-461.

Elliott, S., Chu, S., and Erickson, D.J. (2007). Contours of simulated marine dimethyl sulfide distributions under variation in a Gabric mechanism. *Environ. Modeling and Software,* 22, 349-358.

Elliott, S., Chu, S., Dean, C., and Erickson, D.J. (2008). *Tracegas_Mod: Geochemical processing for low concentration volatiles in the CCSM ocean*. In: Zannetti, P., Rouson, D., and S. Elliott (Eds.), Environmental Science and Environmental Computing Volume III, Fiatlux and the Envirocomp Institute. In press.

Ferek, R.J., and Andreae, M.O. (1984). Photochemical production of carbonyl sulphide in marine surface waters. *Nature*, 307, 148-150.

Gabric, A.J., Ayers, G.P., Murray, C.N., and Parslow, J. (1996). Use of remote sensing and mathematical modeling to predict flux of dimethyl sulfide to the atmosphere of the Southern Ocean. *Adv. Space Res*, 18, 117-128.

IPCC. (1996). *Climate Change* (1995). The Science of Climate Change. Cambridge University Press.

Johnson, J.B., and Bates, T.S. (1996). Sources and sinks of carbon monoxide in the mixed layer of the tropical South Pacific Ocean. *Global Biogeochem. Cycles*, 10, 347-359.

Jumars, P. (1993). *Concepts in Biological Oceanography*. Oxford University Press.

Kettle, A.J. (2000). Comparison of dynamic models for the concentration of a photochemical tracer in the upper ocean. *Mar. Freshwater Res*, 51, 289-304.

Leifer, A. (1988). *The Kinetics of Environmental Aquatic Photochemistry: Theory and Practice*. American Chemical Society.

Liss, P.S. and Duce, R.A. (1997). *The Sea Surface and Global Change*. Cambridge University Press.

Logan, J.A., Prather, M.J., Wofsy, S.C., and McElroy, M.B. (1981). Tropospheric chemistry –a global perspective. *J. Geophys. Res*, 86, 7210-7254.

Miller, W.L., and Fichot, C. (2006). Estimating ultraviolet radiation in the surface ocean with SeaUV. *Solas Newsletter*, 3, 3-4.

Moore, J.K., Doney, S.C., Kleypas, J., Glover, D.M., and Fung, I.Y. (2002). An intermediate complexity ecosystem model for the global domain. *Deep Sea Res*, II 49, 403-462.

Moore, J.K., Doney, S.C., and Lindsay, K. (2004). Upper ocean ecosystem dynamics and iron cycling in a global model. *Global Biogeochem. Cycles,* 18: 10.1029/2004GB002220.

Mopper, K., and Kieber, D.J. (2002). *Photochemistry and the cycling of carbon, sulfur, nitrogen and phosphorus*. In: Hansell, D.A., and C.A. Carlson (Eds.), Biogeochemistry of Marine Dissolved Organic Matter. Academic Press, New York, USA.

Mopper, K., Zhou, X.L., Kieber, R.J., Kieber, D.J., Sikorski, R.J., and Jones, R.D. (1991). Photochemical degradation of dissolved organic carbon and its impact on the oceanic carbon cycle. *Nature*, 353, 60-62.

Morel, F.M.M., and Price, N.M. (2003). The biogeochemical cycles of trace metals in the ocean. *Science*, 300, 944-947.

Najjar, R.G., Erickson, D.J., and Madronich, S. (1995). *Modeling the air-sea fluxes of gases formed from the decomposition of dissolved organic matter*. In: Zepp, R.G., and C. Sonntag (Eds.), The Role of Nonliving Organic Matter in the Earth's Carbon Cycle. John Wiley and Sons, New York, USA.

Nelson, N.B., and Siegel, D.A. (2002). *Chromophoric DOM in the open ocean*. In: Hansell, D.A., and C.A. Carlson (Eds.), Biogeochemistry of Marine Dissolved Organic Matter. Academic Press, New York, USA.

Plane, J.M., Zika, R.G., Zepp, R.G., and Burns, L.A. (1987). *Modeling applied to natural waters*. In: Zika, R.G. (Ed.), Photochemistry of Environmental Aquatic Systems. ACS Advances in Chemistry Series, 327, 250-267.

Plass Dulmer, C., Koppmann, R., Ratte, M., and Rudolph, J. (1995). Light nonmethane hydrocarbons in seawater. *Global Biogeochem. Cycles*, 9, 79-100.

Sarmiento, J.L., and Gruber, N. (2006). *Ocean Biogeochemical Dynamics*. Princeton University Press.

Seinfeld, J.H., and Pandis, S.N. (2006). *Atmospheric Chemistry and Physics: From Air Pollution to Global Change*. John Wiley and Sons, New York, USA.

Sikorski, R.J., and Zika, R.G. (1993). Modeling mixed layer photochemistry of H_2O_2: Optical and chemical modeling of production. *J. Geophys. Res*, 98, 2315-2328.

Simo, R. (2003). From cells to globe: Approaching the dynamics of DMS(P) at multiple scales. *Can. J. Fish. Aquat. Sci*, 61, 673-684.

Toole, D., Kieber, D.J., Kiene, R.P., White, E.M., Bisgrove, J., Del Valle, D.A., and Slezak, D. (2004). High dimethylsulfide photolysis rates in nitrate-rich Antarctic waters. *Geophys. Res. Lett*, 31, doi:10.1029/2004GL019863.

Toole, D., Siegel, D.A., and Doney, S.C. (2008). A light driven, one dimensional dimethyl sulfide cycling model for the Sargasso Sea. *J. Geophys. Res*, in press.

Van Hobe, M., Najjar, R.G., Kettle, A.J., and Andreae, M.O. (2003). Photochemical and physical modeling of carbonyl sulfide in the ocean. *J. Geophys. Res*, 108, doi:10.1029/2000JC000712.

Wells, M.L., Price, N.M., and Bruland, K.W. (1995). Iron chemistry in seawater and its relationship to phytoplankton. *Mar. Chem*, 48, 157-182.

Zafiriou, O.C., Andrews, S.S., and Wang, W. (2003).Concordant estimates of oceanic carbon monoxide source/sink processes in the Pacific yield a balanced global "blue water" budget. *Global Biogeochem. Cycles*, 17, doi:10.1029/2001GB001638.

Zepp, R.G., and Cline, D.M. (1977). Rates of direct photolysis in the aquatic environment. *Environ. Sci. Tech*, 11, 359-366.

In: Photochemistry Research Progress
Editors: A. Sánchez, S. J. Gutierrez

ISBN: 978-1-60456-568-3

Chapter 14

PHOTOCATALYTIC REDUCTION OF METALS AND OXIDATION OF ORGANIC COMPOUNDS IN TITANIUM DIOXIDE SUSPENSION

Zoltán Zsilák, Erzsébet Szabó-Bárdos, Attila Horváth and Ottó Horváth*

ABSTRACT

Both various organic compounds (such as surfactants, industrial intermediates, and amino acids) and heavy-metal ions are characteristic pollutants of wastewaters, due to their extended industrial and domestic use. Heavy-metal ions and many organic contaminations are generally toxic, thus their reliable and cost-effective removal or degradation is the focus of numerous investigations. Photocatalytic procedures offer efficient solutions for this problem. One of the most promising directions of research in this area is based on photoactive semiconductors. Colloidal titanium dioxide proved to be applicable for both reductive deposition of metals and (even simultaneously) mineralization of organic compounds. In this chapter, we survey and interpret the results achieved in photocatalytic procedures based mostly on our experiments with titanium dioxide suspensions.

Although heavy-metal ions are mostly toxic, the corresponding metals are generally precious because of their application in the industry and everyday life as well. Thus, photoreductive deposition of these metals such as mercury, silver, and copper is also important from economic points of view. Furthermore,, in several cases, photocatalytic efficiency of colloidal titanium dioxide in mineralization of organic contaminants can be significantly enhanced by silver and copper deposition. The effects of such metal depositions on the mechanism of the photoinduced degradation of various organic compounds are discussed in details.

---

* University of Pannonia, Institute of Chemistry, Department of General and Inorganic Chemistry, H-8201 Veszprém, P.O.B. 158, email: otto@vegic. uni-pannon.hu

In addition, simultaneous photoreduction of metal ions can considerably increase the photocatalytic oxidative mineralization of organic contaminants such as surfactants, and vice versa. This synergism strongly depends on the salt content and pH of the systems due to the effects of surface charge of the colloidal photocatalyst particles as well as complex and ion-pair formations. Detailed interpretation of these effects are presented in this chapter for the preparation of practical application of this heterogeneous photocatalytic method in the treatment of various wastewaters.

1. INTRODUCTION

A significant part of the man-made pollutants getting into our natural environment is toxic and/or biologically non-degradable. Some of these materials reach the upper layers of the atmosphere, destroying the ozone shield, while others (such as pesticides and herbicides) upset the balance of the biosphere in the soils and natural waters, or spoil the resources of drinking water [1a].

Microorganisms relatively fast destroy natural, organic materials, with rather few exceptions (e.g., lignins). The situation is quite different in the case of numerous synthetic organic compounds. The microorganisms lack the enzymes necessary for their degradation, thus they are microbially irreducible, which considerably increase their dangerous impact on our natural environment. Also organic pollutants of natural origin (such as proteins) may cause serious environmental problems if their concentration is too high to be degraded in wastewater plants within a reasonable period of time.

Some of the heavy metals are among the most harmful pollutants and are of particular concern because of their toxicities to humans [1b]. They are highly persistent in the body because they are combined with specific proteins called metallo-thioneines. Since heavy metals, deviating from organic compounds, cannot be degraded, the treatment of waters polluted by them can only be based on their removal or transfer into a non-toxic oxidation state (if their is any).

Photocatalytic methods have been proved to offer efficient solutions for treatment of waters contaminated with organic pollutants and/or heavy-metal ions [2-5]. One of the most promising type of procedures is based on photoactive semiconductors. Titanium dioxide is the most frequently applied semiconducting material for photocatalytic purposes [6, 7].

2. ORGANIC AND INORGANIC POLLUTANTS TO BE TREATED

2.1. Non- or poorly degradable organic pollutants

Production of innumerable poorly degradable materials support the contionuously developing technologies of our modern life style. They may get into the environment already at the point of production, or during their transportation, usage, or subsequently. Increasing their quantity, they become a more and more serious environmental problem.

Oil and petroleum products belong to this category. Oil mostly consists of aliphatic hydrocarbons, but it may contain numerous alicyclic and aromatic ones. Besides, some oxidized compounds, e.g., aldehydes, ketones and organic acids, may also occur in petroleum, although in small quantities. Oil can enter the environent in several ways: at drilling oil wells, accidents or damages of tankers, breaks in pipelines, in refineries, or at removal or drainage of used oil from machines and motor vehicles.

On the water surfaces hydrophobic oil forms a thin film and an emulsion that strongly inhibit the gas exchange between the air and the aqueous phase. Thus, decreasing concentration of dissolved oxygen causes water animals to suffocate slowly, and enhanced amounts of carbon dioxide may result in acidosis in the cells. As an outer effect, oil sticks the feathers of sea birds together, ceasing their water-resistance and capacity to swim. Besides, water-soluble oxidized petroleum derivatives can be rather toxic.

Although several bacteria in the environment are able to degrade oil, this process is rather slow. Thus, in the meantime, evaporation of the lighter components and oxidation of the heavier particles can occur. The products of the latter process can agglomerate and sink, causing long-lasting pollution. In order to avoid this problem, oil spills can be successfully treated by TiO_2 mediated photoassisted oxidation utilizing solar energy [8].

Nowadays, among aromatic compounds, phenols as toxic chemicals mean a less dangerous pollution because their concentration in natural waters is relatively low and in the presence of appropriate microorganisms, UV rays and oxygen they can degrade 96-97% within 7 days. This process is much slower under anaerobic conditions. Nevertheless, smaller quantities of chlorophenols are enough for spoiling the taste of water. These compounds can form in the presence of strongly chlorinated drinking water. Chlorophenols belong to a class of toxic organic compounds listed by EPA as priority pollutants [9]. These contaminants also undergo TiO_2 based photocatalytic degradation [10].

In general, due to their chlorine content, organo-chloro compounds are mostly resistant to biological degrading processes [1c]. Several chlorinated hydrocarbons, e.g., organic solvents, polychlorinated biphenyls, and various pesticides belong to the poorly degradable chemicals that require more than days to degrade. Reactions by hydrolysis are characteristic in aqueous systems. Hydrolysis usually detoxifies the degraded materials as indicated by the example of the insecticide heptachlor.

Heptachlor

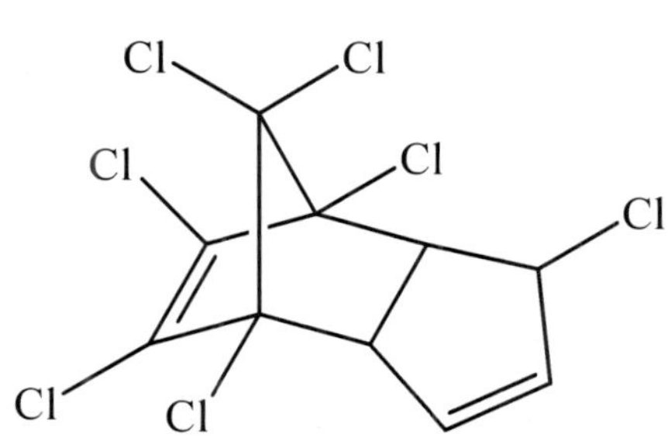

Nevertheless, this detoxifying effect is not as characteristic in the case of halogenated compounds as for esters. Cyclic compounds occasionally produe hydroxy-rings which can subsequently be opened. Under appropriate conditions both abiotic and biotic oxidation is possible. However, in several cases, the oxidation products, e.g., epoxides may be as similarly persistent and toxic as the starting material. In the muddy sediments of water, under anaerobic conditions, mostly reductive reactions take place, e.g., partial dehalogenation as in the case of DDT (1,1,1-trichloro-2,2-bis(4-chlorophenyl)ethane).

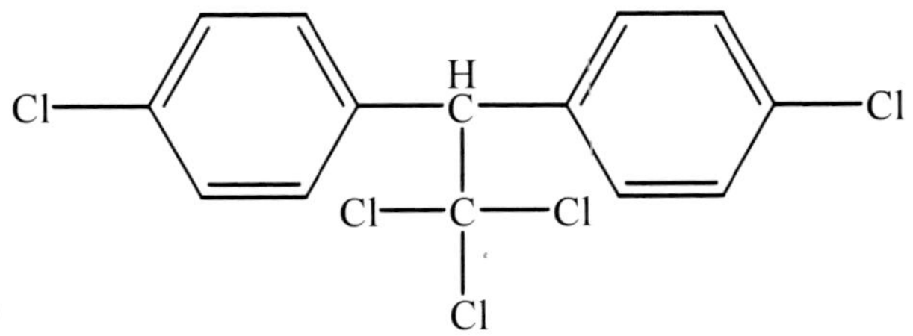

DDT

A variety of pesticides can be completely mineralized by photocatalytic procedures [11].

Lignin-sulphonic acids are also among the poorly degradable pollutants in water. They are actually lignin hydrogen sulphides, i.e., sulphide forming an ester on the propane residue of lignin. These compounds are formed when wood is treated with calcium hydrogen sulphide under elevated pressure and temperature during the manufacture of cellulose [12]. They are rather unpleasant, increasing the viscosity of water and causing taste, odour, and colour problems. Compared to the hemicelluloses and sugar, lignin-sulphonic acids degrade more gradually, and, thus, slowly. Their degradation takes several weeks, hence they are a pollution factor of long duration.

Several detergents belong to the contaminants of our environment due to their industrial and domestic use. The key ingredient of detergents is the surfactant or surface-active agent, which makes water "wetter" and a better cleaning agent. These compounds are not directly toxic, they hinder both the dissolution of atmospheric oxygen into natural waters and the sedimentation of floating particles. They concentrate at interfaces of water with gases (air), solids (dirt), and immiscible liquids (oil). Surfactants are compounds that decrease the surface tension of water and frequently produce foam due to their amphiphilic structure involving a hydrophilic and a hydrophobic molecule group [1c]. Substances of various chemical make-up can possess such a feature.

In alkyl-sulphonic acids a residue of sulphuric acid serves as the hydrophilic part. Until the early 1960s, alkyl benzene sulphonates (ABS) were the most common surfactants used in detergents.

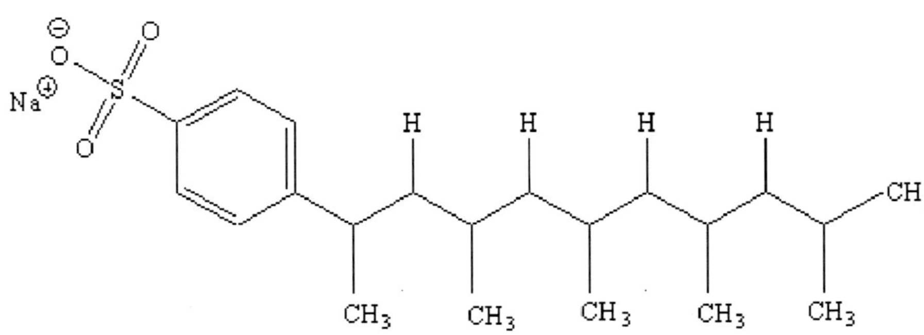

ABS

Since this type of tensids was, however, very slowly biodegradable, due to its branched-chain structure, it was replaced by linear alkyl sulphonates (LAS), decreasing the levels of the anionic surface-active agents in natural waters.

LAS

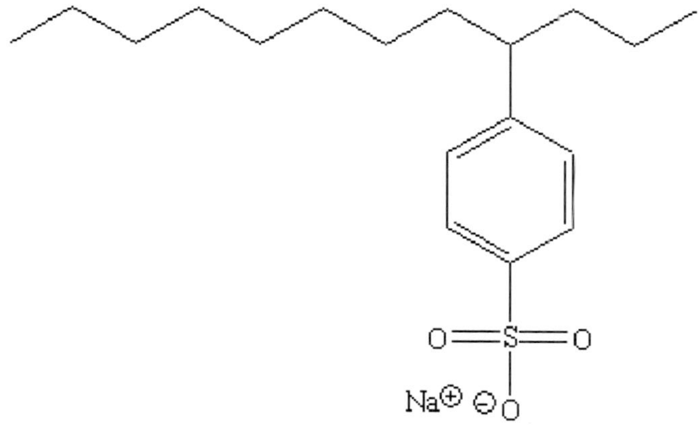

Notably, not only alkyl but also aryl sulphonates, such as naphthaline sulphonate can cause serious foam-formation on rivers.

In alkyl-ammonium compounds the polar component is a positively charged, tertiary ammonium group:

$R_1\text{-}N^+(CH_3)_2\text{-}R_2$

Hence, they are called invert soaps and mostly used in cosmetics.

In non-ionic polyoxyethenes, the hydrophilic molecule parts are alcoholic OH groups:

$R\text{-}(CH_2\text{-}CH_2O)_nH$

They can receive an ester from a fatty acid residue or an ether from a fatty alcohol residue (designated by R).

Although the danger of fish-poisoning and foaming on rivers and ponds have been mostly diminished, rather low surfactant concentrations (e.g. 0.05-0.1 mg/l) are enough to mobilize toxic materials adsorbed in the sediments.

Photocatalysis is an efficient and simple method also for surfactant decontamination [13-17].

2.2. Heavy metals

These elements are, mostly, the transition metals, and some of the representative elements, such as lead and tin, in the lower right-hand corner of the periodic table [1c]. Most of them have an extremely strong affinity for sulphur, and disrupt the original physiological function of different organic compounds (such enzymes) in the body. Protein carboxylic acid ($-CO_2H$) and amino ($-NH_2$) groups are also chemically bound to heavy metals. Some of them bind to cell membranes, hindering transport processes through the cell wall. Heavy metals may precipitate phosphate biocompounds too or catalyze their decomposition.

The main problem with heavy metals that they are toxic not only in their chemically combined forms, but also in the elemental forms, especially in the case of mercury. Since a number of industries emit heavy metals, they are among the most complicated causes of water pollution. Some household wastes can also contribute to this problem. Fortunately, considerable part of heavy metal content in natural waters are adsorbed by mineral or organic sediments, or precipitated as carbonates, sulphites, and sulphides. However, saturation of sediments can increase the concentration of dissolved heavy metals. Besides, chelate-forming materials such as ethylene diamine tetraacetate (EDTA) or nitrile-triacetate (NTA), which may be contained in household cleaners, can mobilize heavy metal ions [12]. Some heavy metals undergo microbial alkylation, becoming able to get into the food chains. For example, Hg^{2+} can be methylated in two steps:

$$Hg^{2+} \rightarrow CH_3Hg^{+} \rightarrow (CH_3)_2Hg$$

Methyl-mercury, similarly to organic derivatives of other heavy metals, is mutagenic too: it can occupy different positions on the nucleic-acid base adenine, hindering its connection to the sugar component of the nucleic acid or to thymine or uracil in the synthesis of nucleic acid. The mutations emerging from these effects can cause cancer by changing the activity of the genome of the eukaryote cells or activating virus-DNA.

Mercury(II) and lead(II) and some other heavy metals, forming alkyl derivatives, can enter the nervous system and cause neurotoxic symptoms such as damage of senses (sight, hearing, or touch) and memory loss. Mercury was the first metal the bio-accumulation of which we experienced in the case of the Minamata sickness in 1953 in Japan. Heterogeneous photocatalysis has been proposed as a convenient tool for mercury removal utilizing even solar light [18, 19].

Toxic effects of the heavy metals are caused by their chelation and sulphide formation with biologically active materials, suppressing their original physiological function. Compounds of Pb^{2+} and Cd^{2+} can enter the bones. In the bone marrow lead hinders the synthesis of blood by inhibition of the enzyme 5-amino-levulin-acid-dehydrate, leading lead anaemia. Cadmium(II) excludes zinc(II) from certain enzymes, and can also replace Ca^{2+} in the bones. The latter effect causes very painful bone deformations known as "Itai-itai sickness" from the cry of pain. Successful photocatalytic procedures for removal of cadmium(II) and lead(II) dissolved in water were also realized [20-22].

Chromium compounds, especially CrO_4^{2-}, can easily enter the cells, mostly because its structure is similar to that of SO_4^{2-}. Being a very oxidative agent, when Cr(VI) penetrates into the core of the cell, it can also change nucleic base by oxidation. But as soon as a reduction to Cr(III) occurs, the transfer through the membrane (the wall) of the cells ceases. The suspected carcinogenic chromium(VI) species is present in a great variety of industrial wastes. It is difficult to be removed from water by physical or chemical methods. Contrary, Cr(III), another environmentally stable oxidation state of chromium, is relatively noncarcinogenic and less toxic. It is not as mobile as Cr(VI) and can be easily removed from alkaline medium. Thus, photocatalytic reduction of Cr(VI) to Cr(III) offers a feasible solution for the treatment of chromium contamination of waters [23, 24].

3. Generation of Radicals Using Various Methods

Conventional water treatment plants generally utilize the processes of mechanical removal, filtration, activated sludge, and chlorination to clean wastewaters containing pollutants of different origin. Unfortunately, these processes are typically not effective in remediating industrial organic contaminants such as substituted aromatic compounds like chlorinated benzene, phenol and benzoic acid derivatives. In addition chlorination which is a widely used procedure to oxidize organic containing effluents often results in a significant increase in the toxicity of organic compounds due to the formation of chlorinated hydrocarbons and aromatic compounds. Considerable efforts have been devoted to develop alternative remediation strategies that provide effective cleaning of the waters contaminated by anthropogenic chemicals. Because of chemical treatment of organic compound using chlorine can led to the formation of substances equally or more toxic than the parent compounds, effective cleaning processes must also address the need for transformation to non toxic materials or mineralization of the contaminants to nature friendly inorganic substances. To realize this approach detailed researches have been focussed on advanced oxidation processes (AOPs) that utilize non chlorine based oxidants of high reactivity. Definitely, advanced oxidation processes are technologies that employ the highly reactive hydroxyl radical as the main oxidizing species for mineralization of a great variety of toxic organic compounds. The hydroxyl radical can be generated by various methods in aqueous systems, such as high energy radiation either by γ rays or high-energy electrons, high frequency ultrasound waves, irradiation by UV light the solution containing H_2O_2, O_3 containing systems using UV light or/and H_2O_2 to promote hydroxyl radical formation, the Fenton reaction (H_2O_2/Fe^{3+}) and photo Fenton reaction, semiconductor particles (SP), such as TiO_2 ZnO, CdS etc., and UV light [25]. Among these processes the last proved to be very effective in various conditions.

3.1. On semiconductor based photocatalysis

Using semiconductor particles (SP), to generate hydroxyl radicals is a practically quite simple method. These particles are readily excited by photons of higher energy than the band gap between the valence and conduction band. The photon absorption leads to form electron-hole pair inside the particle. These species can be recombined within the semiconductor particle or can migrate to the surface of the particle where, in the aerated aqueous reaction mixture, the primary electron transfer reactions compete with the recombination resulting in highly oxidizing radicals:

$SP + h\nu \rightarrow SP\{h^+:e^-\}$	(generation of $\{h^+:e^-\}$pair)	(1)
$SP\{h^+:e^-\} \rightarrow SP$	(recombination)	(2)
$SP\{h^+:e^-\} + O_{2(aq)} \rightarrow SP\{h^+\} + O_2^{-\bullet}$	(electron transfer from CB)	(3)
$SP\{h^+\} + H_2O \rightarrow SP + {}^\bullet OH + H^+$	(electron transfer to VB)	(4)

One of the most important aspects of environmental application of the SPs is that some of these materials are close to the ideal photocatalysts possessing the following characteristics:

chemical and biological stability, non toxicity, relatively low band gap energy, the photogenerated holes are highly oxidizing, and the photogenerated surface electrons are reducing enough to produce superoxide ions in the presence of oxygen, and relatively low cost. It has been turned out that TiO_2 is excellent for photocatalytic mineralization of great variety of organic compounds, because its properties fulfill the requirements above. The energy band diagram (Figure 1) for TiO_2 particle suspended into aqueous solution of pH=7 clearly shows that the redox potential of photogenerated holes (h^+_{VB}) is +2.53 V versus standard hydrogen electrode (SHE). Its reaction with water (4) can produce hydroxyl radical ($^{\bullet}OH$), whose redox potential is only slightly decreased (+2.27 V). On the other hand the reduction potential for conduction band electrons (e^-_{CB}) is -0.52 V, which is enough to evolve hydrogen from water, but these species can become trapped and lose some of their reducing power.

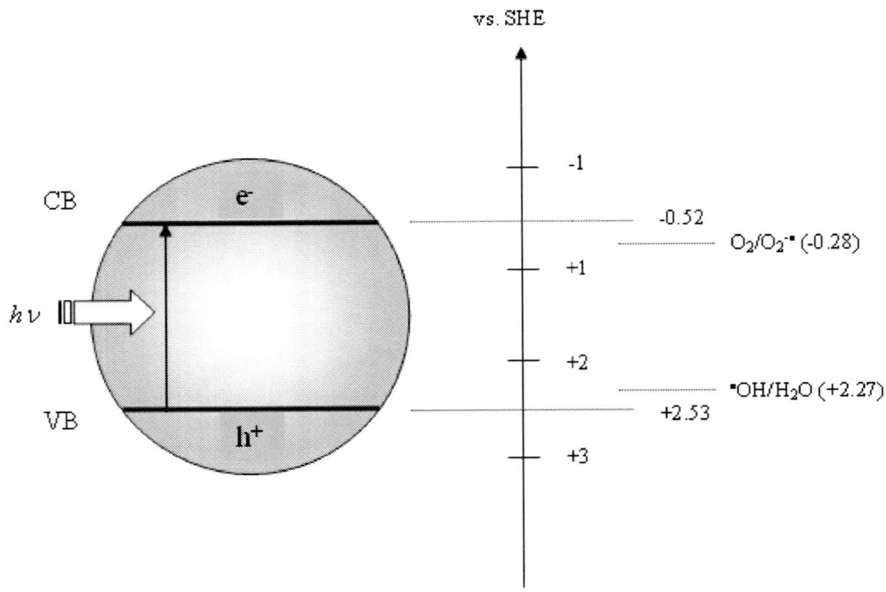

Figure 1. Schematic diagram showing the potentials for dominant redox processes occurring on the surface of the excited TiO_2 particle in aqueous suspension at aerobic conditions.

However, even after trapping, a significant number are still able to reduce oxygen to superoxide anion ($O_2^{-\bullet}$), or to hydrogen peroxide. There are of organic compounds, dominantly those that readily adsorbed at the surface of SP particles and hence can directly react with holes. These compounds are generally oxidized with high efficiency in the presence of molecules or ions readily accepting electrons from CB.

The degradation efficiency of a given organic compound over light induced TiO_2 particles strongly depends on the nature of the SP particles itself. In this regard the most important characteristics are the crystalline structure, the size with the size distribution and the specific surface area of the semiconductor particles. TiO_2 has different crystalline structures; rutile, anatase and brookit. The rutile structure is based on a slightly distorted hexagonal close packing (hcp) of oxygen atoms with half the octahedral interstices being occupied by titanium atoms. Anatase and brookit are both based on cubic rather than hcp of oxygen atoms. These crystals have slightly different bad gap energies (e.g $E_{bg(rutile)}$=3.0 V and $E_{bg(anathase)}$=3.2 V). Since the energy level of their valence band are equals, the threshold energy is higher for anatase, and the electron promoted to the conduction band of this TiO_2 has higher reducing power than that of excited to CB of the rutile particle. Hence the driving force for electron transfer from CB to a given substrate molecule is expected to be higher for anatase than for rutile. The specific surface area has crucial importance. It determines the partial mole fraction of the surface complex of the substrate, and thus the efficiency of the primary electron transfer processes too.

It is important to note that the quantum efficiencies of the photocatalytic mineralization of organic pollutants on commercial bare titania is relatively low, due to the rapid recombination of the photogenerated hole-electron pair within the semiconductor particle. So, the improvement of photocatalytic activity of TiO_2 is one of the most important task for technical applications of heterogeneous photocatalysis in the future. Therefore, in recent years a large number of investigations on the enhancement of photocatalytic activity have been carried out. To achieve significant increase in the efficiency of photoreactions, two principle strategies have been preferred. On the one hand, the procedures for preparation of pure TiO_2 particles have been successfully developed and optimized, and thus in the last decade many new and highly active photocatalysts were produced [26-29]. On the other hand, the modification of bare TiO_2 has been intensively investigated too. These projects include the coupling TiO_2 with other oxide or sulfide semiconductors [30], the covering of the surface with dyes to extend the light absorption in the visible range [31-33] and the doping with metal and non-metal ions [34-38] and the partial depositing with nano-sized noble and transition metal particles [39-43].

4. Application of TiO_2

4.1. Metal reduction and deposition

While utilization of semiconductor for photocatalytic degradation of different organic compounds has been thoroughly studied, it was scarcely investigated for the purpose of decontamination of heavy-metal polluted waters [34, 44-49]. Nevertheless, under appropriate conditions quantitative removal of heavy metal ions has been proved to be possible with irradiated TiO_2 semiconductor particles [50-57]. Mostly, the metal ions are deposited on the surface of the semiconductor particles in a reduced, elemental form. The efficiency of this process depends on the standard reduction potentials of the metal ions, the Gibbs free energy change of the overall reaction, the presence or absence of oxygen, the concentration of the sacrificial electron donor, the irradiation time and the pH of the system. Surfactants as potential electron donors may also affect the photocatalized reduction of heavy metal ions.

As mentioned in Section 3.1., metal or metallate ions with a reduction potential more positive than the conduction band of the photocatalyst can, in principle, serve as electron acceptor. For example, with a toxic metal ion M^{2+}:

$$2M^{2+} + 4e^- \xrightarrow[h\nu]{TiO_2} 2M \quad (5)$$

In this instance, the following overall reaction can take place in the presence of titanium dioxide photocatalyst irradiated:

$$2M^{2+} + 2H_2O \rightarrow 2M + O_2 + 4H^+ \quad (6)$$

If the Gibbs free energy change for this reaction is negative ($\Delta G^0<0$), the radiant energy simply serves to overcome the activation barrier for the process [58]. In most cases, however, Eq (6) involves a positive ΔG^0, i.e., the light drives the system in the thermodynamically up-hill direction, except for metal ions with very positive potentials, i.e., when they have standard reduction potentials that are more positive relative to the O_2/H_2O redox couple. But these reactions too can be made thermodynamically favourable with the addition of appropriate hole scavengers, resulting in a negative free energy change for the overall process.

4.1.1. Reductive deposition of mercury and lead

For the mercury(II) ion (Hg^{2+}) the standard reduction potential more positive than that of the O_2/H_2O pair at pH=7 (0.85 V *vs* 1.229-0.059pH = 0.816 V [59]). This is, however, just a tiny difference between the redox potentials. Moreover, mercury(II) occurs aqueous systems mostly as coordination compounds, e.g., $HgCl_2$, which sparingly dissociate [60]. Complex formation in this case decreases the redox potential (e.g., for the $HgCl_2/Hg$ couple is 0.41 V [61]), making the photoreductive deposition of mercury thermodynamically unfavourable. Thus, in an air-saturated solution less than 10 % of the initial concentration (1×10^{-3} M) of Hg(II) could be reduced after 10-min irradiation. Deoxygenation of the system increased the efficiency of reduction to 40%. For a much more effective deposition, however, a sacrificial electron donor had to be applied. Addition of 5×10^{-4} M ethanol (as a hole scavenger) dramatically increased the efficiency (over 85%) [62]. It clearly indicates that the overall reaction

$$HgCl_2 + CH_3CH_2OH \xrightarrow[h\nu]{TiO_2} Hg + 2\,HCl + CH_3CHO \quad (7)$$

is thermodynamically favourable. Figure 2 shows the dependence of the reduction efficiency vs. ethanol concentration plot.

Another very toxic heavy metal is lead, thus its removal from wastewater is of basic importance. The standard redox potential of the couple Pb^{2+}/Pb is, however, rather low (-0.125 V [63]), moreover that of the couple Pb^{2+}/Pb^+ is even lower (-1.0 V [64]). Hence, for its efficient reduction extremely high ethanol concentration is necessary (40 v/v % $\approx$ 7 M). Even so, a maximum of 84% efficiency could be reached (Figure 2) in deoxygenated system containing 1×10^{-3} M Pb(II).

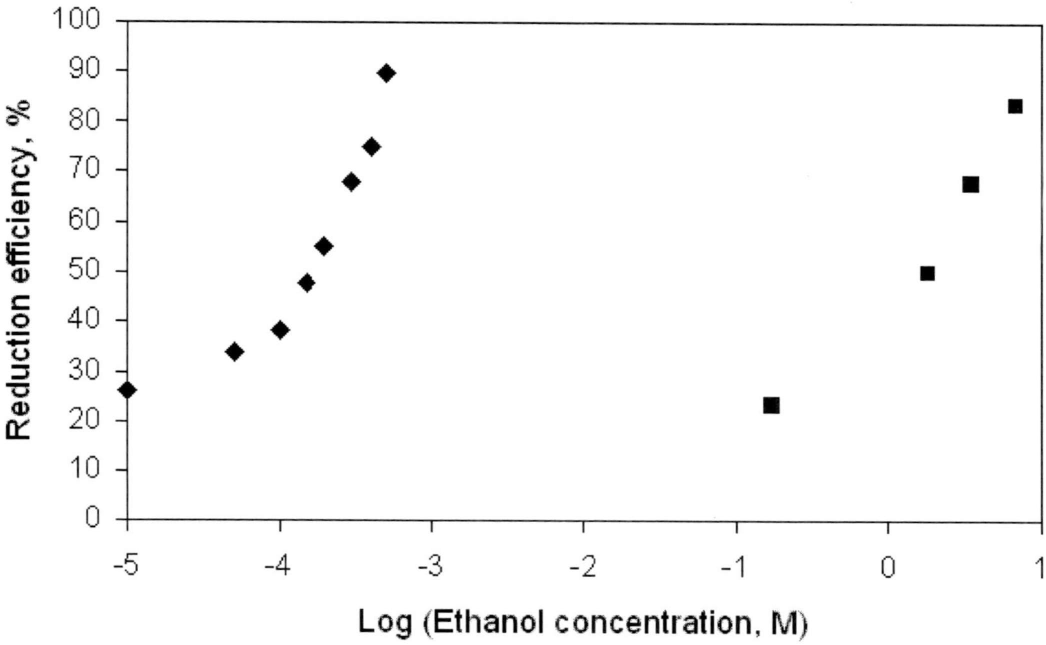

Figure 2. Efficiency of Hg(II) (◆) and Pb(II) (■) reduction versus logarithm of ethanol concentration. Irradiation time is 10 minutes.

2-propanol has been proved to be a more effective electron donor for the photocatalytic reduction of lead(II) than ethanol, due to the stronger reducing ability of the 1-hydroxyalkyl radicals [65]. Notably, a more efficient, almost complete removal of lead(II) was reached by deposition in the form of PbO_2 in photoinduced oxidation on Pt-loaded TiO_2 particles [65, 66]. Metal-free TiO_2, however, proved to be ineffective for this purpose.

While photoassisted reduction of lead(II) is extremely oxygen sensitive, even at higher ethanol concentration, no similar phenomenon can be experienced with mercury at relatively lower concentrations (> 0.01 M) of EtOH. Thus, practically 100% deposition can be reached in the presence of 0.03 M ethanol in air-saturated system.

Effects of surfactants and chloride

Not only ethanol, but other organic compounds such as different surfactants occuring in wastewaters can serve as electron donors in the photocatalytic reduction of heavy metal ions. Thus, sodium dodecyl sulphate (SDS) as an anionic tensid and cetyltrimethylammonium bromide (CTAB) as a cationic surfactant were tested as potential electron donors [62, 67]. Since the degradation of SDS on TiO_2 is an oxidation reaction [68], hence, this process was suspected to promote the photoreduction of Hg(II). In fact, it proved to be a significantly better hole scavenger than ethanol. Faster could be reached practically 100% deposition in the presence of 0.03 M SDS than with ethanol of the same concentration. Moreover, addition of ethanol to the system containing SDS decreased the efficiency of the photoassisted reduction of Hg(II) [62].

Deviating from SDS, CTAB (0.01 M) only slightly promoted the photoreduction of mercury(II). This surfactant proved to be less efficient electron donor than ethanol. This phenomenon might be associated with the charge of the electron donors. The positive charge of the surfactant probably hinders the efficient reaction with the holes on the surface of the TiO_2 particles. For the same reason, the correspondig reaction with the anionic dodecyl sulphate species is favoured. This observation suggested that the charges of the reactants in the photocatalytic redox processes strongly influence the overall efficiency, thus the effects of complex formation and pH were also studied in this system [67].

Addition of chloride ions (0.1 M NaCl) to the air-saturated system containing 5×10^{-4} M SDS as sacrificial electron donor resulted in a dramatic decrease in the efficiency of the photocatalytic reduction of mercury(II) (1.0×10^{-3} M). While in the absence of Cl^- ions nearly 100% deposition could be reached within 10-min irradiation, in salt solution the reduction efficiency did not exceed 40 %. The negative charge of the $HgCl_3^-$ and $HgCl_4^{2-}$ complexes formed hinders the reduction of the metal center by the conduction-band electrons, even in the presence of added hole-scavenger. Moreover, further irradiation resulted in a considerable decrease in the efficiency. This phenomenon suggests that complex formation, as a chemical driving force, promotes the back-oxidation of the deposited mercury in air-saturated system. Besides, large amount of chloride ions in solutions strongly hinders the adsorption of the Hg(II) species and may result in the blocking of active surface sites on TiO_2. In the presence of formate, an efficient photocatalytic separation of metals could be realized in this system containing Cr(VI)/Hg(II) mixture [69]. The important role of the adsorption in the photoassisted reductive deposition of mercury has been demonstrated by application of 2-aminothiazole-modified titania [70]. In this case, the presence of heteroatoms, sulfur and nitrogen, with free electron pairs, promotes the coordinative adsorption of Hg(II), enhancing the efficiency of its photocatalytic reduction.

Effects of pH

According to the expectations, similarly to the complex formation, pH also strongly affects the reduction efficiency of mercury(II) both in the absence and in the presence of SDS as hole scavenger [67]. As it is shown in Figure 3, upon increasing pH, a dramatic increase in the efficiency occurred in the range of pH 5-6.5.

This might be associated with the surface charge of the semiconductor particles, because the zero point of charge (pH_{zpc}) of TiO_2 (both anatase and rutil) is in the same range [71]. Besides, in the oxidation of water molecule (as a hole scavenger) protons are formed, which react with hydroxide ions at higher pH. The slight decrease of the efficiency in the basic solutions may be caused by the formation of anionic mixed ligand halo-hydroxomercurate(II) complexes, hindering the approach to the negatively charged surface of the titanium dioxide particles.

In the presence of SDS hole scavenger (0.003 M) a similar pH dependence of the reduction efficiency was observed (Figure 3), due to the same effects. In this case the decrease of the efficiency in the basic range is much steeper than without SDS. This phenomenon can be attributed to the repulsion between the negatively charged catalyst surface and dodecyl sulphate ions.

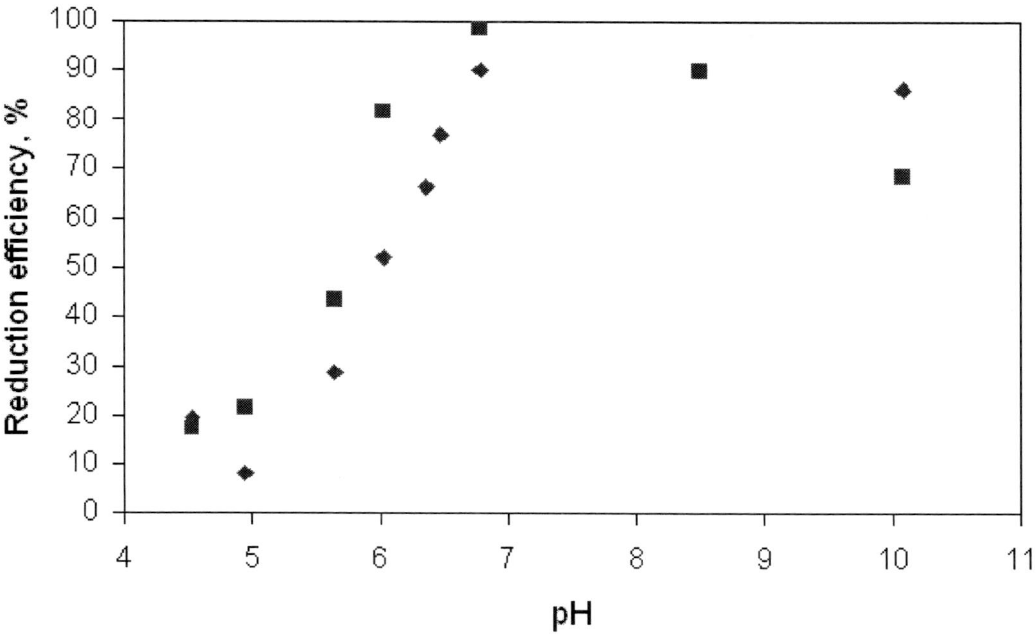

Figure 3. Efficiency of Hg(II) reduction versus initial pH at 3-min irradiation time in air-saturated system in the absence (◆) and in the presence (■) of SDS (3×10^{-3} M).

4.1.2. Photocatalytic reduction of chromate

Depending on the pH, chromate exists in monomeric ($HCrO_4^-/CrO_4^{2-}$) or dimeric ($Cr_2O_7^{2-}$) forms. Accordingly, the pH of the system strongly influences the mechanism of its reduction.

In acidic media:

$$HCrO_4^- + 7\,H^+ + 3\,e^- \leftrightarrow Cr^{3+} + 4\,H_2O \quad E^0 = 1{,}19\text{ V (SHE)} \tag{8}$$

$$Cr_2O_7^{2-} + 14\,H^+ + 6\,e^- \leftrightarrow 2\,Cr^{3+} + 7\,H_2O \quad E^0 = 1{,}23\text{ V (SHE)} \tag{9}$$

In neutral or basic media:

$$CrO_4^{2-} + 4\,H_2O + 3\,e^- \rightarrow Cr(OH)_{3(ad)} + 5\,OH^- \; E^0 = -0{,}13\text{ V (SHE)} \tag{10}$$

Due to the higher reduction potential, the reaction is much faster in acidic than in neutral or basic systems [46, 72, 51, 73-76]. In the latter cases, however, precipitation of $Cr(OH)_3$ may promote the removal of chromium [77]. Besides, the surface charge of titanium dioxide particles is also very pH dependent. One order of magnitude more chromate ions are adsorbed at pH 2.5 than at pH 6 in a system containing 1 g/dm^3 TiO_2 [73]. Above pH_{zpc} (about 6.5 [23]), adsorption of chromate ions is negligible [74]. Moreover, in neutral or alkaline media dissolved oxygen becomes an efficient competitor for the conduction-band electrons [77, 78]. Observations in acidic systems (at pH 2) indicated that, similarly to the case of mercury(II) reduction, water is a rather weak hole scavenger in the photocatalytic redox process [79]. Deviating, however, from the photoassisted deposition of mercury, addition of dodecyl sulphate (4×10^{-4} M) signficantly diminished the reduction efficieny (Figure 4).

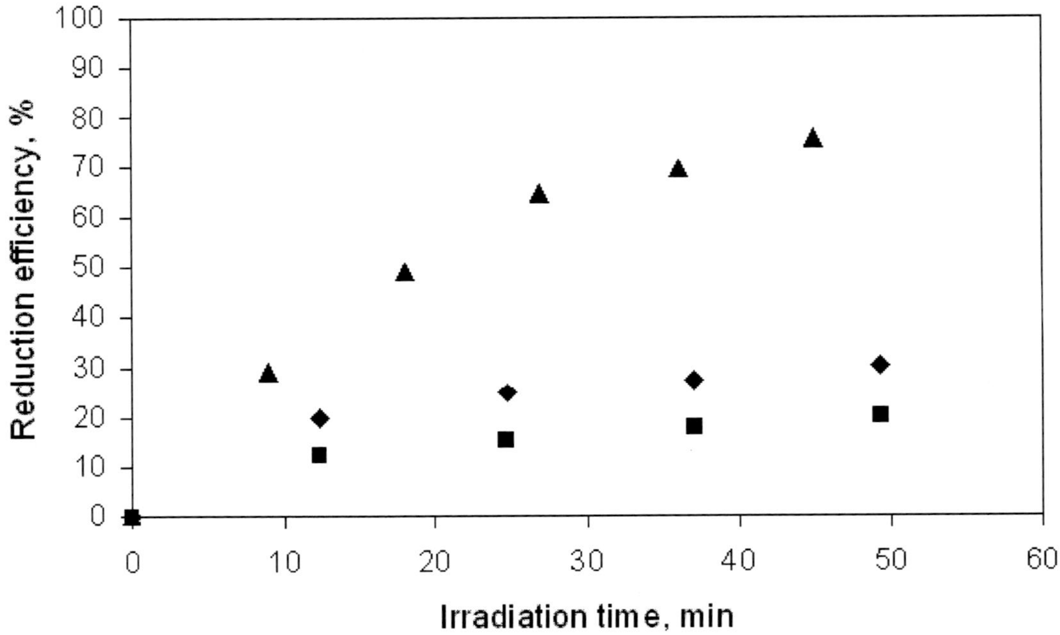

Figure 4. The effect of detergents on the reduction of Cr(VI) to Cr(III): (◆) with no additive, (◆) with SDS, (▲) with CTAB (pH=2.5).

This phenomenon clearly indicates the determining role of the competitive adsorption of the anionic surfactant, while its electron donor ability is less dominant in this case. Accordingly, addition of cationic surfactant (cetyltrimethylammonium ion) dramatically increased the conversion of chromate. This hole scavenger is not adsorbed to the similarly charged particle surface. Notably, it forms photosensitive ion-pair with dichromate, promoting the light-driven charge transfer reaction. Citric acid also proved to be an efficient electron donor in this system [79, 80], while, surprisingly, salicylic and oxalic acids considerably decelerated the reduction of chromate [79].

4.2. Mineralization of organic compounds

4.2.1. Natural organic compounds - of biological importance

Photodegradation of some organic compound of biological importance

Biochemical applications of TiO_2-mediated photodegradation have been proved to be promising for elimination of microorganisms, e. g. bacteria, fungi, and virus, among others. In vivo suppression of cancer cells by illumination of titania was also reported. In living organisms proteins play enormous variety of roles. Considering the multiplicity of their functions, proteins are extremely complex polymer molecules. The monomers that combine to make them are α-amino acids. So to understand the photooxidation of biological system the study of amino acid degradation over light excited SPs is of crucial importance, especially from mechanistic aspects. Investigations on the reactions of various amino acids with hydroxyl radicals and solvated electrons generated by pulsed and continuous radiolysis [81-

87] provided several salient mechanistic features regarding the modes of attack by these radicals that appears as primary species of high reactivity over the excited TiO_2 surfaces.

Decomposition of a series of amino acids of different structures in aqueous titania dispersions under illumination with UV-A and UV-B was carefully studied by Hidaka et al [88]. The decomposition has been analyzed by measuring the final products, such as NH_4^+, NO_3^- and CO_2. The ratios of evolved nitrogen containing products (NH_4^+, NO_3^-) and the rate constants of the formation of these species and that of CO_2 were compared. The mechanistic features were also discussed in this article and it had been concluded that the formation of NH_3 implicated both TiO_2 surface electrons and surface $^{\bullet}OH$ radicals while the reaction with $^{\bullet}OH$, $HO_2^{\bullet}$ radicals lead to formation of CO_2 in various pathways. A more detailed study focussed to the photooxidation of L-analine, L-serine and L-phenylalanine and involved molecular orbital calculations confirmed the proposed mechanism [89]. This would assist in understanding of the deterioration of biological functions which has not been elucidated so far.

In a recent study L-α-alanine was employed as an amino acid and its light induced photodecomposition over various kinds of TiO_2 was investigated. To characterize some of the key intermediates ^{1}H-NMR studies have been performed on photolyzed system [90]. L-alanine was dissolved into D_2O containing 1 mg cm^{-3} TiO_2. The concentration of amino acid and the pH of the solution to be irradiated by a high pressure Hg lamp was adjusted to 10 mM and 5.6±0.3 respectively in NMR sample tube. It was demonstrated that without semiconductor particles no change in the NMR spectrum of amino acid was followed. As it was expected, upon illumination of TiO_2 containing samples with decrease of peak intensities of L-analine signals, several photo-decomposition species appeared. Comparing the chemical shifts, coupling constants, and coupling patterns with those of the authentic reagents some of the intermediates have been identified. Besides the main components of acetic acid, acetaldehyde, paraldehyde, small amount of pyruvic acid, trimer of pyruvic acid, acetamide, formaldehyde, and formic acid were identified. Two different decomposition routes have been revealed. In addition it was also pointed out that the efficiency of these routes depends on the characteristics of the TiO_2 particles, especially on the crystalline structure as it is demonstrated by the next schemes.

$$CH_3CH(NH_3^+)(COO^-) \xrightarrow{TiO_2+h\nu} CH_3COH \quad \text{(acetaldehyde)} \qquad (11)$$

$$CH_3CH(NH_3^+)(COO^-) \xrightarrow{TiO_2+h\nu} CH_3CO(COOH) \quad \text{(pyruvic acid)} \qquad (12)$$

$$CH_3COH \xrightarrow{TiO_2+h\nu} CH_3COOH \quad \text{(acetic acid)} \qquad (13)$$

$$CH_3CO(COOH) \xrightarrow{TiO_2+h\nu} CH_3COOH \quad \text{(acetic acid)} \qquad (14)$$

Scheme 1. Main reaction route occurring over rutile.

$$CH_3CH(NH_3^+)(COO^-) \xrightarrow{TiO_2+h\nu} CH_3COH \quad \text{(acetaldehyde)} \tag{15}$$

$$CH_3COH + NH_3 \xrightarrow{TiO_2+h\nu} CH_3CONH_2 \quad \text{(acetamide)} \tag{16}$$

$$CH_3COH \xrightarrow{TiO_2+h\nu} CH_3COOH \quad \text{(acetic acid)} \tag{17}$$

$$CH_3CONH_2 \xrightarrow{TiO_2+h\nu} CH_3COOH \quad \text{(acetic acid)} \tag{18}$$

Scheme 2. Main reaction route occurring over anatase.

The final two process of both reaction routes are the followings:

$$CH_3COOH \xrightarrow{TiO_2+h\nu} HCHO \quad \text{(formaldehyde)} \tag{19}$$

$$HCHO \xrightarrow{TiO_2+h\nu} HCOOH \quad \text{(formic acid)} \tag{20}$$

So acetic acid is a stable major intermediate, which is produced via an intermediate of acetaldehyde using either rutile or anatase. Pyruvic acid is formed dominantly over rutile particle, on which the formation of hydroperoxo species has been reported, while acetamide is the first detectable intermediate on anatase surface. It has been suggested that the reaction of acetaldehyde and the ammonia, which is released by the primary electron transfer leads to the formation of acetamide. It was deduced by considering some incubation time characterizing the appearance of this species.

A detailed investigations on photodegradation of aspartic acid over UV irradiated bare and silver deposited titanium dioxide of various structure and particle size have been performed recently [91]. The results of this study clearly demonstrated that the nitrogen content of the amino acid is converted dominantly into NH_3 (detected as NH_4^+). The evolution of CO_2 was also measured. Both of these processes resulted in a considerable increase of pH in the liquid phase of the reaction mixture providing significant change in the reaction mechanism. Under prolonged irradiation NH_4^+ is oxidized through NO_2^- intermediate to NO_3^-. (Figure 5).

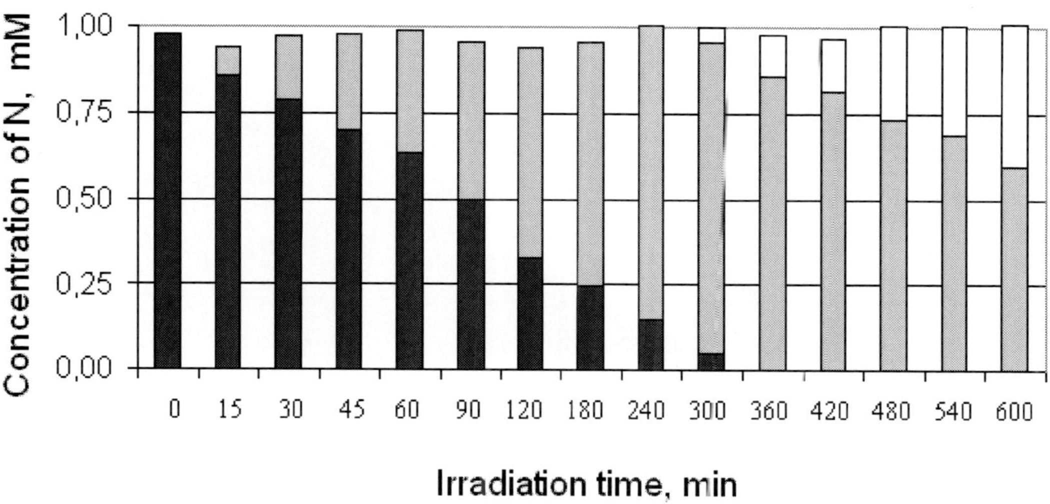

Figure 5. Nitrogen balance of the reaction mixture of 1x10^{-3} M aspartic acid and 1 g dm^{-3} TiO_2 under irradiation: concentration of aspartic acid (black), concentration of NH_4^+ (gray) and concentration of NO_3^- (white).

The primary steps of the mechanism of photodegradation are analyzed and described considering the results of competitive kinetic studies performed by using oxalic acid as efficient hole scavenger, coumarin as $^{\bullet}$OH scavenger and various bare and silver deposited titanium dioxide. The disappearance of the amino function under photolysis using P25 TiO_2 occurs rapidly, while the evolution of ammonium ion (acidic conditions) is rather slow over both anatase (Aldrich) and rutile (Fluka). The silver deposition to the surface of these semiconductor particles resulted in significant increase in the photomineralization rates, especially in the case of anatase and rutile. To obtain further information on the effect of the photodeposited silver the formation of $^{\bullet}$OH radical over the photoexcited bare and silver modified titatania has been investigated and compared. These experiments, in which coumarin and oxygen were used as scavengers for $^{\bullet}$OH radical [92, 93] and electron respectively, revealed that among the investigated bare titania it is the Aldrich catalyst that produces hydroxyl radicals with the lowest efficiency. On the other hand, the silver deposited form of this catalyst produces most efficiently the hydroxyl radicals. The results of these radical scavenging experiments have been compared with those obtained for photodecomposition of aspartic acid. The data suggest that the silver deposit on TiO_2 surface provides a better separation of charges generated by photons $\{h^+:e^-\}$ because the silver clusters act as electron traps. Thus the electron transfer from water to hole competes with the $\{h^+:e^-\}$ recombination (reaction 2) more effectively using catalyst modified by silver deposit than the bare titania. Competitive kinetic studies have been performed using various bare and silver modified titanium dioxide and oxalic acid as efficient hole scavenger. These experiments have clearly indicated that the photodecomposition of aspartic acid over bare TiO_2 particles is inhibited by oxalic acid (Figure 6).

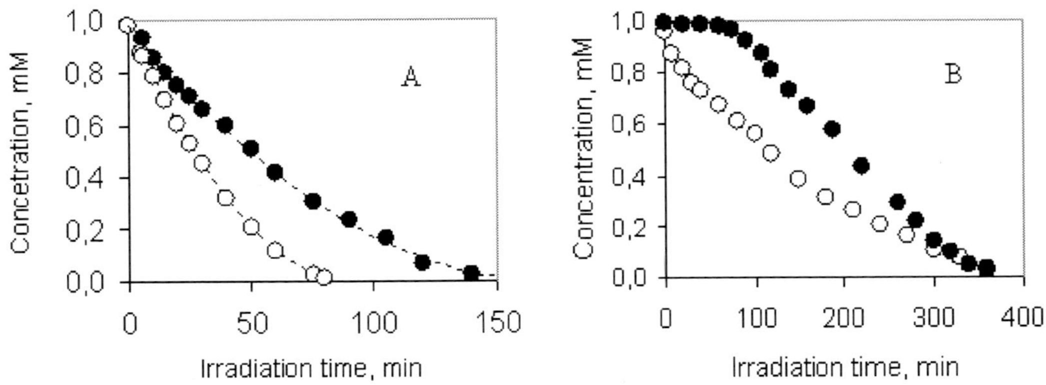

Figure 6. Competitive photodegradation of oxalic and asparctic acids over Degussa P25 photocatalyst: A.) Photooxidation of oxalic acid at $[Ox]_0=1x10^{-3}$ M (○), photooxidation of oxalic acid at $[Ox]_0=1x10^{-3}$ M and $[Asp]_0=1x10^{-3}$ M (●); B.) Photodecomposition of aspartic acid at $[Asp]_0=1x10^{-3}$ M (○), photodecomposition of aspartic acid at $[Asp]_0=1x10^{-3}$ M and $[Ox]_0=1x10^{-3}$ M (●).

Moreover, there is a significant decrease in the photooxidation rate of the oxalic acid in the presence of aspartic acid. These findings have been explained by a mechanism in which the key reaction step is the reaction of $^{\bullet}COOH$ radical, which is the primary product of the electron transfer from oxalate ion to the hole, with the radical species obtained by the electron transfer from conduction band to the aspartic acid. The silver clusters deposited on the surface of semiconductor particles resulting in the efficient charge separation providing a relatively large distance between the primary radicals. Hence a considerable decrease in the efficiency of their reaction is followed, that leads to minimize the inhibition of the photodecomposition of aspartic acid by oxalic acid. So this study has provided further evidence that combination of the silver deposition with using suitable hole scavenger can be applied as an efficient tool for deepening our knowledge on the elementary reaction steps occurring over the surface of the photoexcited semiconductor particles.

4.2.2. Degradation of surfactants

Since both anionic and cationic surfactants proved to be efficient hole scavengers in the TiO_2-mediated photoassisted reduction of Hg(II), one can expect that they can be oxidatively degraded in this photocatalytic system. In fact, SD^- undergoes photoinduced oxidation in air-saturated aqueous TiO_2 suspension with no additives (Figure 7) [67, 68]. Similarly to the reductive deposition of mercury, also this process is rather sensitive to pH. At pH 2 the oxidative conversion of this surfactant approched 100% after 25-min irradiation. In the same system at pH 6, the efficiency was much lower [62], indicating the determining effect of the adsorption in this case too (Figure 7).

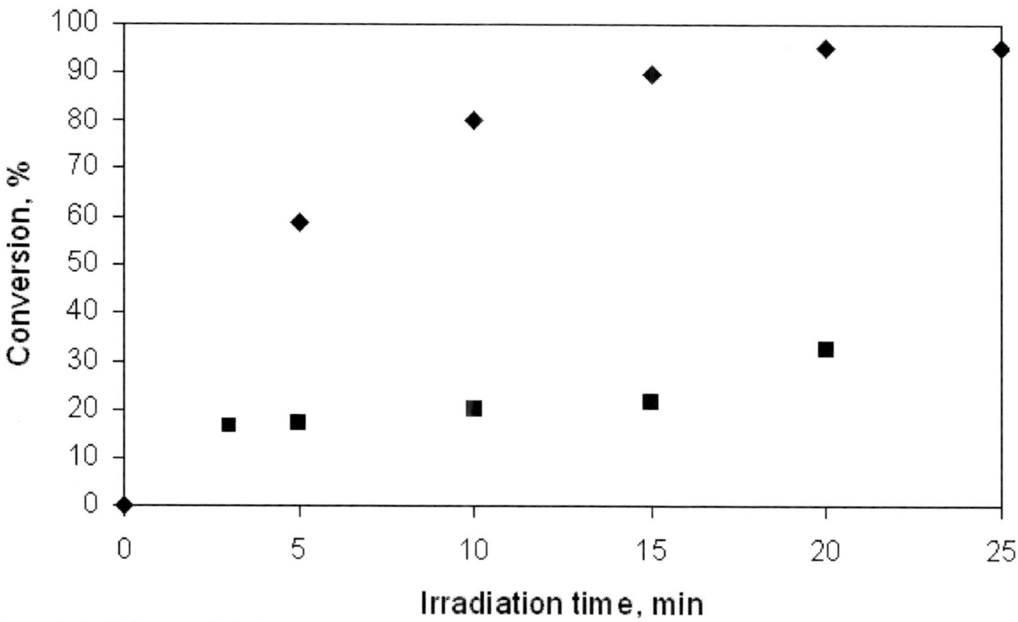

Figure 7. Efficiency of dodecyl sulfate degradation versus irradiation time at pH=2 (◆) and 6 (■).

The surface charge of the catalyst particles is positive at pH 2 promoting the adsorption of the anionic detergent, while pH 6 is very close to the zero point of charge of the titanium dioxide. It should, however, be mentioned that in the presence of mercury(II) as electron scavenger a slight increase in the efficiency of the conversion of SD^- was observed, which can be attributed to the formation of proton in the oxidation process (Eq. 21).

$$-CH_2- + 6h^+ + 2H_2O \longrightarrow CO_2 + 6H^+ \qquad (21)$$

In Eq. 21 $-CH_2-$ designates one unit of the reducing hydrocarbon chain of dodecyl sulphate. The decrease of pH in this system (not buffered) during the irradiation from the initial values of 4.5-8.5 to 2.8-3.0 supports the interpretation above [67].

Electron acceptors can be expected to enhance the efficiency of the oxidative degradation of organic hole scavengers. Accordingly, in the presence of 1×10^{-3} M $HgCl_2$, the average rate of the SD^- conversion in this system was about three times higher than that without additives [67]. Similar effect was observed when Cu^{2+} ions was applied as a positively charged oxidant. The presence of chromate ions ($HCrO_4^-/Cr_2O_7^{2-}$) in acidic media, however, significantly inhibited the photoassisted oxidation of dodecyl sulphate [79]. In this case the negatively charged hole scavenger competes with the surfactant ions for the active sites on the TiO_2 surface.

Cetyltrimethylammonium cationic surfactant as an electron donor could also be efficiently degraded in this photocatalytic system at pH 6. The rate of its degradation was about twice as high as that of the dodecyl sulphate ions under similar circumstances [67]. Due to its positive charge, however, the effects of pH and added electron acceptors strongly deviate from those observed in the case of the anionic tensid. Decreasing pH to the range of 2.0-3.5 considerably slow down the photoinduced oxidation of CTA^+, due to the impeded adsorption of the cationic detergent on the surface of TiO_2 particles. Surprisingly, addition of a cationic electron acceptor (Cu^{2+}) spectacularly increased the conversion in this system [79]. This phenomenon indicated that redox activity was more dominant in this case than adsorption probably because both reactants were repelled by the positively charged surface of photocatalyst particles. At pH 6, however, close to the zero point of charge of the titanium dioxide, the presence of $HgCl_2$ dramatically decreased the efficiency of the photodestruction of CTA^+, indicating the determining role of competitive adsorption [67]. An effect on the photoassisted oxidation of cetyltrimethylammonium was observed upon addition of chromate to the acidic system (pH 2.0-3.5). Although this anionic electron acceptor is adsorbed to the surface of the catalyst, it also forms ion-pairs with the oppositely charged surfactant. Thus, a dramatic enhancement in the efficiency of the oxidative degradation of CTA^+ (and, as mentioned in section 4.1.2., the reduction of the chromate ion too) [79].

4.2.3. Photocatalytic removal of agricultural residuals

The use of great variety of herbicides and pesticide in agriculture has been generally employed to increase the yield of crop. These materials exhibit some degree of water solubility, hence their massive use in the various fields of the industrialized agriculture led to measurable levels of these compound and their degradation products in natural waters, many of which supply the human population with fresh drinking water. An additional concern is the undesirable and often more toxic and/or persistent intermediates formed in the natural degradation pathways of these compounds that can accumulate in the ecosystem.

Because of the rather intense application of herbicides worldwide the continuous monitoring of their level in natural waters is required. It is important to note that when herbicides reach the water supply, natural degradation processes, such as UV light degradation or biodegradation from microorganizms, may initiate the breakdown of the compounds. One of the most heavily used herbicides in the world is 2,4-dichlorophenoxyacetic (2,4-D) acid, which is belongs to a large class of this type of compounds known as pnenoxyl herbicides, 2,4-dichlorophenoxy ester (2,4-DME) and 2-(2,4-dichlorophenoxyl) propionic acid (2,4-DP). The major degradation products of (2,4-D) are 2,4-dichlorophenol (2,4-DCP) and 2,4,6-trichlorophenol (2,4,6-TCP). A careful degradation study of these 2,4-D derivatives using the AOPs demonstrated that the combination the TiO_2 based photocatalysis with sonolysis is a promising technique for mineralization of 2,4-DP and its derivatives [94].

4.3. Enhanced photoactivity

4.3.1. Enhanced photoactivity of noble metal deposited TiO_2 particles

Recent studies have revealed that metal deposition on semiconductor enhances the efficiency of some photocatalytic processes. In particular, the deposition of silver on titania has been of considerable interest for both mechanistic and applicability reasons, since silver clusters can act as electron traps aiding the separation of the hole-electron pair. Metallisation of TiO_2 was reported to significantly increase the photocatalytic degradation of some organic compounds, such us formic acid, acetic acid, oxalic acid, phenol, 2-chlorophenol, sucrose etc.. However, other studies clearly indicate that using silver deposited TiO_2 particles as photocatalysts is not universal. Insignificant effects have been observed for malic acid and salicylic acid. To understand the observed substrate selective enhancement by silver clusters during photooxidation, a series of carefully selected organic molecules of different structure have been investigated and the rate of their photodegradation over bare and silver deposited TiO_2 have been compared (Table 1).

Table 1. Photocatalytic enhancement provided by silver deposition on TiO_2 for decomposition of various organic compounds

Numbering	Organic compounds	Relative enhancement	Reference
1.	Formic acid	2	[95. 96]
2.	Acetic acid	0.67	[43. 95. 95]
3.	Oxalic acid	3.7	[95. 96]
4.	Maleic acid	4.0	[95. 96]
5.	Malonic acis	1.8	[95. 96]
6.	Fructose	8.0	[95. 96]
7.	Glucose	7.2	[95. 96]
8.	Sucrose	5.7	[95. 96]
9.	Phenol	No	[95. 96]
10.	Dichloroacetic acid	2.0	[95. 96]

In has been concluded that the enhancing effect of silver deposits on TiO_2 is more probable for molecules possessing fewer C-H bonds and /or more C=O and C-O bonds [95, 96]. Another approach suggests that the photodegradation rate of organic compounds forming surface complex and readily transferring electron to the photogenerated hole considerable increases in the presence of silver cluster on the semiconductor particle, because this metal deposit acts as efficient electron trap and thus delays the hole-electron recombination.

Quantum sized effect of metal cluster deposited

The increased catalytic activity of TiO_2 photocatalysts due to its modification by gold nanoparticles has been also demonstrated [97-101]. The rate of photodegradation of methyl *tert*-butyl eter using gold-modified Degussa P25 catalyst significantly depends on the amount and size of Au loaded to the semiconductor surface. The optimum corresponds to a mean size of ≤ 3 nm at which point the gold may no longer be metallic. The size-dependent shift in the apparent Fermi level (E_F^*) was measured by using C_{60}/C_{60}^- redox couple probe under steady state and time resolved excitation of TiO_2-Au composite particles [97]. The smaller gold particles induced greater negative shift in E_F^* than the larger particles. Using 3-nm-diameter gold nanoparticles –60 mV shift in flat band potential was observed which is significantly higher than that was measured for 5-nm (-40 MV) and 8-nm (-20 mV) metal particles. Such a shift in the Fermi level improves the energetics of the composite system and enhances the efficiency of interfacial charge transfer process resulting in the increase of the photodecomposition rate of the substrate.

4.3.2. Simultaneous reduction and oxidation

As it has been proved in sections 4.1 and 4.2.2., the efficiencies of light-driven reduction of metal or metallate ions and oxidative degradation of organic pollutants (such as surfactants) in photocatalytic systems based on TiO_2 can be significantly enhanced by addition of appropriate sacrificial hole scavengers and electron donors, respectively. Accordingly, mutual promoting effects on the corresponding redox processes were observed in the systems containing both Hg(II) species and dodecyl sulphate ions compared to the cases where only one of these two contaminants existed alone in the system (and, thus, water and dissolved oxygen played only the role of the electron donor and acceptor, respectively) [67]. Since the pH range of 6-9 was the most favourable for the reduction of Hg(II) and the oxidation rate of DS^- showed a slight increase even a wider range of pH, neutral or slightly basic reaction mixture ist most preferable for simultaneous photocatalytic decontamination of wastewaters containing these pollutants. Besides, nearly neutral solutions are favorable in environmental respects too.The reduction of the positively charged Cu^{2+} ions also promoted the simultaneous oxidation of dodecyl sulphate, while the presence of anionic chromate considerably hindered the degradation of this surfactant [79].

In the case of the cationic cetyltrimethylammonium ions the effects of the added electron acceptors were significantly different from those observed with the anionic surfactant. While also in this system (at lower pH) copper(II) ions strongly speeded up the oxidative degradation, mercury(II) (in neutral solution) dramatically decreased the conversion of CTA^+. Anionic chromate, however, spectacularly enhanced the destruction of this tensid, and, *vice versa*, the efficiency of the photocatalytic reduction of $Cr_2O_7^{2-}$ and $HCrO_4^-$ ions could be multipled by the presence of cetyltrimethylammonium ions in acidic system.

A similar synergism in the simultaneous conversion of heavy metal ions (e.g., silver(I), chromate) and organic pollutants (such as dichlorvos and various dyes) was also observed in aqueous TiO_2 suspensions under illumination [102, 103]. Thus, taking such combinations, i.e., applying simultaneous photocatalytic redox reactions of these contaminants, the mutual promotion effects can be utilized for wastewater treatment.

5. Conclusion

Photocatalytic method based on TiO_2 semiconductor proved to be a promising procedure for both oxidative degradation of organic contaminants and reductive conversion and deposition of metal pollutants in wastewaters. Detailed studies indicate that, in numerous cases, the oxidation rate of various organic compounds can be significantly increased by metallisation of the TiO_2 particles, applying silver, copper, platinum or gold deposition. This promoting effect is attributed to the inhibition of the electron-hole recombination and/or the formation of surface complexes between the metals and the organic ligands to be degraded. Reductive removal of toxic metal ions such as mercury(II), however, can be considerably hindered by complex formation in the bulk solution. Simiraly, pH can also dramatically influence the efficiency of the photocatalytic redox processes in this system, affecting the surface charge of the semiconductor particles and the speciation of the potential hole scavengers and electron donors. Accodingly, the synergic effects of the simultaneous oxidative degradation of organic pollutants such as surfactants and reduction of metal or metallate ions are strongly determined by the charge of the reactants. While similarly charged reactants generally displayed mutual inhibition, mostly due to the competition for the active sites on the surface of the catalyst particles, oppositely charged tensid and metal ions promoted the redox change of both components. Taking these mutual effects into consideration, efficient, cost-effective and environmental friendly technologies using semiconductor-mediated photocatalyses can be designed and realized for decontamination of various polluted waters.

ACKNOWLEDGMENT

Support for this work by the Hungarian Research Fund (OTKA K63494) is gratefully acknowledged.

REFERENCES

[1] Manahan, S. E. (2000). *Fundamentals of Environmental Chemistry, 2nd ed.* Lewis Publishers: Boca Raton, FL, pp (a) 721-758, (b) 807-847, (c) 375-418.
[2] Kanki, T., Yoneda, H., Sano, N., Toyoda, A., Nagai, C. (2004). *Chem. Eng. J,* 97, 77-81.
[3] Kajitvichyanukul, P., Ananpattarachai, J., Pongpom, S. (2005). *Sci. Technol. Adv. Mater,* 6, 344-347.
[4] Kajitvichyanukul, P., Ananpattarachai, J., Pongpom, S. (2005). *Sci. Technol. Adv. Mater,* 6, 352-358.
[5] Gkika, E., Troupis, A., Hiskia, A., Papaconstantinou, E. (2006). *Appl. Catal, B,* 62, 28-34.
[6] Fujishima, A., Rao, T. N., Tryk, D. A. (2000). *J. Photochem. Photobiol, A,* 1, 1-21.
[7] Patsoura, A., Kondarides, D. I., Verykios, E. X. (2007). *Catal. Today,* 124, 94-102.
[8] Heller A., Brock, J. R. (1994). In *Aquatic and Surface Photochemistry*. Helz, G. R., Zepp, R. G., Crosby, D. G., Ed. Lewis Publishers: Boca Raton, FL, pp 427-436.
[9] U.S. Environmental Protection Agency. (1980). *Ambient Water quality criteria for chlorinated phenols*. EPA-440/5-80-032.
[10] Fabbri, D., Bianco Prevot, A., Pramauro, E. (2006). *Appl. Catal, B,* 62, 21-27.
[11] Devipriya, S., Yesodharan, S. (2005). *Sol. Energy Mater. Sol. Cells,* 86, 309-348.
[12] Fellenberg, G. (2000).*The Chemistry of Pollution*.Wiley: Chichester, pp 67-99.
[13] Zhang, R., Gao, L., Zhang, Q. (2004). *Chemosphere,* 54, 405-411.
[14] Lizama, C., Bravo, C., Caneo, C., Ollino, M. (2005). *Environ. Technol, 26*, 909-914.
[15] Kimura, T., Yoshikawa, N., Matsumura, N., Kawase, Y. (2004). *J. Environ. Sci. Health, A,* 39, 2867-2881.
[16] Zhang, H., Quan, X., Chen, S., Zhao, H., Zhao, Y. (2006). *Appl. Surf. Sci,* 252, 8598-8604.
[17] Zhang, T., Oyama, T., Horikoshi, S., Zhao, J., Serpone, N., Hidaka, H. (2003). *Appl. Catal, B,* 42, 13-24.
[18] De La Fournière, E. M., Leyva, A. G., Gautier, E. A., Litter, M. I. (2007). *Chemosphere,* 69, 682-688.

[19] Botta, S. G., Rodríguez, D. J., Leyva, A. G., Litter, M. I. (2002). *Catal. Today,* 76, 247-258.

[20] Chenthamarakshan, C. R., Rajeshwar, K. (2000).*Electrochem. Commun,* 2, 527-530.

[21] Nguyen, V. N. H., Amal, R., Beydoun, D. (2003). *Chem. Eng. Sci,* 58, 4429-44439.

[22] Serpone, N., Borgarello, E., Pelizetti, E. (1988). In *Photocatalysis and environment, trends and applications*. Schiavello, M. Ed. Kluwer Academic Publisher: Dordrecht, pp 527-565.

[23] Sun, B., Reddy, E. P., Smirniotis, P. G. (2005). *Environ. Sci. Technol,* 39, 6251-6259.

[24] Testa, J. J., Grela, M. A., Litter, M. I. (2004). *Environ. Sci. Technol,* 38, 1589-1594.

[25] U.S. Environmental Protection Agency. (1998). *Advanced Photochemical Oxidation Processes*. EPA/625/R-98/004.

[26] Anderson, C., Bard, A. J. (1997). *J. Phys. Chem, B,* 101, 2611-2616.

[27] Guillard, C., Disdier, J., Herrmann, J. M., Lehaut, C., Chopin, T., Malato, S., Blanco, J. (1999). *Catal. Today,* 54, 217-228.

[28] Yu, J. G., Yu, J. C., Ho, W. K., Jiang, Z. T. (2002). *New J. Chem,* 26, 607-613.

[29] Yu, J. G., Yu, J. C. (2003). *Chin. J. Chem,* 21, 994-997.

[30] Linsebigler, A. L., Lu, G., Yates Jr. J. T. (1995). *Chem Rev,* 95, 735-758.

[31] Hodak, J., Quinteros, C., Litter, M. I., San Román, E. (1996). *J. Chem. Soc. Faraday Trans,* 92, 5081-5088.

[32] He, J., Zhao, J., Shen, T., Hidaka, H., Serpone, J. (1997).*Phys. Chem, B,* 101, 9027-9034.

[33] Stylidi, M., Kondarides, D. J., Verykios, X. E. (2003). *Appl. Catal, B,* 40, 271-286.

[34] Litter, M. I. (1999). *Appl. Catal, B,* 23, 89-114.

[35] Asahi, R., Morikawa, T., Ohwaki, T., Aoki, K., Taga, Y. (2001). *Science,* 293, 269-271.

[36] Anpo, M., Takeuchi, M., Ikeue, K., Dohshi, S. (2002).*Curr. Opin. Solid State Matter Sci,* 6, 381-388.

[37] Yu, J. G., Yu, J. C., Cheng, B., Zhao, X. J. (2002). *Sol-Gel Sci. Tech,* 24, 39-48.

[38] Yao, W. F., Wang, H., Xu, X. H., Yang, X. N., Zhang, Y., Shang, S. X., Wang, M. (2003). *Appl. Catal, A,* 251, 235-239.

[39] Hufschmidt, D., Bahnemann, D., Testa, J. J., Emilio, C. A., Litter, M. I. (2002). *J. Photochem. Photobiol, A,* 148, 223-231.

[40] Henglein, A. (1995). *Ber. Bunsen-Ges. Phys. Chem,* 99, 903-913.

[41] Dawson, A., Kamat, P. V. (2001). *J. Phys. Chem. B,* 105, 960-966.

[42] Kamat, P. V., Flumiani, M., Dawson, A. (2002).*Colloids Surf, A*, 269-279.

[43] Szabó-Bárdos, E., Czili, H., Horváth, A. (2003). *J. Photochem. Photobiol, A,* 154, 195-201.

[44] Curran, J. S., Domenech, J., Jaffrezic-Renault, N., Philippe, R. (1985). *J. Phys. Chem,* 89, 957-963.

[45] Borgarello, E., Serpone, N., Emo, G., Harris, R., Pelizetti, E. (1986). *Inorg. Chem,* 25, 4499-4503.

[46] Domenech, J., Munoz, J. (1987). *Electrochim. Acta,* 32, 1383-1386.

[47] Herrmann, J.M., Disdier, J., Pichat, P. (1988). *J. Catal,* 113, 72-81.

[48] Herrmann, J.M., Guillard, C., Pichat, P. (1993). *Catal. Today,* 17, 7-20.

[49] Lau, L. D., Rodriguez, R., Henery, S., Manuel, D. (1998). *Environ. Sci.Technol,* 32, 670-675.

[50] Serpone, N., Ah You, Y. K., Tran, T. P., Harris, R., Pelizetti, E., Hidaka, H. (1987). *Sol. Energy,* 139, 491-498.

[51] Prairie, M. R., Evans, L. R., Stange, B. M., Martinez, S. L. (1993).*Environ. Sci.Technol,* 27, 1176-1182.
[52] Tennakone, K., Wijayantha, K. G. U. (1997). *J. Photochem. Photobiol. A,113*, 89-92.
[53] Fox , M. A. (1983). *Acc. Chem. Res,* 16, 314-321.
[54] Hsiao, C., Lee, C., Ollis, D. F. (1983). *J. Catal,* 82, 418-423.
[55] Ahmed, S., Ollis, D. F. (1984). *Sol. Energy,* 32, 597-601.
[56] Goering, S., Chenthamarakshan, C. R., Rajeshwar, K. (2001). *Electrochem. Commun,* 3, 290-292.
[57] Kanki, T., Yoneda, H., Sano, N., Toyoda, A., Nagai, C. (2004). *Chem. Eng. J, 97*, 77-81.
[58] Rajeshwar, K., Ibanez, J. G. (1995). *J. Chem. Educ,* 72, 1044-1049.
[59] Greenwood, N. N., Earnshaw, A. (1984). *Chemistry of the Elements*. Pergamon Press: Oxford, pp 736, 1409.
[60] Latimer, W. L. (1952).*Oxidation Potentials. 2nd ed*. Prentice-Hall: Englewood Cliffs, NJ, USA.
[61] Bard, A. J. (ed). (1982). *Encyclopedia of Electrochemistry of the Elements. 4th ed.* Marcel Dekker: New York, NY, USA, Vol. 9
[62] Horváth, O., Hegyi, J. (2001).*Prog. Colloid. Polym. Sci,* 117, 211-216.
[63] Shriver, D. F., Atkins, P. W. (2002). *Inorganic Chemistry. 3rd ed.* Oxford University Press: Oxford, pp 696-711.
[64] Breitenkamp, M., Henglein, A, Lilie, J. (1976).*Ber. Bunsen-Ges. Phys. Chem,* 80, 973-979.
[65] Murruni, L., Leyva, G., Litter, M. I. (2007). *Catal. Today,* 129, 127-135.
[66] Tanaka, K., Harada, K., Murata, S. (1986). *Sol. Energy,* 36, 159-161.
[67] Hegyi, J., Horváth, O.(2004). *Prog. Colloid. Polym. Sci,* 125, 10-16.
[68] Lea, J., Adesina, A. A. (1998). *J. Photochem. Photobiol, A,* 118, 111-122.
[69] Wang, Xiaoling., Pehkonen, S. O., Ray, A. K. (2004).*Electrochim. Acta,* 49, 1435-1444.
[70] Cristante, V. M., Araujo, A. B., Jorge, S. M. A., Florentino, A. O., Valente, J. P. S., Padilha, P. M. (2006), *J. Braz. Chem. Soc,* 17, 453-457.
[71] Aguado, M. A., Cervera-March, S., Giménez, J. (1995).*Chem. Eng. Sci,* 50, 1561-1569.
[72] Doménech, X., Muñoz, J. (1990). *J. Chem. Technol. Biotechnol,* 47, 101-107.
[73] Fu, H., Lu, G., Li, S. (1998). *J. Photochem. Photobiol, A*, 114, 81-88.
[74] Ku, Y., Jung, I.L. (2001). *Water Res,* 35, 135-142.
[75] Chenthamarakshan, C. R., Rajeshwar, K., Wolfrum, E. J. (2000). *Langmuir,* 16, 2715-2721.
[76] Giménez, J., Aguado, M. A., Cervera-March, S. (1996). *J. Mol. Catal, A,* 105, 67-78.
[77] Lin, W. Y., Wei, C., Rajeshwar, K. (1993). *J. Electrochem. Soc,* 140, 2477-2482.
[78] Muñoz, J., Doménech, X. (1990). *J. App. Electrochem,* 20, 518-521.
[79] Horváth, O., Bodnár, E., Hegyi, J. (2005). *Colloids Surf, A,* 265, 135-140.
[80] Meichtry, J. M., Brusa, M., Mailhot, G., Grela, M. A., Litter, M. I. (2007). *Appl. Catal, B,*71, 101-107.
[81] Houée-Levin, C., Sicard-Roselli, C. (2001). In *Radiation Chemistry, Present Status and Future Trends*. Jonah, C. D., Rao, B. S. M., Ed. Studies in Physical and Theoretical Chemistry 87, Elsevier: Amsterdam, pp. 553-584.
[82] Neta, P., Simic, M., Hayon, E. (1970). *J. Phys. Chem,* 74, 1214–1220.
[83] Hiller, K.O., Masloch, B., Gobl, M., Asmus, K.D. (1981). *J. Am. Chem. Soc*, 103, 2734-2743.

[84] Hand, V. C., Snyder, M. P., Margerum, D. W. (1983). *J. Am. Chem. Soc,* 105, 4022-4025.

[85] Mönig, J., Chapman, R., Asmus, K.D. (1985). *J. Phys. Chem,* 89, 3139-3144.

[86] Asmus, K.D., Gobl, M., Hiller, K.O., Mahling, S., Mönig, J. (1985). *J. Chem. Soc. Perkin Trans II,* 641-646.

[87] Mönig, J., Gobl, M., Asmus, K.D. (1985). *J.Chem. Soc. Perkin Trans II,* 647-651.

[88] Hidaka, H., Hirokoshi, S., Ajisaka, K., Zhao, J., Serpone, N. (1997). *J. Photochem. Photobiol, A,* 108, 197-205.

[89] Horikoshi, S., Serpone, N., Zhao, J., Hidaka, H. (1998). *J. Photochem.Photobiol, A,* 118, 123-129.

[90] Matsushita, M., Tran, T. H., Nosaka, A.Y., Nosaka, Y. (2006). *Catal.Today,* 120, 240-244.

[91] Szabó-Bárdos, E., Pétervári, E., El-Zein, V., Horváth, A. (2006). *J. Photochem. Photobiol, A,* 184, 221-227.

[92] Ishibashi, K., Fujishima, A., Watanabe, T., Hashimoto, K. (2000). *Electrochem. Commun,* 3, 207-221.

[93] Louit, G., Foley, S., Cabillic, J., Coffigny, H., Tara, F., Valleix, A., Renault, J.P., Pin, S. (2005). *Radiat. Phys. Chem,* 72, 119-124.

[94] Peller, J., Wiest, O., Kamat, P. V. (2004). *J. Phys. Chem,* 108, 10925-10933.

[95] Tran, H., Chiang, K., Scott, J., Amal, R. (2005). *Photochem. Photobiol. Sci,* 4, 565-567.

[96] Tran, H., Scott, J., Chiang, K., Amal, R. (2006). *J. Photochem. Photobiol, A,*183, 41-52.

[97] Subramanian, V., Wolf, E. E., Kamat, P. V. (2004). *J. Am. Chem. Soc,* 126, 4943-4950.

[98] Orlov, A., Jefferson, D. A., Macleod, N., Lambert, R. M. (2004). *Catal. Lett,* 92, 41-47.

[99] Kamat, P. V., Flumiani, M., Dawson, A. (2002). *Colloids Surf,* 202, 269-279.

[100] Valden, M., Lai, X., Goodman, D. W. (1998). *Science,* 281, 1647-1650.

[101] Orlov, A., Jefferson, D. A., Tikhov, M., Lambert, R. M. (2007). *Catal. Commun,*8, 821-824.

[102] Kyung, H., Lee, J., Choi, W. Y. (2005).*Environ. Sci. Technol,* 39, 2376-2382.

[103] Shifu, C., Gengyu, C. (2005). *Chemosphere,* 60, 1308-1315.

In: Photochemistry Research Progress
Editors: A. Sánchez, S. J. Gutierrez

ISBN: 978-1-60456-568-3

Chapter 15

PHOTOCHEMISTRY IN IONIC LIQUIDS

Belen Ferrer and Hermenegildo Garcia
Universidad Politécnica de Valencia, Spain

ABSTRACT

Room temperature ionic liquids (RTILs) have recently emerged as one of the preferred "green" solvents. These liquids are composed of ions and are fluid at room temperature, but still can dissolve a wide range of organic molecules. The use of RTILs as novel media for various photophysical and photochemical studies is a topic of recent interest since the influence of the solvent in molecular organic photochemistry is well established. This chapter comments on some aspects of the photochemistry in RTILs, including solvent relaxation dynamics, electron and energy transfer kinetics, photoisomerization, solvatochromic absorption and fluorescence of dipolar molecules, unconventional wavelength-dependent fluorescence, empirical polarity parameters based on dye molecules, transparency of RTILs in the UV-visible region and the effect of viscosity and/or polarity of RTILs on the mechanism, dynamics (rate constant) and yield of photoinduced electron transfer (PET) processes.

INTRODUCTION

Inorganic ionic compounds and the vast majority of organic ionic compounds are high melting point solids. Thus, even though they were known long before as molten salts, ionic liquids (ILs) are unique in the sense that they are constituted by anions and cations, but they are liquid at low temperature. As defined by Welton, ionic liquids include those ionic compounds that are liquid at temperatures below 100 °C, the most important ones being those that are liquids at room temperature (RTILs).[1]

Structurally most ionic liquids have in common a low-symmetry, conformationally flexible organic cation combined with a non coordinating anion [1-3]. These three features are responsible for the low tendency of RTILs to crystallize. The solid crystal state is characterized by a highly ordered structure that is disfavoured by conformational freedom and dissymmetry of the molecules. In spite of being constituted by ions, RTILs can dissolve a large number of organic molecules and chemical reagents and have essentially zero vapour pressure, so they are non volatile solvents. Examples of RTILs are given in Figure 1.

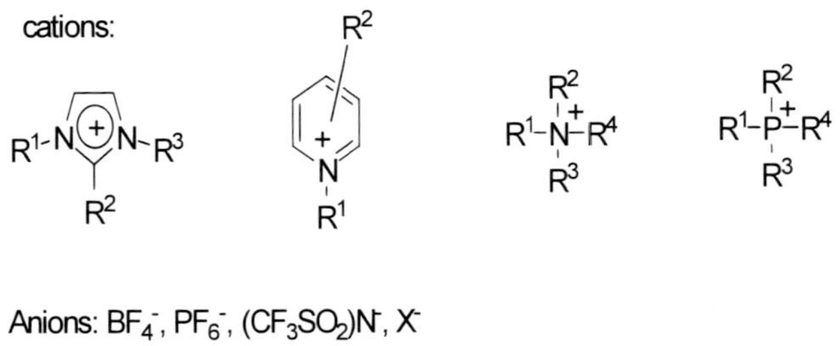

Figure 1. Structure of some common RTILs.

The present importance and primary interest in RTILs comes from the field of Green Chemistry. The aim of Green Chemistry is to develop new processes and provide products using reagents and technology that are benign for the environment. In this context, one source of environmental unfriendliness is the use of conventional volatile organic solvents such as chlorinated compounds, hydrocarbons and ethers. Volatile organic solvents interfere negatively in the natural atmospheric processes and are believed to be largely responsible for ozone depletion and climate change. In addition to the need to find suitable replacements for conventional solvents, Green Chemistry also intends to develop catalytic processes in which the catalyst can be easily separated from the products and reused indefinitely.[4,5] Both goals can be easily achieved using RTILs and for these reasons, RTILs are key solvents in homogeneous catalysis allowing reusability.

The lack of vapour pressure is the feature that gives to these liquids the label of green solvents, since they do not play adverse effects with respect to the atmosphere. However, it is known that some ionic liquids exhibit toxicity and, accordingly, they are not perfectly green.[6,7] In addition, RTILs also fulfil other necessary requirements as solvents, such as thermal and chemical stability and no flammability. The use as new solvents for homogeneous catalysis is based on the fact that even though RTILs can dissolve many organic compounds and catalysts, at the end of the reaction the products can be separated by precipitation, distillation or extraction. Extraction is based on the fact that RTILs are immiscible with alkanes and ethers, what allows recovering the products, while the catalyst remains dissolved in the RTIL and this liquid phase can be reused in a consecutive reaction. Figure 2 summarizes common procedure when using RTILs in catalysis.

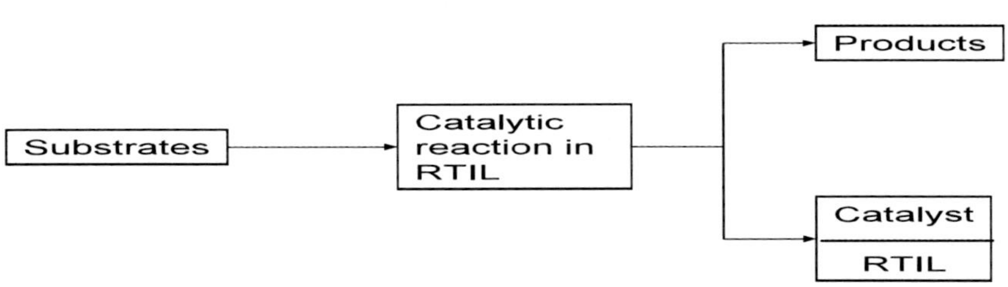

Figure 2. Use of RTILs in homogeneous catalysis to develop reusable systems.

IONIC LIQUIDS AND PHOTOCHEMISTRY

Initially, the photochemistry of organic compounds was studied in solvents as inert as possible with the aim to determine the intrinsic photochemical properties of organic functional groups.[8] Soon it became evident that the molecular photochemistry, i.e., the intrinsic photochemical events triggered in the molecule by light absorption and that primarily involve electron excitation from ground to excited states, can be controlled and altered in a significant extent by the solvent.[9] The solvent can determine the fate of the electronic excitation and the resulting product distribution in many different ways, but there are two general properties, namely viscosity and polarity, that always play an effect. Other possible effects of the solvent include hydrogen atom donor, electron donor or acceptor properties, acidity and basicity, etc. (Figure 3).

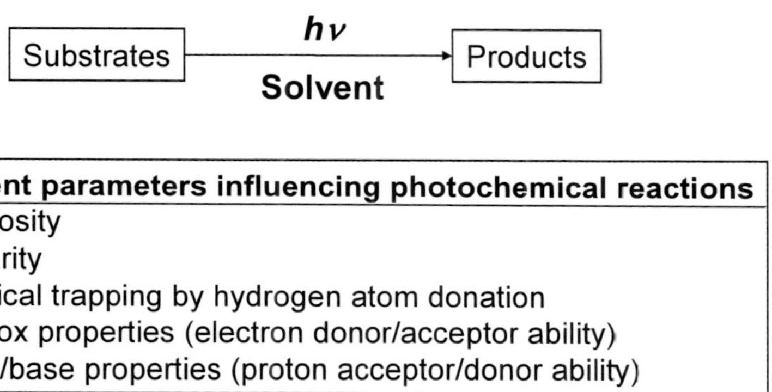

Figure 3. Influence of solvent parameters in the outcome of photochemical reactions.

Viscosity affects dramatically in those cases in which diffusion is important in controlling the product distribution and this is the case when two geminate species are formed in the same photochemical event. There are many photochemical reactions in which a chemical bond is cleaved, either homolytically or heterolytically. When a bond is broken two reactive intermediates are generated simultaneously. There is a possibility for these geminate species to recombine reverting to the parent molecule or forming an isomer or derivative. With the general term solvent cage it is denoted that the solvent molecules are impeding the escape of the geminate species and are promoting the recombination of geminate species. Figure 4 shows a paradigmatic example of variation of the product distribution that has been proposed to measured the cage effect. In this case, photolysis of disymmetrically substituted dibenzyl ketones leads primarily to benzyl and benzoyl radicals that after loosing CO form a geminate radical pair. When escape does not occur the only product is the dissymmetric 1,2-diarylethane, but the formation of symmetrically substituted 1,2-diarylethane indicates the extent of out-of-cage recombination.

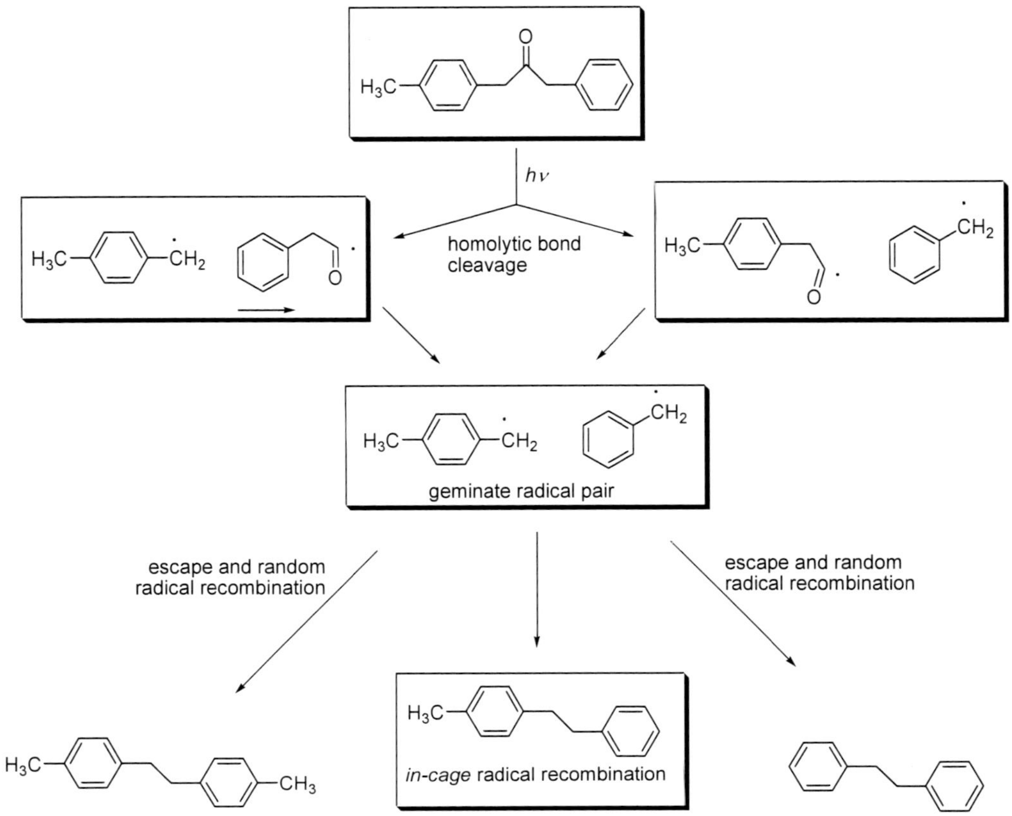

Figure 4. The solvent cage effect is related to the ease in which species can diffuse in the solvent and depends on the viscosity of the medium. It can be determined by the percentage of in-cage radical recombination vs. the total percentage of diarylethanes.

The cage effect is low for low viscosity solvents and the contrary applies for highly viscous liquids. Ionic liquids are about two/three orders or magnitude more viscous than conventional volatile organic solvents and, therefore, diffusion of molecules and reaction intermediates are accordingly much slower than in common organic solvents. To give a reference, the viscosity of methanol and water is 0.544 and 0.89 cP, while that of 1-butyl-3-methylimidazolium tetrafluoroborate is about 115 cP. Not surprisingly, the expected diffusion in ionic liquids is very slow and in cage recombination of geminal radicals much enhanced.

In addition to viscosity controlling the diffusion of species in the medium, polarity is another key factor influencing the outcome of a photochemical reaction. Excited electronic states reached upon photochemical excitation are more easily oxidized or reduced than ground states (Figure 5). If oxidation or reduction occurs through photoinduced electron transfer (PET) from electron donors to electron acceptors, then radical ions are obtained. Photoinduced electron transfer is a general phenomenon that some times competes with energy transfer. In these cases, the ratio between energy and electron transfer is largely affected by the polarity of the medium. Due to Coulombic interactions stabilizing ionic intermediates, highly polar solvents favour photoinduced electron transfer versus energy transfer. Thus, considering that they are constituted by ions, RTILs should favour PET vs. energy transfer.

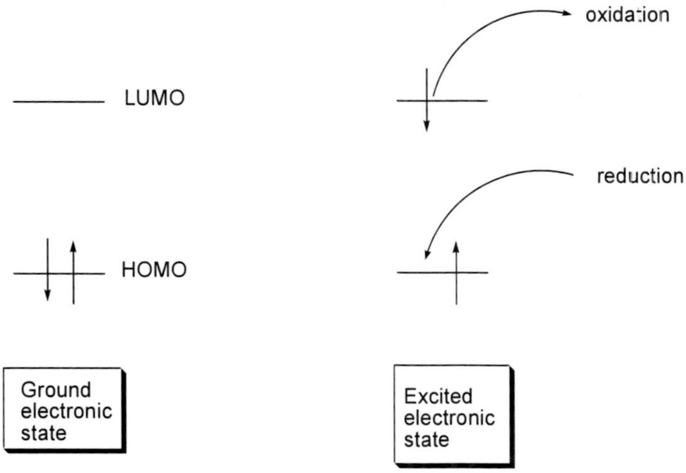

Figure 5. Light absorption generates excited electronic states that are easier oxidizable and reducible species than the ground electronic state.

Transparency Of RTILs in the UV-Visible Region

It is interesting examining the electronic absorption and fluorescence behavior of RTILs when exploring the suitability of several RTILs in optical applications. RTILs are generally regarded as optically transparent materials, possessing no absorption in the UV-visible region above 250 nm for non-aromatic and above 330 nm for imidazolium ILs. The current literature has attributed any absorption beyond this wavelength to the presence of impurities.

By using control experiments with imidazolium cations, Samanta et al. have reported that the weak absorption at 300 nm extending to longer wavelength had nothing to do with the impurity in the RTILs, but was clearly due to the imidazolium salts [10-12]. They have also established that the imidazolium ionic liquids studied, Bmim-PF_6, Bmim-BF_4, Emim-BF_4, Bmim-Cl, Bmim-$(CF_3SO_2)_2N$ and Emim-$(CF_3SO_2)_2N$ are characterized by significant absorption in the entire UV region and a long absorption tail that extends into the visible region (370 nm). They have attributed these absorption characteristics to the imidazolium moiety and its various associated structures.

The same authors have also reported that when exciting in the UV or in the early part of the visible region, these imidazolium ionic liquids exhibit a broad fluorescence, which covers a large part of the visible region. This emission was found to exhibit a very interesting and unusual behaviour, namely, the fluorescence peak position of these imidazolium ionic liquids is strongly dependent on the excitation wavelength. The excitation wavelength dependent shift of the fluorescence maximum has been rationalized taking into consideration the presence of a distribution of energetically different associated structures of the ionic liquids consisting in strongly bound anions and cations forming a cluster. By the obtained results, they have afforded the limitations that these ionic liquids may have in optical studies of the dissolved solutes, thus having serious drawbacks in some of the optical studies. In general it can be considered that RTILs, even with imidazolium moieties, are transparent beyond 350 nm. This is very convenient since it allows excitation of dissolved molecules using the third harmonic of a Nd:YAG laser as well as visible light.

Empirical Polarity Parameters Based on Dye Molecules

Several studies have focused in recent years on the empirical determination of the RTILs polarity by means of a variety of solvatochromic probe dyes. These dyes exhibit absorption and/or emission spectra whose wavelength maxima shift according to the polarity of the medium in which they are dissolved.

Concerning determination of polarity by fluorescence measurements two previous issues is to determine fluorescence lifetimes and to show that the emission does not depend on the excitation wavelength. Since solvation is a slow process in RTILs, only those dipolar systems having fluorescence lifetime longer than the solvent reorganization time can ensure that the maximum wavelength corresponds to the state of solvation equilibrium, therefore qualifying as polarity indicators. In this context, it has been observed that emission lifetimes of highly fluorescent aluminium complexes can be longer than in conventional solvents.[13]

Given that the measured emission maxima of dipolar systems are routinely used for the estimation of the polarity of the RTILs, it is necessary to confirm that the emission maximum of the selected probe does not depend on the excitation wavelength.

Aki et al. were among the first who estimated the polarity of some RTILs from the fluorescence maxima of two aminophthalimide derivatives AP and DAP (Figure 6). [14] The fluorescence properties of these two relatively small and neutral molecules are known to be highly sensitive to the polarity of the media. They have concluded that the RTILs investigated, Bmim-PF_6, 1-methyl-3-octylimidazolium hexafluorophosphate (C_8mim-PF_6), Bmim-NO_3 and N-butylpyryidinium tetrafluoroborate (N-bupy-BF_4), are more polar than the most polar aprotic solvents, such as acetonitrile, but less polar than methanol. They have also observed that the replacement of the counter ion PF_6^- by NO_3^- did not change the apparent polarity of the medium.

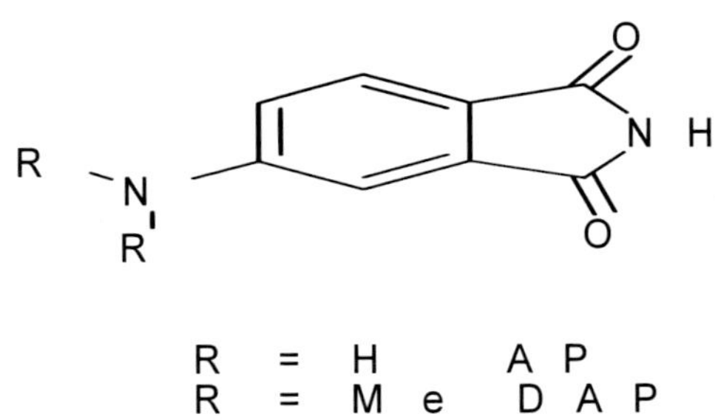

Figure 6. Molecular structure of the two aminophthalimide derivatives AP and DAP.

Most of the studies on polarity of RTILs are based on absorption rather than on emission spectroscopy. Reichardt [15] has employed a betaine dye, namely 2,6-diphenyl-4-(2,4,6-triphenylpyridinium-1-yl) phenolate known as Reichardt's dye or betaine no. 30 (Figure 7), to empirically determine the solvent polarity of RTILs. This zwitterionic betaine dye is a negative solvatochromic absorption probe, i.e. its long-wavelength intramolecular charge-transfer (CT) absorption band shifts to shorter wavelengths (higher energies) when increasing the polarity of the media. Reichardt's dye has been used directly or indirectly to characterize ionic liquids by the solvent polarity parameter $E_T(30)$. This $E_T(30)$ value is simply defined as the molar transition energies of the standard betaine dye no. 30 (Reichardt's dye), measured in solvents of different polarity at room temperature (25 °C) and normal pressure (1 bar), according to the equation [16]:

$E_T(30)$/ kcal mol^{-1} = $hc_{max}N_A$

By using this $E_T(30)$ empirical scale of solvent polarity, the polarity of a large number of RTILs such as ammonium and phosphonium salts, imidazolium and pyridinium based ionic liquids with different alkyl substituents has been determined. [15] Reichardt et al. [15] have investigated the influence of added water on the $E_T(30)$ values of different RTILs, the influence of the length of the alkyl substituents on the imidazolium cation and the effect of the substitution on the C(2)-H atom of the imidazolium cation. The experimental $E_T(30)$ values measured in RTILs range from 42 to 63 kcal mol^{-1}. These values fit quite well into the existing empirical solvent polarity scale of conventional organic solvents. Their polarity is comparable to that of molecular dipolar non-hydrogen bonding and dipolar hydrogen bonding solvents. Thus, in contrast to the ionic nature of RTILs, the scale based on $E_T(30)$ indicates that RTILs do not behave as superpolar as anticipated but as quite normal polar solvents.

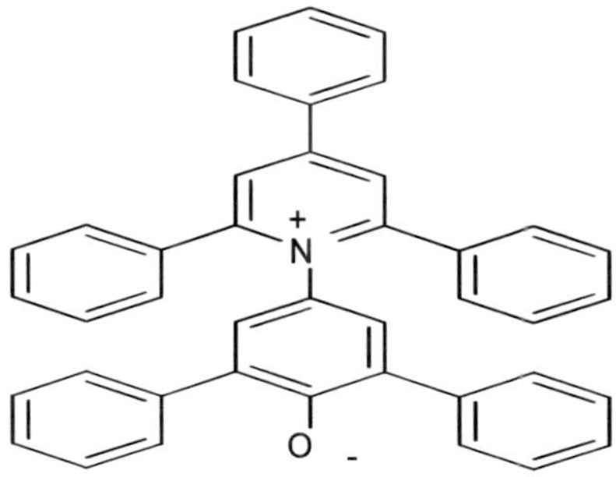

Figure 7. Molecular structure of the negatively solvatochromic standard betaine dye no. 30, 2,6-diphenyl-4-(2,4,6-triphenylpyridinium-1-yl) phenolate, used for the determination of ET(30) values.

Another study, in which the absorption of the solvatochromic dye Nile Red has been investigated for the estimation of the solvent polarity of the RTILs, has been undertaken by Seddon et al. [17] Nile red (Figure 8) is a positive solvatochromic absorption dye, i.e., when dissolved in increasingly polar media, the wavelength of its visible absorption maximum, $_{max}$, moves to longer wavelengths (lower energies). In fact, amongst positively solvatochromic dyes, Nile Red display one of the largest bathochromic shifts known. On changing the solvent from water to pentane a change in $_{max}$ of 110 nm is observed. [18] Seddon and coworkers have found that the 1-alkyl-3-methylimidazolium ionic liquids studied are in the same polarity region as the lower alcohols, such as 2-aminoethanol, methanol, ethanol and 1-butanol.

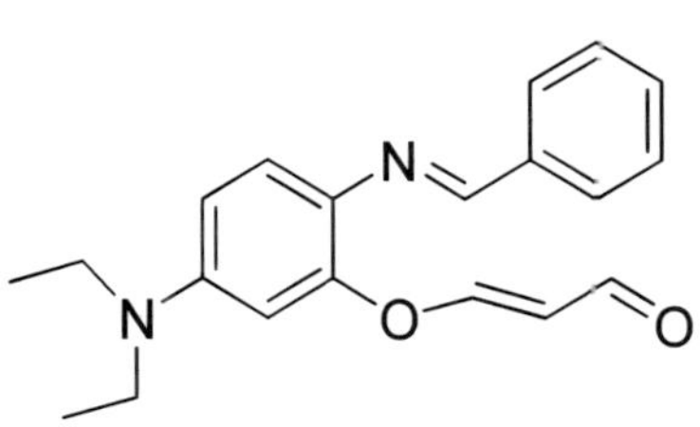

Figure 8. Molecular structure of the Nile Red.

They have also reported the effects on polarity on changing anion and 1-alkyl group. The polarity decreases through the anion series of nitrite, nitrate, tetrafluoroborate and hexafluorophosphate. This decrease in polarity correlates with increasing anion size. For 1-alkyl-3-methylimidazolium ionic liquids, the polarity is apparently determined by the anion for those containing short 1-alkyl groups and by the cation for those containing long 1-alkyl groups. [17] Analogous conclusions were also established using mixed valence compounds as probes (Figure 9). [19]

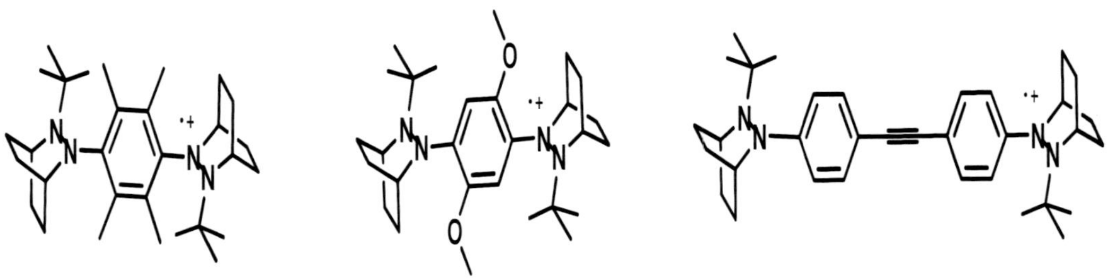

Figure 9. Mixed valence compounds (note the radical cation in one of the nitrogen atom terminus and not in the opposite) used as polarity probes (based on ref.19).

Pandey et al. have reported that mixtures of RTILs often exhibit emergent properties not readily apparent from those of the individual components. [20] They suggest that mixing ionic liquids may be a viable approach in the design and modification of these green solvent systems towards specific ends.

The fluorescence spectra of the various dipolar molecules studied in different imidazolium and pyrrolidinium ionic liquids has been found to be very similar, despite the large difference in the structures of the constituent ions. [21] In all cases, the microenvironment experienced by these molecules is similar to that in short-chain alcohols. The similar polarity determined in the different ionic liquids has been proposed to be a reflection of the hydrogen bonding interaction of the constituent ions of the RTILs with the probe molecules.

Given the complexity of the extended three-dimensional liquid structure of ionic liquids, with a supramolecular structure influenced by hydrogen bonds in which several positive and negative ions form strongly associated clusters, it is not surprising that the use of probes does not yield definitive answers on the question of polarity and raises many interesting questions for the future.

Solvent Relaxation Dynamics in Photoinduced Processes

Light absorption gives rise to electronic excited states whose electron distribution can differ considerably from that of the ground states. An extreme case in where a dramatic change in electron distribution occurs is when light excitation gives rise to a state of charge separation. When photon absorption produces a remarkable charge distribution of a molecule in solution, the solvent that is solvating the ground state has to relocate to minimize the energy of the excited state (Figure 10). Solvent relocation takes place in the submicrosecond time scale. The lifetime of the excited state has to be longer than the time required for solvent relocation.

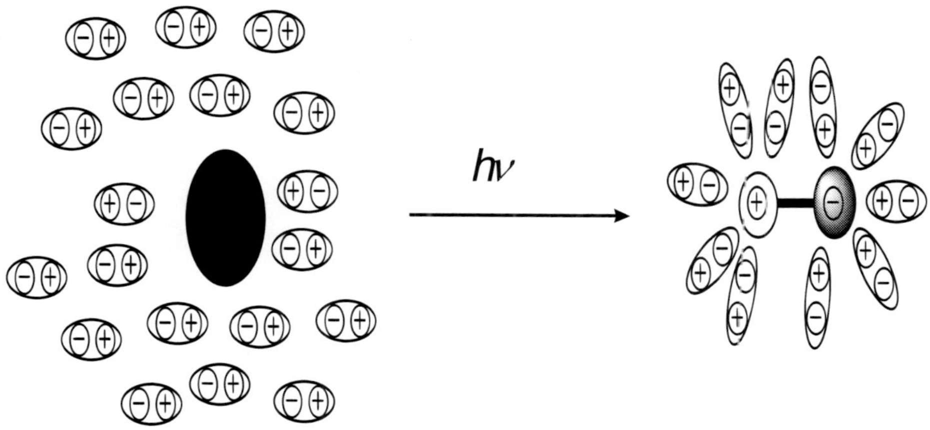

Figure 10. Pictorial illustration of the dynamics of a polar solvent around an apolar molecule that undergoes intramolecular photoinduced charge separation.

Solvation dynamics, i.e., the time required for the solvent molecules to relocate with respect to the probe molecule upon photoexcitation, has been studied in a wide variety of polar solvents employing fluorescent probe molecules which undergo instantaneous charge separation. The temporal evolution of fluorescence can provide useful information on how the solvent molecules reorganize after instantaneous separation of charge in the fluorophore upon absorption of photons. As RTILs are sufficiently polar media, the time-dependent fluorescence data of the dipolar systems provide information about the dynamics and mechanism of solvation processes in these media, which differ considerably from the conventional solvents.

Aki et al. [14] have investigated the polarity of four imidazolium and pyridinium based IL using UV-Vis absorption and fluorescence spectroscopy. Depending on the cations present the polarity of these RTILs range in polarity in between methanol and acetonitrile.

Different solutes have been used as probe molecules for studying the solvation dynamics in RTILs. The list of probes includes (Figure 11):

A) Coumarin-153 (C153) having a rigid structure, the charge separation following electronic excitation is formed instantaneously and displays a significant shift of its emission maximum as a function of the solvent polarity, thus, providing spectroscopic information on polarity. The solvation dynamics of C153 has been extensively investigated in several other media, there being sufficient information for comparing RTILs with conventional organic solvents.

B) Coumarin 102 (C102) has a rigid structure, very similar to that of the C153, but it is somewhat more sensitive to hydrogen bonding

C) Coumarin 152, (C152), has a conformationally flexible structure and as C153 has been widely used as solvatochromic probe to predict the polarity and solvation dynamics of various media. The position of the emission maximum points at shorter wavelength comparing with C153 and the average solvation time is shorter lived compared with C153.

D) 6-Propionyl-2-(dimethylamino)naphthalene (PRODAN) shows a red shift of 130 nm on the position of the fluorescence maximum when the medium is changed from cyclohexane to water. PRODAN emits from a locally excited state and hydrogen bonding plays a major role on dictating the wavelength of the fluorescence maximum in alcoholic or aqueous media.

E) 4-Aminophthalamide (4AP) is a solvatochromic probe often used in dynamic Stokes shift measurements and, as the case of PRODAN, its steady-state fluorescence spectrum is significantly influenced by hydrogen bonding.

F) 4-(Dimethylamino)-4'-cyanostilbene (DCS) is less commonly used as a solvatochromic probe, but, despite the fact that it isomerizes on a nanosecond time scale, DCS exhibits many properties that make it useful to measure the solvation dynamics in RTILs. The excited states involved in the emission of this probe molecule have been discussed by several groups. [22,23]

G) 9-Phenyl-10-methylacridinium (PAc^+) that is a probe having a positive charge able to establish Coulombic interactions with RITLs. [24]

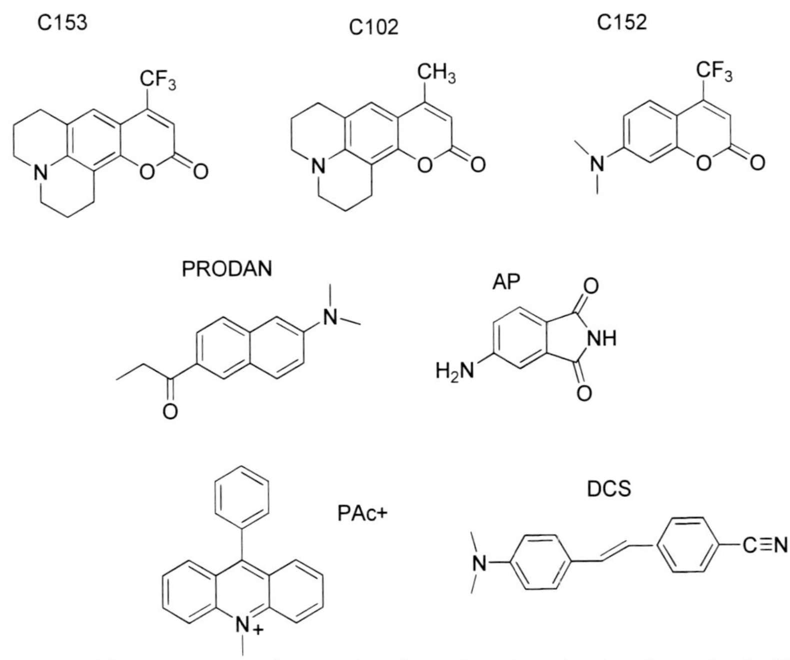

Figure 11. Structure of fluorescence probes used to determine the solvation dynamics in RTILs.

Among the RTILs in where solvation dynamics has been studied the list includes:

A) 1-Butyl-3-methylimidazolium tetrafluoroborate (Bmim-BF_4) and 1-Ethyl-3-methylimidazolium tetrafluoroborate (Emim-BF_4) have been chosen for the study of solvation dynamics because the tetrafluoroborate salts are reported to be air and water stable. [25,26] Even though the BF_4^- salts exhibit a higher miscibility with water compared to analogous PF_6^- or Cl^- salts, they are more stable, less hygroscopic and less corrosive than the halides. Emim-BF_4 possesses an imidazolium cation with a short alkyl chain and is miscible in all proportions with water even at low temperature. The miscibility of Bmim-BF_4 is comparatively low with ice-cold water.

B) Bmim-$[(CF_3SO_2)_2N]$ and Emim-$[(CF_3SO_2)_2N]$ have been chosen because the relaxation dynamics is expected to be much faster in less viscous RTILs. The relatively low viscosity of the $[(CF_3SO_2)_2N]^-$-based ionic liquids of the imidazolium salts compared to others is attributed to the weak hydrogen bonding ability of this anion. RTILs having $(CF_3SO_2)_2N^-$ as counteranion can be prepared conveniently in high yield, they are highly thermally stable and the trace amounts of water can be removed from the liquids at elevated temperatures. The halide anions of the precursor compound can be simply removed by washing these hydrophobic ionic liquids with water.

C) Bmim-PF_6 and 1-hexyl-3-methylimidazolium hexafluorophosphate (Hmim-PF_6) are hydrophobic RTILs. In contrast to BF_4^--based ionic liquids, they are not miscible with water. The melting point of the PF_6^- salt is higher than that of the BF_4^- analogues and similar to that of the chloride salts. Both RTILs have very high viscosity. Hmim-PF_6 is more viscous than Bmim-PF_6 due to the longer alkyl chain present in the C-1 of the imidazolium ring.

With the above fluorescent probes, Samanta et al. have concluded that in comparison to conventional organic solvents the average solvation time of RTILs is very long and depending on the probe used. [27,28] Moreover, the time dependent evolution of C153 and PRODAN fluorescence spectra indicate a biphasic solvation dynamics for Bmim-BF_4 and Emim-BF_4, solvent reorganization occurring in the picosecond and nanosecond timescales for each of the two phases. The biphasic solvation dynamics in the two RTILs is proposed to be the result of a relatively fast initial response of the smaller species, the anion, and a slow collective diffusion of the cations and the anions.

The same solvation mechanism has also been proposed for Bmim-$[(CF_3SO_2)_2N]$ and Emim-$[(CF_3SO_2)_2N]$ using C153, PRODAN and AP as probes. [29] It was found that the solvation dynamics in $(CF_3SO_2)_2N^-$ RTILs is biphasic in nature though significantly faster compared to that in BF_4^--based ionic liquids. This has been attributed to the lower viscosity of RTIL having $(CF_3SO_2)_2N^-$ as counteranion.

Maroncelli et al. have addressed the issue of probe dependency of the solvation dynamics in Bmim-PF_6 using a wide variety of probe molecules, C153, C102, PRODAN, 4-AP, DCS and PAc^+. [30] They found that the solvation times vary by a factor of 2 among the series of solvatochromic probes studied and then they concluded that it should be recognized that solvation times in RTILs can not be assumed to be completely independent of the solutes used to measure this dynamics. Importantly, they have also observed a biphasic solvation in which the ultrafast component (< 5 ps) was attributed to the translational adjustments of the cations and the anions in close proximity with the solute and the slow component to the large scale spatial rearrangement of the ions. [30,31]

Sarkar et al. [32] have measured the emission temporal profile of C152 and C153 in Bmim-PF_6 and Hmim-PF_6. They also observed two well defined time regimes for the solvent relaxation in these RTILs that occurs in a considerably longer time scale compared to conventional solvents of similar polarity. They assumed that the motion of the ions is responsible for the slow solvation in these ionic liquids. The fact that the solvation time in Hmim-PF_6 is longer than in Bmim-PF_6 supports their conjecture that the high viscosity of these ionic liquids could also be responsible for the slow solvation time.

From the above reports, there are clear disagreements not only about the origin of the two temporal components of the solvation dynamics in RTILs, but also on how fast is the fastest component of solvent relaxation. In this regard, several molecular dynamics simulations have been performed, but none of the calculated motions appear to be responsible for the dynamic

Stokes shift observed for dipolar molecules on the time-resolved fluorescence measurements. [33-35] Recently, femtosecond Kerr-gated emission spectroscopy and picosecond time-correlated single photon counting have been used to measure the emission dynamics on the 100 fs-100 ps time scale of C153, PRODAN and 4-AP as probe molecules in N-butyl-N-methylpyrrolidinium bis(trifluoromethanesulfonyl)imide ionic liquids. [36] Using this technique, data with sufficient high temporal resolution to complete the experimental picture of solvation dynamics in ionic liquids has been obtained. Thus, it was measured that the solvent relaxation consists of a short (hundreds of ps) and a long (1 ns) component. As in the case of the imidazolium ionic liquids, the average solvation time is consistent with the viscosity of this ionic liquid and the dynamics of solvation is dependent on the probe molecule.

As summary of the present situation it can be concluded that the mechanism of solvation dynamics in RTILs is far from clear due to the complex nature of the problem and the lack of convincing theoretical model to rationalize the experimental results.

Unconventional Excitation-Wavelength-Dependent Fluorescence of Dipolar Molecules

The fluorescence of the dipolar molecules employed for the study of the dynamics of solvation in RTILs do not show dependence with the excitation wavelength. In contrast, the fluorescence of several different dipolar molecules exhibit in RTILs unexpectedly relatively strong dependence with the excitation wavelength. Thus, the wavelength of the maximum emission intensity varies as a function of the excitation wavelength. Samanta et al. were the first who observed this unconventional behavior during the course of their studies on solvation dynamics in RTILs using the dipolar fluorescent probe 2-amino-7-nitrofluorene (ANF). [37] Using the same ANF probe, Margulis et al. have also studied the microheterogeneity in RTILs. [38]

To understand the origin of the excitation wavelength dependent fluorescence, Samanta et al. have undertaken an investigation on several dipolar probes in three different RTILs of different viscosities. [37] They have reported that some dipolar molecules like C153, C102, PRODAN, 4-AP and three nitrobenzoxadiazole (NBD) derivatives (Figure 12), 4NBD, ENBD and 6NBD, do not exhibit an excitation wavelength dependence whereas other dipolar molecules like ANF, 1-Anilinonaphthalene-8 sulphonate, ANS, and the nitrobenzoxadiazole derivative HNBD (Figure 12) show an unusual excitation-wavelength-dependent fluorescence in RTILs. The excitation wavelength dependent shift of the fluorescence maximum has been measured to be between 10 and 35 nm.

The red shift of the emission wavelength maximum of these dipolar molecules when excited at the long-wavelength edge of the first absorption band has been previously observed in highly viscous or frozen media such as low-temperature glasses, polymer matrices and organized assemblies such as micelles, vesicles, proteins and membranes.[39-41] This phenomenon is termed the "red-edge effect" (REE). [42,43] Samanta et al. have concluded that the observation of the REE in RTILs suggests the existence of a distribution of ground-state molecules differing in their interaction energies with the RTILs and implies that the excited state relaxation processes such as solvation and/or energy transfer are significantly slower in

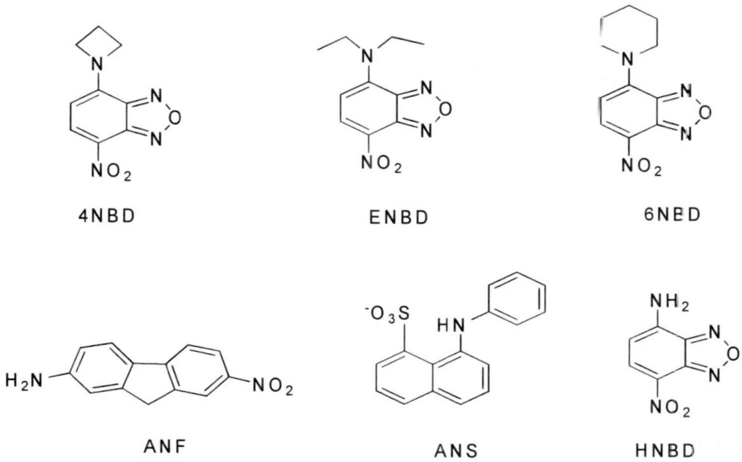

Figure 12. Molecular structures of the three nitrobenzoxadiazole (NBD) derivatives that do not exhibit excitation wavelength dependence and of the ANF, ANS and the HNBD derivative showing an unusual excitation-wavelength-dependent fluorescence behavior.

these media. Although the ground state heterogeneity responsible for this anomaly of influence of the excitation wavelength on the emission is also present for conventional fluid media, in where a strong solute-solvent interaction occurs, what is unique in RTILs is the low rate of the excited state vibration relaxation. In other words, the heterogeneous distribution in the ground state interaction should not be translated into a heterogeneous distribution of excited states if they relax fast to a common excited state-solvent species of minimum energy.

This apparently is not the case in RTILs. Thus, it has been proposed that the excitation wavelength-dependent fluorescence of ANF in RTILs can be a consequence of the short fluorescence lifetime of this system compared to the relatively long solvation time. [21] However, unlike ANF, HNBD and ANS have a fluorescence lifetime longer than the average solvation time in RTILs and, therefore, they do exhibit excitation wavelength dependence of the fluorescence. Then, Samanta et al. have assessed that additional factors such as hydrogen bonding, in the case of HNBD, and electrostatic interactions, in the case of ANS, are playing a role on the observation of this REE effect.

Photoisomerization in RTILs

The photoisomerization of different molecules in Bmim-PF_6 ionic liquid is one of the few photochemical reactions undertaken in RTILs. There are only two reports regarding the study of photoisomerization dynamics in RTILs (Figure 13). Hamaguchi et al. [44] have investigated the photosiomerization of trans-stilbene in Bmim-PF_6 and more recently, Sarkar et al. have used the photophysics of 3,3'-diethyloxadicarbocyanine iodide (DODCI) to determine how the rate of photoisomerization of this molecule is affected by an ionic liquid as a solvent. [45] The photophysics of DODCI is dominated by a very fast isomerization process from the first singlet excited state. Due to its complicated structure, the isomerization of

DODCI involves a tremendous molecular movement and a process with activation barrier, thus it has been influenced by temperature and medium viscosity.

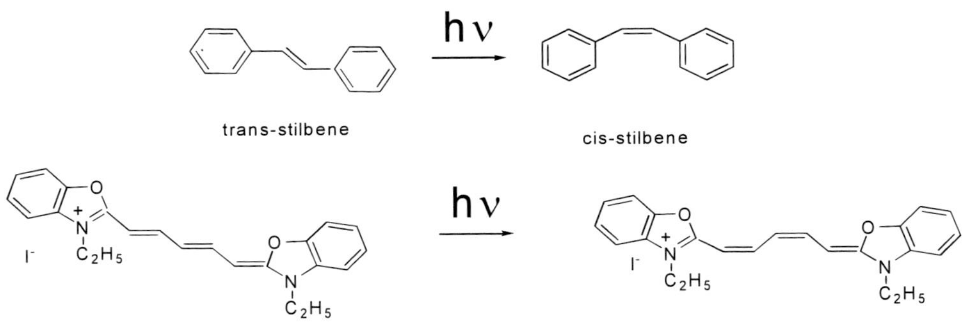

DODCI

Figure 13. Photoisomerization of trans-stilbene and DODCI.

Both reports conclude that photoisomerization indeed proceeds in Bmim-PF_6, but at a rate considerably smaller than organic solvents of similar polarity. This slow isomerization rate was attributed to the high viscosity of neat Bmim-PF_6. The high viscosity of the ionic liquid should inhibit the twisting motion of the conjugate double bonds. To support this proposal, Sarkar et al. have shown that addition of water to neat Bmim-PF_6 increases gradually the photoisomerization rate. This is due to the decreased viscosity of the water-RTIL mixtures as compared to pure Bmim-PF_6. The decrease in viscosity offers less friction to the atom displacement around the C=C double bond.

PHOTOINDUCED ELECTRON TRANSFER IN RTILS

As mentioned earlier, electronic excited states are better oxidizing and reducing species than the ground state. For this reason, single electron transfer between an electron acceptor and an electron donor promoted by light is a general phenomenon (Figure 14). Considering that the result of the electron transfer are positively and negatively charged radical ions, those factors that stabilize charged species are important to increase the lifetime of the charge separated state. In this context, solvent polarity has been shown to be of considerable importance in the stabilization of the charge separate state through electrostatic interaction. Considering that RTILs are constituted by ions and therefore they represent the extreme case of polar solvents, it was anticipated that RTILs will be suitable solvents to promote PET. The data available shows that this is the case. [46]

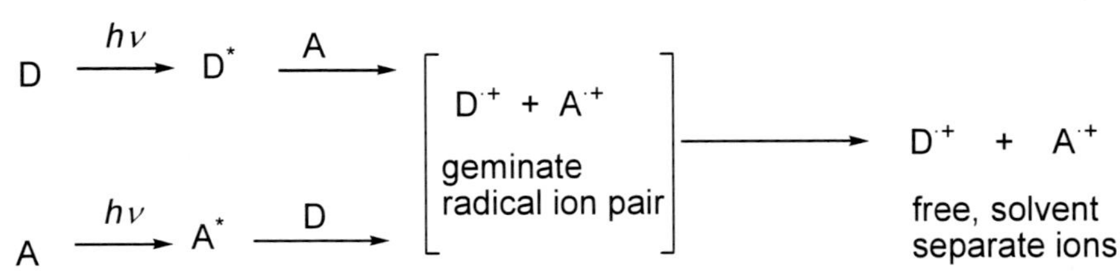

Figure 14. PET among an excited electronic state and an electron donor or electron acceptor molecule.

One interesting observation is that the rate of PET in RTILs can be about two orders of magnitude higher than diffusion. Since the common mechanism for PET requires contact between the donor and the acceptor, one of them in its excited electronic state, the maximum possible rate is the diffusion rate in this medium that according to the Smoluchowski equation is governed by the viscosity of the medium. Thus, the diffusion rate in Bmim-PF_6 and Omim-PF_6 should be about 1×10^7 $M^{-1}\times s^{-1}$, this being the limit for the maximum reaction rate in this solvent for diffusion controlled processes. In contrast to this general rule about the reaction rate limit and following the lead of McLean et al. for energy transfer between benzophenone and naphthalene, [47] Skrzypczak and Neta observed a rate constant for the electron transfer between N-butylpyridinyl radical and duroquinone about one order of magnitude higher than the values estimated from the viscosity.[48] Moreover, the discrepancy between the diffusion rate and the PET rate increases as the length and flexibility of the aliphatic chains increases. [48]

In a parallel study, Vieira and Falvey using the PET quenching of 9,10-dicyanoanthracene fluorescence by a series of electron donor quenchers showed that the quenching rate constants can be about $1.57\text{-}1.70\times10^9$ $M^{-1}\times s^{-1}$. [49] To solve this inconsistency, it has been proposed that the probe molecules reside in and diffuse through nonpolar microdomains in the imidazolium ionic liquids and that PET occurs in these regions in which the local diffusion rate should be much higher than that of the bulk ionic liquid. Similar conclusion with rate constants for PET from 2-4 times higher than the diffusion rate has been reached by Paul and Samanta studying the quenching of pyrene excited state by N,N-dimethylaniline.[50] Quantification of the yield of solvent separated radical ions, i.e., those that have escaped from the solvent cage that are long-lived, by laser flash photolysis has revealed that even in the less viscous RTILs this yield is very low (0.015 ± 0.005), being negligible in more viscous Bmim-PF_6 RTILs.[50] This low yield of free pyrene radical cation indicates that the rate of escape out of the cage is about 2-3 orders of magnitude lower than the back electron transfer and therefore, the cage effect in RTIL must be very high.

Analogous conclusion about the long lifetime of the free radical ions has also been reached for the PET q uenching of 2,4,6-triphenylpyrylium triplets by dicyclopentadiene.[51] An interesting idea is that the photochemistry in RTILs has many similarities with that performed in zeolites that are microporous hosts with a polyelectrolyte structure.[51]

PET between triplet excited state of polypyridyl ruthenium (II) [$Ru(bpy)_3^{2+}$] as donor and viologens (V^{++}) as acceptor is one of the favourite model systems whose behaviour has been studied in a large variety of homogeneous and heterogeneous media, [52] including RTILs.[53,54] In this context, a photocatalytic system that is able to effect the acid- and base-free Meerwein-Ponndorf-Verley-type reduction of carbonyl by isopropanol based on $Ru(bpy)_3^{2+}/V^{++}$ has been reported.[55,56] This photocatalytic system operates in Bmim-PF_6. Upon light absorption, a PET between $Ru(bpy)_3^{2+}$ triplet excited state and V^{++} in its ground state yielding $Ru(bpy)_3^{3+}$ and $V^{\cdot+}$. The latter species abstract a hydrogen atom from isopropanol forming a Meisenheimer intermediate $[V\text{-}H]^+$ analogous to NADH that is able to reduce the ketone. The cycle is closed when $Ru(bpy)_3^{3+}$ is reduced by the ketyl radical regenerating the starting $Ru(bpy)_3^{2+}$ complex. Scheme 15 summarizes the proposed operation cycle of this photocatalytic system.

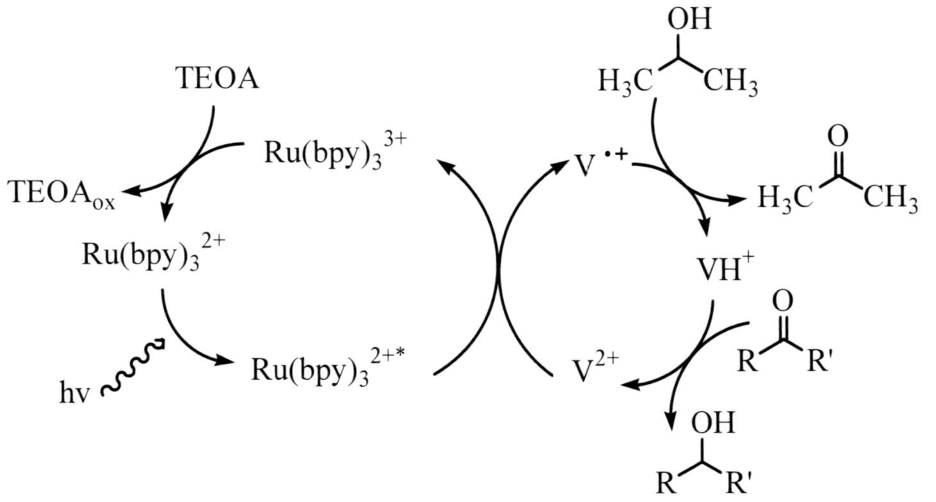

Figure 15. Photocatalytic cycle to effect the Meerwein-Ponndorf-Verley reduction of a carbonyl group by isopropanol.

Concluding Remarks and Future Prospects

RTILs have consolidated as green solvents to conduct catalytic reactions. The special features of this novel media makes them also attractive for performing photochemical reactions. Up to now most of the studies have been limited to simple ionic liquids. Most of the effects observed arise from the high viscosity of these liquids rather than from the polarity that turns out to be in the range of alcohols and polar organic solvents. Some points that need further clarification are the origin of the higher PET constant rate compared to diffusion, understanding of the structure-activity for the observation of excitation wavelength dependence of the emission, solvent reorganization, among other effects.

It can be expected that following the lead of homogeneous catalysis, the interest of RTILs in photochemistry will continue to grow in the near future. As in catalysis, task-specific ionic liquids exhibiting the target photochemical properties will be designed. Also more fundamental studies are needed, other novel fields based at the interphase between photochemistry and nanoscience will develop in the near future. In this sense, it is worth mentioning that ionic liquids can be used in photovoltaic cells to replace volatile electrolyte solutions.[57,58] Also RTILs can serve to enhance the emission efficiency in electroluminescence devices. [59] It can be anticipated that RTILs will play a key role in fields like photocatalysis, photosplitting of water and alternative energy resources. Considering the importance of nanoscience and energy, it is very likely that RTILs will become in the near future one of the key media for photochemistry.

Acknowledgments

Financial support by the Spanish DGI (CTQ2006-5687) is gratefully acknowledged.

REFERENCES

[1] (Bradley, D., Dyson, P., Welton, T. (2000). *Chemistry Review (Deddington, United Kingdom)*, 9, 18-21.
[2] Wasserscheid, P., Keim, W. (2000). *Angew. Chem., Int. Ed*, 39, 3772-3789.
[3] Holbrey, J. D. (2004). *Chimica Oggi*, 22, 35-37.
[4] Dupont, J., de Souza, R. F., Suarez, P. A. Z. (2002). *Chemical Reviews*, 102, 3667-3691.
[5] Corma, A., Garcia, H. (2003). *Chem. Rev*, 103, 4307-4365.
[6] Zhao, D. B., Liao, Y. C., Zhang, Z. D. (2007). *Clean-Soil Air Water*, 35, 42-48.
[7] Andrade, C. K. Z., Alves, L. M. (2005). *Current Organic Chemistry*, 9, 195-218.
[8] Gilbert, A., Baggott, J. (1990). *Essentials of Organic Photochemisry*, Blackwell: Oxford.
[9] Ramamurthy, V. Ed. (1991). *Photochemistry in Organized and Constrained Media*; VCH: New York.
[10]Paul, A., Kumar Mandal, P., Samanta, A. (2005). *J. Phys. Chem. B*, 109, 9148-9153.
[11]Paul, A., Kumar Mandal, P., Samanta, A. (2005). *Chem. Phys. Lett.*, 402, 375-379.
[12]Paul, A., Samanta, A. (2006). *J. Chem. Sci*, 118, 335-340.
[13]Marquis, S., Ferrer, B., Alvaro, M., Garcia, H. (2007). *Journal of Physical Chemistry A*, 111, 6034-6038.
[14]Aki, S. N. V. K., Brennecke, J. F., Samanta, A. (2001). *Chem. Commun*, 413-414.
[15]Reichardt, C. (2005). *Green Chemistry* 7, 339-351.
[16]Dimroth, K., Reichardt, C., Siepmann, T., Bohlmann, F. (1963). *Justus Liebigs Ann. Chem*, 661, 1-37.
[17]Carmichael, A. J., Seddon, K. R. (2000). *J. Phys. Org. Chem*, 13, 591-595.
[18]Deye, J. F., Berger, T. A., Anderson, A. G. (1990). *Anal. Chem*, 62, 615-622.
[19]Ferrer, B., Garcia, H., Schultz, K. P., Nelsen, S. F. (2007). *J. Phys. Chem. B*, 111, 13967-13970.
[20]Fletcher, K. A., Baker, S. N., Baker, G. A., Pandey, S. (2003). *New J. Chem*, 27, 1706-1712.
[21]Samanta, A. (2006). *J. Phys. Chem. B*, 110, 13704-13716.
[22]Kovalenko, S. A., Schanz, R., Senyushkina, T. A., Ernsting, N. P. (2002). *Phys. Chem. Chem. Phys*, 4, 703.
[23]Pines, D., Pines, E., Rettig, W. (2003). *J. Phys. Chem. A*, 107, 236.
[24]Horng, M. L., Dahl, K., Jones, G. I., Maroncelli, M. (1999). *Chem. Phys. Lett*, 315, 363.
[25]Holbrey, J. D., Seddon, K. R. (1999). *J. Chem. Soc., Dalton Trans*, 2133.
[26]Wilkes, J. S., Zaworotko, M. J. (1992). *J. Chem. Soc. , Chem. Commun*, 965.
[27]Karmakar, R., Samanta, A. (2002). *J. Phys. Chem. A*, 106, 4447-4452.
[28]Karmakar, R., Samanta, A. (2002). *J. Phys. Chem. A*, 106, 6670-6675.
[29]Karmakar, R., Samanta, A. (2003). *J. Phys. Chem. A*, 107, 7340-7346.
[30]Ito, N., Arzhantsev, S., Maroncelli, M. (2004). *Chem. Phys. Lett*, 396, 83-91.
[31]Ingram, J. A., Moog, R. S., Ito, N., Biswas, R., Maroncelli, M. (2003). *J. Phys. Chem. B*, 107, 5926.
[32]Chakrabarty, D., Hazra, P., Chakraborty, A., Seth, D., Sarkar, N. (2003). *Chem. Phys. Lett*, 381, 697-704.
[33]Shim, Y., Duan, J. S., Choi, M. Y., Kim, H. J. (2003). *J. Chem. Phys*, 119, 6411.
[34]Kobrak, M. N., Znamenskiy, V. (2004). *Chem. Phys. Lett*, 395, 127.

[35]Arzhantsev, S., Jin, H., Ito, N., Maroncelli, M. (2006). *Chem. Phys. Lett*, 417, 524-529.
[36]Mandal, P. K., Samanta, A. (2005). *J. Phys. Chem. B*, 109, 15172-15177.
[37]Mandal, P. K., Sarkar, M., Samanta, A. (2004). *J. Phys. Chem. A*, 108, 9048-9053.
[38]Hu, Z., Margulis, C. J. (2006). *Proc. Nat. Ac. Sci*, 103, 831-836.
[39]Lakowicz, J. R., Nakamoto, S. K. (1984). *Biochemistry*, 23, 3013.
[40]Demchenko, A. P. (1991). *Topics in Fluorescence Spectroscopy*; Plenum Press: New York, Vol. 3.
[41]Demchenko, A. P. (2002). *Luminescence*, *17*, 19.
[42]Weber, G., Shinitzky, M. (1970). *Proc. Nat. Ac. Sci. USA*, 65, 823.
[43]Valeur, B., Weber, G. (1977). *Chem. Phys. Lett*, 45, 140.
[44]Ozawa, R., Hamaguchi, H. (2001). *Chem. Lett*, 736.
[45]Chakrabarty, D., Chakraborty, A., Hazra, P., Seth, D., Sarkar, N. (2004). *Chem. Phys. Lett*, 397, 216-221.
[46]Alvaro, M., Ferrer, B., Garcia, H., Narayana, M. (2002). *Chem. Phys. Lett*, 362, 435-440.
[47]McLean, A. J., Muldoon, M. J., Gordon, C. M., Dunkin, I. R. (2002). *Chemical Communications*, 1880-1881.
[48]Skrzypczak, A., Neta, P. (2003). *Journal of Physical Chemistry A*, 107, 7800-7803.
[49]Vieira, R. C., Falvey, D. E. (2007). *Journal of Physical Chemistry B*, 111, 5023-5029.
[50]Paul, A., Samanta, A. (2007). *Journal of Physical Chemistry B*, 111, 1957-1962.
[51]Marquis, S., Ferrer, B., Alvaro, M., Garcia, H., Roth, H. D. (2006). *Journal of Physical Chemistry B*, 110, 14956-14960.
[52]Garcia, H., Roth, H. D. (2002). *Chem Rev* , 102, 3947-4008.
[53]Gordon, C. M., McLean, A. J. (2000). *Chemical Communications*, 1395-1396.
[54]Gordon, C. M., McLean, A. J., Muldoon, M. J., Dunkin, I. R. (2002). In *Ionic Liquids*, Vol. 818, pp 428-443.
[55]Herance, J. R., Ferrer, B., Bourdelande, J. L., Marquet, J., Garcia, H. (2006). *Chemistry-a European Journal*, 12, 3890-3895.
[56]Alvaro, M., Carbonell, E., Ferrer, B., Garcia, H., Herance, J. R. (2006). *Photochemistry and Photobiology*, 82, 185-190.
[57]Gratzel, M. (2003). *Journal of Photochemistry and Photobiology C-Photochemistry Reviews*, 4, 145-153.
[58]Gratzel, M. (2004). *Journal of Photochemistry and Photobiology a-Chemistry*, 164, 3-14.
[59]Yang, C. H., Sun, Q. J., Qiao, J., Li, Y. F. (2003). *Journal of Physical Chemistry B*, 107, 12981-12988.

In: Photochemistry Research Progress
Editors: A. Sánchez, S. J. Gutierrez

ISBN: 978-1-60456-568-3

Chapter 16

PHOTOINITIATED DISPERSION POLYMERIZATION FOR THE PREPARATION OF POLYMER MICROSPHERES WITH NARROW SIZE DISTRIBUTION

Zhaohua Zeng[1], Jia Chen and Jianwen Yang

ABSTRACT

Photoinitiated free radical polymerization is a very rapid procedure in which photoinitiator decomposes and then produces free radicals very quickly under UV or visible light irradiation. The disadvantage of photopolymerization is that the light source is necessary for initiating and maintaining the reaction. For this reason, photopolymerization is generally used in photo-imaging or photolithography technology and surface application such as coating and printing, but not suitable for the preparation of materials. Recently, we have tried to introduce the procedure of photopolymerization to the preparation of polymeric microspheres. In this chapter, we report on the photoinitiated dispersion polymerization of methyl methacrylate which was carried out in a mixture of ethanol and water as dispersion medium in the presence of poly(N-vinylpyrrolidone) (PVP) as the steric stabilizer and Darocur 1173 as photoinitiator. within 30 mins of UV irradiation at room temperature, 93.7% of conversion was achieved, and microspheres with 0.94μm number-average diameter and 1.04 polydispersity index (PDI) were obtained. Effects of reaction parameters on particle size and distribution of the resulted microspheres have been investigated. It is shown that by photoinitiated dispersion polymerization, polymeric microspheres with narrow size distribution can be prepared in a very short time, whereas around 20 hours is normally needed in a traditional thermal initiated dispersion polymerization.

[1] School of Chemistry and Chemical Engineering, Sun Yat-sen University, Guangzhou 510275, China. E-mail: ceszzh@mail.sysu.edu.cn

INTRODUCTION

Photoinitiated free radical polymerization is a very rapid procedure in which a photoinitiator decomposes and then produces free radicals very quickly under UV or visible light irradiation. The disadvantage is that the light source is necessary for initiating and maintaining the reaction. For this reason, photopolymerization is generally used in photo-imaging or photolithography technology and surface application such as coating and printing, but not suitable for the preparation of bulk materials.

Dispersion polymerization is an attractive method for preparing monodisperse polymeric microspheres, and has been given great interest over the past two decades[1] . The system of dispersion polymerization is transparent at the beginning and turns to translucent with the formation of microspheres, and keeps high fluidity throughout the reaction process. These characteristics make it possible to introduce photopolymerization into the procedure, as it has been done in other microspheres generating procedures, such as microemulsion polymerization[2] , emulsion polymerization[3] and precipitation polymerization[4] . Another advantage of photoinitiated polymerization in addition to the high reaction rate is that the reaction can be carried out at room temperature or lower. This characteristic is in favor of the preparation of heat sensitive functional microspheres. In this chapter, we report photoinitiated dispersion polymerization of methyl methacrylate in ethanol/water mixture medium in the presence of PVP as stabilizer with 2,2-dimethyl-2-phenyl-acetophenone as photoinitiator. Uniform microspheres have been successfully obtained with monomer conversion over 90% within 30mins of UV irradiation. Effects of reaction parameters on particle size and distribution of the resulted microspheres have been investigated.

EXPERIMENTAL

Materials

Methyl methacrylate (MMA, Tianjing Kermel Chemical Reagents Development Centre) was purified by distilling under reduced pressure and stored in a refrigerator prior to use. 2,2-dimethyl-2-phenyl-acetophenone (Darocur 1173, Ciba company) and Poly(N-vinylpyrrolidone) (PVP, K30, number-average molecular weight 10,000, Shanghai BoAo Biologic technology Ltd.) were used as the photoinitiator and stabilizer, respectively, without further purification. Dispersion medium was a mixture of ethanol (Guangzhou chemical reagents factory) and deionized water.

Photoinitiated Dispersion Polymerization

The apparatus for photoinitiated dispersion polymerization is shown in Figure 1.

A constant temperature in polymerization was maintained by using a jacketed reaction cell and thermostated distilled water circulating inside the jacket. In a typical experiment, ethanol/water mixture with weight ratio of 40/60 was introduced into the reactor as reaction media, and 10 wt% of MMA relative to the system, 15wt% of PVP relative to MMA and 2wt.% of Darocur 1173 relative to MMA were dissolved into the media. The reaction mixture was gently purged with nitrogen for 15mins in order to remove oxygen prior to exposure and then the reactor was sealed. A 400W high-pressure mercury lamp was employed to irradiate the mixture from the top of the reaction cell. UV intensity was measured by a UV-radiometer (type UV-A, from Photoelectric Instrument Factory, Beijing Normal University) which is sensitive in the wavelength range of 320~400nm. Typical light intensity used in the experiment was $1.0 mW\cdot cm^{-1}$. The reaction proceeded for 30mins at 30°C. The product was precipitated by centrifugation and rinsed with water/ethanol mixture and centrifuged repeatedly in order to remove residual PVP, then dried in a vacuum oven for 24hrs to gain fine powder.

Characterization of the PMMA Microspheres

Morphology analysis was carried out on a JSM-6330F Field Emission Scanning Electron Microscope(FMSEM) at 10kV. Samples were gold-coated after being dispersed in water and dropped onto glass slides.

All the clear microspheres for each sample in FESEM images, 500~1000 counts in most cases, were measured in software named Image-Pro Plus 5.1(Media Cybernetics, Inc., USA), and the number-average diameter (D_n), weight-average diameter (D_w), and polydispersity index (PDI) were calculated as follows:

$$D_n = \sum_{i=1}^{n} n_i D_i / n \quad (1)$$

$$D_w = (\sum_{i=1}^{n} D_i^4)/(\sum_{i=1}^{n} D_i^3) \quad (2)$$

$$\mathrm{PDI} = D_w / D_n \quad (3)$$

Molecular weight and distribution were characterized using gel permeation chromatography (Breeze GPC, Waters Corp.) with tetrahydrofuran (Guangzhou chemical reagents factory, distilled before use) as the eluent at 40°C using Styragel HR1, HR3, HR4 columns in series, a 1515 HPLC pump, and a 2417 differential refractive index detector. Calibration was based on nineteen polystyrene narrow standard samples with molecular weight ranging from 1,310 to $1,570,000 g\cdot mol^{-1}$. Both the calibration and the sample analysis were performed at a flow rate of $1.0 mL\cdot min^{-1}$.

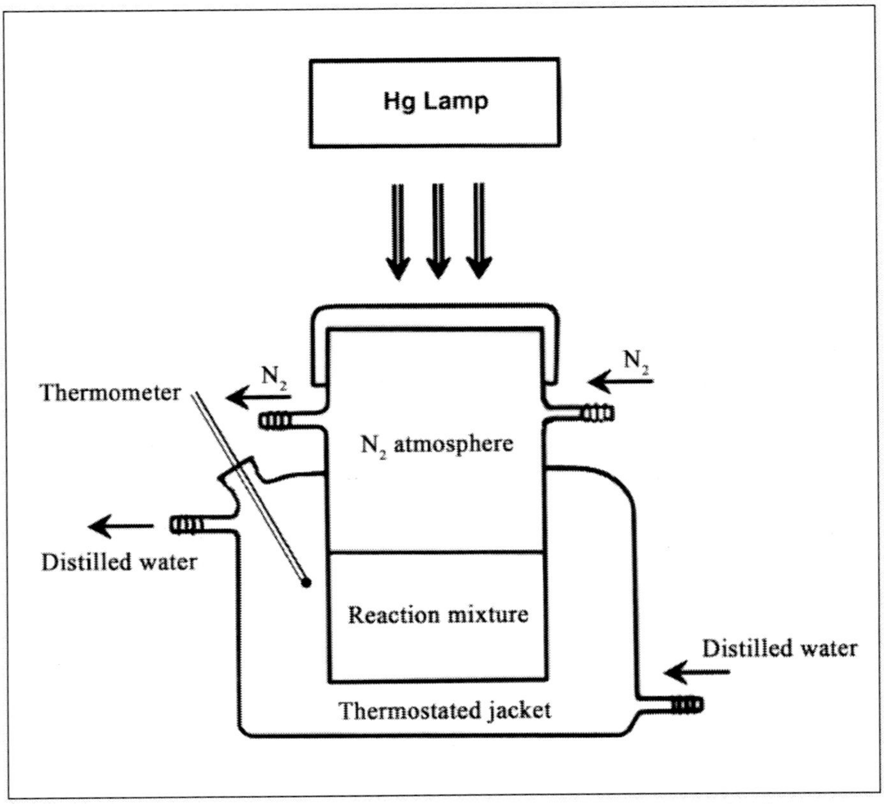

Figure 1. Schematic diagram of apparatus for photoinitiated dispersion polymerization.

Results And Discussion

Photoinitiated dispersion polymerization of MMA was carried out in a mixture of ethanol and water as the dispersion medium in the presence of poly(N-vinylpyrrolidone) (PVP) as the steric stabilizer and Darocur 1173 as photoinitiator. The reaction mixture was clear at the beginning and became opalescent after about twenty seconds of UV irradiation, and turned into a white, stable dispersion as the reaction proceeded.

Figure 2 shows the FESEM images of PMMA microspheres obtained at different reaction time. PMMA microspheres were formed in the polymerization. The size of the particles increased with increasing irradiation time, and the weight-average diameter reached to 0.98μm with PDI of 1.04 after 60min irradiation.

Figure 3 shows that the monomer conversion increased quickly after UV irradiation, and reached to 93.70% at 30mins and 96.57% at 60mins. The diameter of the microspheres increases with time in the same manner as conversion. Number-average molecular weight(M_n), Weight-average molecular weight(M_w) and molecular weight distribution(MWD) of the final polymer are 6.05×10^4kg/kmol, 1.77×10^5kg/kmol and 2.92, respectively.

In a thermal initiated free radical dispersion polymerization, reaction period up to 24h is commonly needed to achieve high monomer conversion due to the long half life of the thermal-initiator. Photoinitiator decomposes and then produces free radicals very quickly under UV irradiation (Scheme 1). As a result, the photoinitiated dispersion polymerization reaches high conversion in a shot time.

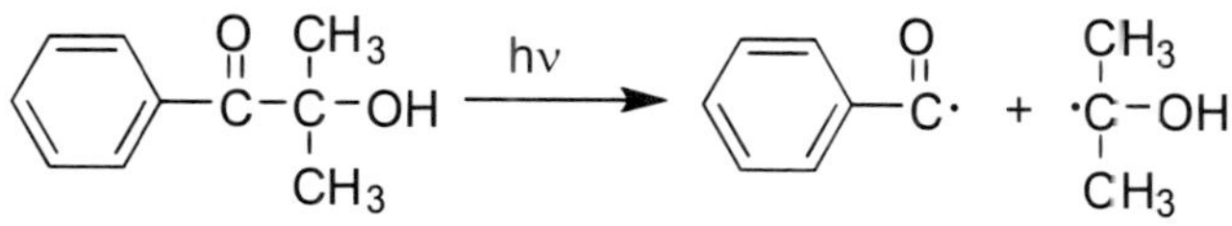

Scheme 1. Initiation mechanism of Darocur 1173

Figure 2. FESEM images of PMMA microspheres obtained by photoinitiated dispersion polymerization of MMA at reaction time of (a) 0.3min, (b) 4mins, (c) 12mins and (d) 60mins with condition described in Experimental section.

Effect of steric stabilizer

In dispersion polymerization, primary particles are generated from the precipitation and coagulation of oligomeric species, and subsequently grown by absorbing oligomers and monomers from the medium. The particles are stabilized by absorbing stabilizer from the reaction medium. Stabilizer plays a crucial role in dispersion polymerization. Typical stabilizers include soluble homopolymer [5-7] ,block copolymer [8-11] and macromonomer [12-14] Poly(N-vinylpyrrolidone) (PVP), a homopolymer soluble in water and ethanol, was used as stabilizer in this work.

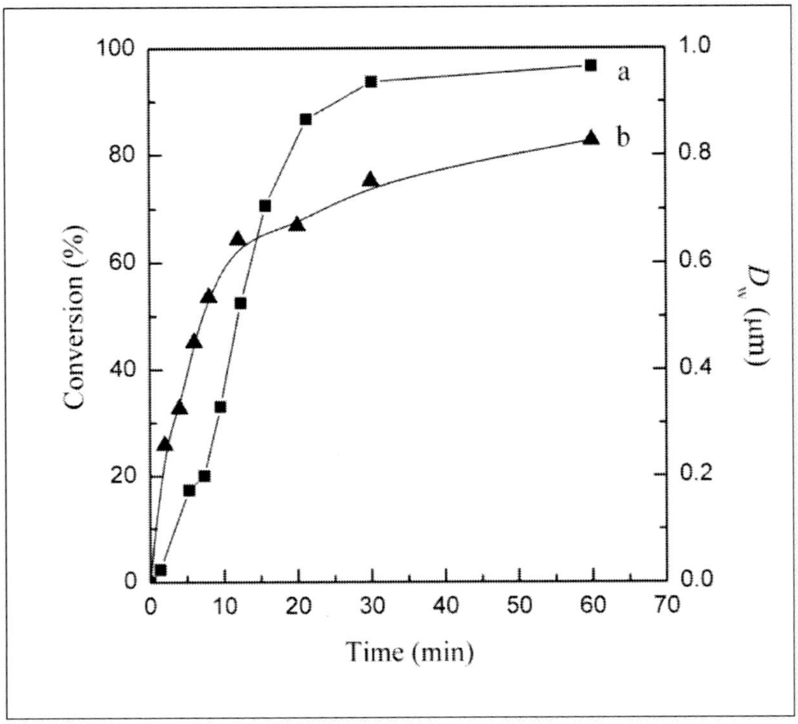

Figure 3. Plots of (a) conversion of MMA and (b) diameter of the microspheres vs. irradiation time for photoinitiated dispersion polymerization of MMA with condition described in Experimental section.

Figure 4 is the dual logarithmic plot of particle diameter vs PVP concentration. It shows a linear relationship with a slope of -0.44 at PVP concentration between 2wt% and 15wt%. The decrease of particle diameter with increasing stabilizer concentration is a normal aspect for a conventional thermal initiated dispersion polymerization. Higher stabilizer concentration induces faster adsorption of grafted stabilizer, which reduces the aggregation of the particle nuclei and thus final microspheres of smaller size are obtained. An exponent of -0.5 for the dependence of particle diameter on stabilizer concentration was predicted by Paine's theoretical analysis [15] . The exponent of -0.44 obtained in the work is very close to the theoretical predicted value.

The wide size distribution at lower PVP concentration is due to the coagulation of the growing particles which were not sufficiently stabilized, as shown in Figure 5(a).

Increasing PVP concentration may reduce the coagulation and thus improve the uniformity of the ultimate particles (Figure 4). Both particle size and uniformity trend to constant as PVP concentration reaches to 15wt%. Extra stabilizer has no effect on the final particle size and distribution.

Molecular weight and distribution of PMMA microspheres with different stabilizer concentration are listed in Table 2. It is showed that PVP concentration had no obvious effect on the molecular weight and distribution of the polymers.

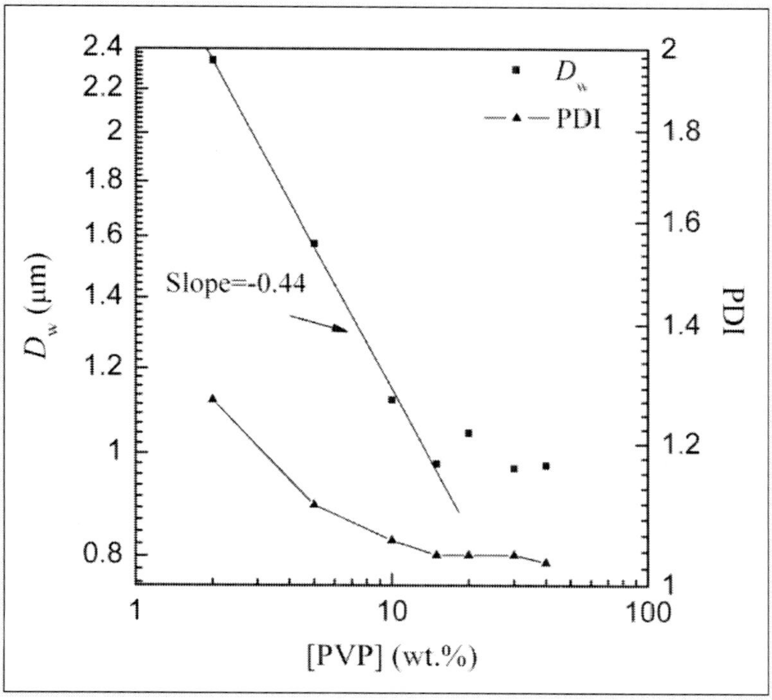

Figure 4. Dual logarithmic plots of weight-average diameter and PDI vs PVP concentration for PMMA microspheres obtained from photoinitiated dispersion polymerization of MMA with reaction parameters described in Experimental section.

Figure 5. FESEM images of PMMA microspheres obtained by photoinitiated dispersion polymerization of MMA with different PVP concentration(wt%): (a)0; (b) 2; (c) 5; (d) 10; (e) 15; (f) 30.

Table 2. Relationship of average molecular weights and molecular weight distributions of PMMA microspheres with different stabilizer concentration

Stabilizer concentration/ % (wt)	$M_n \times 10^{-4}$ (Daltons)	$M_w \times 10^{-4}$ (Daltons)	MWD
0	3.41	21.40	6.27
2	6.16	20.33	3.30
5	6.60	21.06	3.19
10	5.92	20.06	3.39
15	6.05	17.69	2.92
30	7.36	22.92	3.12

Effect of reaction medium

Reaction medium in dispersion polymerization is generally a poor solvent for the polymer chains generated and a solvent for stabilizer, and its effect on the diameter of resulted particles is attributed to its solvency toward polymer chains or stabilizer. Figure 6 shows that particle diameter steadily increases with increasing ethanol content. This phenomenon is generally explained by the solvency of the reaction medium toward the growing polymer chains, which controls the critical chain length above which the polymer precipitates during the nucleation stage[16-18] . Thus, the critical chain length increases with increasing ethanol content in the mixture medium because of the higher solubility of PMMA in ethanol than in water, resulting in a decrease in the number of nuclei and, correspondingly, an increase in particle size. While in another article[19] , the effect of dispersion medium on particle size was attributed to the solvency of the medium toward the stabilizer. It was suggested that the adsorption of stabilizer will be relatively difficult from its good solvent, so that the primary stabilization leads to larger but fewer particles and hence to greater ultimate size.

In photoinitiated polymerization, the photoinitiator decomposes very quickly under the irradiation of UV light, leading to the formation of large amount of nuclei or unstable particles in a short time. The interparticle collision should be more frequent than in thermal-initiated dispersion polymerization. In this case, the adsorption of stabilizer is crucial. Rising the content of ethanol and hence increasing the solvency of the medium toward PVP will decrease the chance of nuclei to absorb PVP from the medium, leading to a less number of primary stabilized particles and hence a larger ultimate particle size. Poor stabilization due to high solvency of the medium toward the stabilizer may induce coagulation of large particles, resulting in the formation of nonspherical particles as the content of ethanol is high enough, as shown in Figure 7 (c). Particle size distribution becomes broader correspondently because of the coagulation (Figure 6).

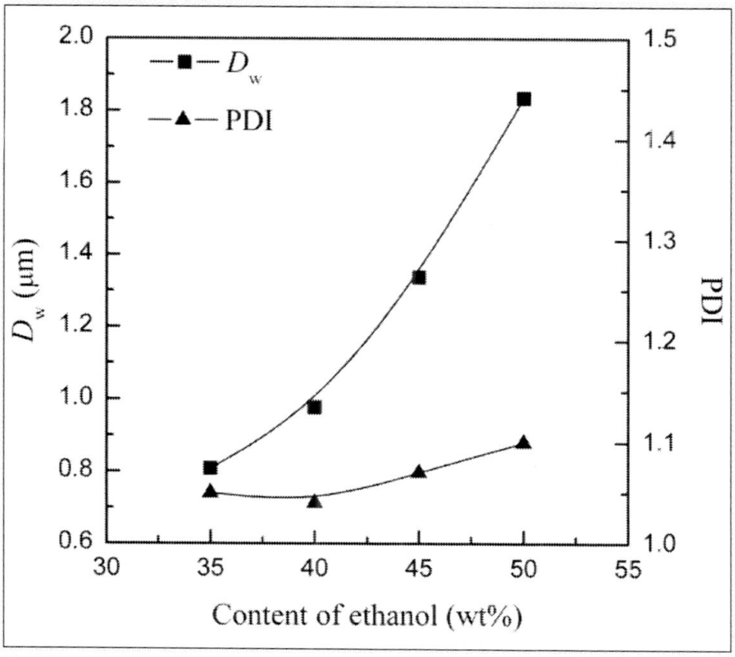

Figure 6. Plots of weight-average diameter and PDI vs ethanol content for PMMA microspheres obtained from photoinitiated dispersion polymerization of MMA with reaction parameters described in Experimental section.

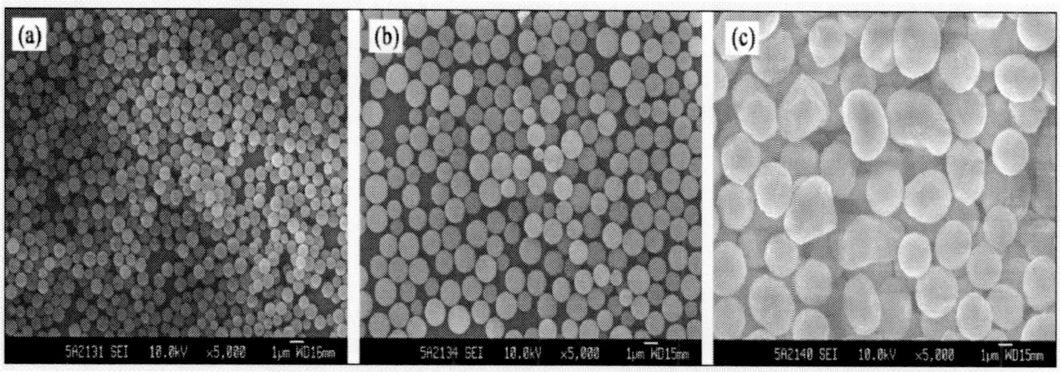

Figure 7. FESEM images of PMMA microspheres obtained by photoinitiated dispersion polymerization of MMA with different ethanol content(wt %): (a) 35; (b) 45; (c) 55.

Effect of initial monomer concentration

The particle diameter increases with increasing MMA concentration, as shown in Figure 8. An increase in monomer concentration may increase the solvency of the continuous phase, resulting in longer critical chain length, lower adsorption rate of the stabilizer on the surface of particles and higher swelling of the particles by the monomer[20-21] . As mentioned above, the adsorption of stabilizer is crucial in photoinitiated dispersion polymerization. High MMA concentration is disadvantageous for the adsorption of MMA-grafted PVP from the medium, for MMA is a good solvent of its polymer. This results in an increase in the coagulation of unstable particles, leading to larger particles with wider side distribution. The wide size distribution at MMA initial concentration of 5wt% is probably due to the insufficient growth of the particles at low monomer content. Narrow size distribution was obtained at MMA initial concentration of 10wt% and 15wt%, as shown in Figure 8.

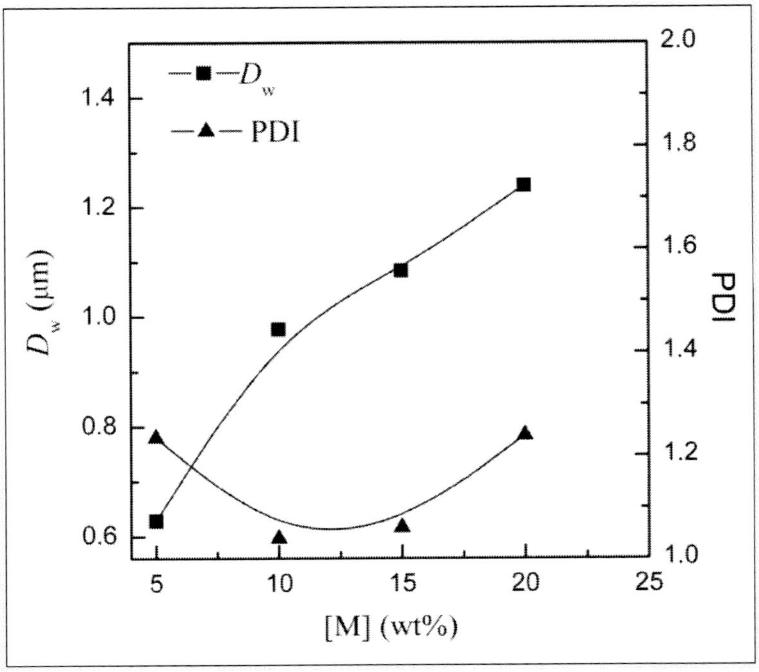

Figure 8. Plots of weight-average diameter and uniformity *vs* initial monomer concentration[M] for PMMA microspheres obtained from photoinitiated dispersion polymerization of MMA with reaction parameters described in Experimental section.

Effect of photoinitiator concentration

The particle size increased with increasing photoinitiator concentration, as shown in Figure 9.

Kim et al.[5] suggested that a higher initiator concentration generates large amount of oligomeric species, which results in a low molecular weight and prefer to stay in the medium, and hence give the reduced number of primary particles during the nucleation step. On the other hand, Paine et al.[7] believed that a greater initiator concentration would lead to grafts of shorter chain lengths which have greater solubility in the reaction medium. The adsorption of the grafted stabilizer on the particles would therefore become more difficult, leading to larger size particles. Both opinions can explain the phenomena. In the case of photoinitiated dispersion polymerization, stabilizer adsorption should be the factor to be taken into account first, as mentioned above.

The broader particles size distribution at high initiator concentration is attributed to the secondary nucleation during the particles growth stage because of the high concentration of the growing oligomers. This induces to the formation of small particles accompanying with large particles, as shown in Figure 10(c).

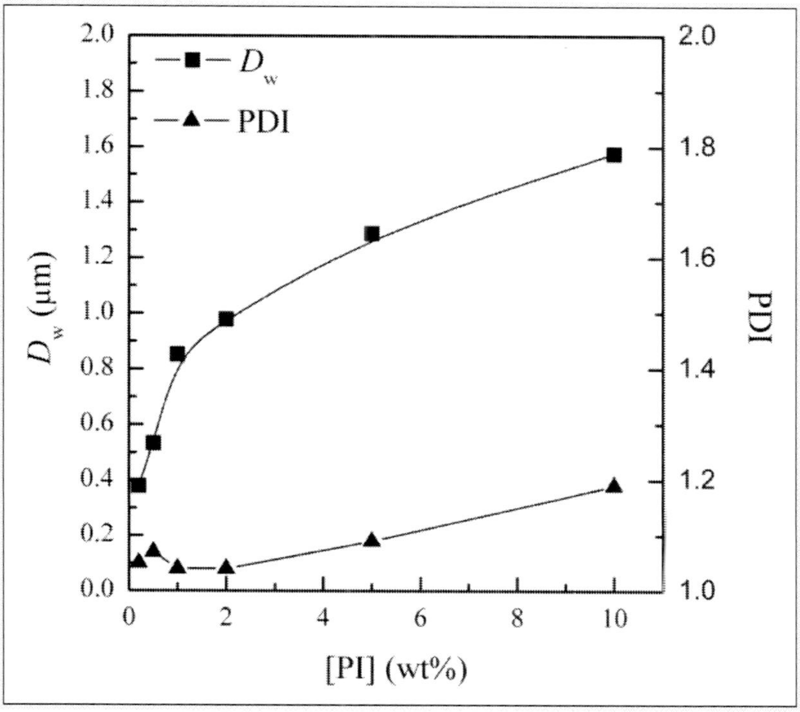

Figure 9. Plots of weight-average diameter and uniformity *vs* photoinitiator concentration[PI] for PMMA microspheres obtained from photoinitiated dispersion polymerization of MMA with reaction parameters described in Experimental section.

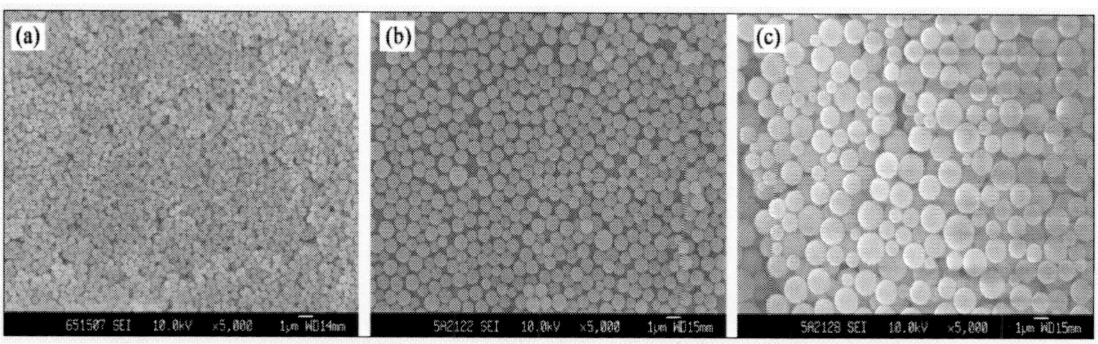

Figure 10. FESEM images of PMMA microspheres obtained by photoinitiated dispersion polymerization of MMA with different photoinitiator concentration related to monomer(wt%): (a) 0.2; (b) 1; (c) 10.

Molecular weight and distribution of PMMA microspheres with different photoinitiator concentration are listed in Table 3. The number-average molecular weight is ranged from 8.08×10^4 to 2.53×10^4 when the photoinitiator concentration changes from 1% (wt) to 10% (wt) relative to monomer. As the photoinitiator Darocur 1173 concentration increases, both the number and weight-average molecular weight, as well as the polydispersity value, tends to decline. This is consistent with the normal free radical polymerization. For a free radical polymerization, the kinetic chain length(v) is inversely dependent on the initiator concentration[I], as shown in the following equation:

$$\nu = \frac{k_p[\mathrm{M}]}{2(fk_d k_t[\mathrm{I}])^{1/2}} \quad (4)$$

Figure 11 shows that the number-average molecular weight of PMMA microspheres decreased with increasing photoinitiator concentration with an exponent of -0.51, in agreement with equation (4).

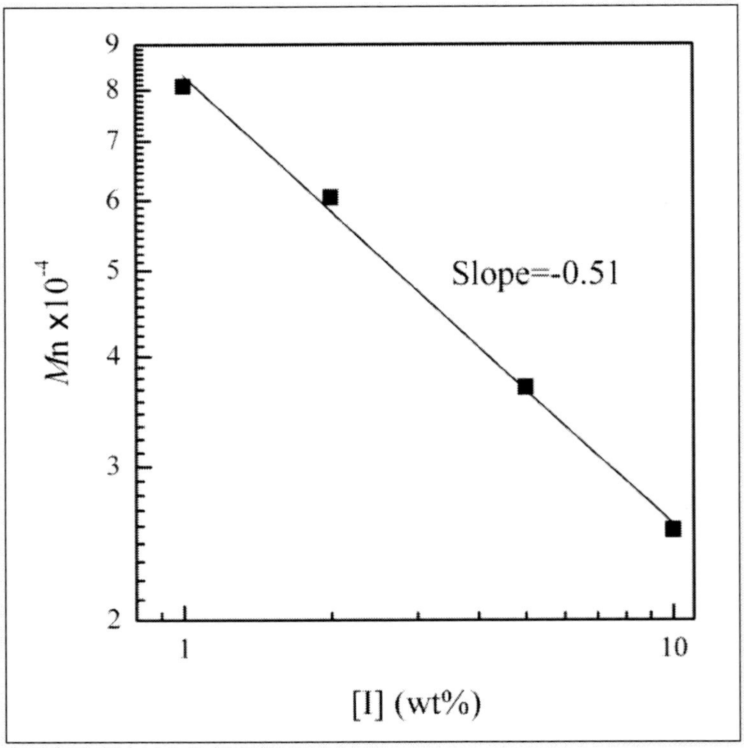

Figure 11. Dual logarithmic plots of number-average molecular weight vs photoinitiator concentration [I] for PMMA microspheres obtained from photoinitiated dispersion polymerization of MMA with reaction parameters described in the Experimental section.

Table 3. Relationship of average molecular weights and molecular weight distributions of PMMA microspheres with different photoinitiator concentration relative to monomer

Photoinitiator concentration / % (wt)	$M_n\times10^{-4}$ (Daltons)	$M_w\times10^{-4}$ (Daltons)	Polydispersity
1	8.08	27.84	3.45
2	6.05	17.69	2.92
5	3.69	10.67	2.89
10	2.53	6.62	2.62

CONCLUSION

PMMA microspheres with narrow size distribution have been successfully synthesized by photoinitiated dispersion polymerization within 30mins of UV irradiation. The morphology, size and size distribution of the microspheres were affected by stabilizer concentration, medium composition, monomer concentration and initiator concentration. Polydispersity index of 1.04 can be achieved under optimum conditions. It seems difficult to generate monodisperse microspheres during such a short reaction period. Secondary nucleation and coagulation of large particles must be avoided for obtaining uniform particles. Solvency of reaction medium toward stabilizer, which affects the adsorption of stabilizer on the particle surface, was shown to be crucial due to the high rate in the formation and growth of particles in photoinitiated dispersion polymerization. In-depth investigation of the mechanism of nucleation and particles growing in photoinitiated dispersion polymerization, and the design or selection of suitable stabilizer for obtained microspheres with improved uniformity are included in the further studies.

This work was supported by the National Natural Science Foundation of China (Grant No. 20774110), which is gratefully acknowledged.

REFERENCES

[1] Kawaguchi, S., Ito, K. (2005). Dispersion polymerization. *Adv Polym Sci,* 175, 299-328.

[2] Kuo, P., Turro, N. J. (1987). Photoinitiated Polymerization of Styrene in Microemulsions. *Macromolecules,* 20, (6), 1216-1221.

[3] Liu, L. Y., Yang, W. T.(2004). Photoinitiated, inverse emulsion polymerization of acrylamide: Some mechanistic and kinetic aspects. *J Polym Sci Pol Chem,* 42, (4), 846-852.

[4] Lime, F., Irgum, K. (2007).Monodisperse polymeric particles by photoinitiated precipitation polymerization. *Macromolecules,* 40, (6), 1962-1968.

[5] Kim, O. H., Lee, K., Kim, K., Lee, B. H., Choe, S. (2006). Effect of PVA in dispersion polymerization of MMA. *Polymer,* 47, (6), 1953-1959.

[6] Shim, S. E., Oh, S., Chang, Y. H., Jin, M. J., Choe, S. (2004). Solvent effect on TEMPO-mediated living free radical dispersion polymerization of styrene. *Polymer,*45, (14), 4731-4739.

[7] Paine, A. J., Luymes, W., McNulty, J. (1990). Dispersion Polymerization of Styrene in Polar Solvents. 6. Influence of Reaction Parameters on Particle Size and Molecular Weight in Poly(N-vinylpyrrolidone)-Stabilized Reactions. *Macromolecules,* 23, (12), 3104-3109.

[8] Jakubowski, W., Lutz, J. F., Slomkowski, S., Matyjaszewski, K. (2005).Block and random copolymers as Surfactants for dispersion polymerization. I. Synthesis via atom transfer radical polymerization and ring-opening polymerization. *J Polym Sci Pol Chem,* 43, (7), 1498-1510.

[9] Riess, G., Labbe, C. (2004). Block copolymers in emulsion and dispersion polymerization. *Macromol Rapid Comm,* 25, (2), 401-435.

[10] Gibanel, S., Heroguez, V., Forcada, J., Gnanou, Y. (2002). Dispersion polymerization of styrene in ethanol-water mixture using polystyrene-b-poly(ethylene oxide) macromonomers as stabilizers. *Macromolecules,* 35, (7), 2467-2473.

[11] Ma, Y. H., Yang, W. T. (2004).Nonaqueous dispersion polymerization of styrene in methanol with the ionomer block copolymer poly[(4-methylstyrene)-co-(4-vinyltriethylbenzyl ammonium bromide)]-b-polyisobutene as a stabilizer. *J Polym Sci Pol Chem,*42, (11), 2678-2685.

[12] Lee, K., Jung, H., Kim, S. Y., Lee, B. H., Choe, S. (2006). Synthesis and characterization of cross-type vinylurethane macromonomer (C-VUM) and their application in the dispersion polymerization of styrene. *Polymer,* 47, (6), 1830-1836.

[13] Kim, S. Y., Lee, K., Jung, H., Shim, S. E., Lee, B. H., Choe, S. (2005).Macromonomers having different molecular weights of polyethylene glycol and end group functionalities in dispersion polymerization of styrene. *Polymer,* 46, (19), 7974-7981.

[14] Jung, H., Kim, S. Y., Lee, K., Lee, B. H., Shim, S. E., Choe, S. (2005). High molecular weight monodisperse polystyrene microspheres prepared by dispersion polymerization, usine a novel bifunctional macromonomer. *J Polym Sci Pol Chem,* 43, (16), 3566-3573.

[15] Paine, A. J. (1990).Dispersion polymerization of styrene in polar solvents. 7. A simple mechanistic model to predict particle size. *Macromolecules,* 23, (12), 3109-3117.

[16] Ugelstad, J., Berge, A., Ellingsen, T., Schmid, R., Nilsen, T., Mork, P. C., Stenstad, P., Hornes, E., Olsvik, O. (1992).Preparation and application of new monosized polymer particles. *Prog Polym Sci,* 17, (1), 87-161.

[17] Horak, D. (1999).Effect of reaction parameters on the particle size in the dispersion polymerization of 2-hydroxyethyl methacrylate. *J Polym Sci Pol Chem,* 37, (20), 3785-3792.

[18] Mackova, H., Horak, D. (2006). Effects of the reaction parameters on the properties of thermosensitive poly(N-isopropylacrylamide) microspheres prepared by precipitation and dispersion polymerization. *J Polym Sci Pol Chem,* 44, (2), 968-982.

[19] Ray, B., Mandal, B. M. (1997). Dispersion Polymerization of Acrylamide. *Langmuir,*13, (8), 2191-2196.

[20] Bamnolker, H., Margel, S. (1996).Dispersion polymerization of styrene in polar solvents: Effect of reaction parameters on microsphere surface composition and surface properties, size and size distribution, and molecular weight. *Journal of Polymer Science: Part A Polymer Chemistry,* 34, 1857-1871.

[21] Xu, Z. S., Deng, Z. W., Hu, X. X., Li, L., Yi, C. F. (2005). Monodisperse polystyrene microspheres prepared by dispersion polymerization with microwave irradiation. *J Polym Sci Pol Chem,* 43, (11), 2368-2376.

INDEX

3

A

B

C

D

E

F

G

H

I

M

N

O

P

Q

R

S

T

U

V

W

X

Y

Z